A Modern Introduction to
Linear Algebra

A Modern Introduction to
Linear Algebra

HENRY RICARDO

CRC Press
Taylor & Francis Group
Boca Raton London New York

CRC Press is an imprint of the
Taylor & Francis Group, an **informa** business

A CHAPMAN & HALL BOOK

CRC Press
Taylor & Francis Group
6000 Broken Sound Parkway NW, Suite 300
Boca Raton, FL 33487-2742

First issued in paperback 2019

© 2010 by Taylor and Francis Group, LLC
Chapman & Hall/CRC is an imprint of Taylor & Francis Group, an Informa business

No claim to original U.S. Government works

ISBN-13: 978-0-367-38504-0

Library of Congress Cataloging-in-Publication Data

Ricardo, Henry.
    A modern introduction to linear algebra / Henry Ricardo.
        p. cm.
    Includes bibliographical references and index.

      1. Algebras, Linear. I. Title.

QA184.2.R53 2010
512'.5--dc22
                                         2009013704

Visit the Taylor & Francis Web site at
http://www.taylorandfrancis.com

and the CRC Press Web site at
http://www.crcpress.com

For my wife, Catherine—"Age cannot wither her, nor custom stale her infinite variety."—
and for

Henry and Marta Ricardo   *Cathy and Mike Corcoran*
Tomás and Nicholas Ricardo   *Christopher Corcoran*

Christine and Greg Gritmon

Contents

An asterisk * marks optional sections, those discussions not needed for later work.

Author

Henry Ricardo received his BS from Fordham College and his MA and PhD from Yeshiva University.

Dr. Ricardo has taught at Manhattan College and has worked at IBM in various technical and business positions. He is currently professor of mathematics at Medgar Evers College of the City University of New York, where he was presented with the 2008 Distinguished Service Award by the School of Science, Health and Technology. In 2009, he was given the Distinguished Service Award by the Metropolitan New York Section of the MAA. The second edition of his previous book, *A Modern Introduction to Differential Equations*, was published by Academic Press in 2009. The first edition, published by Houghton Mifflin, has been translated into Spanish.

Professor Ricardo is a member of Phi Beta Kappa and Sigma Xi, as well as of the AMS, MAA, NCTM, SIAM, and the International Linear Algebra Society. He is currently the governor of the Metropolitan New York Section of the MAA.

Introduction

I.1 Rationale

Linear algebra has always been an important foundation course for higher mathematics. Traditionally, this has been the course that introduces students to the concept of **mathematical proof**, although in recent years there have been alternative paths to this enlightenment. More recently, especially with the availability of computers and powerful handheld calculators, linear algebra has emerged as an essential prerequisite for many areas of application—even penetrating more deeply than calculus.

This book is for a one-semester or a two-semester course at the sophomore–junior level intended for a wide variety of students: mathematics majors, engineering students, and those in other scientific areas such as biology, chemistry, and computer science. It stresses **proofs**, is **matrix-oriented**, and has examples, applications, and exercises reflecting some of the many disciplines that use linear algebra. Although it is assumed that the student has had at least two semesters of calculus, those remarks, examples, and exercises invoking calculus can easily be skipped.

If the instructor chooses to deemphasize proofs and avoid the more theoretical topics, he or she can even use this book for students in business, economics, and the social sciences. To echo the last page of a recent book* in another area of mathematics, the subject is LINEAR ALGEBRA and NOT Calculus, Numerical Analysis, Graph Theory, Physics, Computer Science, Economics, or Sociology. I provide indications of some of the applications of linear algebra in many disciplines, but I delegate the burden of more extensive expositions to the instructors in these areas. While providing some applied examples, as well as discussions of numerical aspects, **I have tried to focus on those beautiful and useful concepts that are at the heart of linear algebra**. I hope that this book will be read as a textbook and then kept as a reference for linear algebra concepts and results.

I.2 Pedagogy

In teaching linear algebra, I have been guided by my experiences as a student and teacher and (more recently) by the efforts and recommendations of the Linear Algebra Curriculum Study Group (LACSG), the ATLAST[†] project, and the Linear Algebra Modules Project (LAMP). In 1997 I was a participant in an ATLAST workshop at the University of Wisconsin (Madison). Based on their collective research, teaching experience, and consultations with users of linear algebra in industry, these groups are unanimous in advocating that the first course in linear algebra be taught from a matrix-oriented point of view and that software be used to enhance teaching and learning.[‡] The content of my text reflects the recommendations cited; and although I have been using software (primarily *Maple*®) in my courses for several years, I have written my text in a "pencil and paper" manner, but with opportunities for calculator or computer usage.

---

* D.A. Sánchez. *Ordinary Differential Equations: A Brief Eclectic Tour* (Washington, DC: MAA, 2002).

† An acronym for Augmenting the Teaching of Linear Algebra through the use of Software Tools.

‡ See, for example, the *Special Issue on Linear Algebra, Coll. Math. J.* **24**(1), (January 1993), *Resources for Teaching Linear Algebra* (MAA Notes, No. 42, 1997), and *Linear Algebra Gems: Assets for Undergraduate Mathematics* (MAA Notes, No. 59, 2002).

My writing style is informal, even conversational at points. I provide **proofs** for virtually all results, leaving some others (or parts of others) as exercises. I have generally provided more details and more "between the lines" explanation than is customary in mathematical proofs, even in textbooks. I have tried to find the "best" proofs from the point of view of the reader, usually accompanying them with motivating examples and illuminating discussions.

Student access to computers and calculators makes the analysis of theorems and algorithms an important part of the learning process, but I do not impose technology on the reader. Virtually all linear algebra calculations can be done routinely using graphing calculators and computer algebra systems. The individual instructor should determine the extent to which his or her students make use of technology.

I believe in teaching linear algebra computationally (numerically, algorithmically), algebraically, and geometrically, with an emphasis on the algebraic aspects. Concrete, easy-to-understand examples motivate the theory. My typical pedagogical cycle is **motivating concrete example(s)** → **analysis** → **general principle (theorem and proof)** → **additional concrete example(s)**. I also rely on a spiral approach, by which I introduce an idea in a very simple situation and then revisit it often, building through successive sections and chapters to the full glory of the concept. For example, the concept of *linear transformation* is introduced briefly in the context of matrix–vector multiplication (Section 3.2), discussed more generally in Example 5.2.4 (the range as a subspace of a vector space), and developed fully in Chapter 6.

Each chapter begins with an **introductory text** that motivates the topics to come. These opening remarks link the material that has been treated previously with the content of the present chapter. There are **Exercises** at the end of each section, divided into A, B, and C problems. I have found such a triage useful in the past. In this book, there are over **1200 numbered exercises**, many with multiple parts. Generally, "A" exercises are drill problems, providing practice with routine calculations and symbol manipulations. "B" exercises are typically more theoretical, asking for simple proofs (sometimes the completion of proofs in the text proper) or demanding more elaborate calculations. Among "B" exercises are requests for examples or counterexamples and exercises encouraging the recognition of patterns and generalizations. "C" exercises are meant to challenge student understanding of the concepts and calculations taught in the text. These problems may require more complex calculations, manipulations, and/or proofs. A typical "C" exercise will be broken up into several parts and may have more generous hints given for its solution. Overall, many of the exercises involve concepts and applications that do not appear in the book, thus enriching the student's view of linear algebra. These exercises also introduce some concepts that will be developed further in later sections.

I have included problems for those who have access to graphing calculators and computers, but these exercises are not platform-specific. I have labeled certain exercises with the symbol *𝒩* to suggest that technology be used; but, in fact, an instructor may choose to allow students to use technology for many exercises involving calculation—perhaps asking for a hand computation followed by a check via calculator or computer.

There are many **figures**, both algebraic and geometric, to illustrate various results.

I end each chapter with a **Summary** of key definitions and results.

Finally, the **Appendices** review or introduce certain basic mathematical tools. In particular, there is an exposition of mathematical induction, which is used throughout the text.

There will be an **Instructor's Solutions Manual**, containing worked-out solutions to all exercises.

I.3 Content

The content of my book reflects the recommendations of the groups mentioned in the last section: LACSG, ATLAST, and LAMP. An asterisk (*) marks optional sections, those topics not essential for later work.

Although most of the drama takes place in the context of finite-dimensional real vector spaces, there are roles for complex spaces and cameo appearances by infinite-dimensional spaces.

Chapter 1 introduces the reader to vectors, the fundamental units of linear algebra. The space $\mathbb{R}^n$ is introduced naturally and its algebraic and geometric properties are explored. Using many examples in $\mathbb{R}^2$ and $\mathbb{R}^3$, the concepts of linear independence, spans, bases, and subspaces are discussed. In particular, the calculations required to determine the linear independence/dependence of sets of vectors serve as motivation for the work in Chapter 2.

Systems of linear equations are discussed thoroughly in **Chapter 2**, from geometric, algebraic, and numerical points of view. Gaussian elimination and reduced row echelon forms are at the heart of this chapter. The connection between the rank and the nullity of a matrix is established and is shown to have important consequences in the analysis of systems of m linear equations in n unknowns.

Chapter 3 provides a basic minicourse in the theory of matrices. Introduced as notational devices in Chapter 2, matrices now develop a personality of their own. The explanation of matrix–vector multiplication provides a glimpse of the matrix as a dynamic object, operating on vectors to transform them linearly. Inverses of matrices are handled via elementary matrices and Gaussian elimination. The important LU and PA = LU factorizations are explained and illustrated. In the next chapter, *the determinant of a square matrix is defined in terms of these decompositions*.

Even though the role of determinants in linear algebra has diminished, there is still some need for this concept. In a unique way, **Chapter 4** *defines* the determinant of a matrix A in terms of the PA = LU factorization. Fundamental properties are then proved easily and traditional methods for hand calculation of determinants are introduced. There follows a treatment of similarity and diagonalization of matrices. The Cayley–Hamilton Theorem is proved for diagonalizable matrices and the *minimal polynomial* is introduced.

Chapter 5 introduces the abstract idea of a *real vector space*. The connection between the concepts introduced in Chapter 1 in the context of $\mathbb{R}^n$ and the more abstract ideas is stressed. In particular, this chapter emphasizes that *every n-dimensional real vector space is algebraically identical to* $\mathbb{R}^n$. There is some discussion of the concept of an *infinite-dimensional vector space*.

Chapter 6 provides a comprehensive coverage of *linear transformations*, including details of the algebra of linear transformations and the matrix representation of linear transformations with respect to bases. Invertibility, isomorphisms, and similarity are also treated.

Chapter 7 introduces complex vector spaces and expands the treatment of the dot product defined in Chapter 1. General inner product spaces over $\mathbb{R}$ and $\mathbb{C}$ are discussed. Orthonormal bases and the Gram–Schmidt process are explained and illustrated. The chapter also treats unitary and orthogonal matrices, the Schur factorization, the Cayley–Hamilton Theorem for general square matrices, the QR factorization, orthogonal complements, and projections.

Chapter 8 introduces the concept of a linear functional and the adjoint of an operator. Discussions of Hermitian and normal matrices (including spectral theorems) follow, with a digression on quadratic forms. The book concludes with the Singular Value Decomposition and the Polar Decomposition.

Appendices A through D provide the basics of **set theory**, an explanation of **summation and product notation**, an exposition of **mathematical induction**, and a quick survey of **complex numbers**.

I.4 Using This Book

Depending on the requirements of the course and the sophistication of the students, this book contains enough material for a one- or two-semester course. Chapters 1 through 4 form the core of a one-semester course for an average class. This material is the heart of linear algebra and prepares a student for future abstract courses as well as applied courses having linear algebra as a prerequisite. A class with a strong background may want to investigate parts of Chapters 5 and 6.

Alternatively, if the instructor has the luxury of a two-semester linear algebra course, he or she may want to cover the relatively concrete Chapters 1 through 4 in the first semester and explore the more abstract Chapters 5 through 7 in the second semester. Additional material from Chapter 8 may be used to complete the student's introduction to linear algebraic concepts.

Overall, an instructor may skip or deemphasize Chapter 4 and go directly to Chapters 5 through 8 from the first three chapters. The second half of the book, more abstract than the first, requires only a nodding acquaintance with determinants.

MATLAB® is a registered trademark of The MathWorks, Inc. For product information, please contact:

The MathWorks, Inc.
3 Apple Hill Drive
Natick, MA 01760-2098 USA
Tel: 508 647 7000
Fax: 508-647-7001
E-mail: info@mathworks.com
Web: www.mathworks.com

Acknowledgments

Dr. Michael Aissen was my first linear algebra teacher and was responsible for introducing me to other beautiful and interesting areas of mathematics, as my classroom instructor and as my mentor in undergraduate research projects. I thank him for all this.

Students of my generation were inspired by the classic linear algebra texts of Paul Halmos and Hoffman/Kunze. (Halmos also inspired my later PhD work in operator theory.) In writing this book, I find myself indebted to many fine contemporary researchers and expositors, notably Carl Meyer and Gilbert Strang. My online review for the MAA of *Handbook of Linear Algebra*, edited by Leslie Hogben (Chapman & Hall/CRC, Boca Raton, FL, 2007), did not do justice to the richness of information in this authoritative tome.

At Medgar Evers College, it has been my pleasure to have Darius Movasseghi as my supportive Chair and to have inspiring colleagues, among them Tatyana Flesher, Mahendra Kawatra, Umesh Nagarkatte, Mahmoud Sayrafiezadeh, and Raymond Thomas. Several generations of students have endured my efforts to demonstrate the beauty of linear algebra, among them Patrice Williams, Reginald Dorcely, Jonathan Maitre, Frantz Voltaire, Yeku Oladapo, Andre Robinson, and Barry Gregory.

I am grateful to Lauren Schultz Yuhasz for suggesting that I write a book on linear algebra. I thank David Grubbs, my editor at CRC Press, for his encouragement and his enthusiastic support of my project. Amy Blalock, the project coordinator, deserves my appreciation for guiding me through the production process, as does Joette Lynch, the project editor.

I thank my reviewers for their insightful comments and helpful suggestions, especially Hugo J. Woerdeman (Drexel University) and Matthias K. Gobbert (University of Maryland).

Above all, I owe my wife a great deal for her support of my work on this book.

Of course, any errors that remain are my own responsibility. I look forward to your comments, suggestions, and questions.

Henry Ricardo
henry@mec.cuny.edu

Vectors

We begin our investigation of linear algebra by studying *vectors*, the basic components of much of the theory and applications of linear algebra. Historically, vectors were not the first objects in linear algebra to be studied intensively. The solution of systems of linear equations provided much of the early motivation for the subject, with determinants arising as an early topic of interest. Anyone reading this book has probably worked with linear systems and may even have encountered determinants in this context. We will investigate these topics fully in Chapters 2 and 4.

Now that the subject has matured to some extent—although both the theory and applications continue to be developed—we can begin our study logically with vectors, the fundamental mathematical objects which form the foundation of this magnificent structure known as linear algebra. Vectors are not only important in their own right, but they link linear algebra and geometry in a vital way.

1.1 Vectors in $\mathbb{R}^n$

At some time in our earlier mathematical studies, we learned the concept of *ordered pairs* (x, y) of real numbers, then *ordered triplets* (x, y, z), and perhaps *ordered n-tuples* $(x_1, x_2, \ldots, x_n)$. This idea and these notations are simple and natural. In addition to being more important in theoretical applications (in defining a *function*, for example), *these n-tuples serve to hold data and to carry information.** For example, the ordered 4-tuple (123456789, 35, 70, 145) might describe a person with Social Security number 123-45-6789 who is 35 years old, $5'10''$ tall, and weighs 145 lb. In interpreting such data, it is important that we understand the *order* in which the information is given.

Despite the generality of the idea we have just been discussing, we will consider only real numerical data for most of our study of linear algebra. Eventually, beginning in Chapter 5, we will illustrate the flexibility and wider applicability of the vector concept by allowing complex numbers and functions as data elements.

---

* The word *vector* is the Latin word for "carrier." In pathology and epidemiology, this word refers to an organism (tick, mosquito, or infected human) that carries a disease from one host to another.

Definition 1.1.1

A **vector** is an ordered finite list of real numbers. A **row vector**, denoted by $[x_1 \quad x_2 \quad \cdots \quad x_n]$, is an ordered finite list of numbers written horizontally. A **column vector**, denoted by $\begin{bmatrix} x_1 \\ x_2 \\ \vdots \\ x_n \end{bmatrix}$, is an ordered finite list of real numbers written vertically. Each number x_i making up a vector is called a **component** of the vector.

---

*From now on, we will use lowercase **bold** letters to denote vectors and all other lowercase letters (usually italicized) to represent real numbers.* For example,

$$\mathbf{y} = \begin{bmatrix} y_1 \\ y_2 \\ y_3 \end{bmatrix}$$

denotes a (column) vector $\mathbf{y}$ whose components (real numbers) are $y_1, y_2,$ and y_3. (An alternative, especially prevalent in applied courses, is to use arrows to denote vectors: $\vec{v}$. This form is also used in classroom presentations because it is difficult to indicate boldface with chalk or markers.)

It is useful to introduce the concept of a *transpose* here. If we have a column vector $\mathbf{x} = \begin{bmatrix} x_1 \\ x_2 \\ \vdots \\ x_n \end{bmatrix}$, then its **transpose**, denoted by $\mathbf{x}^T$, is the row vector $[x_1 \quad x_2 \quad \cdots \quad x_n]$. Similarly, the **transpose** of a row vector, $\mathbf{w} = [w_1 \quad w_2 \quad \cdots \quad w_n]$, denoted by $\mathbf{w}^T$, is the column vector $\begin{bmatrix} w_1 \\ w_2 \\ \vdots \\ w_n \end{bmatrix}$.

The form of vector we use will depend on the context of a problem and will change to suit various theoretical and applied considerations—sometimes just for typographical convenience.

A simple data table can illustrate the use of both row and column vectors.

Example 1.1.1: Column Vectors and Row Vectors

Consider the following grade sheet for an advanced math class.

Student	Test 1	Test 2	Test 3
Albarez, Javier	78	84	87
DeLuca, Matthew	85	76	89
Farley, Rosemary	93	88	94
Nguyen, Bao	89	94	95
VanSlambrouck, Jennifer	62	79	87

We can form various vectors from this table. For example, the

column vector $\begin{bmatrix} 84 \\ 76 \\ 88 \\ 94 \\ 79 \end{bmatrix}$, formed from the second column of the

table, gives all the grades on Test 2. The position (row) of a number associates the grade with a particular student. Similarly, the row vector $[89 \quad 94 \quad 95]$ represents all Bao Nguyen's test marks, the position (column) of a number indicating the number of the test. Note that in this illustration, the vectors can have different shapes and numbers of components depending on the meaning of the vectors.

In working with vectors, it is important that the components be *ordered*, that is, each component represents a specific piece of information and any rearrangement of the data represents a different vector. A more precise way of saying this is that *two vectors*, say $\mathbf{x} = [x_1 \quad x_2 \quad \cdots \quad x_n]$ and $\mathbf{y} = [y_1 \quad y_2 \quad \cdots \quad y_n]$, are **equal** *if and only if* $x_1 = y_1$, $x_2 = y_2, \ldots, x_n = y_n$. Of course, two column vectors are said to be equal if the same component-by-component equalities hold.

1.1.1 Euclidean *n*-Space

From this point on, for reasons that will be made clear in Chapters 2 and 3, we show a preference for *column* vectors, meaning that unless otherwise indicated, the term *vector* will refer to a column vector.

Definition 1.1.2

For a positive integer n, a set of all vectors $\begin{bmatrix} x_1 \\ x_2 \\ \vdots \\ x_n \end{bmatrix}$ with components

consisting of real numbers is called **Euclidean *n*-space**, and is denoted by $\mathbb{R}^n$:

$$\mathbb{R}^n = \left\{ \begin{bmatrix} x_1 \\ x_2 \\ \vdots \\ x_n \end{bmatrix} : x_i \in \mathbb{R} \text{ for } i = 1, 2, \ldots, n \right\}.$$

$$= \left\{ [x_1 \quad x_2 \quad \cdots \quad x_n]^{\mathrm{T}} : x_i \in \mathbb{R} \text{ for } i = 1, 2, \ldots, n \right\}$$

(If necessary, see Appendix A for a discussion of set notation.) The positive integer n is called the **dimension** of the space. For instance, the column vectors in Example 1.1.1 are elements of Euclidean 5-space, or $\mathbb{R}^5$, whereas the row vectors derived from the data table are the transposes of elements of $\mathbb{R}^3$. We will develop a deeper understanding of "dimension" in Section 1.5 and in Chapter 5.

Of course, $\mathbb{R}^1$, or just $\mathbb{R}$, is the set of all real numbers, whereas $\mathbb{R}^2$ and $\mathbb{R}^3$ are familiar geometrical entities: the sets of all points in two-dimensional and three-dimensional "space," respectively. For values of $n > 3$, we lose the geometric interpretation, but we can still deal with $\mathbb{R}^n$ algebraically.

1.1.2 Vector Addition/Subtraction

The next example suggests a way of combining two vectors and motivates the general definition that follows it.

Example 1.1.2: Addition of Vectors

Suppose the components of vector $\mathbf{v}_1 = \begin{bmatrix} 322 \\ 283 \\ 304 \\ 292 \end{bmatrix}$ in $\mathbb{R}^4$ repre-

sent the revenue (in dollars) from sales of a certain item in a store during weeks 1, 2, 3, and 4, respectively, and vector $\mathbf{v}_2 = \begin{bmatrix} 187 \\ 203 \\ 194 \\ 207 \end{bmatrix}$ represents the revenue from sales of a different

item during the same time period. If we combine ("add") the two vectors of data in a component-by-component way, we get the total revenue generated by the two items in the same 4 week period:

$$\mathbf{v}_1 + \mathbf{v}_2 = \begin{bmatrix} 322 \\ 283 \\ 304 \\ 292 \end{bmatrix} + \begin{bmatrix} 187 \\ 203 \\ 194 \\ 207 \end{bmatrix} = \begin{bmatrix} 322 + 187 \\ 283 + 203 \\ 304 + 194 \\ 292 + 207 \end{bmatrix} = \begin{bmatrix} 509 \\ 486 \\ 498 \\ 499 \end{bmatrix}.$$

Definition 1.1.3

If $\mathbf{x} = \begin{bmatrix} x_1 \\ x_2 \\ \vdots \\ x_n \end{bmatrix}$ and $\mathbf{y} = \begin{bmatrix} y_1 \\ y_2 \\ \vdots \\ y_n \end{bmatrix}$ are two vectors in $\mathbb{R}^n$, then the **sum** of $\mathbf{x}$ and $\mathbf{y}$ is the vector in $\mathbb{R}^n$, defined by

$$\mathbf{x} + \mathbf{y} = \begin{bmatrix} x_1 + y_1 \\ x_2 + y_2 \\ \vdots \\ x_n + y_n \end{bmatrix},$$

and the **difference** of $\mathbf{x}$ and $\mathbf{y}$ is the vector in $\mathbb{R}^n$, defined by

$$\mathbf{x} - \mathbf{y} = \begin{bmatrix} x_1 - y_1 \\ x_2 - y_2 \\ \vdots \\ x_n - y_n \end{bmatrix}.$$

Clearly, we can combine vectors only if they have the same number of components, that is, only if both vectors come from $\mathbb{R}^n$ for the same value of n. Because the sum (or difference) of two vectors in $\mathbb{R}^n$ is again a vector in $\mathbb{R}^n$, we say that $\mathbb{R}^n$ is **closed under addition** (or **closed under subtraction**). In other words, we do not find ourselves outside the space $\mathbb{R}^n$ when we add or subtract two vectors in $\mathbb{R}^n$.

Example 1.1.3: Addition and Subtraction of Vectors

In $\mathbb{R}^5$,

$$\begin{bmatrix} -5 \\ 7 \\ 0 \\ -4 \\ 11 \end{bmatrix} + \begin{bmatrix} 3 \\ -2 \\ 5 \\ 4 \\ 6 \end{bmatrix} = \begin{bmatrix} -5+3 \\ 7+(-2) \\ 0+5 \\ -4+4 \\ 11+6 \end{bmatrix} = \begin{bmatrix} -2 \\ 5 \\ 5 \\ 0 \\ 17 \end{bmatrix}$$

and

$$\begin{bmatrix} -5 \\ 7 \\ 0 \\ -4 \\ 11 \end{bmatrix} - \begin{bmatrix} 3 \\ -2 \\ 5 \\ 4 \\ 6 \end{bmatrix} = \begin{bmatrix} -5-3 \\ 7-(-2) \\ 0-5 \\ -4-4 \\ 11-6 \end{bmatrix} = \begin{bmatrix} -8 \\ 9 \\ -5 \\ -8 \\ 5 \end{bmatrix}.$$

1.1.3 Scalar Multiplication

Another operation we need is that of *multiplying a vector by a number.* We can motivate the process in a natural way.

Example 1.1.4: Multiplication of a Vector by a Number

Suppose the vector $\begin{bmatrix} 18.50 \\ 24 \\ 15.95 \end{bmatrix}$ in $\mathbb{R}^3$ represents the prices of three items in a store.

If the store charges $7\frac{1}{4}$% sales tax, then we can compute the tax on the three items as follows, producing a "sales tax vector":

$$\text{Sales tax} = 0.0725 \begin{bmatrix} 18.50 \\ 24 \\ 15.95 \end{bmatrix} = \begin{bmatrix} 0.0725(18.50) \\ 0.0725(24) \\ 0.0725(15.95) \end{bmatrix} = \begin{bmatrix} 1.34 \\ 1.74 \\ 1.16 \end{bmatrix}.$$

Now we can find the "total cost" vector:

$$\text{Total cost} = \text{Price} + \text{Sales tax} = \begin{bmatrix} 18.50 \\ 24 \\ 15.95 \end{bmatrix} + 0.0725 \begin{bmatrix} 18.50 \\ 24 \\ 15.95 \end{bmatrix}$$

$$= \begin{bmatrix} 18.50 \\ 24 \\ 15.95 \end{bmatrix} + \begin{bmatrix} 1.34 \\ 1.74 \\ 1.16 \end{bmatrix} = \begin{bmatrix} 19.84 \\ 25.74 \\ 17.11 \end{bmatrix}.$$

We can generalize this multiplication in an obvious way.

Definition 1.1.4

If k is a real number and $\mathbf{x} = \begin{bmatrix} x_1 \\ x_2 \\ \vdots \\ x_n \end{bmatrix}$, then **scalar multiplication** of the vector $\mathbf{x}$ by the number k is defined by $k\mathbf{x} = \begin{bmatrix} kx_1 \\ kx_2 \\ \vdots \\ kx_n \end{bmatrix}$. The number k in this case is called a **scalar** (or **scalar quantity**) to distinguish it from a vector.

Because a scalar multiple of a vector in $\mathbb{R}^n$ is again a vector in $\mathbb{R}^n$, we say that $\mathbb{R}^n$ is **closed under scalar multiplication**.

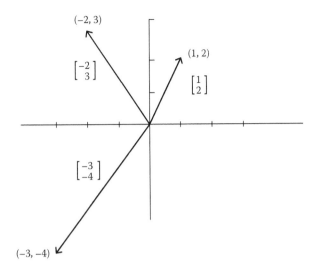

Figure 1.1 Vectors in $\mathbb{R}^2$.

1.1.4 Geometric Vectors in $\mathbb{R}^2$ and $\mathbb{R}^3$

Even though the concept of a vector is simple, vectors are very important in scientific applications. To a physicist or engineer, a vector is a quantity that has both *magnitude* (size) and *direction*, for example, velocity, acceleration, and other forces. In $\mathbb{R}^2$, we can view vectors themselves, vector addition, and scalar multiplication in a nice geometric way that is consistent with the physicist's view. A vector $\begin{bmatrix} x \\ y \end{bmatrix}$ is interpreted as a directed line segment, or arrow, from the origin to the point (x, y) in the usual Cartesian coordinate plane (Figure 1.1):*

The *magnitude* of a vector, a nonnegative quantity, is indicated by the length of the arrow. The *direction* of such a geometric vector is determined by the angle θ which the arrow makes with the positive x-axis (measured in a counterclockwise direction).

The addition of vectors is carried out according to the **Parallelogram Law**, and the sum of two vectors is usually referred to as their **resultant** (vector) (Figure 1.2).

Multiplication of a vector by a positive scalar k does not change the direction of the vector, but affects its magnitude by a factor of k. Multiplication of a vector by a negative scalar *reverses* the vector's direction and affects its magnitude by a factor of $|k|$ (Figure 1.3).

Similarly, in $\mathbb{R}^3$, a vector $\begin{bmatrix} x \\ y \\ z \end{bmatrix}$ can be interpreted as an arrow connecting the origin $(0, 0, 0)$ to a point (x, y, z) in the usual x–y–z plane. The

---

* In vector analysis courses, it is usual to consider vectors that emanate from points other than the origin. See, for example, H. Davis and A. Snider, *Introduction to Vector Analysis*, 7th edn. (Dubuque, IA: Wm. C. Brown, 1995). However, we will use our definition of vectors to illustrate key linear algebra concepts, which will be generalized in this chapter.

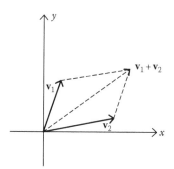

Figure 1.2 Parallelogram law.

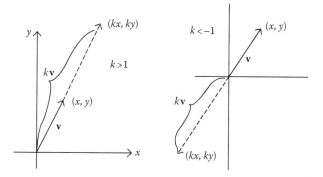

Figure 1.3 Scalar multiplication.

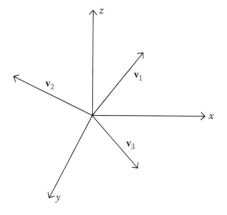

Figure 1.4 Vectors in $\mathbb{R}^3$.

operations of addition and scalar multiplication have the same geometric meaning as in $\mathbb{R}^2$ (Figure 1.4). In Chapters 4, 5, and 8, we will return to these geometrical interpretations for motivation.

1.1.5 Algebraic Properties

Because the components of vectors are real numbers, we should expect the vector operations we have defined to follow the usual rules of algebra.

Theorem 1.1.1: Properties of Vector Operations

If $\mathbf{x}$, $\mathbf{y}$, and $\mathbf{z}$ are any elements of $\mathbb{R}^n$ and k, k_1, and k_2 are real numbers, then

(1) $\mathbf{x} + \mathbf{y} = \mathbf{y} + \mathbf{x}$. [*Commutativity of Addition*]

(2) $\mathbf{x} + (\mathbf{y} + \mathbf{z}) = (\mathbf{x} + \mathbf{y}) + \mathbf{z}$. [*Associativity of Addition*]

(3) There is a **zero vector**, denoted by $\mathbf{0}$, such that $\mathbf{x} + \mathbf{0} = \mathbf{x} = \mathbf{0} + \mathbf{x}$ for every $\mathbf{x} \in \mathbb{R}^n$. [*Additive Identity*]

(4) For any vector $\mathbf{x}$, there exists a vector denoted by $-\mathbf{x}$ such that $\mathbf{x} + (-\mathbf{x}) = \mathbf{0}$, where $\mathbf{0}$ denotes the zero vector. [*Additive Inverse of* $\mathbf{x}$]

(5) $(k_1 k_2)\mathbf{x} = k_1(k_2 \mathbf{x})$. [*Associative Property of Scalar Multiplication*]

(6) $k(\mathbf{x} + \mathbf{y}) = k\mathbf{x} + k\mathbf{y}$. [*Distributivity of Scalar Multiplication over Vector Addition*]

(7) $(k_1 + k_2)\mathbf{x} = k_1\mathbf{x} + k_2\mathbf{x}$. [*Distributivity of Scalar Multiplication over Scalar Addition*]

(8) $1 \cdot \mathbf{x} = \mathbf{x}$. [*Identity Element for Scalar Multiplication*]

We will prove some of these properties and leave the rest as exercises.

Proof of (1)   Suppose that $\mathbf{x} = \begin{bmatrix} x_1 \\ x_2 \\ \vdots \\ x_n \end{bmatrix}$ and $\mathbf{y} = \begin{bmatrix} y_1 \\ y_2 \\ \vdots \\ y_n \end{bmatrix}$. Then

$$\mathbf{x} + \mathbf{y} = \begin{bmatrix} x_1 + y_1 \\ x_2 + y_2 \\ \vdots \\ x_n + y_n \end{bmatrix} = \begin{bmatrix} y_1 + x_1 \\ y_2 + x_2 \\ \vdots \\ y_n + x_n \end{bmatrix}$$

[because all components are real numbers and real numbers commute]

$$= \begin{bmatrix} y_1 \\ y_2 \\ \vdots \\ y_n \end{bmatrix} + \begin{bmatrix} x_1 \\ x_2 \\ \vdots \\ x_n \end{bmatrix} = \mathbf{y} + \mathbf{x}.$$

Proof of (3)   In $\mathbb{R}^n$, define $\mathbf{0} = \left. \begin{bmatrix} 0 \\ 0 \\ \vdots \\ 0 \end{bmatrix} \right\} n$ zeros. Then, for any $\mathbf{x} = \begin{bmatrix} x_1 \\ x_2 \\ \vdots \\ x_n \end{bmatrix} \in \mathbb{R}^n$,

$$\mathbf{x} + \mathbf{0} = \begin{bmatrix} x_1 + 0 \\ x_2 + 0 \\ \vdots \\ x_n + 0 \end{bmatrix} = \begin{bmatrix} x_1 \\ x_2 \\ \vdots \\ x_n \end{bmatrix} = \mathbf{x}.$$

Proof of (4)   Given any vector $\mathbf{x} = \begin{bmatrix} x_1 \\ x_2 \\ \vdots \\ x_n \end{bmatrix}$, define $-\mathbf{x}$ to be the vector

$\begin{bmatrix} -x_1 \\ -x_2 \\ \vdots \\ -x_n \end{bmatrix}$. It follows that $\mathbf{x} + (-\mathbf{x}) = \mathbf{0}$, the zero vector. This property

gives us another way to view vector subtraction: $\mathbf{x} - \mathbf{y} = \mathbf{x} + (-\mathbf{y}) = \mathbf{x} + (-1)\mathbf{y}$. In other words, subtraction can be interpreted as the addition of an additive inverse.

Note that in the proof of property (3), there are two distinct uses of the symbol zero: the scalar 0 and the zero vector $\mathbf{0}$.

Exercises 1.1

A.

1.   If $\mathbf{u} = \begin{bmatrix} 2 \\ 1 \\ 8 \\ 6 \end{bmatrix}$, $\mathbf{v} = \begin{bmatrix} 3 \\ -5 \\ 2 \\ 0 \end{bmatrix}$, and $\mathbf{w} = \begin{bmatrix} 6 \\ 6 \\ -6 \\ -6 \end{bmatrix}$, compute each of the following vectors.

   a. $\mathbf{u} + \mathbf{w}$ b. $5\mathbf{w}$ c. $\mathbf{v} - \mathbf{u}$ d. $3\mathbf{u} + 7\mathbf{v} - 2\mathbf{w}$ e. $\frac{1}{2}\mathbf{w} + \frac{3}{4}\mathbf{u}$
   f. $\mathbf{u} - \mathbf{w} - \mathbf{v}$ g. $-2\mathbf{u} + 3\mathbf{v} - 100\mathbf{w}$

2.   If $\mathbf{v}$ and $\mathbf{w}$ are vectors such that $2\mathbf{v} - \mathbf{w} = \mathbf{0}$, what is the relationship between the components of $\mathbf{v}$ and those of $\mathbf{w}$?

3.   Find three vectors $\mathbf{u}$, $\mathbf{v}$, and $\mathbf{w}$ in $\mathbb{R}^3$ such that $\mathbf{w} = 3\mathbf{u}$, $\mathbf{v} = 2\mathbf{u}$, and $2\mathbf{u} + 3\mathbf{v} + 4\mathbf{w} = \begin{bmatrix} 20 \\ 10 \\ -25 \end{bmatrix}$.

4.   a. Show that the vector equation $x\begin{bmatrix} 5 \\ 7 \end{bmatrix} + y\begin{bmatrix} 3 \\ -10 \end{bmatrix} = \begin{bmatrix} -16 \\ 77 \end{bmatrix}$
      represents two simultaneous linear equations in the two variables x and y.
      b. Solve the simultaneous equations from part (a) for x and y and substitute into the vector equation above to check your answer.

5.   Suppose that we associate with each person a vector in $\mathbb{R}^3$ having the following components: age, height, and weight. Would it make sense to add the vectors associated with two different persons? Would it make sense to multiply one of these vectors by a scalar?

6.   Each day, an investment analyst records the high and low values of the price of Microsoft stock. The analyst stores the data for a given week in two vectors in $\mathbb{R}^5$: $\mathbf{v}_H$, giving the daily high values, and $\mathbf{v}_L$, giving the daily low values. Find a vector expression that provides the average daily values of the price of Microsoft stock for the entire 5 day week.

B.

1.   Find nonzero real numbers a, b, and c such that

$$a\mathbf{u} + b(\mathbf{u} - \mathbf{v}) + c(\mathbf{u} + \mathbf{v}) = \mathbf{0}$$

for every pair of vectors $\mathbf{u}$ and $\mathbf{v}$ in $\mathbb{R}^2$.

2.   Prove property (2) of Theorem 1.1.1.

3.   Prove property (6) of Theorem 1.1.1.

4.   Let $\mathbf{x} = \begin{bmatrix} x_1 \\ x_2 \end{bmatrix}$. Define $\mathbf{x} \geq \mathbf{0}$ to mean $x_1 \geq 0$ and $x_2 \geq 0$. Define $\mathbf{x} \leq \mathbf{0}$ analogously. If $(x_1 + x_2)\mathbf{x} \geq \mathbf{0}$, what must be true of $\mathbf{x}$?

5.   Using the definition in the previous exercise, define $\mathbf{u} \geq \mathbf{v}$ to mean $\mathbf{u} - \mathbf{v} \geq \mathbf{0}$, where $\mathbf{u}$ and $\mathbf{v}$ are vectors having the same number of components. Consider the following vectors:

$$\mathbf{x} = \begin{bmatrix} 3 \\ 5 \\ -1 \end{bmatrix}, \mathbf{y} = \begin{bmatrix} 6 \\ 5 \\ 6 \end{bmatrix}, \mathbf{u} = \begin{bmatrix} 0 \\ 0 \\ -2 \end{bmatrix}, \mathbf{v} = \begin{bmatrix} 4 \\ 2 \\ 0 \end{bmatrix}.$$

a. Show that $\mathbf{x} \geq \mathbf{u}$.
b. Show that $\mathbf{v} \geq \mathbf{u}$.
c. Is there any relationship between $\mathbf{x}$ and $\mathbf{v}$?
d. Show that $\mathbf{y} \geq \mathbf{x}, \mathbf{y} \geq \mathbf{u}$, and $\mathbf{y} \geq \mathbf{v}$.

6.   Extending the definitions in Exercises B4 and B5 in the obvious way, prove the following results for vectors $\mathbf{x}, \mathbf{y}, \mathbf{z}$, and $\mathbf{w}$ in $\mathbb{R}^n$:
a. If $\mathbf{x} \leq \mathbf{y}$ and $\mathbf{y} \leq \mathbf{z}$, then $\mathbf{x} \leq \mathbf{z}$.
b. If $\mathbf{x} \leq \mathbf{y}$ and $\mathbf{z} \leq \mathbf{w}$, then $\mathbf{x} + \mathbf{z} \leq \mathbf{y} + \mathbf{w}$.
c. If $\mathbf{x} \leq \mathbf{y}$ and λ is a nonnegative real number, then $\lambda\mathbf{x} \leq \lambda\mathbf{y}$. If $\mathbf{x} \leq \mathbf{y}$ and μ is a negative real number, then $\mu\mathbf{x} \geq \mu\mathbf{y}$.

*A set of vectors, $S \subseteq \mathbb{R}^n$, is a **convex set** if $\mathbf{x} \in S$ and $\mathbf{y} \in S$ imply $t\mathbf{x} + (1 - t)\mathbf{y} \in S$ for all real numbers t such that $0 \leq t \leq 1$. Geometrically, a set in $\mathbb{R}^n$ is convex if, whenever it contains two points (vectors), it also contains the line segment joining them. The **convex hull** of a set of vectors, $\{\mathbf{v}_1, \mathbf{v}_2, \ldots, \mathbf{v}_k\}$, is the set of all vectors of the form $a_1\mathbf{v}_1 + a_2\mathbf{v}_2 + \cdots + a_k\mathbf{v}_k$, where $a_i \geq 0$, $i = 1, 2, \ldots, k$, and $a_1 + a_2 + \cdots + a_k = 1$. Use these definitions in Exercises C1 through C6.*

C.

1. Prove that the intersection of two convex sets S_1 and S_2 is either convex or empty.

2. Show by example that the union of two convex sets does not have to be convex.

3. Is the complement of a set $\{\mathbf{x}\}$ with a single vector a convex set? Explain your answer.

4. Does a convex set in $\mathbb{R}^3$ that contains the vectors
$\begin{bmatrix} 1 \\ 2 \\ 3 \end{bmatrix}, \begin{bmatrix} -1 \\ -3 \\ 4 \end{bmatrix}$, and $\begin{bmatrix} 0 \\ 1 \\ -7 \end{bmatrix}$ necessarily contain the vector
$\begin{bmatrix} 0 \\ 0 \\ 0 \end{bmatrix}$? Justify your answer.

5. Suppose that S is a finite subset of $\mathbb{R}^n$. Prove the following statements.
   a. S is a subset of its convex hull.
   b. The convex hull of S is the smallest convex set containing S. [*Hint*: Suppose that $\mathcal{H}$ is the convex hull of S and $S \subseteq T \subseteq \mathcal{H} \subseteq \mathbb{R}^n$, where T is a convex set. Then show $\mathcal{H} \subseteq T$.]

1.2 The Inner Product and Norm

In Example 1.1.4, we dealt with three quantities (prices), each of which had to be multiplied by the same number (a sales tax rate). There are situations in which each of several quantities must be multiplied by a different number. The next example shows how to handle such problems using vectors.

Example 1.2.1: A Product of Vectors

Suppose that the vector $[\,1235 \quad 985 \quad 1050 \quad 3460\,]^{\mathsf{T}}$ holds four prices expressed in Euros, British pounds, Australian dollars, and Mexican pesos, respectively. On a particular day, we know that 1 Euro $= \$1.46420$, 1 British pound $= \$1.83637$, 1 Australian dollar $= \$0.83580$, and 1 Mexican peso $= \$0.09294$. How can we use vectors to find the total of the four prices in U.S. dollars?

Arithmetically, we can calculate the total U.S. dollars as follows:

$$Total = 1235(1.46420) + 985(1.83637) + 1050(0.83580)$$
$$+ 3460(0.09294)$$
$$= \$4816.27 \text{ (rounding to the nearest cent at the end).}$$

We can interpret this answer as the result of combining two vectors—one holding the original prices and the other carrying the currency conversion rates—in a way that multiplies the vectors' corresponding components and then adds the resulting products:

$$\overbrace{\begin{bmatrix} 1235 \\ 985 \\ 1050 \\ 3460 \end{bmatrix}}^{\text{Price vector}} \cdot \overbrace{\begin{bmatrix} 1.46420 \\ 1.83637 \\ 0.83580 \\ 0.09294 \end{bmatrix}}^{\text{Conversion rate vector}}$$

$$= 1235(1.46420) + 985(1.83637) + 1050(0.83580)$$
$$+ 3460(0.09294)$$

$$= \overbrace{\$4816.27}^{\text{Total price (a scalar)}} \; .$$

A problem such as the one given in the last example naturally leads to a new vector operation.

Definition 1.2.1

If $\mathbf{x} = \begin{bmatrix} x_1 \\ x_2 \\ \vdots \\ x_n \end{bmatrix}$ and $\mathbf{y} = \begin{bmatrix} y_1 \\ y_2 \\ \vdots \\ y_n \end{bmatrix}$, then the **(Euclidean) inner product**

(or **dot product**) of $\mathbf{x}$ and $\mathbf{y}$, denoted $\mathbf{x} \cdot \mathbf{y}$, is defined as follows:

$$\mathbf{x} \cdot \mathbf{y} = x_1 y_1 + x_2 y_2 + \cdots + x_n y_n.$$

The dot product of two vectors is a *scalar* quantity.

In more sophisticated mathematical terms, we can describe the dot product as a function from the set $\mathbb{R}^n \times \mathbb{R}^n$ into the set $\mathbb{R}$. (A set such as $\mathbb{R}^n \times \mathbb{R}^n$ is called a *Cartesian product*. See Appendix A.)

Dot product notation is valuable in ways that have nothing to do with numerical calculations. For example, $-3x + 4y = 5$, the equation of a straight line in $\mathbb{R}^2$, can be written using the dot product: $\begin{bmatrix} -3 \\ 4 \end{bmatrix} \cdot \begin{bmatrix} x \\ y \end{bmatrix} = 5$.

More generally, any linear equation, $a_1 x_1 + a_2 x_2 + \cdots + a_n x_n = b$, can be represented as $\mathbf{a} \cdot \mathbf{x} = b$, where $\mathbf{a} = \begin{bmatrix} a_1 \\ a_2 \\ \vdots \\ a_n \end{bmatrix} \in \mathbb{R}^n$ is a **vector of**

coefficients, $\mathbf{x} = \begin{bmatrix} x_1 \\ x_2 \\ \vdots \\ x_n \end{bmatrix}$ is a **vector of variables** (or a **vector of unknowns**),

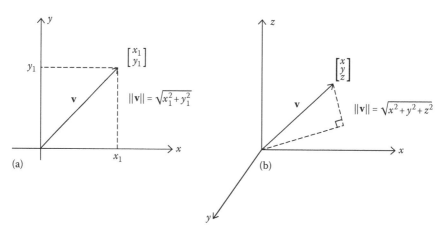

Figure 1.5 The norm of a vector in (a) $\mathbb{R}^2$ and (b) $\mathbb{R}^3$.

and b is a scalar. This representation sets the stage for important developments in Chapters 2 and 3.

Returning to the geometrical interpretation of vectors in Section 1.1, we see that if $\mathbf{v} = \begin{bmatrix} x \\ y \end{bmatrix}$ is a vector in $\mathbb{R}^2$, then its **length**, often called the **norm** of $\mathbf{v}$ and written as $\|\mathbf{v}\|$, corresponds to the length of the hypotenuse of a right triangle and is given by the Pythagorean theorem as $\sqrt{x^2 + y^2}$ (Figure 1.5a).

For the same reason, the length of $\mathbf{v} = \begin{bmatrix} x \\ y \\ z \end{bmatrix}$ in $\mathbb{R}^3$ is $\sqrt{x^2 + y^2 + z^2}$ (Figure 1.5b).

We see that these vector lengths are related to dot products:

$$\text{In } \mathbb{R}^2: \|\mathbf{v}\| = \left\| \begin{bmatrix} x \\ y \end{bmatrix} \right\| = \sqrt{x^2 + y^2} = \sqrt{\mathbf{v} \cdot \mathbf{v}}$$

$$\text{In } \mathbb{R}^3: \|\mathbf{v}\| = \left\| \begin{bmatrix} x \\ y \\ z \end{bmatrix} \right\| = \sqrt{x^2 + y^2 + z^2} = \sqrt{\mathbf{v} \cdot \mathbf{v}}$$

.

Example 1.2.2: Norms of Vectors in $\mathbb{R}^2$ and $\mathbb{R}^3$

If $\mathbf{v} = \begin{bmatrix} -2 \\ -3 \end{bmatrix}$, then $\|\mathbf{v}\| = \sqrt{(-2)^2 + (-3)^2} = \sqrt{\mathbf{v} \cdot \mathbf{v}} = \sqrt{13}$;

if $\mathbf{w} = \begin{bmatrix} 1 \\ -2 \\ 3 \end{bmatrix}$, then $\|\mathbf{w}\| = \sqrt{1^2 + (-2)^2 + 3^2} = \sqrt{\mathbf{w} \cdot \mathbf{w}} = \sqrt{14}$.

This definition of vector length makes perfect geometric sense in $\mathbb{R}^2$ and in $\mathbb{R}^3$, and we can generalize this idea to vectors in any space $\mathbb{R}^n$.

Definition 1.2.2

If $\mathbf{x} = \begin{bmatrix} x_1 \\ x_2 \\ \vdots \\ x_n \end{bmatrix}$ is an element of $\mathbb{R}^n$, then we define the (**Euclidean**)

norm (or **length**) of $\mathbf{x}$ as follows:

$$\| \mathbf{x} \| = \sqrt{x_1^2 + x_2^2 + \cdots + x_n^2} = \sqrt{\mathbf{x} \bullet \mathbf{x}}.$$

The norm is a *nonnegative scalar quantity*.

The (Euclidean) norm of a vector is a consistent and useful measure of the magnitude of a vector. Furthermore, expressions such as $\sqrt{x_1^2 + x_2^2 + \cdots + x_n^2}$ occur often in mathematics and its applications, and can be related to vector concepts. In the exercises following this section, we will see other ways to define a product of vectors and a norm of a vector, but Definitions 1.2.1 and 1.2.2 are the most commonly used because of their geometric interpretations in $\mathbb{R}^2$ and $\mathbb{R}^3$.

Example 1.2.3: Norms of Vectors in $\mathbb{R}^4$ and $\mathbb{R}^5$

For $\mathbf{v} = \begin{bmatrix} 1 \\ -2 \\ 3 \\ 4 \end{bmatrix} \in \mathbb{R}^4$, we have $\|\mathbf{v}\| = \sqrt{1^2 + (-2)^2 + 3^2 + 4^2} = \sqrt{30}$;

if $\mathbf{w} = \begin{bmatrix} 5 \\ -1 \\ 2 \\ 0 \\ 4 \end{bmatrix}$ in $\mathbb{R}^5$, then $\|\mathbf{w}\| = \sqrt{5^2 + (-1)^2 + 2^2 + 0^2 + 4^2} = \sqrt{46}$.

The inner product and the norm have some important algebraic properties that we state in the following theorem.

Theorem 1.2.1: Properties of the Inner Product and the Norm

If $\mathbf{v}$, $\mathbf{v}_1$, $\mathbf{v}_2$, and $\mathbf{v}_3$ are any elements of $\mathbb{R}^n$ and k is a real number, then

(a) $\mathbf{v}_1 \bullet \mathbf{v}_2 = \mathbf{v}_2 \bullet \mathbf{v}_1$
(b) $\mathbf{v}_1 \bullet (\mathbf{v}_2 + \mathbf{v}_3) = \mathbf{v}_1 \bullet \mathbf{v}_2 + \mathbf{v}_1 \bullet \mathbf{v}_3$ and $(\mathbf{v}_1 + \mathbf{v}_2) \bullet \mathbf{v}_3 = (\mathbf{v}_1 \bullet \mathbf{v}_3) + (\mathbf{v}_2 \bullet \mathbf{v}_3)$
(c) $(k\mathbf{v}_1) \bullet \mathbf{v}_2 = \mathbf{v}_1 \bullet (k\mathbf{v}_2) = k(\mathbf{v}_1 \bullet \mathbf{v}_2)$
(d) $\|k\mathbf{v}\| = |k| \|\mathbf{v}\|$

The proofs of these properties follow easily from the basic definitions and the algebraic properties of real numbers, and they appear as exercises at the end of this section.

1.2.1 The Angle between Vectors

Now let us get back to the geometry of $\mathbb{R}^2$, as outlined in Section 1.1. Given two vectors $\mathbf{u} = \begin{bmatrix} u_1 \\ u_2 \end{bmatrix} \neq \mathbf{0}$ and $\mathbf{v} = \begin{bmatrix} v_1 \\ v_2 \end{bmatrix} \neq \mathbf{0}$, we can focus on the angle θ between them (Figure 1.6), where the positive direction of measurement is counterclockwise.

First of all, $\theta = A - B$, so that a standard trigonometric formula gives us

$$\cos \theta = \cos (A - B)$$
$$= \cos A \cos B + \sin A \sin B$$
$$= \frac{v_1}{\|\mathbf{v}\|} \times \frac{u_1}{\|\mathbf{u}\|} + \frac{v_2}{\|\mathbf{v}\|} \times \frac{u_2}{\|\mathbf{u}\|}$$
$$= \frac{u_1 v_1 + u_2 v_2}{\|\mathbf{u}\| \|\mathbf{v}\|} = \frac{\mathbf{u} \cdot \mathbf{v}}{\|\mathbf{u}\| \|\mathbf{v}\|}.$$

Thus, if $\mathbf{u}, \mathbf{v} \in \mathbb{R}^2$ and θ is the angle between these vectors, we have $\cos \theta = \frac{\mathbf{u} \cdot \mathbf{v}}{\|\mathbf{u}\| \|\mathbf{v}\|}$. It follows that $-1 \leq \frac{\mathbf{u} \cdot \mathbf{v}}{\|\mathbf{u}\| \|\mathbf{v}\|} \leq 1$, or $|\mathbf{u} \cdot \mathbf{v}| \leq \|\mathbf{u}\| \|\mathbf{v}\|$.

This last inequality is a special case of the **Cauchy–Schwarz Inequality**,* whose validity for vectors in $\mathbb{R}^n$ for $n > 2$ will allow us to extend the concept of the angle between vectors.

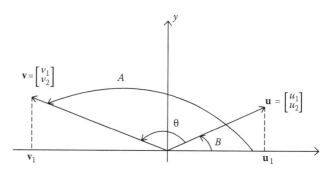

Figure 1.6 The angle between vectors in $\mathbb{R}^2$.

---

* This inequality is named for the French mathematician Augustin-Louis Cauchy (1789–1857) and the German mathematician Hermann Amandus Schwarz (1843–1921). This result is sometimes called the Cauchy–Bunyakovsky–Schwarz (CBS) Inequality, acknowledging an extension of Cauchy's work by the Russian V.J. Bunyakovsky (1804–1889).

Theorem 1.2.2: The Cauchy–Schwarz Inequality

If $\mathbf{x} = [x_1 \ x_2 \ \ldots \ x_n]^T$ and $\mathbf{y} = [y_1 \ y_2 \ \ldots \ y_n]^T$ are vectors in $\mathbb{R}^n$, then

$$\boxed{|\mathbf{x} \cdot \mathbf{y}| \leq \|\mathbf{x}\| \|\mathbf{y}\|}.$$

[In courses such as advanced analysis and statistics, the Cauchy–Schwarz inequality is often used in the form

$$|x_1 y_1 + x_2 y_2 + \cdots + x_n y_n| \leq (x_1^2 + x_2^2 + \cdots + x_n^2)^{1/2} (y_1^2 + y_2^2 + \cdots + y_n^2)^{1/2}.]$$

Proof We give a proof by induction on the dimension of $\mathbb{R}^n$ (see Appendix C). We have already established the inequality for $n = 2$. Now we assume the inequality is valid for some $n > 2$ and show that the result is true for $n + 1$.

Consider

$$\mathbf{x} = [x_1 \ x_2 \ \cdots \ x_n \ x_{n+1}]^T, \mathbf{y} = [y_1 \ y_2 \ \cdots \ y_n \ y_{n+1}]^T \in \mathbb{R}^{n+1}.$$

Then

$$\begin{aligned}
|\mathbf{x} \cdot \mathbf{y}| &= |x_1 y_1 + x_2 y_2 + \cdots + x_n y_n + x_{n+1} y_{n+1}| \\
&= |(x_1 y_1 + x_2 y_2 + \cdots + x_n y_n) + x_{n+1} y_{n+1}| \\
&\leq |x_1 y_1 + x_2 y_2 + \cdots + x_n y_n| + |x_{n+1} y_{n+1}| \\
&\leq \sqrt{x_1^2 + x_2^2 + \cdots + x_n^2} \sqrt{y_1^2 + y_2^2 + \cdots + y_n^2} + |x_{n+1} y_{n+1}|,
\end{aligned}$$

where we have used the inductive hypothesis that the inequality holds for vectors in $\mathbb{R}^n$.

If we apply the Cauchy–Schwarz inequality to the vectors $\mathbf{X} = \begin{bmatrix} \sqrt{x_1^2 + x_2^2 + \cdots + x_n^2} \\ |x_{n+1}| \end{bmatrix}$ and $\mathbf{Y} = \begin{bmatrix} \sqrt{y_1^2 + y_2^2 + \cdots + y_n^2} \\ |y_{n+1}| \end{bmatrix}$ in $\mathbb{R}^2$, we find that $|\mathbf{X} \cdot \mathbf{Y}| \leq \|\mathbf{X}\| \|\mathbf{Y}\| = \sqrt{x_1^2 + x_2^2 + \cdots + x_n^2 + x_{n+1}^2}$ $\sqrt{y_1^2 + y_2^2 + \cdots + y_n^2 + y_{n+1}^2}$. Thus

$$\begin{aligned}
|\mathbf{x} \cdot \mathbf{y}| &\leq \sqrt{x_1^2 + x_2^2 + \cdots + x_n^2} \sqrt{y_1^2 + y_2^2 + \cdots + y_n^2} + |x_{n+1} y_{n+1}| = |\mathbf{X} \cdot \mathbf{Y}| \\
&\leq \sqrt{x_1^2 + x_2^2 + \cdots + x_n^2 + x_{n+1}^2} \sqrt{y_1^2 + y_2^2 + \cdots + y_n^2 + y_{n+1}^2} = \|\mathbf{x}\| \|\mathbf{y}\|.
\end{aligned}$$

Thus the Cauchy–Schwarz inequality is valid in all Euclidean spaces $\mathbb{R}^n$ ($n \geq 2$).

Equality holds in the Cauchy–Schwarz inequality if and only if one of the vectors is a scalar multiple of the other: $\mathbf{x} = c\mathbf{y}$ or $\mathbf{y} = k\mathbf{x}$ for scalars c and k (Exercise B11).

As a consequence of the Cauchy–Schwarz inequality, the formula for the cosine of the angle between two vectors can be extended in a natural way to $\mathbb{R}^3$ and generalized—without the geometric visualization—to the case of any two vectors in $\mathbb{R}^n$. Because the graph of $y = \cos \theta$ for

$0 \leq \theta \leq \pi$ shows that for any real number $r \in [-1, 1]$, there is a unique real number θ such that $\cos \theta = r$, we see that there is a unique real number θ such that $\cos \theta = \frac{\mathbf{u} \cdot \mathbf{v}}{\|\mathbf{u}\| \|\mathbf{v}\|}, 0 \leq \theta \leq \pi$, for any nonzero vectors $\mathbf{u}$ and $\mathbf{v}$ in $\mathbb{R}^n$.

Definition 1.2.3

If $\mathbf{u}$ and $\mathbf{v}$ are nonzero elements of $\mathbb{R}^n$, then we define the **angle** θ **between u and v** as the unique angle between 0 and π rad inclusive, satisfying

$$\cos \theta = \frac{\mathbf{u} \cdot \mathbf{v}}{\|\mathbf{u}\| \|\mathbf{v}\|}.$$

Example 1.2.4: The Angle between Two Vectors in $\mathbb{R}^3$

Suppose we want to find the angle between the two vectors in $\mathbb{R}^3$

$$\mathbf{u} = \begin{bmatrix} 1 \\ \sqrt{3} \\ 2 \end{bmatrix} \quad \text{and} \quad \mathbf{v} = \begin{bmatrix} 2\sqrt{3} \\ 2 \\ \sqrt{3} \end{bmatrix}.$$

We have

$$\cos \theta = \frac{\mathbf{u} \cdot \mathbf{v}}{\|\mathbf{u}\| \|\mathbf{v}\|} = \frac{\begin{bmatrix} 1 \\ \sqrt{3} \\ 2 \end{bmatrix} \cdot \begin{bmatrix} 2\sqrt{3} \\ 2 \\ \sqrt{3} \end{bmatrix}}{\left\| \begin{bmatrix} 1 \\ \sqrt{3} \\ 2 \end{bmatrix} \right\| \left\| \begin{bmatrix} 2\sqrt{3} \\ 2 \\ \sqrt{3} \end{bmatrix} \right\|}$$

$$= \frac{2\sqrt{3} + 2\sqrt{3} + 2\sqrt{3}}{\sqrt{1^2 + (\sqrt{3})^2 + 2^2} \sqrt{(2\sqrt{3})^2 + 2^2 + (\sqrt{3})^2}}$$

$$= \frac{6\sqrt{3}}{\sqrt{8} \cdot \sqrt{19}} = \frac{3\sqrt{3}}{\sqrt{38}} = \frac{3\sqrt{114}}{38},$$

and $\theta = \arccos\left(\frac{3\sqrt{114}}{38}\right) \approx 0.5681$ rad or $32.55°$.

We can write the formula in Definition 1.2.3 as

$$\boxed{\mathbf{u} \cdot \mathbf{v} = \|\mathbf{u}\| \|\mathbf{v}\| \cos \theta}.$$

If vectors $\mathbf{u}$ and $\mathbf{v}$ in $\mathbb{R}^2$ or $\mathbb{R}^3$ are *perpendicular* to each other, then $\theta = \pi/2$ rad and $\cos \theta = 0$, so $\mathbf{u} \cdot \mathbf{v} = 0$. On the other hand, if $\mathbf{u}$ and $\mathbf{v}$ are nonzero, it is obvious that $\|\mathbf{u}\| > 0$ and $\|\mathbf{v}\| > 0$, so the only way for $\mathbf{u} \cdot \mathbf{v}$ to equal 0 is if $\cos \theta = 0$, that is, if $\theta = \pi/2$ and $\mathbf{u}$ and $\mathbf{v}$ are therefore perpendicular. (Definition 1.2.3 requires $0 \leq \theta \leq \pi$.) In later discussions, a generalization of the concept of perpendicularity will be important, so we make a formal definition.

Definition 1.2.4

Two vectors $\mathbf{u}$ and $\mathbf{v}$ in $\mathbb{R}^n$ are called **orthogonal*** (or **perpendicular** if $n = 2$, 3) if $\mathbf{u} \cdot \mathbf{v} = 0$. In this case, we write $\mathbf{u} \perp \mathbf{v}$ (pronounced "u perp v").

Example 1.2.5: Orthogonal Vectors in $\mathbb{R}^5$

In $\mathbb{R}^5$, if $\mathbf{u} = \begin{bmatrix} 1 \\ -2 \\ 3 \\ -4 \\ 5 \end{bmatrix}$ and $\mathbf{v} = \begin{bmatrix} 10 \\ -4 \\ 1 \\ -1 \\ -5 \end{bmatrix}$, then $\mathbf{u} \perp \mathbf{v}$ because

$$\begin{bmatrix} 1 \\ -2 \\ 3 \\ -4 \\ 5 \end{bmatrix} \cdot \begin{bmatrix} 10 \\ -4 \\ 1 \\ -1 \\ -5 \end{bmatrix} = 10 + 8 + 3 + 4 - 25 = 0.$$

Vector algebra is often used to prove theorems in geometry.

Example 1.2.6: A Geometry Theorem

Let us prove that *the diagonals of a rhombus are perpendicular.* We recall that a rhombus is a parallelogram with all four sides having the same length (Figure 1.7).

We may assume that the rhombus is the figure OABC, positioned so that its lower left-hand vertex is at the origin and its

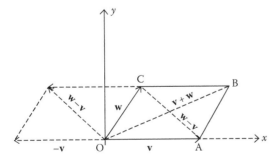

Figure 1.7 A rhombus.

---

* The word *orthogonal* comes from Greek words meaning "straight" and "angle."

bottom side lies along the x-axis. Let $\mathbf{v}$ denote the vector from O to A and let $\mathbf{w}$ denote the vector from the origin to C.

Then the diagonal $\overline{OB}$ represents $\mathbf{v} + \mathbf{w}$ (by the Parallelogram Law). Because the side $\overline{AB}$ has the same direction and length as $\mathbf{w}$, this side can also represent the vector $\mathbf{w}$.

We can see that the diagonal $\overline{AC}$ has the same magnitude and direction as $\mathbf{w} - \mathbf{v} = \mathbf{w} + (-\mathbf{v})$ (again, by the Parallelogram Law). Then, using Definition 1.2.2 and Theorem 1.2.1,

$$(\mathbf{w} - \mathbf{v}) \bullet (\mathbf{w} + \mathbf{v}) = (\mathbf{w} - \mathbf{v}) \bullet \mathbf{w} + (\mathbf{w} - \mathbf{v}) \bullet \mathbf{v}$$

$$= \mathbf{w} \bullet \mathbf{w} - \mathbf{v} \bullet \mathbf{w} + \mathbf{w} \bullet \mathbf{v} - \mathbf{v} \bullet \mathbf{v}$$

$$= \|\mathbf{w}\|^2 - \|\mathbf{v}\|^2.$$

Because all sides of a rhombus are equal, it follows that $\|\mathbf{w}\| = \|\mathbf{v}\|$ and so $(\mathbf{w} - \mathbf{v}) \bullet (\mathbf{w} + \mathbf{v}) = 0$. Thus vectors $\mathbf{w} - \mathbf{v}$ and $\mathbf{w} + \mathbf{v}$ are orthogonal, indicating that diagonals $\overline{OB}$ and $\overline{AC}$ must be perpendicular, which is what we wanted to prove.

Exercises 1.2

A.

1. Let $\mathbf{u} = \begin{bmatrix} 2 \\ -1 \\ 2 \end{bmatrix}$, $\mathbf{v} = \begin{bmatrix} 5 \\ 0 \\ 2 \end{bmatrix}$, $\mathbf{x} = \begin{bmatrix} -7 \\ 1 \\ 2 \end{bmatrix}$, and $\mathbf{y} = \begin{bmatrix} 0 \\ 8 \\ 3 \end{bmatrix}$.

    Compute the following.
    a. $(\mathbf{u} + \mathbf{v}) \bullet (\mathbf{x} + \mathbf{y})$ b. $((3\mathbf{u} \bullet \mathbf{x})\mathbf{v}) \bullet \mathbf{y}$ c. $\mathbf{u} \bullet \mathbf{x} - 4\mathbf{v} \bullet \mathbf{y}$
    d. $\mathbf{u} \bullet \mathbf{x} + 3\mathbf{u} \bullet \mathbf{y} - \mathbf{v} \bullet \mathbf{y}$ e. $(2(\mathbf{v} + \mathbf{u}) \bullet \mathbf{y}) - 5\mathbf{u} \bullet \mathbf{y}$
    f. $4\mathbf{u} \bullet \mathbf{x} + 6[\mathbf{v} \bullet (3\mathbf{x} - \mathbf{y})]$

2. If $\mathbf{v}_1$, $\mathbf{v}_2$, and $\mathbf{v}_3$ are in $\mathbb{R}^n$, does the "law"

    $$\mathbf{v}_1 \bullet (\mathbf{v}_2 \bullet \mathbf{v}_3) = (\mathbf{v}_1 \bullet \mathbf{v}_2) \bullet \mathbf{v}_3$$

    make sense? Explain.

3. Calculate $\|\mathbf{u}\|$ for each of the following vectors.

    a. $\mathbf{u} = \begin{bmatrix} -3 \\ 5 \end{bmatrix}$ b. $\mathbf{u} = \begin{bmatrix} 1 \\ 0 \\ -7 \end{bmatrix}$ c. $\mathbf{u} = \begin{bmatrix} 2 \\ 2 \\ 2 \\ 2 \end{bmatrix}$

    d. $\mathbf{u} = \begin{bmatrix} 1 \\ -1 \\ 2 \\ -2 \\ 3 \end{bmatrix}$ e. $\mathbf{u} = \begin{bmatrix} 1 \\ 2 \\ 3 \\ 4 \\ 5 \\ 6 \end{bmatrix}$ f. $\mathbf{u} = \begin{bmatrix} -2 \\ 3 \\ 1 \\ \sqrt{10} \\ 4 \\ -3 \end{bmatrix}$

4. What is the length of the vector $\begin{bmatrix} \cos a \cos b \\ \cos a \sin b \\ \sin a \end{bmatrix}$, where a and b are any real numbers? (Your answer should be independent of a and b.)

5. For each pair of vectors $\mathbf{u}$, $\mathbf{v}$, calculate the angle between $\mathbf{u}$ and $\mathbf{v}$. Express your answer in both radians and degrees.

a. $\mathbf{u} = \begin{bmatrix} 3 \\ 4 \end{bmatrix}$, $\mathbf{v} = \begin{bmatrix} 5 \\ 12 \end{bmatrix}$ b. $\mathbf{u} = \begin{bmatrix} 4 \\ 6 \end{bmatrix}$, $\mathbf{v} = \begin{bmatrix} -3 \\ 2 \end{bmatrix}$

c. $\mathbf{u} = \begin{bmatrix} 2 \\ 2 \\ -1 \end{bmatrix}$, $\mathbf{v} = \begin{bmatrix} 5 \\ -4 \\ 2 \end{bmatrix}$ d. $\mathbf{u} = \begin{bmatrix} 2 \\ 2 \\ -1 \end{bmatrix}$, $\mathbf{v} = \begin{bmatrix} 5 \\ -3 \\ 2 \end{bmatrix}$

e. $\mathbf{u} = \begin{bmatrix} 0 \\ 1 \\ 1 \end{bmatrix}$, $\mathbf{v} = \begin{bmatrix} 1 \\ 2 \\ -3 \end{bmatrix}$ f. $\mathbf{u} = \begin{bmatrix} -1 \\ 2 \\ 5 \end{bmatrix}$, $\mathbf{v} = \begin{bmatrix} 3 \\ 4 \\ -1 \end{bmatrix}$

g. $\mathbf{u} = \begin{bmatrix} 1 \\ 2 \\ 3 \\ 4 \end{bmatrix}$, $\mathbf{v} = \begin{bmatrix} -2 \\ 1 \\ -3 \\ 4 \end{bmatrix}$ h. $\mathbf{u} = \begin{bmatrix} 2 \\ 0 \\ 1 \\ 0 \\ 3 \end{bmatrix}$, $\mathbf{v} = \begin{bmatrix} -1 \\ 4 \\ -2 \\ 3 \\ 0 \end{bmatrix}$

6. The angle between the vectors $\begin{bmatrix} 1 \\ 7 \\ b \end{bmatrix}$ and $\begin{bmatrix} -2 \\ 2 \\ 1 \end{bmatrix}$ is given by $\arccos(1/3)$. Find b.

7. Suppose that $\mathbf{u}$ and $\mathbf{v}$ are nonzero vectors in $\mathbb{R}^n$ and α, β are positive real numbers. Show that the angle between $\alpha\mathbf{u}$ and $\beta\mathbf{v}$ is the same as the angle between $\mathbf{u}$ and $\mathbf{v}$.

8. Find the angles between the sides and a diagonal of a 1×2 rectangle.

9. Suppose that $\mathbf{u}, \mathbf{v} \in \mathbb{R}^n$. Find a real number α so that $\mathbf{u} - \alpha\mathbf{v}$ is orthogonal to $\mathbf{v}$, given that $\|\mathbf{v}\| = 1$.

10. Given two vectors $\mathbf{u} = \begin{bmatrix} u_1 \\ u_2 \\ u_3 \end{bmatrix}$ and $\mathbf{v} = \begin{bmatrix} v_1 \\ v_2 \\ v_3 \end{bmatrix}$ in $\mathbb{R}^3$, the **cross product** (also called the **vector product**), $\mathbf{u} \times \mathbf{v}$, is the vector defined as

$$\mathbf{u} \times \mathbf{v} = \begin{bmatrix} u_2 v_3 - u_3 v_2 \\ u_3 v_1 - u_1 v_3 \\ u_1 v_2 - u_2 v_1 \end{bmatrix}^*.$$

---

* The cross product is a useful concept in many physics and engineering applications—for example, in discussing *torque*. See, for example, D. Zill and M. Cullen, *Advanced Engineering Mathematics*, 7th edn. (Sudbury, MA: Jones and Bartlett, 2000). In contrast to the dot product, the cross product of two vectors in $\mathbb{R}^3$ results in a vector.

Calculate $\mathbf{u} \times \mathbf{v}$ in each of the following cases.

a. $\mathbf{u} = \begin{bmatrix} 1 \\ 2 \\ 3 \end{bmatrix}$ and $\mathbf{v} = \begin{bmatrix} -1 \\ 2 \\ -3 \end{bmatrix}$

b. $\mathbf{u} = \begin{bmatrix} 2 \\ -1 \\ 3 \end{bmatrix}$ and $\mathbf{v} = \begin{bmatrix} 1 \\ -2 \\ -1 \end{bmatrix}$

c. $\mathbf{u} = \begin{bmatrix} -1 \\ 2 \\ -3 \end{bmatrix}$ and $\mathbf{v} = \begin{bmatrix} 1 \\ 2 \\ 3 \end{bmatrix}$

d. $\mathbf{u} = \begin{bmatrix} 1 \\ 2 \\ 3 \end{bmatrix}$ and $\mathbf{v} = \begin{bmatrix} 1 \\ 2 \\ 3 \end{bmatrix}$

11. Let $\mathbf{i} = \begin{bmatrix} 1 \\ 0 \\ 0 \end{bmatrix}$, $\mathbf{j} = \begin{bmatrix} 0 \\ 1 \\ 0 \end{bmatrix}$, and $\mathbf{k} = \begin{bmatrix} 0 \\ 0 \\ 1 \end{bmatrix}$. Using the definition of the *cross product* given in Exercise A10, fill in the entries of the following "multiplication table."

$\times$	$\mathbf{i}$	$\mathbf{j}$	$\mathbf{k}$
$\mathbf{i}$			
$\mathbf{j}$			
$\mathbf{k}$			

B.

1. Prove property (a) of Theorem 1.2.1 for vectors in $\mathbb{R}^3$.

2. Prove property (b) of Theorem 1.2.1 for vectors in $\mathbb{R}^3$.

3. Prove property (c) of Theorem 1.2.1 for vectors in $\mathbb{R}^3$.

4. Prove property (d) of Theorem 1.2.1 for vectors in $\mathbb{R}^3$.

5. For what values of a are the vectors $\begin{bmatrix} -6 \\ a \\ 2 \end{bmatrix}$ and $\begin{bmatrix} a \\ a^2 \\ a \end{bmatrix}$ orthogonal?

6. For what values of a is the angle between the vectors $\begin{bmatrix} 1 \\ 2 \\ 1 \end{bmatrix}$ and $\begin{bmatrix} 1 \\ 0 \\ a \end{bmatrix}$ equal to $60°$?

7. Let $\mathbf{v}$ and $\mathbf{w}$ be vectors of length 1 along the sides of a given angle. Show that the vector $\mathbf{v} + \mathbf{w}$ bisects the angle.

8. If a is a positive real number, find the cosines of the angles between the vector $\mathbf{v} = \begin{bmatrix} a \\ a \\ \vdots \\ a \end{bmatrix}$ and the vectors

$\mathbf{e}_1 = \begin{bmatrix} 1 \\ 0 \\ \vdots \\ 0 \end{bmatrix}, \mathbf{e}_2 = \begin{bmatrix} 0 \\ 1 \\ \vdots \\ 0 \end{bmatrix}, \ldots, \mathbf{e}_n = \begin{bmatrix} 0 \\ 0 \\ \vdots \\ 1 \end{bmatrix}$ in $\mathbb{R}^n$. What are the actual angles in $\mathbb{R}^2$ and $\mathbb{R}^3$?

9. In $\mathbb{R}^3$, find the angle between a diagonal of a cube and one of its edges.

10. Find and prove a formula for $\|\mathbf{u}\|$, where

    a. $\mathbf{u} = \begin{bmatrix} \sqrt{1} \\ \sqrt{2} \\ \sqrt{3} \\ \vdots \\ \sqrt{n} \end{bmatrix} \in \mathbb{R}^n$ b. $\mathbf{u} = \begin{bmatrix} 1 \\ 2 \\ 3 \\ \vdots \\ n \end{bmatrix} \in \mathbb{R}^n$.

    (You may want to consult the discussion of mathematical induction in Appendix C.)

11. Prove that equality holds in the Cauchy–Schwarz inequality (i.e., $|\mathbf{x} \bullet \mathbf{y}| = \|\mathbf{x}\| \|\mathbf{y}\|$) if and only if $\mathbf{x} = c\mathbf{y}$ or $\mathbf{y} = k\mathbf{x}$, where c and k are scalars.

12. Under what conditions on two vectors $\mathbf{u}$ and $\mathbf{v}$ is it true that

    $$\|\mathbf{u} + \mathbf{v}\|^2 = \|\mathbf{u}\|^2 + \|\mathbf{v}\|^2?$$

    Justify your answer and interpret the equation geometrically in $\mathbb{R}^2$ and $\mathbb{R}^3$.

13. If we know that $\mathbf{u} \bullet \mathbf{v} = \mathbf{u} \bullet \mathbf{w}$ for vectors $\mathbf{u}, \mathbf{v}, \mathbf{w} \in \mathbb{R}^n$, does it follow that $\mathbf{v} = \mathbf{w}$? If it does, give a proof. Otherwise, find a set of vectors $\mathbf{u}, \mathbf{v}, \mathbf{w}$ in some space $\mathbb{R}^n$, for which $\mathbf{u} \bullet \mathbf{v} = \mathbf{u} \bullet \mathbf{w}$ but $\mathbf{v} \neq \mathbf{w}$.

14. Prove that if $\mathbf{u}$ is orthogonal to both $\mathbf{v}$ and $\mathbf{w}$, then $\mathbf{u}$ is orthogonal to $a\mathbf{v} + b\mathbf{w}$ for all scalars a and b.

15. Prove that $\|\mathbf{u} + \mathbf{v}\| = \|\mathbf{u} - \mathbf{v}\|$ if and only if $\mathbf{u}$ and $\mathbf{v}$ are orthogonal.

16. Suppose that $\mathbf{u}$, $\mathbf{v}$, $\mathbf{w}$, and $\mathbf{x}$ are vectors in $\mathbb{R}^n$ such that $\mathbf{u} + \mathbf{v} + \mathbf{w} + \mathbf{x} = \mathbf{0}$. Prove that $\mathbf{u} + \mathbf{v}$ is orthogonal to $\mathbf{u} + \mathbf{x}$ if and only if

$$\|\mathbf{u}\|^2 + \|\mathbf{w}\|^2 = \|\mathbf{v}\|^2 + \|\mathbf{x}\|^2.*$$

[*Hint:* Show that $\mathbf{w} \bullet \mathbf{w} = \|\mathbf{v}\|^2 + \|\mathbf{x}\|^2 - \|\mathbf{u}\|^2 + 2(\mathbf{u} \bullet \mathbf{u} + \mathbf{u} \bullet \mathbf{v} + \mathbf{u} \bullet \mathbf{x} + \mathbf{x} \bullet \mathbf{v}).$]

17. If $\mathbf{u} = \begin{bmatrix} u_1 \\ u_2 \\ \vdots \\ u_n \end{bmatrix} \in \mathbb{R}^n$, define $\mathbf{u} \geq \mathbf{0}$ to mean that $u_i \geq 0$ for $i = 1, 2, \ldots, n$; also define $\mathbf{u} \geq \mathbf{v}$ to mean $\mathbf{u} - \mathbf{v} \geq \mathbf{0}$, where $\mathbf{u}, \mathbf{v} \in \mathbb{R}^n$. (See Exercises 1.1, B4 through B6.) Now let $\mathbf{x}, \mathbf{y}$, and $\mathbf{z}$ be vectors in $\mathbb{R}^n$, with $\mathbf{x} > \mathbf{0}$, $\mathbf{y} \geq \mathbf{z}$, and $\mathbf{y} \neq \mathbf{z}$. Prove that $\mathbf{x} \bullet \mathbf{y} > \mathbf{x} \bullet \mathbf{z}$.

18. Define the **distance** d between two vectors $\mathbf{u}$ and $\mathbf{v}$ in $\mathbb{R}^n$ as

$$d(\mathbf{u},\mathbf{v}) = \|\mathbf{u} - \mathbf{v}\|.$$

Prove that
a. $d(\mathbf{u},\mathbf{v}) \geq 0$.
b. $d(\mathbf{u},\mathbf{v}) = 0$ if and only if $\mathbf{u} = \mathbf{v}$.
c. $d(\mathbf{u},\mathbf{v}) = d(\mathbf{v},\mathbf{u})$.

19. Using the definition of the *cross product* given in Exercise A10, prove that
a. $\mathbf{u} \times \mathbf{v} = -(\mathbf{v} \times \mathbf{u})$ b. $\mathbf{u} \times (\mathbf{v} + \mathbf{w}) = \mathbf{u} \times \mathbf{v} + \mathbf{u} \times \mathbf{w}$.
c. $k(\mathbf{u} \times \mathbf{v}) = (k\mathbf{u}) \times \mathbf{v} = \mathbf{u} \times (k\mathbf{v})$, where k is a scalar.
d. $\mathbf{u} \times \mathbf{u} = 0$ for every $\mathbf{u} \in \mathbb{R}^3$.

20. Using the definition of the *cross product* given in Exercise A10, prove that $\mathbf{u} \times (\mathbf{v} \times \mathbf{w}) = (\mathbf{u} \bullet \mathbf{w})\mathbf{v} - (\mathbf{u} \bullet \mathbf{v})\mathbf{w}$.

21. Using the definition of the *cross product* given in Exercise A10, prove that $\mathbf{u} \times \mathbf{v}$ is orthogonal to both $\mathbf{u}$ and $\mathbf{v}$.

22. Using the definition of the *cross product* given in Exercise A10, prove *Lagrange's identity*: $\|\mathbf{u} \times \mathbf{v}\|^2 = \|\mathbf{u}\|^2\|\mathbf{v}\|^2 - (\mathbf{u} \bullet \mathbf{v})^2$.

---

* This result has geometric significance in $\mathbb{R}^2$. I am indebted to my colleague, Mahmoud Sayrafiezadeh, for calling it to my attention.

C.

1. In 1968, Shmuel Winograd published* a more efficient method to calculate inner products of vectors in $\mathbb{R}^n$. Efficiency in this situation means fewer multiplications in certain types of problems. If x is a real number, the notation $[[x]]$ denotes the largest integer less than or equal to x, for example, $[[\pi]] = 3$ and $[[-\pi]] = -4$. For two vectors

$$\mathbf{x} = \begin{bmatrix} x_1 \\ x_2 \\ \vdots \\ x_n \end{bmatrix} \text{ and } \mathbf{y} = \begin{bmatrix} y_1 \\ y_2 \\ \vdots \\ y_n \end{bmatrix} \text{ in } \mathbb{R}^n,$$

$$\mathbf{x} \bullet \mathbf{y} = \begin{cases} \sum_{j=1}^{[[n/2]]} (x_{2j-1} + y_{2j})(x_{2j} + y_{2j-1}) - \alpha - \beta, \text{ for } n \text{ even,} \\ \sum_{j=1}^{[[n/2]]} (x_{2j-1} + y_{2j})(x_{2j} + y_{2j-1}) - \alpha - \beta + x_n y_n, \text{ for } n \text{ odd,} \end{cases}$$

where $\alpha = \sum_{j=1}^{[[n/2]]} x_{2j-1} x_{2j}, \quad \beta = \sum_{j=1}^{[[n/2]]} y_{2j-1} y_{2j}.$

    The cleverness of this method lies in the fact that if you have to compute the inner products of a vector $\mathbf{x}$ with several other vectors, the calculation of α (or β) in Winograd's formula can be done just once (and stored) instead of using the vector $\mathbf{x}$ each time.

    Suppose we want to calculate each of the 16 inner products, $\mathbf{u}_i \bullet \mathbf{v}_j$, for $i, j = 1, 2, 3, 4$, where each $\mathbf{u}_i$ and $\mathbf{v}_j$ is a vector in $\mathbb{R}^4$.

    a. If you use Definition 1.2.1, how many multiplications and how many additions/subtractions are needed to calculate the 16 inner products?

    b. If you use Winograd's formula, how many multiplications and how many additions/subtractions will you need to calculate the 16 inner products? (Remember that you do not have to repeat calculations for α or β if you have already done them.)

    Compare your answers with those in part (a).

2. Prove the **Cauchy–Schwarz Inequality** (Theorem 1.2.2) for vectors $\mathbf{x}$ and $\mathbf{y}$ in $\mathbb{R}^n$ as follows:

    a. Assume $\mathbf{y} \neq \mathbf{0}$ and let t be an arbitrary parameter (a real number). Show that $0 \leq \|\mathbf{x} - t\mathbf{y}\|^2 = \|\mathbf{x}\|^2 - 2t(\mathbf{x} \bullet \mathbf{y}) + t^2 \|\mathbf{y}\|^2$.

    b. Let $t = (\mathbf{x} \bullet \mathbf{y})/\|\mathbf{y}\|^2$ and show that $0 \leq \|\mathbf{x}\|^2 - \frac{|\mathbf{x} \bullet \mathbf{y}|^2}{\|\mathbf{y}\|^2}$.

    c. Use the result of part (b) to establish that $|\mathbf{x} \bullet \mathbf{y}| \leq \|\mathbf{x}\| \, \|\mathbf{y}\|$.

---

* S. Winograd, A new algorithm for inner product, *IEEE Trans. Comput.* **17** (1968): 693–694.

3. Prove the **Triangle Inequality** for vectors **u** and **v** in $\mathbb{R}^n$:

$$\|\mathbf{u} + \mathbf{v}\| \leq \|\mathbf{u}\| + \|\mathbf{v}\|.$$

[*Hint*: Consider $\|\mathbf{u} + \mathbf{v}\|^2$ and use the Cauchy–Schwarz inequality.]

4. Prove that $\|\mathbf{u} - \mathbf{v}\| \geq \|\mathbf{u}\| - \|\mathbf{v}\|$ for all vectors **u** and **v** in $\mathbb{R}^n$. [*Hint*: Use the result of Exercise C3.]

5. Prove the **Parallelogram Law** for vectors **u** and **v** in $\mathbb{R}^n$:

$$\|\mathbf{u} + \mathbf{v}\|^2 + \|\mathbf{u} - \mathbf{v}\|^2 = 2\|\mathbf{u}\|^2 + 2\|\mathbf{v}\|^2.$$

6. Prove the **Polarization Identity** for vectors **u** and **v** in $\mathbb{R}^n$:

$$\mathbf{u} \bullet \mathbf{v} = \frac{1}{4}\|\mathbf{u} + \mathbf{v}\|^2 - \frac{1}{4}\|\mathbf{u} - \mathbf{v}\|^2 .$$

*If **u** and **v** are vectors in $\mathbb{R}^n$ and $\mathbf{u} \neq \mathbf{0}$, then the **orthogonal projection of **v** onto **u** is the vector $\mathbf{proj}_\mathbf{u}(\mathbf{v})$ defined by $\mathbf{p} = \mathbf{proj}_\mathbf{u}(\mathbf{v}) = \left(\frac{\mathbf{u} \bullet \mathbf{v}}{\mathbf{u} \bullet \mathbf{u}}\right) \cdot \mathbf{u}$. Here's an illustration of the orthogonal projection of a vector **v** onto another vector **u** in $\mathbb{R}^2$:*

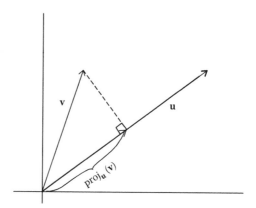

7. Prove that **u** is orthogonal to $\mathbf{v} - \mathbf{proj}_\mathbf{u}(\mathbf{v})$ for all vectors **u** and **v** in $\mathbb{R}^n$, where $\mathbf{u} \neq \mathbf{0}$.

8. Prove that $\mathbf{proj}_\mathbf{u}(\mathbf{v} - \mathbf{proj}_\mathbf{u}(\mathbf{v})) = \mathbf{0}$.

9. Prove that $\mathbf{proj}_\mathbf{u}(\mathbf{proj}_\mathbf{u}(\mathbf{v})) = \mathbf{proj}_\mathbf{u}(\mathbf{v})$.

10. Define the **sum norm**, or ℓ_1 **norm**, of a vector $\mathbf{u} = \begin{bmatrix} u_1 \\ u_2 \\ \vdots \\ u_n \end{bmatrix}$ as follows:

$$\|\mathbf{u}\|_1 = |u_1| + |u_2| + \cdots + |u_n|.$$

Assuming the *Triangle Inequality* for real numbers x and y, $|x + y| \leq |x| + |y|$, prove that $\|\mathbf{u} + \mathbf{v}\|_1 \leq \|\mathbf{u}\|_1 + \|\mathbf{v}\|_1$.

11. Define the **max norm**, or ℓ_∞ **norm**, of a vector $\mathbf{u} = \begin{bmatrix} u_1 \\ u_2 \\ \vdots \\ u_n \end{bmatrix}$ as

follows:

$$\|\mathbf{u}\|_\infty = \max\{|u_1|, |u_2|, \ldots, |u_n|\}$$
$$= \text{the largest of the numbers } |u_i|.$$

Assuming the Triangle Inequality (see Exercise C3) for real numbers, prove that

$$\|\mathbf{u} + \mathbf{v}\|_\infty \leq \|\mathbf{u}\|_\infty + \|\mathbf{v}\|_\infty.$$

1.3 Spanning Sets

Given a set of vectors in some Euclidean space $\mathbb{R}^n$, we know how to combine these vectors using the operations of vector addition and scalar multiplication. The set of all vectors resulting from such combinations is important in both theory and applications.

Definition 1.3.1

Given a nonempty finite set of vectors $S = \{\mathbf{v}_1, \mathbf{v}_2, \ldots, \mathbf{v}_k\}$ in $\mathbb{R}^n$, a **linear combination** of these vectors is any vector of the form $a_1\mathbf{v}_1 + a_2\mathbf{v}_2 + \cdots + a_k\mathbf{v}_k$, where $a_1, a_2, \ldots, a_k$ are scalars.

In each space $\mathbb{R}^n$, there are special sets of vectors that play an important role in describing the space. For example, in $\mathbb{R}^2$ the vectors $\mathbf{e}_1 = \begin{bmatrix} 1 \\ 0 \end{bmatrix}$ and $\mathbf{e}_2 = \begin{bmatrix} 0 \\ 1 \end{bmatrix}$ have the significant property that *any* vector $\mathbf{v} = \begin{bmatrix} x \\ y \end{bmatrix}$ can be written as a linear combination of vectors $\mathbf{e}_1$ and $\mathbf{e}_2$: $\mathbf{v} = \begin{bmatrix} x \\ y \end{bmatrix} = x\mathbf{e}_1 + y\mathbf{e}_2$. In physics and engineering, the symbols $\mathbf{i}$ and $\mathbf{j}$ (or $\vec{\mathbf{i}}$ and $\vec{\mathbf{j}}$) are often used for $\mathbf{e}_1$ and $\mathbf{e}_2$, respectively. If we extend (stretch) the vectors $\mathbf{e}_1$ and $\mathbf{e}_2$ using multiplication by positive and negative scalars, we get the usual x- and y-axes in the Euclidean plane (Figure 1.8).

Similarly, any vector $\mathbf{w} = \begin{bmatrix} x \\ y \\ z \end{bmatrix}$ in $\mathbb{R}^3$ can be expressed uniquely in terms of the vectors $\mathbf{e}_1 = \begin{bmatrix} 1 \\ 0 \\ 0 \end{bmatrix}$, $\mathbf{e}_2 = \begin{bmatrix} 0 \\ 1 \\ 0 \end{bmatrix}$, and $\mathbf{e}_3 = \begin{bmatrix} 0 \\ 0 \\ 1 \end{bmatrix}$:

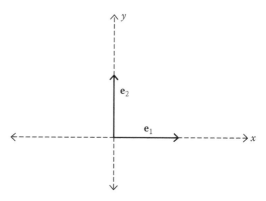

Figure 1.8 The vectors $\mathbf{e_1}$ and $\mathbf{e_2}$ in $\mathbb{R}^2$.

$\mathbf{w} = \begin{bmatrix} x \\ y \\ z \end{bmatrix} = x\mathbf{e}_1 + y\mathbf{e}_2 + z\mathbf{e}_3$. In applied courses, these three vectors are

often denoted by $\mathbf{i}, \mathbf{j}$, and $\mathbf{k}$ (or by $\vec{\mathbf{i}}, \vec{\mathbf{j}}$, and $\vec{\mathbf{k}}$), respectively.

Generalizing, any vector $\mathbf{x} = \begin{bmatrix} x_1 \\ x_2 \\ \vdots \\ x_n \end{bmatrix}$ in $\mathbb{R}^n$ can be written as a linear

combination of the n vectors $\mathbf{e}_1 = \begin{bmatrix} 1 \\ 0 \\ \vdots \\ 0 \end{bmatrix}, \mathbf{e}_2 = \begin{bmatrix} 0 \\ 1 \\ \vdots \\ 0 \end{bmatrix}, \ldots, \mathbf{e}_n = \begin{bmatrix} 0 \\ 0 \\ \vdots \\ 1 \end{bmatrix}$:

$\mathbf{v} = x_1\mathbf{e}_1 + x_2\mathbf{e}_2 + \cdots + x_n\mathbf{e}_n$.

The special set $\{\mathbf{e}_1, \mathbf{e}_2, \ldots, \mathbf{e}_n\}$ will be used often in our analysis of $\mathbb{R}^n$. This set of vectors is a particular example of a **spanning set** for a particular space $\mathbb{R}^n$—appropriately named because such a set reaches across (spans) the entire space when all linear combinations of vectors in the set are formed.

Definition 1.3.2

Given a nonempty set of vectors $S = \{\mathbf{v}_1, \mathbf{v}_2, \ldots, \mathbf{v}_k\}$ in $\mathbb{R}^n$, the **span** of S, denoted by $span(S)$, is the set of all linear combinations of vectors from S:

$$span(S) = \{a_1\mathbf{v}_1 + a_2\mathbf{v}_2 + \cdots + a_k\mathbf{v}_k | a_i \in \mathbb{R}, \mathbf{v}_i \in S, i = 1, 2, \ldots, k\}.$$

(Note that some or all of the scalars may be zero.)

A nonempty set S of vectors in $\mathbb{R}^n$ **spans** $\mathbb{R}^n$ (or is a **spanning set** for $\mathbb{R}^n$) if every vector in $\mathbb{R}^n$ is an element of $span(S)$, that is, if every vector in $\mathbb{R}^n$ is a linear combination of vectors in S.

Note that $span(S)$ is an infinite set of vectors unless $S = \{\mathbf{0}\}$, in which case $span(S) = S = \{\mathbf{0}\}$.

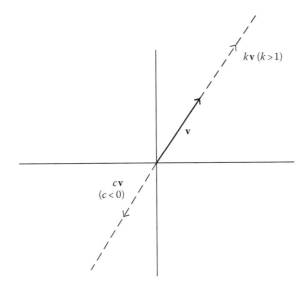

Figure 1.9 The span of a nonzero vector in $\mathbb{R}^2$.

Given a single nonzero vector $\mathbf{v} = \begin{bmatrix} x \\ y \end{bmatrix}$ in $\mathbb{R}^2$, the span of $\{\mathbf{v}\}$ is the set of all scalar multiples of $\mathbf{v}$: $span(\{\mathbf{v}\}) = \left\{ a\mathbf{v} = \begin{bmatrix} ax \\ ay \end{bmatrix} : a \in \mathbb{R} \right\}$. Because $a = 0$ is a possibility, the origin is in the span and we can interpret the span of $\{\mathbf{v}\}$ as *the set of all points on a straight line through the origin* (see Figure 1.9). The slope of this line depends on the original coordinates of $\mathbf{v}$: Slope $= ay/ax = y/x$, assuming that $a \neq 0$ and $x \neq 0$. If $x = 0$, the span is just the y-axis, whereas if $y = 0$, the span is the x-axis.

Suppose we have two nonzero vectors $\mathbf{v}$ and $\mathbf{w}$ in $\mathbb{R}^2$. If one of these vectors is a scalar multiple of the other, then span$(\{\mathbf{v}, \mathbf{w}\}) = span(\{\mathbf{v}\}) = span(\{\mathbf{w}\})$, again a straight line through the origin. On the other hand, if the vectors do not lie on the same straight line through the origin, their span is *all of* $\mathbb{R}^2$ (see Figure 1.10), a fact we will prove after Theorem 1.4.2.

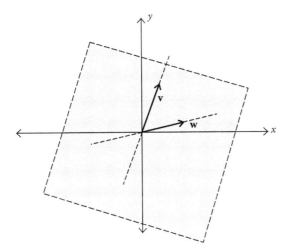

Figure 1.10 The span of linearly independent vectors in $\mathbb{R}^2$.

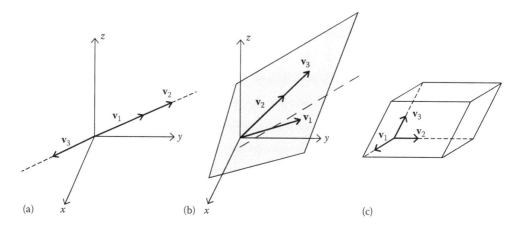

Figure 1.11 Spans of three nonzero vectors in $\mathbb{R}^3$.

In $\mathbb{R}^3$ the span of a single nonzero vector is a line through the origin in a three-dimensional space. Given two nonzero vectors in $\mathbb{R}^3$, their span is either a line through the origin or a plane passing through the origin, depending on whether the vectors are scalar multiples of each other.

Finally, three nonzero vectors in $\mathbb{R}^3$ span a line, a plane, or all of $\mathbb{R}^3$, depending on whether all three of the vectors lie on the same straight line through the origin, only two of the vectors are collinear, or no two vectors lie on the same straight line, respectively (Figure 1.11a through c, respectively).

Example 1.3.1: The Span of a Set of Vectors in $\mathbb{R}^2$

(a) Let us determine the span of the three vectors, $\begin{bmatrix} 1 \\ -2 \end{bmatrix}, \begin{bmatrix} -3 \\ 6 \end{bmatrix}, \begin{bmatrix} 4 \\ -8 \end{bmatrix}$. According to Definition 1.3.2, the span of this set of vectors is the set

$$\left\{ a\begin{bmatrix} 1 \\ -2 \end{bmatrix} + b\begin{bmatrix} -3 \\ 6 \end{bmatrix} + c\begin{bmatrix} 4 \\ -8 \end{bmatrix} : a, b, c \in \mathbb{R} \right\}$$

$$= \left\{ \begin{bmatrix} a - 3b + 4c \\ -2a + 6b - 8c \end{bmatrix} : a, b, c \in \mathbb{R} \right\}.$$ If we examine the vectors in the span carefully, we see that no matter how the scalars a, b, and c vary, the second component is always -2 times the first component. This means that we can express the span as $\left\{ \begin{bmatrix} x \\ -2x \end{bmatrix} : x \in \mathbb{R} \right\}$, or as $\left\{ x\begin{bmatrix} 1 \\ -2 \end{bmatrix} : x \in \mathbb{R} \right\}$. Thus the span of our set of three vectors consists of all scalar multiples of one of them, that is, the span is a straight line through the origin with slope equal to -2. (Going back to the original set of vectors, we notice that all three vectors lie on the same straight line through the origin.)

(b) Now let us consider the span of the set $S = \left\{ \begin{bmatrix} 1 \\ -2 \end{bmatrix}, \begin{bmatrix} -3 \\ 6 \end{bmatrix}, \begin{bmatrix} 2 \\ 1 \end{bmatrix} \right\}$. We have $span(S) = \left\{ a \begin{bmatrix} 1 \\ -2 \end{bmatrix} + b \begin{bmatrix} -3 \\ 6 \end{bmatrix} + c \begin{bmatrix} 2 \\ 1 \end{bmatrix} : a, b, c \in \mathbb{R} \right\} = \left\{ \begin{bmatrix} a - 3b + 2c \\ -2a + 6b + c \end{bmatrix} : a, b, c \in \mathbb{R} \right\}$, and there is no apparent connection between the components of the vectors in the span.

However, we can write

$$span(S) = \left\{ \begin{bmatrix} a - 3b \\ -2a + 6b \end{bmatrix} + \begin{bmatrix} 2c \\ c \end{bmatrix} : a, b, c \in \mathbb{R} \right\}$$

$$= \left\{ k_1 \begin{bmatrix} 1 \\ -2 \end{bmatrix} + k_2 \begin{bmatrix} 2 \\ 1 \end{bmatrix} : k_1, k_2 \in \mathbb{R} \right\},$$

showing that $span \left\{ \begin{bmatrix} 1 \\ -2 \end{bmatrix}, \begin{bmatrix} -3 \\ 6 \end{bmatrix}, \begin{bmatrix} 2 \\ 1 \end{bmatrix} \right\} = span \left\{ \begin{bmatrix} 1 \\ -2 \end{bmatrix}, \begin{bmatrix} 2 \\ 1 \end{bmatrix} \right\}$. (We can see, for example, that $\begin{bmatrix} -3 \\ 6 \end{bmatrix} = -3 \begin{bmatrix} 1 \\ -2 \end{bmatrix}$, so $\begin{bmatrix} -3 \\ 6 \end{bmatrix}$ does not contribute to the spanning capability of the original set of vectors.) The vectors $\begin{bmatrix} 1 \\ -2 \end{bmatrix}$ and $\begin{bmatrix} 2 \\ 1 \end{bmatrix}$ do not lie on the same straight line, and we claim that $span \left\{ \begin{bmatrix} 1 \\ -2 \end{bmatrix}, \begin{bmatrix} 2 \\ 1 \end{bmatrix} \right\} = \mathbb{R}^2$.

To see this, suppose that $\begin{bmatrix} x \\ y \end{bmatrix}$ is any vector in $\mathbb{R}^2$. We show that there are real numbers k_1 and k_2 such that $\begin{bmatrix} x \\ y \end{bmatrix} = k_1 \begin{bmatrix} 1 \\ -2 \end{bmatrix} + k_2 \begin{bmatrix} 2 \\ 1 \end{bmatrix} = \begin{bmatrix} k_1 + 2k_2 \\ -2k_1 + k_2 \end{bmatrix}$. This vector equation is equivalent to the system of two equations in two unknowns, $k_1 + 2k_2 = x$, $-2k_1 + k_2 = y$; we can solve this system by substitution or elimination to conclude that $k_1 = (x - 2y)/5$ and $k_2 = (2x + y)/5$. (For instance, if $\begin{bmatrix} x \\ y \end{bmatrix} = \begin{bmatrix} 2 \\ -7 \end{bmatrix}$, then $k_1 = [2 - 2(-7)]/5 = 16/5$, $k_2 = [2(2) + (-7)]/5 = -3/5$, and $\begin{bmatrix} 2 \\ -7 \end{bmatrix} = \frac{16}{5} \begin{bmatrix} 1 \\ -2 \end{bmatrix} - \frac{3}{5} \begin{bmatrix} 2 \\ 1 \end{bmatrix}$.)

Example 1.3.2: The Span of a Set of Vectors in $\mathbb{R}^3$

Consider the set $S = \left\{ \begin{bmatrix} 1 \\ -3 \\ 2 \end{bmatrix}, \begin{bmatrix} 2 \\ -4 \\ -1 \end{bmatrix}, \begin{bmatrix} 1 \\ -5 \\ 7 \end{bmatrix} \right\}$. Then

$$span(S) = \left\{ a \begin{bmatrix} 1 \\ -3 \\ 2 \end{bmatrix} + b \begin{bmatrix} 2 \\ -4 \\ -1 \end{bmatrix} + c \begin{bmatrix} 1 \\ -5 \\ 7 \end{bmatrix} : a, b, c \in \mathbb{R} \right\}$$

$$= \left\{ \begin{bmatrix} a + 2b + c \\ -3a - 4b - 5c \\ 2a - b + 7c \end{bmatrix} : a, b, c \in \mathbb{R} \right\}$$

$$= \left\{ \begin{bmatrix} x \\ y \\ z \end{bmatrix} \in \mathbb{R}^3 : 11x + 5y + 2z = 0 \right\}.$$

Although not obvious, this last description of the span is easily verified, and it indicates that the span is a plane through the origin.

An examination of the original vectors indicates that they are not collinear. However, there is a relationship among them, for example, $\begin{bmatrix} 1 \\ -5 \\ 7 \end{bmatrix} = 3 \begin{bmatrix} 1 \\ -3 \\ 2 \end{bmatrix} - \begin{bmatrix} 2 \\ -4 \\ -1 \end{bmatrix}$. Finally, we can conclude that the span is not all of $\mathbb{R}^3$ because (for example) the components of the vector, $\begin{bmatrix} x \\ y \\ z \end{bmatrix} = \begin{bmatrix} 1 \\ 0 \\ 0 \end{bmatrix}$, do not satisfy the
equation $11x + 5y + 2z = 0$ and so this vector cannot be in the span. (Example 1.3.4 provides a more detailed explanation of nonspanning.)

Example 1.3.3: A Spanning Set for $\mathbb{R}^3$

We show that the set $S = \left\{ \begin{bmatrix} 1 \\ 1 \\ 1 \end{bmatrix}, \begin{bmatrix} 0 \\ 1 \\ 1 \end{bmatrix}, \begin{bmatrix} 0 \\ 0 \\ 1 \end{bmatrix} \right\}$ spans $\mathbb{R}^3$ by
demonstrating that for any vector $\mathbf{v} = \begin{bmatrix} x \\ y \\ z \end{bmatrix}$ in $\mathbb{R}^3$, we can find
scalars a, b, and c, such that $\mathbf{v}$ can be written as

$$\begin{bmatrix} x \\ y \\ z \end{bmatrix} = a \begin{bmatrix} 1 \\ 1 \\ 1 \end{bmatrix} + b \begin{bmatrix} 0 \\ 1 \\ 1 \end{bmatrix} + c \begin{bmatrix} 0 \\ 0 \\ 1 \end{bmatrix}, \quad \text{or} \quad \begin{bmatrix} x \\ y \\ z \end{bmatrix} = \begin{bmatrix} a \\ a + b \\ a + b + c \end{bmatrix}.$$

This last vector equation is equivalent to the linear system

$$a = x$$
$$a + b = y$$
$$a + b + c = z$$

with the solutions $a = x$, $b = y - a = y - x$ and $c = z - a - b = z - x - (y - x) = z - y$.

Therefore,

$$\begin{bmatrix} x \\ y \\ z \end{bmatrix} = x\begin{bmatrix} 1 \\ 1 \\ 1 \end{bmatrix} + (y - x)\begin{bmatrix} 0 \\ 1 \\ 1 \end{bmatrix} + (z - y)\begin{bmatrix} 0 \\ 0 \\ 1 \end{bmatrix},$$

and we have shown that S is a spanning set for $\mathbb{R}^3$. (For instance, if $\begin{bmatrix} x \\ y \\ z \end{bmatrix} = \begin{bmatrix} 1 \\ -2 \\ 3 \end{bmatrix}$, then $\begin{bmatrix} 1 \\ -2 \\ 3 \end{bmatrix} = (1)\begin{bmatrix} 1 \\ 1 \\ 1 \end{bmatrix} +$

$(-3)\begin{bmatrix} 0 \\ 1 \\ 1 \end{bmatrix} + 5\begin{bmatrix} 0 \\ 0 \\ 1 \end{bmatrix}$.)

Example 1.3.4: A Nonspanning Set for $\mathbb{R}^3$

Let us examine the vectors $\mathbf{u} = \begin{bmatrix} 1 \\ 1 \\ 0 \end{bmatrix}$, $\mathbf{v} = \begin{bmatrix} 2 \\ 5 \\ 3 \end{bmatrix}$ and $\mathbf{w} = \begin{bmatrix} 0 \\ 1 \\ 1 \end{bmatrix}$

to see if they span $\mathbb{R}^3$. We will take an arbitrary vector $\mathbf{x} = \begin{bmatrix} x_1 \\ x_2 \\ x_3 \end{bmatrix}$

in $\mathbb{R}^3$ and try to find scalars a, b, and c such that $\mathbf{x}$ can be written as $a\mathbf{u} + b\mathbf{v} + c\mathbf{w}$. This problem is equivalent to solving the system

(1) $a + 2b = a + 2b = x_1$.
(2) $a + 5b + c = a + 5b + c = x_2$.
(3) $3b + c = 3b + c = x_3$.

Subtracting equation (3) from equation (2) yields $a + 2b = x_2 - x_3$. Comparing this result to the original equation (1), we are forced to conclude that $x_1 = x_2 - x_3$. In other words, the only way we can find the scalars a, b, and c is if the vector $\mathbf{x}$ has the

form $\begin{bmatrix} x_2 - x_3 \\ x_2 \\ x_3 \end{bmatrix}$. This means that any vector *not* of this form

will not be "reached" by the vectors $\mathbf{u}$, $\mathbf{v}$, and $\mathbf{w}$. For example,

the vector $\begin{bmatrix} 1 \\ 2 \\ 3 \end{bmatrix}$ cannot be written as a linear combination of

these three vectors,* although the vector $\begin{bmatrix} -1 \\ 2 \\ 3 \end{bmatrix}$ *can* be

expressed this way. Thus $A = \left\{ \begin{bmatrix} 1 \\ 1 \\ 0 \end{bmatrix}, \begin{bmatrix} 2 \\ 5 \\ 3 \end{bmatrix}, \begin{bmatrix} 0 \\ 1 \\ 1 \end{bmatrix} \right\}$ is not a

---

* Just finding such a counterexample constitutes a proof that the set does not span $\mathbb{R}^3$.

spanning set for $\mathbb{R}^3$. We note that there is some redundancy in

set A, for example, $\begin{bmatrix} 2 \\ 5 \\ 3 \end{bmatrix} = 2 \begin{bmatrix} 1 \\ 1 \\ 0 \end{bmatrix} + 3 \begin{bmatrix} 0 \\ 1 \\ 1 \end{bmatrix}$.

Exercises 1.3

A.

1. Give a geometric interpretation of the span of $\left\{ \begin{bmatrix} 1 \\ 1 \\ 1 \end{bmatrix}, \begin{bmatrix} 0 \\ 0 \\ 0 \end{bmatrix} \right\}$

   in $\mathbb{R}^3$.

2. What is the span of the set $\left\{ \begin{bmatrix} 1 \\ 1 \\ 1 \end{bmatrix}, \begin{bmatrix} 0 \\ -1 \\ 1 \end{bmatrix}, \begin{bmatrix} 0 \\ 0 \\ -1 \end{bmatrix} \right\}$ in $\mathbb{R}^3$?

3. Let $\mathbf{u} = \begin{bmatrix} 2 \\ 1 \\ 1 \end{bmatrix}$ and $\mathbf{v} = \begin{bmatrix} 2 \\ t \\ 2t \end{bmatrix}$. Find all values of t (if any) for
   which $\mathbf{u}$ and $\mathbf{v}$ span $\mathbb{R}^3$.

4. Let $S = \left\{ \begin{bmatrix} 1 \\ 2 \\ -3 \end{bmatrix}, \begin{bmatrix} -2 \\ 1 \\ 1 \end{bmatrix} \right\}$. Describe $span(S)$. Is the vector

   $\begin{bmatrix} -5 \\ 2 \\ 3 \end{bmatrix}$ in $span(S)$?

5. Show that $span \left\{ \begin{bmatrix} 2 \\ -1 \\ 6 \end{bmatrix}, \begin{bmatrix} -3 \\ 4 \\ 1 \end{bmatrix} \right\} = span \left\{ \begin{bmatrix} -1 \\ 3 \\ 7 \end{bmatrix}, \begin{bmatrix} 8 \\ -9 \\ 4 \end{bmatrix} \right\}$.

6. Find a vector $\mathbf{u} \in \mathbb{R}^3$ *not* in the span of
   $S = \left\{ \begin{bmatrix} 1 \\ 2 \\ 2 \end{bmatrix}, \begin{bmatrix} 1 \\ 1 \\ 1 \end{bmatrix}, \begin{bmatrix} -1 \\ 0 \\ 0 \end{bmatrix} \right\}$.

7. Let $\mathbf{u}_1 = \begin{bmatrix} 2 \\ 4 \\ 3 \end{bmatrix}$, $\mathbf{u}_2 = \begin{bmatrix} 2 \\ 7 \\ 6 \end{bmatrix}$, $\mathbf{v}_1 = \begin{bmatrix} 2 \\ 1 \\ 0 \end{bmatrix}$, and $\mathbf{v}_2 = \begin{bmatrix} -2 \\ 2 \\ 3 \end{bmatrix}$.
   Show that $span\{\mathbf{u}_1, \mathbf{u}_2\} \subseteq span\{\mathbf{v}_1, \mathbf{v}_2\}$.

8. In $\mathbb{R}^3$, determine the graph of

a. $span\left\{\begin{bmatrix} 1 \\ 3 \\ 2 \end{bmatrix}, \begin{bmatrix} 2 \\ 6 \\ 4 \end{bmatrix}, \begin{bmatrix} -3 \\ -9 \\ -6 \end{bmatrix}\right\}$ b. $span\left\{\begin{bmatrix} 2 \\ -4 \\ 6 \end{bmatrix}, \begin{bmatrix} -6 \\ 12 \\ -18 \end{bmatrix}, \begin{bmatrix} 1 \\ -2 \\ 3 \end{bmatrix}\right\}$.

9. Find the span of
$$S = \left\{[1 \quad 4 \quad 0 \quad 0]^{\mathrm{T}}, [2 \quad 0 \quad 4 \quad 0]^{\mathrm{T}}, [0 \quad 0 \quad 4 \quad 1]^{\mathrm{T}}\right\}.$$

10. Is the vector $\begin{bmatrix} 3 \\ -1 \\ 0 \\ -1 \end{bmatrix}$ in the span of the vectors $\begin{bmatrix} 2 \\ -1 \\ 3 \\ 2 \end{bmatrix}$,

$\begin{bmatrix} -1 \\ 1 \\ 1 \\ -3 \end{bmatrix}$, and $\begin{bmatrix} 1 \\ 1 \\ 9 \\ 5 \end{bmatrix}$ in $\mathbb{R}^4$?

B.

1. Prove that if $S \subseteq T \subseteq \mathbb{R}^n$, then $span(S) \subseteq span(T)$.

2. Prove that if S and T are nonempty finite sets such that S is a set of vectors in $\mathbb{R}^n$, then $span(span(S)) = span(S)$.

3. If U and V are nonempty finite subsets of $\mathbb{R}^n$, show that $span(U) + span(V) \subseteq span(U \cup V)$. [If A and B are sets, then $A + B = \{x | x = a + b, \text{ where } a \in A \text{ and } b \in B\}$.]

4. For any two nonempty finite subsets U and V of $\mathbb{R}^n$, show that
   a. $span(U) \cup span(V) \subseteq span(U \cup V)$.
   b. $span(U \cap V) \subseteq span(U) \cap span(V)$.

5. Let $S = \{\mathbf{v}_1, \mathbf{v}_2, \ldots, \mathbf{v}_k\}$ and $T = \{\mathbf{v}_1, \mathbf{v}_2, \ldots, \mathbf{v}_k, \mathbf{w}\}$ be two sets of vectors from $\mathbb{R}^n$ where $\mathbf{w} \neq \mathbf{0}$. Prove that $span(S) = span(T)$ if and only if $\mathbf{w} \in span(S)$.

6. Prove that if $\mathbf{u}$ is orthogonal to every vector in the set $S = \{\mathbf{v}_1, \mathbf{v}_2, \ldots, \mathbf{v}_k\}$, then $\mathbf{u}$ is orthogonal to every vector in $span(S)$. (This is a generalization of Problem B14, Exercises 1.2.)

C.

1. Suppose that $\mathbf{v} \in span\{\mathbf{v}_1, \mathbf{v}_2, \ldots, \mathbf{v}_k\}$ and each $\mathbf{v}_i \in span\{\mathbf{w}_1, \mathbf{w}_2, \ldots, \mathbf{w}_r\}$.
   a. Show that $\mathbf{v} \in span\{\mathbf{w}_1, \mathbf{w}_2, \ldots, \mathbf{w}_r\}$.
   b. State conditions for $span\{\mathbf{v}_1, \mathbf{v}_2, \ldots, \mathbf{v}_k\} = span\{\mathbf{w}_1, \mathbf{w}_2, \ldots, \mathbf{w}_r\}$.

1.4 Linear Independence

Let us look at the set of vectors $S = \{\mathbf{v}_1, \mathbf{v}_2, \mathbf{v}_3\}$ in $\mathbb{R}^3$, where $\mathbf{v}_1 = \begin{bmatrix} -1 \\ 2 \\ 0 \end{bmatrix}$, $\mathbf{v}_2 = \begin{bmatrix} -8 \\ 4 \\ 9 \end{bmatrix}$, and $\mathbf{v}_3 = \begin{bmatrix} 2 \\ 0 \\ -3 \end{bmatrix}$. If we analyze the relationships among the vectors in S (perhaps for a long time), we might discover that $\mathbf{v}_2$ is a linear combination of vectors $\mathbf{v}_1$ and $\mathbf{v}_3$: $\begin{bmatrix} -8 \\ 4 \\ 9 \end{bmatrix} = 2\begin{bmatrix} -1 \\ 2 \\ 0 \end{bmatrix} - 3\begin{bmatrix} 2 \\ 0 \\ -3 \end{bmatrix}$, or $\mathbf{v}_2 = 2\mathbf{v}_1 - 3\mathbf{v}_3$. This algebraic relationship can also be written as $-2\mathbf{v}_1 + \mathbf{v}_2 + 3\mathbf{v}_3 = \mathbf{0}$ (the zero vector). (See Examples 1.3.1(a), 1.3.2, and 1.3.4 for other instances of this kind of redundancy.)

We say that this set of vectors S is *linearly dependent* because one vector can be expressed as a linear combination of other vectors in the set. Let us define this concept precisely in $\mathbb{R}^n$.

Definition 1.4.1

A nonempty finite set of vectors $S = \{\mathbf{v}_1, \mathbf{v}_2, \ldots, \mathbf{v}_k\}$ in $\mathbb{R}^n$ is called **linearly independent** if the only way that $a_1\mathbf{v}_1 + a_2\mathbf{v}_2 + \cdots + a_k\mathbf{v}_k = \mathbf{0}$, where $a_1, a_2, \ldots, a_k$ are scalars, is if $a_1 = a_2 = \cdots = a_k = 0$. Otherwise, S is called **linearly dependent**.

[**Note that any set containing the zero vector cannot be linearly independent** (Exercise B12).]

If we speak somewhat loosely and say that vectors $\mathbf{v}_1, \mathbf{v}_2, \ldots, \mathbf{v}_k$ are linearly independent (or dependent), we mean that the set $S = \{\mathbf{v}_1, \mathbf{v}_2, \ldots, \mathbf{v}_k\}$ is linearly independent (or dependent).

Example 1.4.1: A Linearly Independent Set in $\mathbb{R}^3$

We consider the set $S = \left\{ \begin{bmatrix} 1 \\ 2 \\ -1 \end{bmatrix}, \begin{bmatrix} -1 \\ 1 \\ 0 \end{bmatrix}, \begin{bmatrix} 1 \\ 3 \\ -1 \end{bmatrix} \right\}$ in $\mathbb{R}^3$.

To determine the linear independence or dependence of S, we have to investigate the solutions a_1, a_2, a_3 to the equation $a_1\begin{bmatrix} 1 \\ 2 \\ -1 \end{bmatrix} + a_2\begin{bmatrix} -1 \\ 1 \\ 0 \end{bmatrix} + a_3\begin{bmatrix} 1 \\ 3 \\ -1 \end{bmatrix} = \begin{bmatrix} 0 \\ 0 \\ 0 \end{bmatrix}$, or

$$\begin{bmatrix} a_1 - a_2 + a_3 \\ 2a_1 + a_2 + 3a_3 \\ -a_1 - a_3 \end{bmatrix} = \begin{bmatrix} 0 \\ 0 \\ 0 \end{bmatrix}.$$

This vector equation is equivalent to the system of equations

$$a_1 - a_2 + a_3 = 0$$
$$2a_1 + a_2 + 3a_3 = 0$$
$$-a - a_3 = 0$$

Solving this system, we can add the first and third equations to conclude that $a_2 = 0$. Continuing in this way, we find that a_1 and a_3 must be zero as well. Therefore, the set of vectors S is *linearly independent*.

As we saw in the last example, we have to be able to solve a system of linear equations to determine if a set of vectors is linearly independent. Historically, the solution of systems of linear equations was a major motivating factor in the development of linear algebra,* and we will focus on this topic in the next chapter. For now, we continue to learn more about vectors.

We can characterize linear dependence (and hence independence) in an alternative way.

Theorem 1.4.1

A set $S = \{\mathbf{v}_1, \mathbf{v}_2, \ldots, \mathbf{v}_k\}$ in $\mathbb{R}^n$ that contains at least two vectors is linearly dependent if and only if some vector $\mathbf{v}_j$ (with $j > 1$) is a linear combination of the remaining vectors in the set.

Proof   First, suppose that some vector $\mathbf{v}_j$ is a linear combination of the remaining vectors in the set:

$$\mathbf{v}_j = a_1 \mathbf{v}_1 + a_2 \mathbf{v}_2 + \cdots + a_{j-1} \mathbf{v}_{j-1} + a_{j+1} \mathbf{v}_{j+1} + \cdots + a_k \mathbf{v}_k.$$

Then

$$a_1 \mathbf{v}_1 + a_2 \mathbf{v}_2 + \cdots + a_{j-1} \mathbf{v}_{j-1} + (-1)\mathbf{v}_j + a_{j+1} \mathbf{v}_{j+1} + \cdots + a_k \mathbf{v}_k = \mathbf{0}.$$

Because the coefficient of $\mathbf{v}_j$ in this linear combination is not zero, the set S is linearly dependent.

On the other hand, suppose that we know S is linearly dependent. Then we can find scalars $a_1, a_2, \ldots, a_k$ such that $\sum_{i=1}^{k} a_i \mathbf{v}_i = \mathbf{0}$ with at least one scalar, say a_j, not zero. This implies that

$$a_j \mathbf{v}_j = -a_1 \mathbf{v}_1 - a_2 \mathbf{v}_2 - \cdots - a_{j-1} \mathbf{v}_{j-1} - a_{j+1} \mathbf{v}_{j+1} - \cdots - a_k \mathbf{v}_k,$$

so that $\mathbf{v}_j = \left(\frac{-a_1}{a_j} \right) \mathbf{v}_1 + \left(\frac{-a_2}{a_j} \right) \mathbf{v}_2 + \cdots + \left(\frac{-a_{j-1}}{a_j} \right) \mathbf{v}_{j-1} + \left(\frac{-a_{j+1}}{a_j} \right) \mathbf{v}_{j+1} + \cdots + \left(\frac{-a_k}{a_j} \right) \mathbf{v}_k.$ Thus $\mathbf{v}_j$ can be written as a linear combination of the remaining vectors in S.

---

* See, for example, V. Katz, Historical ideas in teaching linear algebra, in *Learn from the Masters!*, F. Swetz et al., eds. (Washington, DC: MAA, 1995).

Example 1.4.2: Linearly Dependent Sets of Vectors in $\mathbb{R}^4$ and $\mathbb{R}^5$

a. The vectors $[1 \;\; -1 \;\; 1 \;\; 2 \;\; 1]^T$, $[4 \;\; -1 \;\; 6 \;\; 6 \;\; 2]^T$, $[-4 \;\; -2 \;\; -3 \;\; -4 \;\; -2]^T$, and $[-2 \;\; -1 \;\; 1 \;\; -2 \;\; -2]^T$ are linearly dependent in $\mathbb{R}^5$ because (for example) $[-4 \;\; -2 \;\; -3 \;\; -4 \;\; -2]^T = 2[1 \;\; -1 \;\; 1 \;\; 2 \;\; 1]^T - [4 \;\; -1 \;\; 6 \;\; 6 \;\; 2]^T + [-2 \;\; -1 \;\; 1 \;\; -2 \;\; -2]^T$.

b. The vectors $[1 \;\; 1 \;\; 0 \;\; 0]^T$, $[1 \;\; 0 \;\; 1 \;\; 0]^T$, $[1 \;\; 0 \;\; 0 \;\; 1]^T$, $[0 \;\; 1 \;\; 1 \;\; 0]^T$, and $[0 \;\; 0 \;\; 1 \;\; 1]^T$ are linearly dependent in $\mathbb{R}^4$ because $[1 \;\; 1 \;\; 0 \;\; 0]^T = [1 \;\; 0 \;\; 0 \;\; 1]^T + [0 \;\; 1 \;\; 1 \;\; 0]^T - [0 \;\; 0 \;\; 1 \;\; 1]^T$, or equivalently, $(1)[1 \;\; 1 \;\; 0 \;\; 0]^T + (0)[1 \;\; 0 \;\; 1 \;\; 0]^T + (-1)[1 \;\; 0 \;\; 0 \;\; 1]^T + (-1)[0 \;\; 1 \;\; 1 \;\; 0]^T + (1)[0 \;\; 0 \;\; 1 \;\; 1]^T = [0 \;\; 0 \;\; 0 \;\; 0]^T$.

Geometrically, in $\mathbb{R}^2$ or $\mathbb{R}^3$, a set of vectors is linearly independent if and only if no two vectors (viewed as emanating from the origin) lie on the same straight line (Figure 1.12a through c).

At this point, we investigate the connection between linearly independent sets and spanning sets. For any space $\mathbb{R}^n$ the possible relationships are indicated by Figure 1.13. The letters refer to the components of

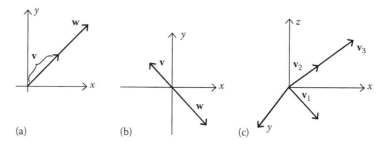

(a)   (b)   (c)

Figure 1.12 Linearly dependent vectors in $\mathbb{R}^2$ and $\mathbb{R}^3$.

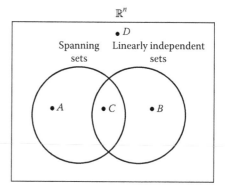

Figure 1.13 Spanning sets and linearly independent sets in $\mathbb{R}^n$.

Example 1.4.3, which provides specifics in $\mathbb{R}^3$. The verification of these statements is required in Exercises A2 through A5.

Example 1.4.3: Spanning Sets and Linearly Independent Sets in $\mathbb{R}^3$

a. The set $A = \left\{ \begin{bmatrix} 1 \\ 1 \\ 1 \end{bmatrix}, \begin{bmatrix} 0 \\ 1 \\ 1 \end{bmatrix}, \begin{bmatrix} 0 \\ 0 \\ 1 \end{bmatrix}, \begin{bmatrix} 1 \\ 0 \\ 0 \end{bmatrix} \right\}$ spans $\mathbb{R}^3$ but is not linearly independent.

b. The set $B = \left\{ \begin{bmatrix} 1 \\ -3 \\ 2 \end{bmatrix}, \begin{bmatrix} 2 \\ -4 \\ -1 \end{bmatrix} \right\}$ is linearly independent but does not span $\mathbb{R}^3$.

c. The set $C = \left\{ \begin{bmatrix} 1 \\ 1 \\ 1 \end{bmatrix}, \begin{bmatrix} 0 \\ 1 \\ 1 \end{bmatrix}, \begin{bmatrix} 0 \\ 0 \\ 1 \end{bmatrix} \right\}$ is linearly independent and also spans $\mathbb{R}^3$.

d. The set $D = \left\{ \begin{bmatrix} 1 \\ 1 \\ 1 \end{bmatrix}, \begin{bmatrix} 0 \\ 1 \\ 1 \end{bmatrix}, \begin{bmatrix} 1 \\ 0 \\ 0 \end{bmatrix} \right\}$ is linearly dependent and does not span $\mathbb{R}^3$.

Despite the possibilities shown in Example 1.4.3, we *can* establish a relationship between linearly independent sets and spanning sets in any Euclidean space $\mathbb{R}^n$: **The size of any linearly independent set is always less than or equal to the size of any spanning set.** (In Example 1.4.3, notice that the linearly independent sets have two or three vectors, whereas any spanning set has three or four vectors.) The proof of this important result is a bit intricate, using what is sometimes called the **Steinitz* replacement** (or **exchange**) **technique**; but the next example should illuminate the basic idea. The name of the process comes from the fact that vectors of a spanning set are replaced by (exchanged for) vectors of a linearly independent set.

Example 1.4.4: A Taste of Theorem 1.4.2

We know that $\mathbf{e}_1 = \begin{bmatrix} 1 \\ 0 \\ 0 \end{bmatrix}$, $\mathbf{e}_2 = \begin{bmatrix} 0 \\ 1 \\ 0 \end{bmatrix}$, and $\mathbf{e}_3 = \begin{bmatrix} 0 \\ 0 \\ 1 \end{bmatrix}$ span $\mathbb{R}^3$.

Suppose we believe that $\mathbf{w}_1 = \begin{bmatrix} 1 \\ 2 \\ 3 \end{bmatrix}$, $\mathbf{w}_2 = \begin{bmatrix} 3 \\ 4 \\ 5 \end{bmatrix}$,

---

* This technique was named for Ernst Steinitz (1871–1928), who gave the first abstract definition of the algebraic structure known as a *field* and made other contributions to algebra.

$\mathbf{w}_3 = \begin{bmatrix} 5 \\ 6 \\ 7 \end{bmatrix}$, and $\mathbf{w}_3 = \begin{bmatrix} 7 \\ 8 \\ 9 \end{bmatrix}$ are linearly independent vectors in $\mathbb{R}^3$.

We can write $\mathbf{w}_1 = \mathbf{e}_1 + 2\mathbf{e}_2 + 3\mathbf{e}_3$. Then $\mathbf{e}_1$ can be expressed as a linear combination of $\mathbf{w}_1, \mathbf{e}_2$, and $\mathbf{e}_3$: $\mathbf{e}_1 = \mathbf{w}_1 + (-2)\mathbf{e}_2 + (-3)\mathbf{e}_3$. Thus the set $\{\mathbf{w}_1, \mathbf{e}_2, \mathbf{e}_3\}$ spans $\mathbb{R}^3$ because any linear combination of $\mathbf{e}_1, \mathbf{e}_2$, and $\mathbf{e}_3$ is actually a combination of $\mathbf{w}_1, \mathbf{e}_2$, and $\mathbf{e}_3$. Note that we have replaced one of the spanning vectors by a vector from the alleged linearly independent set.

Now repeat this process by writing $\mathbf{w}_2 = 3\mathbf{w}_1 + (-2)\mathbf{e}_2 + (-4)\mathbf{e}_3$. Therefore, we have $\mathbf{e}_2 = \frac{3}{2}\mathbf{w}_1 + (-\frac{1}{2})\mathbf{w}_2 + (-2)\mathbf{e}_3$, so the set $\{\mathbf{w}_2 1, \mathbf{w}_2, \mathbf{e}_3\}$ spans $\mathbb{R}^3$. Proceeding one step further, we can write $\mathbf{w}_3$ in terms of $\mathbf{w}_1, \mathbf{w}_2$, and $\mathbf{e}_3$, and then express $\mathbf{e}_3$ as a linear combination of $\mathbf{w}_1, \mathbf{w}_2$, and $\mathbf{w}_3$. Thus we have shown that $\{\mathbf{w}_1, \mathbf{w}_2, \mathbf{w}_3\}$ spans $\mathbb{R}^3$. But then $\mathbf{w}_4$ must be a linear combination of $\mathbf{w}_1, \mathbf{w}_2$, and $\mathbf{w}_3$—a contradiction of the linear independence of $\{\mathbf{w}_1, \mathbf{w}_2, \mathbf{w}_3, \mathbf{w}_4\}$ (by Theorem 1.4.1). [From this analysis, we might suspect that no set of four or more vectors in $\mathbb{R}^3$ can be linearly independent. In this example, note that $\mathbf{w}_4 = (-1)\mathbf{w}_2 + 2\mathbf{w}_3$.]

Now we should be ready to tackle the relationship between linearly independent sets and spanning sets in any Euclidean space.

Theorem 1.4.2

Suppose that the vectors $\mathbf{v}_1, \mathbf{v}_2, \dots, \mathbf{v}_m$ span $\mathbb{R}^n$ and that the nonzero vectors $\mathbf{w}_1, \mathbf{w}_2, \dots, \mathbf{w}_k$ in $\mathbb{R}^n$ are linearly independent. Then $k \le m$. (That is, the size of any linearly independent set is always less than or equal to the size of any spanning set.)

Proof Because $\mathbf{v}_1, \mathbf{v}_2, \dots, \mathbf{v}_m$ span $\mathbb{R}^n$, every vector in $\mathbb{R}^n$ can be written as a linear combination of the vectors $\mathbf{v}_i$. In particular,

$$\mathbf{w}_1 = a_1\mathbf{v}_1 + a_2\mathbf{v}_2 + \cdots + a_m\mathbf{v}_m.$$

Because $\mathbf{w}_1 \ne \mathbf{0}$, not all the coefficients a_i are equal to 0. After renumbering $\mathbf{v}_1, \mathbf{v}_2, \dots, \mathbf{v}_m$ if necessary, we can say that $a_1 \ne 0$. Then $\mathbf{v}_1$ can be expressed as a linear combination of $\mathbf{w}_1$ and the remaining vectors $\mathbf{v}_i$: $\mathbf{v}_1 = \left(\frac{1}{a_1}\right)\mathbf{w}_1 + \left(\frac{-a_2}{a_1}\right)\mathbf{v}_2 + \cdots + \left(\frac{-a_m}{a_1}\right)\mathbf{v}_m$. Therefore, the set $\{\mathbf{w}_1, \mathbf{v}_2, \dots, \mathbf{v}_m\}$, consisting of the $\mathbf{v}_i$'s with $\mathbf{v}_1$ replaced by $\mathbf{w}_1$, spans $\mathbb{R}^n$. We continue this process, replacing the $\mathbf{v}$'s by $\mathbf{w}$'s and relabeling the vectors $\mathbf{v}_i$, if necessary. The key claim, which we will prove by mathematical induction, is that for every $i = 1, 2, \dots, m-1$, the set $\{\mathbf{w}_1, \mathbf{w}_2, \dots \mathbf{w}_i, \mathbf{v}_{i+1}, \dots \mathbf{v}_m\}$ spans $\mathbb{R}^n$.

We have proved the claim for $i = 1$. Now, assuming that for a specific i, $\{\mathbf{w}_1, \mathbf{w}_2, \ldots \mathbf{w}_i, \mathbf{v}_{i+1}, \ldots \mathbf{v}_m\}$ spans $\mathbb{R}^n$, we can write

$$\mathbf{w}_{i+1} = (a_1 \mathbf{w}_1 + a_2 \mathbf{w}_2 + \cdots + a_i \mathbf{w}_i) + (a_{i+1} \mathbf{v}_{i+1} + a_{i+2} \mathbf{v}_{i+2} + \cdots + a_m \mathbf{v}_m).$$

Not all a_i's in the second set of parentheses can be 0 because this would imply that $\mathbf{w}_{i+1}$ is a linear combination of other $\mathbf{w}_k$'s, so that the set of $\mathbf{w}_k$'s is linearly dependent. Say that $a_{i+1} \neq 0$ (if not, just rearrange or relabel the terms in the second set of parentheses). Then

$$\mathbf{v}_{i+1} = \frac{\mathbf{w}_{i+1}}{a_{i+1}} - \left\{ \left(\frac{a_1}{a_{i+1}}\right) \mathbf{w}_1 + \cdots + \left(\frac{a_i}{a_{i+1}}\right) \mathbf{w}_i \right\}$$
$$- \left\{ \left(\frac{a_{i+2}}{a_{i+1}}\right) \mathbf{v}_{i+2} + \cdots + \left(\frac{a_m}{a_{i+1}}\right) \mathbf{v}_m \right\},$$

which implies that $\{\mathbf{w}_1, \mathbf{w}_2, \ldots, \mathbf{w}_{i+1}, \mathbf{v}_{i+2}, \ldots, \mathbf{v}_m\}$ spans $\mathbb{R}^n$. Thus we have proved the claim by induction.

If $k > m$, eventually the $\mathbf{v}_i$'s disappear, having been replaced by the $\mathbf{w}_j$'s. Furthermore, span $\{\mathbf{w}_1, \mathbf{w}_2, \ldots, \mathbf{w}_m\} = \mathbb{R}^n$. Then each of the vectors $\mathbf{w}_{m+1}, \mathbf{w}_{m+2}, \ldots, \mathbf{w}_k$ is a linear combination of $\mathbf{w}_1, \mathbf{w}_2, \ldots,$ $\mathbf{w}_m$—contradicting the assumed linear independence of the $\mathbf{w}_i$'s. Therefore, k must be less than or equal to m.

We can use Theorem 1.4.2 to make a stronger statement about linearly independent/dependent sets of vectors.

Theorem 1.4.3

Suppose that $A = \{\mathbf{v}_1, \mathbf{v}_2, \ldots, \mathbf{v}_k\}$ is a set of vectors in $\mathbb{R}^n$. If $k > n$, that is, if the number of vectors in A exceeds the dimension of the space, then A is linearly dependent.

Proof We have observed that $\mathbb{R}^n$ has a spanning set consisting of the n vectors $\mathbf{e}_1, \mathbf{e}_2, \ldots, \mathbf{e}_n$, where $\mathbf{e}_i$ has 1 as component i and zeros elsewhere. Theorem 1.4.2 implies that if A is a set of k linearly independent vectors in $\mathbb{R}^n$, then $k \leq n$. Therefore, any set of $k > n$ vectors must be dependent.

Another way of stating Theorem 1.4.3 is that **if** $A = \{\mathbf{v}_1, \mathbf{v}_2, \ldots, \mathbf{v}_k\}$ **is a linearly independent subset of** $\mathbb{R}^n$, **then** $k \leq n$. In the language of logic, this last statement is just the *contrapositive* of Theorem 1.4.3 and so is an equivalent statement. Note that this does *not* say that $k \leq n$ guarantees linear independence.

Now let $\mathbf{v}$ and $\mathbf{w}$ be two nonzero linearly independent vectors in $\mathbb{R}^2$. We want to prove our earlier claim (Section 1.3) that the span of these two vectors is all of $\mathbb{R}^2$. Suppose that there is a vector $\mathbf{z} \in \mathbb{R}^2$ such that $\mathbf{z} \notin \text{span}\{\mathbf{v}, \mathbf{w}\}$. Exercise B1 implies that if $\mathbf{z} \notin \text{span}(\{\mathbf{v}, \mathbf{w}\})$, then $\{\mathbf{v}, \mathbf{w}, \mathbf{z}\}$ must be linearly independent. But this contradicts Theorem 1.4.3

since we would have three vectors in a two-dimensional space. Therefore, $\mathbf{z} \in span\{\mathbf{v}, \mathbf{w}\} = \mathbb{R}^2$. In this same way, we can show that three linearly independent vectors in $\mathbb{R}^3$ must span $\mathbb{R}^3$.

In the next section, we investigate the consequences of having a set of vectors in $\mathbb{R}^n$ that are linearly independent and that span $\mathbb{R}^n$ at the same time.

Exercises 1.4

A.

1. Determine whether each of the following sets of vectors is linearly independent in the appropriate space $\mathbb{R}^n$.

   a. $\begin{bmatrix} -1 \\ 2 \end{bmatrix}, \begin{bmatrix} 1 \\ -2 \end{bmatrix}$ b. $\begin{bmatrix} 1 \\ 1 \end{bmatrix}, \begin{bmatrix} 0 \\ 1 \end{bmatrix}$ c. $\begin{bmatrix} 1 \\ 2 \end{bmatrix}, \begin{bmatrix} 0 \\ -1 \end{bmatrix}, \begin{bmatrix} 3 \\ 4 \end{bmatrix}$

   d. $\begin{bmatrix} -3 \\ 1 \\ 5 \end{bmatrix}, \begin{bmatrix} -6 \\ -3 \\ 15 \end{bmatrix}$ e. $\begin{bmatrix} 1 \\ 1 \\ 0 \end{bmatrix}, \begin{bmatrix} 0 \\ 1 \\ 1 \end{bmatrix}, \begin{bmatrix} 0 \\ 1 \\ 0 \end{bmatrix}$

   f. $\begin{bmatrix} 2 \\ -3 \\ 1 \end{bmatrix}, \begin{bmatrix} 3 \\ -1 \\ 5 \end{bmatrix}, \begin{bmatrix} 1 \\ -4 \\ 3 \end{bmatrix}$

   g. $\begin{bmatrix} 2 \\ -3 \\ 1 \end{bmatrix}, \begin{bmatrix} 3 \\ -1 \\ 5 \end{bmatrix}, \begin{bmatrix} 1 \\ -4 \\ 3 \end{bmatrix}, \begin{bmatrix} 0 \\ -1 \\ 4 \end{bmatrix}$ h. $\begin{bmatrix} 1 \\ 2 \\ 3 \\ 0 \end{bmatrix}, \begin{bmatrix} 2 \\ 4 \\ 6 \\ 0 \end{bmatrix}$

   i. $\begin{bmatrix} 1 \\ 2 \\ 1 \\ 2 \end{bmatrix}, \begin{bmatrix} 2 \\ 3 \\ 2 \\ 3 \end{bmatrix}, \begin{bmatrix} 1 \\ 2 \\ 3 \\ 4 \end{bmatrix}, \begin{bmatrix} 1 \\ 1 \\ 1 \\ 1 \end{bmatrix}$ j. $\begin{bmatrix} 1 \\ 0 \\ 1 \\ 0 \end{bmatrix}, \begin{bmatrix} 0 \\ 1 \\ 0 \\ 1 \end{bmatrix}, \begin{bmatrix} 1 \\ 0 \\ 0 \\ 0 \end{bmatrix}, \begin{bmatrix} 0 \\ 1 \\ 0 \\ 0 \end{bmatrix}$

   k. $\begin{bmatrix} -1 \\ 2 \\ -3 \\ 4 \end{bmatrix}, \begin{bmatrix} 0 \\ 2 \\ 0 \\ 3 \end{bmatrix}, \begin{bmatrix} 1 \\ 2 \\ 3 \\ 4 \end{bmatrix}, \begin{bmatrix} 2 \\ 1 \\ 0 \\ 5 \end{bmatrix}, \begin{bmatrix} 1 \\ 2 \\ 3 \\ 4 \end{bmatrix}$

   l. $\begin{bmatrix} 1 \\ 1 \\ 1 \\ 1 \\ 1 \end{bmatrix}, \begin{bmatrix} 0 \\ 1 \\ 1 \\ 1 \\ 1 \end{bmatrix}, \begin{bmatrix} 0 \\ 0 \\ 1 \\ 1 \\ 1 \end{bmatrix}, \begin{bmatrix} 0 \\ 0 \\ 0 \\ 1 \\ 1 \end{bmatrix}, \begin{bmatrix} 0 \\ 0 \\ 0 \\ 0 \\ 1 \end{bmatrix}$

2. Show that the set A in Example 1.4.3 spans $\mathbb{R}^3$ but is not linearly independent.

3. Show that the set B in Example 1.4.3 is linearly independent but does not span $\mathbb{R}^3$.

4. Show that the set C in Example 1.4.3 is linearly independent and also spans $\mathbb{R}^3$.

5. Show that the set D in Example 1.4.3 is linearly dependent and does not span $\mathbb{R}^3$.

6. a. Show that the vectors $\begin{bmatrix} a \\ b \end{bmatrix}$ and $\begin{bmatrix} c \\ d \end{bmatrix}$ in $\mathbb{R}^2$ are linearly dependent if and only if $ad - bc = 0$.

    b. For what values of r are the vectors $\begin{bmatrix} r \\ 1 \end{bmatrix}$ and $\begin{bmatrix} r+2 \\ r \end{bmatrix}$ linearly independent?

7. Suppose that $U = \{\mathbf{u}_1, \mathbf{u}_2\}$ and $V = \{\mathbf{v}_1, \mathbf{v}_2\}$ are linearly independent subsets of $\mathbb{R}^3$. Give a geometrical description of the intersection $span(U) \cap span(V)$.

8. Express a general vector $\begin{bmatrix} x \\ y \\ z \end{bmatrix}$ in $\mathbb{R}^3$ as a linear combination of the vectors $\begin{bmatrix} 1 \\ 2 \\ 1 \end{bmatrix}, \begin{bmatrix} 1 \\ 0 \\ -1 \end{bmatrix}$, and $\begin{bmatrix} 1 \\ -2 \\ 1 \end{bmatrix}$.

9. Let $\mathbf{u} = \begin{bmatrix} 2 \\ 1 \\ 1 \end{bmatrix}$ and $\mathbf{v} = \begin{bmatrix} 2 \\ t \\ 2t \end{bmatrix}$. Find all values of t (if any), for which $\mathbf{u}$ and $\mathbf{v}$ are linearly dependent.

10. Determine a *maximal* set of linearly independent vectors from the set $S = \left\{ [1 \ -1 \ -4 \ 0]^{\mathrm{T}}, [1 \ 1 \ 2 \ 4]^{\mathrm{T}}, \ [2 \ 1 \ 1 \ 6]^{\mathrm{T}}, [2 \ -1 \ -5 \ 2]^{\mathrm{T}} \right\}$, that is, find a linearly independent subset of S such that adding any other vector from S to it renders the subset linearly dependent.

11. Show that the set $\mathcal{E}_n = \{\mathbf{e}_1, \mathbf{e}_2, \dots, \mathbf{e}_n\}$ is linearly independent in $\mathbb{R}^n$. (See Section 1.3 for the definition of the vectors $\mathbf{e}_i, i = 1, 2, \dots, n$.)

B.

1. Let S be a linearly independent set of vectors in $\mathbb{R}^n$. Suppose that $\mathbf{v}$ is a vector in $\mathbb{R}^n$ that is not in the span of S. Prove that the set $S \cup \{\mathbf{v}\}$ is linearly independent.

2. If $\mathbf{u}_1$ and $\mathbf{u}_2$ are linearly independent in $\mathbb{R}^n$ and $\mathbf{w}_1 = a\mathbf{u}_1 + b\mathbf{u}_2$, $\mathbf{w}_2 = c\mathbf{u}_1 + d\mathbf{u}_2$, show that $\mathbf{w}_1$ and $\mathbf{w}_2$ are linearly independent if and only if $ad \neq bc$.

3. Prove that if v_1, v_2, and v_3 are linearly independent in $\mathbb{R}^n$, then so are the vectors $x = v_1 + v_2, y = v_1 + v_3$, and $z = v_2 + v_3$.

4. Suppose that $S = \{v_1, v_2, v_3\}$ is a linearly independent set of vectors in $\mathbb{R}^n$. Are the vectors $v_1 - v_2, v_2 - v_3$, and $v_3 - v_1$ linearly independent? Explain.

5. Suppose that $S = \{v_1, v_2, v_3\}$ is a linearly independent set of vectors in $\mathbb{R}^n$. Show that $T = \{w_1, w_2, w_3\}$ is also linearly independent, where $w_1 = v_1 + v_2 + v_3$, $w_2 = v_2 + v_3$, and $w_3 = v_3$.

6. Suppose that $S = \{v_1, v_2, v_3\}$ is a linearly dependent set of vectors in $\mathbb{R}^n$. Is $T = \{w_1, w_2, w_3\}$, where $w_1 = v_1$, $w_2 = v_1 + v_2$, and $w_3 = v_1 + v_2 + v_3$, linearly dependent or linearly independent? Justify your answer.

7. For what values of the scalar a are the vectors
$$\begin{bmatrix} a \\ 1 \\ 0 \end{bmatrix}, \begin{bmatrix} 1 \\ a \\ 1 \end{bmatrix}, \text{ and } \begin{bmatrix} 0 \\ 1 \\ a \end{bmatrix}$$ linearly independent, and for what values of a are they linearly dependent? Explain your reasoning.

8. Find an integer k so that the following set is linearly dependent:
$$\left\{ \begin{bmatrix} k \\ -1 \\ 1 \end{bmatrix}, \begin{bmatrix} 1 \\ k \\ -1 \end{bmatrix}, \begin{bmatrix} -1 \\ 1 \\ k \end{bmatrix} \right\}.$$

9. Find three vectors in $\mathbb{R}^3$ that are linearly dependent, but are such that any two of them are linearly independent.

10. a. Let S_1 and S_2 be finite subsets of vectors in $\mathbb{R}^n$ such that $S_1 \subseteq S_2$. If S_2 is linearly dependent, show by examples in $\mathbb{R}^3$ that S_1 may be either linearly dependent or linearly independent.
    b. Let S_1 and S_2 be finite subsets of vectors in $\mathbb{R}^n$ such that $S_1 \subseteq S_2$. If S_1 is linearly independent, show by examples in $\mathbb{R}^3$ that S_2 may be either linearly dependent or linearly independent.

11. Let $S = \{v_1, v_2, \ldots, v_k\}$ be a set of vectors in $\mathbb{R}^n$. Show that if one of these vectors is the zero vector, then S is linearly dependent.

12. a. Prove that any subset of a linearly independent set of vectors is linearly independent.
    b. Prove that any set of vectors containing a linearly dependent subset is linearly dependent.

13.   Prove that any linearly independent set $S = \{\mathbf{v}_1, \mathbf{v}_2, \ldots, \mathbf{v}_n\}$ consisting of n vectors in $\mathbb{R}^n$ must span $\mathbb{R}^n$. [*Hint*: Use the result of Problem B1 of Exercises 1.4.]

C.

1.   If

$$\mathbf{v}_1 = \mathbf{u}_1 + \mathbf{u}_2 + \mathbf{u}_3$$
$$\mathbf{v}_2 = \mathbf{u}_1 + \alpha\mathbf{u}_2$$
$$\mathbf{v}_3 = \mathbf{u}_2 + \beta\mathbf{u}_3,$$

where $\mathbf{u}_1$, $\mathbf{u}_2$, and $\mathbf{u}_3$ are given linearly independent vectors, find the conditions that must be satisfied by α and β to ensure that $\mathbf{v}_1$, $\mathbf{v}_2$, and $\mathbf{v}_3$ are linearly independent.

1.5 Bases

Figure 1.13 indicates that there need be no connection between the concepts of linear independence and spanning. However, a set that is both linearly independent and spanning provides valuable insights into the structure of a Euclidean space.

Definition 1.5.1

A nonempty set $S = \{\mathbf{v}_1, \mathbf{v}_2, \ldots, \mathbf{v}_k\}$ of vectors in $\mathbb{R}^n$ is called a **basis** for $\mathbb{R}^n$ if both the following conditions hold:

(1) The set S is linearly independent.
(2) The set S spans $\mathbb{R}^n$.

In $\mathbb{R}^n$, the vectors $\mathbf{e}_1 = \begin{bmatrix} 1 \\ 0 \\ \vdots \\ 0 \end{bmatrix}$, $\mathbf{e}_2 = \begin{bmatrix} 0 \\ 1 \\ \vdots \\ 0 \end{bmatrix}, \ldots, \mathbf{e}_3 = \begin{bmatrix} 0 \\ 0 \\ \vdots \\ 1 \end{bmatrix}$ constitute a special basis called the **standard basis** for $\mathbb{R}^n$. As we have noted before, these vectors span $\mathbb{R}^n$. The proof that these vectors are linearly independent constituted Exercise A11 in Section 1.4. From now on, we will denote the standard basis in $\mathbb{R}^n$ by $\mathcal{E}_n$: $\mathcal{E}_n = \{\mathbf{e}_1, \mathbf{e}_2, \ldots, \mathbf{e}_n\}$. We note that the vectors in the standard basis for $\mathbb{R}^n$ are **mutually orthogonal**: $\mathbf{e}_i \cdot \mathbf{e}_j = 0$ if $i \neq j$. Also, $\|\mathbf{e}_i\| = 1$, $i = 1, 2, \ldots, n$. A set of vectors that are mutually orthogonal and of unit length is called an **orthonormal set**. In particular, a basis consisting of vectors that are mutually orthogonal and of unit length is called an **orthonormal basis**. (See Exercises B5 through B7 for some properties of orthonormal bases.) We will discuss orthonormal sets and orthonormal bases more thoroughly in Chapters 7 and 8.

Example 1.5.1: A Basis for $\mathbb{R}^3$

We show that the set $B = \left\{ \begin{bmatrix} 2 \\ 1 \\ 1 \end{bmatrix}, \begin{bmatrix} 1 \\ 7 \\ 7 \end{bmatrix}, \begin{bmatrix} 4 \\ -1 \\ 0 \end{bmatrix} \right\}$ is a basis for $\mathbb{R}^3$.

First of all, B spans $\mathbb{R}^3$. If $\mathbf{x} = \begin{bmatrix} x_1 \\ x_2 \\ x_3 \end{bmatrix}$ is any vector in $\mathbb{R}^3$, then

the equation $a \begin{bmatrix} 2 \\ 1 \\ 1 \end{bmatrix} + b \begin{bmatrix} 1 \\ 7 \\ 7 \end{bmatrix} + c \begin{bmatrix} 4 \\ -1 \\ 0 \end{bmatrix} = \begin{bmatrix} x_1 \\ x_2 \\ x_3 \end{bmatrix}$ is equiva-

lent to the system of linear equations:

(1) $2a + b + 4c = x_1$.
(2) $a + 7b - c = x_2$.
(3) $a + 7b = x_3$.

Equation (2) $-$ equation (3) yields $-c = x_2 - x_3$, or $c = x_3 - x_2$. Equation (1) $-$ twice equation (3) gives us $-13b + 4c = x_1 - 2x_3$, or (replacing c by $x_3 - x_2$) $b = -\frac{1}{13}(x_1 - 6x_3 + 4x_2)$. Finally, equation (3) implies that $a = x_3 - 7 b = x_3 + \frac{7}{13}(x_1 - 6x_3 + 4x_2)$. Thus any vector in $\mathbb{R}^3$ can be expressed as a linear combination of vectors in B.

From the formulas for a, b, and c we have just derived, we see that if $\begin{bmatrix} x_1 \\ x_2 \\ x_3 \end{bmatrix} = \begin{bmatrix} 0 \\ 0 \\ 0 \end{bmatrix}$, then $a = b = c = 0$. This says that B is a linearly independent set of vectors. As a linearly independent spanning set of vectors in $\mathbb{R}^3$, B is a basis for $\mathbb{R}^3$.

The next example shows that $\mathbb{R}^2$ has an orthonormal basis other than the standard basis. The validity and generalization of the constructive process demonstrated in this example is shown in Section 7.4. The crucial concept is that of an *orthogonal projection* (see the explanation given between Problems C6 and C7 of Exercises 1.2).

Example 1.5.2: Constructing an Orthonormal Basis for $\mathbb{R}^2$

It is easy to show that the set $B = \{\mathbf{v}_1, \mathbf{v}_2\} = \left\{ \begin{bmatrix} 1 \\ 2 \end{bmatrix}, \begin{bmatrix} -3 \\ 4 \end{bmatrix} \right\}$ is a basis for $\mathbb{R}^2$. However, B is not an orthonormal basis: The vectors in B are not mutually orthogonal and neither vector in B has unit length. However, we illustrate a process (algorithm) that transforms the vectors of B into an orthonormal basis for $\mathbb{R}^2$.

Let $\mathbf{w}_1 = \mathbf{v}_1 = \begin{bmatrix} 1 \\ 2 \end{bmatrix}$ and calculate the vector

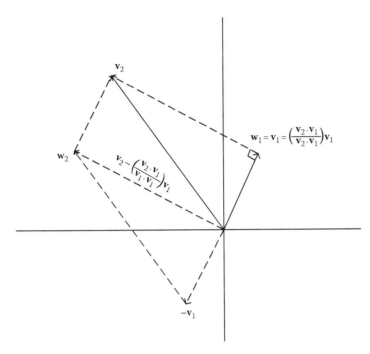

Figure 1.14 Constructing an orthonormal basis.

$$\mathbf{w}_2 = \mathbf{v}_2 - \left(\frac{\mathbf{v}_2 \bullet \mathbf{v}_1}{\mathbf{v}_1 \bullet \mathbf{v}_1}\right)\mathbf{v}_1 = \begin{bmatrix} -3 \\ 4 \end{bmatrix} - \left(\frac{\begin{bmatrix} -3 \\ 4 \end{bmatrix} \bullet \begin{bmatrix} 1 \\ 2 \end{bmatrix}}{\begin{bmatrix} 1 \\ 2 \end{bmatrix} \bullet \begin{bmatrix} 1 \\ 2 \end{bmatrix}}\right)\begin{bmatrix} 1 \\ 2 \end{bmatrix}$$

$$= \begin{bmatrix} -3 \\ 4 \end{bmatrix} - \frac{5}{5}\begin{bmatrix} 1 \\ 2 \end{bmatrix} = \begin{bmatrix} -4 \\ 2 \end{bmatrix}.$$

The vector $\left(\frac{\mathbf{v}_2 \bullet \mathbf{v}_1}{\mathbf{v}_1 \bullet \mathbf{v}_1}\right)\mathbf{v}_1$ is the *orthogonal projection* of $\mathbf{v}_2$ on $\mathbf{v}_1$, and $\mathbf{w}_2 = \mathbf{v}_2 - \left(\frac{\mathbf{v}_2 \bullet \mathbf{v}_1}{\mathbf{v}_1 \bullet \mathbf{v}_1}\right)\mathbf{v}_1$ is a vector orthogonal to $\mathbf{v}_1 = \mathbf{w}_1$ (Figure 1.14).

Algebraically, we have $\mathbf{w}_1 \bullet \mathbf{w}_2 = \begin{bmatrix} 1 \\ 2 \end{bmatrix} \bullet \begin{bmatrix} -4 \\ 2 \end{bmatrix} = -4 + 4 = 0.$

We can verify that $\{\mathbf{w}_1, \mathbf{w}_2\} = \left\{\begin{bmatrix} 1 \\ 2 \end{bmatrix}, \begin{bmatrix} -4 \\ 2 \end{bmatrix}\right\}$ is a basis for $\mathbb{R}^2$. All that is left is to "normalize" these vectors by dividing each by its length:

$$\hat{\mathbf{w}}_1 = \frac{\mathbf{w}_1}{\|\mathbf{w}_1\|} = \frac{\begin{bmatrix} 1 \\ 2 \end{bmatrix}}{\sqrt{5}} = \begin{bmatrix} 1/\sqrt{5} \\ 2/\sqrt{5} \end{bmatrix},$$

$$\hat{\mathbf{w}}_2 = \frac{\mathbf{w}_2}{\|\mathbf{w}_2\|} = \frac{\begin{bmatrix} -4 \\ 2 \end{bmatrix}}{\sqrt{20}} = \begin{bmatrix} -2/\sqrt{5} \\ 1/\sqrt{5} \end{bmatrix}.$$

The result is that $\{\hat{\mathbf{w}}_1, \hat{\mathbf{w}}_2\} = \left\{ \begin{bmatrix} 1/\sqrt{5} \\ 2/\sqrt{5} \end{bmatrix}, \begin{bmatrix} -2/\sqrt{5} \\ 1/\sqrt{5} \end{bmatrix} \right\}$ is an orthonormal basis for $\mathbb{R}^2$.

Examples 1.5.1 and 1.5.2 imply that, for any positive integer value of n, the space $\mathbb{R}^n$ may have many bases (the plural of *basis*). In fact, any space $\mathbb{R}^n$ has *infinitely many* bases: It is easy to verify that if the vectors $\mathbf{v}_1, \mathbf{v}_2, \ldots, \mathbf{v}_m$ constitute a basis, then the vectors $\alpha\mathbf{v}_1, \mathbf{v}_2, \ldots, \mathbf{v}_m$, where α is any scalar, also form a basis. However, the next theorem reveals that *any basis for a given $\mathbb{R}^n$ must have the same number of elements*. Theorem 1.4.2 implies that this number is less than or equal to n. An important consequence of the next theorem is that **the number of vectors in any basis for $\mathbb{R}^n$ must be exactly n**.

Theorem 1.5.1

Any two bases for $\mathbb{R}^n$ must have the same number of vectors.

Proof   Suppose $B = \{\mathbf{u}_1, \mathbf{u}_2, \ldots, \mathbf{u}_k\}$ and $T = \{\mathbf{v}_1, \mathbf{v}_2, \ldots, \mathbf{v}_r\}$ are two bases for $\mathbb{R}^n$, where n is a given positive integer. We will show that $k = r$.

As bases, both B and T are linearly independent and span $\mathbb{R}^n$. First let us focus on the fact that B is a spanning set for $\mathbb{R}^n$ and that the vectors $\mathbf{v}_i$ in T are linearly independent. Then we can apply Theorem 1.4.2 to conclude that $r \leq k$. Reversing the roles of B and T—so that T is viewed as spanning $\mathbb{R}^n$ and B is linearly independent—we use Theorem 1.4.2 again to see that $k \leq r$. The only way we can have both $r \leq k$ and $k \leq r$ is if $k = r$.

Corollary 1.5.1

Any basis for $\mathbb{R}^n$ must have n elements.

Proof   The standard basis $\{\mathbf{e}_1, \mathbf{e}_2, \ldots, \mathbf{e}_n\}$ for $\mathbb{R}^n$ contains n elements, and therefore, by Theorem 1.5.1, *every* basis for $\mathbb{R}^n$ contains n vectors.

Later, in Chapter 5, when we discuss the more general (and more abstract) notion of a *vector space* V (instead of just $\mathbb{R}^n$), we will define the *dimension* of such a space as the number of elements in a basis for V.

Although any basis for $\mathbb{R}^n$ must contain n elements, not every set of n vectors in $\mathbb{R}^n$ is a basis—see Example 1.4.3(d), for instance. However, as we will see shortly, **every set of n linearly independent vectors in $\mathbb{R}^n$ must span $\mathbb{R}^n$ and so must be a basis**.

Theorem 1.4.2 states that the number of vectors in any linearly independent subset of $\mathbb{R}^n$ is less than or equal to the number of vectors in any spanning set for $\mathbb{R}^n$. The next result indicates how we can fill in this "gap" between linearly independent sets and spanning sets.

Theorem 1.5.2

If S is a finite spanning set for some Euclidean space $\mathbb{R}^n$ and if I is a linearly independent subset of $\mathbb{R}^n$, such that $I \subseteq S$, then there is a basis B of $\mathbb{R}^n$, such that $I \subseteq B \subseteq S$.

Proof Suppose that I, a subset of a spanning set S, is a linearly independent set. We note first that if $span(I) = \mathbb{R}^n$, then I must be a basis by definition. Now suppose that $span(I) \neq \mathbb{R}^n$. This implies that $span(I) \subset \mathbb{R}^n$ and that $I \subset S$, or else I would be a spanning set. Then there exists at least one element $\mathbf{s}_1 \in S\backslash I$ such that $\mathbf{s}_1 \notin span(I)$: If this last condition were not true, then every element of $S\backslash I$ would belong to $span(I)$, so that $\mathbb{R}^n = span(S) \subseteq span(I)$ and $\mathbb{R}^n = span(I)$, a contradiction.

   Now we can see that $I \cup \{\mathbf{s}_1\}$ is a linearly independent set. Otherwise, $\mathbf{s}_1 \in span(I)$, another contradiction (see Exercises 1.4, Problem B1). If $I \cup \{\mathbf{s}_1\}$ spans $\mathbb{R}^n$, then it is a basis and we can let $B = I \cup \{\mathbf{s}_1\}$. If $I \cup \{\mathbf{s}_1\}$ does not span $\mathbb{R}^n$, we can repeat our argument to find an element $\mathbf{s}_2 \in S\backslash(I \cup \{\mathbf{s}_1\})$ with $I \cup \{\mathbf{s}_1, \mathbf{s}_2\}$ linearly independent. Because S is a finite set by hypothesis, if we continue this process, we see that for some value m the set, $B = I \cup \{\mathbf{s}_1, \mathbf{s}_2, \dots, \mathbf{s}_m\}$, is a basis of $\mathbb{R}^n$ with $I \subset B \subseteq S$.

The following consequence of Theorem 1.5.2 is very useful when working with linearly independent vectors, and will be invoked often in later sections. The corollary asserts that we can add enough vectors to any linearly independent subset to produce a basis, that is, an expanded version of the linearly independent set that also spans the space.

Corollary 1.5.2

Every linearly independent subset I, of a Euclidean space $\mathbb{R}^n$, can be extended to form a basis.

Proof Suppose that I is a linearly independent subset of $\mathbb{R}^n$ that is not a basis for $\mathbb{R}^n$, and take $S = I \cup B$, where B is any basis for $\mathbb{R}^n$. Then S is a spanning set and, by Theorem 1.5.2, there is a basis $\hat{B}$ such that $I \subseteq \hat{B} \subseteq I \cup B$. This basis $\hat{B}$ is an extension of I.

   Of course, the extension method described in Corollary 1.5.2 is not unique. There are other ways to form a basis by extending a linearly independent set.

   The *Steinitz replacement technique*, used in the proof of Theorem 1.4.2, actually extended a linearly independent set of vectors to a basis, but we chose not to emphasize this fact at that time.

Corollary 1.5.3

Any set of n linearly independent vectors in $\mathbb{R}^n$ is a basis of $\mathbb{R}^n$.

Proof Suppose that I is a set of n linearly independent vectors. If I is not a basis, then it can be extended to form a basis for $\mathbb{R}^n$. But then the number of elements in the extended set is greater than or equal to $n+1$, and the set must be linearly dependent by Theorem 1.4.3, a contradiction. Therefore, I must already be a basis.

Example 1.5.3: Extending a Linearly Independent Set to a Basis

Suppose that we are given the set $S = \left\{ \begin{bmatrix} 1 \\ -1 \\ 1 \end{bmatrix}, \begin{bmatrix} 2 \\ 0 \\ 3 \end{bmatrix} \right\}$ in $\mathbb{R}^3$.

It is easy to check that S is linearly independent. However, by Corollary 1.5.1, S cannot be a basis for $\mathbb{R}^3$. Because any three linearly independent vectors in $\mathbb{R}^3$ constitute a basis for $\mathbb{R}^3$ (Corollary 1.5.3), S needs one more appropriate vector.

First we calculate $span(S) = \left\{ \begin{bmatrix} a+2b \\ -a \\ a+3b \end{bmatrix} : a, b \in \mathbb{R} \right\}$. Now we find any vector that does not belong to the span of S. For example, $\begin{bmatrix} 1 \\ 0 \\ 1 \end{bmatrix}$ does not fit the pattern of the vectors in $span(S)$. (If $a=0$, the third component would have to be 3/2 of the first component.) Then $\hat{S} = \left\{ \begin{bmatrix} 1 \\ -1 \\ 1 \end{bmatrix}, \begin{bmatrix} 2 \\ 0 \\ 3 \end{bmatrix}, \begin{bmatrix} 1 \\ 0 \\ 1 \end{bmatrix} \right\}$ is a linearly independent set by Problem B1 of Exercises 1.4. (Alternatively, we can consider a linear combination of the three vectors that equals the zero vector and see that all the scalar coefficients must be zero.) Thus $\hat{S}$, as a linearly independent set of three vectors in $\mathbb{R}^3$, must be a basis.

Any vector in $\mathbb{R}^n$ can be represented as a linear combination of basis vectors, and it is important to realize that any such representation is *unique*. The next result has far-reaching consequences; the converse of this theorem is also true (Exercise B1).

Theorem 1.5.3

If $B = \{\mathbf{v}_1, \mathbf{v}_2, \ldots, \mathbf{v}_n\}$ is a basis for $\mathbb{R}^n$, then every vector $\mathbf{v}$ in $\mathbb{R}^n$ can be written as a linear combination $\mathbf{v} = a_1\mathbf{v}_1 + a_2\mathbf{v}_2 + \cdots + a_n\mathbf{v}_n$ in one and only one way.

Proof We give a proof by contradiction. First of all, because B is a basis, any vector $\mathbf{v}$ in $\mathbb{R}^n$ can be written as a linear combination of vectors in B:

$$\mathbf{v} = a_1\mathbf{v}_1 + a_2\mathbf{v}_2 + \cdots + a_n\mathbf{v}_n.$$

Now suppose that $\mathbf{v}$ has another representation in terms of the vectors in B:

$$\mathbf{v} = b_1\mathbf{v}_1 + b_2\mathbf{v}_2 + \cdots + b_n\mathbf{v}_n.$$

Then

$$0 = \mathbf{v} - \mathbf{v} = a_1\mathbf{v}_1 + a_2\mathbf{v}_2 + \cdots + a_n\mathbf{v}_n - (b_1\mathbf{v}_1 + b_2\mathbf{v}_2 + \cdots + b_n\mathbf{v}_n)$$
$$= (a_1 - b_1)\mathbf{v}_1 + (a_2 - b_2)\mathbf{v}_2 + \cdots + (a_n - b_n)\mathbf{v}_n.$$

But the linear independence of the vectors $\mathbf{v}_i$ implies that $(a_1 - b_1) = (a_2 - b_2) = \cdots = (a_n - b_n) = 0$, that is, $a_1 = b_1, a_2 = b_2, \ldots, a_n = b_n$, so $\mathbf{v}$ has a unique representation as a linear combination of basis vectors.

We should notice that the proofs of Theorems 1.4.1 through 1.4.3 and 1.5.1 through 1.5.3 do not make explicit use of the fact that the vectors are n-tuples. These demonstrations use only definitions and the algebraic properties of vectors contained in Theorem 1.1.1. We shall see this abstract algebraic approach in its full glory beginning in Chapter 5.

Example 1.5.4: RGB Color Space

In terms of the perception of light by the human eye, experiments have shown that every color can be duplicated by using mixtures of the three primary colors, red, green, and blue.* There is, however, a difference between mixtures of *light* (as used on a computer screen) and mixtures of *pigment* an artist uses.[†]

In studies of color matching, involving the projection of light onto a screen, the standard basis vectors, $\mathbf{e}_1 = \begin{bmatrix} 1 \\ 0 \\ 0 \end{bmatrix}$,

$\mathbf{e}_2 = \begin{bmatrix} 0 \\ 1 \\ 0 \end{bmatrix}$, and $\mathbf{e}_3 = \begin{bmatrix} 0 \\ 0 \\ 1 \end{bmatrix}$, represent the projection of one foot-candle of **red, green**, and **blue**, respectively, onto a screen. The zero vector represents the color **black**, whereas the vector $[1 \quad 1 \quad 1]^\mathsf{T}$ represents **white**. The scalar multiple $c\mathbf{e}_i$ represents the projection of c foot-candles of the color represented by $\mathbf{e}_i$, that is, the *intensity* or *brightness* of the light is changed

---

* The theory was presented by Thomas Young in 1802 and developed further by Hermann von Helmholtz in the 1850s. The theory was finally proved in a 1983 experiment by Dartnall, Bowmaker, and Mollon.
[†] See, for example, R.G. Gonzalez and R.E. Woods, *Digital Image Processing*, 2nd edn., Chap. 6 (Upper Saddle River, NJ: Prentice-Hall, 2002).

by scalar multiplication. Then a nonzero vector with nonnegative components,

$$\begin{bmatrix} x \\ y \\ z \end{bmatrix} = x\mathbf{e}_1 + y\mathbf{e}_2 + z\mathbf{e}_3,$$

describes the light produced by projecting x foot-candles of pure red, y foot-candles of pure green, and z foot-candles of pure blue onto a screen. Adding two vectors $\begin{bmatrix} x_1 \\ x_2 \\ x_3 \end{bmatrix}$ and $\begin{bmatrix} y_1 \\ y_2 \\ y_3 \end{bmatrix}$

yields $\begin{bmatrix} x_1 + y_1 \\ x_2 + y_2 \\ x_3 + y_3 \end{bmatrix}$, which represents the light produced by projecting $x_1 + y_1$, $x_2 + y_2$, and $x_3 + y_3$ foot-candles of red, green, and blue, respectively, onto a screen simultaneously. The color

purple, for example, can be represented as $\begin{bmatrix} c \\ 0 \\ c \end{bmatrix}$, a mixture of

red and blue in which the colors are equally bright.

In summary, we have described the *CIE-RGB* color space*, a subset of $\mathbb{R}^3$, to be the set $V = \{c_1\mathbf{e}_1 + c_2\mathbf{e}_2 + c_3\mathbf{e}_3 | c_i \geq 0, \ i = 1,2,3\}$. If we restrict the coefficients so that $0 \leq c_i \leq 1$, then the coefficients represent the percentage of each pure color in the mixture. Figure 1.15

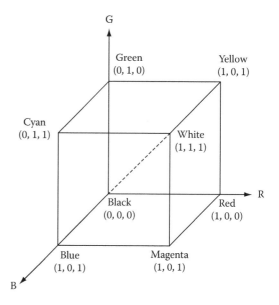

Figure 1.15 The RGB color cube.

* CIE stands for the *Commission Internationale de l'Éclairage* (International Commission on Illumination).

shows the *RGB color cube*, in which the dotted line represents shades of gray.

1.5.1 Coordinates/Change of Basis

If $B = \{\mathbf{v}_1, \mathbf{v}_2, \ldots, \mathbf{v}_n\}$ is a basis for $\mathbb{R}^n$, and vector $\mathbf{v}$ in $\mathbb{R}^n$ is written as a linear combination of these basis vectors, $\mathbf{v} = a_1\mathbf{v}_1 + a_2\mathbf{v}_2 + \cdots + a_n\mathbf{v}_n$, then the uniquely defined scalars a_i (Theorem 1.5.3) are called the **coordinates of the vector v relative to** (or **with respect to) the basis B**. Thus, whenever we use a particular basis in $\mathbb{R}^n$, we are establishing a coordinate system according to which any vector in $\mathbb{R}^n$ is identified uniquely by the n-tuple of its coordinates $(a_1, a_2, \ldots, a_n)$ with respect to the basis vectors $\mathbf{v}_1, \mathbf{v}_2, \ldots, \mathbf{v}_n$. The scalar multiples of these basis vectors serve as **coordinate axes** for $\mathbb{R}^n$. We will use the notation $[\mathbf{v}]_B$ to denote the unique **coordinate vector** corresponding to vector $\mathbf{v}$ relative to basis B: $[\mathbf{v}]_B = \begin{bmatrix} a_1 \\ a_2 \\ \vdots \\ a_n \end{bmatrix}$. To be precise, we must have an *ordered* basis (i.e., we should think of our basis vectors in a particular order) so that the coordinate vector with respect to this basis is unambiguous.

If basis B for a space $\mathbb{R}^n$ is the standard basis $\mathcal{E}_n$ and $\mathbf{v} \in \mathbb{R}^n$, then $\mathbf{v}$ is its own coordinate vector relative to B: $\mathbf{v} = [\mathbf{v}]_{\mathcal{E}_n}$. For example, take the vector $\mathbf{v} = \begin{bmatrix} 1 \\ -5 \\ 24 \end{bmatrix}$ in $\mathbb{R}^3$. Then $\mathbf{v} = \begin{bmatrix} 1 \\ -5 \\ 24 \end{bmatrix} = (1)\mathbf{e}_1 + (-5)\mathbf{e}_2 + (24)\mathbf{e}_3$, so the coordinate vector corresponding to $\mathbf{v}$ is $[\mathbf{v}]_{\mathcal{E}_3} = \begin{bmatrix} 1 \\ -5 \\ 24 \end{bmatrix} = \mathbf{v}$. Thus we should always assume that **a vector in $\mathbb{R}^n$, given without explicit reference to a basis, may be taken as the coordinate vector relative to $\mathcal{E}_n$.**

Example 1.5.5: Two Coordinate Systems for $\mathbb{R}^3$

If we adopt the standard ordered basis $\mathcal{E}_3 = \{\mathbf{e}_1, \mathbf{e}_2, \mathbf{e}_3\}$ for $\mathbb{R}^3$, then every vector $\mathbf{v}$ in $\mathbb{R}^3$ is represented uniquely as $x\mathbf{e}_1 + y\mathbf{e}_2 + z\mathbf{e}_3$, or by the coordinate vector $\begin{bmatrix} x \\ y \\ z \end{bmatrix}$.

On the other hand, if we use the basis $B = \left\{ \begin{bmatrix} 2 \\ 1 \\ 1 \end{bmatrix}, \begin{bmatrix} 1 \\ 7 \\ 7 \end{bmatrix}, \begin{bmatrix} 4 \\ -1 \\ 0 \end{bmatrix} \right\}$ of Example 1.5.1, then a vector we think of as $\begin{bmatrix} x \\ y \\ z \end{bmatrix}$, with respect to the standard basis $\mathcal{E}_3$, is

represented by the coordinate vector $\begin{bmatrix} z + \frac{7}{13}(x - 6z + 4y) \\ -\frac{1}{13}(x - 6z + 4y) \\ z - y \end{bmatrix}$.

For example, the vector, whose representation with respect to the standard basis is $\begin{bmatrix} 13 \\ 0 \\ 0 \end{bmatrix}$, has coordinate vector,

$\begin{bmatrix} 0 + \frac{7}{13}(13 - 6(0) + 4(0)) \\ -\frac{1}{13}(13 - 6(0) + 4(0)) \\ 0 - 0 \end{bmatrix} = \begin{bmatrix} 7 \\ -1 \\ 0 \end{bmatrix}$,   relative   to   the

basis B.

An equation of the form $\begin{bmatrix} x \\ y \\ z \end{bmatrix}_{B_1} = \begin{bmatrix} u \\ v \\ w \end{bmatrix}_{B_2}$ will indicate that a vector,

whose coordinates with respect to an ordered basis B_1 are x, y, and z, has coordinates u, v, and w with respect to the ordered basis B_2. Thus,

in Example 1.5.4, we can write $\begin{bmatrix} 13 \\ 0 \\ 0 \end{bmatrix}_{\mathcal{E}_3} = \begin{bmatrix} 7 \\ -1 \\ 0 \end{bmatrix}_B$. We will devote

more time to this idea of changing bases in later chapters, starting with Chapter 4.

Exercises 1.5

A.

1. Determine whether each of the following sets of vectors is a basis for the appropriate space $\mathbb{R}^n$ and explain your answer.

a. $\left\{ \begin{bmatrix} 2 \\ -3 \end{bmatrix} \right\}$ b. $\left\{ \begin{bmatrix} 1 \\ 1 \end{bmatrix}, \begin{bmatrix} 0 \\ -1 \end{bmatrix} \right\}$ c. $\left\{ \begin{bmatrix} 1 \\ 2 \end{bmatrix}, \begin{bmatrix} 3 \\ 5 \end{bmatrix}, \begin{bmatrix} -6 \\ 4 \end{bmatrix} \right\}$

d. $\left\{ \begin{bmatrix} -1 \\ 0 \\ 2 \end{bmatrix}, \begin{bmatrix} 0 \\ 0 \\ 0 \end{bmatrix}, \begin{bmatrix} 5 \\ -2 \\ 7 \end{bmatrix} \right\}$ e. $\left\{ \begin{bmatrix} 2 \\ 1 \\ 1 \end{bmatrix}, \begin{bmatrix} 1 \\ 7 \\ 7 \end{bmatrix}, \begin{bmatrix} 4 \\ -1 \\ 0 \end{bmatrix} \right\}$

f. $\left\{ \begin{bmatrix} 0 \\ 2 \\ 3 \end{bmatrix}, \begin{bmatrix} -1 \\ 0 \\ 4 \end{bmatrix}, \begin{bmatrix} 3 \\ -2 \\ 7 \end{bmatrix}, \begin{bmatrix} 3 \\ 4 \\ -7 \end{bmatrix} \right\}$ g. $\left\{ \begin{bmatrix} 1 \\ 1 \\ 0 \end{bmatrix}, \begin{bmatrix} 1 \\ 2 \\ 3 \end{bmatrix}, \begin{bmatrix} -1 \\ 0 \\ 1 \end{bmatrix} \right\}$

h. $\left\{ \begin{bmatrix} -3 \\ 1 \\ 4 \end{bmatrix}, \begin{bmatrix} 0 \\ 2 \\ 9 \end{bmatrix}, \begin{bmatrix} 5 \\ 3 \\ 6 \end{bmatrix}, \begin{bmatrix} 1 \\ 2 \\ 3 \end{bmatrix} \right\}$

i. $\left\{ \begin{bmatrix} 1 \\ 1 \\ 1 \\ 1 \end{bmatrix}, \begin{bmatrix} 2 \\ 5 \\ 6 \\ 4 \end{bmatrix}, \begin{bmatrix} 1 \\ 2 \\ 3 \\ 2 \end{bmatrix}, \begin{bmatrix} 2 \\ 6 \\ 8 \\ 5 \end{bmatrix} \right\}$ j. $\left\{ \begin{bmatrix} 1 \\ -1 \\ 2 \\ 3 \end{bmatrix}, \begin{bmatrix} 4 \\ 3 \\ 1 \\ 2 \end{bmatrix}, \begin{bmatrix} 0 \\ 1 \\ 0 \\ 2 \end{bmatrix} \right\}$

2. Show that $\mathbf{u}_1 = \begin{bmatrix} \frac{1}{3\sqrt{2}} \\ \frac{1}{3\sqrt{2}} \\ \frac{-4}{3\sqrt{2}} \end{bmatrix}$, $\mathbf{u}_2 = \begin{bmatrix} \frac{2}{3} \\ \frac{2}{3} \\ \frac{1}{3} \end{bmatrix}$, and $\mathbf{u}_3 = \begin{bmatrix} \frac{1}{\sqrt{2}} \\ \frac{-1}{\sqrt{2}} \\ 0 \end{bmatrix}$ consti-

tute an orthonormal basis for $\mathbb{R}^3$.

3. Find a basis for $\mathbb{R}^3$ that includes the vectors

    a. $\begin{bmatrix} 1 \\ 0 \\ 2 \end{bmatrix}$ and $\begin{bmatrix} 0 \\ 1 \\ 3 \end{bmatrix}$, b. $\begin{bmatrix} 1 \\ 0 \\ 2 \end{bmatrix}$. [*Hint*: Use the technique of

    Example 1.5.3.]

4. Find a basis for $\mathbb{R}^4$ that includes the vectors $\begin{bmatrix} 1 \\ 0 \\ 1 \\ 0 \end{bmatrix}$ and $\begin{bmatrix} 0 \\ 1 \\ -1 \\ 0 \end{bmatrix}$.

    [*Hint*: Use the hint given in the previous exercise.]

5. In Example 1.5.4, the pure colors red, green, and blue are represented by the linearly independent vectors $\mathbf{e}_1, \mathbf{e}_2$, and $\mathbf{e}_3$. What is the physical significance of linear independence here?

6. Consider the ordered basis B for $\mathbb{R}^2$ consisting of the vectors $\begin{bmatrix} 1 \\ 2 \end{bmatrix}$ and $\begin{bmatrix} 3 \\ 4 \end{bmatrix}$. Suppose we know that $[\mathbf{x}]_B = \begin{bmatrix} 7 \\ 11 \end{bmatrix}$ for a particular vector $\mathbf{x} \in \mathbb{R}^2$. Find $\mathbf{x}$ with respect to $\mathcal{E}_2$.

7. Find a basis B for $\mathbb{R}^2$ such that $\begin{bmatrix} 1 \\ 2 \end{bmatrix}_B = \begin{bmatrix} 3 \\ 5 \end{bmatrix}_{\mathcal{E}_2}$ and $\begin{bmatrix} 3 \\ 4 \end{bmatrix}_B = \begin{bmatrix} 2 \\ 3 \end{bmatrix}_{\mathcal{E}_2}$.

8. Suppose that $B = \{\mathbf{u}, \mathbf{v}\}$ is a basis for $\mathbb{R}^2$, where $\mathbf{u}$ and $\mathbf{v}$ are pictured below.

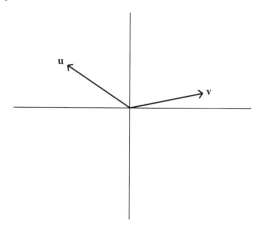

On a reproduction of this diagram, sketch the vector $\mathbf{x}$ with $[\mathbf{x}]_B = \begin{bmatrix} -1 \\ 2 \end{bmatrix}$.

9. a. Show that the vectors $\mathbf{v}_1 = \begin{bmatrix} 1 \\ 0 \\ -1 \end{bmatrix}$, $\mathbf{v}_2 = \begin{bmatrix} 1 \\ 2 \\ 1 \end{bmatrix}$, and $\mathbf{v}_3 = \begin{bmatrix} 0 \\ -3 \\ 2 \end{bmatrix}$ form a basis for $\mathbb{R}^3$.

   b. Express each of the standard basis vectors $\mathbf{e}_1, \mathbf{e}_2$, and $\mathbf{e}_3$ as a linear combination of $\mathbf{v}_1, \mathbf{v}_2$, and $\mathbf{v}_3$.

10. Find the coordinate vector of $\mathbf{u} = \begin{bmatrix} 1 \\ -1 \\ 1 \end{bmatrix}$ relative to the basis consisting of the vectors $\mathbf{v}_1 = \begin{bmatrix} 1 \\ 0 \\ 0 \end{bmatrix}$, $\mathbf{v}_2 = \begin{bmatrix} 1 \\ 1 \\ 0 \end{bmatrix}$, and $\mathbf{v}_3 = \begin{bmatrix} 1 \\ 1 \\ 1 \end{bmatrix}$.

11. Given the basis B, consisting of the vectors

$$\mathbf{v}_1 = \begin{bmatrix} 1 \\ 1 \\ 1 \\ 1 \end{bmatrix}, \mathbf{v}_2 = \begin{bmatrix} 0 \\ 1 \\ 1 \\ 1 \end{bmatrix}, \mathbf{v}_3 = \begin{bmatrix} 0 \\ 0 \\ 1 \\ 1 \end{bmatrix}, \mathbf{v}_4 = \begin{bmatrix} 0 \\ 0 \\ 0 \\ 1 \end{bmatrix},$$

write the coordinates of each vector $\mathbf{e}_1, \mathbf{e}_2, \mathbf{e}_3, \mathbf{e}_4$ of the standard basis for $\mathbb{R}^4$ relative to the basis B. Likewise, write the coordinates of each $\mathbf{v}_i$ with respect to the standard basis.

B.

1. Let $S = \{\mathbf{v}_1, \mathbf{v}_2, \ldots, \mathbf{v}_n\}$ be a set of n nonzero vectors in $\mathbb{R}^n$ such that every vector in $\mathbb{R}^n$ can be written in one and only one way as a linear combination of the vectors in S. Show that S is a basis for $\mathbb{R}^n$. (This is the *converse* of Theorem 1.5.3.)

2. Suppose that $B = \{\mathbf{v}_1, \mathbf{v}_2, \ldots, \mathbf{v}_n\}$ is a basis for $\mathbb{R}^n$ and let $a_1, a_2, \ldots, a_n$ be arbitrary nonzero scalars. Prove that $\tilde{B} = \{a_1\mathbf{v}_1, a_2\mathbf{v}_2, \ldots, a_n\mathbf{v}_n\}$ is a basis for $\mathbb{R}^n$.

3. Find a vector $\mathbf{v}$ in $\mathbb{R}^3$ such that the vectors $\begin{bmatrix} -1 \\ 2 \\ 3 \end{bmatrix}$, $\begin{bmatrix} 7 \\ 2 \\ 1 \end{bmatrix}$, and $\mathbf{v}$ are mutually orthogonal.

4. In $\mathbb{R}^n$, suppose the vectors $\mathbf{v}_1, \mathbf{v}_2, \ldots, \mathbf{v}_k$, where $k \leq n$ and no vector is the zero vector, are mutually orthogonal. Show that these vectors form a linearly independent set.

   [*Hint*: Suppose that you can find scalars a_i such that $\sum_{i=1}^{k} a_i\mathbf{v}_i = \mathbf{0}$. Now consider the dot product of each vector $\mathbf{v}_i$ with the sum $\sum_{i=1}^{k} a_i\mathbf{v}_i$.]

5. a. Prove that if $\mathbf{v}_1, \mathbf{v}_2, \ldots, \mathbf{v}_n$ is an orthonormal basis for $\mathbb{R}^n$ and $\mathbf{v} \in \mathbb{R}^n$, then

$$\mathbf{v} = \sum_{i=1}^{n} (\mathbf{v} \bullet \mathbf{v}_i)\mathbf{v}_i.$$

   b. Let $\mathbf{x} = \begin{bmatrix} 1 \\ 1 \\ 1 \end{bmatrix}$. Use part (a) to write $\mathbf{x}$ as a linear combination of the orthonormal basis vectors $\mathbf{u}_1, \mathbf{u}_2,$ and $\mathbf{u}_3$ defined in Exercise A2.

6. a. Suppose that $\{\mathbf{u}_1, \mathbf{u}_2, \ldots, \mathbf{u}_n\}$ is an orthonormal basis for $\mathbb{R}^n$. If $\mathbf{u} = \sum_{i=1}^{n} a_i\mathbf{u}_i$ and $\mathbf{v} = \sum_{i=1}^{n} b_i\mathbf{u}_i$, show that $\mathbf{u} \bullet \mathbf{v} = \sum_{i=1}^{n} a_ib_i$.

   b. If $\{\mathbf{u}_1, \mathbf{u}_2, \ldots, \mathbf{u}_n\}$ is an orthonormal basis for $\mathbb{R}^n$ and $\mathbf{v} = \sum_{i=1}^{n} c_i\mathbf{u}_i$, show that $\|\mathbf{v}\|^2 = \sum_{i=1}^{n} c_i^2$ [**Parseval's Formula**].

7. Let $\{\mathbf{u}_1, \mathbf{u}_2, \mathbf{u}_3\}$ be an orthonormal basis for $\mathbb{R}^3$. If $\mathbf{x} = c_1\mathbf{u}_1 + c_2\mathbf{u}_2 + c_3\mathbf{u}_3$ is a vector with the properties $\|\mathbf{x}\| = 5$, $\mathbf{u}_1 \bullet \mathbf{x} = 4$, and $\mathbf{x} \perp \mathbf{u}_2$, then what are the possible values of c_1, c_2, and c_3? [*Hint*: You may want to use the result of part (b) of the previous exercise.]

C.

1. Find all values of a for which $\left\{ \begin{bmatrix} a^2 \\ 0 \\ 1 \end{bmatrix}, \begin{bmatrix} 0 \\ a \\ 2 \end{bmatrix}, \begin{bmatrix} 1 \\ 0 \\ 1 \end{bmatrix} \right\}$ is a basis for $\mathbb{R}^3$.

2. Example 1.5.3 suggests that all colors can be represented by a subset V of $\mathbb{R}^3$. If $\begin{bmatrix} x \\ y \\ z \end{bmatrix}$ represents a color, the usual Euclidean norm (Definition 1.2.2) does not correspond to the *brightness* of the color. Define a norm $\|\cdot\|$ on this subset of $\mathbb{R}^3$ that would give a measure of brightness and satisfy (1) $\|c\mathbf{u}\| = |c|\|\mathbf{u}\|$ and (2) $\|\mathbf{u} + \mathbf{v}\| \leq \|\mathbf{u}\| + \|\mathbf{v}\|$ for all vectors $\mathbf{u}, \mathbf{v} \in V$ and all scalars c.

3. a. Show that the vectors

$$\boldsymbol{\varepsilon}_1 = \begin{bmatrix} 1 \\ 0 \\ 0 \\ \vdots \\ 0 \end{bmatrix}, \boldsymbol{\varepsilon}_2 = \begin{bmatrix} 1 \\ 1 \\ 0 \\ \vdots \\ 0 \end{bmatrix}, \ldots, \boldsymbol{\varepsilon}_n = \begin{bmatrix} 1 \\ 1 \\ 1 \\ \vdots \\ 1 \end{bmatrix}$$

form a basis for $\mathbb{R}^n$.

 b. If a vector $\mathbf{v}$ in $\mathbb{R}^n$ has the coordinates $(1, 2, \ldots, n)$, with respect to the standard basis $\mathbf{e}_1, \mathbf{e}_2, \ldots, \mathbf{e}_n$, what are the coordinates of $\mathbf{v}$ with respect to the basis $\boldsymbol{\varepsilon}_1, \boldsymbol{\varepsilon}_2, \ldots, \boldsymbol{\varepsilon}_n$?

1.6 Subspaces

Suppose we consider the set U of all vectors in $\mathbb{R}^3$ that are orthogonal to a fixed vector $\mathbf{x}$: $U = \{\mathbf{u} | \mathbf{u} \in \mathbb{R}^3$ and $\mathbf{u} \cdot \mathbf{x} = 0\}$. For example, if $\mathbf{x}$ were $\begin{bmatrix} 1 \\ 2 \\ 3 \end{bmatrix}$, then $\mathbf{u} = \begin{bmatrix} -7 \\ 2 \\ 1 \end{bmatrix}$ would be one of the vectors in U. Notice that the zero vector is also an element of U.

If $\mathbf{v}$ and $\mathbf{w}$ are in U, we see that $(\mathbf{v} + \mathbf{w}) \cdot \mathbf{x} = \mathbf{v} \cdot \mathbf{x} + \mathbf{w} \cdot \mathbf{x} = 0 + 0 = 0$, so $\mathbf{v} + \mathbf{w} \in U$. Also, if c is a scalar, then $(c\mathbf{v}) \cdot \mathbf{x} = c(\mathbf{v} \cdot \mathbf{x}) = c(0) = 0$, showing that $c\mathbf{v} \in U$. Thus this subset U of $\mathbb{R}^3$ is closed under the addition and scalar multiplication defined for $\mathbb{R}^3$. Furthermore, U "inherits" all the properties of vector operations stated in Theorem 1.1.1, and so we can think of U as a subset of $\mathbb{R}^3$ that has the same algebraic structure as the entire space $\mathbb{R}^3$. We call U a *subspace* of $\mathbb{R}^3$.

Definition 1.6.1

A nonempty subset W of $\mathbb{R}^n$ is called a **subspace** of $\mathbb{R}^n$ if, for every pair of vectors $\mathbf{u}$ and $\mathbf{v}$ in W and each scalar c,

(1) $\mathbf{u} + \mathbf{v} \in W$.
(2) $c\mathbf{u} \in W$.

It is worth repeating that if W is a subspace of $\mathbb{R}^n$, then W is a subset of $\mathbb{R}^n$ that has the same algebraic behavior as $\mathbb{R}^n$ itself. One immediate deduction from the definition is that any subspace must have the zero vector as an element—just let $c = 0$ in property (2). (Some treatments of subspaces make the presence of the zero vector part of the definition.) It should be clear that $\mathbb{R}^n$ and $\{\mathbf{0}\}$ qualify as subspaces of $\mathbb{R}^n$, usually referred to as **improper subspaces**. All other subspaces are called **proper subspaces**.

As we might expect, a **basis** for a subspace W of $\mathbb{R}^n$ is a linearly independent subset of W that spans W. The number of elements in any basis for W is called the **dimension** of the subspace W. (For a given subspace, we can establish that this number is unique by a method similar to that used in proving Theorem 1.5.1.) It is customary to take the dimension of the subspace $\{\mathbf{0}\}$ to be zero.

Example 1.6.1: A Subspace of $\mathbb{R}^3$

In $\mathbb{R}^3$, consider the subset $U = \left\{ \begin{bmatrix} x_1 \\ x_2 \\ x_3 \end{bmatrix} : x_3 = x_1 + x_2 \right\}$, the collection of all vectors in $\mathbb{R}^3$ whose third component is the sum of the first two components. Note that $\mathbf{0} \in U$. If

$$\mathbf{u} = \begin{bmatrix} u_1 \\ u_2 \\ u_3 \end{bmatrix} \text{ and } \mathbf{v} = \begin{bmatrix} v_1 \\ v_2 \\ v_3 \end{bmatrix} \text{ are in } U, \text{ then } \mathbf{u}+\mathbf{v} = \begin{bmatrix} u_1 \\ u_2 \\ u_3 \end{bmatrix} + \begin{bmatrix} v_1 \\ v_2 \\ v_3 \end{bmatrix}$$

$$= \begin{bmatrix} u_1+v_1 \\ u_2+v_2 \\ u_3+v_3 \end{bmatrix} = \begin{bmatrix} u_1+v_1 \\ u_2+v_2 \\ (u_1+u_2)+(v_1+v_2) \end{bmatrix} = \begin{bmatrix} u_1+v_1 \\ u_2+v_2 \\ (u_1+v_1)+(u_2+v_2) \end{bmatrix} \in U$$

and $c\mathbf{u} = \begin{bmatrix} cu_1 \\ cu_2 \\ cu_3 \end{bmatrix} = \begin{bmatrix} cu_1 \\ cu_2 \\ c(u_1+u_2) \end{bmatrix} = \begin{bmatrix} cu_1 \\ cu_2 \\ cu_1+cu_2 \end{bmatrix} \in U.$ Therefore, U

is a subspace of $\mathbb{R}^3$. (Geometrically, U is a plane through the origin, a two-dimensional object in Euclidean 3-space. See Figure 1.16.)

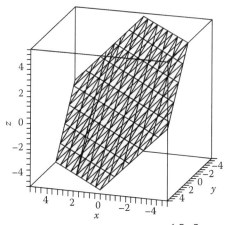

Figure 1.16 The two-dimensional subspace $U = \left\{ \begin{bmatrix} x \\ y \\ z \end{bmatrix} : z = x + y \right\}$, $-5 \le x, y, z \le 5$.

Example 1.6.2: A Subset of $\mathbb{R}^n$ That Is Not a Subspace

Suppose we consider the subset U of vectors in $\mathbb{R}^n$ whose Euclidean norms do not exceed 1: $U = \{\mathbf{u} \in \mathbb{R}^n : \|\mathbf{u}\| \le 1\}$. Once we think about it, it becomes almost painfully obvious that U is not a subspace. For instance, if $\mathbf{e}_1$ and $\mathbf{e}_2$ are the first two vectors of the standard basis for $\mathbb{R}^n$, then $\|\mathbf{e}_1\| = 1 = \|\mathbf{e}_2\|$, so that $\mathbf{e}_1, \mathbf{e}_2 \in U$; but $\mathbf{e}_1 + \mathbf{e}_2 = [1 \quad 1 \quad 0 \quad \cdots \quad 0]^\mathsf{T}$ and $\|\mathbf{e}_1 + \mathbf{e}_2\| = \sqrt{1^2 + 1^2 + 0^2 + \cdots + 0^2} = \sqrt{2} > 1$, so that U is not closed under addition. Also, if $\mathbf{u} \in U$ ($\mathbf{u} \ne \mathbf{0}$) and c is a scalar, then $\|c\mathbf{u}\| = |c|\|\mathbf{u}\|$, showing that $c\mathbf{u} \notin U$ if $|c| > \frac{1}{\|\mathbf{u}\|}$.

A useful and important fact is that **the span of any finite subset of $\mathbb{R}^n$ is a subspace of $\mathbb{R}^n$** (Exercise B1).

We will discuss subspaces in more detail in Chapter 5. For now, we will consider an important example that will motivate our way into Chapter 2.

Example 1.6.3: A Special Subspace of $\mathbb{R}^n$

Let V denote the set of all solutions of a homogeneous system of m linear equations in n unknowns

$$
\begin{aligned}
a_{11}x_1 + a_{12}x_2 + \cdots + a_{1n}x_n &= 0 \\
a_{21}x_1 + a_{22}x_2 + \cdots + a_{2n}x_n &= 0 \\
\vdots \qquad \vdots \qquad\qquad \vdots \qquad \vdots \\
a_{m1}x_1 + a_{m2}x_2 + \cdots + a_{mn}x_n &= 0
\end{aligned}
\tag{$*$}
$$

where the a_{ij}'s are constants. At one time we might have thought of a solution as an ordered n-tuple $(x_1, x_2, \ldots, x_n)$, but now we can interpret

a solution as a vector $\begin{bmatrix} x_1 \\ x_2 \\ \vdots \\ x_n \end{bmatrix}$ in $\mathbb{R}^n$. The zero vector is an obvious

element of V.

If $\mathbf{x} = \begin{bmatrix} x_1 \\ x_2 \\ \vdots \\ x_n \end{bmatrix}$ and $\mathbf{y} = \begin{bmatrix} y_1 \\ y_2 \\ \vdots \\ y_n \end{bmatrix}$ are two solutions of (*), then we

can substitute their sum, $\begin{bmatrix} x_1 + y_1 \\ x_2 + y_2 \\ \vdots \\ x_n + y_n \end{bmatrix}$, into the left side of (*),

obtaining m expressions of the form

$$a_{i1}(x_1 + y_1) + a_{i2}(x_2 + y_2) + \cdots + a_{in}(x_n + y_n),$$

or

$$[a_{i1}x_1 + a_{i2}x_2 + \cdots + a_{in}x_n] + [a_{i1}y_1 + a_{i2}y_2 + \cdots + a_{in}y_n].$$

But each expression in brackets equals 0 because $\mathbf{x}$ and $\mathbf{y}$ individually are solutions of (*). This says that $\mathbf{x} + \mathbf{y}$ is a solution of the system (*). Similarly, substituting $c\mathbf{x}$ into the left side of (*), we get m expressions of the form

$$a_{i1}(cx_1) + a_{i2}(cx_2) + \cdots + a_{in}(cx_n), \quad \text{or}$$
$$c[a_{i1}x_1 + a_{i2}x_2 + \cdots + a_{in}x_n],$$

each of which is 0. Thus $c\mathbf{x}$ is a solution of (*) for any scalar c.

We have shown that V is a subspace of $\mathbb{R}^n$. We will analyze the algebraic and geometric characteristics of this kind of subspace (a **null space**) in Chapters 2, 5, and 6.

Exercises 1.6

A.

1. Prove that $B = \left\{ \begin{bmatrix} 1 \\ 0 \\ 1 \end{bmatrix}, \begin{bmatrix} 0 \\ 1 \\ 1 \end{bmatrix} \right\}$ is a basis for the subspace U
    in Example 1.6.1.

2. Find a basis for the subspace spanned by each of the following
    set of vectors and indicate the dimension of each space.

a. $\left\{ \begin{bmatrix} 1 \\ 0 \\ 1 \\ 0 \end{bmatrix}, \begin{bmatrix} 1 \\ 0 \\ 1 \\ 1 \end{bmatrix} \right\}$ b. $\left\{ \begin{bmatrix} 0 \\ 1 \\ 1 \\ 1 \end{bmatrix}, \begin{bmatrix} 1 \\ 1 \\ 1 \\ 0 \end{bmatrix}, \begin{bmatrix} 1 \\ 1 \\ 0 \\ 1 \end{bmatrix} \right\}$

c. $\left\{ \begin{bmatrix} 0 \\ 1 \\ 0 \\ 1 \end{bmatrix}, \begin{bmatrix} 1 \\ 0 \\ 1 \\ 0 \end{bmatrix}, \begin{bmatrix} 0 \\ 0 \\ 1 \\ 1 \end{bmatrix} \right\}$

3.   Suppose that V is a two-dimensional subspace of $\mathbb{R}^3$ having the bases

$$B_1 = \{\mathbf{v}_1, \mathbf{v}_2\} = \left\{ \begin{bmatrix} 1 \\ -1 \\ 1 \end{bmatrix}, \begin{bmatrix} 0 \\ 1 \\ 1 \end{bmatrix} \right\} \text{ and}$$

$$B_2 = \{\mathbf{w}_1, \mathbf{w}_2\} = \left\{ \begin{bmatrix} 1 \\ 0 \\ 2 \end{bmatrix}, \begin{bmatrix} 1 \\ -2 \\ 0 \end{bmatrix} \right\}. \text{ Find the coordinate vec-}$$

tors for $\mathbf{w}_1$ and $\mathbf{w}_2$ relative to basis B_1.

4.   Suppose $V \neq \{\mathbf{0}\}$ is a k-dimensional subspace of $\mathbb{R}^n$ and $B = \{\mathbf{v}_1, \mathbf{v}_2, \ldots, \mathbf{v}_k\}$ is a basis for V. What is the coordinate vector of $\mathbf{v}_i$ with respect to B $(i = 1, 2, \ldots, k)$?

5.   Prove that the set $W = \left\{ \mathbf{u} = \begin{bmatrix} x \\ y \end{bmatrix} : y = 5x \right\}$ is a subspace of $\mathbb{R}^2$. Interpret W geometrically. What is the dimension of W?

6.   Prove that the set $W = \left\{ \begin{bmatrix} x \\ y \\ z \end{bmatrix} : x = y = z \right\}$ is a subspace of $\mathbb{R}^3$. Interpret W geometrically. What is the dimension of W?

7.   Is the following subset of $\mathbb{R}^3$ a subspace of $\mathbb{R}^3$? If not, what property is violated? If so, find a basis.

$$S = \left\{ \begin{bmatrix} x \\ y \\ z \end{bmatrix} : xy = 0, \, z \text{ arbitrary} \right\}$$

8.   Is $\mathbb{R}^3$ a subspace of $\mathbb{R}^4$? Explain your answer.

9.   Consider the set $S = \{[x_1 \, x_2 \ldots x_n]^T \in \mathbb{R}^n : x_i \geq 0, \, i = 1, 2, \ldots, n\}$. Is S a subspace of $\mathbb{R}^n$? Explain your answer.

10.   Suppose V is a subspace of $\mathbb{R}^n$ with basis $B = \{\mathbf{v}_1, \mathbf{v}_2, \ldots, \mathbf{v}_k\}$. If $\mathbf{x} \in V$ is orthogonal to each vector $\mathbf{v}_i \in B$, show that $\mathbf{x} = \mathbf{0}$.

B.

1.   Prove that the span of any finite subset of vectors in $\mathbb{R}^n$ is a subspace of $\mathbb{R}^n$. Why is the subset V of Example 1.5.4 not a subspace of $\mathbb{R}^3$?

2. Suppose that W is a subspace of $\mathbb{R}^n$. The **orthogonal comple-ment**, $W^\perp$, of W is the set of those vectors $\mathbf{v}$ in $\mathbb{R}^n$ that are orthogonal to all vectors in W:

$$W^\perp = \{\mathbf{v} \mid \mathbf{v} \in \mathbb{R}^n \text{ and } \mathbf{v} \bullet \mathbf{w} = 0 \text{ for all } \mathbf{w} \in W\}.$$

Show that $W^\perp$ is a subspace of $\mathbb{R}^n$. [$W^\perp$ is pronounced "W perp."]

3. Consider the subspace V of $\mathbb{R}^5$ spanned by the vector $\begin{bmatrix} 1 \\ 0 \\ -2 \\ 3 \\ 4 \end{bmatrix}$.

Find a basis for $V^\perp$. (See the previous exercise for the defin-ition of $V^\perp$, the **orthogonal complement** of V.)

4. Suppose that U is a subspace of V. Prove that the dimension of U is strictly less than the dimension of V if and only if $U \neq V$.

5. Give an example of two subspaces U and V of $\mathbb{R}^2$ which shows that having the dimension of U equal to the dimension of V does not imply that $U = V$.

C.

1. If B is a basis for a subspace W of $\mathbb{R}^n$, show that
    a. $[\mathbf{x} + \mathbf{y}]_B = [\mathbf{x}]_B + [\mathbf{y}]_B$ for $\mathbf{x}, \mathbf{y} \in W$
    b. $[k\mathbf{x}]_B = k[\mathbf{x}]_B$ for $\mathbf{x} \in W$ and k a scalar.

2. Let B be a basis of a subspace W of $\mathbb{R}^n$ and let $\mathbf{w}_1, \mathbf{w}_2, \ldots, \mathbf{w}_k$ be in W.
    a. Show that $\mathbf{w} \in W$ is a linear combination of $\mathbf{w}_1, \mathbf{w}_2, \ldots, \mathbf{w}_k$ if and only if $[\mathbf{w}]_B$ is a linear combination of $[\mathbf{w}_1]_B$, $[\mathbf{w}_2]_B, \ldots, [\mathbf{w}_k]_B$. (Assume the results of the previous exer-cise.)
    b. Show that $\{\mathbf{w}_1, \mathbf{w}_2, \ldots, \mathbf{w}_k\}$ is linearly independent if and only if $\{[\mathbf{w}_1]_B, [\mathbf{w}_2]_B, \ldots, [\mathbf{w}_k]_B\}$ is linearly independent.

3. If W_1 and W_2 are subspaces of the same Euclidean space $\mathbb{R}^n$, prove that $W_1 \cap W_2$ is a subspace of $\mathbb{R}^n$.

4. Suppose that W_1 and W_2 are nonempty proper subspaces of $\mathbb{R}^n$. Show that there exists an element $\mathbf{v} \in \mathbb{R}^n$ such that $\mathbf{v} \notin W_1$ and $\mathbf{v} \notin W_2$. Is this true for more than two subspaces?

5. If W_1 and W_2 are nonempty proper subspaces of $\mathbb{R}^n$, prove that $W_1 \cup W_2$ cannot be a subspace if $W_1 \not\subset W_2$ and $W_2 \not\subset W_1$.

6. Let p and q be fixed positive integers. Consider the pq row vectors $\mathbf{x}_{ij}$ such that $\mathbf{x}_{ij}^T \in \mathbb{R}^{p+q+1}$, $i = 1, 2, \ldots, p$ and

$j = 1, 2, \ldots, q$, and $\mathbf{x}_{ij}$ has 1's in the first, the $(i+1)$th, and the $(p+j+1)$th positions and 0's elsewhere. Show that the dimension of the span of these vectors is $p + q - 1$. (Exercise B1 establishes that such a span is a subspace.)

1.7 Summary

The dominant mathematical object in this chapter is **Euclidean n-space**, defined as

$$\mathbb{R}^n = \left\{ \begin{bmatrix} x_1 & x_2 & \cdots & x_n \end{bmatrix}^{\mathrm{T}} : x_i \in \mathbb{R} \text{ for } i = 1, 2, \ldots, n \right\},$$

consisting of **column vectors**, the **transposes** of **row vectors**. The positive integer n is called the **dimension** of the space. Euclidean n-space is endowed with both an algebraic and a geometric structure. Vectors in $\mathbb{R}^n$ can be added and subtracted in a component-by-component way and a vector can be multiplied by a **scalar**. These operations obey familiar algebraic laws, such as commutativity and associativity of addition and associativity and distributivity of scalar multiplication. In any space $\mathbb{R}^n$, there is a universal additive identity, and each vector has an additive inverse. Geometrically, we can define the **(Euclidean) inner product** (or **dot product**), a scalar quantity, from which we can derive the **length**, or **norm**, of a vector. The **Cauchy–Schwarz Inequality** for vectors $\mathbf{u}, \mathbf{v} \in \mathbb{R}^n$, $|\mathbf{u} \bullet \mathbf{v}| \leq \|\mathbf{u}\| \, \|\mathbf{v}\|$, enables us to define the **angle θ between nonzero vectors** $\mathbf{u}$ and $\mathbf{v}$ as the unique angle between 0 and π rad inclusive, satisfying $\cos \theta = \frac{\mathbf{u} \bullet \mathbf{v}}{\|\mathbf{u}\| \|\mathbf{v}\|}$. Vectors $\mathbf{v}$ and $\mathbf{w}$ are **orthogonal** (**perpendicular** in $\mathbb{R}^2$ or $\mathbb{R}^3$), denoted $\mathbf{v} \perp \mathbf{w}$, if $\mathbf{v} \bullet \mathbf{w} = 0$.

This chapter introduces the idea of **linear combination** of a nonempty finite set of vectors and the important related concept of **linearly independent** (**linearly dependent**) sets. A nonempty set S of vectors in $\mathbb{R}^n$ **spans** $\mathbb{R}^n$ (or is a **spanning set** for $\mathbb{R}^n$) if every vector in $\mathbb{R}^n$ is an element of span(S), that is, if every vector in $\mathbb{R}^n$ is a linear combination of vectors in S. The size of any linearly independent set is always less than or equal to the size of any spanning set. If the number of vectors in any subset A of $\mathbb{R}^n$ exceeds n, then A must be linearly dependent.

A nonempty set S of vectors in $\mathbb{R}^n$ is called a **basis** for $\mathbb{R}^n$ if S is a linearly independent set that spans $\mathbb{R}^n$. The vectors $\mathbf{e}_1 = \begin{bmatrix} 1 \\ 0 \\ \vdots \\ 0 \end{bmatrix}$,

$\mathbf{e}_2 = \begin{bmatrix} 0 \\ 1 \\ \vdots \\ 0 \end{bmatrix}, \ldots, \mathbf{e}_n = \begin{bmatrix} 0 \\ 0 \\ \vdots \\ 1 \end{bmatrix}$ constitute a special basis $\mathcal{E}_n$ called the **standard basis** for $\mathbb{R}^n$. This basis consists of vectors that are mutually orthogonal and of unit length, and is an example of an **orthonormal basis**.

Although $\mathbb{R}^n$ has infinitely many bases, the number of vectors in any basis for $\mathbb{R}^n$ must be n. Any vector in $\mathbb{R}^n$ can be written *uniquely* as a linear combination of basis vectors. This uniqueness leads to the concept of a **coordinate system**, according to which any vector in $\mathbb{R}^n$ is identified uniquely by its coordinates.

An important concept is that of a **subspace** of $\mathbb{R}^n$, a nonempty subset of $\mathbb{R}^n$ that has the same algebraic structure as $\mathbb{R}^n$ itself. We can speak of a **basis** for a subspace and, consequently, the **dimension** of a subspace.

Systems of Equations

In Chapter 1, we studied vectors and some of their applications, introducing significant geometric and algebraic concepts that will be generalized in this chapter and in later chapters.

In examining the linear independence or linear dependence of vectors, we saw the important role played by systems of linear equations and their solutions. In virtually every field of study to which linear algebra can be applied, systems of linear equations occur and their solutions are important.

For example, in analyzing the electrical circuit

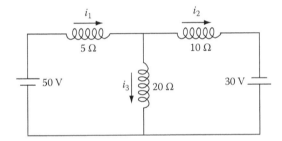

Kirchhoff's law* provides the following system of linear equations

$$i_1 - i_2 - i_3 = 0$$
$$5i_1 + 20i_3 = 50$$
$$10i_2 - 20i_3 = 30$$
$$5i_1 + 10i_2 = 80$$

where i_1, i_2, and i_3 represent the currents indicated in various parts of the circuit diagram.

This system is not difficult to solve manually by simple techniques; but a system of linear equations arising in some industrial application, for example, may involve thousands of variables and equations.[†] Solving such problems requires accurate and efficient methods, algorithms that can be implemented on computers in a way that minimizes computing time, and/or the use of memory.

---

* Named after the German physicist Gustav Robert Kirchhoff, who formulated several laws relating the voltages, resistances, and currents of a circuit in 1847.

† See N.J. Higham, *Accuracy and Stability of Numerical Algorithms*, 2nd edn., p. 191 (Philadelphia, PA: SIAM, 2002) for a list of large linear systems solved by computer. As of 2001, Higham cites a $525,000 \times 525,000$ system as the record holder.

In this chapter, we analyze systems of linear equations and their solutions geometrically, algebraically, and computationally. Along the way, we will introduce a significant extension of the vector concept—the *matrix*—which we will explore further in Chapter 3 and in succeeding chapters.

2.1 The Geometry of Systems of Equations in $\mathbb{R}^2$ and $\mathbb{R}^3$

There are three standard elementary methods of solving a system of two linear equations in two unknowns

$$ax + by = c$$
$$dx + ey = f$$

These are (1) graphing, (2) substitution, and (3) elimination.

Analytic geometry, the "marriage of algebra and geometry"* that Rene Descartes brought about, tells us that if we are trying to solve such a system, there are only three geometric (graphical) possibilities—and therefore only three equivalent algebraic possibilities:

I. The lines representing the equations intersect at a single point ↔ the system is **consistent** and has a **unique** solution (Figure 2.1a).
II. The lines are parallel and therefore do not intersect ↔ the system is **inconsistent** and has **no** solution (Figure 2.1b).
III. One line is a disguised, therefore equivalent, form of the other ↔ the system is **dependent** and there are **infinitely many** solutions (Figure 2.1c).

Algebraically, the **elimination method** for a system of two equations in two variables involves adding a multiple of one or both equations in the

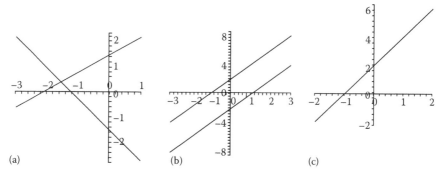

(a)   (b)   (c)

Figure 2.1 Linear systems in $\mathbb{R}^2$.

---

* See, for example, R. Mankiewicz, *The Story of Mathematics*, Chap. 11 (Princeton, NJ: Princeton University Press, 2001).

system to the other equation in the system or to each other. The goal is to eliminate one of the variables, so only one equation in one unknown remains. The **substitution method** requires solving one equation for one of the variables and then substituting that expression in the other equation, yielding one equation in one unknown.

Example 2.1.1: Elimination and Substitution: Two Equations, Two Variables

Suppose that we have the system of linear equations

$$5x + 3y = 9$$
$$2x - 4y = 14$$

If we multiply the first equation by 4 and the second equation by 3, we get

$$20x + 12y = 36$$
$$6x - 12y = 42$$

Adding the new equations *eliminates* y, and we find that $26x = 78$, so $x = 3$. Substituting this value of x in the original first equation, for example, we find that $y = -2$.

Alternatively, we could solve the second equation for x (for arithmetic/algebraic convenience) and then *substitute* for x in the first equation. This yields $5(7 + 2y) + 3y = 9$, so $y = -2$. Finally, we use this value of y in the formula $x = 7 + 2y$ to find that $x = 3$.

In dealing with a system having three equations in three unknowns, the possibilities are similar, but this time a linear equation has the form

$$ax + by + cz = d$$

and represents a *plane* in three-dimensional space (Figure 2.2).

Corresponding to this geometric interpretation, we see that a system of three linear equations in three unknowns represents three planes, and solutions correspond to points common to all three planes. Even

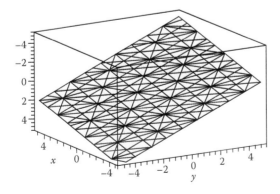

Figure 2.2 The plane $x + 2y + 3z = 1$ $(-5 \leq x, y, z \leq 5)$ in $\mathbb{R}^3$.

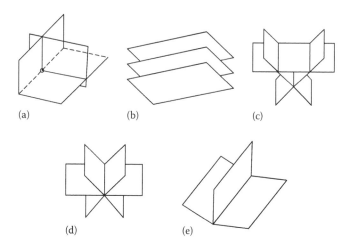

Figure 2.3 Linear systems in $\mathbb{R}^3$.

though the geometry may be more complicated, the solution possibilities follow a familiar pattern:

I. The planes representing the equations intersect at a single point $\leftrightarrow$ the system is **consistent** and has a **unique** solution (e.g., Figure 2.3a).

II. There is no common intersection of the three planes $\leftrightarrow$ the system is **inconsistent** and has **no** solution (e.g., Figure 2.3b or c).

III. The planes intersect in a line or in a plane $\leftrightarrow$ the system is **dependent** and there are **infinitely many** solutions. (e.g., Figure 2.3d or e).

Algebraically, we can use the same methods of elimination and substitution, but the manipulations become more complicated.

Example 2.1.2: Elimination and Substitution: Three Equations, Three Variables

Suppose we have the system

$$x - 2y + 3z = 9$$
$$-x + 3y = -4$$
$$2x - 5y + 5z = 17$$

Adding the first two equations and then adding -2 times the original first equation to the third equation eliminates the variable x:

$$y + 3z = 5$$
$$-y - z = -1$$

If we add these last two equations, we eliminate y and find that $z = 2$.

Substituting $z = 2$ into either of the two equations, we obtain $y = -1$. Finally, we substitute the values of y and z into the second original equation to get $x = 1$. (We could have eliminated variables in different ways to get the same unique solution. See the following picture of the three planes intersecting at a point.)

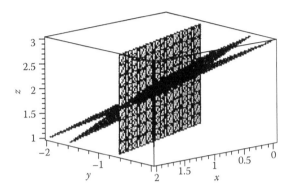

To use the method of substitution, we could, for example, solve the second original equation to obtain $x = 3y + 4$. Substituting this expression for x into the remaining equations gives us a system of two equations in y and z. We can continue in this way, solving one equation for one of the variables y and z in terms of the other and then substituting in the other equation so that we determine the value of one of the variables. Substituting values found in previous equations eventually leads to the complete solution.

As indicated in pattern III above, sometimes the planes representing the equations in a system intersect at infinitely many points.

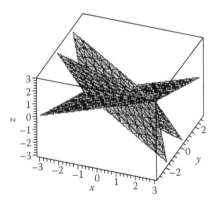

Example 2.1.3: Infinitely Many Solutions: Free Variables and Basic Variables

Suppose that we have the system

$$-x + y + 2z = 0$$
$$3x + 4y + z = 0$$
$$2x + 5y + 3z = 0$$

Substituting $x = y + 2z$ from the first equation into each of the remaining equations yields the single relation $y + z = 0$. Geo-metrically, the infinite solution set $\left\{ \begin{bmatrix} x \\ y \\ z \end{bmatrix} \in \mathbb{R}^3 : y + z = 0 \right\}$ represents the fact that the three planes intersect in the line $y + z = 0$.

We may solve for both y and x in terms of z,

$$y = -z, \quad x = y + 2z = z$$

and represent the solution set in terms of the **free variable** (or **parameter**) z, so-called because it is an independent variable, free to take on any value in the interval $(-\infty, \infty)$: $\left\{ \begin{bmatrix} z \\ -z \\ z \end{bmatrix} : z \in \mathbb{R} \right\}$, or $\left\{ z \begin{bmatrix} 1 \\ -1 \\ 1 \end{bmatrix} : z \in \mathbb{R} \right\}$.

(The variables that are not free are called **basic variables**.) In the following sections of this chapter, we will see that free variables and basic variables play an important role in analyzing systems of linear equations.

The same basic techniques can be tried if the number of equations does not equal the number of variables. In the next section, we begin a systematic study of arbitrary systems of linear equations, including theory, solution methods, and discussions of computational accuracy and efficiency.

Exercises 2.1

These exercises should be regarded as a review of basic algebraic/graphical techniques learned in high school or in college algebra courses. However, some may require a higher level of thought.

A.

*In Exercises 1–6, solve each system **graphically**.*

1. $x + 2y = 3$
   $x - y = -3$

2. $2x + 3y = 4$
   $3x - 2y = -7$

3. $2x + y = 8$
   $4x + 2y = 12$

4. $x - 3y = -1$
   $-2x + 6y = 3$

5. $x - 3y = 4$
   $-3x + 9y = -12$

6. $-12x + 2y = -3$
   $2x - \frac{1}{3}y = \frac{1}{2}$

7. Solve the system in Exercise 1 by *substitution*.

8. Solve the system in Exercise 4 by *substitution*.

9. Solve the system in Exercise 6 by *substitution*.

10. Solve the system in Exercise 2 by *elimination*.

11. One of the earliest recorded problems involving a system of linear equations is the following, which appeared in the Chinese work (ca. 200 BC) *Chiu-chang Suan-shu* (*Nine Chapters on Arithmetic*):*

>   Three sheafs of a good crop, two sheafs of a mediocre crop, and one sheaf of a bad crop are sold for 39 dou. Two sheafs of good, three mediocre, and one bad are sold for 34 dou; and one good, two mediocre, and three bad are sold for 26 dou. What is the price received for each sheaf of a good crop, each sheaf of a mediocre crop, and each sheaf of a bad crop?

     Solve the problem.

12. On Saturday night, the manager of a shoe store evaluates the receipts of the previous week's sales. Two hundred forty pairs of two different styles of tennis shoes were sold. One style sold for $66.95 and the other sold for $84.95. The total receipts were $17,652. The cash register that was supposed to record the number of each type of shoe sold malfunctioned. Can you recover the information? If so, how many shoes of each type were sold?

13. A farmer used four test plots to determine the relationship between wheat yield (in bushels per acre) and the amount of fertilizer (in hundreds of pounds per acre).

     The results are shown in the following table.

Fertilizer, x	1.0	1.5	2.0	2.5
Yield, y	32	41	48	53

     a. Find the **least-squares regression line**[†] $y = ax + b$ by solving the following system for the coefficients a and b.

$$4b + 7.0a = 174$$
$$7b + 13.5a = 322$$

     b. On the same set of axes, plot the data in the table and the least squares regression line.

---

* See V.J. Katz, Historical ideas in teaching linear algebra, in *Learn from the Masters!*, F. Swetz et al., eds., pp. 189–192 (Washington, DC: MAA, 1995) or G.G. Joseph, *The Crest of the Peacock: The Non-European Roots of Mathematics*, revised edition, pp. 173–176 (Princeton, NJ: Princeton University Press, 2000).

† A measure of how well a linear equation "fits" a set of data points $\{(x_1, y_1), (x_2, y_2), \ldots, (x_n, y_n)\}$ is the sum of the squares of the differences between the actual y values and the values given by the linear equation. The line that yields the smallest sum of squared differences is the **least-squares regression line**.

14. Find the equation of the parabola $y = ax^2 + bx + c$ that passes through the points $(-1, 6)$, $(1, 4)$, and $(2, 9)$.

15. If a, b, c, and d are real numbers, with at least one of them not equal to 0, then the solution set of the equation $ax + by + cz = d$ is a plane through the origin, a subspace of $\mathbb{R}^3$. Must the solution set of a system of two linear equations in three unknowns always represent a line in three dimensions? Explain.

16. Explain why you cannot possibly find two equations in three unknowns having exactly one solution.

17. Draw diagrams like those in Figure 2.1 to describe what can happen if you have three equations in two unknowns. (You should have *seven* visually distinct situations to describe.)

18. Solve the system

$$a + 2b - 4c = -4$$
$$5a + 11b - 21c = -22$$
$$3a - 2b + 3c = 11$$

by *elimination*.

19. Solve the system (manually)

$$-x + y + z = 0$$
$$x - y + z = 0$$
$$3x - 3y - z = 0$$

using any method you wish.

20. Solve the system representing an electrical circuit on the first page of this chapter.

21. Let $I_A, I_B,$ and I_C represent the current in parts A, B, and C, respectively, of an electrical circuit. Suppose that the relationships among the currents are given by the system

$$I_A + I_B + I_C = 0$$
$$-8I_B + 10I_C = 0$$
$$4I_A - 8I_B = 6$$

Determine the current in each part of the circuit.

22. An investor has a portfolio totaling $500,000 that is invested in certificates of deposit (CDs), municipal bonds, blue-chip stocks, and growth or speculative stocks. The CDs pay 10%

annually and the municipal bonds pay 8% annually. Over a 5 year period, the investor expects the blue-chip stocks to return 12% annually and the growth stocks to return 13% annually. The investor wants a combined (total) annual return of 10% as well as only one-fourth of the portfolio invested in stocks. How much should he or she put in each type of investment?

B.

1. Consider the system

$$ax + by = e$$
$$cx + dy = f$$

If $ad - bc = 0$, show that the vectors $\begin{bmatrix} a \\ b \end{bmatrix}$ and $\begin{bmatrix} c \\ d \end{bmatrix}$ in $\mathbb{R}^2$ are linearly dependent. [*Hint*: Consider several cases, such as $a = 0$, $b \neq 0$.]

2. Does the system

$$x + y = 5$$
$$2x - y = 1$$
$$3x + 2y = 5$$

have a solution? Justify your answer.

3. Show that the system

$$4x - 4y + az = c$$
$$3x - 2y + bz = d$$

has a solution for all values of a, b, c, and d. [*Hint*: Assume that z is any real number and show that you can solve for x and y.]

4. Is the vector $\begin{bmatrix} 6 \\ 10 \\ 16 \\ 4 \end{bmatrix} \in \mathbb{R}^4$ contained in span

$$\left\{ \begin{bmatrix} 1 \\ 2 \\ 3 \\ 1 \end{bmatrix}, \begin{bmatrix} 2 \\ -1 \\ 1 \\ -3 \end{bmatrix}, \begin{bmatrix} 1 \\ 3 \\ 4 \\ 2 \end{bmatrix} \right\}?$$

5. Determine a value of k that makes the following vectors linearly dependent:

$$\begin{bmatrix} 1 \\ 2 \\ k \end{bmatrix}, \begin{bmatrix} 0 \\ 1 \\ k-1 \end{bmatrix}, \begin{bmatrix} 3 \\ 4 \\ 3 \end{bmatrix}.$$

6. For what value of the constant k does the following system have a unique solution? Find the solution in this case. What happens if k does not equal this value?

$$2x + 4z = 6$$
$$3x + y + z = -1$$
$$2y - z = -2$$
$$x - y + kz = -5$$

7. Find conditions on a, b, and c such that the system

$$3x + 2y = a$$
$$-4x + y = b$$
$$2x + 5y = c$$

has a solution.

8. Consider the following system of linear equations:

$$x - y = a$$
$$w + z = b$$
$$y - z = c$$
$$x + w = d$$

a. For what conditions on a, b, c, and d will this system have a solution?

b. Give a set of values for a, b, c, and d for which the system does *not* have a solution.

c. Show that if the system has one solution, then it has infinitely many solutions.

C.

1. Solve the system

$$x_1 + 2x_2 + 3x_3 + 4x_4 = 1$$
$$x_2 + 2x_3 + 3x_4 + 4x_1 = 2$$
$$x_3 + 2x_4 + 3x_1 + 4x_2 = 3$$
$$x_4 + 2x_1 + 3x_2 + 4x_3 = 4$$

[*Hint*: Letting $s = x_1 + x_2 + x_3 + x_4$ may simplify the problem a bit.]

2. Consider the following system of equations containing a parameter a:

$$x + y - z = 2$$
$$x + 2y + z = 3$$
$$x + y + (a^2 - 5)z = a$$

a. For what value(s) of a will the system have a *unique* solution?

b. For what value(s) of a will the system have *no* solution?

c. For what value(s) of a will the system have *infinitely many* solutions?

3. Investigate the following system and describe its solution in
    relation to the value of the parameter λ.

$$x + y + \lambda z = 1$$
$$x + \lambda y + z = 1$$
$$\lambda x + y + z = 1$$

2.2 Matrices and Echelon Form

The most widely used method of solving systems of linear equations is
elimination—or, in the systematic form we shall describe in the next
section, *Gaussian elimination*—which converts a given system of equations into a new system that can be solved easily. Although this method is
attributed to the great German mathematician Karl Friedrich Gauss (1777–
1855), apparently it was known to the Chinese over 2000 years ago.*
There are slightly different versions of the method in the literature, but a
common feature is that this conversion is accomplished in a finite
sequence of steps. The important point is that *the final converted system
should have exactly the same solutions (if any) as the original system*. We
say that the original and converted systems are **equivalent**. (Theorem
2.2.2 describes and proves this equivalence.)

The possible steps that enable us to convert any system of linear
equations to an equivalent (but more easily solved) system are the following:

   I. Interchange any two equations.
  II. Replace any equation by a nonzero multiple of itself.
 III. Replace any equation by the sum of itself and a nonzero multiple
     of any other equation.

2.2.1 The Matrix of Coefficients

If we examine Example 2.1.2, for instance, we note that the calculations
involve the *coefficients* of the equations and the *constants* on the right-hand
sides of the equations, not the variables themselves. The variables x, y, and z
function only as placeholders in each equation and play no direct role in the
elimination process.

Let us look more closely at the system of equations in Example 2.1.2
and focus on the coefficients of the unknown quantities:

$$\mathbf{1}x + (\mathbf{-2})y + \mathbf{3}z = 9$$
$$(\mathbf{-1})x + \mathbf{3}y + \mathbf{0}z = -4$$
$$\mathbf{2}x + (\mathbf{-5})y + \mathbf{5}z = 17$$

---

* See, for example, G.G. Joseph, *The Crest of the Peacock*, pp. 173–176 (Princeton, NJ:
Princeton University Press, 2000). Gauss's interest grew from his need to find approximate
solutions to problems in astronomy and geodesy. See G.W. Stewart, Gauss, statistics, and
Gaussian elimination, *J. Comput. Graph. Stat.* **4** (1995): 1–11.

Written in the form of a square array

$$\begin{bmatrix} 1 & -2 & 3 \\ -1 & 3 & 0 \\ 2 & -5 & 5 \end{bmatrix},$$

these numbers constitute the **matrix of coefficients** of the system. In general, a **matrix*** (plural: *matrices*) is any rectangular array of elements, called **entries**—for now, real numbers—held together by large brackets [ ] or sometimes by large parentheses ( ). We describe a matrix as having m **rows** and n **columns**, or we just call it an $m \times n$ **matrix** ("m by n matrix"). Two matrices are **equal** if they have the same size—that is, the same number of rows and the same number of columns—and if corresponding entries of the two matrices are equal.

For example, the matrix $\begin{bmatrix} -1 & 0 & 2 & 3 & 4 \\ 5 & -3 & 7 & 1 & 0 \\ 4 & 3 & 2 & 1 & -2 \end{bmatrix}$ is a 3×5 matrix.

Its rows

$$[-1\ 0\ 2\ 3\ 4], \quad [5\ -3\ 7\ 1\ 0], \quad [4\ 3\ 2\ 1\ -2]$$

can be considered as row vectors in $\mathbb{R}^5$ (more properly, as the transposes of vectors in $\mathbb{R}^5$) and its columns

$$\begin{bmatrix} -1 \\ 5 \\ 4 \end{bmatrix}, \quad \begin{bmatrix} 0 \\ -3 \\ 3 \end{bmatrix}, \quad \begin{bmatrix} 2 \\ 7 \\ 2 \end{bmatrix}, \quad \begin{bmatrix} 3 \\ 1 \\ 1 \end{bmatrix}, \quad \begin{bmatrix} 4 \\ 0 \\ -2 \end{bmatrix}$$

can be interpreted as vectors in $\mathbb{R}^3$. Alternatively, we can think of a row of an $m \times n$ matrix as a $1 \times n$ **row matrix** and a column of the matrix as an $m \times 1$ **column matrix**. An $n \times n$ matrix (one with the same number of columns as rows) is called a **square matrix**. A system of linear equations having the same number of equations as unknowns will have a square matrix of coefficients.

In this and the next section, we will be using matrices (the plural of *matrix*) primarily as holders of data. But there is much more depth to the use of matrices in linear algebra, and we will begin to uncover this deeper significance in Chapter 3.

In the system analyzed in Example 2.1.2, we also have the **solution vector** (or **vector of inputs**), $\mathbf{x} = \begin{bmatrix} x \\ y \\ z \end{bmatrix}$. Finally, we have the vector comprised of the right-hand sides of the equations, $\mathbf{b} = \begin{bmatrix} 9 \\ -4 \\ 17 \end{bmatrix}$, which we can think of as the **vector of constants** or the **vector of outputs**. As an abbreviation for the system

---

* One meaning of the Latin word *matrix* is "womb." For now we interpret our mathematical terminology as referring to the fact that *a matrix holds data*. Later we will view a matrix more dynamically as a generator, a transformer.

$$x - 2y + 3z = 9$$
$$-x + 3y = -4$$
$$2x - 5y + 5z = 17$$

we will write

$$\begin{bmatrix} 1 & -2 & 3 \\ -1 & 3 & 0 \\ 2 & -5 & 5 \end{bmatrix} \begin{bmatrix} x \\ y \\ z \end{bmatrix} = \begin{bmatrix} 9 \\ -4 \\ 17 \end{bmatrix},$$

or $A\mathbf{x} = \mathbf{b}$, where $A = \begin{bmatrix} 1 & -2 & 3 \\ -1 & 3 & 0 \\ 2 & -5 & 5 \end{bmatrix}$, $\mathbf{x} = \begin{bmatrix} x \\ y \\ z \end{bmatrix}$, and $\mathbf{b} = \begin{bmatrix} 9 \\ -4 \\ 17 \end{bmatrix}$. For
the moment, this matrix–vector equation is just a shorthand way of
representing the system in terms of its significant components.

In general, if we are given any system of m equations in n unknowns,

$$a_{11}x_1 + a_{12}x_2 + \cdots + a_{1n}x_n = b_1$$
$$a_{21}x_1 + a_{22}x_2 + \cdots + a_{2n}x_n = b_2$$
$$\vdots \qquad \vdots \qquad \qquad \vdots \quad \vdots$$
$$a_{m1}x_1 + a_{m2}x_2 + \cdots + a_{mn}x_n = b_m$$

we can express it in the matrix–vector form

$$\begin{bmatrix} a_{11} & a_{12} & \cdots & a_{1n} \\ a_{21} & a_{22} & \cdots & a_{2n} \\ \vdots & \vdots & \vdots & \vdots \\ a_{m1} & a_{m2} & \cdots & a_{mn} \end{bmatrix} \begin{bmatrix} x_1 \\ x_2 \\ \vdots \\ x_n \end{bmatrix} = \begin{bmatrix} b_1 \\ b_2 \\ \vdots \\ b_m \end{bmatrix}.$$

Defining $A = \begin{bmatrix} a_{11} & a_{12} & \cdots & a_{1n} \\ a_{21} & a_{22} & \cdots & a_{2n} \\ \vdots & \vdots & \vdots & \vdots \\ a_{m1} & a_{m2} & \cdots & a_{mn} \end{bmatrix}$, the matrix of coefficients,

$\mathbf{x} = \begin{bmatrix} x_1 \\ x_2 \\ \vdots \\ x_n \end{bmatrix}$, and $\mathbf{b} = \begin{bmatrix} b_1 \\ b_2 \\ \vdots \\ b_m \end{bmatrix}$, we can simplify the system to $A\mathbf{x} = \mathbf{b}$.

Furthermore, we can abbreviate any matrix A as $A = [a_{ij}]$, where a_{ij}
denotes the entry in row i, column j of the matrix ($i = 1, 2, \ldots, m$; $j = 1, 2, \ldots, n$). Sometimes we will refer to a_{ij} as the **(i, j) entry** of A. In terms
of a system of equations, a_{ij} denotes the coefficient of the variable x_j in
equation i. Thus in Example 2.1.2, a_{32} ($=-5$) is the coefficient of the
second variable, y, in the third equation.

2.2.2 Elementary Row Operations

Because the basic system conversion steps needed for elimination affect
only the coefficients of the variables and the constants on the right-hand
side of the equations, we can focus our attention on various *matrices*
associated with the system of equations and restate operations I–III given

above in *matrix* form. Now the possible steps—operations on the rows of a matrix this time—that enable us to convert a system of linear equations to an equivalent system are the following **elementary row operations**:

    I. Interchange any two rows of a matrix.

    II. Replace any row of a matrix by a nonzero multiple of itself.

    III. Replace any row of a matrix by the sum of itself and a nonzero multiple of any other row.

We can define **elementary column operations** by changing the word *row* or *rows* in steps I–III to *column* or *columns*, although we will not use these column processes right now.

In working with a system of equations in the form $A\mathbf{x} = \mathbf{b}$, to ensure that we are performing the same operations on the entries of the vector of outputs $\mathbf{b}$ as we are performing on the rows of A, we enlarge the original matrix of coefficients to include the vector of outputs as the last column. In terms of Example 2.1.2, we form the 3×4 matrix

$$\begin{bmatrix} 1 & -2 & 3 & 9 \\ -1 & 3 & 0 & -4 \\ 2 & -5 & 5 & 17 \end{bmatrix}$$

sometimes written in the *partitioned* form

$$\begin{bmatrix} 1 & -2 & 3 & \vdots & 9 \\ -1 & 3 & 0 & \vdots & -4 \\ 2 & -5 & 5 & \vdots & 17 \end{bmatrix}$$

and carry out the appropriate steps I–III on this new matrix. This enlarged matrix is called the *augmented matrix* of the system.

Definition 2.2.1

Given the system of m equations in n unknowns

$$a_{11}x_1 + a_{12}x_2 + \cdots + a_{1n}x_n = b_1$$
$$a_{21}x_1 + a_{22}x_2 + \cdots + a_{2n}x_n = b_2$$
$$\vdots \qquad \vdots \qquad \qquad \vdots \qquad \vdots$$
$$a_{m1}x_1 + a_{m2}x_2 + \cdots + a_{mn}x_n = b_m$$

the $m \times (n+1)$ matrix

$$\begin{bmatrix} a_{11} & a_{12} & \cdots & a_{1n} & b_1 \\ a_{21} & a_{22} & \cdots & a_{2n} & b_2 \\ \vdots & \vdots & \vdots & \vdots & \vdots \\ a_{m1} & a_{m2} & \cdots & a_{mn} & b_m \end{bmatrix}$$

is called the **augmented matrix** of the system of equations and is denoted by $[A \mid \mathbf{b}]$.

Symbolically, let R_i denote row i of the augmented matrix. Then the elementary row operation I is denoted by $R_i \leftrightarrow R_j$, indicating that row i of the matrix has been replaced by row j and vice versa. The expression $R_i \to cR_i$ means that row i has been replaced by cR_i, where c is a nonzero constant. Finally, a type III elementary row operation can be expressed as $R_i \to R_i + aR_k$, where $k \neq i$ and $a \neq 0$. (Column operations will use the notation C_i for column i.)

The next example shows how to use this simplifying notation to keep track of the steps involved in solving a linear system.

Example 2.2.1: Elimination Using Matrices

If we want to solve the system

$$x + y + z = 2$$
$$x + 3y + 2z = 1$$
$$y + 3z = 2$$

we first express it in matrix form:

$$\begin{bmatrix} 1 & 1 & 1 \\ 1 & 3 & 2 \\ 0 & 1 & 3 \end{bmatrix} \begin{bmatrix} x \\ y \\ z \end{bmatrix} = \begin{bmatrix} 2 \\ 1 \\ 2 \end{bmatrix}.$$

Then we append the vector of constants to the matrix of coefficients to form the augmented matrix

$$\begin{bmatrix} 1 & 1 & 1 & 2 \\ 1 & 3 & 2 & 1 \\ 0 & 1 & 3 & 2 \end{bmatrix}$$

and proceed to convert our original system to an equivalent but more readily solved system, using a particular "top-down" process that will be explained fully later:

$$\begin{bmatrix} 1 & 1 & 1 & 2 \\ 1 & 3 & 2 & 1 \\ 0 & 1 & 3 & 2 \end{bmatrix} \xrightarrow{R_2 \to R_2 - R_1} \begin{bmatrix} 1 & 1 & 1 & 2 \\ 0 & 2 & 1 & -1 \\ 0 & 1 & 3 & 2 \end{bmatrix} \xrightarrow{R_2 \leftrightarrow R_3} \begin{bmatrix} 1 & 1 & 1 & 2 \\ 0 & 1 & 3 & 2 \\ 0 & 2 & 1 & -1 \end{bmatrix}$$

$$\xrightarrow{R_3 \to R_3 - 2R_2} \begin{bmatrix} 1 & 1 & 1 & 2 \\ 0 & 1 & 3 & 2 \\ 0 & 0 & -5 & -5 \end{bmatrix} \xrightarrow{R_3 \to -\frac{1}{5}R_3} \begin{bmatrix} 1 & 1 & 1 & 2 \\ 0 & 1 & 3 & 2 \\ 0 & 0 & 1 & 1 \end{bmatrix}.$$

This last matrix represents the converted system

$$x + y + z = 2$$
$$y + 3z = 2$$
$$z = 1$$

Substituting $z = 1$ in the first two equations, we get

$$x + y = 1 \quad \text{and} \quad y = -1.$$

Therefore, $x = 2$, $y = -1$, and $z = 1$, so that the solution of the system is the vector $\begin{bmatrix} 2 \\ -1 \\ 1 \end{bmatrix}$.

The last matrix in Example 2.2.2 is said to be in *echelon* (or "steplike") *form*. The process of solving an equation and then substituting in earlier equations is called **back substitution**.

Definition 2.2.2

An $m \times n$ matrix M is in **echelon form** (or **row echelon** form) if

(1) Any rows consisting entirely of zeros are grouped together at the bottom of the matrix.
(2) The first nonzero entry in any row, called the **leading entry*** (**pivotal element**, **pivot**), is to the right of the first nonzero entry in any higher row.
(3) All entries in a column below a leading entry are zeros.

Here is the last matrix in Example 2.2.2 with the leading entries boxed:

$$\begin{bmatrix} \boxed{1} & 1 & 1 & 2 \\ 0 & \boxed{1} & 3 & 2 \\ 0 & 0 & \boxed{1} & 1 \end{bmatrix}$$

Let us look at some additional examples to understand this very useful form of a matrix.

Example 2.2.2: Echelon Form

The matrices

$$\begin{bmatrix} \boxed{2} & 3 & 4 \\ 0 & \boxed{-1} & 5 \\ 0 & 0 & \boxed{3} \end{bmatrix}, \quad \begin{bmatrix} \boxed{1} & 3 & -2 & 4 \\ 0 & \boxed{2} & 5 & 7 \end{bmatrix}, \quad \text{and} \quad \begin{bmatrix} \boxed{4} & 3 & 2 & 1 \\ 0 & 0 & \boxed{-3} & 5 \\ 0 & 0 & 0 & 0 \\ 0 & 0 & 0 & 0 \end{bmatrix}$$

are in echelon form, with the leading entries highlighted; whereas the matrices

---

* Some authors require that the leading entry of any row be 1.

$$A = \begin{bmatrix} 2 & 4 & 6 \\ 0 & 0 & 0 \\ 0 & 0 & -1 \end{bmatrix}, \quad B = \begin{bmatrix} 1 & 3 & 5 \\ 0 & -3 & 4 \\ 0 & 2 & 6 \end{bmatrix}, \quad \text{and}$$

$$C = \begin{bmatrix} 1 & 2 & 3 & 4 \\ 5 & 6 & 7 & 8 \end{bmatrix}$$

are *not* in echelon form. Matrix A does not satisfy condition (1) of Definition 2.2.3, B violates conditions (2) and (3), and C also violates conditions (2) and (3).

We can make this concept of transforming one matrix into another via elementary operations more formal.

Definition 2.2.3

Two matrices are **row equivalent** if one matrix may be obtained from the other matrix by means of a finite sequence of elementary row operations.

If A is row equivalent to B, we write $A \overset{R}{\sim} B$.

Example 2.2.3: Row Equivalent Matrices

Suppose $A = \begin{bmatrix} 1 & 1 & 1 \\ 1 & 3 & 2 \\ 2 & 1 & -1 \end{bmatrix}$. Then A is row equivalent to

$B = \begin{bmatrix} 3 & 3 & 3 \\ 1 & 3 & 2 \\ 2 & 1 & -1 \end{bmatrix}$ because $A = \begin{bmatrix} 1 & 1 & 1 \\ 1 & 3 & 2 \\ 2 & 1 & -1 \end{bmatrix} \xrightarrow{R_1 \to 3R_1} \begin{bmatrix} 3 & 3 & 3 \\ 1 & 3 & 2 \\ 2 & 1 & -1 \end{bmatrix} = B.$

Similarly, A is row equivalent to $C = \begin{bmatrix} 2 & 1 & -1 \\ 3 & 5 & 4 \\ 1 & 1 & 1 \end{bmatrix}$:

$A = \begin{bmatrix} 1 & 1 & 1 \\ 1 & 3 & 2 \\ 2 & 1 & -1 \end{bmatrix} \xrightarrow{R_2 \to R_2 + 2R_1} \begin{bmatrix} 1 & 1 & 1 \\ 3 & 5 & 4 \\ 2 & 1 & -1 \end{bmatrix} \xrightarrow{R_1 \leftrightarrow R_3} \begin{bmatrix} 2 & 1 & -1 \\ 3 & 5 & 4 \\ 1 & 1 & 1 \end{bmatrix} = C.$

Obviously a given matrix may be row equivalent to many other matrices of the same size.

It is important to realize that if $A \overset{R}{\sim} B$, then $B \overset{R}{\sim} A$. Just apply "inverse" row operations on B to get A. This technique is shown in Example 2.2.5. Also, if $A \overset{R}{\sim} B$ and $A \overset{R}{\sim} C$, then $B \overset{R}{\sim} C$ [Exercise C2(b)].

Example 2.2.4: Inverse Row Operations

We can use Example 2.2.2 to illustrate the idea of *inverse row operations*. Additions (subtractions) become subtractions (additions) and simple multiplications (divisions)

become divisions (multiplications). Row interchanges remain the same:

[Original matrix]   **[Original matrix]**

$$\begin{bmatrix} 1 & 1 & 1 & 2 \\ 1 & 3 & 2 & 1 \\ 0 & 1 & 3 & 2 \end{bmatrix} \qquad \begin{bmatrix} 1 & 1 & 1 & 2 \\ 1 & 3 & 2 & 1 \\ 0 & 1 & 3 & 2 \end{bmatrix}$$

$\downarrow R_2 \rightarrow R_2 - R_1$ $\uparrow R_2 \rightarrow R_2 + R_1$

$$\begin{bmatrix} 1 & 1 & 1 & 2 \\ 0 & 2 & 1 & -1 \\ 0 & 1 & 3 & 2 \end{bmatrix} \qquad \begin{bmatrix} 1 & 1 & 1 & 2 \\ 0 & 2 & 1 & -1 \\ 0 & 1 & 3 & 2 \end{bmatrix}$$

$\downarrow R_2 \leftrightarrow R_3$ $\uparrow R_2 \leftrightarrow R_3$

$$\begin{bmatrix} 1 & 1 & 1 & 2 \\ 0 & 1 & 3 & 2 \\ 0 & 2 & 1 & -1 \end{bmatrix} \qquad \begin{bmatrix} 1 & 1 & 1 & 2 \\ 0 & 1 & 3 & 2 \\ 0 & 2 & 1 & -1 \end{bmatrix}$$

$\downarrow R_3 \rightarrow R_3 - 2R_2$ $\uparrow R_3 \rightarrow R_3 + 2R_2$

$$\begin{bmatrix} 1 & 1 & 1 & 2 \\ 0 & 1 & 3 & 2 \\ 0 & 0 & -5 & -5 \end{bmatrix} \qquad \begin{bmatrix} 1 & 1 & 1 & 2 \\ 0 & 1 & 3 & 2 \\ 0 & 0 & -5 & -5 \end{bmatrix}$$

$\downarrow R_3 \rightarrow -\dfrac{1}{5}R_3$ $\uparrow R_3 \rightarrow -5R_3$

$$\begin{bmatrix} 1 & 1 & 1 & 2 \\ 0 & 1 & 3 & 2 \\ 0 & 0 & 1 & 1 \end{bmatrix} \qquad \begin{bmatrix} 1 & 1 & 1 & 2 \\ 0 & 1 & 3 & 2 \\ 0 & 0 & 1 & 1 \end{bmatrix}.$$

[Echelon form]   **[Echelon form]**

The next result is important in both theory and practice.

Theorem 2.2.1

Every matrix is row equivalent to a matrix in echelon form.

The proof of this assertion will be given in the next section, which introduces an important computational algorithm.

The following theorem establishes the equivalence of two systems connected by elementary row operations. This result will be particularly useful in Section 2.6.

Theorem 2.2.2

If $[A \mid \mathbf{b}]$ and $[C \mid \mathbf{d}]$ are row equivalent augmented matrices of two systems of linear equations, then the two systems have the same solution sets—that is, any solution of one system is a solution of the other.

Proof   Let us start with the system $A\mathbf{x} = \mathbf{b}$, or

$$
\begin{aligned}
a_{11}x_1 + a_{12}x_2 + \cdots + a_{1n}x_n &= b_1 \\
a_{21}x_1 + a_{22}x_2 + \cdots + a_{2n}x_n &= b_2 \\
\vdots \qquad \vdots \qquad\quad \vdots \qquad \vdots \\
a_{m1}x_1 + a_{m2}x_2 + \cdots + a_{mn}x_n &= b_m
\end{aligned}
\tag{$*$}
$$

Letting $\mathbf{R}_i$ denote row i of the matrix of coefficients A, $i = 1, 2, \ldots, m$, we can use the dot product to write this system in terms of a set of vector equations

$$
\mathbf{R}_1^{\mathrm{T}} \bullet \mathbf{x} = b_1, \ \mathbf{R}_2^{\mathrm{T}} \bullet \mathbf{x} = b_2, \ldots, \ \mathbf{R}_m^{\mathrm{T}} \bullet \mathbf{x} = b_m.
\tag{$**$}
$$

I.   Clearly, exchanging any two equations of $(*)$—or, rows of $[A \mid \mathbf{b}]$—does not change the solution set of the system $(*)$.

II.   If we multiply equation i of $(*)$ by a nonzero scalar k, the vector equations $(**)$ do not change except for $\mathbf{R}_i^{\mathrm{T}} \bullet \mathbf{x} = b_i$, which becomes $k\mathbf{R}_i^{\mathrm{T}} \bullet \mathbf{x} = kb_i$. But we know that multiplying both sides of an equation by a nonzero scalar does not change the solution set of that equation. In other words, we can multiply a row of the augmented matrix $[A \mid \mathbf{b}]$ by $k \neq 0$ and the resulting system has the same solution as $A\mathbf{x} = \mathbf{b}$.

III.   Now suppose we replace equation i of $(*)$ by the linear combination (equation i) $+ k$ (equation j). This is the same as changing only the vector equation $\mathbf{R}_i^{\mathrm{T}} \bullet \mathbf{x} = b_i$, which becomes $\mathbf{R}_i^{\mathrm{T}} \bullet \mathbf{x} + k\mathbf{R}_j^{\mathrm{T}} \bullet \mathbf{x} = b_i + kb_j$. If a vector $\mathbf{u}$ is a solution of the system $(**)$, then (in particular) $\mathbf{R}_i^{\mathrm{T}} \bullet \mathbf{u} = b_i$ and $\mathbf{R}_j^{\mathrm{T}} \bullet \mathbf{u} = b_j$, so that $\mathbf{R}_i^{\mathrm{T}} \bullet \mathbf{u} + k\mathbf{R}_j^{\mathrm{T}} \bullet \mathbf{u} = b_i + kb_j$—that is, $\mathbf{u}$ is a solution of the new system as well. (The vector $\mathbf{u}$ satisfies all the original equations and the changed vector equation.) Conversely, suppose that $\mathbf{u}$ is a solution of the changed system $\mathbf{R}_1^{\mathrm{T}} \bullet \mathbf{x} = b_1, \mathbf{R}_2^{\mathrm{T}} \bullet \mathbf{x} = b_2, \ldots, \mathbf{R}_i^{\mathrm{T}} \bullet \mathbf{x} + k\mathbf{R}_j^{\mathrm{T}} \bullet \mathbf{x} = b_i + kb_j$, $\ldots, \mathbf{R}_j^{\mathrm{T}} \bullet \mathbf{x} = b_j, \ldots, \mathbf{R}_m^{\mathrm{T}} \bullet \mathbf{x} = b_m$. Then $\mathbf{u}$ is a solution of every equation $\mathbf{R}_k^{\mathrm{T}} \bullet \mathbf{x} = b_k$, except possibly for $k = i$. Because $\mathbf{u}$ is assumed to be a solution of $\mathbf{R}_j^{\mathrm{T}} \bullet \mathbf{x} = b_j (j \neq i)$ and of $\mathbf{R}_i^{\mathrm{T}} \bullet \mathbf{x} + k\mathbf{R}_j^{\mathrm{T}} \bullet \mathbf{x} = b_i + kb_j$, we have $\mathbf{R}_i^{\mathrm{T}} \bullet \mathbf{u} + k\mathbf{R}_j^{\mathrm{T}} \bullet \mathbf{u} = \mathbf{R}_i^{\mathrm{T}} \bullet \mathbf{u} + kb_j = b_i + kb_j$. Subtracting kb_j from both sides of this last equation, we see that $\mathbf{R}_i^{\mathrm{T}} \bullet \mathbf{u} = b_i$—that is, $\mathbf{u}$ is a solution of *every* equation $\mathbf{R}_k^{\mathrm{T}} \bullet \mathbf{x} = b_k$ of the original system.

As we have stated before, two systems that have the same solution sets are called **equivalent**.

Example 2.2.5: Equivalent Systems

Consider the two systems

$$x + 2y = 0 \qquad\qquad 2x + 4y = 0$$
$$(1)\ 2x + y = 2 \quad \text{and} \quad (2) \quad x - y = 2$$
$$x - y = 2 \qquad\qquad 4x - y = 6.$$

The augmented matrices of these systems are

$$A = \begin{bmatrix} 1 & 2 & 0 \\ 2 & 1 & 2 \\ 1 & -1 & 2 \end{bmatrix} \quad \text{and} \quad B = \begin{bmatrix} 2 & 4 & 0 \\ 1 & -1 & 2 \\ 4 & -1 & 6 \end{bmatrix},$$

respectively.

First we note that $A \overset{R}{\sim} B$ by performing the following elementary row operations on A to get B: $R_2 \to R_2 + 2R_3$, $R_2 \leftrightarrow R_3$, $R_1 \to 2R_1$. Next, with very little effort, we see that systems (1) and (2) have precisely the same solution, $(\frac{4}{3}, -\frac{2}{3})$, or $\begin{bmatrix} \frac{4}{3} \\ -\frac{2}{3} \end{bmatrix} = \frac{2}{3}\begin{bmatrix} 2 \\ -1 \end{bmatrix}$.

Exercises 2.2

A.

Write each of the systems in Exercises 1–8 in the form $A\mathbf{x} = \mathbf{b}$, where A is a matrix and $\mathbf{x}, \mathbf{b}$ are vectors. Do not solve the systems.

1. $-3x + 4y = 2$
    $x - 2y = -1$

2. $4x - 5y = 0$
    $-5x + 4y = 1$

3. $x + y + z = 1$
    $x + 2y + 3z = 2$
    $2x + 3y + 4z = 3$

4. $x + z = -2$
    $x - 3y + z = 4$
    $2y - z = 5$

5. $a - b + c - d = 0$
    $2a + 3b - c + 4d = -2$
    $3a - 2b + 4c - d = 1$
    $-2a + b + 2c + d = -3$

6. $x + 3z - w = 4$
    $2x - 3y + z = 7$
    $2y + 3w = -2$
    $-x + y - z + w = 0$

7. $x + y + z = -2$
    $3x - 4y - 3z = 4$

8. $x + y = 3$
    $-3x + 4y = 0$
    $4x - 3y = -2$

9. Let

$$A = \begin{bmatrix} 0 & 1 & -2 & 3 \\ 5 & -1 & 4 & 2 \\ 4 & 3 & 2 & 0 \end{bmatrix}.$$

Find the individual matrices obtained by performing the following elementary row operations on A.
a. Interchanging the first and third rows
b. Multiplying the third row by 4
c. Adding (-3) times the second row to the third row

10. Let

$$A = \begin{bmatrix} 1 & 2 & 3 \\ -3 & 2 & -1 \\ 0 & 4 & 5 \\ 3 & -5 & 2 \end{bmatrix}.$$

Find the individual matrices obtained by performing the following elementary row operations on A.
a. Interchanging the second and fourth rows
b. Multiplying the second row by (-1)
c. Adding 2 times the first row to the third row

11. Find three matrices that are row equivalent to

$$A = \begin{bmatrix} 2 & -1 & 3 & 4 \\ 0 & 1 & 2 & -1 \\ 5 & 2 & -3 & 4 \end{bmatrix}.$$

12. Find three matrices that are row equivalent to

$$A = \begin{bmatrix} 0 & -1 & 2 \\ -3 & 4 & 5 \\ 1 & -2 & 3 \\ 3 & -4 & -5 \end{bmatrix}.$$

In Exercises 13–16, the augmented matrices for systems of linear equations have been reduced to the given matrices in echelon form and the variables (unknowns) indicated. Solve each system.

13. $\begin{bmatrix} 1 & 0 & 2 \\ 0 & 1 & -1 \end{bmatrix}; x, y$ 14. $\begin{bmatrix} 1 & 2 & -1 & 3 \\ 0 & -3 & 3 & -6 \\ 0 & 0 & 5 & 5 \end{bmatrix}; x_1, x_2, x_3$

15. $\begin{bmatrix} -2 & 4 & 6 \\ 0 & 3 & 5 \\ 0 & 0 & 0 \end{bmatrix}; a, b$ 16. $\begin{bmatrix} 1 & 2 & 3 & 4 & 5 \\ 0 & 6 & 7 & 8 & 9 \\ 0 & 0 & -1 & 2 & -3 \\ 0 & 0 & 0 & 3 & 4 \end{bmatrix}; x, y, z, w$

17.   Suppose that the augmented matrix for a system of linear equations has been row reduced to

$$\begin{bmatrix} 1 & 2 & 3 \\ 0 & 4 & 5 \\ 0 & 0 & 6 \end{bmatrix}.$$

What does this information tell you about the solution of the system?

18.   How many solutions does a consistent linear system of three equations and four unknowns have? Why?

B.

1.   Find all vectors $\mathbf{x} \in \mathbb{R}^4$ that are orthogonal to both $\begin{bmatrix} 1 \\ 0 \\ 1 \\ 1 \end{bmatrix}$ and

$\begin{bmatrix} 0 \\ 1 \\ -1 \\ 2 \end{bmatrix}.$

2.   Find all vectors $\mathbf{x} \in \mathbb{R}^3$ such that $\|\mathbf{x}\| = 1$ and $\mathbf{x}$ makes an angle of $\pi/3$ with each of the vectors $\begin{bmatrix} 1 \\ 0 \\ -1 \end{bmatrix}$ and $\begin{bmatrix} 0 \\ 1 \\ 1 \end{bmatrix}.$

3.   Find all solutions a, b, c, d, e of the system

$$e + b = ya$$
$$a + c = yb$$
$$b + d = yc$$
$$c + e = yd$$
$$d + a = ye$$

where y is a parameter. [*Hint*: Consider the following cases: (1) $y = 2$; (2) $y \neq 2$, $y^2 + y - 1 \neq 0$; and (3) $y \neq 2$, $y^2 + y - 1 = 0$.]

4.   Describe the elementary row operation that "undoes" each of the three elementary row operations $R_i \leftrightarrow R_j$, $R_i \rightarrow cR_i$ $(c \neq 0)$, and $R_i \rightarrow R_i + aR_k$, where $k \neq i$ and $a \neq 0$.

5. Suppose that 100 insects are distributed in an enclosure consisting of four chambers with passageways between them as shown below.*

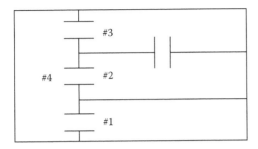

At the end of 1 min, the insects have redistributed themselves. Assume that a minute is not enough time for an insect to visit more than one chamber and that at the end of a minute 40% of the insects in each chamber have not left the chamber they occupied at the beginning of the minute. The insects that leave a chamber disperse uniformly among the chambers that are directly accessible from the one they initially occupied—for example, from #3, half move to #2 and half move to #4.

a. If at the end of 1 min there are 12, 25, 26, and 37 insects in chambers #1, #2, #3, and #4, respectively, determine what the initial distribution had to be.

b. If the initial distribution is 20, 20, 20, and 40, what is the distribution at the end of 1 min?

6. Prove that the nonzero rows of a matrix in echelon form are linearly independent.

7. Prove that if a matrix is in echelon form, the columns containing pivot elements are linearly independent.

8. Regarding the rows (actually, their transposes) of an $m \times n$ matrix A as vectors in $\mathbb{R}^n$, define the *row space* V as the span of the rows of A. Prove that the following matrices have the same row space V and find a basis for V:

$$\begin{bmatrix} 2 & 3 & 4 \\ 0 & 1 & 1 \\ 2 & -1 & 0 \end{bmatrix}, \quad \begin{bmatrix} 4 & 2 & 4 \\ 2 & 0 & 1 \\ 2 & 7 & 8 \end{bmatrix}.$$

9. Regarding the columns of an $m \times n$ matrix A as vectors in $\mathbb{R}^m$, define the *column space* V as the span of the columns of A. Prove that the following matrices have the same column space V and find a basis for V:

$$\begin{bmatrix} 1 & 1 & 0 \\ 0 & 1 & 1 \\ -1 & 0 & 1 \end{bmatrix}, \quad \begin{bmatrix} 2 & 1 \\ 1 & 2 \\ -1 & 1 \end{bmatrix}.$$

* From C.D. Meyer, *Matrix Analysis and Applied Linear Algebra*, p. 13 (Philadelphia, PA: SIAM, 2000).

10. Consider the matrix

$$A = \begin{bmatrix} 1 & 0 & 1 \\ 3 & 1 & 4 \\ 7 & 2 & 9 \end{bmatrix}.$$

a. Determine the number of linearly independent rows of A.
b. Determine the number of linearly independent columns of A.
c. What do you notice about your answers to parts (a) and (b)?

11. Consider the matrix

$$B = \begin{bmatrix} 3 & 1 & 3 & 7 \\ -1 & -3 & -1 & -5 \\ 7 & 0 & 2 & 9 \\ 0 & 1 & 3 & 4 \\ 2 & 0 & 1 & 3 \end{bmatrix}.$$

a. Determine the number of linearly independent rows of B.
b. Determine the number of linearly independent columns of B.
c. What do you notice about your answers to parts (a) and (b)?

12. Assuming that the first nonzero entry in every nonzero row is 1 (if there is such a row), display all seven possible (distinct) configurations for a 2×3 matrix that is in echelon form. (For example, one of the seven is $\begin{bmatrix} 1 & * & * \\ 0 & 1 & * \end{bmatrix}$, where the entries marked $*$ can be zero or nonzero.)

13. Assuming that the first nonzero entry in every nonzero row is 1 (if there is such a row), display all seven possible (distinct) configurations for a 3×3 matrix that is in echelon form. (For example, one of the seven is $\begin{bmatrix} 0 & 1 & * \\ 0 & 0 & 1 \\ 0 & 0 & 0 \end{bmatrix}$, where the entries marked $*$ can be zero or nonzero.)

14. In a certain biological system, there are n species of animals and m sources of food. Let x_j represent the population of the jth species, for each $j = 1, \ldots, n$; b_i represent the available daily supply of the ith food; and a_{ij} represent the amount of the ith food consumed on the average by a member of the jth species. The linear system

$$a_{11}x_1 + a_{12}x_2 + \cdots + a_{1n}x_n = b_1$$
$$a_{21}x_1 + a_{22}x_2 + \cdots + a_{2n}x_n = b_2$$
$$\vdots \qquad \vdots \qquad\quad \vdots \qquad \vdots$$
$$a_{m1}x_1 + a_{m2}x_2 + \cdots + a_{mn}x_n = b_m$$

represents an *equilibrium* where there is a daily supply of food to precisely meet the average daily consumption of each species.
a. Let

$$A = \begin{bmatrix} 1 & 2 & 0 & 3 \\ 1 & 0 & 2 & 2 \\ 0 & 0 & 1 & 1 \end{bmatrix}, \quad \mathbf{x} = \begin{bmatrix} 1000 \\ 500 \\ 350 \\ 400 \end{bmatrix}, \quad \text{and} \quad \mathbf{b} = \begin{bmatrix} 3500 \\ 2700 \\ 900 \end{bmatrix}.$$

Is there sufficient food to satisfy the average daily consumption?

b. What is the maximum number of animals of each species that could be individually added to the system with the supply of food still meeting the consumption?

c. If species 1 became extinct, how much of an individual increase of each of the remaining species could be supported?

d. If species 2 became extinct, how much of an individual increase of each of the remaining species could be supported?

C.

1. Show that the three types of elementary row operations discussed in this section are not independent by showing that the interchange operation (I) can be accomplished by a sequence of the other two types (II and III) of row operations.

2. If $A \overset{R}{\sim} B$ means that matrix A is row equivalent to matrix B, prove that

a. $A \overset{R}{\sim} A$.

b. If $A \overset{R}{\sim} B$ and $A \overset{R}{\sim} C$, then $B \overset{R}{\sim} C$.

c. If $A \overset{R}{\sim} B$ and $B \overset{R}{\sim} C$, then $A \overset{R}{\sim} C$.

3. Determine the unique coefficients of the nth degree polynomial $p(x) = a_0 + a_1 x + a_2 x^2 + \cdots + a_n x^n$ that satisfies the conditions $p(a) = c_0, p'(a) = c_1, p''(a) = c_2, \ldots, p^{(n)}(a) = c_n$ for some real number a and constants $c_0, c_1, c_2, \ldots, c_n$. [The symbol $p^{(k)}(a)$ denotes the kth derivative of p, evaluated at a.]

2.3 Gaussian Elimination

Theorem 2.2.1 stated that every matrix is equivalent to one in echelon form. The validity of this claim is established by seeing that we can carry out the following algorithm (procedure), called **Gaussian elimination**, to reduce any matrix to row echelon form:

Reducing a Matrix to Echelon Form (Gaussian Elimination)

STEP 1. Find the leftmost column that does not consist entirely of zeros.

STEP 2. By interchanging rows if necessary, obtain a nonzero entry, say a in the first row of the column found in STEP 1. This entry is called the leading entry (pivotal element, pivot).

STEP 3. Add suitable multiples of the first row to the rows below, so that all the entries below the leading a become 0.

STEP 4. Ignore the first row of the matrix and repeat the above procedure on the matrix that remains. Continue in this way until the entire matrix is in echelon form.

At each stage of Gaussian elimination, the strategy is to focus on the pivot (the first nonzero entry in a row) and to eliminate all terms below this position using the steps listed above. If we encounter a zero coefficient in what should be a pivot position, we must interchange this row with a row *below* it to produce a nonzero element to serve as the pivot—if possible. When a matrix is in echelon form, a pivot is a leading entry in a nonzero row. In an echelon matrix, a column (row) containing a pivot is called a **pivot column (row)** of the original matrix A.

From now on, unless otherwise stated, we will use the Gaussian elimination as our basic algorithm for examining systems of linear equations.

Let us implement this algorithm in specific examples.

Example 2.3.1: Linear Independence/ Dependence of Vectors

If we want to determine the linear independence/dependence

of the vectors $\begin{bmatrix} 1 \\ 2 \\ 1 \\ 3 \\ 2 \end{bmatrix}$, $\begin{bmatrix} 1 \\ 3 \\ 3 \\ 5 \\ 3 \end{bmatrix}$, $\begin{bmatrix} 3 \\ 8 \\ 7 \\ 13 \\ 8 \end{bmatrix}$, $\begin{bmatrix} 1 \\ 4 \\ 6 \\ 9 \\ 7 \end{bmatrix}$, and $\begin{bmatrix} 5 \\ 13 \\ 13 \\ 25 \\ 19 \end{bmatrix}$ in $\mathbb{R}^5$, we

have to find the coefficients of a linear combination of these vectors that equals the zero vector. In matrix terms, this is equivalent to solving the system

$$\begin{bmatrix} 1 & 1 & 3 & 1 & 5 \\ 2 & 3 & 8 & 4 & 13 \\ 1 & 3 & 7 & 6 & 13 \\ 3 & 5 & 13 & 9 & 25 \\ 2 & 3 & 8 & 7 & 19 \end{bmatrix} \begin{bmatrix} c_1 \\ c_2 \\ c_3 \\ c_4 \\ c_5 \end{bmatrix} = \begin{bmatrix} 0 \\ 0 \\ 0 \\ 0 \\ 0 \end{bmatrix}.$$

We should form the augmented matrix, but since the appended column consists entirely of zeros and would not change with row operations, we will just reduce the matrix of coefficients to an echelon form, indicating the new pivot element at each stage of the reduction:

$$\begin{bmatrix} \boxed{1} & 1 & 3 & 1 & 5 \\ 2 & 3 & 8 & 4 & 13 \\ 1 & 3 & 7 & 6 & 13 \\ 3 & 5 & 13 & 9 & 25 \\ 2 & 3 & 8 & 7 & 19 \end{bmatrix} \xrightarrow[\substack{R_4 \to R_4 - 3R_1 \\ R_5 \to R_5 - 2R_1}]{\substack{R_2 \to R_2 - 2R_1 \\ R_3 \to R_3 - R_1}} \begin{bmatrix} 1 & 1 & 3 & 1 & 5 \\ 0 & \boxed{1} & 2 & 2 & 3 \\ 0 & 2 & 4 & 5 & 8 \\ 0 & 2 & 4 & 6 & 10 \\ 0 & 1 & 2 & 5 & 9 \end{bmatrix}$$

$$\xrightarrow[\substack{R_4 \to R_4 - 2R_2 \\ R_5 \to R_5 - R_2}]{R_3 \to R_3 - 2R_2} \begin{bmatrix} 1 & 1 & 3 & 1 & 5 \\ 0 & 1 & 2 & 2 & 3 \\ 0 & 0 & 0 & \boxed{1} & 2 \\ 0 & 0 & 0 & 2 & 4 \\ 0 & 0 & 0 & 3 & 6 \end{bmatrix} \xrightarrow[R_5 \to R_5 - 3R_3]{R_4 \to R_4 - 2R_3} \begin{bmatrix} 1 & 1 & 3 & 1 & 5 \\ 0 & 1 & 2 & 2 & 3 \\ 0 & 0 & 0 & 1 & 2 \\ 0 & 0 & 0 & 0 & 0 \\ 0 & 0 & 0 & 0 & 0 \end{bmatrix}$$

The echelon form reveals that there are infinitely many solutions, so the set of vectors is linearly dependent. (Remembering that there should be a last column of zeros, we see that the last nonzero row of the echelon form is equivalent to the equation $c_4 + 2c_5 = 0$, or $c_4 = -2c_5$, indicating that c_5 can be considered a *free variable* (see Example 2.1.3). Similarly, c_3 can be taken as a free variable, and all other solutions can be expressed in terms of c_3 and c_5 by the process of back substitution. Note that these free variables correspond to nonpivot columns.

Examining the echelon form more carefully, we see that the nonzero rows are linearly independent and the columns containing pivot elements are linearly independent as vectors in $\mathbb{R}^5$. (See Exercises B12 and B13.)

Theorem 2.2.1 does *not* say that the row echelon form of a matrix is *unique*. As Figure 2.4 and the next example show, different sequences of elementary row operations applied to a matrix may yield different echelon forms for the same matrix; and two different matrices may have the same echelon form.

Example 2.3.2: Echelon Form Is Not Unique

Consider the matrix $\begin{bmatrix} 3 & 4 & 5 \\ 0 & 1 & 2 \\ -1 & 0 & 6 \end{bmatrix}$.

Each of the following elementary row operations produces the same echelon form $\begin{bmatrix} 3 & 4 & 5 \\ 0 & 1 & 2 \\ 0 & 0 & 5 \end{bmatrix}$: $R_3 \rightarrow R_3 + \frac{1}{3}R_1$,

$R_3 \rightarrow R_3 - \frac{4}{3}R_2$; whereas the sequence of operations $R_1 \rightarrow \frac{1}{3}R_1$, $R_3 \rightarrow R_3 + R_1$, $R_3 \rightarrow R_3 - \frac{4}{3}R_2$ yields the row

echelon form $\begin{bmatrix} 1 & \frac{4}{3} & \frac{5}{3} \\ 0 & 1 & 2 \\ 0 & 0 & 5 \end{bmatrix}$. On the other hand, $A = \begin{bmatrix} 3 & 4 & 5 \\ 0 & 1 & 2 \\ -1 & 0 & 6 \end{bmatrix}$

and $B = \begin{bmatrix} 0 & 1 & 2 \\ 3 & 4 & 5 \\ -1 & 0 & 6 \end{bmatrix}$ are different matrices, yet each can be

reduced to the same echelon form $\begin{bmatrix} 3 & 4 & 5 \\ 0 & 1 & 2 \\ 0 & 0 & 5 \end{bmatrix}$. (We can

interchange the first two rows of B and then apply the same elementary operations in the same order to both A and the new version of B.)

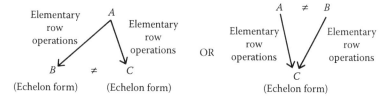

Figure 2.4 Echelon form is not unique.

Exercises 2.3

A.

*In Exercises 1–6, solve each system of equations **manually** by using the Gaussian elimination.*

1. $2x_1 - x_2 - x_3 = 4$
    $3x_1 + 4x_2 - 2x_3 = 11$
    $3x_1 - 2x_2 + 4x_3 = 11$

2. $a + b + 2c = -1$
    $2a - b + 2c = -4$
    $4a + b + 4c = -2$

3. $3x + 2y + z = 5$
    $2x + 3y + z = 1$
    $2x + y + 3z = 11$

4. $x + 2y + 4z = 31$
    $5x + y + 2z = 29$
    $3x - y + z = 10$

5. $x + y + 2z + 3w = 1$
    $3x - y - z - 2w = -4$
    $2x + 3y - z - w = -6$
    $x + 2y + 3z - w = -4$

6. $x + 2y + 3z - 2w = 6$
    $2x - y - 2z - 3w = 8$
    $3x + 2y - z + 2w = 4$
    $2x - 3y + 2z + w = -8$

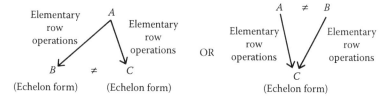 7. Solve the following linear system:

$$x + \tfrac{1}{2}y + \tfrac{1}{3}z + \tfrac{1}{4}w = \tfrac{1}{8}$$
$$\tfrac{1}{2}x + \tfrac{1}{3}y + \tfrac{1}{4}z + \tfrac{1}{5}w = \tfrac{1}{9}$$
$$\tfrac{1}{3}x + \tfrac{1}{4}y + \tfrac{1}{5}z + \tfrac{1}{6}w = \tfrac{1}{10}$$
$$\tfrac{1}{4}x + \tfrac{1}{5}y + \tfrac{1}{6}z + \tfrac{1}{7}w = \tfrac{1}{11}$$

8. Solve the following system:

$$x + 0.50000y + 0.33333z + 0.25000w = 0.12500$$
$$0.50000x + 0.33333y + 0.25000z + 0.20000w = 0.11111$$
$$0.33333x + 0.25000y + 0.20000z + 0.16667w = 0.10000$$
$$0.25000x + 0.20000y + 0.16667z + 0.14286w = 0.09091$$

9. a. What is the largest possible number of pivots a 4×6 matrix can have? Why?
    b. What is the largest possible number of pivots a 6×4 matrix can have? Why?

10. Suppose the matrix of coefficients corresponding to a linear system is 4×6 and has three pivot columns. How many pivot columns does the augmented matrix have if the linear system is inconsistent?

B.

1. Suppose you have a system of three linear equations in four variables and the Gaussian elimination reduces the coefficient matrix A of the system to

$$R = \begin{bmatrix} 1 & 2 & 0 & -3 \\ 0 & 0 & 1 & 1 \\ 0 & 0 & 0 & 0 \end{bmatrix}.$$

    a. Describe all solutions of $A\mathbf{x} = \mathbf{0}$.
    b. But you were never told the entries of A! What theorem justifies your answer to part (a)?
    c. Now suppose that the Gaussian elimination required no row switches to reach R. Find a vector $\mathbf{b}$ such that $A\mathbf{x} = \mathbf{b}$ has no solution.

2. Row operations and the Gaussian elimination can be used even when the matrix entries are not numbers.
    a. Assuming that $x \neq -1$ or 2, reduce the following matrix to echelon form in such a way that no x's appear in the final matrix:

$$\begin{bmatrix} 8-x & 9 & 9 \\ 3 & 2-x & 3 \\ -9 & -9 & -10-x \end{bmatrix}$$

    b. What are correct echelon forms if $x = -1$ or $x = 2$?

3. Consider the following three systems, where the coefficients are the same for each system but the right-hand sides are different:

$$\begin{matrix} 4x - 8y + 5z = 1 & 0 & 0 \\ 4x - 7y + 4z = 0 & 1 & 0 \\ 3x - 4y + 2z = 0 & 0 & 1 \end{matrix}$$

    Solve all three systems at one time by performing the Gaussian elimination on an augmented matrix of the form $[A \mid \mathbf{b}_1 \mid \mathbf{b}_2 \mid \mathbf{b}_3]$.

4. Use the method indicated in the previous exercise to solve both of the following systems:

$$x - y + z = -1 \quad x - y + z = 1$$
$$2x + y - z = 0 \quad 2x + y - z = -1$$

5. Find angles α, β, and γ such that

$$2 \sin \alpha - \cos \beta + 3 \tan \gamma = 3$$
$$4 \sin \alpha + 2 \cos \beta - 2 \tan \gamma = 2$$
$$6 \sin \alpha - 3 \cos \beta + \tan \gamma = 9$$

where $0 \le \alpha \le 2 \pi, 0 \le \beta \le 2 \pi$, and $0 \le \gamma < \pi$.

6. Let A be the $(2n + 1) \times (2n + 1)$ matrix

$$\begin{bmatrix} 1 & 0 & 0 & \cdots & 0 & 0 & 1 \\ 0 & 1 & 0 & \cdots & 0 & 1 & 0 \\ 0 & 0 & 1 & \cdots & 1 & 0 & 0 \\ \vdots & \vdots & \vdots & 1 & \vdots & \vdots & \vdots \\ 0 & 0 & 1 & \cdots & 1 & 0 & 0 \\ 0 & 1 & 0 & \cdots & 0 & 1 & 0 \\ 1 & 0 & 0 & \cdots & 0 & 0 & 1 \end{bmatrix}.$$

(The 1's form an X.) Prove that A has $n + 1$ linearly independent rows (or columns). Each row or column is taken as a vector in $\mathbb{R}^{2n+1}$.

C.

1. Suppose that we have a system of n equations in n unknowns

$$a_{11}x_1 + a_{12}x_2 + \cdots + a_{1n}x_n = b_1$$
$$a_{21}x_1 + a_{22}x_2 + \cdots + a_{2n}x_n = b_2$$
$$\vdots \qquad \vdots \qquad \qquad \vdots \qquad \vdots$$
$$a_{n1}x_1 + a_{n2}x_2 + \cdots + a_{nn}x_n = b_n$$

If the columns of the matrix of coefficients, regarded as vectors in $\mathbb{R}^n$, are linearly independent, prove that the system has a (unique) solution.

2. Suppose that we have a system of n equations in n unknowns

$$a_{11}x_1 + a_{12}x_2 + \cdots + a_{1n}x_n = 0$$
$$a_{21}x_1 + a_{22}x_2 + \cdots + a_{2n}x_n = 0$$
$$\vdots \qquad \vdots \qquad \qquad \vdots \qquad \vdots$$
$$a_{n1}x_1 + a_{n2}x_2 + \cdots + a_{nn}x_n = 0$$

Prove that the system has a solution $\ne \mathbf{0}$ if and only if the columns of the matrix of coefficients, regarded as vectors in $\mathbb{R}^n$, are linearly dependent.

3. Suppose **x** is a solution vector of the system in the previous exercise and **y** is a solution vector of the system in Exercise C1. Show that **x** + **y** is a solution of the system in Exercise C1.

4. Suppose that the echelon form of the $p \times q$ matrix A has p linearly independent rows and that matrix B is $p \times r$. Show that the echelon form of the $p \times (q+r)$ augmented matrix $[A \mid B]$ has p linearly independent rows. (Form $[A \mid B]$ by just "gluing" B onto the right side of A.)

5. Suppose that $[A \mid \mathbf{b}]$ is the augmented matrix associated with a linear system. We know that performing row operations on $[A \mid \mathbf{b}]$ does not change the solution of the system. However, it turns out that *column operations* can alter the solution.
    a. Describe the effect on the solution of a linear system when columns j and k of $[A \mid \mathbf{b}]$ are interchanged.
    b. Describe the effect when column j is replaced by $c \cdot (\text{column } j)$ for $c \neq 0$.
    c. Describe the effect when column j is replaced by $(\text{column } j) + c \cdot (\text{column } k)$ for $c \neq 0$ and $k \neq j$. [*Hint*: Experiment with a 2×2 or 3×3 system.]

6. Discuss the solutions of the system

$$ax_1 + bx_2 + 2x_3 = 1$$
$$ax_1 + (2b - 1)x_2 + 3x_3 = 1$$
$$ax_1 + bx_2 + (b + 3)x_3 = 2b - 1$$

[*Hint*: Row reduce and then consider various cases focusing on the values ± 1 and 5 for b and the value 0 for a.]

*2.4 Computational Considerations—Pivoting

Sometimes in trying to implement the Gaussian elimination a nonzero pivot does not appear automatically. The next example emphasizes the occasional need for STEP 2 in the algorithm for reducing a matrix to echelon form, not necessarily at the beginning of the reduction process. The term **pivoting** is used to describe the strategy of choosing "good" pivot elements, usually by the strategic interchange of rows (and perhaps of columns). Generally, this means ensuring that there are nonzero elements in the pivot position. However, as we will see in Example 2.4.2, from the point of view of computer calculation, this may also mean avoiding very small pivots.

Example 2.4.1: A Need for Pivoting

Consider the linear system

$$-2x + 4y - 2z - 6w = 4$$
$$3x - 6y + 6z + 10w = -1$$
$$-2x + 6y - z + w = 1$$
$$2x - 5y + 4z + 8w = -3$$

A reasonable first step in reducing the augmented matrix to echelon form is

$$\begin{bmatrix} -2 & 4 & -2 & -6 & 4 \\ 3 & -6 & 6 & 10 & -1 \\ -2 & 6 & -1 & 1 & 1 \\ 2 & -5 & 4 & 8 & -3 \end{bmatrix} \xrightarrow[\substack{R_2 \to R_2 + \frac{3}{2}R_1 \\ R_3 \to R_3 - R_1 \\ R_4 \to R_4 + R_1}]{} \begin{bmatrix} -2 & 4 & -2 & -6 & 4 \\ 0 & 0 & 3 & 1 & 5 \\ 0 & 2 & 1 & 7 & -3 \\ 0 & -1 & 2 & 2 & 1 \end{bmatrix}.$$

We would like to use the number 3 in the second row as a pivot, but there is no way that any number in the second row can eliminate the numbers below it in the second column. Therefore, we are forced to interchange rows 2 and 3 or rows 2 and 4. For example, we could continue the reduction process as follows, where we indicate the pivot at each stage:

$$\begin{bmatrix} \boxed{-2} & 4 & -2 & -6 & 4 \\ 0 & 0 & 3 & 1 & 5 \\ 0 & 2 & 1 & 7 & -3 \\ 0 & -1 & 2 & 2 & 1 \end{bmatrix} \xrightarrow{R_2 \leftrightarrow R_4} \begin{bmatrix} -2 & 4 & -2 & -6 & 4 \\ 0 & \boxed{-1} & 2 & 2 & 1 \\ 0 & 2 & 1 & 7 & -3 \\ 0 & 0 & 3 & 1 & 5 \end{bmatrix}$$

$$\xrightarrow{R_3 \to R_3 + 2R_2} \begin{bmatrix} -2 & 4 & -2 & -6 & 4 \\ 0 & -1 & 2 & 2 & 1 \\ 0 & 0 & \boxed{5} & 11 & -1 \\ 0 & 0 & 3 & 1 & 5 \end{bmatrix}$$

$$\xrightarrow{R_4 \to R_4 - \frac{3}{5}R_3} \begin{bmatrix} -2 & 4 & -2 & -6 & 4 \\ 0 & -1 & 2 & 2 & 1 \\ 0 & 0 & 5 & 11 & -1 \\ 0 & 0 & 0 & \boxed{-\frac{28}{5}} & \frac{28}{5} \end{bmatrix}.$$

It follows that the solution of the system is $x = 1$, $y = 1$, $z = 2$, and $w = -1$.

Clearly a matrix such as

$$\begin{bmatrix} 0 & 0 & 0 & 1 \\ 0 & 0 & 2 & 3 \\ 0 & 4 & 5 & 6 \\ 7 & 8 & 9 & 10 \end{bmatrix}$$

would require several interchanges of rows before being reduced to echelon form.

Now let us look at what happens at any stage of the reduction process when we have found a nonzero pivot a_{ij} in row i and column j, a number that can be used to eliminate all entries below it in column j. Algebraically, we are using the following elementary operation on any row below row i: $R_k \rightarrow R_k - \left(\frac{a_{kj}}{a_{ij}}\right) R_i$, $a_{ij} \neq 0$, $k > i$. Even though a_{ij} is nonzero, if $|a_{ij}|$ is small compared to $|a_{kj}|$, the term a_{kj}/a_{ij} (called the **multiplier**) will be large, so the modified matrix (system) will have entries (coefficients) that are many times larger than those in the original system. This could distort the final solution. Looked at another way, if the pivot element at a particular step of the elimination process is very small, it is possible that this number should be 0 and appears to be nonzero only because of roundoff errors introduced in previous steps. Using a multiplier having this small number as its denominator may compound the error.

Partial pivoting refers to the strategy of searching the entries on and below the pivotal position for the pivot of *maximum magnitude* and making this maximal entry the pivot, interchanging rows if necessary. In practice, guaranteeing that no multiplier is greater than 1 reduces (but does not completely eliminate) the possibility of creating relatively large numbers that can overwhelm the significance of smaller numbers. (**Complete pivoting** requires the examination of every entry below or to the right of the pivotal position for the entry of maximum magnitude and may involve exchanges of *columns* as well as rows. We will not explore this technique.*) The following example, although burdened by admittedly unrealistic restrictions, illustrates the philosophy of pivoting very well.

Example 2.4.2: Partial Pivoting

Let us consider the system

$$0.0001x + y = 1$$
$$x - y = 0$$

Furthermore, to illustrate how roundoff errors can affect a numerical algorithm, suppose that we assume that our calculator or computer has a *word length* of three digits—that is, numbers are represented in the form $0.abc \times 10^e$. We also assume that our calculator rounds the results of calculations. (For example, the coefficient of x in the first equation of our system is represented as 0.100×10^{-3}, or just 0.1×10^{-3}; and a number such as 21.23675 becomes 0.212×10^2.) Thus whatever solution we get by using the Gaussian elimination will only be an *approximation* of the exact solution.

We start with the augmented matrix for our system:

$$\begin{bmatrix} 0.0001 & 1 & 1 \\ 1 & -1 & 0 \end{bmatrix}.$$

Proceeding with the basic (some would say "naïve") Gaussian algorithm, we choose the nonzero entry 0.0001 as

---

* The terms **partial pivoting** and **complete pivoting** were introduced by J.H. Wilkinson in a 1961 paper. See N.J. Higham, *Accuracy and Stability of Numerical Algorithms*, 2nd edn., p. 188 (Philadelphia, PA: SIAM, 2002).

our pivot and use the row operation $R_2 \rightarrow R_2 - \left(\frac{1}{0.0001}\right)$ $R_1 = R_2 - (0.1 \times 10^5)R_1$. The augmented matrix is now $\begin{bmatrix} 0.0001 & 1 & 1 \\ 0 & (-0.1 \times 10^1) - (0.1 \times 10^5)(0.1 \times 10^1) & 0 \end{bmatrix}$

$= \begin{bmatrix} 0.0001 & 1 & 1 \\ 0 & -0.1 \times 10^5 & -0.1 \times 10^5 \end{bmatrix}$. We see that the exact value of the entry in the second row and second column is $-10001 = -0.10001 \times 10^5$, but our hypothetical computer represents this as -0.100×10^5, the first and only instance of roundoff error in our calculation.

Working backward from the matrix in echelon form, we conclude that

$$y = \frac{-0.1 \times 10^5}{-0.1 \times 10^5} = 0.1 \times 10^1 = 1 \quad \text{and}$$

$$x = \frac{(0.1 \times 10^1) - (0.1 \times 10^1)}{0.1 \times 10^{-3}} = 0.$$

The exact solution is $x = y = \frac{10000}{10001} = 0.999900009999\ldots$, so our approximate solution is good for y but terrible for x. The heart of the problem is that in rounding the calculated entry in the second row and second column, we eliminated the quantity -0.00001×10^5, which was the coefficient of y in our original system. In effect, using the Gaussian algorithm on our hypothetical computer has given us a solution of the system

$$0.0001x + y = 1$$
$$x = 0$$

Now let us interchange the equations of our original system, so that the natural pivot is 1, not 0.0001:

$$x - y = 0$$
$$0.0001x + y = 1$$

This time we use the operation

$$R_2 \rightarrow R_2 - \left(\frac{0.1 \times 10^{-3}}{0.1 \times 10^1}\right)R_1 = R_2 - (0.1 \times 10^{-3})R_1,$$

so the augmented matrix is transformed into

$$\begin{bmatrix} 1 & -1 & 0 \\ 0 & (0.1 \times 10^1) - (0.1 \times 10^{-3})(-0.1 \times 10^1) & 0.1 \times 10^1 \end{bmatrix}$$

$$= \begin{bmatrix} 1 & -1 & 0 \\ 0 & 0.1 \times 10^1 & 0.1 \times 10^1 \end{bmatrix}.$$

Therefore, after partial pivoting, the solution is

$$y = \frac{0.1 \times 10^1}{0.1 \times 10^1} = 0.1 \times 10^1 = 1 \quad \text{and}$$

$$x = \frac{0 - (-0.1 \times 10^1)}{0.1 \times 10^1} = 0.1 \times 10^1 = 1,$$

a very good approximation to the exact solution.

Despite the success of partial pivoting in the last example, this strategy of avoiding small pivots does not always work, as the next example shows.

Example 2.4.3:* Partial Pivoting—What Happened?

We will use the Gaussian elimination with partial pivoting to solve the system

$$-10x + 10^5 y = 10^5$$
$$x + y = 2$$

assuming (as in Example 2.3.2) that our calculator or computer has a word length of three digits and that the calculator rounds.

Because $|-10| > 1$, we can use -10 as the pivot without interchanging rows:

$$\begin{bmatrix} -10 & 10^5 & 10^5 \\ 1 & 1 & 2 \end{bmatrix} \xrightarrow{R_2 \to R_2 + 10^{-1}R_1} \begin{bmatrix} -10 & 10^5 & 10^5 \\ 0 & 10^4 & 10^4 \end{bmatrix}$$

(The row operation indicated yields the $(2,2)$ entry $1 + 10^4 = 0.10001 \times 10^5$, and the word length and rounding assumptions reduce this to $0.100 \times 10^5 = 10^4$. Similarly, our hypothetical device gives us 10^4 as the $(2,3)$ entry.)

Back substitution yields $x = 0$ and $y = 1$. However, solving the system manually gives us the exact solution $x = 1/1.0001$ and $y = 1.0002/1.0001$. Thus, despite partial pivoting, the computed solution for x is a very bad approximation to the exact solution. Although the *absolute error*, |exact answer − approximation|, is small, the *relative error* is |exact answer − approximation|/(exact answer) = $|1 - 0|/1 = 1$, representing a 100% error!

The multiplier in this example is not the problem; but, rather, the *relative sizes of the coefficients in each equation*.

---

* Taken from C.D. Meyer, *Matrix Analysis and Applied Linear Algebra*, p. 26 (Philadelphia, PA: SIAM, 2000).

The coefficients of the first equation of our original system are much larger than the coefficients in the second equation. If we *rescale* the system by multiplying the first equation by 10^{-5}, we get the equivalent system

$$-10^{-4}x + y = 1$$
$$x + y = 2$$

which yields a very good approximation to the exact solution via partial pivoting.

An extensive discussion of the various strategies for pivoting, with scaling and without scaling, lies beyond the range of this text. For further information, see a numerical analysis text such as *Numerical Mathematics and Computing (Fifth Edition)* by W. Cheney and D. Kincaid (Belmont, CA: Brooks/Cole, 2004) or a more specialized book such as *Accuracy and Stability of Numerical Algorithms (Second Edition)* by Nicholas J. Higham (Philadelphia, PA: SIAM, 2002).

Exercises 2.4

A.

1. Use the Gaussian elimination with partial pivoting *manually* to reduce the following matrix to row echelon form:

$$\begin{bmatrix} 1 & 0 & 0 & 0 & 1 \\ -1 & 1 & 0 & 0 & 1 \\ -1 & -1 & 1 & 0 & 1 \\ -1 & -1 & -1 & 1 & 1 \\ -1 & -1 & -1 & -1 & 1 \end{bmatrix}.$$

2. On a hypothetical computer with a word length of three digits and truncation (as in Example 2.4.2), compute the solution of

$$-3x + y = -2$$
$$10x - 3y = 7$$

   a. Without partial pivoting
   b. With partial pivoting
   c. Exactly

3. Solve the system

$$x - y + z = 1$$
$$2x + 3y - z = 4$$
$$-3x + y + z = -1$$

a. Without partial pivoting
b. With partial pivoting

4. Assuming a computer with a word length of three digits and truncation, use the Gaussian elimination to solve

$$x + y = 1$$
$$x + 1.0001y = 0$$

a. Without partial pivoting
b. With partial pivoting
c. Exactly

B.

1. Consider the following system:

$$0.001x - y = 1$$
$$x + y = 0$$

a. Use three-digit arithmetic with no pivoting to solve this system.
b. Find a system that is exactly satisfied by your solution from part (a), and note how close this system is to the original system.
c. Now use partial pivoting and three-digit arithmetic to solve the original system.
d. Find a system that is exactly satisfied by your solution from part (c), and note how close this system is to the original system.
e. Use exact arithmetic to obtain the solution to the original system, and compare the exact solution with the results of parts (a) and (c).
f. Round the exact solution to three significant digits and compare the result with those of parts (a) and (c).

2. a. Using two-digit arithmetic, use the Gaussian elimination without pivoting to solve the system

$$0.98x + 0.43y = 0.91$$
$$-0.61x + 0.23y = 0.48$$

b. Compare your answer with both the true solution $x^* = 0.005946\ldots, y^* = 2.102727\ldots$ and the best possible two-digit approximation $x' = 0.0059, y' = 2.1$.
c. Find a system from which the Gaussian elimination in perfect arithmetic would produce x' and y' as the exact solution. [Note that the small absolute errors in the numerical answers actually represent large relative errors and may not be acceptable.]

C.

1. Suppose that A is an $n \times n$ matrix of real numbers that has been scaled so that each entry satisfies $|a_{ij}| \leq 1$, and consider reducing A to echelon form using the Gaussian elimination with partial pivoting. Show that after k steps of the process, no entry can have a magnitude that exceeds 2^k. [*Hint*: What is the maximum size of the multiplier at each step? Use the triangle inequality and induction.]

2.5 Gauss–Jordan Elimination and Reduced Row Echelon Form

The Gaussian elimination technique introduced in Section 2.3 proceeds by performing elementary row operations to produce zeros below the pivot elements of a matrix to reduce it to *echelon form* (or *row echelon form*).

In this section, we will look at a modification of the Gaussian elimination, called **Gauss–Jordan elimination**,* that reduces (or eliminates entirely) the computations involved in back substitution by performing additional row operations to transform the matrix from echelon form to something called *reduced row echelon form* (or *reduced echelon form*, rref). To see the differences between the methods, let us revisit Example 2.2.2.

The system

$$
\begin{aligned}
x + y + z &= 2 \\
x + 3y + 2z &= 1 \\
y + 3z &= 2
\end{aligned}
$$

was represented in terms of an augmented matrix and then row reduced as follows:

$$
\begin{bmatrix} 1 & 1 & 1 & 2 \\ 1 & 3 & 2 & 1 \\ 0 & 1 & 3 & 2 \end{bmatrix} \rightarrow \begin{bmatrix} 1 & 1 & 1 & 2 \\ 0 & 2 & 1 & -1 \\ 0 & 1 & 3 & 2 \end{bmatrix}
$$

$$
\rightarrow \begin{bmatrix} 1 & 1 & 1 & 2 \\ 0 & 1 & 3 & 2 \\ 0 & 2 & 1 & -1 \end{bmatrix} \rightarrow \begin{bmatrix} 1 & 1 & 1 & 2 \\ 0 & 1 & 3 & 2 \\ 0 & 0 & -5 & -5 \end{bmatrix} \rightarrow \begin{bmatrix} 1 & 1 & 1 & 2 \\ 0 & 1 & 3 & 2 \\ 0 & 0 & 1 & 1 \end{bmatrix}.
$$

Finally, back substitution led to the solution $x = 2$, $y = -1$, and $z = 1$.

---

* For some history, see V.J. Katz, Who is the Jordan of Gauss–Jordan? *Math. Mag.* **61** (1988): 99–100. Also see S.C. Althoen and R. McLaughlin, Gauss–Jordan reduction: A brief history, *Amer. Math. Monthly* **94** (1987): 130–142.

Now instead of back substituting, we are going to take the last augmented matrix (in echelon form) and use an appropriate multiple of the third row to eliminate the second nonzero entry, 3, from the second row:

$$\begin{bmatrix} 1 & 1 & 1 & 2 \\ 0 & 1 & 3 & 2 \\ 0 & 0 & 1 & 1 \end{bmatrix} \xrightarrow{R_2 \to R_2 - 3R_3} \begin{bmatrix} 1 & 1 & 1 & 2 \\ 0 & 1 & 0 & -1 \\ 0 & 0 & 1 & 1 \end{bmatrix}.$$

Next, we use the last row to eliminate the third nonzero entry from the first row:

$$\begin{bmatrix} 1 & 1 & 1 & 2 \\ 0 & 1 & 0 & -1 \\ 0 & 0 & 1 & 1 \end{bmatrix} \xrightarrow{R_1 \to R_1 - R_3} \begin{bmatrix} 1 & 1 & 0 & 1 \\ 0 & 1 & 0 & -1 \\ 0 & 0 & 1 & 1 \end{bmatrix}.$$

Finally, we eliminate the second nonzero entry from the first row:

$$\begin{bmatrix} 1 & 1 & 0 & 1 \\ 0 & 1 & 0 & -1 \\ 0 & 0 & 1 & 1 \end{bmatrix} \xrightarrow{R_1 \to R_1 - R_2} \begin{bmatrix} 1 & 0 & 0 & 2 \\ 0 & 1 & 0 & -1 \\ 0 & 0 & 1 & 1 \end{bmatrix}.$$

This final matrix is said to be in reduced row echelon form (*rref*).

Definition 2.5.1

A matrix is in rref if it satisfies the following conditions:

  (1) Any rows consisting entirely of zeros are grouped together at the bottom of the matrix.
  (2) The first nonzero entry in each row is 1 (called a *leading* 1), and it appears in a column to the right of the first nonzero entry of any preceding row.
  (3) The first nonzero entry in each nonzero row is the *only* nonzero entry in its column.

Of course, conditions (1) and (2) describe a matrix in echelon form (see Definition 2.2.2), with all leading entries equal to 1. Thus a matrix in rref is a matrix in a particular echelon form that also satisfies condition (3).

Example 2.5.1: Reduced Row Echelon Form

The following matrices are in reduced echelon form:

$$\begin{bmatrix} 1 & 0 & 0 & -5 \\ 0 & 1 & 0 & 6 \\ 0 & 0 & 1 & 3 \end{bmatrix}, \begin{bmatrix} 1 & 0 \\ 0 & 1 \\ 0 & 0 \end{bmatrix}, \begin{bmatrix} 0 & 1 & 0 & -3 \\ 0 & 0 & 1 & 1 \end{bmatrix}, \begin{bmatrix} 1 & 0 & 0 \\ 0 & 1 & 0 \\ 0 & 0 & 1 \end{bmatrix}.$$

These matrices are *not* in reduced echelon form:

$$\begin{bmatrix} 0 & 1 & 0 \\ 0 & 0 & 1 \\ 1 & 0 & 0 \end{bmatrix}$$ Condition (2) is not satisfied.

$$\begin{bmatrix} 0 & 0 & 0 \\ 1 & 0 & 0 \\ 0 & 1 & 0 \end{bmatrix}$$ Condition (1) is not satisfied.

$$\begin{bmatrix} 0 & 1 & 2 & 3 \\ 0 & 0 & 1 & 2 \\ 0 & 0 & 0 & 1 \end{bmatrix}$$ Condition (3) is not satisfied.

An observation that will be important later, in the proof of Theorem 2.5.1, is that deleting the last column (or the last k columns) of a matrix in rref results in a matrix that is still in rref. Because the rref pattern proceeds from left to right, eliminating the last k columns does not change the basic pattern. For example, consider the following deletions performed on an rref matrix:

$$\begin{bmatrix} 1 & 0 & -3 & 0 & 5 & 4 \\ 0 & 1 & 2 & 0 & -1 & 0 \\ 0 & 0 & 0 & 1 & 2 & 3 \end{bmatrix} \xrightarrow{\text{Delete the last column}} \begin{bmatrix} 1 & 0 & -3 & 0 & 5 \\ 0 & 1 & 2 & 0 & -1 \\ 0 & 0 & 0 & 1 & 2 \end{bmatrix}$$

$\downarrow$ Delete the last three columns

$$\begin{bmatrix} 1 & 0 & -3 \\ 0 & 1 & 2 \\ 0 & 0 & 0 \end{bmatrix}.$$

With slight modifications, the algorithm used in Section 2.3 can be used to find the rref of a matrix A, denoted rref(A).

Reducing a Matrix to Reduced Row Echelon Form (Gauss–Jordan Elimination)

STEP 1. Reduce the matrix to echelon form using the algorithm of Section 2.2.

STEP 2. Begin with the bottommost row that contains a leading entry and work *upward* through the rows, adding appropriate multiples of each row to the rows above it to introduce zeros in the column *above* the leading entry.

STEP 3. Multiply each row by the appropriate reciprocal, so each leading entry becomes 1.

We should note that the algorithm just given may not be the most efficient way to obtain the rref. There may be shortcuts that eliminate some steps and make the arithmetic easier. For example, for ease of calculation, we may want to force a leading entry to be 1 earlier in the process of reduction. As we will see in Theorem 2.5.1, no matter what sequence of elementary row operations we perform, the final rref will be the same. Recall that Example 2.3.2 showed that the basic echelon form of a matrix is *not* unique.

In Gaussian elimination, we work from the top down, using a particular row (equation) to eliminate in the rows (equations) below that row (equation). Gauss–Jordan elimination continues where the Gaussian elimination leaves off, eliminating entries (variables) *above* that row (equation). Essentially, the Gauss–Jordan procedure performs the back substitution of the Gaussian elimination as it goes along, rather than waiting until the end.

The procedures for both the Gaussian elimination and the Gauss–Jordan elimination are sometimes described with slight modifications; but it is agreed that of all the procedures for obtaining the rref of a matrix, the Gaussian elimination requires the fewest arithmetic operations and is therefore the preferred method for computer solution of systems of linear equations. For large values of n, the Gaussian elimination on an $n \times n$ system requires approximately $n^3/3$ multiplications (which were once considered more "costly" than additions/subtractions*), whereas the Gauss–Jordan method needs approximately $n^3/2$ multiplications.[†] So if we are solving an $n \times n$ system and n is large, the Gauss–Jordan procedure requires about 50% more effort $\left(\frac{n^3}{2}/\frac{n^3}{3} = 1.5\right)$ than the Gaussian elimination. When $n = 1000$, for example, the Gauss–Jordan procedure needs about $1000^3/2 = 500,000,000$ multiplications, compared to $1000^3/3 = 333,333,333$ for the Gaussian elimination.

However, the Gauss–Jordan method has some theoretical advantages, as we shall see in the next section. For now, let us consider additional examples of the Gauss–Jordan procedure.

Example 2.5.2: Gauss–Jordan Elimination

To illustrate the flexibility we have in using the Gauss–Jordan algorithm, we will convert the matrix

$$\begin{bmatrix} 4 & 3 & 1 & 6 \\ 0 & 2 & 5 & -3 \\ 1 & 0 & -3 & 4 \end{bmatrix}$$

---

* On modern computers, floating point multiplications have become comparable to floating point additions. See N.J. Higham *Accuracy and Stability of Numerical Algorithms*, 2nd edn., p. 434 (Philadelphia, PA: SIAM, 2002) or the footnote of C.D. Meyer, *Matrix Analysis and Applied Linear Algebra*, p. 10 (Philadelphia, PA: SIAM, 2000).

[†] See, for example, C.D. Meyer, *Matrix Analysis and Applied Linear Algebra*, p. 10, 16 (Philadelphia, PA: SIAM, 2000).

to rref in two different ways. First,

$$\begin{bmatrix} 4 & 3 & 1 & 6 \\ 0 & 2 & 5 & -3 \\ 1 & 0 & -3 & 4 \end{bmatrix} \xrightarrow{R_3 \to R_3 - \frac{1}{4}R_1} \begin{bmatrix} 4 & 3 & 1 & 6 \\ 0 & 2 & 5 & -3 \\ 0 & -\frac{3}{4} & -\frac{13}{4} & \frac{5}{2} \end{bmatrix}$$

$$\xrightarrow[\text{[row echelon form]}]{R_3 \to R_3 + \frac{3}{8}R_2} \begin{bmatrix} 4 & 3 & 1 & 6 \\ 0 & 2 & 5 & -3 \\ 0 & 0 & -\frac{11}{8} & \frac{11}{8} \end{bmatrix}$$

$$\xrightarrow[R_3 \to -\frac{8}{11}R_3]{R_1 \to \frac{1}{4}R_1, R_2 \to \frac{1}{2}R_2} \begin{bmatrix} 1 & \frac{3}{4} & \frac{1}{4} & \frac{3}{2} \\ 0 & 1 & \frac{5}{2} & -\frac{3}{2} \\ 0 & 0 & 1 & -1 \end{bmatrix}$$

$$\xrightarrow[R_1 \to R_1 - \frac{1}{4}R_3]{R_2 \to R_2 - \frac{5}{2}R_3} \begin{bmatrix} 1 & \frac{3}{4} & 0 & \frac{7}{4} \\ 0 & 1 & 0 & 1 \\ 0 & 0 & 1 & -1 \end{bmatrix}$$

$$\xrightarrow[\text{[reduced row echelon form]}]{R_1 \to R_1 - \frac{3}{4}R_2} \begin{bmatrix} 1 & 0 & 0 & 1 \\ 0 & 1 & 0 & 1 \\ 0 & 0 & 1 & -1 \end{bmatrix}.$$

Next, we decide to start the process by interchanging rows:

$$\begin{bmatrix} 4 & 3 & 1 & 6 \\ 0 & 2 & 5 & -3 \\ 1 & 0 & -1 & 4 \end{bmatrix} \xrightarrow{R_1 \leftrightarrow R_3} \begin{bmatrix} 1 & 0 & -3 & 4 \\ 0 & 2 & 5 & -3 \\ 4 & 3 & 1 & 6 \end{bmatrix}$$

$$\xrightarrow{R_3 \to R_3 - 4R_1} \begin{bmatrix} 1 & 0 & -3 & 4 \\ 0 & 2 & 5 & -3 \\ 0 & 3 & 13 & -10 \end{bmatrix}$$

$$\xrightarrow{R_2 \to R_2 - R_3} \begin{bmatrix} 1 & 0 & -3 & 4 \\ 0 & -1 & -8 & 7 \\ 0 & 3 & 13 & -10 \end{bmatrix}$$

$$\xrightarrow[\text{[row echelon form]}]{R_3 \to R_3 + 3R_2} \begin{bmatrix} 1 & 0 & -3 & 4 \\ 0 & -1 & -8 & 7 \\ 0 & 0 & -11 & 11 \end{bmatrix}$$

$$\xrightarrow[R_3 \to -\frac{1}{11}R_3]{R_2 \to -R_2} \begin{bmatrix} 1 & 0 & -3 & 4 \\ 0 & 1 & 8 & -7 \\ 0 & 0 & 1 & -1 \end{bmatrix}$$

$$\xrightarrow[\text{[reduced row echelon form]}]{\substack{R_1 \to R_1 + 3R_3 \\ R_2 \to R_2 - 8R_3}} \begin{bmatrix} 1 & 0 & 0 & 1 \\ 0 & 1 & 0 & 1 \\ 0 & 0 & 1 & -1 \end{bmatrix}.$$

Notice how the arithmetic was simpler the second time. Although we can always achieve an rref by following the steps of the algorithm slavishly, thinking a few steps ahead may suggest a better way of reaching the goal.

Before coming to the theoretical high point of this section, let us look at an application of linear systems and their solution by the Gauss–Jordan method.

Example 2.5.3: Balancing Chemical Equations

A chemical reaction occurs, for example, when molecules of one substance A are combined under certain conditions with molecules of another substance B to produce a quantity of a third substance C: $A + B \rightarrow C$. The term *stoichiometry*, from Greek words meaning "element" and "measurement," refers to the study of quantitative relationships in chemical reactions. A major aspect of this study is the writing of "balanced" chemical equations.

For example, we can describe the fact that glucose ($C_6H_{12}O_6$) reacts with oxygen gas (O_2) to form carbon dioxide and water: $\overbrace{C_6H_{12}O_6 + O_2}^{\text{Reactants}} \rightarrow \overbrace{CO_2 + H_2O}^{\text{Products}}$.

Stoichiometry demands a balance. For instance, there are eight atoms of oxygen on the left ($O_6 + O_2 = O_8$), but only three atoms of oxygen on the right. Because of the basic principle that matter is neither created nor destroyed in a reaction, we must try to find a chemical equation that accounts for all quantities involved in the reaction. Mathematically, we look for the smallest positive integers a, b, c, d such that

$$a(C_6H_{12}O_6) + b(O_2) = c(CO_2) + d(H_2O).$$

Comparing the number of atoms of carbon (C), hydrogen (H), and oxygen (O) on each side of the reaction, we get the system

Carbon: $6a = c$ $6a - c = 0$
Hydrogen: $12a = 2d$ or $12a - 2d = 0$
Oxygen: $6a + 2b = 2c + d$ $6a + 2b - 2c - d = 0$

We form the augmented matrix

$$\begin{bmatrix} 6 & 0 & -1 & 0 & \vdots & 0 \\ 12 & 0 & 0 & -2 & \vdots & 0 \\ 6 & 2 & -2 & -1 & \vdots & 0 \end{bmatrix},$$

which has the rref

$$\begin{bmatrix} 1 & 0 & 0 & -\frac{1}{6} & \vdots & 0 \\ 0 & 1 & 0 & -1 & \vdots & 0 \\ 0 & 0 & 1 & -1 & \vdots & 0 \end{bmatrix}.$$

From this last matrix, we find that the system has an infinite number of solutions of the form $\begin{bmatrix} d/6 \\ d \\ d \\ d \end{bmatrix}$, where d is a free variable. Choosing $d = 6$ gives us the solution with the smallest positive integer values for a, b, c, and d. Thus $a = 1$, $b = 6$, $c = 6$, and $d = 6$, so that the balanced chemical equation for our reaction is

$$C_6H_{12}O_6 + 6O_2 \rightarrow 6CO_2 + 6H_2O.$$

We could have obtained this answer by careful guessing, but it is important to develop methodology that can deal with more complicated reactions.

As we have noted, one consequence of the next theorem is that the rref of A is independent of the particular elimination method (sequence of row operations) used.

Theorem 2.5.1

The rref of a matrix A is unique.

*Proof** Let A be an $m \times n$ matrix. We will prove the theorem by mathematical induction on n.

If A is a matrix with only one column (i.e., $n = 1$), then there are just two possibilities for $R = \text{rref}(A)$. The entry in the first row of R is either 0 or 1, and all other entries are 0:

$$R = \begin{bmatrix} 0 \\ 0 \\ \vdots \\ 0 \end{bmatrix} \quad \text{or} \quad R = \begin{bmatrix} 1 \\ 0 \\ \vdots \\ 0 \end{bmatrix}.$$

Because any row operation applied to the first of these column matrices leaves it unchanged, we cannot use row operations to transform the first matrix into the second (nor can we change the second into the first). This means that any matrix A with just one column is row equivalent to one of the matrices R above but not to both of them.

Now suppose that $n > 1$ and that the theorem is true for all matrices with *fewer* than n columns. (This is our inductive hypothesis.) If B and C are two rrefs of A, let A', B', and C' be the matrices obtained by deleting the last columns from A, B, and C, respectively. Then B' and C' are row equivalent to A', and, moreover, B' and C' are in rref (see the comment right after Example 2.5.1). But A' has fewer columns than A and so, by the inductive hypothesis, $B' = C'$. This in turn implies that B and C can differ in the nth column only.

Now assume that $B \neq C$. Then there is an integer j such that row j of B is not equal to row j of C. Let $\mathbf{u}$ be any column vector such that $B\mathbf{u} = \mathbf{0}$. Then $C\mathbf{u} = \mathbf{0}$ by Theorem 2.2.2 and hence $(B - C)\mathbf{u} = \mathbf{0}$.[†]

---

* Based on T. Yuster, The reduced row echelon form of a matrix is unique: A simple proof, *Math. Mag.* **57** (1984): 93–94.
[†] The (i, j) entry of the matrix $B - C$ is $d_{ij} = b_{ij} - c_{ij}$.

The first $n-1$ columns of $B-C$ are zero columns, and we get the matrix equation

$$\begin{bmatrix} 0 & 0 & \cdots & 0 & b_{1n}-c_{1n} \\ \vdots & \vdots & \cdots & \vdots & \vdots \\ 0 & 0 & \cdots & 0 & b_{jn}-c_{jn} \\ \vdots & \vdots & \cdots & \vdots & \vdots \\ 0 & 0 & \cdots & 0 & b_{mn}-c_{mn} \end{bmatrix} \begin{bmatrix} u_1 \\ \vdots \\ u_j \\ \vdots \\ u_n \end{bmatrix} = \begin{bmatrix} 0 \\ \vdots \\ 0 \\ \vdots \\ 0 \end{bmatrix},$$

which is equivalent to

$$(b_{1n}-c_{1n})u_n = (b_{2n}-c_{2n})u_n = \cdots = (b_{jn}-c_{jn})u_n$$
$$= \cdots = (b_{mn}-c_{mn})u_n = 0.$$

In this case, because we have assumed that $b_{jn} \neq c_{jn}$, we conclude that $u_n = 0$.

It follows that the nth columns of both B and C must contain leading 1's, for otherwise those columns would be "free" columns, and we could choose the value of u_n arbitrarily. But because the first $n-1$ columns of B and C are identical, the row in which this leading 1 must appear must be the same for B and C, namely, the row which is the first zero row of rref (A'). Because the remaining entries in the nth columns of B and C must all be zero, we have $B = C$, which is a contradiction.

Despite the computational inefficiency of going through extra steps to produce the rref, the uniqueness of this form makes it very useful for theoretical purposes, as we shall see in the rest of this chapter.

Exercises 2.5

A.

1. Which of the following matrices are in rref? If a matrix is not in this form, explain which part of the definition is violated.

a. $\begin{bmatrix} 1 & 0 & 3 & -2 \\ 0 & 1 & 5 & 6 \end{bmatrix}$ b. $\begin{bmatrix} 0 & 0 & 1 \\ 0 & 1 & 0 \\ 1 & 0 & 0 \end{bmatrix}$ c. $\begin{bmatrix} 0 & 0 & 1 & 0 \\ 1 & 0 & 0 & 1 \\ 0 & 1 & 0 & 0 \\ 0 & 0 & 0 & 0 \end{bmatrix}$

d. $\begin{bmatrix} 2 & 0 & 1 & 0 \\ 0 & 1 & 0 & 0 \\ 0 & 0 & 0 & 1 \\ 0 & 0 & 0 & 0 \end{bmatrix}$ e. $\begin{bmatrix} 1 & 2 & 3 \\ 0 & 1 & 0 \\ 0 & 0 & 1 \end{bmatrix}$ f. $\begin{bmatrix} 1 & 0 & 0 & 0 \\ 0 & 0 & 0 & 1 \\ 0 & 1 & 0 & 0 \\ 0 & 0 & 1 & 0 \end{bmatrix}$

g. $\begin{bmatrix} 1 & 0 & 3 & 1 \\ 0 & 1 & 2 & 4 \end{bmatrix}$ h. $\begin{bmatrix} 1 & 0 & 0 & 3 \\ 0 & 1 & 1 & 2 \\ 0 & 0 & 0 & 3 \\ 0 & 0 & 0 & 0 \end{bmatrix}$ i. $\begin{bmatrix} 1 & 3 & 0 & 2 & 0 \\ 1 & 0 & 2 & 2 & 0 \\ 0 & 0 & 0 & 0 & 1 \\ 0 & 0 & 0 & 0 & 0 \end{bmatrix}$

2. Transform each of the following matrices to rref:

    a. $\begin{bmatrix} 1 & 2 \\ 3 & 4 \end{bmatrix}$
    b. $\begin{bmatrix} 1 & 1 & 3 & 7 \\ 0 & 1 & 6 & 2 \\ 0 & 0 & 1 & 5 \end{bmatrix}$
    c. $\begin{bmatrix} 1 & 1 & 0 \\ 0 & 1 & 0 \\ 0 & 0 & 0 \end{bmatrix}$

    d. $\begin{bmatrix} 1 & 2 & 3 & 4 \\ 5 & 6 & 7 & 8 \\ 9 & 10 & 11 & 12 \end{bmatrix}$
    e. $\begin{bmatrix} 2 & -4 & 8 \\ 3 & 5 & 8 \\ -6 & 0 & 4 \end{bmatrix}$

    f. $\begin{bmatrix} 0 & -1 & 1 \\ 2 & 4 & 3 \\ 0 & 6 & -2 \end{bmatrix}$
    g. $\begin{bmatrix} 2 & -7 \\ 3 & 5 \\ 4 & -3 \end{bmatrix}$

3. Reduce the matrix

    $$\begin{bmatrix} 2 & 1 & 3 \\ 0 & -2 & 7 \\ 3 & 4 & 5 \end{bmatrix}$$

    to rref without introducing fractions as matrix entries at any intermediate stage.

4. For each part, suppose that the augmented matrix for a system of linear equations has been transformed by row operations to the given rref. Solve the system in each part, assuming that the variables are $x_1, x_2, \ldots$ from left to right.

    a. $\begin{bmatrix} 0 & 1 & 0 & -3 \\ 0 & 0 & 1 & 1 \end{bmatrix}$
    b. $\begin{bmatrix} 1 & 0 \\ 0 & 1 \\ 0 & 0 \end{bmatrix}$
    c. $\begin{bmatrix} 1 & 0 & 0 & -5 \\ 0 & 1 & 0 & 6 \\ 0 & 0 & 1 & 3 \end{bmatrix}$

    d. $\begin{bmatrix} 1 & 0 & 0 & 2 \\ 0 & 1 & 0 & -1 \\ 0 & 0 & 1 & 1 \end{bmatrix}$
    e. $\begin{bmatrix} 1 & 2 & 0 & 3 & 1 \\ 0 & 0 & 1 & 3 & 4 \end{bmatrix}$

    f. $\begin{bmatrix} 1 & -3 & 0 & 0 \\ 0 & 0 & 1 & 0 \\ 0 & 0 & 0 & 1 \end{bmatrix}$
    g. $\begin{bmatrix} 1 & -6 & 0 & 0 & 3 & -2 \\ 0 & 0 & 1 & 0 & 4 & 7 \\ 0 & 0 & 0 & 1 & 5 & 8 \\ 0 & 0 & 0 & 0 & 0 & 0 \end{bmatrix}$

5. Suppose you have a linear system whose augmented matrix has the form

    $$\begin{bmatrix} 1 & 2 & 1 & \vdots & 1 \\ -1 & 4 & 3 & \vdots & 2 \\ 2 & -2 & k & \vdots & 3 \end{bmatrix}.$$

    For what values of k will the system have a *unique* solution?

6. Consider a linear system whose augmented matrix has the form

    $$\begin{bmatrix} 1 & 2 & 1 & \vdots & 0 \\ 2 & 5 & 3 & \vdots & 0 \\ -1 & 1 & p & \vdots & 0 \end{bmatrix}.$$

    a. Is it possible for the system to have *no* solution?
    b. For what values of p will the system have *infinitely many* solutions?

7. Consider a linear system whose augmented matrix has the form

$$\begin{bmatrix} 1 & 1 & 3 & \vdots & 2 \\ 1 & 2 & 4 & \vdots & 3 \\ 1 & 3 & a & \vdots & b \end{bmatrix}.$$

   a. For what values of a and b will the system have *infinitely many* solutions?
   b. For what values of a and b will the system have *no* solution?

8. Solve each of the following systems (if possible) by using Gauss–Jordan elimination:

   a. $2x + y - 3z = 1$ b. $x + 2y - 2z = -1$
       $5x + 2y - 6z = 5$ $3x - y + 2z = 7$
       $3x - y - 4z = 7$ $5x + 3y - 4z = 2$

   c. $x + 2y + 3z = 9$ d. $x + 2y + 3z + 4w = 5$
       $2x - y + z = 8$ $x + 3y + 5z + 7w = 11$
       $3x - z = 3$ $x - z - 2w = -6.$

   e. $x + y + 2z - 5w = 3$
       $2x + 5y - z - 9w = -3$
       $2x + y - z + 3w = -11$
       $x - 3y + 2z + 7w = -5$

   f. $a + 2b - 3d + e = 2$
       $a + 2b + c - 3d + e + 2f = 3$
       $a + 2b + 3d + 2e + f = 4$
       $3a + 6b + c - 9d + 4e + 3f = 9$

9. The general form of the equation of a circle in $\mathbb{R}^2$ is

$$x^2 + y^2 + Ax + By + C = 0.$$

   Find the equation of the circle passing through the three (noncollinear) points

$$(1, 2),\ (-2, 1),\ \text{and } (-3, -1).$$

10. Find values of the coefficients a and b so that $y = a \sin 2x + b \cos 2x$ is a solution of the differential equation $y'' + 3y' - y = \sin 2x.$

11. An investor remarks to a stockbroker that all her stock holdings are in three companies, IBM, Coca Cola, and Gillette, and that 2 days ago the value of her stocks went down $350 but yesterday the value increased by $600. The broker recalls that 2 days ago the price of IBM and Coca Cola stocks dropped by $1 and $1.50 a share, respectively, but the price of Gillette stock rose by $0.50. The broker also remembers that yesterday the price of IBM stock rose by $1.50, there was a further drop of $0.50 a share in Coca Cola, and the price of Gillette stock rose by $1. Show that

the broker does not have enough information to calculate the number of shares the investor owns of each company's stock, but that when the investor says that she owns 200 shares of Gillette stock, the broker can calculate the number of shares of IBM and Coca Cola. [*Hint*: Take the prices three days ago as your base.]

12. In each part, determine the smallest positive integer values a, b, c, and d that balance the equation, where B = boron, C = carbon, H = hydrogen, N = nitrogen, O = oxygen, and S = sulfur (see Example 2.5.3).

  a. $a(C_6H_6) + b(O_2) \rightarrow c(C) + d(H_2O)$

  b. $a(B_3S_3) + b(H_3N) \rightarrow c(B_3N_2) + d(S_3H_4)$

13. The process of *photosynthesis* can be described by the reaction equation

$$a(CO_2) + b(H_2O) \rightarrow c(O_2) + d(C_6H_{12}O_6).$$

Find the values of a, b, c, and d that balance the equation.

14. Find the values of a, b, c, d, and e that balance the equation

$$a(AgNO_3) + b(H_2O) \rightarrow c(Ag) + d(O_2) + e(HNO_3),$$

where Ag = silver and N = nitrogen.

15. Balance the following chemical reaction:

$$PbN_6 + CrMn_2O_8 \rightarrow Cr_2O_3 + MnO_2 + Pb_3O_4 + NO,$$

where Pb represents lead, Cr represents chromium, and Mn denotes magnesium.

16. Use a calculator or computer algebra system (CAS) to solve the system

$$\tfrac{1}{5}x + \tfrac{1}{4}y + \tfrac{1}{2}z = \tfrac{37}{120}$$
$$\tfrac{1}{3}x + \tfrac{1}{7}y + \tfrac{1}{4}z = \tfrac{93}{336}$$
$$\tfrac{1}{4}x + \tfrac{1}{6}y + \tfrac{1}{3}z = \tfrac{43}{180}$$

B.

1. Suppose the rref of A is

$$\begin{bmatrix} 1 & 0 & 2 & 0 & -2 \\ 0 & 1 & -5 & 0 & -3 \\ 0 & 0 & 0 & 1 & 6 \end{bmatrix}.$$

Determine A if the first, second, and fourth columns of A are

$$\begin{bmatrix} 1 \\ -1 \\ 3 \end{bmatrix}, \quad \begin{bmatrix} 0 \\ -1 \\ 1 \end{bmatrix}, \quad \text{and} \quad \begin{bmatrix} 1 \\ -2 \\ 0 \end{bmatrix},$$

respectively.

2. Suppose that the rref of A is

$$\begin{bmatrix} 1 & -3 & 0 & 4 & 0 & 5 \\ 0 & 0 & 1 & 3 & 0 & 2 \\ 0 & 0 & 0 & 0 & 1 & -1 \\ 0 & 0 & 0 & 0 & 0 & 0 \end{bmatrix}.$$

Determine A if the first, third, and sixth columns of A are

$$\begin{bmatrix} 1 \\ -2 \\ -1 \\ 3 \end{bmatrix}, \quad \begin{bmatrix} -1 \\ 1 \\ 2 \\ -4 \end{bmatrix}, \quad \text{and} \quad \begin{bmatrix} 3 \\ -9 \\ 2 \\ 5 \end{bmatrix},$$

respectively.

3. Determine the condition on b_1, b_2, and b_3 so that the following system has a solution.

$$x + 2y + 6z = b_1$$
$$2x - 3y - 2z = b_2$$
$$3x - y + 4z = b_3$$

4. Show that if $ad - bc \neq 0$, then the rref of $\begin{bmatrix} a & b \\ c & d \end{bmatrix}$ is $\begin{bmatrix} 1 & 0 \\ 0 & 1 \end{bmatrix}$.

5. Regard two row reduced matrices as having the same echelon pattern if the leading 1's occur in the same positions. Describe all possible distinct rrefs for 2×2 matrices. [*Hint*: There are four of them.]

6. If $[A \,|\, \mathbf{b}]$ is in rref, prove that A is also in reduced row echelon form.

7. The solution of the system with the augmented matrix given below yields a coded message that can be understood as follows: Each letter of the alphabet is numbered by its alphabetical position ($A = 1$, $B = 2$, and so forth). What is the message?

$$\begin{bmatrix} 1 & -1 & -1 & -1 & 0 & 0 & 0 & -27 \\ 0 & 0 & -1 & 0 & -1 & 0 & -1 & -27 \\ 0 & -1 & 0 & 1 & -1 & 0 & 0 & -21 \\ 1 & 1 & 1 & 0 & -1 & 0 & 0 & 27 \\ 1 & 1 & 0 & 1 & -1 & 0 & 0 & 16 \\ 0 & 0 & 1 & 1 & -1 & -1 & -1 & -8 \\ 0 & 1 & 0 & 0 & 0 & 0 & -1 & 13 \end{bmatrix}$$

(You should be able to solve the system by hand as well.)

C.

1. Write down the possible reduced row echelon patterns for a 2×3 matrix (see Exercise B5). [*Hint*: There are seven.]

2. Describe all possible rrefs of the matrix (see Exercise B5).

$$\begin{bmatrix} a & b & c \\ d & e & f \\ g & h & i \end{bmatrix}.$$

3. How many reduced row echelon patterns are possible for a 3×5 matrix? (see Exercise B5). [*Hint*: There are 26.]

4. Suppose a system of nine linear equations in seven variables has an augmented matrix which, when put in rref, has four leading 1's.
    a. Describe the solution set if one of the leading 1's is in the rightmost column.
    b. Describe the solution set if none of the leading 1's is in the rightmost column.

5. Suppose a system of m linear equations in n variables has an augmented matrix which, when put in rref, has r leading 1's.
    a. Describe the solution set if one of the leading 1's is in the rightmost column.
    b. Describe the solution set if none of the leading 1's is in the rightmost column.

*2.6 Ill-Conditioned Systems of Linear Equations

So far in this chapter, we have seen a number of techniques and theorems that may have created the impression that we have a practical understanding of the solution of systems of linear equations. In particular, we believe we have some insight into the algorithms used by calculators and computers to solve such systems. It is precisely because computers are used to solve most of the real-world problems involving linear algebra that we addressed some computational considerations in Section 2.4.

However, at this point we may still have an unconscious bias toward the *continuity* of mathematical processes—that is, we may expect a small change in an input to result in a small change in the output. Specifically, given a linear system $A\mathbf{x} = \mathbf{b}$, we may feel that a slight change ("perturbation") in the matrix A should result in a slight change in the solution $\mathbf{x}$ or that a slight change in $\mathbf{b}$ causes only a slight change in $\mathbf{x}$. Unfortunately, the situation is more complicated than this.

As an example, let us look at the three systems:

$$(1) \quad \begin{array}{l} x + y = 1 \\ x + 1.01y = 1 \end{array} \qquad (2) \quad \begin{array}{l} x + y = 1 \\ x + 1.01y = 1.01 \end{array} \qquad (3) \quad \begin{array}{l} x + y = 1 \\ 0.99x + y \end{array}$$

The solution of system (1) is $x = 1$ and $y = 0$, whereas the solution of system (2) is $x = 0$ and $y = 1$. A change of 0.01 in the right-hand side of the second equation has produced a surprising change in the solution. Finally, the solution of the third system is $x = -1$ and $y = 2$, so that in going from system (1) to system (3) a change of 0.01 in two of the coefficients has changed the solution even more dramatically. (The rrefs of the three augmented matrices are significantly different.)

It is a common practice to measure changes in a system *relative* to the size (norm) of the object being perturbed, although sometimes it turns out to be more appropriate to measure change on an *absolute* scale (see Example 2.4.3).

Definition 2.6.1

Let $\hat{x}$ denote an approximation of the real number x. Then the **absolute error** of the approximation is defined as $|\hat{x} - x|$ and the **relative error** is given by $\frac{|\hat{x}-x|}{|x|}$, $x \neq 0$. (If $\hat{x}$ and x are vectors, the absolute value symbol may be replaced by an appropriate norm symbol $\|\cdot\|$.)

As a simple numerical example, let $x_1 = 1.31$ and $\hat{x}_1 = 1.30$, $x_2 = 0.12$, and $\hat{x}_2 = 0.11$. Then $|\hat{x}_1 - x_1| = 0.01 = |\hat{x}_2 - x_2|$, which seems to indicate that $\hat{x}_1$ is as close to x_1 as $\hat{x}_2$ is to x_2. However, to four decimal places, $\frac{|\hat{x}_1-x_1|}{|x_1|} = 0.0076$, whereas $\frac{|\hat{x}_2-x_2|}{|x_2|} = 0.0833$, showing that $\hat{x}_1$ is closer to x_1 than $\hat{x}_2$ is to x_2 (relatively speaking).

Some insight into the behavior of the systems

$$(1) \quad \begin{array}{l} x + y = 1 \\ x + 1.01y = 1 \end{array} \qquad (2) \quad \begin{array}{l} x + y = 1 \\ x + 1.01y = 1.01 \end{array} \qquad (3) \quad \begin{array}{l} x + y = 1 \\ 0.99x + y = 1.01 \end{array}$$

can be gained by thinking about the graphical representation of each system. Each of the six straight lines making up systems (1)–(3) has a slope that is equal to or nearly equal to -1. If we try to graph these lines using a graphing calculator or computer software, we would have a difficult time distinguishing one line from another. What this means geometrically is that the two lines forming each system are nearly parallel and so intersect at a very small angle. This, in turn, means that a small movement of a line parallel to itself moves the point of intersection (the solution of the system) a distance along the line that is great compared with the normal distance moved by the line (Figure 2.5).

A system of linear equations $A\mathbf{x} = \mathbf{b}$ that is overly sensitive to perturbations in A or $\mathbf{b}$ is called **ill-conditioned**. More precisely, an ill-conditioned system is one in which a significant change in the solution can be produced by a small change in the values of the matrix of coefficients or in the output vector $\mathbf{b}$. A system is said to be **well-conditioned** if relatively small changes in A or $\mathbf{b}$ result in relatively small changes in the solution of the system.

Here is an example of a larger ill-conditioned system.

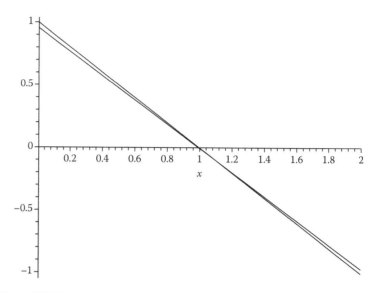

Figure 2.5 Two lines intersecting at a small angle.

Example 2.6.1: An Ill-Conditioned System

We consider the system $A\mathbf{x} = \mathbf{b}$, where

$$A = \begin{bmatrix} 0.932 & 0.443 & 0.417 \\ 0.712 & 0.915 & 0.887 \\ 0.632 & 0.514 & 0.493 \end{bmatrix}.$$

We define different output vectors $\mathbf{b}_1 = \begin{bmatrix} 1 \\ 1 \\ -1 \end{bmatrix}$ and

$\mathbf{b}_2 = \begin{bmatrix} 1.01 \\ 1.01 \\ -1.01 \end{bmatrix}$ and solve the resulting systems $A\mathbf{x} = \mathbf{b}_1$ and

$A\mathbf{x} = \mathbf{b}_2$. A CAS (in this case, *Maple*) gives the solution of $A\mathbf{x} = \mathbf{b}_1$ as

$$\begin{bmatrix} 106.395 \\ -4909.194 \\ 4979.886 \end{bmatrix},$$

rounded to three decimal places to match the precision of the coefficients, and the solution of $A\mathbf{x} = \mathbf{b}_2$ as

$$\begin{bmatrix} 107.459 \\ -4958.286 \\ 5029.685 \end{bmatrix}.$$

Using the Euclidean norm (Definition 1.2.2), we see that the difference (distance) between $\mathbf{b}_1$ and $\mathbf{b}_2$ (see Problem B16 of Exercises 1.2) is

$$\|\mathbf{b}_1 - \mathbf{b}_2\| = \sqrt{(1 - 1.01)^2 + (1 - 1.01)^2 + (-1 + 1.01)^2} = 0.17,$$

whereas there is quite an absolute distance between the *solutions* of the two systems:

$$\sqrt{(106.395 - 107.459)^2 + (-4909.194 + 4958.286)^2 + (4979.886 - 5029.685)^2}$$
$$= 69.94.$$

The system $A\mathbf{x} = \mathbf{b}$ is ill-conditioned.

The geometrical interpretation of the last example is that the system $A\mathbf{x} = \mathbf{b}$ represents the intersection of three planes in $\mathbb{R}^3$ in a single point, but the planes intersect at small angles. Thus any slight perturbation (disturbance) of the output vector $\mathbf{b}$ results in moving the point of intersection further out, relatively speaking.

There is another way that a system can be ill-conditioned. A calculator or computer engaged in performing row reductions will round off entries, and this can cause difficulties—as the next example illustrates.

Example 2.6.2: An Ill-Conditioned System

Suppose that we start with the system

$$2.693x + 1.731y = 1$$
$$7.083x + 4.552y = 2$$

A CAS gives us the solution $\begin{bmatrix} -510.061 \\ 794.104 \end{bmatrix}$ (rounded to three decimal places). If we drop one significant digit from each of our coefficients, we get the system

$$2.69x + 1.73y = 1$$
$$7.08x + 4.55y = 2$$

so that the original matrix of coefficients is perturbed slightly. The solution of this new system is $\begin{bmatrix} -122.472 \\ 191.011 \end{bmatrix}$, so that once again we have an ill-conditioned system.

Linear systems often arise in analyzing the results of experiments. Ill-conditioned systems can originate from experimental errors in data or from rounding errors introduced in the calculator or computer solution of linear systems. Sometimes it is possible to reformulate the linear model or at least solve the system in ways that do not accentuate the ill-conditioning. There are modifications of Gaussian and Gauss–Jordan elimination that minimize rounding errors. For example, a combination of *pivoting* and *scaling* (as seen in Section 2.4) is often effective.

Exercises 2.6

A.

1. Consider the equations

$$28x + 25y = 30$$
$$19x + 17y = 20$$

   a. Find the unique solution of this system manually.
   b. Substitute the values $x = 18$ and $y = -19$ in the system. Does it look as if these values should approximate the exact solution? Compare these values with the solution found in part (a).
   c. Change the 20 on the right-hand side of the system to 19, leaving everything else unchanged. Now solve the new system manually and comment on the difference between this solution and the solution found in part (a).

2. Consider the linear system

$$100a + 99b = 398$$
$$99a + 98b = 394$$

   a. Find the exact solution of the system by hand.
   b. Now change the value 394 on the right-hand side of the system to 393.98 and solve the resulting system. Comment on the sensitivity of the solution to changes in the output vector.

3. Consider the linear system

$$\begin{bmatrix} 1000 & 999 \\ 999 & 998 \end{bmatrix} \begin{bmatrix} x_1 \\ x_2 \end{bmatrix} = \begin{bmatrix} 1999 \\ 1997 \end{bmatrix}.$$

   a. Find the exact solution of the system by hand.
   b. Change the input vector to $\begin{bmatrix} 1998.99 \\ 1997.01 \end{bmatrix}$ and solve the resulting system.

   Comment on the sensitivity of the solution to changes in the output vector.

4. Consider the system $A\mathbf{x} = \mathbf{b}$, with
   $\mathbf{b} = \begin{bmatrix} \frac{137}{60} & \frac{29}{20} & \frac{153}{140} & \frac{743}{840} & \frac{1879}{2520} \end{bmatrix}^{\mathrm{T}}$ and

   a. $A = \begin{bmatrix} 1 & \frac{1}{2} & \frac{1}{3} & \frac{1}{4} & \frac{1}{5} \\ \frac{1}{2} & \frac{1}{3} & \frac{1}{4} & \frac{1}{5} & \frac{1}{6} \\ \frac{1}{3} & \frac{1}{4} & \frac{1}{5} & \frac{1}{6} & \frac{1}{7} \\ \frac{1}{4} & \frac{1}{5} & \frac{1}{6} & \frac{1}{7} & \frac{1}{8} \\ \frac{1}{5} & \frac{1}{6} & \frac{1}{7} & \frac{1}{8} & \frac{1}{9} \end{bmatrix}$ b. $A = \begin{bmatrix} 1 & \frac{1}{2} & \frac{1}{3} & \frac{1}{4} & \frac{1}{5} \\ \frac{1}{2} & \frac{1}{3} & \frac{1}{4} & \frac{1}{5} & \frac{1}{6} \\ \frac{1}{3} & \frac{1}{4} & \frac{1}{5} & \frac{1}{6} & \frac{1}{7} \\ \frac{1}{4} & \frac{1}{5} & \frac{1}{6} & \frac{1}{7} & \frac{1}{8} \\ .20001 & \frac{1}{6} & \frac{1}{7} & \frac{1}{8} & \frac{1}{9} \end{bmatrix}.$

   Comment on the sensitivity of the solution to changes in the matrix of coefficients.

B.

1. Consider the two systems

    $$x + y = 4$$
    $$1.01x + y = 3.02$$

    and

    $$x + y = 4$$
    $$1.06125x + y = 3.02$$

    Solve each system manually and comment on the sensitivity of the solutions to slight variations.

2. Using an ordinary calculator for the arithmetic, determine the solution of

    $$1.985x - 1.358y = 2.212$$
    $$0.953x - 0.652y = b_2$$

    correct to three decimal places
    a. When $b_2 = 1.062$.
    b. When $b_2 = 1.063$.

3. Consider the system

    $$x + y = 5$$
    $$x - y = 1$$

    a. Solve the system exactly.
    b. Change the system to

    $$x + y = 5$$
    $$x - y = 1 + h$$

    and solve the system in terms of h. If h is small, does the change in the output vector from $\mathbf{b} = \begin{bmatrix} 5 \\ 1 \end{bmatrix}$ to $\hat{\mathbf{b}} = \begin{bmatrix} 5 \\ 1 + h \end{bmatrix}$ change the solution much in terms of absolute error?
    c. Letting $\Delta\mathbf{b} = \mathbf{b} - \hat{\mathbf{b}}$, define the *relative change* in $\mathbf{b}$ to be the scalar $\|\Delta\mathbf{b}\|/\|\mathbf{b}\|$, where $\|\cdot\|$ denotes the Euclidean norm in $\mathbb{R}^2$. A similar definition holds for the relative change in the solution $\mathbf{x} = \begin{bmatrix} x \\ y \end{bmatrix}$. Now calculate $\|\Delta\mathbf{b}\|/\|\mathbf{b}\|$ and $\|\Delta\mathbf{x}\|/\|\mathbf{x}\|$. Is the system well-conditioned or ill-conditioned?

C.

In trying to solve a system $A\mathbf{x} = \mathbf{b}$, the **residual vector** *$\mathbf{r} = A\mathbf{x} - \mathbf{b}$ gives a measure of how close any vector $\mathbf{x}$ is to a solution. In more detail, if $\mathbf{x} = \begin{bmatrix} \xi_1 \\ \xi_2 \\ \vdots \\ \xi_n \end{bmatrix}$ is a calculated solution for an $n \times n$ system*

$$a_{11}x_1 + a_{12}x_2 + \cdots + a_{1n}x_n = b_1$$
$$a_{21}x_1 + a_{22}x_2 + \cdots + a_{2n}x_n = b_2$$
$$\vdots \qquad \vdots \qquad \qquad \vdots \qquad \vdots$$
$$a_{n1}x_1 + a_{n2}x_2 + \cdots + a_{nn}x_n = b_n$$

then the numbers

$$r_i = a_{i1}\xi_1 + a_{i2}\xi_2 + \cdots + a_{in}\xi_n - b_i$$

for $i = 1, 2, \ldots, n$, *n are called the **residuals**. Use these concepts in Exercises 1–2.*

1. Consider the system

$$5x + 10y = 15$$
$$5.00001x + 10y = 15.00001$$

   a. Solve the system exactly by hand.
   b. Find the residual vector corresponding to the "approximate" solution $\tilde{\mathbf{x}} = \begin{bmatrix} 3 \\ 0 \end{bmatrix}$.
   c. How does the size of the components of the residual vector found in part (b) compare to the closeness between your answer to part (a) and the vector $\tilde{\mathbf{x}} = \begin{bmatrix} 3 \\ 0 \end{bmatrix}$. Does a small residual vector guarantee the accuracy of an approximate solution?

2. Suppose that for the ill-conditioned system

$$0.835x + 0.667y = 0.168$$
$$0.333x + 0.266y = 0.067$$

   we somehow compute a solution to be $\xi_1 = -666$ and $\xi_2 = 834$.
   a. Compute the residuals r_1 and r_2. What do you believe you can say about the computed solution $\begin{bmatrix} \xi_1 \\ \xi_2 \end{bmatrix}$?
   b. Solve the system exactly by hand.
   c. Think about your answers to parts (a) and (b). What do you conclude about the process of "checking the answer" by substituting a calculated solution back into the left-hand side of the original system of equations?

3. Consider the system

$$x + y = 3$$
$$x + 1.00001y = 3.00001$$

   a. Solve the system exactly by hand.
   b. Change the system to

$$x + y = 3$$
$$x + 1.00001y = 3.00001 + h$$

and solve the system in terms of h. If h is very small, does the change in the output vector from $\mathbf{b} = \begin{bmatrix} 3 \\ 3.00001 \end{bmatrix}$ to $\hat{\mathbf{b}} = \begin{bmatrix} 3 \\ 3.00001 + h \end{bmatrix}$ change the solution much in terms of absolute error?

c. As in Exercise B3, calculate $\|\Delta\mathbf{b}\|/\|\mathbf{b}\|$ and $\|\Delta\mathbf{x}\|/\|\mathbf{x}\|$. Is the system well-conditioned or ill-conditioned?

2.7 Rank and Nullity of a Matrix

2.7.1 Row Spaces and Column Spaces

If A is an $m \times n$ matrix, $A = \begin{bmatrix} a_{11} & a_{12} & \cdots & a_{1n} \\ a_{21} & a_{22} & \cdots & a_{2n} \\ \vdots & \vdots & \vdots & \vdots \\ a_{m1} & a_{m2} & \cdots & a_{mn} \end{bmatrix}$, we may think of the columns of A,

$$\begin{bmatrix} a_{1j} \\ a_{2j} \\ \vdots \\ a_{mj} \end{bmatrix} \quad j = 1, 2, \ldots, n,$$

as elements of the Euclidean space $\mathbb{R}^m$ and the transposes of the rows of A, $[a_{i1} \ a_{i2} \ \cdots \ a_{in}]^T$, $i = 1, 2, \ldots, m$, as elements of $\mathbb{R}^n$.

At this point, it is useful to introduce the **transpose** of a matrix A, denoted by A^T—the matrix you get by interchanging rows and columns of A: If $A = [a_{ij}]$, then $A^T = [b_{ij}]$, where $b_{ij} = a_{ji}$. Thus column i of A^T is row i of A. We use the transpose of a matrix sparingly in this chapter and investigate it in more detail in Section 3.1.

Definition 2.7.1

Given an $m \times n$ matrix A, the **row space** of A is the subspace of $\mathbb{R}^n$ spanned by the (transposes of the) rows of A. Similarly, the **column space** of A is the subspace of $\mathbb{R}^m$ spanned by the columns of A.

(Problem B1 of Exercises 1.6 states that the span of any set of vectors in a Euclidean space $\mathbb{R}^n$ is a subspace. The concepts of row space and column space first appeared in Problems B8 and B9 of Exercises 2.2.) Note that **the row space of A is the column space of A^T and the column space of A is the row space of A^T.**

Example 2.7.1: A Row Space and a Column Space

Consider the matrix

$$A = \begin{bmatrix} 1 & 0 & 1 & 1 \\ 1 & 1 & 2 & 3 \\ -1 & 1 & 0 & 1 \end{bmatrix}.$$

The row space of A is the subspace of $\mathbb{R}^4$ defined as

$$span\{[1\ 0\ 1\ 1]^T, [1\ 1\ 2\ 3]^T, [-1\ 1\ 0\ 1]^T\}$$
$$= \{[x+y-z \quad y+z \quad x+2y \quad x+3y+z]^T : x, y, z \in \mathbb{R}\}$$

and the column space of A is the subspace of $\mathbb{R}^3$ defined as

$$span\left\{\begin{bmatrix} 1 \\ 1 \\ -1 \end{bmatrix}, \begin{bmatrix} 0 \\ 1 \\ 1 \end{bmatrix}, \begin{bmatrix} 1 \\ 2 \\ 0 \end{bmatrix}, \begin{bmatrix} 1 \\ 3 \\ 1 \end{bmatrix}\right\}$$
$$= \left\{\begin{bmatrix} x+z+w \\ x+y+2z+3w \\ -x+y+w \end{bmatrix} : x, y, z, w \in \mathbb{R}\right\}.$$

The column space plays a fundamental role in determining the solutions (if any) of a system of linear equations. For example, consider the system

$$\begin{aligned} x_1 + x_2 + 3x_4 &= 4 \\ 2x_1 + x_2 - x_3 + x_4 &= 1 \\ 3x_1 - x_2 - x_3 + 2x_4 &= -3 \\ -x_1 + 2x_2 + 3x_3 - x_4 &= 4 \end{aligned} \tag{$*$}$$

This is equivalent to the vector equation

$$x_1 \begin{bmatrix} 1 \\ 2 \\ 3 \\ -1 \end{bmatrix} + x_2 \begin{bmatrix} 1 \\ 1 \\ -1 \\ 2 \end{bmatrix} + x_3 \begin{bmatrix} 0 \\ -1 \\ -1 \\ 3 \end{bmatrix} + x_4 \begin{bmatrix} 3 \\ 1 \\ 2 \\ -1 \end{bmatrix} = \begin{bmatrix} 4 \\ 1 \\ -3 \\ 4 \end{bmatrix}.$$

Saying that system ($*$) has a solution is equivalent to saying that the vector

of outputs $\begin{bmatrix} 4 \\ 1 \\ -3 \\ 4 \end{bmatrix}$ can be expressed as a linear combination of the

columns of the matrix of coefficients

$$\begin{bmatrix} 1 & 1 & 0 & 3 \\ 2 & 1 & -1 & 1 \\ 3 & -1 & -1 & 2 \\ -1 & 2 & 3 & -1 \end{bmatrix}.$$

We can state and prove this observation as a general theorem.

Theorem 2.7.1

A system of m linear equations in n unknowns $A\mathbf{x} = \mathbf{b}$ has a solution if and only if the vector $\mathbf{b}$ can be expressed as a linear combination of the columns of A. (In other words, the system can be solved if and only if $\mathbf{b}$ belongs to the column space of A.)

Proof   First suppose the vector $\mathbf{b} = \begin{bmatrix} b_1 \\ b_2 \\ \vdots \\ b_m \end{bmatrix}$ can be written as a linear combination of the columns of A:

$$x_1 \begin{bmatrix} a_{11} \\ a_{21} \\ \vdots \\ a_{m1} \end{bmatrix} + x_2 \begin{bmatrix} a_{12} \\ a_{22} \\ \vdots \\ a_{m2} \end{bmatrix} + \cdots + x_n \begin{bmatrix} a_{1n} \\ a_{2n} \\ \vdots \\ a_{mn} \end{bmatrix} = \begin{bmatrix} b_1 \\ b_2 \\ \vdots \\ b_m \end{bmatrix}$$

for some real numbers $x_1, x_2, \ldots, x_n$. Then (by the definitions of vector equality and scalar multiplication in Section 1.1) we can write

$$a_{11}x_1 + a_{12}x_2 + \cdots + a_{1n}x_n = b_1$$
$$a_{21}x_1 + a_{22}x_2 + \cdots + a_{2n}x_n = b_2$$
$$\vdots \qquad \vdots \qquad \qquad \vdots \qquad \vdots$$
$$a_{m1}x_1 + a_{m2}x_2 + \cdots + a_{mn}x_n = b_m$$

which says that $\mathbf{x} = \begin{bmatrix} x_1 \\ x_2 \\ \vdots \\ x_n \end{bmatrix}$ is a solution of the system $A\mathbf{x} = \mathbf{b}$, where

$$A = \begin{bmatrix} a_{11} & a_{12} & \cdots & a_{1n} \\ a_{21} & a_{22} & \cdots & a_{2n} \\ \vdots & \vdots & \vdots & \vdots \\ a_{m1} & a_{m2} & \cdots & a_{mn} \end{bmatrix} \quad \text{and} \quad \mathbf{b} = \begin{bmatrix} b_1 \\ b_2 \\ \vdots \\ b_m \end{bmatrix}.$$

Now suppose the system $A\mathbf{x} = \mathbf{b}$ has a solution, where

$$A = \begin{bmatrix} a_{11} & a_{12} & \cdots & a_{1n} \\ a_{21} & a_{22} & \cdots & a_{2n} \\ \vdots & \vdots & \vdots & \vdots \\ a_{m1} & a_{m2} & \cdots & a_{mn} \end{bmatrix} \quad \text{and} \quad \mathbf{b} = \begin{bmatrix} b_1 \\ b_2 \\ \vdots \\ b_m \end{bmatrix},$$

This is equivalent to having a vector $\mathbf{x} = \begin{bmatrix} x_1 \\ x_2 \\ \vdots \\ x_n \end{bmatrix}$ such that

$$a_{11}x_1 + a_{12}x_2 + \cdots + a_{1n}x_n = b_1$$
$$a_{21}x_1 + a_{22}x_2 + \cdots + a_{2n}x_n = b_2$$
$$\vdots \qquad \vdots \qquad\qquad \vdots \qquad \vdots$$
$$a_{m1}x_1 + a_{m2}x_2 + \cdots + a_{mn}x_n = b_m$$

This last system can be written in the vector form

$$x_1 \begin{bmatrix} a_{11} \\ a_{21} \\ \vdots \\ a_{m1} \end{bmatrix} + x_2 \begin{bmatrix} a_{12} \\ a_{22} \\ \vdots \\ a_{m2} \end{bmatrix} + \cdots + x_n \begin{bmatrix} a_{1n} \\ a_{2n} \\ \vdots \\ a_{mn} \end{bmatrix} = \begin{bmatrix} b_1 \\ b_2 \\ \vdots \\ b_m \end{bmatrix},$$

which states that $\mathbf{b}$ is a linear combination of the columns of A—that is, $\mathbf{b}$ belongs to the column space of A.

It is important to note that if A is an $m \times n$ matrix and $\mathbf{x} = [x_1 \ x_2 \ \dots \ x_n]^T$ is any vector in $\mathbb{R}^n$, Theorem 2.7.1 provides us with an interpretation of the expression $A\mathbf{x}$ as $x_1\mathbf{a}_1 + x_2\mathbf{a}_2 + \cdots + x_n\mathbf{a}_n$, where $\mathbf{a}_i$ is column i of A. As a linear combinations of vectors, $A\mathbf{x}$ is again a vector. We note that this interpretation of $A\mathbf{x}$ gives us an alternative way to view the column space of A: The column space of $A = \{A\mathbf{x} | \mathbf{x} \in \mathbb{R}^n\}$.

In Chapter 3, we will reinterpret the juxtaposition $A\mathbf{x}$, formally defining the product of a vector by a matrix in this way (Definition 3.2.1(a)) and extending this definition to include the product of matrices. For now we need some immediate consequences of this interpretation.

For example, if $\mathbf{x} = [x_1 \ x_2 \ \dots \ x_n]^T$ and $\mathbf{y} = [y_1 \ y_2 \ \dots \ y_n]^T$, we see that

$$A(\mathbf{x} + \mathbf{y}) = A([x_1 + y_1 \ x_2 + y_2 \ \dots \ x_n + y_n]^T) = (x_1 + y_1)\mathbf{a}_1$$
$$+ (x_2 + y_2)\mathbf{a}_2 + \cdots + (x_n + y_n)\mathbf{a}_n$$
$$= (x_1\mathbf{a}_1 + x_2\mathbf{a}_2 + \cdots + x_n\mathbf{a}_n) + (y_1\mathbf{a}_1 + y_2\mathbf{a}_2 + \cdots + y_n\mathbf{a}_n)$$
$$= A\mathbf{x} + A\mathbf{y}.$$

In general, if $\mathbf{x}_1, \mathbf{x}_2, \dots, \mathbf{x}_m$ are vectors in $\mathbb{R}^n$ and $c_1, c_2, \dots, c_m$ are any scalars, we can use mathematical induction to show that (see Exercise B4).

$$A\left(\sum_{k=1}^{m} c_k\mathbf{x}_k\right) = \sum_{k=1}^{m} A(c_k\mathbf{x}_k) = \sum_{k=1}^{m} c_k(A\mathbf{x}_k). \tag{2.7.1}$$

Definition 2.7.2

The **row rank** of an $m \times n$ matrix A is the dimension of its row space as a subspace of $\mathbb{R}^n$. The **column rank** of an $m \times n$ matrix A is the dimension of its column space as a subspace of $\mathbb{R}^m$.

It is important to realize that the **row (column) rank is the maximum number of linearly independent rows (columns) of A.** These linearly independent rows or columns serve as basis vectors for the appropriate space. By convention, the row rank and the column rank of an $m \times n$ zero matrix are zero. Note that **the row rank of A is the column rank of A^T** and **the column rank of A is the row rank of A^T**.

To find a basis for the row space of a matrix A, just calculate rref(A). Then the nonzero rows of rref(A) form a basis for the row space of A. Thus the row rank of A equals the number of nonzero rows in rref(A).

The boxed statements are based on the next theorem.

Theorem 2.7.2

If A and B are row equivalent matrices, then their row spaces are identical.

Proof   Suppose that A is an $m \times n$ matrix and B is obtained from A by a *single elementary row operation*. Let R_A and R_B denote the row spaces of A and B, respectively. We will show that $R_A = R_B$.

Let $\mathbf{r}_1, \mathbf{r}_2, \ldots, \mathbf{r}_k$ be the nonzero rows of A and suppose that B results from A by the row operation $R_i \to cR_i$, where c is a nonzero scalar. Then each vector $\mathbf{v} \in R_B$ is of the form

$$\mathbf{v} = \alpha_1 \mathbf{r}_1 + \cdots + \alpha_i(k\mathbf{r}_i) + \cdots + \alpha_k \mathbf{r}_k$$
$$= \alpha_1 \mathbf{r}_1 + \cdots + (\alpha_i k)\mathbf{r}_i + \cdots + \alpha_k \mathbf{r}_k$$

and is therefore also in the row space of A. Thus $\boxed{R_B \subseteq R_A}$.

Conversely, because $c \neq 0$, we may use the inverse of the elementary operation to see that each vector $\mathbf{w} \in R_A$ is of the form

$$\mathbf{w} = \alpha_1 \mathbf{r}_1 + \cdots + \alpha_i \mathbf{r}_i + \cdots + \alpha_k \mathbf{r}_k$$
$$= \alpha_1 \mathbf{r}_1 + \cdots + \left(\frac{\alpha_i}{k}\right)(k\mathbf{r}_i) + \cdots + \alpha_k \mathbf{r}_k$$

and is also in the row space of B. Thus, $\boxed{R_A \subseteq R_B}$ and the two statements $R_B \subseteq R_A$ and $R_A \subseteq R_B$ together show that $R_A = R_B$. In a similar way, $R_A = R_B$ can be shown for the other two types of row operations, but the proofs are left as exercises (see Problem B8).

Now any matrix B that is row equivalent to A is the result of a finite number of elementary row operations applied to A. Because we have proved that each such row operation leaves the row space unchanged, we have

$$R_A = R_{B_1} = \cdots = R_{B_k} = R_B,$$

where B_{i+1} is the matrix obtained from B_i by a single elementary row operation.

Note that an immediate consequence of Theorem 2.7.2 is that **row equivalent matrices have the same row rank**.

Corollary 2.7.1

The nonzero rows of rref (A) constitute a basis for the row space of A.

Proof Clearly the nonzero rows of rref(A), because of their steplike structure, are linearly independent (see Problem B6 in Exercises 2.2) and so form a basis for the span of the rows of rref(A). Because Theorem 2.7.2 implies that $span$(rows of A) $= span$[rows of rref(A)], we see that the nonzero rows of rref(A) also serve as a basis for the row space of A.

Example 2.7.2: A Basis for a Row Space

As we saw in Example 2.7.1, the subspace generated by the
rows of matrix $A = \begin{bmatrix} 1 & 0 & 1 & 1 \\ 1 & 1 & 2 & 3 \\ -1 & 1 & 0 & 1 \end{bmatrix}$ is

$$\{ [x+y-z \quad y+z \quad x+2y \quad x+3y+z]^T : x, y, z \in \mathbb{R} \}$$

If we make the substitutions $X = x+y-z$ and $Y = y+z$, the row space can be written simply as $\{ [X \quad Y \quad X+Y \quad X+2Y]^T : X, Y \in \mathbb{R} \}$.

The rref of A is

$$\begin{bmatrix} 1 & 0 & 1 & 1 \\ 0 & 1 & 1 & 2 \\ 0 & 0 & 0 & 0 \end{bmatrix},$$

so the row space of rref (A) is $span\{[1 \ 0 \ 1 \ 1]^T,$ $[0 \ 1 \ 1 \ 2]^T\} = \{ [x \quad y \quad x+y \quad x+2y]^T : x, y \in \mathbb{R} \} =$ the row space of A, as Theorem 2.7.2 claims. Furthermore, Corollary 2.7.1 indicates that $[1 \ 0 \ 1 \ 1]^T$ and $[0 \ 1 \ 1 \ 2]^T$, the nonzero rows of rref (A), are basis vectors for the row space of A, and so the row rank of A is 2.

There is another way to determine a basis for the span of the rows of A. The row space of A can be written as

$$\{ [x+y-z \quad y+z \quad x+2y \quad x+3y+z]^T : x, y, z \in \mathbb{R} \}$$
$$= \{ (x+y-z)[1 \ 0 \ 1 \ 1]^T + (y+z)[0 \ 1 \ 1 \ 2]^T : x, y, z \in \mathbb{R} \}$$
$$= \{ a[1 \ 0 \ 1 \ 1]^T + b[0 \ 1 \ 1 \ 2]^T : a, b \in \mathbb{R} \},$$

so $[1 \ 0 \ 1 \ 1]^T$ and $[0 \ 1 \ 1 \ 2]^T$ are basis vectors for the row space of A.

The situation for the column space is just a bit more complicated. We get rref(A) by using elementary row operations on A; but, unfortunately, a sequence of elementary row operations applied to A does not necessarily produce a matrix with the same column space as A—in particular, A *and* $rref(A)$ *will generally have different column spaces.*

Example 2.7.3: Row Equivalent Matrices with Different Column Spaces

If we define $A = \begin{bmatrix} 1 & 0 & -3 & 5 & 0 \\ 0 & 1 & 2 & -1 & 0 \\ 0 & 0 & 0 & 0 & 1 \\ 0 & 0 & 0 & 0 & 0 \end{bmatrix}$ and

$$B = \begin{bmatrix} 1 & 3 & 3 & 2 & -9 \\ -2 & -2 & 2 & -8 & 2 \\ 2 & 3 & 0 & 7 & 1 \\ 3 & 4 & -1 & 11 & -8 \end{bmatrix}, \text{ then } A \text{ is row}$$

equivalent to B—in fact, $A = rref(B)$.

However, any vector in the column space of A must have a zero fourth entry, whereas there is no such restriction on the vectors making up the column space of B—for instance, $[1 \ -2 \ 2 \ 3]^T$ is in the column space of B, but is not in the column space of A. Thus the column space of A is not equal to the column space of B.

However, despite the possible changes in the column space of a matrix caused by row operations, it turns out that these row operations do preserve the column *rank* of a matrix. The following lemma establishes this fact by showing that the columns of two row equivalent matrices satisfy the same linear dependence relations. We will use this information to determine a basis for the column space of a matrix.

Lemma 2.7.1

If an $m \times n$ matrix A with columns $\mathbf{a}_1, \mathbf{a}_2, \ldots, \mathbf{a}_n$ is row equivalent to matrix B with columns $\mathbf{b}_1, \mathbf{b}_2, \ldots, \mathbf{b}_n$, then $c_1\mathbf{a}_1 + c_2\mathbf{a}_2 + \cdots + c_n\mathbf{a}_n = \mathbf{0}$ if and only if $c_1\mathbf{b}_1 + c_2\mathbf{b}_2 + \cdots + c_n\mathbf{b}_n = \mathbf{0}$.

Proof Suppose that A is row equivalent to B. Then the systems $A\mathbf{x} = \mathbf{0}$ and $B\mathbf{x} = \mathbf{0}$ have the same solutions by Theorem 2.2.2. Therefore, $A\mathbf{c} = \mathbf{0}$ if and only if $B\mathbf{c} = \mathbf{0}$. If $\mathbf{c} = [c_1 \quad c_2 \quad \ldots \quad c_n]^T$, the last equivalence becomes

$$c_1\mathbf{a}_1 + c_2\mathbf{a}_2 + \cdots + c_n\mathbf{a}_n = \mathbf{0} \text{ if and only if } c_1\mathbf{b}_1 + c_2\mathbf{b}_2 + \cdots + c_n\mathbf{b}_n = \mathbf{0}$$

by Theorem 2.7.1.

We emphasize that Lemma 2.7.1 implies that for row equivalent matrices A and B any set of columns of A is linearly dependent (or independent) if and only if the corresponding set of columns of B is linearly dependent (or independent).

Theorem 2.7.3

If A is an $m \times n$ matrix and the pivots (leading 1's) of rref(A) occur in columns $i_1, i_2, \ldots, i_k$, where $\{i_1, i_2, \ldots, i_k\} \subseteq \{1, 2, \ldots, n\}$, then columns $i_1, i_2, \ldots, i_k$ of A form a basis for the column space of A.

Proof In rref(A), because of the unique "staircase" configuration of the entries, the columns containing leading 1's must be linearly independent and thus form a basis for the column space of rref(A). But Lemma 2.7.1 implies that these same pivot columns of A also serve as a basis for the column space of A.

We can restate Theorem 2.7.3 in a more algorithmic way.

> To find a basis for the column space of a matrix A, just calculate rref(A). Then the pivot columns of A form a basis for the column space of A. Thus the column rank of A equals the number of pivot columns in rref(A).

It is important to realize that if A is row equivalent to the rref B, it is not necessarily true that the pivot columns of B form a basis for the column space of A (although they *do* constitute a basis for the column space of B). It is just the *numbers* (labels) of the pivot columns of B that we can use to identify a set of columns of A that is a basis for the column space of A.

Of course, we could have used elementary *column* operations to get the **reduced column echelon form*** of A. Then if we discard any zero columns, the remaining columns will form a basis for the column space of A. Clearly the column space of A is spanned by the nonzero columns of the reduced column echelon form of A. Furthermore, these nonzero columns are linearly independent because of the arrangement of 0's and 1's in this form. (It turns out that elementary column operations leave the row rank unchanged, but we do not need this fact here.)

Example 2.7.4: A Basis for a Column Space

If $A = \begin{bmatrix} 2 & 2 & 0 & 0 \\ 1 & 0 & 1 & -2 \\ 2 & 1 & 1 & -2 \end{bmatrix}$, then rref($A$) $= \begin{bmatrix} 1 & 0 & 1 & -2 \\ 0 & 1 & -1 & 2 \\ 0 & 0 & 0 & 0 \end{bmatrix}$.

The first and second columns of rref(A) are the pivot columns of A, so Theorem 2.7.2 tells us that columns 1 and 2 of A, $\begin{bmatrix} 2 \\ 1 \\ 2 \end{bmatrix}$

---

* In Definition 2.5.1, exchange the words "row" and "column" and replace "bottom" with "right side." More simply, transpose A, compute rref(A^T), and transpose again—that is, finish with $[\text{rref}(A^T)]^T$.

and $\begin{bmatrix} 2 \\ 0 \\ 1 \end{bmatrix}$, serve as basis vectors for the column space of A as a subspace of $\mathbb{R}^3$.

Alternatively, we can see that the column space of A is the set

$$\left\{ x\begin{bmatrix} 2 \\ 1 \\ 2 \end{bmatrix} + y\begin{bmatrix} 2 \\ 0 \\ 1 \end{bmatrix} + z\begin{bmatrix} 0 \\ 1 \\ 1 \end{bmatrix} + w\begin{bmatrix} 0 \\ -2 \\ -2 \end{bmatrix} : x, y, z, w \in \mathbb{R} \right\}$$

$$= \left\{ \begin{bmatrix} 2x + 2y \\ x + z - 2w \\ 2x + y + z - 2w \end{bmatrix} : x, y, z, w \in \mathbb{R} \right\}$$

$$= \left\{ (x + z - 2w)\begin{bmatrix} 2 \\ 1 \\ 2 \end{bmatrix} + (y - z + 2w)\begin{bmatrix} 2 \\ 0 \\ 1 \end{bmatrix} : x, y, z, w \in \mathbb{R} \right\}$$

$$= \left\{ a\begin{bmatrix} 2 \\ 1 \\ 2 \end{bmatrix} + b\begin{bmatrix} 2 \\ 0 \\ 1 \end{bmatrix} : a, b \in \mathbb{R} \right\}.$$

In this last example, the reduced column echelon form of A is

$$\begin{bmatrix} 1 & 0 & 0 & 0 \\ 0 & 1 & 0 & 0 \\ 1/2 & 1 & 0 & 0 \end{bmatrix}$$ (see Exercise A12), so a basis for the column space

of A is $\left\{ \begin{bmatrix} 1 \\ 0 \\ 1/2 \end{bmatrix}, \begin{bmatrix} 0 \\ 1 \\ 1 \end{bmatrix} \right\}$, or $\left\{ \begin{bmatrix} 2 \\ 0 \\ 1 \end{bmatrix}, \begin{bmatrix} 0 \\ 1 \\ 1 \end{bmatrix} \right\}$ (eliminating the fractional

component of the first vector).

The various results and examples in this section lead us to the following important conclusion.

Theorem 2.7.4

The row rank of a matrix A equals the column rank of A.

Proof First of all, Corollary 2.7.1 indicates that the row rank of A equals the number of nonzero rows of rref(A). But each such row corresponds to a leading 1, so the rank of A is the same as the number of pivotal columns of A. Furthermore, Theorem 2.7.3 tells us that the column rank of A also equals the number of pivot columns of A. Thus the row rank and column rank of A are identical.

One consequence of Theorem 2.7.3 is that we can now use the term **rank** to refer to either the row rank or the column rank of a matrix A, whichever is more appropriate, and write rank(A) or $r(A)$ for this common value. If A is an $m \times n$ matrix, we can see that $r(A) \leq m$ and $r(A) \leq n$. We can write this more compactly as

$$\boxed{r(A) \leq \min(m, n)},$$

where $\min(m, n)$ denotes the smaller of the two numbers m and n if they are not equal and their common value if $m = n$.

Because we know that the row rank of A equals the column rank of A^T (see the paragraph right after Definition 2.7.1), we can draw the following conclusion from Theorem 2.7.4.

Corollary 2.7.2

If A is an $m \times n$ matrix, then $\operatorname{rank}(A) = \operatorname{rank}(A^T)$.

2.7.2 The Null Space

Example 1.6.4 provided an informal definition of the null space of a matrix A as a special subspace. Now we define this concept more precisely.

Definition 2.7.3

If A is an $m \times n$ matrix, the **null space** of A (sometimes called the **kernel** of A) is the subspace of all solutions of $A\mathbf{x} = \mathbf{0}$—that is, the subspace $\{\mathbf{x} \in \mathbb{R}^n : A\mathbf{x} = \mathbf{0}\}$. The dimension of the null space is called the **nullity** of A, denoted nullity (A).

We know that the row space and the column space of an $m \times n$ matrix A have the same dimension, the rank of A. The next theorem, a fundamental result of linear algebra, shows the close connection between rank and nullity.

Theorem 2.7.5

For any matrix A having n columns, nullity$(A) = n - \operatorname{rank}(A)$—or, in the usual form of the statement,

$$\boxed{\operatorname{rank}(A) + \operatorname{nullity}(A) = n.}$$

Proof Let $B = \{\mathbf{x}_1, \mathbf{x}_2, \ldots, \mathbf{x}_k\}$ be a basis for the null space of A, so nullity$(A) = k$. Because B is a linearly independent subset of $\mathbb{R}^n$, we know by Corollary 1.5.2 that B can be extended to a basis $\{\mathbf{x}_1, \mathbf{x}_2, \ldots, \mathbf{x}_k, \mathbf{x}_{k+1}, \ldots, \mathbf{x}_n\}$ of $\mathbb{R}^n$. Now we will show that the vectors $A\mathbf{x}_{k+1}, A\mathbf{x}_{k+2}, \ldots, A\mathbf{x}_n$ form a basis of the column space of A.

These vectors clearly belong to the column space. Take any vector $\mathbf{z}$ in the column space of A. Then, because $\mathbf{z}$ is a linear combination of the columns of A, $\mathbf{z} = A\mathbf{x}$ for some $\mathbf{x} \in \mathbb{R}^n$. But $\mathbf{x}$ can be written as

$\mathbf{x} = \sum_{i=1}^{n} a_i \mathbf{x}_i$ for some scalars $a_1, a_2, \ldots, a_n$. Then, by the property shown in Equation 2.7.1,

$$\mathbf{z} = A\mathbf{x} = A\left(\sum_{i=1}^{n} a_i \mathbf{x}_i\right) = \sum_{i=1}^{n} a_i(A\mathbf{x}_i) = \sum_{i=k+1}^{n} a_i(A\mathbf{x}_i)$$

because $A\mathbf{x}_1 = \mathbf{0}$, $A\mathbf{x}_2 = \mathbf{0}, \ldots, A\mathbf{x}_k = \mathbf{0}$. Thus the vectors $A\mathbf{x}_{k+1}$, $A\mathbf{x}_{k+2}, \ldots, A\mathbf{x}_n$ span the column space of A. To prove linear independence, suppose that $b_1(A\mathbf{x}_{k+1}) + b_2(A\mathbf{x}_{k+2}) + \cdots + b_{n-k}(A\mathbf{x}_n) = \mathbf{0}$. But this last equation can be written as $A(b_1\mathbf{x}_{k+1} + b_2\mathbf{x}_{k+2} + \cdots + b_{n-k}\mathbf{x}_n) = \mathbf{0}$, so that $b_1\mathbf{x}_{k+1} + b_2\mathbf{x}_{k+2} + \cdots + b_{n-k}\mathbf{x}_n$ is in the null space of A. Hence

$$b_1\mathbf{x}_{k+1} + b_2\mathbf{x}_{k+2} + \cdots + b_{n-k}\mathbf{x}_n = c_1\mathbf{x}_1 + c_2\mathbf{x}_2 + \cdots + c_k\mathbf{x}_k$$

for some scalars $c_1, c_2, \ldots, c_k$, or

$$c_1\mathbf{x}_1 + c_2\mathbf{x}_2 + \cdots + c_k\mathbf{x}_k - b_1\mathbf{x}_{k+1} - b_2\mathbf{x}_{k+2} - \cdots - b_{n-k}\mathbf{x}_n = \mathbf{0}$$

Because the set $\{\mathbf{x}_1, \mathbf{x}_2, \ldots, \mathbf{x}_n\}$, as a basis for $\mathbb{R}^n$, is linearly independent, we must have $b_1 = b_2 = \cdots = b_{n-k} = 0$. Thus the vectors $A\mathbf{x}_{k+1}$, $A\mathbf{x}_{k+2}, \ldots, A\mathbf{x}_n$ are linearly independent and so form a basis for the column space of A. Therefore, rank(A) = the dimension of the column space of $A = n - k = n$–the nullity of A, or rank(A) + nullity(A) = n.

Example 2.7.5: Theorem 2.7.5 Illustrated

The matrix

$$A = \begin{bmatrix} 0 & 1 & 0 & 0 \\ 0 & 0 & 2 & 0 \\ 0 & 0 & 0 & 3 \\ 0 & 0 & 0 & 0 \end{bmatrix}$$

clearly has rank equal to 3. The null space of A is easily found to be the set of all scalar multiples of the vector $[1\ 0\ 0\ 0]^{\mathrm{T}}$, a subspace of dimension 1. Thus rank(A) + nullity(A) = $3 + 1 = 4 =$ the number of columns of A.

Theorem 2.7.5 can be considered "a kind of conservation law"—as the amount of "stuff" in the row or column space of A increases, the amount of "stuff" in the null space of A must decrease and vice versa.* As we will see in the next section, Theorem 2.7.5 has important consequences in the analysis of systems of linear equations.

---

*C.D. Meyer, *Matrix Analysis and Applied Linear Algebra*, p. 200 (Philadelphia, PA: SIAM, 2000).

Exercises 2.7

A.

1. Find the row space and row rank of

$$A = \begin{bmatrix} 1 & -1 & 0 \\ 2 & 3 & 1 \\ 3 & -2 & 4 \end{bmatrix}.$$

2. Find the column space and column rank of the matrix A in the previous exercise.

3. Find the row rank and the column rank of

$$A = \begin{bmatrix} 1 & 0 & 1 & 2 \\ 2 & 1 & 0 & 3 \\ 1 & -1 & 3 & 3 \end{bmatrix}.$$

4. Find bases for the row and column spaces of the following matrices:

a. $\begin{bmatrix} 2 & -3 & 9 \\ 4 & -5 & 36 \end{bmatrix}$ b. $\begin{bmatrix} -1 & 6 & 1 & 5 \\ 3 & 1 & 1 & 5 \\ 1 & 13 & 3 & 15 \end{bmatrix}$

5. Find a basis for the row space of

$$A = \begin{bmatrix} 1 & -1 & 3 & 0 & -2 \\ -2 & 2 & -6 & 0 & 4 \\ 0 & 2 & 5 & -1 & 0 \\ 2 & -6 & -4 & 2 & -4 \end{bmatrix}.$$

6. Find a basis for the row space of the matrix in Example 2.7.1.

7. Find a basis for the column space of the matrix in Example 2.7.1.

8. Find a basis for the column space of

$$A = \begin{bmatrix} 1 & -2 & 1 & 1 & 2 \\ -1 & 3 & 0 & 2 & -2 \\ 0 & 1 & 1 & 3 & 4 \\ 1 & 2 & 5 & 13 & 5 \end{bmatrix}.$$

9. Find a basis for the column space of

$$A = \begin{bmatrix} 1 & 2 & 4 & 0 & -3 \\ 0 & 1 & -5 & -1 & 2 \\ 0 & 1 & 3 & -1 & 0 \\ 2 & 0 & -1 & -1 & 0 \end{bmatrix}.$$

10. If A is a 4×5 matrix, what is the largest possible value for rank(A)?

11. If A is a 9×6 matrix, what is the largest possible value for rank(A)?

12. Show that the *reduced column echelon form* of
$$A = \begin{bmatrix} 2 & 2 & 0 & 0 \\ 1 & 0 & 1 & -2 \\ 2 & 1 & 1 & -2 \end{bmatrix} \text{ is } \begin{bmatrix} 1 & 0 & 0 & 0 \\ 0 & 1 & 0 & 0 \\ 1/2 & 1 & 0 & 0 \end{bmatrix}.$$

13. Find the null space of $\begin{bmatrix} 1 & -2 & 1 & 1 \\ -1 & 2 & 0 & 1 \\ 2 & -4 & 1 & 0 \end{bmatrix}$.

14. Find the null space of $\begin{bmatrix} 5 & 0 & -7 & 1 \\ 0 & 1 & 4 & 0 \\ 0 & 0 & 1 & 0 \\ 5 & 0 & 0 & 1 \end{bmatrix}$.

15. Determine the rank and nullity of the matrix
$$A = \begin{bmatrix} 2 & -1 & 1 \\ 1 & 1 & 2 \\ 0 & 1 & -3 \end{bmatrix}.$$

What does your answer say about the solution set of the system?
$$2x - y + z = 0$$
$$x + y + 2z = 0$$
$$y - 3z = 0$$

16. For each matrix A, compute rank(A), find a basis for the null space of A, and verify Theorem 2.7.5.

a. $A = \begin{bmatrix} 1 & 2 & 0 & 5 & 7 & 1 \\ 2 & 1 & 1 & -1 & 0 & 3 \\ -5 & 2 & -4 & 19 & 21 & 9 \\ 1 & -1 & 1 & -6 & -7 & 2 \end{bmatrix}$

b. $A = \begin{bmatrix} 3 & 5 & 5 & 2 & 0 \\ 1 & 0 & 2 & 2 & 1 \\ 1 & 1 & 1 & -2 & -2 \\ 2 & 0 & 4 & 4 & 2 \end{bmatrix}$

17. Suppose that A is an $n \times n$ matrix. If the column space of A is the same as the null space of A, show that n is an even number.

18. Can the null space of a 3×6 matrix have dimension 2? Explain your answer.

B.

1. Let

$$A = \begin{bmatrix} 0 & 1 & 1 & b \\ 1 & 0 & 1 & 4 \\ 1 & -1 & 0 & 1 \end{bmatrix}.$$

   Determine a basis for the row space and a basis for the column space of A. At least one of the answers should depend on the value of b. [*Hint*: Consider the cases $b \neq 3$ and $b = 3$.]

2. Let

$$A = \begin{bmatrix} 2 & 2 & 0 & 0 \\ 1 & 0 & 1 & -2 \\ 2 & 1 & 1 & -2 \end{bmatrix}.$$

   a. Find a basis for the row space of A.
   b. Find a basis for the column space of A.
   c. Find a basis for the null space of A.
   d. What are the rank and nullity of A?

3. Let a, b, and c be real numbers. If a, b, and c are not all equal,
   a. Show that the matrix

$$A = \begin{bmatrix} 1 & a & b+c \\ 1 & b & a+c \\ 1 & c & a+b \end{bmatrix}$$

   has rank 2.
   b. Find a basis for the null space of A.

4. Use mathematical induction to prove Equation 2.7.1.

5. Show that if matrix A is row equivalent to matrix B, then the null space of A equals the null space of B.

6. If A is an $m \times n$ matrix, prove directly that every vector $\mathbf{v}$ in the null space of A^T is orthogonal to every column of A. (Review Definition 1.2.4 if necessary.)

7. Suppose the columns of an $n \times k$ matrix M are linearly independent and form a basis for the null space of matrix A. Show that $\text{rank}(A) = n - k$.

8. Prove Theorem 2.7.2 for the elementary row operations $R_i \leftrightarrow R_j$ and $R_i \rightarrow R_i + cR_j$ $(c \neq 0,\ i \neq j)$.

9. Suppose that A is an $m \times n$ matrix whose null space is $\{\mathbf{0}\}$. If $\mathbf{v}_1, \mathbf{v}_2, \ldots, \mathbf{v}_k$ are linearly independent vectors in $\mathbb{R}^n$, prove that $A\mathbf{v}_1, A\mathbf{v}_2, \ldots, A\mathbf{v}_k$ are linearly independent vectors in $\mathbb{R}^n$.

C.

1. Suppose that A and B are matrices that have the same number of columns. The notation $\begin{bmatrix} A \\ \cdots \\ B \end{bmatrix}$ denotes the new matrix formed by stacking B directly below A.

    Show that the null space of $\begin{bmatrix} A \\ \cdots \\ B \end{bmatrix}$ equals (the null space of A) $\cap$ (the null space of B).

2. If A is an $m \times n$ matrix with rank r, we can construct a basis for the null space of A as follows. Assume, without loss of generality, that the first r columns of A form a basis for the column space of A. Then for $j = r+1, r+2, \ldots, n$, we have $\mathbf{a}_j = x_{j1}\mathbf{a}_1 + x_{j2}\mathbf{a}_2 + \cdots + x_{jr}\mathbf{a}_r$ (*) for some scalars $x_{j1}, x_{j2}, \ldots, x_{jr}$.

    Now for $j = r+1, r+2, \ldots, n$, consider the vectors

    $$\mathbf{x}_j = [x_{j1} \quad \ldots \quad x_{jr} \quad 0 \quad \ldots \quad 0 \quad -1 \quad 0 \quad \ldots \quad 0]^{\mathrm{T}} \in \mathbb{R}^n,$$

    where the -1 appears as component j.
    a. Prove that $\mathbf{x}_{r+1}, \mathbf{x}_{r+2}, \ldots, \mathbf{x}_n$ belong to the null space of A.
    b. Prove that the vectors $\mathbf{x}_{r+1}, \mathbf{x}_{r+2}, \ldots, \mathbf{x}_n$ are linearly independent.
    c. Prove that these vectors span the null space of A.

2.8 Systems of m Linear Equations in n Unknowns

We are now ready to state and prove some general results about systems of linear equations. The basic facts about two- and three-dimensional systems discussed in Section 2.1 hold true for systems of any size (any number of equations, any number of unknowns). For a system of m linear equations in n unknowns, there are only three possibilities:

(1) The system has no solution.
(2) The system has exactly one solution.
(3) The system has infinitely many solutions.

From now on, any system that has at least one solution will be called **consistent**. If a system has no solution, it will be called **inconsistent**.

In this section, we will apply our knowledge of reduced echelon form to the coefficient matrix and the augmented matrix of a system of linear equations in order to determine conditions under which each of the three possibilities listed above is realized.

The *rank* of the matrix A, introduced in Section 2.7, plays a key role in our study of the solutions of a system of linear equations $A\mathbf{x} = \mathbf{b}$. The discussion in Section 2.7 gives us the following three equivalent ways to view $r(A)$: (1) The maximum number of linearly independent columns of

A; (2) the number of nonzero rows in any (ordinary) row echelon form of A; and (3) the number of leading 1's in the rref of A.

2.8.1 The Solution of Linear Systems

Theorem 2.7.1 established the fundamental fact that a system of m linear equations in n unknowns $A\mathbf{x} = \mathbf{b}$ has a solution if and only if the vector $\mathbf{b}$ belongs to the column space of A.

Now suppose we have a system of the form $A\mathbf{x} = \mathbf{0}$.

Definition 2.8.1

A system of linear equations of the form $A\mathbf{x} = \mathbf{0}$ or, equivalently,

$$a_{11}x_1 + a_{12}x_2 + \cdots + a_{1n}x_n = 0$$
$$a_{21}x_1 + a_{22}x_2 + \cdots + a_{2n}x_n = 0$$
$$\vdots \qquad \vdots \qquad \qquad \vdots \quad \vdots$$
$$a_{m1}x_1 + a_{m2}x_2 + \cdots + a_{mn}x_n = 0$$

is called a **homogeneous** system. Otherwise, the system is called **nonhomogeneous**.

It is obvious that the zero vector is a solution of any homogeneous system. This solution is called the **trivial solution**. A solution different from $\mathbf{0}$ is called a **nontrivial solution**.

As we have seen, the set of all solutions of a homogeneous $m \times n$ system $A\mathbf{x} = \mathbf{0}$ is a subspace of $\mathbb{R}^n$ called the **null space** of A with dimension nullity(A).

Recall that rank(A) + nullity(A) = n (Theorem 2.7.5).

Theorem 2.8.1

If A is an $m \times n$ matrix, then the homogeneous system of equations $A\mathbf{x} = \mathbf{0}$ has a nontrivial solution if and only if rank(A) < n.

Proof   First suppose $A\mathbf{x} = \mathbf{0}$ has a nontrivial solution. This means that the null space of A has a nonzero vector in it, so nullity(A) ≥ 1. Theorem 2.7.5 allows us to conclude that rank(A) = $n -$ nullity(A) $\leq n - 1 < n$.

Conversely, suppose rank(A) < n. Then nullity(A) = $n -$ rank(A) $\geq n - (n-1) = 1$, so the null space of A is not $\{\mathbf{0}\}$—that is, $A\mathbf{x} = \mathbf{0}$ has a nontrivial solution.

An immediate consequence of this last theorem is that a homogeneous system of linear equations with more variables than equations has a nontrivial solution.

Corollary 2.8.1

A homogeneous $m \times n$ system of linear equations with $m < n$ has a nontrivial solution.

Proof   In the rref of A, there can be at most m leading 1's—that is, $\text{rank}(A) \leq m < n$. Theorem 2.8.1 tells us that the system has a nontrivial solution.

In fact, an $m \times n$ homogeneous system with $m < n$ has *infinitely many* nontrivial solutions (Exercise B9).

Example 2.8.1: An $m \times n$ Homogeneous System with $m < n$

Consider the system

$$x + 3y - 2z + 5u - 3v = 0$$
$$2x + 7y - 3z + 7u - 5v = 0$$
$$3x + 11y - 4z + 10u - 9v = 0$$

with $m = 3 < 5 = n$. The augmented matrix for this system is

$$[A \,|\, \mathbf{0}] = \begin{bmatrix} 1 & 3 & -2 & 5 & -3 & 0 \\ 2 & 7 & -3 & 7 & -5 & 0 \\ 3 & 11 & -4 & 10 & -9 & 0 \end{bmatrix},$$

and it is easy to calculate the reduced echelon form of $[A \,|\, \mathbf{0}]$:

$$B = \text{rref}[A \,|\, \mathbf{0}] = \begin{bmatrix} 1 & 0 & -5 & 0 & 22 & 0 \\ 0 & 1 & 1 & 0 & -5 & 0 \\ 0 & 0 & 0 & 1 & -2 & 0 \end{bmatrix}.$$

We have $\text{rank}(B) = 3 = \text{rank}(A)$. Working backward from the last row of B, we see that $u - 2v = 0$, $y + z - 5v = 0$, and $x - 5z + 22v = 0$. Letting z and v be free variables, we see that the solutions have the form

$$\begin{bmatrix} 5z - 22v \\ 5v - z \\ z \\ 2v \\ v \end{bmatrix} = z \begin{bmatrix} 5 \\ -1 \\ 1 \\ 0 \\ 0 \end{bmatrix} + v \begin{bmatrix} -22 \\ 5 \\ 0 \\ 2 \\ 1 \end{bmatrix}.$$

Expressed another way, the solution set of the system is the two-dimensional subspace of $\mathbb{R}^5$ spanned by $[5 \;\; -1 \;\; 1 \;\; 0 \;\; 0]^T$ and $[-22 \;\; 5 \;\; 0 \;\; 2 \;\; 1]^T$, an infinite set.

Example 2.8.2: A Homogeneous System Dependent on a Parameter

Consider the following system of equations:

$$3x - z + t = 0$$
$$x + 2y + 4t = 0$$
$$-x + \alpha y + z + (3 + \alpha)t = 0$$

The solution space (as a subspace of $\mathbb{R}^4$) depends on the value of the parameter α. Gaussian elimination can reduce the augmented matrix

$$[A \mid \mathbf{0}] = \begin{bmatrix} 3 & 0 & -1 & 1 & 0 \\ 1 & 2 & 0 & 4 & 0 \\ -1 & \alpha & 1 & 3 + \alpha & 0 \end{bmatrix}$$

to the echelon form

$$\begin{bmatrix} 3 & 0 & -1 & 1 & 0 \\ 0 & 2 & \frac{1}{3} & \frac{11}{3} & 0 \\ 0 & 0 & \frac{2}{3} - \frac{\alpha}{6} & \frac{10}{3} - \frac{5\alpha}{6} & 0 \end{bmatrix}.$$

If $\frac{2}{3} - \frac{\alpha}{6} = 0 = \frac{10}{3} - \frac{5\alpha}{6}$ —that is, if $\alpha = 4$—the last row of the echelon form is a row of zeros, so rank $[A \mid \mathbf{0}] = \text{rank}(A) = 2$ and the system has two free variables, say z and t. Using back substitution, we find that any solution has the form

$$\begin{bmatrix} \frac{z}{3} - \frac{t}{3} \\ -\frac{z}{6} - \frac{11t}{6} \\ z \\ t \end{bmatrix} = \begin{bmatrix} \frac{z}{3} \\ -\frac{z}{6} \\ z \\ 0 \end{bmatrix} + \begin{bmatrix} -\frac{t}{3} \\ -\frac{11t}{6} \\ 0 \\ t \end{bmatrix} = \frac{1}{6} z \begin{bmatrix} 2 \\ -1 \\ 6 \\ 0 \end{bmatrix} - \frac{1}{6} t \begin{bmatrix} 2 \\ 11 \\ 0 \\ -6 \end{bmatrix}$$

$$= c_1 \begin{bmatrix} 2 \\ -1 \\ 6 \\ 0 \end{bmatrix} + c_2 \begin{bmatrix} 2 \\ 11 \\ 0 \\ -6 \end{bmatrix},$$

where z and t (and therefore c_1 and c_2) are arbitrary real numbers. The solution space is a two-dimensional subspace of $\mathbb{R}^4$.

On the other hand, if $\alpha \neq 4$, we can see that rank $[A \mid \mathbf{0}] = \text{rank}(A) = 3$, so there is only one free variable, say t. Back substitution leads to solutions of the form

$$\begin{bmatrix} -2t \\ -t \\ -5t \\ t \end{bmatrix} = -t \begin{bmatrix} 2 \\ 1 \\ 5 \\ -1 \end{bmatrix} = c \begin{bmatrix} 2 \\ 1 \\ 5 \\ -1 \end{bmatrix},$$

where c is an arbitrary real number. The solution set is a line in four-space.

If, at the beginning of our analysis, we had entered the augmented matrix into a CAS and asked for the rref, we might have obtained the result

$$\begin{bmatrix} 1 & 0 & 0 & 2 & 0 \\ 0 & 1 & 0 & 1 & 0 \\ 0 & 0 & 1 & 5 & 0 \end{bmatrix}$$

with no disclaimer, which would have completely obscured the role of the parameter α in the problem. In the event that $\alpha = 4$, the last row of any echelon form and also of the reduced echelon form should be a zero row.

The next theorem provides an important criterion for a *nonhomogeneous* system to have a solution.

Theorem 2.8.2

A nonhomogeneous system of linear equations $A\mathbf{x} = \mathbf{b}$ has a solution if and only if rank $[A \mid \mathbf{b}\,] = \text{rank}(A)$.

Proof Suppose the system has a solution $\mathbf{x} = \begin{bmatrix} x_1 \\ x_2 \\ \vdots \\ x_n \end{bmatrix}$. By Theorem

2.7.1, this is equivalent to $x_1\mathbf{a}_1 + x_2\mathbf{a}_2 + \cdots + x_n\mathbf{a}_n = \mathbf{b}$—that is, $\mathbf{b}$ is linearly dependent on the columns of A. But this last statement is equivalent to saying that column $\mathbf{b}$ does not change the number of linearly independent columns of A, so we have rank $[A \mid \mathbf{b}] = \text{rank}(A)$.

Example 2.8.3: A System with No Solution

Consider the system

$$2a + 2b - 2c = 5$$
$$7a + 7b + c = 10$$
$$5a + 5b - c = 5$$

The augmented matrix for this system is

$$[A \mid \mathbf{b}] = \begin{bmatrix} 2 & 2 & -2 & 5 \\ 7 & 7 & 1 & 10 \\ 5 & 5 & -1 & 5 \end{bmatrix},$$

which has rref

$$B = \begin{bmatrix} 1 & 1 & 0 & 0 \\ 0 & 0 & 1 & 0 \\ 0 & 0 & 0 & 1 \end{bmatrix}.$$

Here $n = 3$, rank $[A \,|\, \mathbf{b}] = 3$, but rank$(A) = 2$. The last row of B is equivalent to the equation $0 \cdot a + 0 \cdot b + 0 \cdot c = 1$, which has no solution.

In particular, Theorem 2.8.2 states that if we have a nonhomogeneous system of m equations in n unknowns, so that the coefficient matrix A is $m \times n$, then when rank$(A) = m$, the system $A\mathbf{x} = \mathbf{b}$ will be consistent for *all* vectors $\mathbf{b} \in \mathbb{R}^m$. To see this, note that if rank$(A) = m$, there is no row of zeros in rref(A) and hence no possibility of inconsistency— that is, in rref $[A \,|\, \mathbf{b}]$ there will not be an implied equation of the form $0 \cdot x_1 + 0 \cdot x_2 + \cdots + 0 \cdot x_n = c$ for some nonzero scalar c.

We can extend Theorem 2.8.2 to guarantee *uniqueness* of a solution.

Theorem 2.8.3

A nonhomogeneous system $A\mathbf{x} = \mathbf{b}$ of linear equations in n unknowns has a unique solution if and only if rank $[A \,|\, \mathbf{b}] = $ rank$(A) = n$.*

Proof First of all, suppose a system $A\mathbf{x} = \mathbf{b}$ of equations in n unknowns has a unique solution. Then there exist unique numbers $x_1, x_2, \ldots, x_n$ such that

$$x_1\mathbf{a}_1 + x_2\mathbf{a}_2 + \cdots + x_n\mathbf{a}_n = \mathbf{b}. \tag{*}$$

Statement (*) implies that rank $[A \,|\, \mathbf{b}] = $ rank$(A) \leq n$. If rank$(A) < n$, the set of column vectors of A is linearly dependent—that is, we can write

$$\tilde{x}_1\mathbf{a}_1 + \tilde{x}_2\mathbf{a}_2 + \cdots + \tilde{x}_n\mathbf{a}_n = \mathbf{0}, \tag{**}$$

where not all the coefficients $\tilde{x}_i$ are zero. Adding (*) and (**), we have $(x_1 + \tilde{x}_1)\mathbf{a}_1 + (x_2 + \tilde{x}_2)\mathbf{a}_2 + \cdots + (x_n + \tilde{x}_n)\mathbf{a}_n = \mathbf{b}$, contradicting the uniqueness of the original system's solution. Thus rank $[A \,|\, \mathbf{b}] = $ rank$(A) = n$.

On the other hand, if rank $[A \,|\, \mathbf{b}] = $ rank$(A) = n$, then the system has a solution (by Theorem 2.8.2) and the set of columns of A is linearly independent. Now suppose that we have two distinct solutions of $A\mathbf{x} = \mathbf{b}$:

$$x_1\mathbf{a}_1 + x_2\mathbf{a}_2 + \cdots + x_n\mathbf{a}_n = \mathbf{b},$$

and

$$\hat{x}_1\mathbf{a}_1 + \hat{x}_2\mathbf{a}_2 + \cdots + \hat{x}_n\mathbf{a}_n = \mathbf{b}.$$

---

* See K. Hardy, K.S. Williams, and B.K. Spearman, Uniquely determined unknowns in systems of linear equations, *Math. Mag.* **75** (2002): 53–57 for interesting results related to this theorem.

Subtracting, we get $(x_1 - \hat{x}_1)\mathbf{a}_1 + (x_2 - \hat{x}_2)\mathbf{a}_2 + \cdots + (x_n - \hat{x}_n)\mathbf{a}_n = \mathbf{0}$. The linear independence of the columns implies that $x_i - \hat{x}_i = 0$ for $i = 1, 2, \ldots, n$, so any solution must be unique.

Corollary 2.8.2

A nonhomogeneous system of m equations in n unknowns cannot have a unique solution if $m < n$.

Proof   Given an $m \times n$ system $A\mathbf{x} = \mathbf{b}$ with $m < n$, then $\mathrm{rank}(A) \le m < n$.

By Theorem 2.8.3, the system cannot have a unique solution.

Example 2.8.1 analyzes a system with $m < n$. It has infinitely many solutions.

Example 2.8.4: A System with a Unique Solution

The system

$$x + y = 0$$
$$x - y = 1$$
$$4x + 2y = 1$$

has the augmented matrix

$$[A \,|\, \mathbf{b}] = \begin{bmatrix} 1 & 1 & 0 \\ 1 & -1 & 1 \\ 4 & 2 & 1 \end{bmatrix},$$

which can be row reduced to

$$B = \begin{bmatrix} 1 & 0 & \frac{1}{2} \\ 0 & 1 & -\frac{1}{2} \\ 0 & 0 & 0 \end{bmatrix}.$$

Thus rank $[A \,|\, \mathbf{b}] = \mathrm{rank}(A) = 2 = n$ and there is a unique solution that we can read directly from B: $x = \frac{1}{2}$, $y = -\frac{1}{2}$.

Looking back at Example 2.8.2, we see that $n = 5$, $\mathrm{rank}(A) = 3$, and there are 2 $[= n - \mathrm{rank}(A)]$ free variables in the solution of the system. This pattern is always true for a consistent system, as we show in the next theorem.

Theorem 2.8.4

Let $A\mathbf{x} = \mathbf{b}$ be a consistent $m \times n$ system of linear equations. If $\mathrm{rank}(A) = r$, then $n - r$ of the unknowns can be given arbitrary values and the equations can be solved by using these parameters. (Here we assume that the system has a nontrivial solution if it is homogeneous.)

Proof   We start with the augmented matrix $[A \mid \mathbf{b}]$ and transform it to rref. Thus we obtain a matrix of the form $[\tilde{A} \mid \tilde{\mathbf{b}}]$ in which, if $\text{rank}(A) = r$, there are r nonzero rows. Assume that the pivot columns are columns number c_i for $1 \leq i \leq r$. Then $1 \leq c_1 < c_2 < \cdots < c_r \leq n+1$. Also assume that the remaining (i.e., nonpivot) column numbers are c_{r+1}, $c_{r+2}, \ldots, c_n$, where $1 \leq c_{r+1} < c_{r+2} < \cdots < c_n \leq n+1$. The corresponding system of equations $\tilde{A}\mathbf{x} = \tilde{\mathbf{b}}$ is equivalent to the original system (Theorem 2.2.2), and its form allows us to assign $n - r$ of the unknowns as solution parameters in the following way:

Case 1: $r < n$. ... Because there are r pivot columns, we can take the unknowns $x_{c_1}, x_{c_2}, \ldots, x_{c_r}$ corresponding to these pivot columns as *dependent* unknowns and use the unknowns $x_{c_{r+1}}, x_{c_{r+2}}, \ldots, x_{c_n}$ corresponding to the nonpivot columns as *independent* unknowns, which can take on arbitrary values.

The nonzero equations corresponding to the reduced echelon form of A look like

$$x_{c_1} + \tilde{a}_{1,c_{r+1}} x_{c_{r+1}} + \cdots\cdots\cdots\cdots + \tilde{a}_{1c_n} x_{c_n} = \tilde{b}_1$$

$$x_{c_2} + \tilde{a}_{2,c_{r+1}} x_{c_{r+1}} + \cdots\cdots\cdots + \tilde{a}_{2c_n} x_{c_n} = \tilde{b}_2$$

$$\ddots \qquad\qquad \vdots \qquad\qquad \vdots \qquad \vdots$$

$$x_{c_r} + \tilde{a}_{r,c_{r+1}} x_{c_{r+1}} + \cdots + \tilde{a}_{rc_n} x_{c_n} = \tilde{b}_r$$

If we assign values to the variables $x_{c_{r+1}}, x_{c_{r+2}}, \ldots, x_{c_n}$, we can solve each equation for the unknowns $x_{c_1}, x_{c_2}, \ldots, x_{c_r}$.

Case 2: $r = n$. ... In this case, the only form that $[\tilde{A} \mid \tilde{\mathbf{b}}] = \text{rref}[A \mid \mathbf{b}]$ can take is

$$
\begin{bmatrix}
1 & 0 & \ldots & 0 & \tilde{b}_1 \\
0 & 1 & \ldots & 0 & \tilde{b}_2 \\
\vdots & \vdots & \vdots & \vdots & \vdots \\
0 & 0 & \ldots & 1 & \tilde{b}_n \\
0 & 0 & \ldots & 0 & 0 \\
\vdots & \vdots & \vdots & \vdots & \vdots \\
0 & 0 & \ldots & 0 & 0
\end{bmatrix}.
$$

Because $A\mathbf{x} = \mathbf{b}$ and $\tilde{A}\mathbf{x} = \tilde{b}$ have the same solution set, we can just solve the easier (reduced) system by observing the equations that the nonzero rows represent and conclude that $x_1 = \tilde{b}_1, x_2 = \tilde{b}_2, \ldots, x_n = \tilde{b}_n$. There are $n - n = 0$ free variables, meaning that the solution is unique.

If the system in the hypothesis of Theorem 2.8.4 is *homogeneous*, a careful reading of Case 1 of the proof reveals that solving for the first r variables in terms of the last $n - r$ variables indicates that the null space of A has dimension $n - r$. This provides an alternative way of proving Theorem 2.7.4, the relationship between rank and nullity.

To understand Theorem 2.8.4 better, let us look at an example.

Example 2.8.5: A System with Infinitely Many Solutions

The augmented matrix of the system

$$x + 3y - 2z + 5u - 3v = 1$$
$$2x + 7y - 3z + 7u - 5v = 2$$
$$3x + 11y - 4z + 10u - 9v = 3$$

reduces to

$$B = \begin{bmatrix} 1 & 0 & -5 & 0 & 22 & 1 \\ 0 & 1 & 1 & 0 & -5 & 0 \\ 0 & 0 & 0 & 1 & -2 & 0 \end{bmatrix}.$$

Because $\text{rank}(A) = r = 3 < n = 5$, Theorem 2.8.4 leads us to expect infinitely many solutions. The pivot columns are columns 1, 2, and 4, so we take the first, second, and fourth unknowns—x, y, and u, respectively—as dependent variables and the third and fifth unknowns (z and v, respectively) as free variables.

Beginning with the last row, the three equations corresponding to the rows of B are

$$u = 0 - (-2)v$$
$$y = 0 - (1)z - (-5)v$$
$$x = 1 - (-5)z - 22v$$

or

$$u = 2v$$
$$y = -z + 5v$$
$$x = 1 + 5z - 22v$$

For each pair of values we assign to z and v, we obtain definite values of x, y, and u, so that $\begin{bmatrix} x \\ y \\ z \\ u \\ v \end{bmatrix}$ is a solution of the original system. For example, letting $z = 4$ and $v = 1$, we find that $x = 1 + 5(4) - 22(1) = -1$, $y = -4 + 5(1) = 1$, and $u = 2(1) = 2$. Thus the vector $\begin{bmatrix} -1 \\ 1 \\ 4 \\ 2 \\ 1 \end{bmatrix}$ is one solution of the system.

For nonhomogeneous systems, we can summarize the results of this section so far in Figure 2.6.

A homogeneous $m \times n$ system $A\mathbf{x} = \mathbf{0}$ always has the zero vector as a solution, but has an infinite number of solutions if and only if rank$(A) < n$.

Figure 2.6 indicates that column rank is the important criterion for the number of solutions of an $m \times n$ nonhomogeneous system of linear equations. Although Corollaries 2.8.2 and 2.8.3 refer directly to the relationship between m and n, this comparison between the number of equations and the number of unknowns has limited usefulness.

An $m \times n$ nonhomogeneous system of linear equations in which $m < n$—that is, in which there are fewer equations than unknowns—is said to be **underdetermined**. Such systems often (but not always) have infinitely many solutions (see Example 2.8.3). The system $\{x + y + z = 3, 3x + 3y + 3z = 4\}$, however, is inconsistent.

A nonhomogeneous $m \times n$ system in which $m > n$—that is, in which the number of equations exceeds the number of unknowns—is called **overdetermined**. An overdetermined system will often (but not always) be inconsistent. Example 2.8.5 provides a *consistent* overdetermined system.

Finally, as the next theorem shows, to determine all solutions of a nonhomogeneous system, just find *one* solution of the nonhomogeneous system and then solve a related homogeneous system. This result is important in the application of linear algebra to the theory of differential equations.*

If we are given a system $A\mathbf{x} = \mathbf{b}$, $\mathbf{b} \neq \mathbf{0}$, then the homogeneous system $A\mathbf{x} = \mathbf{0}$ is called the **associated system**, or **reduced system**, corresponding to the original nonhomogeneous system.

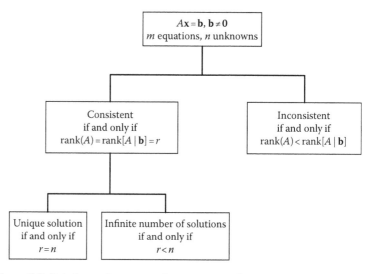

Figure 2.6 Solutions of $m \times n$ nonhomogeneous linear systems.

---

* See, for example, Section 5.6 of H. Ricardo, *A Modern Introduction to Differential Equations*, 2nd edn. (San Diego, CA: Academic Press, 2009).

Theorem 2.8.5

Let $A\mathbf{x} = \mathbf{b}$, $\mathbf{b} \neq \mathbf{0}$, be a consistent $m \times n$ system of linear equations and suppose $\mathbf{x}_1$ is a solution of $A\mathbf{x} = \mathbf{b}$. Then $\mathbf{x}^*$ is a solution of the system if and only if $\mathbf{x}^* = \mathbf{x}_1 + \mathbf{z}$, where $\mathbf{z}$ is some solution of the associated system $A\mathbf{x} = \mathbf{0}$.

Proof   First suppose that $\mathbf{x}^* = \mathbf{x}_1 + \mathbf{z}$, where $\mathbf{x}_1 = \begin{bmatrix} x_1 \\ x_2 \\ \vdots \\ x_n \end{bmatrix}$ is a solution

of $A\mathbf{x} = \mathbf{b}$ and $\mathbf{z} = \begin{bmatrix} z_1 \\ z_2 \\ \vdots \\ z_n \end{bmatrix}$ is a solution of the associated system

$A\mathbf{x} = \mathbf{0}$. Then, by Equation 2.7.1, we have $A\mathbf{x}^* = A(\mathbf{x}_1 + \mathbf{z}) = A\mathbf{x}_1 + A\mathbf{z} = \mathbf{b} + \mathbf{0} = \mathbf{b}^*$, so $\mathbf{x}^*$ is a solution of the nonhomogeneous system.

On the other hand, suppose that $\mathbf{x}^* = \begin{bmatrix} x_1^* \\ x_2^* \\ \vdots \\ x_n^* \end{bmatrix}$ is a solution of the

nonhomogeneous system and let $\mathbf{z} = \mathbf{x}^* - \mathbf{x}_1$. Then $A\mathbf{z} = A(\mathbf{x}^* - \mathbf{x}_1) = A\mathbf{x}^* - A\mathbf{x}_1 = \mathbf{b} - \mathbf{b} = \mathbf{0}$, so $\mathbf{z}$ is a solution of the homogeneous system $A\mathbf{x} = \mathbf{0}$. But the definition of $\mathbf{z}$ tells us that $\mathbf{x}^* = \mathbf{x}_1 + \mathbf{z}$, which is what we want to prove.

In Theorem 2.8.5 we see that $A\mathbf{x} = \mathbf{b}$ has a unique solution of the form $\mathbf{x}^* = \mathbf{x}_1 + \mathbf{z}$ if the only possible value of $\mathbf{z}$ is $\mathbf{0}$.

Example 2.8.6: Theorem 2.8.5 Illustrated

In Example 2.7.6, we saw that one solution of the system

$$x + 3y - 2z + 5u - 3v = 1$$
$$2x + 7y - 3z + 7u - 5v = 2$$
$$3x + 11y - 4z + 10u - 9v = 3$$

is the vector $\begin{bmatrix} -1 \\ 1 \\ 4 \\ 2 \\ 1 \end{bmatrix}$. Now let us look at the associated system

$$x + 3y - 2z + 5u - 3v = 0$$
$$2x + 7y - 3z + 7u - 5v = 0$$
$$3x + 11y - 4z + 10u - 9v = 0$$

The calculations in Example 2.7.6 indicate that a system equivalent to the homogeneous system just given is

$$x = 0 - (-5)z - 22v = 5z - 22v$$
$$y = 0 - (1)z - (-5)v = -z + 5v$$
$$u = 0 - (-2)v = 2v$$

where the unknowns z and v can be given arbitrary real values. Thus the solution of the reduced system is the set of all vectors in $\mathbb{R}^5$ of the form $\begin{bmatrix} 5z - 22v \\ -z + 5v \\ z \\ 2v \\ v \end{bmatrix}$, where z and v are arbitrary.

Theorem 2.8.5 tells us that any solution of the original nonhomogeneous system is given by a vector of the form

$$\begin{bmatrix} -1 \\ 1 \\ 4 \\ 2 \\ 1 \end{bmatrix} + \begin{bmatrix} 5z - 22v \\ -z + 5v \\ z \\ 2v \\ v \end{bmatrix} = \begin{bmatrix} 5z - 22v - 1 \\ 5v - z + 1 \\ z + 4 \\ 2v + 2 \\ v + 1 \end{bmatrix},$$

where z and v are arbitrary. For example, letting $z = 1$ and $v = -1$, we get the vector $\begin{bmatrix} 26 \\ -5 \\ 5 \\ 0 \\ 0 \end{bmatrix}$, which we can check is a solution of the nonhomogeneous system.

Exercises 2.8

A.

Find the rank of each matrix in Exercises 1–12.

1. $\begin{bmatrix} -2 & -6 & 12 \\ 1 & 2 & -3 \\ 3 & 8 & -12 \end{bmatrix}$
2. $\begin{bmatrix} 24 & -7 & -14 & 21 \\ -3 & 1 & 2 & -3 \\ 6 & -2 & -4 & 6 \end{bmatrix}$

3. $\begin{bmatrix} 12 & 18 & 6 \\ 10 & 15 & 5 \\ 34 & 51 & 17 \\ 14 & 21 & 7 \end{bmatrix}$
4. $\begin{bmatrix} 3 & 0 & -1 & 2 \\ 3 & 2 & 4 & -8 \\ 9 & 6 & 10 & -20 \end{bmatrix}$

5. $\begin{bmatrix} 1 & 2 & 4 & -2 \\ -1 & 4 & 8 & -4 \\ 1 & -3 & -6 & 3 \end{bmatrix}$
6. $\begin{bmatrix} 2 & -1 & 3 & -2 & 4 \\ 4 & -2 & 5 & 1 & 2 \\ 2 & -1 & 1 & 8 & 2 \end{bmatrix}$

7. $\begin{bmatrix} 1 & 1 & 1 & 1 \\ 1 & 1 & 1 & 1 \\ 1 & 1 & 1 & 1 \\ 1 & 1 & 1 & 1 \end{bmatrix}$ 8. $\begin{bmatrix} 0 & 0 & 0 & 1 \\ 0 & 0 & 1 & 0 \\ 0 & 1 & 0 & 0 \\ 1 & 0 & 0 & 0 \end{bmatrix}$

9. $\begin{bmatrix} 1 & 3 & 5 & -1 \\ 2 & -1 & -3 & 4 \\ 5 & 1 & -1 & 7 \\ 7 & 7 & 9 & 1 \end{bmatrix}$ 10. $\begin{bmatrix} 0 & 4 & 10 & 1 \\ 4 & 8 & 18 & 7 \\ 10 & 18 & 40 & 17 \\ 1 & 7 & 17 & 3 \end{bmatrix}$

11. $\begin{bmatrix} 1 & -1 & 2 & 3 & 4 \\ 2 & 1 & -1 & 2 & 0 \\ -1 & 2 & 1 & 1 & 3 \\ 1 & 5 & -8 & -5 & -12 \\ 3 & -7 & 8 & 9 & 13 \end{bmatrix}$ 12. $\begin{bmatrix} 2 & 1 & 1 & 1 \\ 1 & 3 & 1 & 1 \\ 1 & 1 & 4 & 1 \\ 1 & 1 & 1 & 5 \\ 1 & 2 & 3 & 4 \\ 1 & 1 & 1 & 1 \end{bmatrix}$

13. Let

$$A = \begin{bmatrix} 1 & 2 & -1 & 0 & 4 & 3 \\ 2 & -2 & 0 & -3 & 1 & 4 \\ -1 & -8 & 3 & -3 & -11 & -5 \\ 0 & -6 & 2 & -3 & -7 & -2 \end{bmatrix}.$$

Determine the rank of A and bases for the row space and column spaces of A.

14. Suppose the matrix in Exercise 5 is the augmented matrix for a system.
    a. Write out the system of equations in full algebraic form.
    b. Solve the system if possible.

15. Suppose the matrix in Exercise 11 is the augmented matrix for a system.
    a. Write out the system of equations in full algebraic form.
    b. Solve the system if possible.

16. If A is the coefficient matrix for a homogeneous system consisting of four equations in eight unknowns and if there are five free variables, what is the rank of A?

17. Suppose that A is the coefficient matrix for a homogeneous system consisting of four equations in six unknowns and suppose that A has at least one nonzero row.
    a. Determine the smallest number of free variables that are possible.
    b. Determine the maximum number of free variables that are possible.

18. Suppose that A is the 4×4 matrix all of whose entries are zero. Describe the solution set of the system $A\mathbf{x} = \mathbf{0}$.

19. Suppose that A is a 3×3 matrix whose rref has two pivot columns.
   a. Does the system $A\mathbf{x} = \mathbf{0}$ have a nontrivial solution? Explain.
   b. Does the equation $A\mathbf{x} = \mathbf{b}$ have at least one solution for every possible $\mathbf{b}$? Explain.

20. Suppose that A is a 2×4 matrix whose rref has two pivot positions.
   a. Does the system $A\mathbf{x} = \mathbf{0}$ have a nontrivial solution? Explain.
   b. Does the equation $A\mathbf{x} = \mathbf{b}$ have at least one solution for every possible $\mathbf{b}$? Explain.

21. Solve the following system and express the solution in the form $\mathbf{x}^* = \mathbf{x}_1 + \mathbf{z}$, as described in Theorem 2.8.5:

$$x + 2y - z - 2w = 2$$
$$2x + y - 2z + 3w = 2$$
$$x + 2y + 3z + 4w = 5$$
$$4x + 5y - 4z - w = 6$$

22. Solve the following system and express the solution in the form $\mathbf{x}^* = \mathbf{x}_1 + \mathbf{z}$, as described in Theorem 2.8.5:

$$x - y - 2z + 3w = 4$$
$$3x + 2y - z + 2w = 5$$
$$-y - 7z + 9w = -2$$

23. Given that

$$A = \begin{bmatrix} 1 & 2 & 2 \\ 2 & 2 & 3 \\ 1 & -1 & 3 \end{bmatrix}, \quad C = \begin{bmatrix} 2 & 1 & 1 \\ 2 & 2 & 1 \\ 1 & 1 & 1 \end{bmatrix}, \quad \mathbf{d} = \begin{bmatrix} 10 \\ 13 \\ 9 \end{bmatrix},$$

and that $C\mathbf{b} = \mathbf{d}$, solve the linear system $A\mathbf{x} = \mathbf{b}$.

24. Find a 3×3 matrix A, not all of whose entries are zero, such that the vector $\begin{bmatrix} 1 \\ -2 \\ 1 \end{bmatrix}$ is a solution of $A\mathbf{x} = \mathbf{0}$.

B.

1. If A is a 4×6 matrix, show that the columns of A are linearly dependent.

2. If A is a 5×3 matrix, show that the rows of A are linearly dependent.

3. Suppose that A is a 6×4 matrix whose rank is 4.
   a. Are the rows of A linearly dependent or linearly independent? Justify your answer.
   b. Are the columns of A linearly dependent or linearly independent? Justify your answer.

4. Let A be an $m \times n$ matrix with $m \neq n$. Show that either the rows or the columns of A are linearly dependent.

5. Show by an example that if an elementary column operation is applied to the augmented matrix of a linear system, the resulting linear system need not be equivalent to the original one.

6. Suppose that A is an $m \times n$ matrix of rank r. Determine the possible values for the rank of the matrix obtained by
   a. Changing exactly one element of A.
   b. Changing two elements of A.

7. Indicate the rank of the following matrix for all possible real values of λ.

$$\begin{bmatrix} 3 & 1 & 1 & 4 \\ \lambda & 4 & 10 & 1 \\ 1 & 7 & 17 & 3 \\ 2 & 2 & 4 & 3 \end{bmatrix}.$$

8. Indicate the rank of the following matrix for different real values of λ:

$$\begin{bmatrix} 1 & \lambda & -1 & 2 \\ 2 & -1 & \lambda & 5 \\ 1 & 10 & -6 & \lambda \end{bmatrix}.$$

9. Show that an $m \times n$ homogeneous system with $m < n$ has *infinitely many* solutions. [*Hint*: If $\mathbf{x}$ is a nontrivial solution, show that $c\mathbf{x}$ is a solution for every scalar c.]

10. Suppose that A is an $m \times n$ matrix, all of whose entries are rational numbers $\frac{a}{b}$ with integers a and b $(b \neq 0)$. Prove or disprove the following statements, giving an example or an explanation if the statement is not true:
    a. There is a row echelon form for A consisting of integers.
    b. The homogeneous linear system $A\mathbf{x} = \mathbf{0}$ has integer solutions.
    c. If $\mathbf{b} \in \mathbb{R}^m$ has integer components, then $A\mathbf{x} = \mathbf{b}$ has an integer solution, provided the system is solvable.

11. For which values of the parameter t does the following homogeneous linear system have nontrivial solutions?

$$6x - y + z = 0$$
$$tx + z = 0$$
$$y + tz = 0$$

12. Suppose that A is the 3×4 matrix

$$\begin{bmatrix} 4 & -20 & -1 & -13 \\ 2 & -10 & 7 & 1 \\ 3 & -15 & -2 & -11 \end{bmatrix}.$$

Let $\boldsymbol{\alpha}_1, \boldsymbol{\alpha}_2, \boldsymbol{\alpha}_3$ denote the row vectors of A and let $\boldsymbol{\beta}_1, \boldsymbol{\beta}_2, \boldsymbol{\beta}_3,$ $\boldsymbol{\beta}_4$ denote the column vectors of A.

a. Show that A has rank 2 and find a basis $\{\mathbf{x}_1, \mathbf{x}_2\}$ for the subspace of $\mathbb{R}^4$ spanned by the rows of A.

b. Find scalars c_{ij} such that $\boldsymbol{\alpha}_i = c_{i1}\mathbf{x}_1 + c_{i2}\mathbf{x}_2$ for $i = 1, 2, 3$.

c. Express each column vector $\boldsymbol{\beta}_j$ as a linear combination of the vectors

$$\boldsymbol{\gamma}_1 = \begin{bmatrix} c_{11} \\ c_{21} \\ c_{31} \end{bmatrix} \quad \text{and} \quad \boldsymbol{\gamma}_2 = \begin{bmatrix} c_{12} \\ c_{22} \\ c_{32} \end{bmatrix}.$$

d. Show that the vector $\boldsymbol{\beta} = \begin{bmatrix} 2 \\ 4 \\ 1 \end{bmatrix}$ can be expressed as a linear combination of $\boldsymbol{\gamma}_1$ and $\boldsymbol{\gamma}_2$. Does the linear system $A\mathbf{x} = \boldsymbol{\beta}$ have a solution? Explain.

13. A certain 4×5 matrix A is known to have rank 3. Furthermore, if $\boldsymbol{\alpha}_1, \boldsymbol{\alpha}_2, \boldsymbol{\alpha}_3, \boldsymbol{\alpha}_4$ are the row vectors of A, it is known that there exist vectors $\boldsymbol{\rho}_1, \boldsymbol{\rho}_2, \boldsymbol{\rho}_3$ such that

$$\boldsymbol{\alpha}_1 = -7\boldsymbol{\rho}_1 + 2\boldsymbol{\rho}_2 - 5\boldsymbol{\rho}_3$$
$$\boldsymbol{\alpha}_2 = 4\boldsymbol{\rho}_1 + 0\boldsymbol{\rho}_2 + \boldsymbol{\rho}_3$$
$$\boldsymbol{\alpha}_3 = 9\boldsymbol{\rho}_1 + \boldsymbol{\rho}_2 - 2\boldsymbol{\rho}_3$$
$$\boldsymbol{\alpha}_4 = -2\boldsymbol{\rho}_1 + \boldsymbol{\rho}_2 + 3\boldsymbol{\rho}_3$$

a. Show that $\{\boldsymbol{\rho}_1, \boldsymbol{\rho}_2, \boldsymbol{\rho}_3\}$ is a basis for the subspace spanned by the rows of A.

b. Find three vectors $\boldsymbol{\gamma}_1, \boldsymbol{\gamma}_2, \boldsymbol{\gamma}_3$ that span the subspace spanned by the columns of A.

c. Determine whether or not the linear system $A\mathbf{x} = \boldsymbol{\beta}$ has a solution, where

$$\boldsymbol{\beta} = \begin{bmatrix} 0 \\ 3 \\ 13 \\ -4 \end{bmatrix}.$$

14. Construct a homogeneous linear system of three equations in four unknowns, $x_1, x_2, x_3,$ and x_4, that has

$$x_2 \begin{bmatrix} -2 \\ 1 \\ 0 \\ 0 \end{bmatrix} + x_4 \begin{bmatrix} -3 \\ 0 \\ 2 \\ 1 \end{bmatrix}$$

as its solution (where x_2 and x_4 can take arbitrary real values). [*Hint*: Look at Theorem 2.8.4 and consider what the rref of the matrix of coefficients might look like.]

15. Construct a nonhomogeneous linear system of three equations in four unknowns, x_1, x_2, x_3, and x_4, that has

$$\begin{bmatrix} 1 \\ 0 \\ 1 \\ 0 \end{bmatrix} + x_2 \begin{bmatrix} -2 \\ 1 \\ 0 \\ 0 \end{bmatrix} + x_4 \begin{bmatrix} -3 \\ 0 \\ 2 \\ 1 \end{bmatrix}$$

as its solution (where x_2 and x_4 can take arbitrary real values).

C.

1. If the matrix

$$\begin{bmatrix} a & 1 & a & 0 & 0 & 0 \\ 0 & b & 1 & b & 0 & 0 \\ 0 & 0 & c & 1 & c & 0 \\ 0 & 0 & 0 & d & 1 & d \end{bmatrix}$$

has rank r, prove that
a. $r > 2$.
b. $r = 3$ if and only if $a = d = 0$ and $bc = 1$.
c. $r = 4$ in all cases other than those mentioned in part (b).

2. a. Show that $\begin{bmatrix} a_1 \\ b_1 \end{bmatrix}, \begin{bmatrix} a_2 \\ b_2 \end{bmatrix}, \ldots, \begin{bmatrix} a_n \\ b_n \end{bmatrix}$ are vectors collinear in $\mathbb{R}^2$ if and only if

$$\mathrm{rank} \begin{bmatrix} 1 & a_1 & b_1 \\ 1 & a_2 & b_2 \\ \vdots & \vdots & \vdots \\ 1 & a_n & b_n \end{bmatrix} \leq 2.$$

b. What is the appropriate generalization of part (a) to vectors

$$\begin{bmatrix} a_1 \\ b_1 \\ c_1 \end{bmatrix}, \begin{bmatrix} a_2 \\ b_2 \\ c_2 \end{bmatrix}, \ldots, \begin{bmatrix} a_n \\ b_n \\ c_n \end{bmatrix} \text{ in } \mathbb{R}^3?$$

3. Suppose that A is an $n \times n$ matrix and that m rows (not necessarily consecutive) of A are selected to form an $m \times n$ "submatrix" B. Prove that $\mathrm{rank}(B) \geq m - n + \mathrm{rank}(A)$.

4. If B is a submatrix of an $m \times n$ matrix A obtained by deleting s rows and t columns from A, then $\mathrm{rank}(A) \leq s + t + \mathrm{rank}(B)$.

5. Consider the system of m equations in n unknowns represented by $A\mathbf{x} = \mathbf{b}$.

    Prove that the system has a solution if and only if every vector in $\mathbb{R}^m$ that is orthogonal to all columns of A is also orthogonal to the vector $\mathbf{b}$.

2.9 Summary

In $\mathbb{R}^2$ and $\mathbb{R}^3$, the solutions of systems of linear equations can be interpreted geometrically as the intersections of lines and of planes, respectively. A system of m equations in n unknowns has no solution, exactly one solution, or an infinite number of solutions.

The most common method of solving systems of linear equations is **Gaussian elimination**. By introducing the concept of a **matrix** and writing a system in the matrix–vector form $A\mathbf{x} = \mathbf{b}$, we can interpret Gaussian elimination as an algorithm that performs **elementary row operations** on A, the **matrix of coefficients**. Two matrices that can be converted to each other by means of elementary row operations are called **row equivalent**. Appending the **vector of outputs b** to the last column of the matrix of coefficients A, we obtain the **augmented matrix** $[A \mid \mathbf{b}]$ of the system. If $[A \mid \mathbf{b}]$ and $[C \mid \mathbf{d}]$ are row equivalent augmented matrices of two systems of linear equations, then the two systems have the same solution sets.

Elementary row operations can be used to obtain the (**row**) **echelon form** of any $m \times n$ matrix M. Every matrix can be reduced to a row echelon form, but not uniquely. [Considerations of computational efficiency demand that we develop strategies for choosing "good" pivot elements. **Partial pivoting** refers to the strategy of searching the entries on and below the pivotal position for the pivot of *maximum magnitude* and making this maximal entry the pivot, interchanging rows if necessary. In practice, guaranteeing that no multiplier is greater than 1 reduces the possibility of creating relatively large numbers that can overwhelm the significance of smaller numbers.]

Gauss–Jordan elimination is a modification of the Gaussian elimination that reduces the computations involved in back substitution by performing additional row operations to transform the matrix from echelon form to something called **rref** (or **reduced echelon form**). The Gauss–Jordan method may require a larger number of arithmetic operations than the basic Gaussian elimination, but the rref of a matrix has the advantage of being *unique*.

[Systems $A\mathbf{x} = \mathbf{b}$ for which small changes in A or $\mathbf{b}$ result in unexpectedly large changes in $\mathbf{x}$ are called **ill-conditioned**. A system is **well-conditioned** if relatively small changes in A or $\mathbf{b}$ yield relatively small changes in the solution of the system.]

Given an $m \times n$ matrix A, the span of the columns of A is a subspace of $\mathbb{R}^m$ called the **column space** of A, whereas the span of the (transposes of the) rows of A is a subspace of $\mathbb{R}^n$ called the **row space** of A. The dimensions of the column space and the row space are called the **column rank** and **row rank** of A, respectively. A system $A\mathbf{x} = \mathbf{b}$ has a solution if and only if $\mathbf{b}$ belongs to the column space of A.

The nonzero rows of rref(A) form a basis for the row space of A. If we determine the pivot columns of A in calculating rref(A), then these pivot columns constitute a basis for the column space of A. Even though the row space and the column space of a matrix A may be different, **the row rank of A equals the column rank of A**. We call this common value the **rank** of A, denoted $r(A)$. If A is an $m \times n$ matrix, then $r(A) \leq \min(m, n)$.

A **homogeneous** $m \times n$ system $A\mathbf{x} = \mathbf{0}$ has a nonzero solution if and only if $\text{rank}(A) < n$. Consequently, an $m \times n$ homogeneous system with $m < n$ must have a nontrivial solution—in fact, infinitely many solutions.

For **nonhomogeneous** systems $A\mathbf{x} = \mathbf{b}$, $\mathbf{b} \neq \mathbf{0}$, the solution possibilities are summarized by Figure 2.6. Given an $m \times n$ nonhomogeneous system $A\mathbf{x} = \mathbf{b}$ with $\text{rank}(A) = r$, we can give $n - r$ of the unknowns arbitrary values and then solve the equations in terms of these values as parameters.

If $\mathbf{x}_1$ is a solution of an $m \times n$ nonhomogeneous system $A\mathbf{x} = \mathbf{b}$, then *any* solution of the system has the form $\mathbf{x}_1 + \mathbf{z}$, where $\mathbf{z}$ is some solution of the **associated system** (or **reduced system**) $A\mathbf{x} = \mathbf{0}$.

3

Matrix Algebra

In Chapter 2, we introduced matrices and saw how matrix notation occurs naturally in representing and solving systems of linear equations. The usefulness and central importance of matrices will become more obvious as we continue our study of linear algebra.

In this chapter, we introduce various matrix operations and explore their properties, building on the ideas discussed in Chapters 1 and 2.

3.1 Addition and Subtraction of Matrices

Let us look at tables (matrices) representing 2 weeks of sales of ties and pairs of socks in a small mall shop:

Week 1 Sales (No. of Items Sold)

	Blue	Gray	Brown
Ties	5	3	1
Socks	10	7	4

Week 2 Sales

	Blue	Gray	Brown
Ties	7	4	1
Socks	10	9	5

If we want to get a 2 weeks' total, by type of item and color of item, we can simply add the two matrices of data, *entry by entry*:

Week $1 +$ Week 2 $\quad = 2$ weeks' total

$$\begin{bmatrix} 5 & 3 & 1 \\ 10 & 7 & 4 \end{bmatrix} + \begin{bmatrix} 7 & 4 & 1 \\ 10 & 9 & 5 \end{bmatrix} = \begin{bmatrix} 5+7 & 3+4 & 1+1 \\ 10+10 & 7+9 & 4+5 \end{bmatrix}$$

$$\begin{array}{ccc} \text{Bl} & \text{Gr} & \text{Br} \end{array}$$
$$= \begin{bmatrix} 12 & 7 & 2 \\ 20 & 16 & 9 \end{bmatrix} \begin{array}{l} \text{Ties} \\ \text{Socks} \end{array}.$$

Thus, for example, the 2 weeks' total for brown ties is 2 and the total for gray socks is 16.

Similarly, if matrix A represents revenue (dollars) from the first week's sales of the indicated items and matrix B represents the corresponding costs to the seller, then *subtracting* the matrices—that is, calculating $A - B$—gives the matrix of *profits*:

Revenue   Costs

$$A = \begin{bmatrix} 150 & 75 & 25 \\ 80 & 56 & 32 \end{bmatrix}, \; B = \begin{bmatrix} 35 & 18 & 6 \\ 30 & 21 & 12 \end{bmatrix}, \text{ and}$$

Profits
Bl Gr Br

$$C = A - B = \begin{bmatrix} 150 - 35 & 75 - 18 & 25 - 6 \\ 80 - 30 & 56 - 21 & 32 - 12 \end{bmatrix} = \begin{bmatrix} 115 & 57 & 19 \\ 50 & 35 & 20 \end{bmatrix} \begin{matrix} \text{Ties} \\ \text{Socks} \end{matrix}.$$

The matrix C tells us, for example, that the first week's profit on blue ties is \$115 and the week's profit on gray socks is \$35.

These simple examples motivate the general definition of *matrix addition* and *matrix subtraction*.

Definition 3.1.1

If A and B are two $m \times n$ matrices, $A = [a_{ij}]$ and $B = [b_{ij}]$, then the sum $A + B$ is defined as the matrix $[s_{ij}]$, where $s_{ij} = a_{ij} + b_{ij}$ for $i = 1, 2, \ldots,$ m and $j = 1, 2, \ldots, n$. In other words, the element in row i, column j of the sum $A + B$ is the sum of the element in row i, column j of A and the element in row i, column j of B. Similarly, the difference $A - B$ is defined as the matrix $[d_{ij}]$, where $d_{ij} = a_{ij} - b_{ij}$.

Example 3.1.1: Matrix Addition and Subtraction

If $A = \begin{bmatrix} -2 & 7 & 0 \\ 13 & 9 & -4 \end{bmatrix}$ and $B = \begin{bmatrix} 10 & -3 & 4 \\ 5 & -9 & 6 \end{bmatrix}$, then

$$A + B = \begin{bmatrix} -2 + 10 & 7 + (-3) & 0 + 4 \\ 13 + 5 & 9 + (-9) & -4 + 6 \end{bmatrix} = \begin{bmatrix} 8 & 4 & 4 \\ 18 & 0 & 2 \end{bmatrix},$$

$$A - B = \begin{bmatrix} -2 - 10 & 7 - (-3) & 0 - 4 \\ 13 - 5 & 9 - (-9) & -4 - 6 \end{bmatrix} = \begin{bmatrix} -12 & 10 & -4 \\ 8 & 18 & -10 \end{bmatrix}, \text{ and}$$

$$B - A = \begin{bmatrix} 10 - (-2) & -3 - 7 & 4 - 0 \\ 5 - 13 & -9 - 9 & 6 - (-4) \end{bmatrix} = \begin{bmatrix} 12 & -10 & 4 \\ -8 & -18 & 10 \end{bmatrix}.$$

Note that for the operations of addition and subtraction to make sense, the matrices involved must have the same shape—that is, both matrices must be $m \times n$ with the same m and n. Because the operations of matrix addition and subtraction are defined entry-by-entry and the entries are real numbers, these operations "inherit" the useful algebraic properties of the real number system.

In algebraic terms, the first sentence of the next theorem emphasizes that the set of all $m \times n$ matrices is *closed* under the operations of addition and subtraction: If you add two $m \times n$ matrices, the result is an $m \times n$ matrix; and if you subtract one $m \times n$ matrix from another, you again get an $m \times n$ matrix. The entry-by-entry definitions of these operations preserve the original (identical) shapes of the matrices involved.

Theorem 3.1.1: Properties of Matrix Addition and Subtraction

If A, B, and C are any $m \times n$ matrices, then the sum or difference of any two such matrices is again an $m \times n$ matrix. Furthermore,

(a) $A + B = B + A$. [*Commutativity of Addition*]
(b) $A + (B + C) = (A + B) + C$. [*Associativity of Addition*]
(c) The $m \times n$ zero matrix O_{mn}, all of whose entries are zero, has the property that $A + O_{mn} = A = O_{mn} + A$. [*Additive Identity*] [An $n \times n$ **zero matrix** will be written simply as O_n. If the size of the zero matrix is clear from the context, we may just write O.]
(d) For an $m \times n$ matrix A, there exists a matrix $-A$ such that $A + (-A) = O_{mn}$. [*Additive Inverse of A*]

Proof of (a)   By the definition of matrix addition, the element in row i, column j of $A + B$ is $s_{ij} = a_{ij} + b_{ij}$. But $a_{ij} + b_{ij} = b_{ij} + a_{ij}$ for $i = 1$, $2, \ldots, m$ and $j = 1, 2, \ldots, n$, because each entry is a real number and real numbers commute. Thus, $A + B = [s_{ij}] = [a_{ij} + b_{ij}] = [b_{ij} + a_{ij}] = B + A$.

Proof of (b) $A + (B + C) = [a_{ij}] + ([b_{ij}] + [c_{ij}]) = [a_{ij}] + [b_{ij} + c_{ij}] = [a_{ij} + (b_{ij} + c_{ij})] = [(a_{ij} + b_{ij}) + c_{ij}] = [a_{ij} + b_{ij}] + [c_{ij}] = (A + B) + C$ <*because addition of real numbers is associative*>.

Proof of (c)   Trivial.

Proof of (d)   For any matrix $A = [a_{ij}]$, define $-A$ as the matrix $[-a_{ij}]$—that is, the matrix each of whose entries is the *negative* of the corresponding entry of A. Then,
    $A + (-A) = [a_{ij}] + [-a_{ij}] = [a_{ij} + (-a_{ij})] = [a_{ij} - a_{ij}] = O_{mn}$.   (There is a subtle point of real number algebra here: subtraction is interpreted as the addition of a negative quantity—that is, if a and b are real numbers, then $a - b = a + (-b)$.)

3.1.1 Scalar Multiplication

Another basic matrix operation allows us to multiply a matrix by a scalar. This is an obvious extension of the scalar multiplication defined for vectors in Section 1.1.

Definition 3.1.2

If $A = [a_{ij}]$ is an $m \times n$ matrix and c is a scalar, then cA is the $m \times n$ matrix obtained by multiplying each entry of A by c: $cA = c[a_{ij}] = [b_{ij}]$, where $b_{ij} = ca_{ij}$.

This matrix is called a **scalar multiple** of A.

Example 3.1.2: Multiplication of a Matrix by a Scalar

Suppose $A = \begin{bmatrix} -1 & 2 & -3 \\ 0 & 4 & 5 \end{bmatrix}$. Then, for example,

$$5A = \begin{bmatrix} 5(-1) & 5(2) & 5(-3) \\ 5(0) & 5(4) & 5(5) \end{bmatrix} = \begin{bmatrix} -5 & 10 & -15 \\ 0 & 20 & 25 \end{bmatrix},$$

$$(-2)A = \begin{bmatrix} -2(-1) & -2(2) & -2(-3) \\ -2(0) & -2(4) & -2(5) \end{bmatrix} = \begin{bmatrix} 2 & -4 & 6 \\ 0 & -8 & -10 \end{bmatrix},$$

and

$$\frac{1}{2}A = \begin{bmatrix} \frac{1}{2}(-1) & \frac{1}{2}(2) & \frac{1}{2}(-3) \\ \frac{1}{2}(0) & \frac{1}{2}(4) & \frac{1}{2}(5) \end{bmatrix} = \begin{bmatrix} -\frac{1}{2} & 1 & -\frac{3}{2} \\ 0 & 2 & \frac{5}{2} \end{bmatrix}.$$

Notice that the concept of a scalar multiple gives us another way to think of matrix subtraction: $A - B = A + (-B) = A + (-1)B$. (Compare the proof of Theorem 3.1.1(d).)

Theorem 3.1.2: Properties of Scalar Multiplication

If A and B are $m \times n$ matrices and c and d are scalars, then the following properties hold:

(a) cA is again an $m \times n$ matrix. [*Closure Property*]
(b) $(cd) A = c(dA)$. [*Associative Property*]
(c) $c(A + B) = cA + cB$. [*Distributivity of Scalar Multiplication Over Matrix Addition*]
(d) $(c + d)A = cA + dA$. [*Distributivity of Scalar Multiplication Over Scalar Addition*]
(e) $1 \cdot A = A$. [*Identity Element for Scalar Multiplication*]

The proofs of these properties of scalar multiplication follow easily from Definition 3.1.2 and the algebraic properties of real numbers.

3.1.2 Transpose of a Matrix

An important matrix operation is that of *transposition*. It is not related to the operations of matrix algebra we have already seen, but the concept was introduced for vectors in Section 1.1 and for matrices (briefly) in Section 2.7.

Definition 3.1.3

The **transpose** of an $m \times n$ matrix A is the $n \times m$ matrix A^T obtained by interchanging the rows and columns in A. Row i of A is *column i of A^T* and column j of A is *row j of A^T*. Focusing on the individual entries, if $A = [a_{ij}]$, then $A^T = [t_{ij}]$, where $t_{ij} = a_{ji}$—that is, the (i,j) entry of the transpose is the (j,i) entry of the original matrix A.

Example 3.1.3: Matrix Transposes

Suppose $A = \begin{bmatrix} 1 \\ 2 \\ 3 \end{bmatrix}$ (regarded as a 3×1 matrix), then $A^T = [1\ 2\ 3]$. If $B = [-1\ 0\ 2]$ (regarded as a 1×3 matrix), then $B^T = \begin{bmatrix} -1 \\ 0 \\ 2 \end{bmatrix}$.

For $C = \begin{bmatrix} 1 & 2 \\ 3 & 4 \end{bmatrix}$, we have $C^T = \begin{bmatrix} 1 & 3 \\ 2 & 4 \end{bmatrix}$. The matrix

$$D = \begin{bmatrix} 1 & 0 & -2 & 3 \\ 0 & 2 & 4 & 6 \\ -4 & 1 & 2 & 3 \end{bmatrix} \text{ has transpose } D^T = \begin{bmatrix} 1 & 0 & -4 \\ 0 & 2 & 1 \\ -2 & 4 & 2 \\ 3 & 6 & 3 \end{bmatrix}.$$

Finally, if $E = \begin{bmatrix} 0 & -1 & 2 \\ -1 & 4 & 6 \\ 2 & 6 & 5 \end{bmatrix}$, then $E^T = \begin{bmatrix} 0 & -1 & 2 \\ -1 & 4 & 6 \\ 2 & 6 & 5 \end{bmatrix} = E$;

and if $F = \begin{bmatrix} 0 & 2 & -1 \\ -2 & 0 & 3 \\ 1 & -3 & 0 \end{bmatrix}$, then $F^T = \begin{bmatrix} 0 & -2 & 1 \\ 2 & 0 & -3 \\ -1 & 3 & 0 \end{bmatrix} = -F.$

The matrices E and F in the last example are interesting because matrices that are equal to their own transposes or to the negatives of their transposes are often used to model physical phenomena.

Definition 3.1.4

A square matrix $A = [a_{ij}]$ is said to be a **symmetric matrix** if $A = A^T$— that is, if $a_{ij} = a_{ji}$ for every i and j. A square matrix is called **skew-symmetric** if $A^T = -A$, so that $a_{ij} = -a_{ji}$ for all i and j.

If we think for a moment, we will realize that only *square* matrices can be symmetric or skew-symmetric.

Example 3.1.4: A Symmetric Matrix

The following matrix represents distances between three American cities:

$$
\begin{array}{c}
 \\
NY \\
CHI \\
LA
\end{array}
\begin{array}{ccc}
NY & CHI & LA \\
\left[\begin{array}{ccc}
0 & 841 & 2797 \\
841 & 0 & 2092 \\
2797 & 2092 & 0
\end{array}\right].
\end{array}
$$

To get the mileage between any two of the cities, look along the row of one city and the column of the other. The number lying at the intersection of the row and column is the mileage we are looking for. For example, the distance between Chicago and Los Angeles is 2092 miles, the number at the intersection of row 2 and column 3. We see that each nonzero distance appears twice in the matrix and that the matrix is symmetric.

Theorem 3.1.3: Properties of the Transpose

If A and B are matrices of the same size and k is a scalar, then

(a) $(A + B)^T = A^T + B^T$.
(b) $(kA)^T = k(A^T)$.

Proof of (a) $\quad (A + B)^T = ([a_{ij}] + [b_{ij}])^T = ([a_{ij} + b_{ij}])^T = [a_{ji} + b_{ji}] = [a_{ji}] + [b_{ji}] = A^T + B^T$.

Proof of (b) $\quad (kA)^T = (k\,[a_{ij}])^T = [ka_{ij}]^T = [ka_{ji}] = k\,[a_{ji}] = k\,(A^T)$.

Exercises 3.1

A.

1. If $A = \begin{bmatrix} 1 & 2 & 3 \\ 4 & 5 & 6 \end{bmatrix}$, $B = \begin{bmatrix} 1 & 2 \\ 3 & 4 \end{bmatrix}$, and $C = \begin{bmatrix} -1 & 0 & 2 \\ 3 & 4 & -5 \end{bmatrix}$,
   calculate the following, if defined. Otherwise, explain your answer.
   a. $A + B$
   b. $2A + C$
   c. $(A - C)^T$
   d. B^T
   e. $B^T - C^T$
   f. $-4B$
   g. $(-A + 3C)^T$
   h. $B + B^T$
   i. $C + C^T$

2. In Chapter 4, we will have to consider matrices of the form
   $A - \lambda I_n$ (or possibly $\lambda I_n - A$), where A is an $n \times n$ matrix,
   $I_n = [b_{ij}]$ is the $n \times n$ matrix with $b_{11} = b_{22} = \cdots = b_{nn} = 1$
   and all other entries zero, and λ is a scalar. For each of the
   following matrix–scalar pairs, write the combination $A - \lambda I_n$
   as a single matrix.
   a. $A = \begin{bmatrix} 2 & 3 \\ -1 & 4 \end{bmatrix}$, $\lambda = 3$
   b. $A = \begin{bmatrix} 1 & 0 \\ 3 & -2 \end{bmatrix}$, $\lambda = -2$
   c. $A = \begin{bmatrix} 2 & 0 & 1 \\ -4 & 6 & 3 \\ 0 & 1 & 2 \end{bmatrix}$, $\lambda = 2$
   d. $A = \begin{bmatrix} 1 & 2 & 3 & 4 \\ 5 & 6 & 7 & 8 \\ -4 & -3 & -2 & -1 \\ 0 & 0 & 0 & 0 \end{bmatrix}$, $\lambda = -1$

3. Let $A = \begin{bmatrix} 3 & 6 & -1 \\ 0 & -2 & 4 \end{bmatrix}$. Find a matrix B, which is a scalar
   multiple of A and which has 2 as its entry in the first row,
   second column.

4. Find x, y, z, and w if $3 \begin{bmatrix} x & y \\ z & w \end{bmatrix} = \begin{bmatrix} x & 6 \\ -1 & 2w \end{bmatrix} + \begin{bmatrix} 4 & x+y \\ z+w & 3 \end{bmatrix}$.

5. Using Theorem 3.1.1, prove that $A + B = B + C$ implies $A = C$.

6. Prove part (c) of Theorem 3.1.2.

7. Find the transpose of each of the following matrices, identifying which (if any) are symmetric:

a. $\begin{bmatrix} -2 \\ 0 \\ 4 \\ 7 \end{bmatrix}$ b. $[0 \quad 2 \quad 4]$ c. $\begin{bmatrix} -3 & 0 \\ 1 & 2 \end{bmatrix}$ d. $\begin{bmatrix} 1 & 2 \\ 0 & 3 \\ -3 & 0 \end{bmatrix}$

e. $\begin{bmatrix} 5 & -2 \\ -2 & 7 \end{bmatrix}$ f. $\begin{bmatrix} -1 & 2 & -3 \\ 0 & 4 & 5 \\ 1 & -5 & 0 \end{bmatrix}$ g. $\begin{bmatrix} 1 & 5 & 9 \\ 5 & 2 & -3 \\ 9 & -3 & 0 \end{bmatrix}$

8. a. If the 5×5 matrix A is symmetric and $a_{34} = 7$, what is a_{43}?
    b. If the 8×8 matrix B is skew-symmetric and $b_{57} = -3$, what is b_{75}?

9. Using Theorem 3.1.3, prove that $(A - B)^T = A^T - B^T$.

B.

1. Using the properties of Theorem 3.1.1, show that $A + 2(A + 3B) = 3A + 6B$, justifying each step.

2. Prove that $A = -(-A)$.

3. Prove that $A - B = -(B - A)$.

4. Prove that $(A - B) - C = A - (B + C)$.

5. Prove that $A - (B - C) = (A - B) + C$.

6. Prove that $(A^T)^T = A$.

7. a. Construct a 4×4 skew-symmetric matrix $A = [a_{ij}]$.
    b. What do you notice about the diagonal entries $a_{11}, a_{22}, a_{33}, a_{44}$ of the matrix constructed in part (a)? Generalize your observation to the case of an $n \times n$ skew-symmetric matrix and prove that you are right.

C.

1. Carry out steps (a)–(f) to show that **any square matrix A can be expressed uniquely as $A = M + S$, where M is symmetric and S is skew-symmetric**.
    a. Show that if $A = M + S$, where M is symmetric and S is skew-symmetric, then $A^T = M - S$.

b. Given that $A = M + S$, where M is symmetric and S is skew-symmetric, show that $A + A^T = 2M$, so that $M = \frac{1}{2}(A + A^T)$.

c. Given that $A = M + S$, where M is symmetric and S is skew-symmetric, show that $A - A^T = 2S$, so that $S = \frac{1}{2}(A - A^T)$.

d. Show that for any square matrix A, $M = \frac{1}{2}(A + A^T)$ is a symmetric matrix.

e. Show that for any square matrix A, $S = \frac{1}{2}(A - A^T)$ is skew-symmetric.

f. Show that $A = M + S$, where the matrices M and S are as in parts (d) and (e).

g. Write the following matrix as the sum of a symmetric matrix and a skew-symmetric matrix:

$$\begin{bmatrix} 1 & 2 & 3 & 4 \\ 5 & 6 & 7 & 8 \\ 9 & 10 & 11 & 12 \\ 13 & 14 & 15 & 16 \end{bmatrix}.$$

3.2 Matrix–Vector Multiplication

In Section 2.2, a matrix was introduced as a shorthand way of representing a system of linear equations in terms of its significant components:

$$
\begin{array}{l}
a_{11}x_1 + a_{12}x_2 + \cdots + a_{1n}x_n = b_1 \\
a_{21}x_1 + a_{22}x_2 + \cdots + a_{2n}x_n = b_2 \\
\quad \vdots \qquad\quad \vdots \qquad \vdots \qquad \vdots \\
a_{m1}x_1 + a_{m2}x_2 + \ldots + a_{mn}x_n = b_m
\end{array}
\Leftrightarrow
\begin{bmatrix}
a_{11} & a_{12} & \cdots & a_{1n} \\
a_{21} & a_{22} & \cdots & a_{2n} \\
\vdots & \vdots & \vdots & \vdots \\
a_{m1} & a_{m2} & \cdots & a_{mn}
\end{bmatrix}
\begin{bmatrix}
x_1 \\ x_2 \\ \vdots \\ x_n
\end{bmatrix}
$$

$$
=
\begin{bmatrix}
b_1 \\ b_2 \\ \vdots \\ b_m
\end{bmatrix}
\Leftrightarrow A\mathbf{x} = \mathbf{b}.
$$

The juxtaposition of matrix and vector in the middle and last equations strongly suggests a *multiplication* of the vector by the matrix. Theorem 2.7.1 ($A\mathbf{x} = \mathbf{b}$ has a solution if and only if $\mathbf{b}$ is a linear combination of the columns of A) provides motivation for a rigorous definition of matrix–vector multiplication.

Definition 3.2.1(a)

If A is an $m \times n$ matrix whose jth column is denoted by $\mathbf{a}_j$ and

$$\mathbf{x} = \begin{bmatrix} x_1 \\ x_2 \\ \vdots \\ x_n \end{bmatrix}, \text{ then we define the } \textbf{product of } A \textbf{ and } \mathbf{x} \text{ (or the } \textbf{matrix–}$$

vector product) $A\mathbf{x}$ as follows: $A\mathbf{x} = x_1\mathbf{a}_1 + x_2\mathbf{a}_2 + \cdots + x_n\mathbf{a}_n$.

In other words, the product $A\mathbf{x}$ is defined to be a vector, the linear combination of the columns of matrix A with the coefficient (or "weight") of column $\mathbf{a}_i$ equal to component x_i of vector $\mathbf{x}$, $i = 1, 2, \ldots, n$. This idea of matrix–vector multiplication is not limited to the representation of systems of linear equations. We can define such a product in general, *provided that the number of columns of the matrix is equal to the number of components of the vector.* $\overset{m \times n}{A} \,\overset{n \times 1}{\mathbf{x}} = \overset{m \times 1}{\mathbf{b}}$. The size of the product will be (the number of rows of A) $\times 1$.

Example 3.2.1: Matrix–Vector Multiplication

Suppose that

$$A = \begin{bmatrix} -1 & 4 \\ 0 & 5 \end{bmatrix}, \quad B = \begin{bmatrix} 1 & 2 & 3 \\ -3 & 2 & -1 \\ 0 & 1 & 2 \end{bmatrix}, \quad C = \begin{bmatrix} 3 & 4 & 2 \\ 5 & 7 & 3 \\ 1 & 2 & 1 \\ 3 & 3 & 3 \end{bmatrix},$$

$$\mathbf{x} = \begin{bmatrix} -2 \\ 3 \end{bmatrix}, \text{ and } \mathbf{y} = \begin{bmatrix} 2 \\ -1 \\ 3 \end{bmatrix}. \text{ Then,}$$

$$A\mathbf{x} = (-2)\begin{bmatrix} -1 \\ 0 \end{bmatrix} + 3\begin{bmatrix} 4 \\ 5 \end{bmatrix} = \begin{bmatrix} 14 \\ 15 \end{bmatrix},$$

$$B\mathbf{y} = 2\begin{bmatrix} 1 \\ -3 \\ 0 \end{bmatrix} + (-1)\begin{bmatrix} 2 \\ 2 \\ 1 \end{bmatrix} + 3\begin{bmatrix} 3 \\ -1 \\ 2 \end{bmatrix} = \begin{bmatrix} 9 \\ -11 \\ 5 \end{bmatrix},$$

and

$$C\mathbf{y} = 2\begin{bmatrix} 3 \\ 5 \\ 1 \\ 3 \end{bmatrix} + (-1)\begin{bmatrix} 4 \\ 7 \\ 2 \\ 3 \end{bmatrix} + 3\begin{bmatrix} 2 \\ 3 \\ 1 \\ 3 \end{bmatrix} = \begin{bmatrix} 8 \\ 12 \\ 3 \\ 12 \end{bmatrix}.$$

A careful examination of Definition 3.2.1(a) and Example 3.2.1 indicates that component i of the matrix–vector product $A\mathbf{x}$ is actually the dot product of the transpose of row i of A (where the transposed row is interpreted as a vector in $\mathbb{R}^n$) and $\mathbf{x}$. This observation provides an alternative (but equivalent) view of the product of a matrix and a vector.

Definition 3.2.1(b)

Given any $m \times n$ matrix $A = \begin{bmatrix} a_{11} & a_{12} & \cdots & a_{1n} \\ a_{21} & a_{22} & \cdots & a_{2n} \\ \vdots & \vdots & \vdots & \vdots \\ a_{m1} & a_{m2} & \cdots & a_{mn} \end{bmatrix}$ and any vector

$\mathbf{x} = \begin{bmatrix} x_1 \\ x_2 \\ \vdots \\ x_n \end{bmatrix}$ in $\mathbb{R}^n$, the product $A\mathbf{x}$ is defined as the vector $\mathbf{b} = \begin{bmatrix} b_1 \\ b_2 \\ \vdots \\ b_m \end{bmatrix}$ in

$\mathbb{R}^m$, where

$$b_i = [a_{i1} \ a_{i2} \ldots a_{in}]^{\mathrm{T}} \bullet \begin{bmatrix} x_1 \\ x_2 \\ \vdots \\ x_n \end{bmatrix} = a_{i1}x_1 + a_{i2}x_2 + \cdots + a_{in}x_n \ \text{ for } \ i =$$

$1, 2, \ldots, m$—that is, component i of the product vector is the dot product of the $n \times 1$ transpose of row i of A with the column vector $\mathbf{x}$.

Example 3.2.2: Matrix–Vector Multiplication

Let us use matrices B and C and vector $\mathbf{y}$ from Example 3.2.1. Then,

$$\begin{bmatrix} 1 & 2 & 3 \\ -3 & 2 & -1 \\ 0 & 1 & 2 \end{bmatrix} \begin{bmatrix} 2 \\ -1 \\ 3 \end{bmatrix} = \begin{bmatrix} b_1 \\ b_2 \\ b_3 \end{bmatrix},$$

where

$$b_1 = [1 \quad 2 \quad 3]^{\mathrm{T}} \bullet \begin{bmatrix} 2 \\ -1 \\ 3 \end{bmatrix} = 1(2) + 2(-1) + 3(3) = 9,$$

$$b_2 = [-3 \quad 2 \quad -1]^{\mathrm{T}} \bullet \begin{bmatrix} 2 \\ -1 \\ 3 \end{bmatrix} = -3(2) + 2(-1) + (-1)(3) = -11,$$

$$b_3 = [0 \quad 1 \quad 2]^{\mathrm{T}} \bullet \begin{bmatrix} 2 \\ -1 \\ 3 \end{bmatrix} = 0(2) + 1(-1) + 2(3) = 5.$$

Thus, we can write

$$\overbrace{\begin{bmatrix} 1 & 2 & 3 \\ -3 & 2 & -1 \\ 0 & 1 & 2 \end{bmatrix}}^{3 \times 3} \overbrace{\begin{bmatrix} 2 \\ -1 \\ 3 \end{bmatrix}}^{3 \times 1} = \overbrace{\begin{bmatrix} 9 \\ -11 \\ 5 \end{bmatrix}}^{3 \times 1}.$$

Similarly, $\begin{bmatrix} 3 & 4 & 2 \\ 5 & 7 & 3 \\ 1 & 2 & 1 \\ 3 & 3 & 3 \end{bmatrix} \begin{bmatrix} 2 \\ -1 \\ 3 \end{bmatrix} = \begin{bmatrix} b_1 \\ b_2 \\ b_3 \\ b_4 \end{bmatrix}$, where

$$b_1 = \begin{bmatrix} 3 & 4 & 2 \end{bmatrix}^{\mathsf{T}} \bullet \begin{bmatrix} 2 \\ -1 \\ 3 \end{bmatrix} = 3(2) + 4(-1) + 2(3) = 8,$$

$$b_2 = \begin{bmatrix} 5 & 7 & 3 \end{bmatrix}^{\mathsf{T}} \bullet \begin{bmatrix} 2 \\ -1 \\ 3 \end{bmatrix} = 5(2) + 7(-1) + 3(3) = 12,$$

$$b_3 = \begin{bmatrix} 1 & 2 & 1 \end{bmatrix}^{\mathsf{T}} \bullet \begin{bmatrix} 2 \\ -1 \\ 3 \end{bmatrix} = 1(2) + 2(-1) + 1(3) = 3,$$

$$b_4 = \begin{bmatrix} 3 & 3 & 3 \end{bmatrix}^{\mathsf{T}} \bullet \begin{bmatrix} 2 \\ -1 \\ 3 \end{bmatrix} = 3(2) + 3(-1) + 3(3) = 12,$$

so $\overbrace{\begin{bmatrix} 3 & 4 & 2 \\ 5 & 7 & 3 \\ 1 & 2 & 1 \\ 3 & 3 & 3 \end{bmatrix}}^{4 \times 3} \overbrace{\begin{bmatrix} 2 \\ -1 \\ 3 \end{bmatrix}}^{3 \times 1} = \overbrace{\begin{bmatrix} 8 \\ 12 \\ 3 \\ 12 \end{bmatrix}}^{4 \times 1}.$

As the next example shows, matrix–vector products are useful in dealing with certain important algebraic expressions.

Example 3.2.3: Quadratic Forms

In analytic geometry, any conic section (ellipse, parabola, hyperbola) or degenerate case of a conic section (for example, a circle) can be represented by a quadratic equation in two variables as

$$ax^2 + bxy + cy^2 + dx + ey + f = 0.$$

The expression $ax^2 + bxy + cy^2$ is called the *quadratic form** associated with the quadratic equation, and it can be represented in terms of matrix–vector multiplication as

$$\begin{bmatrix} x & y \end{bmatrix} \begin{bmatrix} a & b/2 \\ b/2 & c \end{bmatrix} \begin{bmatrix} x \\ y \end{bmatrix},$$

---

* In general, a **quadratic form** in n variables $x_1, x_2, \ldots, x_n$ is a function $Q(x_1, x_2, \ldots, x_n) = \sum_{i=1}^{n} \sum_{j=1}^{n} a_{ij} x_i x_j$, where a_{ij} is a constant for $i = 1, 2, \ldots, n$ and $j = 1, 2, \ldots, n$. In physics, for example, the quadratic form $(m/2)(u^2 + v^2 + w^2)$ expresses the kinetic energy of a moving body in space with three velocity components u, v, and w.

where $[x \quad y] = \begin{bmatrix} x \\ y \end{bmatrix}^{\mathsf{T}}$ is considered a 1×2 (row) matrix and the product is interpreted as

$$\overbrace{[x \quad y]}^{\text{matrix}} \left(\overbrace{\begin{bmatrix} a & b/2 \\ b/2 & c \end{bmatrix} \begin{bmatrix} x \\ y \end{bmatrix}}^{\text{vector}} \right).$$

For example, the graph of the quadratic equation:

$$5x^2 + 6xy + 5y^2 - 16x - 16y + 8 = 0$$

is an ellipse with center $(1, 1)$:

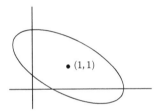

The entire quadratic equation—the quadratic form $5x^2 + 6xy + 5y^2$ plus the linear terms $-16x - 16y + 8$—can be represented in the form:

$$[x \quad y] \begin{bmatrix} 5 & 3 \\ 3 & 5 \end{bmatrix} \begin{bmatrix} x \\ y \end{bmatrix} - [16 \quad 16] \begin{bmatrix} x \\ y \end{bmatrix} + 8 = 0.$$

Similarly, in $\mathbb{R}^3$, $9x_1^2 - x_2^2 + 4x_3^2 + 6x_1x_2 - 8x_1x_3 + 2x_2x_3$ is a quadratic form that we can express as a matrix–vector product $\mathbf{x}^{\mathsf{T}}A\mathbf{x}$:

$$9x_1^2 - x_2^2 + 4x_3^2 + 6x_1x_2 - 8x_1x_3 + 2x_2x_3$$

$$= [x_1 \quad x_2 \quad x_3] \begin{bmatrix} 9 & 3 & -4 \\ 3 & -1 & 1 \\ -4 & 1 & 4 \end{bmatrix} \begin{bmatrix} x_1 \\ x_2 \\ x_3 \end{bmatrix}.$$

We see that the quadratic form $Q(\mathbf{x}) = \mathbf{x}^{\mathsf{T}}A\mathbf{x}$ defines a *scalar-valued function*, $Q: \mathbb{R}^3 \to \mathbb{R}$. (Alternatively, $Q(\mathbf{x})$ can be interpreted as a one-component vector.) Note that the matrix A in this example is symmetric.

Quadratic forms occur frequently in mathematics, statistics, physics, and engineering disciplines. We will continue to work with these important functions in Chapters 4, 7, and (especially) 8.

For any positive integer n, a particularly significant $n \times n$ matrix $[a_{ij}]$ is the one that has all its *main diagonal* entries $a_{11}, a_{22}, \ldots, a_{nn}$ equal to 1 and all other entries equal to 0. This $n \times n$ matrix is called an **identity matrix** and is denoted by I_n. It has the important property that $I_n\mathbf{x} = \mathbf{x}$ for every vector $\mathbf{x} \in \mathbb{R}^n$: According to Definition 3.2.1(b), if $\mathbf{x} = [x_1 \ x_2 \ldots x_n]^T \in \mathbb{R}^n$, component i of the product $I_n\mathbf{x}$ is given

by $\left[\begin{matrix} 0 & 0 & \ldots & \overbrace{1}^{\text{component } i} & 0 & \ldots & 0 \end{matrix}\right]^T \bullet \begin{bmatrix} x_1 \\ x_2 \\ \vdots \\ x_n \end{bmatrix} = 0 \cdot x_1 + 0 \cdot x_2 + \cdots +$

$\overbrace{1 \cdot x_i}^{\text{term } i} + \cdots + 0 \cdot x_n = x_i$, the ith component of $\mathbf{x}$ $(i = 1, 2, \ldots, n)$.

Matrices cI_n that are obtained by multiplying I_n by a real number c are called **scalar matrices**. A scalar matrix is a special case of a **diagonal matrix**, a square matrix whose only nonzero entries lie along the main diagonal. In general, we represent a diagonal matrix by specifying its main diagonal entries as follows:

$$\text{diag}(d_1, \ d_2, \ \ldots, \ d_n) = \begin{bmatrix} d_1 & 0 & 0 & \ldots & 0 \\ 0 & d_2 & 0 & \ldots & 0 \\ \vdots & \vdots & \ddots & \ldots & \vdots \\ 0 & 0 & \ldots & d_{n-1} & 0 \\ 0 & 0 & \ldots & 0 & d_n \end{bmatrix}$$

A square matrix is called an **upper triangular matrix** if all its entries below the main diagonal are zero. An $n \times n$ matrix in echelon or reduced echelon form is an upper triangular matrix. Similarly, a square matrix is called a **lower triangular matrix** if all its entries above the main diagonal are zero.

Example 3.2.4(a): Identity Matrices, Scalar Matrices, and Diagonal Matrices

We have

$$I_2 = \begin{bmatrix} 1 & 0 \\ 0 & 1 \end{bmatrix}, \ I_3 = \begin{bmatrix} 1 & 0 & 0 \\ 0 & 1 & 0 \\ 0 & 0 & 1 \end{bmatrix}, \ I_4 = \begin{bmatrix} 1 & 0 & 0 & 0 \\ 0 & 1 & 0 & 0 \\ 0 & 0 & 1 & 0 \\ 0 & 0 & 0 & 1 \end{bmatrix},$$

and so forth.

We see that $\begin{bmatrix} 1 & 0 & 0 \\ 0 & 1 & 0 \\ 0 & 0 & 1 \end{bmatrix} \begin{bmatrix} -2 \\ 3 \\ -5 \end{bmatrix} = \begin{bmatrix} 1(-2) + 0(3) + 0(-5) \\ 0(-2) + 1(3) + 0(-5) \\ 0(-2) + 0(3) + 1(-5) \end{bmatrix} =$

$\begin{bmatrix} -2 \\ 3 \\ -5 \end{bmatrix}$, for example.

The matrices

$$\begin{bmatrix} -3 & 0 & 0 \\ 0 & -3 & 0 \\ 0 & 0 & -3 \end{bmatrix} = (-3)I_3 \text{ and } \begin{bmatrix} 0.7 & 0 & 0 & 0 & 0 \\ 0 & 0.7 & 0 & 0 & 0 \\ 0 & 0 & 0.7 & 0 & 0 \\ 0 & 0 & 0 & 0.7 & 0 \\ 0 & 0 & 0 & 0 & 0.7 \end{bmatrix} = 0.7 I_5$$

are scalar matrices. Notice what multiplication by a scalar matrix does to a vector:

$$\begin{bmatrix} -3 & 0 & 0 \\ 0 & -3 & 0 \\ 0 & 0 & -3 \end{bmatrix} \begin{bmatrix} x \\ y \\ z \end{bmatrix} = \begin{bmatrix} -3x \\ -3y \\ -3z \end{bmatrix} = -3 \begin{bmatrix} x \\ y \\ z \end{bmatrix}, \text{ or } (-3I_3)\mathbf{x} = (-3)(I_3\mathbf{x})$$

$= (-3)\mathbf{x}$. The matrices

$$\begin{bmatrix} 1 & 0 \\ 0 & -5 \end{bmatrix}, \begin{bmatrix} 3 & 0 & 0 \\ 0 & 0 & 0 \\ 0 & 0 & 7 \end{bmatrix}, \begin{bmatrix} -1 & 0 & 0 & 0 \\ 0 & 2 & 0 & 0 \\ 0 & 0 & \pi & 0 \\ 0 & 0 & 0 & -4 \end{bmatrix}$$

are general diagonal matrices. We can abbreviate the last matrix, for example, as diag$(-1, 2, \pi, -4)$.

Example 3.2.4(b): Upper and Lower Triangular Matrices

The square matrices $\begin{bmatrix} 1 & -2 \\ 0 & 0 \end{bmatrix}, \begin{bmatrix} 1 & 0 & -5 \\ 0 & 2 & 4 \\ 0 & 0 & 3 \end{bmatrix}$, and

$$\begin{bmatrix} 0 & 1 & 2 & 3 \\ 0 & -1 & 7 & -2 \\ 0 & 0 & 0 & 4 \\ 0 & 0 & 0 & 2 \end{bmatrix}$$ are upper triangular matrices, whereas

the matrices $\begin{bmatrix} 1 & 0 \\ 2 & 3 \end{bmatrix}, \begin{bmatrix} -3 & 0 & 0 \\ 2 & -2 & 0 \\ 0 & 5 & -1 \end{bmatrix}$, and $\begin{bmatrix} 1 & 0 & 0 & 0 \\ 2 & 0 & 0 & 0 \\ 3 & 4 & 5 & 0 \\ 6 & 7 & 8 & 9 \end{bmatrix}$

are lower triangular matrices. We should note that some or all the diagonal elements of upper and lower triangular matrices may be zero. All that matters for an upper or a lower triangular matrix is that all the entries *below* or *above* the main diagonal, respectively, should be zero.

Note that if a square matrix is both upper triangular and lower triangular, then it must be a diagonal matrix.

Although for any positive integer n, we described "an" identity matrix I_n, the identity matrix is in fact uniquely determined by the property that it plays the same role in matrix–vector multiplication that the scalar 1 plays in ordinary multiplication.

Theorem 3.2.1

For any positive integer n, the matrix I_n is the *only* $n \times n$ matrix M with the property that $M\mathbf{x} = \mathbf{x}$ for every vector $\mathbf{x} \in \mathbb{R}^n$.

Proof Suppose that $M = [m_{ij}]$ is an $n \times n$ matrix such that $M\mathbf{x} = \mathbf{x}$ for every $\mathbf{x} \in \mathbb{R}^n$. Choose

$$\mathbf{x} = \mathbf{e}_r = \begin{bmatrix} 0 \\ 0 \\ \vdots \\ 1 \\ \vdots \\ 0 \end{bmatrix},$$

where the 1 appears in row r and all other components are 0. Then if $i \neq r$,

$$0 = (\text{component } i \text{ of the product vector } \mathbf{e}_r = M\mathbf{e}_r)$$

$$= [m_{i1} \ m_{i2} \ldots, m_{in}]^{\mathrm{T}} \cdot \begin{bmatrix} 0 \\ 0 \\ \vdots \\ 1 \\ \vdots \\ 0 \end{bmatrix}$$

$$= m_{i1} \cdot 0 + \cdots + m_{i,(r-1)} \cdot 0 + m_{i,r} \cdot 1 + m_{i,(r+1)} \cdot 0 + \cdots + m_{in} \cdot 0 = m_{ir}$$

Similarly, we find that when $i = r$,

$$1 = (\text{component } i \text{ of the product vector } \mathbf{e}_r = M\mathbf{e}_r) = m_{ir}.$$

Because we can let $r = 1, 2, \ldots, n$ in defining $\mathbf{e}_r$, we have shown that
$m_{ij} = \begin{cases} 0 & \text{if } i \neq j \\ 1 & \text{if } i = j \end{cases}$. That is, $M = I_n$.

3.2.1 Matrix–Vector Multiplication as a Transformation

The product of a vector by a matrix has a more *dynamical*, a more action-oriented, interpretation. In particular, multiplying a vector in $\mathbb{R}^n$ by an $n \times n$ matrix has the effect of transforming the vector in geometrical ways—for example, by changing the vector's direction and/or altering its length within $\mathbb{R}^n$. We shall examine this geometric interpretation in greater detail in Chapters 4, 6, and 7.

Example 3.2.5: A Geometric View of Matrix–Vector Multiplication

The vector $\mathbf{x} = \begin{bmatrix} 1 \\ -3 \end{bmatrix}$, whose length is $\sqrt{10}$ and which makes an angle of approximately $71.6°$ with the positive x-axis (by Definition 1.2.3), can be pictured geometrically in $\mathbb{R}^2$ as

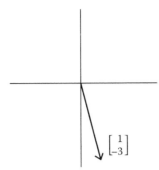

If we multiply this vector by the matrix $A = \begin{bmatrix} 0 & -2 \\ 2 & 0 \end{bmatrix}$, we see

that $\begin{bmatrix} 0 & -2 \\ 2 & 0 \end{bmatrix} \begin{bmatrix} 1 \\ -3 \end{bmatrix} = \begin{bmatrix} 6 \\ 2 \end{bmatrix} = 2 \cdot \begin{bmatrix} 3 \\ 1 \end{bmatrix}$. Geometrically, multi-

plying **x** by A has *transformed* **x** into the vector $\begin{bmatrix} 6 \\ 2 \end{bmatrix}$, which in

this case is the original vector rotated 90° in a counterclockwise direction and doubled in length:

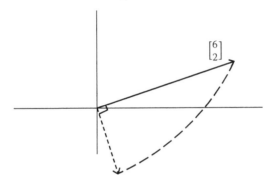

(In this case, the dot product of the original vector and the transformed vector is zero, so the two vectors are orthogonal. Alternatively, we can use Definition 1.2.3 to calculate

that $\begin{bmatrix} 6 \\ 2 \end{bmatrix}$ makes an angle of approximately 18.4° with the

positive x-axis, making the angle between the two vectors $71.6° + 18.4° = 90°$.)

More generally, the multiplication of a vector in $\mathbb{R}^n$ by an $m \times n$ matrix A defines a function, or **transformation**, T_A from $\mathbb{R}^n$ into $\mathbb{R}^m$. For any fixed $m \times n$ matrix A, the function $T_A \colon \mathbb{R}^n \to \mathbb{R}^m$ is defined by

$T_A(\mathbf{v}) = A\mathbf{v}$ for all $\mathbf{v} = \begin{bmatrix} x_1 \\ x_2 \\ \vdots \\ x_n \end{bmatrix} \in \mathbb{R}^n$. If we have a second matrix B, which is

$p \times m$, then we can define a function $T_B \colon \mathbb{R}^m \to \mathbb{R}^p$ by $T_B(\mathbf{x}) = B\mathbf{x}$ for all $\mathbf{x} \in \mathbb{R}^m$. If we start with $\mathbf{v} \in \mathbb{R}^n$, then $A\mathbf{v}$ is a vector in $\mathbb{R}^m$, which we can multiply by B. The final result is a vector $B(A\mathbf{v})$ in $\mathbb{R}^p$. If we write $B(A\mathbf{v}) = (BA)\mathbf{v}$, we gain some insight into what the *product of matrices* might mean—it could represent a *composition of functions* (*transformations*), $T_B \circ T_A$. We can express this graphically as follows:

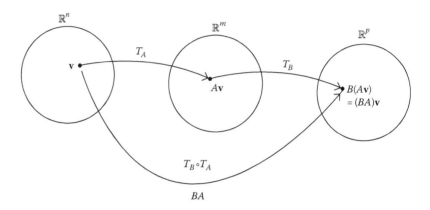

Section 3.3, especially Corollary 3.3.1, continues this line of thought.

The transformations discussed in the last paragraph have important and useful properties.

Theorem 3.2.2

If A is a fixed $m \times n$ matrix and $T_A : \mathbb{R}^n \to \mathbb{R}^m$ is the transformation defined by

$T_A(\mathbf{v}) = A\mathbf{v}$ for all $\mathbf{v} \in \mathbb{R}^n$, then

(1) $T_A(\mathbf{v}_1 + \mathbf{v}_2) = T_A(\mathbf{v}_1) + T_A(\mathbf{v}_2)$ for any vectors $\mathbf{v}_1$ and $\mathbf{v}_2$ in $\mathbb{R}^n$. [*Additivity*]
(2) $T_A(k\mathbf{v}) = kT_A(\mathbf{v})$ for any vector $\mathbf{v}$ in $\mathbb{R}^n$ and any real number k. [*Homogeneity*]

Proof of (1)   Let $\mathbf{v}_1 = \begin{bmatrix} x_1 \\ x_2 \\ \vdots \\ x_n \end{bmatrix}$ and $\mathbf{v}_2 = \begin{bmatrix} y_1 \\ y_2 \\ \vdots \\ y_n \end{bmatrix}$ be arbitrary vectors

in $\mathbb{R}^n$. Then

$$\mathbf{v}_1 + \mathbf{v}_2 = \begin{bmatrix} x_1 + y_1 \\ x_2 + y_2 \\ \vdots \\ x_n + y_n \end{bmatrix}$$ and, by Definition 3.2.1(b), component i of

$T_A(\mathbf{v}_1 + \mathbf{v}_2) = A(\mathbf{v}_1 + \mathbf{v}_2) \in \mathbb{R}^m$ is given by

$$b_i = [a_{i1} \ a_{i2} \ \ldots \ a_{in}]^{\text{T}} \cdot \begin{bmatrix} x_1 + y_1 \\ x_2 + y_2 \\ \vdots \\ x_n + y_n \end{bmatrix} = a_{i1}(x_1 + y_1) + a_{i2}(x_2 + y_2) + \cdots$$

$$+ a_{in}(x_n + y_n) = (a_{i1}x_1 + a_{i2}x_2 + \cdots + a_{in}x_n) + (a_{i1}y_1 + a_{i2}y_2 + \cdots$$

$$+ a_{in}y_n) = [a_{i1} \ a_{i2} \ \ldots \ a_{in}]^{\text{T}} \cdot \begin{bmatrix} x_1 \\ x_2 \\ \vdots \\ x_n \end{bmatrix} + [a_{i1} \ a_{i2} \ \ldots \ a_{in}]^{\text{T}} \cdot \begin{bmatrix} y_1 \\ y_2 \\ \vdots \\ y_n \end{bmatrix}$$

$$= (\text{component } i \text{ of } T_A(\mathbf{v}_1)) + (\text{component } i \text{ of } T_A(\mathbf{v}_1)) \text{ for } i = 1, 2, \ldots, m.$$

Proof of (2)   Similarly, the ith component of $T_A(k\mathbf{v}) = A(k\mathbf{v}) \in \mathbb{R}^m$ is given by

$$b_i = [a_{i1} \ a_{i2} \ \ldots \ a_{in}]^{\text{T}} \cdot \begin{bmatrix} kx_1 \\ kx_2 \\ \vdots \\ kx_n \end{bmatrix} = a_{i1}(kx_1) + a_{i2}(kx_2) + \cdots + a_{in}(kx_n)$$

$$= k\{a_{i1}x_1 + a_{i2}x_2 + \cdots + a_{in}x_n\}$$

$$= k[a_{i1} \ a_{i2} \ \ldots \ a_{in}]^{\text{T}} \cdot \begin{bmatrix} x_1 \\ x_2 \\ \vdots \\ x_n \end{bmatrix}$$

$$= k(\text{component } i \text{ of } T_A(\mathbf{v})).$$

Any function satisfying properties (1) and (2) of Theorem 3.2.2 is called a **linear transformation**. In Chapter 6, we will explore the significance of such transformations in greater depth.

Example 3.2.6: Matrix–Vector Multiplication as a Linear Transformation

Let $A = \begin{bmatrix} 1 & 2 \\ 3 & 4 \end{bmatrix}$, $\mathbf{v}_1 = \begin{bmatrix} -2 \\ 3 \end{bmatrix}$, and $\mathbf{v}_2 = \begin{bmatrix} 4 \\ 5 \end{bmatrix}$. Define $T_A(\mathbf{v}) = A\mathbf{v}$ for all $\mathbf{v} \in \mathbb{R}^2$. Then

$$T_A(\mathbf{v}_1 + \mathbf{v}_2) = T_A\left(\begin{bmatrix} -2 \\ 3 \end{bmatrix} + \begin{bmatrix} 4 \\ 5 \end{bmatrix}\right) = T_A\left(\begin{bmatrix} 2 \\ 8 \end{bmatrix}\right) = A \cdot \begin{bmatrix} 2 \\ 8 \end{bmatrix}$$

$$= \begin{bmatrix} 1 & 2 \\ 3 & 4 \end{bmatrix} \cdot \begin{bmatrix} 2 \\ 8 \end{bmatrix} = \begin{bmatrix} 18 \\ 38 \end{bmatrix},$$

$$T_A(\mathbf{v}_1) = A \cdot \mathbf{v}_1 = \begin{bmatrix} 1 & 2 \\ 3 & 4 \end{bmatrix} \cdot \begin{bmatrix} -2 \\ 3 \end{bmatrix} = \begin{bmatrix} 4 \\ 6 \end{bmatrix}, \text{ and } T_A(\mathbf{v}_2)$$

$$= A \cdot \mathbf{v}_2 = \begin{bmatrix} 1 & 2 \\ 3 & 4 \end{bmatrix} \cdot \begin{bmatrix} 4 \\ 5 \end{bmatrix} = \begin{bmatrix} 14 \\ 32 \end{bmatrix}.$$

We see that $T_A(\mathbf{v}_1 + \mathbf{v}_2) = \begin{bmatrix} 18 \\ 38 \end{bmatrix} = \begin{bmatrix} 4 \\ 6 \end{bmatrix} + \begin{bmatrix} 14 \\ 32 \end{bmatrix}$

$$= T_A(\mathbf{v}_1) + T_A(\mathbf{v}_2). \text{ Also, for}$$

example, $T_A(5 \cdot \mathbf{v}_1) = T_A\left(\begin{bmatrix} -10 \\ 15 \end{bmatrix}\right) = \begin{bmatrix} 1 & 2 \\ 3 & 4 \end{bmatrix} \cdot \begin{bmatrix} -10 \\ 15 \end{bmatrix}$

$$= \begin{bmatrix} 20 \\ 30 \end{bmatrix} = 5 \cdot \begin{bmatrix} 4 \\ 6 \end{bmatrix} = 5 \cdot T_A(\mathbf{v}_1).$$

We can combine properties (1) and (2) of Theorem 3.2.2 to write $T_A(c_1\mathbf{v}_1 + c_2\mathbf{v}_2) = c_1 T_A(\mathbf{v}_1) + c_2 T_A(\mathbf{v}_2)$ for any scalars c_1 and c_2. Furthermore, we can extend this formula to any finite linear combination of vectors in $\mathbb{R}^n$:

$$T_A(c_1\mathbf{v}_1 + c_2\mathbf{v}_2 + \cdots + c_n\mathbf{v}_n) = c_1 T_A(\mathbf{v}_1) + c_2 T_A(\mathbf{v}_2) + \cdots + c_n T_A(\mathbf{v}_n)$$

(See Exercise B9.)

We should note that Theorem 3.2.2 establishes two useful properties of matrix–vector multiplication:

(i) $A(\mathbf{v}_1 + \mathbf{v}_2) = A\mathbf{v}_1 + A\mathbf{v}_2$
(ii) $A(k\mathbf{v}) = kA(\mathbf{v})$

for any $m \times n$ matrix A, any vectors $\mathbf{v}_1$, $\mathbf{v}_2$, and $\mathbf{v}$ in $\mathbb{R}^n$, and $k \in \mathbb{R}$. We can generalize this as we did in the preceding paragraph:

$$A(c_1\mathbf{v}_1 + c_2\mathbf{v}_2 + \cdots + c_n\mathbf{v}_n) = c_1(A\mathbf{v}_1) + c_2(A\mathbf{v}_2) + \cdots + c_n(A\mathbf{v}_n).$$

Exercises 3.2

A.

In Exercises 1–8, find the product $A\mathbf{x}$ if possible. If you cannot find the product, give a reason.

1. $A = \begin{bmatrix} 1 & 2 \\ 3 & 4 \end{bmatrix}, \quad \mathbf{x} = \begin{bmatrix} -3 \\ -1 \end{bmatrix}$

2. $A = \begin{bmatrix} -4 & 0 \\ 1 & -2 \end{bmatrix}, \quad \mathbf{x} = \begin{bmatrix} 0 \\ 5 \end{bmatrix}$

3. $A = \begin{bmatrix} 0 & -1 & 2 \\ -3 & 4 & 1 \\ 5 & 0 & -2 \end{bmatrix}$, $\mathbf{x} = \begin{bmatrix} -2 \\ 5 \end{bmatrix}$

4. $A = \begin{bmatrix} 1 & -2 & 3 \\ 0 & 4 & 5 \\ -3 & 6 & 0 \end{bmatrix}$, $\mathbf{x} = \begin{bmatrix} 1 \\ 2 \\ 3 \end{bmatrix}$

5. $A = \begin{bmatrix} 0 & 1 & 0 & 2 \\ -1 & 3 & -2 & 0 \\ 4 & 0 & 1 & -3 \\ 5 & 4 & 3 & 1 \end{bmatrix}$, $\mathbf{x} = \begin{bmatrix} 1 \\ 2 \\ 3 \\ 4 \end{bmatrix}$

6. $A = \begin{bmatrix} -2 & 0 & 0 & 0 & 0 \\ 0 & -2 & 0 & 0 & 0 \\ 0 & 0 & -2 & 0 & 0 \\ 0 & 0 & 0 & -2 & 0 \\ 0 & 0 & 0 & 0 & -2 \end{bmatrix}$, $\mathbf{x} = \begin{bmatrix} 1 \\ 2 \\ 3 \\ 4 \\ 5 \end{bmatrix}$

7. $A = \begin{bmatrix} 1 & -2 & 4 & 3 & -1 \\ 0 & 2 & 1 & 0 & 4 \\ 0 & 0 & 3 & -3 & 1 \\ 0 & 0 & 0 & 4 & 2 \\ 0 & 0 & 0 & 0 & 5 \end{bmatrix}$, $\mathbf{x} = \begin{bmatrix} -1 \\ 2 \\ -3 \\ 4 \\ -5 \end{bmatrix}$

8. $A = \begin{bmatrix} 1 & -2 & 4 & 3 & -1 \\ 0 & 2 & 1 & 0 & 4 \\ 0 & 0 & 3 & -3 & 1 \\ 0 & 0 & 0 & 4 & 2 \\ 0 & 0 & 0 & 0 & 5 \end{bmatrix}$, $\mathbf{x} = \begin{bmatrix} 1 \\ 0 \\ -5 \\ 3 \end{bmatrix}$

In Exercises 9–14, calculate $\mathbf{x}^\mathrm{T} A \mathbf{x}$ *as* $\mathbf{x}^\mathrm{T}(A\mathbf{x})$.

9. $A = \begin{bmatrix} 1 & 0 \\ -2 & 3 \end{bmatrix}$, $\mathbf{x} = \begin{bmatrix} 2 \\ 4 \end{bmatrix}$

10. $A = \begin{bmatrix} 2 & -4 \\ -4 & 1 \end{bmatrix}$, $\mathbf{x} = \begin{bmatrix} -3 \\ 5 \end{bmatrix}$

11. $A = \begin{bmatrix} 1 & -2 & 3 \\ 4 & 0 & 5 \\ -3 & 2 & -1 \end{bmatrix}$, $\mathbf{x} = \begin{bmatrix} 1 \\ 2 \\ 3 \end{bmatrix}$

12. $A = \begin{bmatrix} 0 & 1 & -2 \\ 1 & 0 & 3 \\ -2 & 3 & -1 \end{bmatrix}$, $\mathbf{x} = \begin{bmatrix} -4 \\ 5 \\ -6 \end{bmatrix}$

13. $A = \begin{bmatrix} 0 & 1 & 0 & 2 \\ -3 & 4 & -1 & 5 \\ 6 & -2 & 0 & -4 \\ 4 & 3 & 2 & 1 \end{bmatrix}$, $\mathbf{x} = \begin{bmatrix} 5 \\ 4 \\ 3 \\ -2 \end{bmatrix}$

14. $A = \begin{bmatrix} 1 & 0 & 0 & 0 \\ 0 & 1 & 0 & 0 \\ 0 & 0 & 1 & 0 \\ 0 & 0 & 0 & 1 \end{bmatrix}$, $\mathbf{x} = \begin{bmatrix} 1 \\ 2 \\ 3 \\ 4 \end{bmatrix}$

In Exercises 15–20, write each quadratic form as $\mathbf{x}^T A \mathbf{x}$ (see Example 3.2.3).

15. $4x^2 + 9y^2$

16. $x_1 x_2 + x_1 x_3 + x_2 x_3$

17. $13x^2 - 10xy + 13y^2$

18. $x_1^2 + x_2^2 + x_3^2$

19. $2x_1^2 - 4x_1 x_2 + 3x_2^2 - 5x_1 x_3 + 10x_3^2$

20. $x^2 + y^2 - z^2 - w^2 + 2xy - 10xw + 4zw$

In Exercises 21–26, plot the vectors $\mathbf{x}$ and $A\mathbf{x}$ on the same set of axes and describe the effect of the multiplication on $\mathbf{x}$ in terms of (a) any rotation (including direction) and (b) any change in length.

21. Use A and $\mathbf{x}$ from Exercise 1.

22. Use A and $\mathbf{x}$ from Exercise 2.

23. $A = \begin{bmatrix} \frac{\sqrt{2}}{2} & -\frac{\sqrt{2}}{2} \\ \frac{\sqrt{2}}{2} & \frac{\sqrt{2}}{2} \end{bmatrix}$, $\mathbf{x} = \begin{bmatrix} 1 \\ 0 \end{bmatrix}$

24. $A = \begin{bmatrix} -1 & 0 \\ 0 & -1 \end{bmatrix}$, $\mathbf{x} = \begin{bmatrix} 1 \\ 1 \end{bmatrix}$

25. $A = \begin{bmatrix} -4 & 0 \\ 0 & -4 \end{bmatrix}$, $\mathbf{x} = \begin{bmatrix} -2 \\ 3 \end{bmatrix}$

26. $A = \begin{bmatrix} 0 & -1 \\ 1 & 0 \end{bmatrix}$, $\mathbf{x} = \begin{bmatrix} -5 \\ -3 \end{bmatrix}$

*Given an $n \times n$ matrix A, if there exists a nonzero vector $\mathbf{x} \in \mathbb{R}^n$ and a scalar λ such that $A\mathbf{x} = \lambda\mathbf{x}$, we call $\mathbf{x}$ an **eigenvector** of A and λ an **eigenvalue** of A. Geometrically, this means that the "image" vector $A\mathbf{x}$ lies on the same straight line as $\mathbf{x}$ (in the opposite direction to $\mathbf{x}$ if $\lambda < 0$) and that the original vector is either stretched or shrunk by a factor of $|\lambda|$: $\|A\mathbf{x}\| = \|\lambda\mathbf{x}\| = |\lambda|\,\|\mathbf{x}\|$. Eigenvectors and eigenvalues will be discussed fully in Chapter 4 and subsequent chapters.*

27.	For each of the following matrices, determine which of the given vectors are eigenvectors. For each eigenvector, find the corresponding eigenvalue.

a.	$\begin{bmatrix} 6 & 0 \\ 1 & 9 \end{bmatrix}$; $\mathbf{x}_1 = \begin{bmatrix} -6 \\ 2 \end{bmatrix}$, $\mathbf{x}_2 = \begin{bmatrix} 1 \\ 0 \end{bmatrix}$, $\mathbf{x}_3 = \begin{bmatrix} 3 \\ -1 \end{bmatrix}$, $\mathbf{x}_4 = \begin{bmatrix} 0 \\ 0 \end{bmatrix}$

b.	$\begin{bmatrix} 2 & 1 \\ 2 & 3 \end{bmatrix}$; $\mathbf{x}_1 = \begin{bmatrix} 2 \\ 3 \end{bmatrix}$, $\mathbf{x}_2 = \begin{bmatrix} 1 \\ 2 \end{bmatrix}$, $\mathbf{x}_3 = \begin{bmatrix} -4 \\ -8 \end{bmatrix}$, $\mathbf{x}_4 = \begin{bmatrix} -1 \\ 2 \end{bmatrix}$

c.	$\begin{bmatrix} 1 & 1 & 0 \\ 1 & 0 & 1 \\ 0 & 1 & 1 \end{bmatrix}$; $\mathbf{x}_1 = \begin{bmatrix} 2 \\ 3 \\ 5 \end{bmatrix}$, $\mathbf{x}_2 = \begin{bmatrix} -2 \\ 0 \\ 2 \end{bmatrix}$, $\mathbf{x}_3 = \begin{bmatrix} 3 \\ 3 \\ 3 \end{bmatrix}$, $\mathbf{x}_4 = \begin{bmatrix} -1 \\ 2 \\ -1 \end{bmatrix}$

28.	If $A = [a_{ij}]$ is any $n \times n$ matrix and $\mathbf{e}_j$ is the column vector that has a 1 in the jth position and 0s everywhere, describe the following products (see the proof of Theorem 3.2.1).

a.	$\mathbf{e}_i \bullet (A\mathbf{e}_j)$

b.	$\mathbf{e}_i^T A \mathbf{e}_i$

c.	$\mathbf{e}_i^T A \mathbf{e}_j$

B.

1.	Find a 2×3 matrix $A \neq O_{23}$ and a vector $\mathbf{x} \neq \mathbf{0}$ in $\mathbb{R}^3$ such that $A\mathbf{x} = \mathbf{0}$.

2.	Let A and B be $m \times n$ matrices. Prove that $A = B$ if and only if

$$A\mathbf{x} = B\mathbf{x} \quad \text{for all } \mathbf{x} \in R^n.$$

3.	Experiment with different vectors and diagonal matrices to determine the effect of multiplying a vector by a diagonal matrix. Prove that your conclusion is correct for any $n \times n$ diagonal matrix and any vector in $\mathbb{R}^n$.

4.	Let A be the $m \times n$ matrix all of whose entries are 1. Describe $A\mathbf{x}$ for any $\mathbf{x} \in \mathbb{R}^n$ and prove that your description is correct for any positive integer values of m and n.

5.	Suppose that the following table indicates the proportion of the output of Industry i (in row i) that is required as input in order to produce one unit of output of Industry j (in column j), where $i, j = 1, 2, 3$.

	Industry 1	Industry 2	Industry 3
Industry 1	$\frac{1}{8}$	$\frac{1}{3}$	$\frac{1}{4}$
Industry 2	$\frac{1}{2}$	$\frac{1}{6}$	$\frac{1}{4}$
Industry 3	$\frac{1}{4}$	$\frac{1}{6}$	$\frac{1}{4}$

Thus, for example, $1/3$ of Industry 1's output is necessary to produce one unit of Industry 2's output. Now let

$$\mathbf{X} = \begin{bmatrix} x_1 \\ x_2 \\ x_3 \end{bmatrix}, \text{ where } x_i \text{ denotes the } total \ output \ of \ Industry \ i,$$

and $\mathbf{Y} = \begin{bmatrix} y_1 \\ y_2 \\ y_3 \end{bmatrix}$, where y_i represents the *output of Industry i sold to satisfy external demand*, $i = 1, 2, 3$. Assume the rule that **the total output for any Industry i equals the total input demand for that product plus the final (external) demand for it.**[*]

a. Express this rule in matrix–vector terms, as an equation of the form:

$$\mathbf{w} = M\mathbf{u} + \mathbf{v}$$

b. If the *final demand vector* is $\mathbf{Y} = \begin{bmatrix} 10 \\ 30 \\ 20 \end{bmatrix}$, use the formula found in part (a) to calculate the vector of industry outputs.

c. If the final demand vector changes to $\mathbf{Y} = \begin{bmatrix} 20 \\ 70 \\ 50 \end{bmatrix}$, calculate the new industry outputs.

6. Show that the quadratic equation $ax^2 + bxy + cy^2 + dx + ey + f = 0$ can be written as $\mathbf{x}^T A \mathbf{x} + \mathbf{z}^T \mathbf{x} + f = 0$ for appropriate vectors $\mathbf{x}$ and $\mathbf{z}$ and a symmetric matrix A. (The term f is a scalar.)

7. If T is a *linear transformation*, as described in Theorem 3.2.2, prove that
    a. $T(\mathbf{0}) = \mathbf{0}$.
    b. $T(\mathbf{u} - \mathbf{v}) = T(\mathbf{u}) - T(\mathbf{v})$ for all $\mathbf{u}$ and $\mathbf{v}$ in $\mathbb{R}^n$.

8. For each of the linear transformations $T_A : \mathbb{R}^n \to \mathbb{R}^n$ described below, find a matrix A such that $T_A(\mathbf{v}) = A\mathbf{v}$ for all $\mathbf{v} \in \mathbb{R}^n$.

    a. $T_A\left(\begin{bmatrix} x \\ y \end{bmatrix} \right) = \begin{bmatrix} x \\ 0 \end{bmatrix}$

    b. $T_A\left(\begin{bmatrix} x \\ y \end{bmatrix} \right) = \begin{bmatrix} 0 \\ y \end{bmatrix}$

    c. $T_A\left(\begin{bmatrix} x \\ y \end{bmatrix} \right) = \begin{bmatrix} y \\ x \end{bmatrix}$

    d. $T_A\left(\begin{bmatrix} x \\ y \\ z \end{bmatrix} \right) = \begin{bmatrix} z \\ y \\ x \end{bmatrix}$

    e. $T_A\left(\begin{bmatrix} x \\ y \\ z \end{bmatrix} \right) = \begin{bmatrix} x \\ -3y \\ 2z \end{bmatrix}$

---

[*] This is a type of **input–output matrix** problem. In 1973, Wassily Leontief (1906–1999) was awarded the Nobel Prize in economics for his development of the input–output method and for its application to important economic problems.

f. $T_A\left(\begin{bmatrix} x \\ y \\ z \\ w \end{bmatrix}\right) = \begin{bmatrix} x - y \\ 2y + z \\ z + w \\ x - y + z - w \end{bmatrix}$

9. If A is an $m \times n$ matrix, use mathematical induction to prove that

$$T_A(c_1\mathbf{v}_1 + c_2\mathbf{v}_2 + \cdots + c_n\mathbf{v}_n) = c_1 T_A(\mathbf{v}_1) + c_2 T_A(\mathbf{v}_2)$$
$$+ \cdots + c_n T_A(\mathbf{v}_n)$$

for all vectors $\mathbf{v}_i \in \mathbb{R}^n$ and all real numbers c_i (see Theorem 3.2.2).

10. Suppose A is a 3×4 matrix satisfying the equations:

$$A\begin{bmatrix} 1 \\ 2 \\ -1 \\ 4 \end{bmatrix} = \begin{bmatrix} 1 \\ 2 \\ 3 \end{bmatrix} \quad \text{and} \quad A\begin{bmatrix} 0 \\ 3 \\ 1 \\ -2 \end{bmatrix} = \begin{bmatrix} 1 \\ 1 \\ 1 \end{bmatrix}.$$

Find a vector $\mathbf{x} \in \mathbb{R}^4$ such that $A\mathbf{x} = \begin{bmatrix} 0 \\ 1 \\ 2 \end{bmatrix}$. Give your reasoning. [*Hint*: You may use the results of Exercise B7.]

C.

1. Explain why a system of linear equations can never have exactly two different solutions.

    Extend your argument to explain the fact that if a system has more than one solution, then it must have *infinitely* many solutions. [*Hint*: Suppose the system is written in matrix form $A\mathbf{x} = \mathbf{b}$ and that $\mathbf{x}_1$ and $\mathbf{x}_2$ are two distinct solutions. Show that for any real number t the vector $\mathbf{x}_1 + t(\mathbf{x}_1 - \mathbf{x}_2)$ is also a solution.]

2. Suppose that $\mathbf{v}_1, \mathbf{v}_2, \ldots, \mathbf{v}_k$ are vectors in $\mathbb{R}^n$. Prove that if T is a linear transformation from $\mathbb{R}^n$ to $\mathbb{R}^m$ (as described in Theorem 3.2.2) and $\{T(\mathbf{v}_1), T(\mathbf{v}_2), \ldots, T(\mathbf{v}_k)\}$ is a linearly independent set in $\mathbb{R}^m$, then $\{\mathbf{v}_1, \mathbf{v}_2, \ldots, \mathbf{v}_k\}$ is a linearly independent set in $\mathbb{R}^n$. (The *converse* of this statement is not true.) [*Hint*: You may want to use one of the results of Exercise B7.]

3. A quadratic form $Q(\mathbf{x}) = \mathbf{x}^T A \mathbf{x}$ is said to be *positive definite* if $Q(\mathbf{x}) > 0$ for all $\mathbf{x} \neq \mathbf{0}$. Transform the matrix A of the quadratic form to echelon form (or reduced echelon form), noting the pivots. The quadratic form is positive definite if and only if each pivot is strictly positive when A is reduced *without interchanging rows*. Determine whether each of the quadratic forms below is positive definite.
    a. $2x_1^2 + x_2^2 + 6x_3^2 + 2x_1x_2 + x_1x_3 + 4x_2x_3$
    b. $2x_1^2 - 2x_1x_2 + 2x_2^2 - 2x_2x_3 + 2x_3^2$

3.3 The Product of Two Matrices

As easy and intuitive as matrix addition, subtraction, and scalar multiplication are, the entry-by-entry calculation we have gotten used to is not the most useful pattern for the *multiplication* of two matrices. As shown in the next example, the multiplication of a matrix by a matrix is best understood in terms of the matrix–vector product defined in Section 3.2.

Example 3.3.1: The Product of Two Matrices

Alpha College and Beta University are planning to buy identical equipment, but in different quantities, for their computer laboratories. Table 3.1 indicates the quantities of each piece of equipment they intend to purchase.

TABLE 3.1 Quantities

	System Unit (incl. 17″ monitor)	Laser Printer	Surge Protector
Alpha College	25	5	20
Beta University	35	3	15

There are two competing vendors involved in the purchase, and their prices are given in Table 3.2.

Each college wants to determine how much it will cost to purchase its equipment from each vendor. This is an easy calculation, but it is the *pattern* that we are interested in:

Alpha College will spend $25(1,286) + 5(399) + 20(39) = \$34,925$ if it buys its equipment from Vendor 1. Note that this calculation is the dot product of row 1 of Table (matrix) 3.1 and column 1 of Table (matrix) 3.2.

Alpha College will spend $25(1,349) + 5(380) + 20(37) = \$36,365$ if it buys its equipment from Vendor 2. We recognize that this number is the dot product of row 1 of Table (matrix) 3.1 and column 2 of Table (matrix) 3.2.

Beta University will spend $35(1,286) + 3(399) + 15(39) = \$46,792$ if it buys its equipment from Vendor 1. This figure is the dot product of row 2 of Table (matrix) 3.1 and column 1 of Table (matrix) 3.2.

TABLE 3.2 Unit Prices in Dollars

	Vendor 1	Vendor 2
System unit (incl. 17″ monitor)	1286	1349
Laser printer	399	380
Surge protector	39	37

Beta University will spend $35(1{,}349) + 3(380) + 15(37) = \$48{,}910$ if it buys its equipment from Vendor 2. The total is the dot product of row 2 of Table (matrix) 3.1 and column 2 of Table (matrix) 3.2.

We can display the results in matrix form as follows:

	Vendor 1	Vendor 2
Alpha College	34,925	36,365
Beta University	46,792	48,910

In pure matrix terms, we can think of this final matrix as the product of a "demand matrix" and a "price matrix":

$$
\begin{array}{c}
\text{Demand} \\
\text{Sys. Pr. SP} \\
\begin{array}{c}\text{Alpha C.}\\ \text{Beta U.}\end{array}
\begin{bmatrix} 25 & 5 & 20 \\ 35 & 3 & 15 \end{bmatrix}
\end{array}
\times
\begin{array}{c}
\text{Price} \\
\begin{array}{cc}V1 & V2\end{array} \\
\begin{bmatrix} 1286 & 1349 \\ 399 & 380 \\ 39 & 37 \end{bmatrix}
\end{array}
=
\begin{array}{c}
\text{Total Cost} \\
\begin{array}{cc}V1 & V2\end{array} \\
\begin{bmatrix} 34925 & 36365 \\ 46792 & 48910 \end{bmatrix}
\begin{array}{c}\text{Alpha C.}\\ \text{Beta U.}\end{array}
\end{array}.
$$

For example, as we have observed earlier, the element in row 1, column 2 of the product matrix, 36365, is found by taking the dot product of row 1 of the demand matrix and column 2 of the price matrix. Note that this process is just an extension of matrix–vector multiplication and for this to make sense, the number of columns of the first matrix (counting from the left) must equal the number of rows of the second matrix.

Taking our cue from the last example, we can generalize this row-by-column operation in a natural way.

Definition 3.3.1

Matrices A and B are said to be **conformable** for multiplication in the order AB if A has exactly as many columns as B has rows—that is, A is $m \times p$ and B is $p \times n$ for some positive integers m, p, and n.

Example 3.3.2: Conformable Matrices

Matrices $A = \begin{bmatrix} 0 & -1 & 2 & -3 \\ 4 & -5 & 6 & -7 \\ 3 & 0 & 0 & 4 \end{bmatrix}$ and $B = \begin{bmatrix} 0 & 5 & -4 & 3 & 2 \\ 1 & -3 & 0 & 0 & 6 \\ -4 & 7 & 2 & 4 & 6 \\ 5 & 4 & 3 & 2 & 1 \end{bmatrix}$

are conformable in the order AB because the number of columns of A, 4, equals the number of rows of B. But these matrices are *not* conformable in the order BA because the number of columns of B, 5, does *not* equal the number of rows of A, 3.

Definition 3.3.2

If $A = [a_{ij}]$ and $B = [b_{ij}]$ are conformable matrices, we regard both the rows of A and the columns of B as vectors and define the **product AB** as follows:

The entry in row i, column j of the product matrix $C = [c_{ij}] = AB$ is the *dot product* of the transpose of row i of A and column j of B—that is,

$$c_{ij} = [\,a_{i1} \quad a_{i2} \quad \dots \quad a_{ip}\,]^{\mathrm{T}} \cdot \begin{bmatrix} b_{1j} \\ b_{2j} \\ \vdots \\ b_{pj} \end{bmatrix} = a_{i1}b_{1j} + a_{i2}b_{2j} + \cdots + a_{ip}b_{pj}.$$

Schematically, we can represent this matrix multiplication $C = AB$ as follows:

$$\begin{bmatrix} c_{11} & c_{12} & \dots & c_{1j} & \dots & c_{1n} \\ c_{21} & c_{22} & \dots & c_{2j} & \dots & c_{2n} \\ \vdots & \vdots & \vdots & \vdots & \vdots & \vdots \\ c_{i1} & c_{i2} & \dots & \boxed{c_{ij}} & \dots & c_{in} \\ \vdots & \vdots & \vdots & \vdots & \vdots & \vdots \\ c_{m1} & c_{m2} & \dots & c_{mj} & \dots & c_{mn} \end{bmatrix}$$

$$= \begin{bmatrix} a_{11} & a_{12} & \dots & a_{1p} \\ a_{21} & a_{22} & \dots & a_{2p} \\ \vdots & \vdots & \vdots & \vdots \\ \boxed{a_{i1}} & \boxed{a_{i2}} & \dots & \boxed{a_{ip}} \\ \vdots & \vdots & \vdots & \vdots \\ a_{m1} & a_{m2} & \dots & a_{mp} \end{bmatrix} \cdot \begin{bmatrix} b_{11} & b_{12} & \dots & \boxed{b_{1j}} & \dots & b_{1n} \\ b_{21} & b_{22} & \dots & \boxed{b_{2j}} & \dots & b_{2n} \\ \vdots & \vdots & \vdots & \vdots & \vdots & \vdots \\ b_{p1} & b_{p2} & \dots & \boxed{b_{pj}} & \dots & b_{pn} \end{bmatrix}$$

By focusing on the sizes of the matrices, it is easy to remember when matrices A and B can be multiplied and what the shape of the product C will be

$$\overbrace{A}^{m \times p} \; \overbrace{B}^{p \times n} = \overbrace{C}^{m \times n}$$

This *row-by-column* view of a matrix product is the traditional one and focuses on the product matrix *one entry at a time*. It is the most widely used definition, although there are others. (See Exercises C3–C5, for example.) At one time, the product was called the *Cayley product* because it was introduced around 1858 by the British mathematician Arthur Cayley (1821–1895), one of the founders of matrix theory.

Example 3.3.3: Matrix Products

$$\begin{bmatrix} 2 & -3 & 0 \\ 4 & 0 & 1 \end{bmatrix} \cdot \begin{bmatrix} 1 & 2 \\ 3 & 4 \\ 5 & 6 \end{bmatrix} = \begin{bmatrix} 2(1) - 3(3) + 0(5) & 2(2) - 3(4) + 0(6) \\ 4(1) + 0(3) + 1(5) & 4(2) + 0(4) + 1(6) \end{bmatrix}$$

$$= \begin{bmatrix} -7 & -8 \\ 9 & 14 \end{bmatrix};$$

$$\begin{bmatrix} \pi & -2 & 6 \\ 0 & 4 & 1 \\ -3 & 5 & 7 \end{bmatrix} \cdot \begin{bmatrix} 2 & -3 & 0 \\ 9 & 2 & -6 \\ 2 & 1 & 4 \end{bmatrix}$$

$$= \begin{bmatrix} 2\pi - 2(9) + 6(2) & -3\pi - 2(2) + 6(1) & \pi(0) - 2(-6) + 6(4) \\ 0(2) + 4(9) + 1(2) & 0(-3) + 4(2) + 1(1) & 0(0) + 4(-6) + 1(4) \\ -3(2) + 5(9) + 7(2) & -3(-3) + 5(2) + 7(1) & -3(0) + 5(-6) + 7(4) \end{bmatrix}$$

$$= \begin{bmatrix} 2\pi - 6 & -3\pi + 2 & 36 \\ 38 & 9 & -20 \\ 53 & 26 & -2 \end{bmatrix}.$$

3.3.1 The Column View of Multiplication

There is another interpretation of matrix multiplication that gives the same result as Definition 3.3.2, but focuses on generating the product matrix *one column at a time*. In considering the matrix product $C = AB$, where $A = [a_{ij}]$ is $m \times p$ and $B = [b_{ij}]$ is $p \times n$, we think of matrix B as *a vector whose components are the columns* of B:

$$B = [\mathbf{b}_1 \ \mathbf{b}_2 \ \cdots \ \mathbf{b}_n], \quad \text{where } \mathbf{b}_1 = \begin{bmatrix} b_{11} \\ b_{21} \\ \vdots \\ b_{p1} \end{bmatrix}, \mathbf{b}_2 = \begin{bmatrix} b_{12} \\ b_{22} \\ \vdots \\ b_{p2} \end{bmatrix}, \dots, \mathbf{b}_n = \begin{bmatrix} b_{1n} \\ b_{2n} \\ \vdots \\ b_{pn} \end{bmatrix}.$$

Somewhat awkwardly, we are saying that

$$B = \begin{bmatrix} \begin{bmatrix} b_{11} \\ b_{21} \\ \vdots \\ b_{p1} \end{bmatrix} \begin{bmatrix} b_{12} \\ b_{22} \\ \vdots \\ b_{p2} \end{bmatrix} \cdots \begin{bmatrix} b_{1n} \\ b_{2n} \\ \vdots \\ b_{pn} \end{bmatrix} \end{bmatrix},$$

but we choose the more compact notation $B = [\mathbf{b}_1 \ \mathbf{b}_2 \dots \mathbf{b}_n]$.

Then the first column of $C = AB$ is obtained by finding the matrix–vector product $A \cdot$ (the first column of B) $= A\mathbf{b}_1$, the second column of C is given by $A \cdot$ (the second column of B) $= A\mathbf{b}_2$, and so forth. Once again, for this process to make sense, *the number of columns of A must equal the number of rows of B*.

Before proving the validity of this process, we will illustrate its use.

Example 3.3.4: The Column View of Matrix Multiplication

Suppose that $A = \begin{bmatrix} 1 & 2 & 3 & 4 \\ 5 & 6 & 7 & 8 \end{bmatrix}$ and $B = \begin{bmatrix} 1 & 0 & 3 \\ 4 & 2 & 1 \\ 0 & 5 & 4 \\ 2 & 1 & 0 \end{bmatrix}$. First

of all, we see that the product AB is defined. (It is not reasonable to try to calculate the product BA because any row of B has only three entries and could not be involved in a dot product with A, whose columns have only two entries.)

If C is the product matrix AB, then we have the following:

The first column of C is $A\mathbf{b}_1 = \begin{bmatrix} 1 & 2 & 3 & 4 \\ 5 & 6 & 7 & 8 \end{bmatrix} \begin{bmatrix} 1 \\ 4 \\ 0 \\ 2 \end{bmatrix}$

$$= \begin{bmatrix} 1(1)+2(4)+3(0)+4(2) \\ 5(1)+6(4)+7(0)+8(2) \end{bmatrix} = \begin{bmatrix} 17 \\ 45 \end{bmatrix}.$$

The second column of C is $A\mathbf{b}_2 = \begin{bmatrix} 1 & 2 & 3 & 4 \\ 5 & 6 & 7 & 8 \end{bmatrix} \begin{bmatrix} 0 \\ 2 \\ 5 \\ 1 \end{bmatrix}$

$$= \begin{bmatrix} 1(0)+2(2)+3(5)+4(1) \\ 5(0)+6(2)+7(5)+8(1) \end{bmatrix} = \begin{bmatrix} 23 \\ 55 \end{bmatrix}.$$

The third column of C is $A\mathbf{b}_3 = \begin{bmatrix} 1 & 2 & 3 & 4 \\ 5 & 6 & 7 & 8 \end{bmatrix} \begin{bmatrix} 3 \\ 1 \\ 4 \\ 0 \end{bmatrix}$

$$= \begin{bmatrix} 1(3)+2(1)+3(4)+4(0) \\ 5(3)+6(1)+7(4)+8(0) \end{bmatrix} = \begin{bmatrix} 17 \\ 49 \end{bmatrix}.$$

We have gone through all the columns of B, so now we can build the product matrix C from the matrix–column vector products we have calculated:

$$C = AB = \begin{bmatrix} 1 & 2 & 3 & 4 \\ 5 & 6 & 7 & 8 \end{bmatrix} \begin{bmatrix} 1 & 0 & 3 \\ 4 & 2 & 1 \\ 0 & 5 & 4 \\ 2 & 1 & 0 \end{bmatrix} = \begin{bmatrix} 17 & 23 & 17 \\ 45 & 55 & 49 \end{bmatrix}.$$

Letting $\mathrm{col}_j(M)$ denote the jth column of matrix M, we can express this alternative procedure for matrix multiplication concisely as a theorem.

Theorem 3.3.1

If $A = [a_{ij}]$ is $m \times p$ and $B = [b_{ij}]$ is $p \times n$, then

$$\boxed{\operatorname{col}_j(AB) = A \cdot \operatorname{col}_j(B).}$$

In words, column j of the product AB is A times column j of B.

Proof By Definition 3.3.2, we see that the entry in row i, column j of AB is given by

$$c_{ij} = a_{i1}b_{1j} + a_{i2}b_{2j} + \cdots + a_{ip}b_{pj}.$$

Therefore, column j of AB is given by the vector

$$\begin{bmatrix} c_{1j} \\ c_{2j} \\ \vdots \\ c_{mj} \end{bmatrix} = \begin{bmatrix} a_{11}b_{1j} + a_{12}b_{2j} + \cdots + a_{1p}b_{pj} \\ a_{21}b_{1j} + a_{22}b_{2j} + \cdots + a_{2p}b_{pj} \\ \vdots \\ a_{m1}b_{1j} + a_{m2}b_{2j} + \cdots + a_{mp}b_{pj} \end{bmatrix}.$$

But, by Definition 3.2.1(a),

$$A \cdot \operatorname{col}_j(B) = A \cdot \begin{bmatrix} b_{1j} \\ b_{2j} \\ \vdots \\ b_{pj} \end{bmatrix} = b_{1j}\operatorname{col}_1(A) + b_{2j}\operatorname{col}_2(A) + \cdots + b_{pj}\operatorname{col}_p(A)$$

$$= b_{1j}\begin{bmatrix} a_{11} \\ a_{21} \\ \vdots \\ a_{m1} \end{bmatrix} + b_{2j}\begin{bmatrix} a_{12} \\ a_{22} \\ \vdots \\ a_{m2} \end{bmatrix} + \cdots + b_{pj}\begin{bmatrix} a_{1p} \\ a_{2p} \\ \vdots \\ a_{mp} \end{bmatrix}$$

$$= \begin{bmatrix} a_{11}b_{1j} + a_{12}b_{2j} + \cdots + a_{1p}b_{pj} \\ a_{21}b_{1j} + a_{22}b_{2j} + \cdots + a_{2p}b_{pj} \\ \vdots \\ a_{m1}b_{1j} + a_{m2}b_{2j} + \cdots + a_{mp}b_{pj} \end{bmatrix} = \operatorname{col}_j(AB).$$

To view Example 3.3.4 differently, we can think of matrix B as *partitioned* into three *blocks* determined by its columns $\mathbf{b}_i$ ($i = 1, 2, 3$):

$$\begin{bmatrix} 1 & \vdots & 0 & \vdots & 3 \\ 4 & \vdots & 2 & \vdots & 1 \\ 0 & \vdots & 5 & \vdots & 4 \\ 2 & \vdots & 1 & \vdots & 0 \end{bmatrix}.$$

Then, we can write the product in abbreviated form (using Theorem 3.3.1) as

$$AB = \left[A\mathbf{b}_1 \;\vdots\; A\mathbf{b}_2 \;\vdots\; A\mathbf{b}_3 \right] = \begin{bmatrix} 17 & \vdots & 23 & \vdots & 17 \\ 45 & \vdots & 55 & \vdots & 49 \end{bmatrix}.$$

A useful alternative formulation of Theorem 3.3.1 is that, with the same assumptions about A and B,

$$\boxed{AB = A[\mathbf{b}_1 \; \mathbf{b}_2 \; \cdots \; \mathbf{b}_n] = [A\mathbf{b}_1 \; A\mathbf{b}_2 \; \cdots \; A\mathbf{b}_n]} \,, \qquad (3.3.1)$$

where $\mathbf{b}_k$ denotes column k of B.

The proof of the following fundamental associativity property of matrix multiplication uses Theorem 3.3.1 in a crucial way.

Corollary 3.3.1

If A is $m \times p$ and B is $p \times n$, then $(AB)\mathbf{x} = A(B\mathbf{x})$ for all $\mathbf{x} \in \mathbb{R}^n$.

Proof

$$(AB)\mathbf{x} = x_1 \mathrm{col}_1(AB) + x_2 \mathrm{col}_2(AB) + \cdots + x_n \mathrm{col}_n(AB)$$
[by Definition 3.2.1(a)]
$$= x_1[A \cdot \mathrm{col}_1(B)] + x_2[A \cdot \mathrm{col}_2(B)] + \cdots + x_n[A \cdot \mathrm{col}_n(B)]$$
[by Theorem 3.3.1]
$$= A \cdot [x_1 \cdot \mathrm{col}_1(B)] + A \cdot [x_2 \cdot \mathrm{col}_2(B)] + \cdots + A \cdot [x_n \cdot \mathrm{col}_n(B)]$$
[by Theorem 3.2.2, (2)]
$$= A \cdot [x_1 \cdot \mathrm{col}_1(B) + x_2 \cdot \mathrm{col}_2(B) + \cdots + x_n \cdot \mathrm{col}_n(B)]$$
[by Theorem 3.2.2, (1)]
$$= A(B\mathbf{x}) \quad \text{[by Definition 3.2.1(a)]}$$

3.3.2 The Row View of Multiplication

In addition to the element-at-a-time and column views of matrix multiplication, there is an interpretation that gives us the individual *rows* of a product matrix. Letting $\mathrm{row}_i(M)$ denote the ith row of a matrix M, we can express this third procedure for matrix multiplication formally, as a theorem.

Theorem 3.3.2

If $A = [a_{ij}]$ is $m \times p$ and $B = [b_{ij}]$ is $p \times n$, then

$$\boxed{\mathrm{row}_i(AB) = \mathrm{row}_i(A)B}\,.$$

In words, row i of the product AB is (row i of A) times B, where $\mathrm{row}_i(A)$ is taken to be a $1 \times p$ matrix.

Proof   According to Definition 3.3.2, the entries in row i of AB are given by

$$c_{ir} = \sum_{k=1}^{p} a_{ik}b_{kr},$$

for $r = 1, 2, \ldots, n$. But

$$\text{row}_i(A) \cdot B = \begin{bmatrix} a_{i1} & a_{i2} & \cdots & a_{ip} \end{bmatrix} \cdot \begin{bmatrix} b_{11} & b_{12} & \cdots & b_{1n} \\ b_{21} & b_{22} & \cdots & b_{2n} \\ \vdots & \vdots & \vdots & \vdots \\ b_{p1} & b_{p2} & \cdots & b_{pn} \end{bmatrix}$$

$$= \begin{bmatrix} \sum_{k=1}^{p} a_{ik}b_{k1} & \sum_{k=1}^{p} a_{ik}b_{k2} & \cdots & \sum_{k=1}^{p} a_{ik}b_{kn} \end{bmatrix},$$

so that $\text{row}_i(AB) = \text{row}_i(A)\, B$.

A useful alternative formulation of Theorem 3.2.2 is that, with the same assumptions about A and B,

$$AB = \begin{bmatrix} \text{row}_1(A) \\ \text{row}_2(A) \\ \vdots \\ \text{row}_m(A) \end{bmatrix} B = \begin{bmatrix} \text{row}_1(A)B \\ \text{row}_2(A)B \\ \vdots \\ \text{row}_m(A)B \end{bmatrix} \tag{3.3.2}$$

we can interpret this product in terms of a partition into blocks determined by the rows of A:

$$AB = \begin{bmatrix} \text{row}_1(A)B \\ \cdots\cdots\cdots \\ \text{row}_2(A)B \\ \cdots\cdots\cdots \\ \vdots \\ \cdots\cdots\cdots \\ \text{row}_m(A)B \end{bmatrix}$$

Example 3.3.5: The Row View of Matrix Multiplication

Let us reuse the matrices of Example 3.3.4 to illustrate Theorem 3.3.2:

$$A = \begin{bmatrix} 1 & 2 & 3 & 4 \\ 5 & 6 & 7 & 8 \end{bmatrix} \quad \text{and} \quad B = \begin{bmatrix} 1 & 0 & 3 \\ 4 & 2 & 1 \\ 0 & 5 & 4 \\ 2 & 1 & 0 \end{bmatrix}.$$

We see that

Row 1 of the product AB is $\text{row}_1(A) \cdot B = \begin{bmatrix} 1 & 2 & 3 & 4 \end{bmatrix} \begin{bmatrix} 1 & 0 & 3 \\ 4 & 2 & 1 \\ 0 & 5 & 4 \\ 2 & 1 & 0 \end{bmatrix}$

$$= [1(1) + 2(4) + 3(0) + 4(2) \quad 1(0) + 2(2) + 3(5) + 4(1) \quad 1(3)$$
$$+ 2(1) + 3(4) + 4(0)]$$
$$= [17 \; 23 \; 17].$$

Row 2 of the product AB is $\text{row}_2(A) \cdot B = [5 \; 6 \; 7 \; 8] \begin{bmatrix} 1 & 0 & 3 \\ 4 & 2 & 1 \\ 0 & 5 & 4 \\ 2 & 1 & 0 \end{bmatrix}$

$$= [5(1) + 6(4) + 7(0) + 8(2) \quad 5(0) + 6(2) + 7(5) + 8(1) \quad 5(3)$$
$$+ 6(1) + 7(4) + 8(0)]$$
$$= [45 \; 55 \; 49].$$

Therefore, the product AB is $\begin{bmatrix} 17 & 23 & 17 \\ 45 & 55 & 49 \end{bmatrix}$, as was shown in Example 3.3.4 using a column approach.

Let us summarize the different ways in which we can interpret the product of matrices.

Matrix Multiplication

Suppose that $A = [a_{ij}]$ is an $m \times p$ matrix and $B = [b_{ij}]$ is a $p \times n$ matrix. Then, the product AB is an $m \times n$ matrix and

(1) The (i, j) entry in the product $C = AB$ is

$$c_{ij} = \sum_{k=1}^{p} a_{ik} b_{kj} = a_{i1} b_{1j} + a_{i2} b_{2j} + \cdots + a_{ip} b_{pj},$$

the dot product of (the transpose of) $\text{row}_i(A)$ and $\text{col}_j(B)$.

(2) Column j of the product $A \cdot B$ is given by

$$\text{col}_j(AB) = A \, \text{col}_j(B).$$

Thus, $AB = A[\text{col}_1(B) \; \text{col}_2(B) \; \ldots \; \text{col}_n(B)] = [A\text{col}_1(B) \; A\text{col}_2(B) \ldots A\text{col}_n(B)]$.

(3) Row i of the product AB is given by
$\text{row}_i(AB) = \text{row}_i(A)B$.

Thus, $AB = \begin{bmatrix} \text{row}_1(A)B \\ \text{row}_2(A)B \\ \vdots \\ \text{row}_m(A)B \end{bmatrix}$.

No matter how we define the product of matrices, a very important algebraic fact about matrix multiplication is that **multiplication is not commutative**—that is, $AB \neq BA$ in general, even if A and B are conformable with respect to both orders of multiplication.

Example 3.3.6: Matrix Multiplication and Commutativity

If $A = \begin{bmatrix} 0 & 4 & 2 \\ -1 & 3 & -2 \end{bmatrix}$ and $B = \begin{bmatrix} 2 & 2 \\ 3 & -1 \\ 1 & 0 \end{bmatrix}$, then $AB = \begin{bmatrix} 14 & -4 \\ 5 & -5 \end{bmatrix}$

and $BA = \begin{bmatrix} -2 & 14 & 0 \\ 1 & 9 & 8 \\ 0 & 4 & 2 \end{bmatrix}$, so that $AB \neq BA$ (which is the usual

case). On the other hand, if $C = \begin{bmatrix} 1 & 2 \\ 3 & 4 \end{bmatrix}$ and $D = \begin{bmatrix} -3 & 0 \\ 0 & -3 \end{bmatrix}$,

then $CD = \begin{bmatrix} -3 & -6 \\ -9 & -12 \end{bmatrix}$ and $DC = \begin{bmatrix} -3 & -6 \\ -9 & -12 \end{bmatrix}$, so that C and D

do commute.

Note that matrix multiplication *may* be commutative in some cases, but commutativity does not hold *all* the time. With the notable exception of commutativity, all the expected algebraic properties of matrix multiplication hold. From now on, whenever a set of matrices is described as "conformable," it will mean that all the matrices have the correct sizes for the additions and/or multiplications being discussed.

Theorem 3.3.3: Properties of Matrix Multiplication

If A, B, and C are conformable for the additions and multiplications indicated below and k is a scalar, then

(a) $A(BC) = (AB)C$. [*Associativity of Matrix Multiplication*]
(b) $k(AB) = (kA)B = A(kB)$.
(c) $A(B + C) = AB + AC$. [*Left Distributivity of Multiplication with Respect to Addition*]
(d) $(B + C)A = BA + CA$. [*Right Distributivity of Multiplication with Respect to Addition*]
(e) if A is $m \times n$, $AI_n = A = I_m A$. [*Right- and Left-Hand Multiplicative Identities*]
(f) $(AB)^T = B^T A^T$. [*Transpose of a Product*]

Proof of (a)   Although this property is not difficult to prove using the row-by-column definition (see Exercise B5), we will use the "column" version of matrix multiplication (Theorem 3.3.1 and Corollary 3.3.1). So suppose that A is $m \times n$, B is $n \times r$, and C is $r \times s$, and let $C = [\mathbf{c}_1 \, \mathbf{c}_2 \cdots \mathbf{c}_s]$. Then, $BC = B[\mathbf{c}_1 \, \mathbf{c}_2 \cdots \mathbf{c}_n] = [B\mathbf{c}_1 \, B\mathbf{c}_2 \cdots B\mathbf{c}_s]$ and $A(BC) = A[B\mathbf{c}_1 \, B\mathbf{c}_2 \cdots B\mathbf{c}_s] = [A(B\mathbf{c}_1) \, A(B\mathbf{c}_2) \ldots A(B\mathbf{c}_s)] = [(AB)\mathbf{c}_1 \, (AB)\mathbf{c}_2 \ldots (AB)\mathbf{c}_s] = (AB)[\mathbf{c}_1 \, \mathbf{c}_2 \cdots \mathbf{c}_s] = (AB)C$.

Proof of (b) Suppose that A is $m \times n$ and B is $n \times r$. Then, by Definition 3.3.2, the (i, j) entry of AB is $\sum_{k=1}^n a_{ik}b_{kj}$, so that the (i, j) entry of $k(AB)$ is $k \sum_{k=1}^n a_{ik}b_{kj} = \sum_{k=1}^n ka_{ik}b_{kj} = \sum_{k=1}^n (ka_{ik})b_{kj}$, which is the (i, j) entry of $(kA)B$.

In the same way, we can show that the (i,j) entry of $k(AB)$ is the (i,j) entry of $A(kB)$.

Proof of (c)   We will use the "row-by-column" method (Definition 3.3.2) here.

(Exercise B6 asks for the "column" definition proof.) First we determine the entry in row i, column j of $A(B+C)$. Because $A(B+C)$ is the product of matrices $[a_{ij}]$ and $[b_{ij}+c_{ij}]$, the (i,j) entry of this product is

$$\sum_{k=1}^{n} a_{ik}(b_{kj}+c_{kj}) = \sum_{k=1}^{n}(a_{ik}b_{kj}+a_{ik}c_{kj}) = \sum_{k=1}^{n}a_{ik}b_{kj} + \sum_{k=1}^{n}a_{ik}c_{kj}.$$

But $\sum_{k=1}^{n} a_{ik}b_{kj}$ is the (i,j) entry of the product AB and $\sum_{k=1}^{n} a_{ik}c_{kj}$ is the (i,j) entry of the product AC. Therefore, $A(B+C)=AB+AC$. The proof of part (d) can be given in a similar way.

Proof of (e)   First we note that for any positive integer n, I_n can be described as $[\delta_{ij}]$, where $\delta_{ij} = \begin{cases} 1 & \text{if } i=j \\ 0 & \text{otherwise} \end{cases}$. The element in row i, column j of AI_n is $c_{ij}=a_{i1}\,\delta_{1j}+a_{i2}\,\delta_{2j}+\cdots a_{in}\,\delta_{nj}$. But $\delta_{ij}=0$ unless $i=j$. Thus, the sum for c_{ij} reduces to the single term $a_{ij}\,\delta_{jj}$, or $a_{ij}\cdot 1 = a_{ij}$. This shows that $AI_n=A$. In the same way, we can show that $I_mA=A$.

Proof of (f):   Let c_{ij} be the (i,j) entry of AB. Then, c_{ji} is the (i,j) entry of $(AB)^T$. But $c_{ij}=\sum_{k=1}^{n} a_{ik}b_{kj}$, so that $c_{ji}=\sum_{k=1}^{n} a_{jk}b_{ki}$. Now the (i,j) entry of B^TA^T is the dot product of row i of B^T and column j of A^T. However, row i of B^T equals column i of B and column j of A^T equals row j of A. Therefore, the (i,j) entry of B^TA^T is

$$b_{1i}a_{j1}+b_{2i}a_{j2}+\cdots+b_{ri}a_{jr} = a_{j1}b_{1i}+a_{j2}b_{2i}+\cdots+a_{jn}b_{ni} = c_{ji},$$

and so $(AB)^T=B^TA^T$.

The fact that matrices do not commute in general causes some algebraic difficulties in calculating with matrices, problems we should keep in mind.

Example 3.3.7: An Algebraic Anomaly

Suppose that A and B are two $n \times n$ matrices and consider the product $(A-B)(A+B)$. Based on our previous experience with algebra, we would expect the result to be A^2-B^2 by the "difference of squares" identity.* However, if we use both the left and right distributive laws [Theorem 3.3.3(c) and (d)], the first time with the factor $(A-B)$ considered as a single quantity, we get

$$(A-B)(A+B) = (A-B)A + (A-B)B = A^2 - BA + AB - B^2.$$

Because we cannot assume that $BA=AB$, we cannot cancel the middle terms of the last expansion to get A^2-B^2 as our final

---

* If M is a square matrix, we denote M times M by M^2.

answer. In ordinary high school or college algebra, A and B would be real numbers, complex numbers, or algebraic expressions, all of which *do* commute.

There are other, more subtle, differences between ordinary algebra and matrix algebra, as our next example shows.

Example 3.3.8: Matrix Cancellation May Not Be Valid

Consider the matrices $A = \begin{bmatrix} 1 & 1 \\ 1 & 1 \end{bmatrix}$, $B = \begin{bmatrix} 2 & 2 \\ 2 & 2 \end{bmatrix}$, and $C = \begin{bmatrix} 3 & 1 \\ 1 & 3 \end{bmatrix}$. Then, $AB = \begin{bmatrix} 4 & 4 \\ 4 & 4 \end{bmatrix} = AC$, but $B \neq C$. That is, we cannot "cancel" the A in the matrix equation $AB = AC$ to conclude that $B = C$, even though $A \neq O_2$. Furthermore, if $D = \begin{bmatrix} 1 & -4 & 2 \\ -1 & 4 & -2 \end{bmatrix}$ and $E = \begin{bmatrix} 2 & 2 \\ 1 & -1 \\ 1 & -3 \end{bmatrix}$, then

$DE = \begin{bmatrix} 0 & 0 \\ 0 & 0 \end{bmatrix} = O_2$, even though neither factor is a zero matrix and neither D nor E has a single zero entry.

3.3.3 Positive Powers of a Square Matrix

If A is an $n \times n$ matrix, we can define positive integer powers of A in a "recursive" way—that is, in such a way that higher powers are defined in terms of lower powers:

$$\begin{aligned} A^0 &= I_n \\ A^1 &= A \\ A^2 &= AA \\ A^3 &= A(A^2) \\ A^4 &= A(A^3) \\ &\vdots \\ A^m &= A(A^{m-1}) \end{aligned}$$

(We should see that calculating a power of a matrix A make sense only if A is a *square* matrix.)

It turns out that the expected laws of exponents hold for powers of matrices:

$$A^m A^n = A^{m+n} \quad \text{and} \quad (A^m)^n = A^{mn}$$

for all nonnegative integers m and n. We leave the proofs of these results as Exercises B20 and B21. The extension of these laws to negative exponents will be discussed in Section 3.4.

Example 3.3.9: Positive Powers of Matrices

Suppose $A = \begin{bmatrix} 1 & -2 \\ 3 & -4 \end{bmatrix}$ and $B = \begin{bmatrix} -2 & -3 & -7 \\ 1 & -1 & -4 \\ 0 & 1 & 3 \end{bmatrix}$. Then,

$$A^2 = \begin{bmatrix} -5 & 6 \\ -9 & 10 \end{bmatrix}$$

and $A^3 = A(A)^2 = \begin{bmatrix} 1 & -2 \\ 3 & -4 \end{bmatrix}\begin{bmatrix} -5 & 6 \\ -9 & 10 \end{bmatrix} = \begin{bmatrix} 13 & -14 \\ 21 & -22 \end{bmatrix}$, whereas

$$B^2 = \begin{bmatrix} 1 & 2 & 5 \\ -3 & -6 & -15 \\ 1 & 2 & 5 \end{bmatrix} \text{ and } B^3 = B(B^2) = \begin{bmatrix} -2 & -3 & -7 \\ 1 & -1 & -4 \\ 0 & 1 & 3 \end{bmatrix}$$

$$\begin{bmatrix} 1 & 2 & 5 \\ -3 & -6 & -15 \\ 1 & 2 & 5 \end{bmatrix} = \begin{bmatrix} 0 & 0 & 0 \\ 0 & 0 & 0 \\ 0 & 0 & 0 \end{bmatrix}. \text{ We can verify that}$$

$$A^2 A^3 = \begin{bmatrix} -5 & 6 \\ -9 & 10 \end{bmatrix}\begin{bmatrix} 13 & -14 \\ 21 & -22 \end{bmatrix} = \begin{bmatrix} 61 & -62 \\ 93 & -94 \end{bmatrix} = A^5. \text{ We can}$$

also confirm that $(A^2)^3 = \begin{bmatrix} -5 & 6 \\ -9 & 10 \end{bmatrix}^3 = \begin{bmatrix} -125 & 126 \\ -189 & 190 \end{bmatrix} = A^6.$

We see that in the last example, a power of B turned out to be the zero matrix, even though B itself was not the zero matrix. Matrices exhibiting such behavior occur in both the theory and applications of linear algebra and so have been given a special name.

Definition 3.3.3

An $n \times n$ matrix N is said to be **nilpotent of index k** if $N^k = O_n$ but $N^{k-1} \neq O_n$—that is, N^{k-1} contains at least one nonzero entry.

Thus, matrix B in Example 3.3.9 is nilpotent of index 3 because $B^3 = O_3$, but $B^2 \neq O_3$.

Sometimes, it is possible to observe patterns in the powers of a matrix, so that we can find and prove a formula for any positive integer power of that matrix.

Example 3.3.10: A Power Pattern

If we start with $A = \begin{bmatrix} 1 & a \\ 0 & 1 \end{bmatrix}$, where a is any real number, then

$$A^2 = \begin{bmatrix} 1 & 2a \\ 0 & 1 \end{bmatrix}, \quad A^3 = \begin{bmatrix} 1 & a \\ 0 & 1 \end{bmatrix}\begin{bmatrix} 1 & 2a \\ 0 & 1 \end{bmatrix} = \begin{bmatrix} 1 & 3a \\ 0 & 1 \end{bmatrix}, A^4 =$$

$$\begin{bmatrix} 1 & a \\ 0 & 1 \end{bmatrix}\begin{bmatrix} 1 & 3a \\ 0 & 1 \end{bmatrix} = \begin{bmatrix} 1 & 4a \\ 0 & 1 \end{bmatrix}, \text{ and so forth. We can } guess$$

that the pattern is $A^n = \begin{bmatrix} 1 & na \\ 0 & 1 \end{bmatrix}$ for any positive integer n. This can be *proved* by mathematical induction (see Appendix C if necessary): Clearly, the formula is true for $n = 1$. Now suppose it is true for $n = k$. Then $A^{k+1} = A(A)^k = \begin{bmatrix} 1 & a \\ 0 & 1 \end{bmatrix}\begin{bmatrix} 1 & ka \\ 0 & 1 \end{bmatrix} = \begin{bmatrix} 1 & (k+1)a \\ 0 & 1 \end{bmatrix}$, so the formula holds for $n = k + 1$. By induction, we know that the formula holds for all positive integers n.

Powers of matrices have interesting and important uses, as the following example shows.

Example 3.3.11: A Computer Network Application

Suppose we have a computer network consisting of four computers. If computer i can communicate directly with computer j, we define $a_{ij} = 1$. Otherwise, $a_{ij} = 0$. Assuming that a computer does not communicate with itself directly, the following matrix could represent a particular network:

$$A = [a_{ij}] = \begin{bmatrix} 0 & 0 & 0 & 1 \\ 1 & 0 & 0 & 1 \\ 1 & 0 & 0 & 0 \\ 0 & 1 & 1 & 0 \end{bmatrix}.$$

This matrix indicates that computers 2 and 4 are the only ones that communicate with each other directly in a reciprocal fashion, whereas computers 2 and 3 do not communicate directly with each other at all.

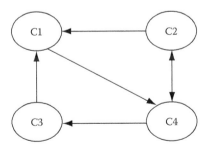

Now suppose we want to know which computers can communicate *indirectly* with the help of intermediaries. For example, computer 2 could communicate with computer k, which in turn communicates with computer 3, so that 2 communicates indirectly with 3. If we look at the entry in the fourth row and first column of A^2, we see something interesting:

$$A^2 = \begin{bmatrix} 0 & 1 & 1 & 0 \\ 0 & 1 & 1 & 1 \\ 0 & 0 & 0 & 1 \\ 2 & 0 & 0 & 1 \end{bmatrix}.$$

This entry b_{41} of the matrix $B = A^2$ equals

$$a_{41}a_{11} + a_{42}a_{21} + a_{43}a_{31} + a_{44}a_{41}.$$

Any one of these four products $a_{4k} a_{k1}$ ($k = 1, 2, 3,$ or 4) is equal to 1 only if both a_{4k} and a_{k1} are 1—that is, only if computer 4 can "speak" to computer k and computer k can in turn communicate with computer 1. Here, we find $b_{41} = 2$, telling us that there are two possible channels of communication by which computer 4 can speak to computer 1 through an intermediary (or by one relay). In fact, computer 4 can speak to either computer 2 or computer 3—that is, $a_{42} a_{21} = 1$ and $a_{43} a_{31} = 1$.

Furthermore, the c_{ij} entry of $C = A^3$ tells us how many channels of communication are open between computer i and computer j using *two* relays. Continuing in this way, we can conclude that the (i, j) entry of A^{n+1} indicates how many channels are open between computer i and computer j using n intermediaries. Also, $A + A^2$ gives the total number of channels that are open between various computers for either zero or one relay. Generalizing,

$$A + A^2 + A^3 + \cdots + A^{n+1}$$

yields the total number of channels of communication that are open between the various computers, with no more than n intermediaries.

Exercises 3.3

A.

1. The following table gives the sizes of five matrices:

A	B	C	D	E
3×7	2×3	3×3	2×2	7×2

    Determine which products are defined and give the sizes of all meaningful products.
    a. BCA b. BD c. EDB d. $A^{\mathrm{T}}C$ e. $E^{\mathrm{T}}A^{\mathrm{T}}CB^{\mathrm{T}}$
    f. CAB g. $BAED$ h. $C^{\mathrm{T}}BA$

2. If $A = \begin{bmatrix} -1 & 0 & 2 \\ 3 & 4 & 5 \end{bmatrix}$, $B = \begin{bmatrix} 0 & 2 & 4 \\ -1 & 3 & 0 \\ 4 & 5 & 6 \end{bmatrix}$, and $C = \begin{bmatrix} 2 & 0 \\ 0 & -1 \\ 3 & 4 \end{bmatrix}$,

   calculate the following, if defined. Otherwise, explain your answer.

   a. AC b. BA c. CA d. $C^{\mathsf{T}}A$ e. $A^{\mathsf{T}}C^{\mathsf{T}}$
   f. ABC g. CAB h. C^2 i. BA^{T} j. $C^{\mathsf{T}}BA^{\mathsf{T}}$
   k. I_2C l. AI_3C m. I_3BCI_2

3. Calculate each of the following products in three ways: (1) by using Definition 3.3.2; (2) by using Theorem 3.3.1; and (3) by using Theorem 3.3.2.

   a. $\begin{bmatrix} -1 & 3 \\ 0 & 4 \end{bmatrix}\begin{bmatrix} 5 & 2 \\ -3 & 1 \end{bmatrix}$ b. $\begin{bmatrix} 1 & 2 \\ 3 & 4 \end{bmatrix}\begin{bmatrix} -5 & 0 & 6 \\ 9 & -4 & 2 \end{bmatrix}$

   c. $\begin{bmatrix} 9 & -8 & 7 \\ -6 & 5 & -4 \\ 3 & -2 & 1 \end{bmatrix}\begin{bmatrix} 0 & 2 & -3 \\ 4 & 8 & 5 \\ -1 & 6 & -2 \end{bmatrix}$

   d. $\begin{bmatrix} 7 & -4 & 0 & 2 \\ 1 & 0 & 5 & 3 \\ -2 & 6 & 9 & 8 \end{bmatrix}\begin{bmatrix} 2 & 4 \\ 6 & 8 \\ 1 & 3 \\ -5 & 7 \end{bmatrix}$

   e. $\begin{bmatrix} -2 & 6 & 8 & 3 & 4 \\ 0 & 9 & -5 & 2 & 0 \\ 5 & -3 & 1 & 6 & 7 \\ 1 & 0 & 1 & 0 & 1 \\ -2 & 4 & -6 & 8 & 0 \end{bmatrix}\begin{bmatrix} 1 & -3 & 5 & -7 & 9 \\ -8 & 6 & -4 & 2 & 0 \\ 2 & 4 & 9 & 8 & -3 \\ 7 & -6 & 0 & 3 & 1 \\ 1 & 1 & 2 & 3 & 5 \end{bmatrix}$

4. Find a, b, c, and d if $\begin{bmatrix} a & b \\ c & d \end{bmatrix}\begin{bmatrix} 1 & 1 \\ 0 & 1 \end{bmatrix} = \begin{bmatrix} 1 & -1 \\ 0 & 1 \end{bmatrix}$.

5. If B is any 2×2 matrix of the form $\begin{bmatrix} a & b \\ a & b \end{bmatrix}$, show that

   $\begin{bmatrix} 1 & -1 \\ 3 & -3 \end{bmatrix}B = O_2$.

6. a. Show that the matrices $3\begin{bmatrix} -5 & 0 & 7 \\ 2 & 1 & 4 \end{bmatrix}$ and

   $\begin{bmatrix} 3 & 0 \\ 0 & 3 \end{bmatrix}\begin{bmatrix} -5 & 0 & 7 \\ 2 & 1 & 4 \end{bmatrix}$ are equal.

   b. Generalize the result in part (a) and express the matrix $c[a_{ij}]$, where c is a scalar, as the product of two matrices.

7. If A is a symmetric $n \times n$ matrix and P is any $m \times n$ matrix, show that PAP^{T} is symmetric. [*Hint*: Use associativity to group factors.]

8. Simplify $(A^{\mathsf{T}}B^{\mathsf{T}} + 3C)^{\mathsf{T}}$ by writing the expression using only one transpose, and give the reason for each step.

9. Verify that $\begin{bmatrix} 1 & 1 \\ -1 & -1 \end{bmatrix}$ is nilpotent of index 2.

10. Verify that $M = \begin{bmatrix} -2 & -3 & -7 \\ 1 & -1 & -4 \\ 0 & 1 & 3 \end{bmatrix}$ is nilpotent of index 3.

11. We can define a *square root* of a square matrix A as any matrix B such that $B^2 = A$.

    Show that for any real number k, the matrix $\begin{bmatrix} k & 1+k \\ 1-k & -k \end{bmatrix}$

    is a square root of I_2, so that I_2 has an *infinite* number of square roots.

B.

1. Let $A = \begin{bmatrix} 1 & -1 \\ -3 & 3 \end{bmatrix}$.

   a. Find a column vector $\mathbf{x}$, $\mathbf{x} \neq \mathbf{0}$, such that $A\mathbf{x} = \mathbf{0}$. Is your answer *unique*?

   b. Find a row vector $\mathbf{y}$, $\mathbf{y} \neq \mathbf{0}$, such that $\mathbf{y}A = \mathbf{0}$. Is your answer *unique*?

2. If B is a 2×2 matrix such that $\begin{bmatrix} 1 & -1 \\ 3 & -3 \end{bmatrix} B = O_2$, show that

   $B = \begin{bmatrix} a & b \\ a & b \end{bmatrix}$ for some numbers a and b. [This is the *converse*

   of the result in Exercise A5.]

3. Find a 2×3 matrix A and a 3×2 matrix B such that $AB = \begin{bmatrix} 1 & 0 \\ 0 & 1 \end{bmatrix}$.

4. Consider the matrix $A = \begin{bmatrix} 1 & 3 \\ 0 & 1 \end{bmatrix}$.

   a. Show that if a and b are real numbers, then any 2×2 matrix of the form $\begin{bmatrix} a & b \\ 0 & a \end{bmatrix}$ commutes with A.

   b. Show that if a matrix B commutes with A, then B must be of the form $\begin{bmatrix} a & b \\ 0 & a \end{bmatrix}$ for some real numbers a and b.

5. Prove part (a) of Theorem 3.3.3 by using the "row-by-column" definition (Definition 3.3.2) of matrix multiplication.

6. Prove part (b) of Theorem 3.3.3 by using the "column" version (Theorem 3.3.1) of matrix multiplication.

7. Prove that the product of two $n \times n$ diagonal matrices is a diagonal matrix.

8. Show that if A is an $n \times n$ symmetric matrix, then A^2 is also symmetric.

9. a. Find an example to show that AB does not have to be symmetric even though A and B are symmetric.
    b. Find an example to show that AB can possibly be symmetric if A and B are both symmetric.

10. Suppose A and B are symmetric. Prove that
    a. If A and B commute, then AB is symmetric.
    b. If AB is symmetric, then A and B commute.

11. Prove that if A, B, and C are conformable, then $(ABC)^{\mathrm{T}} = C^{\mathrm{T}}B^{\mathrm{T}}A^{\mathrm{T}}$.

12. Let A be an $m \times n$ matrix. Show that $\mathbf{x}^{\mathrm{T}} A^{\mathrm{T}} A\mathbf{x} = \mathbf{0}$ if and only if $A\mathbf{x} = \mathbf{0}$.

13. Let $H = I_n - 2\mathbf{x}\mathbf{x}^{\mathrm{T}}$, where $\mathbf{x} \in \mathbb{R}^n$ and $\mathbf{x}^{\mathrm{T}}\mathbf{x} = 1$.* Show that
    a. H is symmetric.
    b. $H^2 = I_n$.
    c. $H^{\mathrm{T}}H = I_n$.

14. Suppose that $A = \begin{bmatrix} 1 & 0 \\ 3 & 1 \end{bmatrix}$.

    a. Calculate $A^2 - 2A + I_2$.
    b. Using the result from part (a), show that $A^3 = 3A - 2I_2$ and $A^4 = 4A - 3I_2$.
    c. Guess a formula for A^n, where n is a positive integer, and use mathematical induction to prove that your formula is correct.

15. Suppose that $A = \begin{bmatrix} 1 & 0 & 2 \\ 0 & 3 & -1 \\ 4 & 5 & 0 \end{bmatrix}$.

    a. Calculate $A^3 - 4A^2 + 19I_3$.
    b. Use the result from part (a) to show that $A^4 = 16A^2 - 19A - 76I_3$.

16. Suppose A and B are 3×3 matrices and $\mathbf{x}$ is a 3×1 column vector. Consider the product $AB\mathbf{x}$. Using Definition 3.3.2, compare the number of computations (scalar additions and multiplications) needed to compute the product in each of the forms $(AB)\mathbf{x}$ and $A(B\mathbf{x})$.

17. Using Definition 3.3.2, how many multiplications of scalars are needed to perform the following matrix operations?
    a. Multiply a 4×5 matrix times a 5×7 matrix.
    b. Square an 8×8 matrix.
    c. Multiply a 7×4 matrix times a 4×7 matrix.

---

* More generally, a matrix of the form $H = I_n - \frac{2\mathbf{x}\mathbf{x}^{\mathrm{T}}}{\mathbf{x}^{\mathrm{T}}\mathbf{x}}$, where $\mathbf{x}$ is a nonzero vector, is called a **Householder matrix** after the American numerical analyst Alston Householder (1904–1993).

d. Cube a 10×10 matrix.

e. Multiply a 2×3 matrix times a 3×4 matrix times a 4×5 matrix.

18. Using the definition of square root given in Exercise A11, show that $\begin{bmatrix} 0 & 1 \\ 0 & 0 \end{bmatrix}$ has no square root.

19. Using the definition of square root given in Exercise A11, show that $\begin{bmatrix} 3 & -4 \\ 1 & -1 \end{bmatrix}$ has only the matrices $\begin{bmatrix} 2 & -2 \\ \frac{1}{2} & 0 \end{bmatrix}$ and $\begin{bmatrix} -2 & 2 \\ -\frac{1}{2} & 0 \end{bmatrix}$ as square roots.

20. Suppose that A is a square matrix. If m and n are positive integers, use mathematical induction to prove that $A^m A^n = A^{m+n}$. (For each fixed value of m, use induction on n.)

21. Suppose that A is a square matrix. If m and n are positive integers, use mathematical induction to prove that $(A^m)^n = A^{mn}$. (For each fixed value of m, use induction on n.)

22. Use mathematical induction to prove that if A is a square matrix and n is a positive integer, then $(A^T)^n = (A^n)^T$.

23. A *semimagic square* is an $n \times n$ matrix with real entries in which every row and every column has the same sum s. Prove that the product of two $n \times n$ semimagic squares is a semimagic square and determine the row and column sum S of the product.

24. If $AB = \lambda B$, where A is $n \times n$, B is $n \times p$, and λ is a scalar, show that $A^m B = \lambda^m B$ for any nonnegative integer m.

25. Define $P(\theta) = \begin{bmatrix} \cos \theta & -\sin \theta \\ \sin \theta & \cos \theta \end{bmatrix}$ for any angle θ.

a. Show that $P(\theta_1)P(\theta_2) = P(\theta_1 + \theta_2)$. [*Hint*: You will need some trigonometric identities.]

b. Show that $[P(\theta)]^n = P(n\theta)$ for any nonnegative integer n.

c. If $\theta = 2\pi/n$ for some positive integer n, show that $[P(\theta)]^n = I_2$.

26. Show that each row of AB is a linear combination of the rows of B.

27. Show that each column of AB is a linear combination of the columns of A.

28. Suppose that A is an $m \times n$ matrix.

a. Show that every vector in the null space of A is orthogonal to every row of A.

b. Show that every vector in the null space of A^T is orthogonal to every column of A.

29. If $\mathbf{a}$ is an $n \times 1$ matrix and $\mathbf{b}$ is a $1 \times m$ matrix, show that $\mathbf{ab}$ is an $n \times m$ matrix of rank at most 1.

30. Prove that multiplication of a matrix by another matrix cannot increase its rank. (That is, if A and B are conformable matrices, then $\text{rank}(AB) \leq \min\{\text{rank}(A), \text{rank}(B)\}$.)
   [*Hint*: Use the results of Exercises B26 and B27.]

31. Find two square matrices A and B such that $\text{rank}(A) = \text{rank}(B)$ but $\text{rank}(A^2) \neq \text{rank}(B^2)$.

32. a. If A is an $m \times n$ matrix, prove that $\text{rank}(AA^T) = \text{rank}(A^T A) = \text{rank}(A)$.
   b. If A is a 17×9 matrix, what is the largest possible rank of $A^T A$?

33. a. If A is $m \times p$ and B is $p \times n$, prove that

   $$AB = \text{col}_1(A)\text{row}_1(B) + \text{col}_2(A)\text{row}_2(B) + \cdots + \text{col}_p(A)\text{row}_p(B).$$

   b. What is the size of each term $\text{col}_k(A)\, \text{row}_k(B)$ in the summation given in part (a)? (Each summand is called an **outer product**.)
   c. Apply the formula in part (a) to calculate the product of

   $$A = \begin{bmatrix} 1 & 0 \\ -3 & 4 \\ 0 & 5 \end{bmatrix} \quad \text{and} \quad B = \begin{bmatrix} 1 & -2 & 3 & -4 \\ 5 & -6 & 7 & -8 \end{bmatrix}.$$

C.

1. Theorem 3.3.3(a) states that $A(BC) = (AB)C$. When B is square and C has fewer columns than A has rows, explain why it is more efficient to compute $A(BC)$ instead of $(AB)C$. [Here "more efficient" means that you need fewer scalar additions and multiplications. Compare with Exercise B17.]

2. If $A = [a_{ij}]$ and $B = [b_{ij}]$ are both $m \times n$ matrices, define the **Hadamard** (or **Schur**)* **product** $A*B$ as

   $$A*B = [a_{ij}b_{ij}] = \begin{bmatrix} a_{11}b_{11} & a_{12}b_{12} & \cdots & a_{1n}b_{1n} \\ a_{21}b_{21} & a_{22}b_{22} & \cdots & a_{2n}b_{2n} \\ \vdots & \vdots & \vdots & \vdots \\ a_{m1}b_{m1} & a_{m2}b_{m2} & \cdots & a_{mn}b_{mn} \end{bmatrix}.$$

---

* Named after the mathematicians Jacques Hadamard (1865–1963) and Issai Schur (1875–1941).

This is just the product obtained by multiplying the corresponding entries in each matrix. Prove the following properties of the Hadamard product.

a. $A*B = B*A$.
b. $A*(B*C) = (A*B)*C$.
c. $A*(B*C) = (A*B)*C$.
d. $A*(B + C) = A*B + A*C$.
e. There is an $n \times n$ "identity matrix" I such that $I*A = A = A*I$ for all $n \times n$ matrices A.

3. If A and B are $n \times n$ matrices, define the **Jordan product**, named after the German physicist Pascual Jordan (1902–1980), as follows:

$$A * B = \frac{AB + BA}{2}.$$

This product is useful in the mathematical treatment of quantum mechanics. Assuming that A, B, and C are $n \times n$ matrices:

a. Show that $A*B = B*A$.
b. Show by example that $A*(B*C)$ is not generally equal to $(A*B)*C$.
c. Show that $A*(B + C) = (A*B) + (A*C)$.
d. Show that $(B + C)*A = (B*A) + (C*A)$.
e. Show that $A*(B*A^2) = (A*B)*A^2$, where $A^2 = A*A = AA$.

4. If A and B are $n \times n$ matrices, define the **Lie product** (pronounced "Lee product"), named after the Norwegian mathematician Marius Sophus Lie (1842–1899), as follows:

$$A \times B = AB - BA.$$

Assuming that A, B, and C are $n \times n$ matrices and k is a scalar:

a. Show that $A \times B = -(B \times A)$.
b. $(kA) \times B = A \times (kB) = k (A \times B)$.
c. $A \times (B + C) = (A \times B) + (A \times C)$.
d. $(B + C) \times A = (B \times A) + (C \times A)$.
e. $A \times (B \times C) + B \times (C \times A) + C \times (A \times B) = O_n$ [the *Jacobi identity*].

5. If $\times$ denotes *Lie multiplication* (as defined in the previous exercise) and A, B are skew-symmetric matrices, prove that $A \times B$ is skew-symmetric.

6. If $*$ denotes *Jordan multiplication* (Exercise 3 previously) and $\times$ denotes *Lie multiplication* (Exercise 4 previously), prove that for any $n \times n$ matrices A, B, and C,

a. $A \times (B*B) = 2 [(A \times B)*B]$.
b. $A \times (B \times C) = 4 [(A*B)*C - (A*C)*B]$.
c. $AB = (A*B) + (A \times B)/2$, where AB is the ordinary (Cayley) product of A and B.

3.4 Partitioned Matrices

Our first look at the concept of partitioned matrices was back in Section 2.2, where we introduced the *augmented matrix*. Also, in discussing the various equivalent definitions of matrix multiplication (Section 3.3), we considered a matrix split into *blocks*, each of which was a column vector or a row vector.

There are good reasons, both theoretical and practical, for examining partitioned matrices. To understand a practical reason, let us consider the product of a 2×3 matrix and a 3×4 matrix. In the final result, there are $(2)(4) = 8$ entries that must be calculated, with each entry requiring three scalar multiplications and two scalar additions. Therefore, altogether there are $(8)(3) = 24$ multiplications and $(8)(2) = 16$ additions to be carried out. In the world of computer calculations, multiplication is traditionally considered a more "expensive" operation than addition, although, as we have already noted in Section 2.5, on modern computers floating point multiplications have become comparable with floating point additions.

In general, the product of an $m \times n$ matrix and an $n \times p$ matrix results in mp entries, each entry requiring n multiplications and $n - 1$ additions, for a total of mpn multiplications and $mp(n - 1)$ additions to be performed. For square matrices with n rows and n columns, this becomes n^3 multiplications and $n \cdot n \cdot (n - 1) = n^3 - n^2$ additions. When n is large, this is a lot of computing! Limited storage space forces scientists to be very efficient in doing matrix computations. Partitioning large matrices is one way of being efficient.

A **submatrix** of a matrix A is a smaller matrix contained within A. We can think of a submatrix of A as the array that results from deleting some (or no) rows and/or columns of A.

Definition 3.4.1

An $m \times n$ matrix A is said to be **partitioned into submatrices** when it is written in the form:

$$A = \begin{bmatrix} A_{11} & A_{12} & \cdots & A_{1s} \\ A_{21} & A_{22} & \cdots & A_{2s} \\ \vdots & \vdots & \cdots & \vdots \\ A_{r1} & A_{r2} & \cdots & A_{rs} \end{bmatrix},$$

where each A_{ij} is an $m_i \times n_j$ submatrix of A with $m_1 + m_2 + \cdots + m_r = m$ and $n_1 + n_2 + \cdots + n_s = n$.

These submatrices A_{ij} are also called **blocks**, and the partitioned form given previously is called a **block matrix**, whose size is $r \times s$.

Example 3.4.1: Partitions of a Matrix

Suppose $A = \begin{bmatrix} 1 & 2 & 3 & 4 & 5 \\ 0 & 1 & -2 & 3 & 6 \\ 4 & 3 & 2 & 1 & 0 \\ 2 & 4 & 1 & 3 & 5 \end{bmatrix}$. We can use dividing lines

to indicate how we want to partition A. For example, we can write

$A = \begin{bmatrix} 1 & 2 & 3 & \vdots & 4 & 5 \\ 0 & 1 & -2 & \vdots & 3 & 6 \\ 4 & 3 & 2 & \vdots & 1 & 0 \\ 2 & 4 & 1 & \vdots & 3 & 5 \end{bmatrix} = [A_1 \; A_2]$, where $A_1 = \begin{bmatrix} 1 & 2 & 3 \\ 0 & 1 & -2 \\ 4 & 3 & 2 \\ 2 & 4 & 1 \end{bmatrix}$ and

$A_2 = \begin{bmatrix} 4 & 5 \\ 3 & 6 \\ 1 & 0 \\ 3 & 5 \end{bmatrix}$. Note that A_1 is the submatrix formed by deleting

columns 4 and 5 of A, whereas submatrix A_2 consists of A with its first three columns removed.

We may choose a more ambitious partition such as

$A = \begin{bmatrix} 1 & 2 & 3 & \vdots & 4 & 5 \\ 0 & 1 & -2 & \vdots & 3 & 6 \\ \cdots & \cdots & \cdots & \cdots & \cdots & \cdots \\ 4 & 3 & 2 & \vdots & 1 & 0 \\ 2 & 4 & 1 & \vdots & 3 & 5 \end{bmatrix} = \begin{bmatrix} A_{11} & A_{12} \\ A_{21} & A_{22} \end{bmatrix}$, where

$A_{11} = \begin{bmatrix} 1 & 2 & 3 \\ 0 & 1 & -2 \end{bmatrix}$, $A_{12} = \begin{bmatrix} 4 & 5 \\ 3 & 6 \end{bmatrix}$, $A_{21} = \begin{bmatrix} 4 & 3 & 2 \\ 2 & 4 & 1 \end{bmatrix}$, and

$A_{22} = \begin{bmatrix} 1 & 0 \\ 3 & 5 \end{bmatrix}$. We can see, for example, that submatrix

A_{22} is obtained by deleting the first two rows and the first three columns of A. There are other possible partitions of A, and we do not necessarily have to delete *consecutive* rows and/or columns to obtain submatrices.

Returning to the question of large-scale computer calculations, let us consider the product of two 100×100 matrices, A and B. There are 10,000 entries in each matrix, and calculating the product of these matrices requires $100^3 = 1,000,000$ scalar multiplications and $100^3 - 100^2 = 990,000$ scalar additions. Instead of straining the storage capacity of our computer (in terms of holding intermediate arithmetic results and so forth), we can partition each matrix into smaller matrices—for example, four 50×50 blocks:

$$A = \begin{bmatrix} a_{11} & a_{12} & \cdots & a_{1,50} & \vdots & a_{1,51} & \cdots & a_{1,100} \\ a_{21} & a_{22} & \cdots & a_{2,50} & \vdots & a_{2,51} & \cdots & a_{2,100} \\ \vdots & \vdots & \vdots & \vdots & \vdots & \vdots & \vdots & \vdots \\ a_{50,1} & a_{50,2} & \cdots & a_{50,50} & \vdots & a_{50,51} & \cdots & a_{50,100} \\ \cdots & \cdots & \cdots & \cdots & \vdots & \cdots & \cdots & \cdots \\ a_{51,1} & a_{51,2} & \cdots & a_{51,50} & \vdots & a_{51,51} & \cdots & a_{51,100} \\ \vdots & \vdots & \vdots & \vdots & \vdots & \vdots & \vdots & \vdots \\ a_{100,1} & a_{100,2} & \cdots & a_{100,50} & \vdots & a_{100,51} & \cdots & a_{100,100} \end{bmatrix} = \begin{bmatrix} A_{11} & A_{12} \\ A_{21} & A_{22} \end{bmatrix}$$

and, similarly, $B = \begin{bmatrix} B_{11} & B_{12} \\ B_{21} & B_{22} \end{bmatrix}$, where A_{ij} is a 50×50 matrix $(i, j = 1, 2)$ and B_{ij} is a 50×50 matrix $(i, j = 1, 2)$.

The payoff in this decomposition is that, because the submatrices are conformable, we can treat the submatrices like scalars and write

$$AB = \begin{bmatrix} A_{11} & A_{12} \\ A_{21} & A_{22} \end{bmatrix} \begin{bmatrix} B_{11} & B_{12} \\ B_{21} & B_{22} \end{bmatrix} = \begin{bmatrix} A_{11}B_{11} + A_{12}B_{21} & A_{11}B_{12} + A_{12}B_{22} \\ A_{21}B_{11} + A_{22}B_{21} & A_{21}B_{12} + A_{22}B_{22} \end{bmatrix}.$$

With these smaller submatrices A_{ij} and B_{ij}, a computer can store the individual products $A_{ik}B_{kj}$ and clear the main storage for the next set of calculations.

Even for hand calculations, this block multiplication can be useful, particularly when there are special patterns in the matrices to be multiplied.

Example 3.4.2: The Product of Partitioned Matrices

In preparation for multiplication, the 4×4 matrices A and B have been partitioned to take advantage of special patterns:

$$A = \begin{bmatrix} 2 & 0 & \vdots & 0 & 0 \\ 0 & 2 & \vdots & 0 & 0 \\ \cdots & \cdots & \vdots & \cdots & \cdots \\ 1 & 0 & \vdots & -3 & 5 \\ 0 & 1 & \vdots & 4 & 6 \end{bmatrix} = \begin{bmatrix} 2I_2 & O_2 \\ I_2 & C \end{bmatrix} \text{ and}$$

$$B = \begin{bmatrix} 1 & 2 & \vdots & 1 & 0 \\ 3 & 4 & \vdots & 0 & 1 \\ \cdots & \cdots & \vdots & \cdots & \cdots \\ -3 & 0 & \vdots & 0 & 0 \\ 0 & -3 & \vdots & 0 & 0 \end{bmatrix} = \begin{bmatrix} D & I_2 \\ -3I_2 & O_2 \end{bmatrix}.$$

Then, $AB = \begin{bmatrix} 2I_2 & O_2 \\ I_2 & C \end{bmatrix} \begin{bmatrix} D & I_2 \\ -3I_2 & O_2 \end{bmatrix}$

$= \begin{bmatrix} 2I_2 D + O_2(-3I_2) & 2I_2 I_2 + O_2 O_2 \\ I_2 D + C(-3I_2) & I_2 I_2 + CO_2 \end{bmatrix}$

$= \begin{bmatrix} 2D & 2I_2 \\ D - 3C & I_2 \end{bmatrix} = \begin{bmatrix} 2 & 4 & \vdots & 2 & 0 \\ 6 & 8 & \vdots & 0 & 2 \\ \cdots & \cdots & \vdots & \cdots & \cdots \\ 10 & -13 & \vdots & 1 & 0 \\ -9 & -14 & \vdots & 0 & 1 \end{bmatrix}$

$= \begin{bmatrix} 2 & 4 & 2 & 0 \\ 6 & 8 & 0 & 2 \\ 10 & -13 & 1 & 0 \\ -9 & -14 & 0 & 1 \end{bmatrix},$

a result we can check by multiplying in one of the usual ways. Note that in this example, the only nontrivial matrix operations needed to calculate the product AB are multiplication by a scalar and subtraction. (The multiplications involving I_2 and O_2 are considered trivial.)

We can generalize all this as follows:

Definition 3.4.2

Suppose that A ($m \times n$) and B ($n \times p$) are partitioned into blocks as indicated:

$$A = \begin{bmatrix} A_{11} & A_{12} & \cdots & A_{1n} \\ A_{21} & A_{22} & \cdots & A_{2n} \\ \vdots & \vdots & \vdots & \vdots \\ A_{m1} & A_{m2} & \cdots & A_{mn} \end{bmatrix}, \quad B = \begin{bmatrix} B_{11} & B_{12} & \cdots & B_{1p} \\ B_{21} & B_{22} & \cdots & B_{2p} \\ \vdots & \vdots & \vdots & \vdots \\ B_{n1} & B_{n2} & \cdots & B_{np} \end{bmatrix}.$$

If each pair (A_{ik}, B_{kj}) is conformable for multiplication, then A and B are said to be **conformably partitioned**.

If matrices A and B are conformably partitioned, we can calculate the product $AB = C$ in the usual row-by-column way (or in any other equivalent way). For example, the (i, j) block in C could be written as

$$A_{i1}B_{1j} + A_{i2}B_{2j} + \cdots + A_{in}B_{nj}.$$

Example 3.4.3: Conformably Partitioned Matrices

Suppose we want to find AB, where

$$A = \begin{bmatrix} 2 & 1 & 3 \\ 0 & -1 & 2 \end{bmatrix} \quad \text{and} \quad B = \begin{bmatrix} 3 & -7 & -7 & 2 \\ -2 & 1 & 4 & 0 \\ 0 & 2 & 4 & 0 \end{bmatrix}.$$

There are several ways to partition these matrices conformably, but we will choose the following partitions by way of example:

$$A = \begin{bmatrix} 2 & \vdots & 1 & 3 \\ 0 & \vdots & -1 & 2 \end{bmatrix} = [\, A_{11} \;\; A_{12} \,], \text{ where } A_{11} = \begin{bmatrix} 2 \\ 0 \end{bmatrix} \text{ and}$$

$$A_{12} = \begin{bmatrix} 1 & 3 \\ -1 & 2 \end{bmatrix} \text{ and}$$

$$B = \begin{bmatrix} 3 & -7 & -7 & 2 \\ \cdots & \cdots & \cdots & \cdots \\ -2 & 1 & 4 & 0 \\ 0 & 2 & 4 & 0 \end{bmatrix} = \begin{bmatrix} B_{11} \\ B_{21} \end{bmatrix}, \text{ where } B_{11} = [3 \; -7 \; -7 \; 2] \text{ and}$$

$$B_{21} = \begin{bmatrix} -2 & 1 & 4 & 0 \\ 0 & 2 & 4 & 0 \end{bmatrix}. \text{ Then,}$$

$$AB = [A_{11} \; A_{12}] \begin{bmatrix} B_{11} \\ B_{21} \end{bmatrix} = [A_{11}B_{11} + A_{12}B_{12}]$$

$$= \left[\begin{bmatrix} 2 \\ 0 \end{bmatrix} [3 \; -7 \; -7 \; 2] + \begin{bmatrix} 1 & 3 \\ -1 & 2 \end{bmatrix} \begin{bmatrix} -2 & 1 & 4 & 0 \\ 0 & 2 & 4 & 0 \end{bmatrix} \right]$$

$$= \left[\begin{bmatrix} 6 & -14 & -14 & 4 \\ 0 & 0 & 0 & 0 \end{bmatrix} + \begin{bmatrix} -2 & 7 & 16 & 0 \\ 2 & 3 & 4 & 0 \end{bmatrix} \right] = \begin{bmatrix} 4 & -7 & 2 & 4 \\ 2 & 3 & 4 & 0 \end{bmatrix}.$$

Exercises 3.4

A.

1.   In each of the following parts, calculate the product AB using the partitions indicated.

   a. $A = \begin{bmatrix} 1 & 2 & 3 & \vdots & 4 & 5 \\ 0 & -1 & -2 & \vdots & 6 & 1 \end{bmatrix}$, $B = \begin{bmatrix} 1 & 2 \\ 3 & 4 \\ 5 & 6 \\ \cdots & \cdots \\ 0 & 3 \\ 4 & 1 \end{bmatrix}$

b. $A = \begin{bmatrix} -1 & 2 & \vdots & 0 \\ 3 & 4 & \vdots & 1 \\ \cdots & \cdots & \vdots & \cdots \\ 0 & 5 & \vdots & 0 \end{bmatrix}$, $B = \begin{bmatrix} 0 & 3 & 2 & \vdots & 1 \\ 4 & 5 & -1 & \vdots & 6 \\ \cdots & \cdots & \cdots & \vdots & \cdots \\ 1 & 0 & 0 & \vdots & 1 \end{bmatrix}$

c. $A = \begin{bmatrix} 1 & 0 & 0 & \vdots & 3 \\ 0 & 1 & 0 & \vdots & 2 \\ 0 & 0 & 1 & \vdots & 1 \end{bmatrix}$, $B = \begin{bmatrix} 1 & 0 & 0 \\ 0 & 1 & 0 \\ 0 & 0 & 0 \\ \cdots & \cdots & \cdots \\ 1 & 2 & 3 \end{bmatrix}$

d. $A = \begin{bmatrix} 1 & 2 & \vdots & 0 & 0 & 0 \\ 3 & 4 & \vdots & 0 & 0 & 0 \\ 5 & 6 & \vdots & 0 & 0 & 0 \\ \cdots & \cdots & \vdots & \cdots & \cdots & \cdots \\ 0 & 0 & \vdots & 5 & 1 & 2 \\ 0 & 0 & \vdots & 3 & 4 & 1 \end{bmatrix}$, $B = \begin{bmatrix} 3 & -2 & \vdots & 0 & 0 \\ 2 & 4 & \vdots & 0 & 0 \\ \cdots & \cdots & \vdots & \cdots & \cdots \\ 0 & 0 & \vdots & 1 & 2 \\ 0 & 0 & \vdots & 2 & -3 \\ 0 & 0 & \vdots & -4 & 1 \end{bmatrix}$

e. $A = \begin{bmatrix} 1 & 2 & \vdots & -1 & 0 \\ \cdots & \cdots & \cdots & \cdots & \cdots \\ 4 & 0 & \vdots & 2 & 1 \\ 2 & -5 & \vdots & 1 & 2 \end{bmatrix}$, $B = \begin{bmatrix} 3 & 1 & \vdots & 2 \\ -4 & 5 & \vdots & -2 \\ \cdots & \cdots & \cdots & \cdots \\ 1 & 0 & \vdots & 3 \\ 2 & 3 & \vdots & -1 \end{bmatrix}$

2. Suppose

$$A = \begin{bmatrix} 1 & 2 & -1 & 0 \\ 4 & 0 & 2 & 1 \\ 2 & -5 & 1 & 2 \end{bmatrix} \quad \text{and} \quad B = \begin{bmatrix} 3 & 1 & 2 \\ -4 & 5 & -2 \\ 1 & 0 & 3 \\ 2 & 3 & -1 \end{bmatrix}.$$

Calculate AB

a. Directly (without using partitions).

b. With the partitions indicated:

$$\begin{bmatrix} 3 & 1 & \vdots & 2 \\ -4 & 5 & \vdots & -2 \\ \cdots & \cdots & \vdots & \cdots \\ 1 & 0 & \vdots & 3 \\ 2 & 3 & \vdots & -1 \end{bmatrix}, \begin{bmatrix} 3 & 1 & \vdots & 2 \\ -4 & 5 & \vdots & -2 \\ \cdots & \cdots & \vdots & \cdots \\ 1 & 0 & \vdots & 3 \\ 2 & 3 & \vdots & -1 \end{bmatrix}$$

c. With a different partition of your own choice.

3. Complete the partitioning for the following product in such a way that each matrix is divided into four submatrices, which are conformable for multiplication:

$$\begin{bmatrix} \times & \times & \times & \times \\ \times & \times & \times & \times \\ \times & \times & \times & \times \\ \cdots & \cdots & \cdots & \cdots \\ \times & \times & \times & \times \end{bmatrix} \begin{bmatrix} \times & \times & \times & \times \\ \cdots & \cdots & \cdots & \cdots \\ \times & \times & \times & \times \\ \times & \times & \times & \times \\ \times & \times & \times & \times \end{bmatrix} = \begin{bmatrix} \times & \vdots & \times & \times & \times \\ \times & \vdots & \times & \times & \times \\ \times & \vdots & \times & \times & \times \\ \times & \vdots & \times & \times & \times \end{bmatrix}$$

4. If A and B are $n \times n$ matrices and C is the $2n \times 2n$ matrix defined by $C = \begin{bmatrix} A & B \\ B & -A \end{bmatrix}$, calculate C^2.

5. Let A be an $n \times n$ matrix and let $B = \begin{bmatrix} A & I_n & O_n \\ I_n & A & I_n \\ O_n & I_n & A \end{bmatrix}$. Calculate B^2 and B^3.

6. If A is $m \times n$, B is $n \times r$, and C is $n \times t$, what is the size of $A \cdot [B\ C] = [AB\ AC]$?

B.

1. In how many ways can a 4×5 matrix be partitioned into a 2×3 block matrix—that is, a partitioned matrix of the form:

$$\begin{bmatrix} A_{11} & A_{12} & A_{13} \\ A_{21} & A_{22} & A_{23} \end{bmatrix},$$

where A_{ij} has size $m_i \times n_j$ $(i = 1, 2;\ j = 1, 2, 3)$ and $m_1 + m_2 = 4$, $n_1 + n_2 + n_3 = 5$?

[*Hint*: In how many ways can you position the horizontal and the vertical dividing lines?]

2. Suppose that a matrix A is partitioned so that $A = \begin{bmatrix} B & C \\ O & O \end{bmatrix}$. Use mathematical induction to show that $A^n = \begin{bmatrix} B^n & B^{n-1}C \\ O & O \end{bmatrix}$.

3. If $A = \begin{bmatrix} A_{11} & A_{12} \\ A_{21} & A_{22} \end{bmatrix}$ is a partitioned matrix, what is A^T? Try to generalize this.

4. Suppose
$$A = \begin{bmatrix} I_2 & O_2 \\ O_2 & -I_2 \end{bmatrix}, \quad B = \begin{bmatrix} -I_2 & G \\ O_2 & I_2 \end{bmatrix}, \quad \text{and } C = \begin{bmatrix} -I_2 & O_2 \\ H & I_2 \end{bmatrix}.$$
a. Show that $A^2 = B^2 = C^2 = I_4$.
b. Show that $AB + BA = AC + CA = -2I_4$.
c. If $GH = HG = O_2$, show that $BC + CD = 2I_4$.

5. If $A = \begin{bmatrix} A_{11} & A_{12} \\ O & A_{22} \end{bmatrix}$, where A_{11} and A_{12} are square matrices, show that $p(A) = \begin{bmatrix} p(A_{11}) & \vdots & \text{(a messy expression)} \\ \cdots & \vdots & \cdots\cdots\cdots\cdots \\ O & \vdots & p(A_{22}) \end{bmatrix}$ for any polynomial $p(x)$. [*Hint*: Start by considering A^k for a nonnegative integer k.]

6. If A is a matrix of rank r, show that $\text{rank}(B) = r$, where
    $$B = \begin{bmatrix} A & O \\ O & O \end{bmatrix}.$$

7. a. Prove that $\text{rank} \begin{bmatrix} A & B \\ O & C \end{bmatrix} \geq \text{rank}(A) + \text{rank}(C)$.

    b. Show that strict inequality can occur in the inequality of part (a).

    c. Explain why the rank of an upper triangular matrix is not less than the number of nonzero diagonal elements.

C.

1. a. Show that an $m \times n$ matrix has $(2^m - 1)(2^n - 1)$ submatrices. [*Hint:* A submatrix corresponds to a subset of rows deleted and/or a subset of columns deleted. Consider the number of possible subsets of the set of m rows and n columns.]

    b. How many submatrices does a 4×5 matrix have?

2. For a partitioned matrix, show that $\begin{bmatrix} A & B \\ C & D \end{bmatrix}^{\mathrm{T}} = \begin{bmatrix} A^{\mathrm{T}} & C^{\mathrm{T}} \\ B^{\mathrm{T}} & D^{\mathrm{T}} \end{bmatrix}$.

3. If $P = \begin{bmatrix} a & b & c & d \\ -b & a & -d & c \\ -c & d & a & -b \\ -d & -c & b & a \end{bmatrix}$, show in a simple way, by appropriate partitioning, that $PP^{\mathrm{T}} = (a^2 + b^2 + c^2 + d^2)I_4$. [*Hint:* You may want to use the result of Exercise C2.]

3.5 Inverses of Matrices

If we have an ordinary (scalar) equation $ax = b$, where $a \neq 0$, we can always use the multiplicative inverse of a to cancel the coefficient a and write the solution as $x = a^{-1}b = b/a$. However, in Example 3.3.8, we saw that cancellation is not always possible when working with matrix equations.

Let us look at a system of n linear equations in n unknowns

$$a_{11}x_1 + a_{12}x_2 + \cdots + a_{1n}x_n = b_1$$
$$a_{21}x_1 + a_{22}x_2 + \cdots + a_{2n}x_n = b_2$$
$$\vdots \qquad \vdots \qquad \qquad \vdots \qquad \vdots \qquad (*)$$
$$a_{n1}x_1 + a_{n2}x_2 + \cdots + a_{nn}x_n = b_n$$

In matrix form, this system becomes $A\mathbf{x} = \mathbf{b}$, where $A = [a_{ij}]$ is the (square) matrix of coefficients, $\mathbf{x} = \begin{bmatrix} x_1 \\ x_2 \\ \vdots \\ x_n \end{bmatrix}$, and $\mathbf{b} = \begin{bmatrix} b_1 \\ b_2 \\ \vdots \\ b_n \end{bmatrix}$. Based on our previous experience with algebra, we would like to solve this system by multiplying both sides of the matrix equation by some kind of *inverse* of the square matrix A, in effect "dividing out" the factor A. Then, we could write the solution as $\mathbf{x} = A^{-1}\mathbf{b}$. There is no general division operation for matrices, but we can come close to this idea, as indicated by the next definition.

Definition 3.5.1

Suppose A is an $n \times n$ matrix. If there is an $n \times n$ matrix B such that $AB = I_n$ and $BA = I_n$, then A is called **nonsingular** (or **invertible**) and B is called an **inverse** of A, denoted by $B = A^{-1}$. (Clearly, we can write $A = B^{-1}$ as well.) If A has no inverse, we call A a **singular** (or **noninvertible**) matrix.

Example 3.5.1: Singular and Nonsingular Matrices

If $A = \begin{bmatrix} 1 & 2 \\ 3 & 4 \end{bmatrix}$, we can see that

$$\begin{bmatrix} 1 & 2 \\ 3 & 4 \end{bmatrix} \begin{bmatrix} -2 & 1 \\ \frac{3}{2} & -\frac{1}{2} \end{bmatrix} = \begin{bmatrix} 1 & 0 \\ 0 & 1 \end{bmatrix} = \begin{bmatrix} -2 & 1 \\ \frac{3}{2} & -\frac{1}{2} \end{bmatrix} \begin{bmatrix} 1 & 2 \\ 3 & 4 \end{bmatrix},$$

so that A is *invertible* and $A^{-1} = \begin{bmatrix} -2 & 1 \\ \frac{3}{2} & -\frac{1}{2} \end{bmatrix}$.

On the other hand, the matrix $\begin{bmatrix} 2 & 5 \\ 4 & 10 \end{bmatrix}$ is *singular*. To see this, suppose that we could find a 2×2 matrix $B = [b_{ij}]$ such that $\begin{bmatrix} 2 & 5 \\ 4 & 10 \end{bmatrix} B = I_2 = B \begin{bmatrix} 2 & 5 \\ 4 & 10 \end{bmatrix}$. If we focus on just the first equation, we see that this equation is equivalent to the system:

$$(1)\ 2b_{11} + 5b_{21} = 1,$$
$$(2)\ 2b_{12} + 5b_{22} = 0,$$
$$(3)\ 4b_{11} + 10b_{21} = 0,$$
$$(4)\ 4b_{12} + 10b_{22} = 1.$$

But we see that the left-hand side of Equation 3 is twice the left-hand side of Equation 1, whereas the corresponding right-hand sides do not have this relationship. Therefore, we know that we cannot find such a matrix B, and so $\begin{bmatrix} 2 & 5 \\ 4 & 10 \end{bmatrix}$ is noninvertible.

It is obvious that the zero matrix O_n is singular for every n, but the last example shows that even a matrix without a single zero entry can be noninvertible. Before we solve the problem of calculating the inverse of a nonsingular matrix—and in fact determining *if* a given matrix is nonsingular—let us focus on an important point we have glided over. Definition 3.5.1 describes "an" inverse rather than "the" inverse. In other words, the definition leaves open the possibility that an invertible matrix could have more than one inverse. Perhaps the inverse we plucked out of our sleeve for A in Example 3.5.1 is only one of the many inverses for A. Before we let our imaginations run wild, let us end the mystery by declaring and proving the uniqueness of inverses.

Theorem 3.5.1

If an $n \times n$ matrix A has an inverse B, then B is the only inverse of A.

Proof We give a proof by contradiction. Suppose we assume that A has two distinct inverses, B_1 and B_2. Then $AB_1 = I_n = B_1A$ and $AB_2 = I_n = B_2A$, so we can write

$$B_2 = B_2I_n = B_2(AB_1) = (B_2A)B_1 = I_nB_1 = B_1.$$

This contradiction shows there can be only one inverse for A.

Now we can speak with confidence about *the* inverse of a nonsingular matrix. It is a fact that **if A is invertible, then A^{-1} is also invertible and *its* inverse is A**: $(A^{-1})^{-1} = A$. (See Exercise A2.)

Furthermore, whereas the definition of a matrix inverse involves a two-sided multiplication, the next theorem points out that we only have to find a *one-sided inverse*. The proof of this result would be more direct if we understood some concepts from the next section, but we can prove the theorem right now, using what we already know.

Theorem 3.5.2

If A is an $n \times n$ matrix for which there exists a matrix B such that $AB = I_n$, then $BA = I_n$ also, so that A is nonsingular and $B = A^{-1}$.

Proof Suppose $A = [\mathbf{a}_1\ \mathbf{a}_2\ \ldots\ \mathbf{a}_n]$ and $B = [\mathbf{b}_1\ \mathbf{b}_2\ \ldots\ \mathbf{b}_n]$. First, we show that $A\mathbf{b}_j = \mathbf{e}_j$, $j = 1, 2, \ldots, n$, where $\mathbf{e}_j$ is the jth standard basis vector in $\mathbb{R}^n$:

$AB = A[\mathbf{b}_1\ \mathbf{b}_2\ \ldots\ \mathbf{b}_n] = [A\mathbf{b}_1\ A\mathbf{b}_2\ \ldots\ A\mathbf{b}_n] = I_n = [\mathbf{e}_1\ \mathbf{e}_2\ \ldots\ \mathbf{e}_n]$

implies $A\mathbf{b}_j = \mathbf{e}_j$ by the definition of matrix equality.

Next, we show that the n columns of B are linearly independent and thus span $\mathbb{R}^n$ (see Section 1.5). Suppose $c_1\mathbf{b}_1 + c_2\mathbf{b}_2 + \cdots + c_n\mathbf{b}_n = \mathbf{0}$. Then,

$A(c_1\mathbf{b}_1 + c_2\mathbf{b}_2 + \cdots + c_n\mathbf{b}_n) = A \cdot \mathbf{0} = \mathbf{0}$, or

$c_1(A\mathbf{b}_1) + c_2(A\mathbf{b}_2) + \cdots + c_n(A\mathbf{b}_n) = c_1\mathbf{e}_1 + c_2\mathbf{e}_2 + \cdots + c_n\mathbf{e}_n = \mathbf{0}$.

Because the set $\{\mathbf{e}_i\}$ is linearly independent, this implies that $c_1 = c_2 = \cdots = c_n = 0$. Therefore, the columns of B span $\mathbb{R}^n$. In particular, there exist scalars x_{ij} such that $x_{1j}\mathbf{b}_1 + x_{2j}\mathbf{b}_2 + \cdots + x_{nj}\mathbf{b}_n = \mathbf{e}_j$ for $i, j = 1, 2, \ldots, n$. We can write this last equation as $B\mathbf{x}_j = \mathbf{e}_j$, where $\mathbf{x}_j = [x_{1j}\ x_{2j} \ldots x_{nj}]^{\mathrm{T}}$. Now $\mathbf{x}_j = I_n\mathbf{x}_j = (AB)\mathbf{x}_j = A(B\mathbf{x}_j) = A\mathbf{e}_j = \mathbf{a}_j$. But then $BA = B[\mathbf{a}_1\ \mathbf{a}_2\ \ldots\ \mathbf{a}_n] = [B\mathbf{a}_1\ B\mathbf{a}_2\ \ldots\ B\mathbf{a}_n] = [B\mathbf{x}_1\ B\mathbf{x}_2 \ldots B\mathbf{x}_n] = [\mathbf{e}_1\ \mathbf{e}_2\ \ldots\ \mathbf{e}_n] = I_n$.

We should see that the conclusion of Theorem 3.5.2 is valid if the hypothesis were $BA = I_n$ for some matrix B. Then, we could conclude that $AB = I_n$ and $B = A^{-1}$.

Getting back to the main problem, we see that we can solve the system:

$$a_{11}x_1 + a_{12}x_2 + \cdots + a_{1n}x_n = b_1$$

$$a_{21}x_1 + a_{22}x_2 + \cdots + a_{2n}x_n = b_2$$

$$\vdots \qquad \vdots \qquad \qquad \vdots \qquad \vdots$$

$$a_{n1}x_1 + a_{n2}x_2 + \cdots + a_{nn}x_n = b_n$$

or $A\mathbf{x} = \mathbf{b}$, by the formula $\mathbf{x} = A^{-1}\mathbf{b}$ if the coefficient matrix A is nonsingular.

In general, if A is nonsingular, we can solve any matrix equation of the form $AX = B$, where A, X, and B are conformable matrices. We just multiply both sides (on the left) by A^{-1} to get the solution $X = A^{-1}B$. Similarly, if an equation has the form $XA = B$ and A is nonsingular, we multiply on the *right* by A^{-1} to obtain the solution $X = BA^{-1}$.*

Example 3.5.2: Solving a System via Matrix Inversion

Consider the system:

$$x + 2z = 1$$
$$2x - y + 3z = 2$$
$$4x + y + 8z = 3$$

which is represented by the matrix equation $A\mathbf{x} = \mathbf{b}$, where

$A = \begin{bmatrix} 1 & 0 & 2 \\ 2 & -1 & 3 \\ 4 & 1 & 8 \end{bmatrix}$, $\mathbf{x} = \begin{bmatrix} x \\ y \\ z \end{bmatrix}$, and $\mathbf{b} = \begin{bmatrix} 1 \\ 2 \\ 3 \end{bmatrix}$. We can verify

that $A^{-1} = \begin{bmatrix} -11 & 2 & 2 \\ -4 & 0 & 1 \\ 6 & -1 & -1 \end{bmatrix}$. Then, $\mathbf{x} = A^{-1}\mathbf{b} =$

$\begin{bmatrix} -11 & 2 & 2 \\ -4 & 0 & 1 \\ 6 & -1 & -1 \end{bmatrix} \begin{bmatrix} 1 \\ 2 \\ 3 \end{bmatrix} = \begin{bmatrix} -1 \\ -1 \\ 1 \end{bmatrix}$.

---

* Because matrix multiplication is not commutative, there is ambiguity in defining matrix division and writing $A \div B$ or A/B: The expression could mean AB^{-1} or $B^{-1}A$.

Once we understand how to find inverses of nonsingular matrices, we can solve more general matrix equations, as our next example shows. However, *except for theoretical purposes, it is seldom necessary to calculate the inverse of a matrix.* Systems of equations are usually solved (with computers) by some variant of Gaussian elimination.

Example 3.5.3: Solving a Matrix Equation

Assuming that we can find the necessary inverse, let us solve the matrix equation:

$$X\begin{bmatrix} 3 & -2 \\ 5 & -4 \end{bmatrix} = \begin{bmatrix} -1 & 2 \\ -5 & 6 \end{bmatrix}.$$

We can verify that $\begin{bmatrix} 3 & -2 \\ 5 & -4 \end{bmatrix}^{-1} = \begin{bmatrix} 2 & -1 \\ \frac{5}{2} & -\frac{3}{2} \end{bmatrix}$, so that multiplying both sides of the original matrix equation by this inverse on the right, we get

$$X\begin{bmatrix} 3 & -2 \\ 5 & -4 \end{bmatrix}\begin{bmatrix} 3 & -2 \\ 5 & -4 \end{bmatrix}^{-1} = \begin{bmatrix} -1 & 2 \\ -5 & 6 \end{bmatrix}\begin{bmatrix} 3 & -2 \\ 5 & -4 \end{bmatrix}^{-1},$$

$$X\begin{bmatrix} 3 & -2 \\ 5 & -4 \end{bmatrix}\begin{bmatrix} 2 & -1 \\ \frac{5}{2} & -\frac{3}{2} \end{bmatrix} = \begin{bmatrix} -1 & 2 \\ -5 & 6 \end{bmatrix}\begin{bmatrix} 2 & -1 \\ \frac{5}{2} & -\frac{3}{2} \end{bmatrix}, \text{ or}$$

$$X = \begin{bmatrix} 3 & -2 \\ 5 & -4 \end{bmatrix}.$$

To get some understanding of what may be involved in determining the invertibility of a matrix and then actually calculating this inverse, we will look at a simple example. We will discuss the general process for finding inverses in Section 3.6.

Example 3.5.4: The Inverse of a 2 × 2 Matrix

Given any 2×2 matrix $A = \begin{bmatrix} a & b \\ c & d \end{bmatrix}$, we want to determine *if* it has an inverse and, if so, what the inverse looks like. First, let us assume that A has an inverse.

This means we are looking for the unique matrix $M = \begin{bmatrix} x & y \\ z & w \end{bmatrix}$ such that $MA = I_2 = AM$. By Theorem 3.5.2, we can focus on the equation $AM = I_2$, or

$$\begin{bmatrix} ax + bz & ay + bw \\ cx + dz & cy + dw \end{bmatrix} = \begin{bmatrix} 1 & 0 \\ 0 & 1 \end{bmatrix}.$$

This matrix equation can be written in the form $C\mathbf{u} = \mathbf{v}$:

$$\begin{bmatrix} a & 0 & b & 0 \\ c & 0 & d & 0 \\ 0 & a & 0 & b \\ 0 & c & 0 & d \end{bmatrix} \begin{bmatrix} x \\ y \\ z \\ w \end{bmatrix} = \begin{bmatrix} 1 \\ 0 \\ 0 \\ 1 \end{bmatrix}.$$

If $a \neq 0$, we can reduce the augmented matrix of this system to the echelon form:

$$\begin{bmatrix} a & 0 & b & 0 & \vdots & 1 \\ 0 & a & 0 & b & \vdots & 0 \\ 0 & 0 & \frac{ad-bc}{a} & 0 & \vdots & -\frac{c}{a} \\ 0 & 0 & 0 & \frac{ad-bc}{a} & \vdots & 1 \end{bmatrix}.$$

Then, if $ad - bc \neq 0$, we can solve by backward substitution to get

$$w = \frac{a}{ad - bc}, \quad z = \frac{-c}{ad - bc}, \quad y = \frac{-b}{ad - bc}, \quad x = \frac{d}{ad - bc},$$

or

$$M = \frac{1}{ad - bc} \begin{bmatrix} d & -b \\ -c & a \end{bmatrix}.$$

It is easy to check that $AM = MA = I_2$, provided that $a \neq 0$ and $ad - bc \neq 0$.

Now if $a = 0$, we can interchange rows to put the augmented matrix into the form:

$$\begin{bmatrix} c & 0 & d & 0 & \vdots & 0 \\ 0 & c & 0 & d & \vdots & 1 \\ 0 & 0 & b & 0 & \vdots & 1 \\ 0 & 0 & 0 & b & \vdots & 0 \end{bmatrix}.$$

We see that neither c nor b can be zero or else rank $[C \mid \mathbf{v}] \neq$ rank C, and the system has no unique solution (Theorem 2.8.3). However, if $c \neq 0$ and $b \neq 0$, we find that

$$M = -\frac{1}{bc} \begin{bmatrix} d & -b \\ -c & 0 \end{bmatrix}.$$

To summarize: (1) matrix A has an inverse if and only if $ad - bc \neq 0$;

(2) if A is invertible, then $A^{-1} = \frac{1}{ad-bc} \cdot \begin{bmatrix} d & -b \\ -c & a \end{bmatrix}$.

In the formula for the inverse of a 2×2 matrix, the critical expression $ad - bc$ is called the *determinant* of the matrix A. Determinants will be discussed further in Chapter 4.

The last theorem of this section states some important algebraic properties of the matrix inverse.

Theorem 3.5.3: Properties of Matrix Inverses

Let A and B be invertible $n \times n$ matrices and k a nonzero scalar. Then

1. $(A^{-1})^{-1} = A$.
2. $(kA)^{-1} = \frac{1}{k}A^{-1}$.
3. $(AB)^{-1} = B^{-1}A^{-1}$.
4. $(A^{T})^{-1} = (A^{-1})^{T}$.

Proof of (1)   Exercise A2.

Proof of (2) $(kA)\frac{1}{k}A^{-1} = \left(k\frac{1}{k}\right)(AA^{-1}) = 1I_n = I_n$, so that $\frac{1}{k}A^{-1} = (kA)^{-1}$ by Theorem 3.5.2.

Proof of (3) $(AB)(B^{-1}A^{-1}) = A(BB^{-1})A^{-1} = A(I_n)A^{-1} = AA^{-1} = I_n$, so that $B^{-1}A^{-1} = (AB)^{-1}$ by Theorem 3.5.2.

Proof of (4) $A^{T}(A^{-1})^{T} = $ [by Theorem 3.3.3(f)] $(A^{-1}A)^{T} = I_n^{T} = I_n$, and so the property is proved.

Example 3.5.5: Properties of Inverses

Let $A = \begin{bmatrix} 1 & -2 \\ 3 & 4 \end{bmatrix}$ and $B = \begin{bmatrix} 0 & 3 \\ -4 & 1 \end{bmatrix}$. Using the formula developed in Example 3.5.4, we find that $A^{-1} = \frac{1}{10}\begin{bmatrix} 4 & 2 \\ -3 & 1 \end{bmatrix}$ and $B^{-1} = \frac{1}{12}\begin{bmatrix} 1 & -3 \\ 4 & 0 \end{bmatrix}$. Then, $(A^{-1})^{-1} = \begin{bmatrix} \frac{4}{10} & \frac{2}{10} \\ -\frac{3}{10} & \frac{1}{10} \end{bmatrix}^{-1} = $

$\frac{1}{\frac{1}{10}}\begin{bmatrix} \frac{1}{10} & -\frac{2}{10} \\ \frac{3}{10} & \frac{4}{10} \end{bmatrix} = \begin{bmatrix} 1 & -2 \\ 3 & 4 \end{bmatrix} = A$. Also, $(5A)^{-1} = \begin{bmatrix} 5 & -10 \\ 15 & 20 \end{bmatrix}^{-1} = $

$\frac{1}{250}\begin{bmatrix} 20 & 10 \\ -15 & 5 \end{bmatrix} = \frac{1}{5}\frac{1}{10}\begin{bmatrix} 4 & 2 \\ -3 & 1 \end{bmatrix} = \frac{1}{5}A^{-1}$. We see that $AB = \begin{bmatrix} 8 & 1 \\ -16 & 13 \end{bmatrix}$, $(AB)^{-1} = \frac{1}{120}\begin{bmatrix} 13 & -1 \\ 16 & 8 \end{bmatrix}$, and $B^{-1}A^{-1} = $

$\frac{1}{12}\begin{bmatrix} 1 & -3 \\ 4 & 0 \end{bmatrix} \cdot \frac{1}{10}\begin{bmatrix} 4 & 2 \\ -3 & 1 \end{bmatrix} = \frac{1}{120}\begin{bmatrix} 13 & -1 \\ 16 & 8 \end{bmatrix} = (AB)^{-1}$. Finally,

we observe that $(A^{T})^{-1} = \begin{bmatrix} 1 & 3 \\ -2 & 4 \end{bmatrix}^{-1} = \frac{1}{10}\begin{bmatrix} 4 & -3 \\ 2 & 1 \end{bmatrix}$ is the

same as $(A^{-1})^{T} = \frac{1}{10}\begin{bmatrix} 4 & -3 \\ 2 & 1 \end{bmatrix}$.

Property (3) of Theorem 3.5.3 can be generalized as follows: if each matrix $A_1, A_2, \ldots, A_k$ is $n \times n$ and invertible, then the product $A_1 A_2 \cdots A_k$ is invertible and $(A_1 A_2 \cdots A_{k-1} A_k)^{-1} = A_k^{-1} A_{k-1}^{-1} \cdots A_2^{-1} A_1^{-1}$ (see Exercise B3).

It is important to realize that the assumption behind property (3) is that *matrices A and B are square and invertible*. As the next example shows, these conditions are critical.

Example 3.5.6: Inverse Properties

Let $A = \begin{bmatrix} 1 & -1 & 1 \\ -2 & 0 & 1 \end{bmatrix}$ and $B = \begin{bmatrix} 1 & -1 \\ 2 & -2 \\ 2 & -1 \end{bmatrix}$. Then $AB = \begin{bmatrix} 1 & 0 \\ 0 & 1 \end{bmatrix} = I_2$, so $(AB)^{-1} = I_2$. But, because A and B are not square, A^{-1} and B^{-1} do not exist. Thus, the existence of the left side of property (3) of Theorem 3.5.3 does not imply the existence of the inverses on the right side.

Furthermore, if we consider the matrices $A = \begin{bmatrix} 1 & 1 \\ 1 & 1 \end{bmatrix}$ and $B = \begin{bmatrix} 3 & 1 \\ 1 & 3 \end{bmatrix}$, we find that A is not invertible (by Example 3.5.4), nor is $AB = \begin{bmatrix} 4 & 4 \\ 4 & 4 \end{bmatrix}$; but B *does* have an inverse.

3.5.1 Negative Powers of a Square Matrix

If A is an $n \times n$ nonsingular matrix, then for any positive integer m, we define A^{-m} as follows: $A^{-m} = (A^{-1})^m$. Exercise B5 asks for a proof that $(A^m)^{-1} = (A^{-1})^m$.

The laws of exponents for square matrices discussed in Section 3.3 can be extended to negative integer powers. Thus, $A^m A^n = A^{m+n}$ (see Exercise B6) and $(A^m)^n = A^{mn}$ hold for *all* integers.

Example 3.5.7: Negative Powers of a Matrix

Suppose $A = \begin{bmatrix} 1 & 2 \\ 0 & 1 \end{bmatrix}$. We can check that $A^{-1} = \begin{bmatrix} 1 & -2 \\ 0 & 1 \end{bmatrix}$.

Then, for example, $A^{-3} = (A^{-1})^3 = \begin{bmatrix} 1 & -2 \\ 0 & 1 \end{bmatrix}^3 = \begin{bmatrix} 1 & -6 \\ 0 & 1 \end{bmatrix}$.

Also, $A^3 = \begin{bmatrix} 1 & 6 \\ 0 & 1 \end{bmatrix}$ and $(A^3)^{-1} = \begin{bmatrix} 1 & -6 \\ 0 & 1 \end{bmatrix}$, so that we have $A^{-3} = (A^3)^{-1}$.

Exercises 3.5

A.

1. Verify each of the following:

   a. $\begin{bmatrix} -5 & 1 \\ -6 & 1 \end{bmatrix}^{-1} = \begin{bmatrix} 1 & -1 \\ 6 & -5 \end{bmatrix}$ b. $\begin{bmatrix} -1 & 5 \\ 3 & 0 \end{bmatrix}^{-1} = \begin{bmatrix} 0 & \frac{1}{3} \\ \frac{1}{5} & \frac{1}{15} \end{bmatrix}$

   c. $\begin{bmatrix} 1 & -2 & 3 \\ 2 & -5 & 10 \\ 0 & 0 & 1 \end{bmatrix}^{-1} = \begin{bmatrix} 5 & -2 & 5 \\ 2 & -1 & 4 \\ 0 & 0 & 1 \end{bmatrix}$

   d. $\begin{bmatrix} 1 & -2 & 3 \\ 2 & -5 & 10 \\ -1 & 2 & -2 \end{bmatrix}^{-1} = \begin{bmatrix} 10 & -2 & 5 \\ 6 & -1 & 4 \\ 1 & 0 & 1 \end{bmatrix}$

   e. $\begin{bmatrix} 1 & 0 & 0 & 0 \\ 3 & -1 & 0 & 0 \\ 0 & 1 & -1 & 0 \\ 2 & 4 & 3 & 1 \end{bmatrix}^{-1} = \begin{bmatrix} 1 & 0 & 0 & 0 \\ 3 & -1 & 0 & 0 \\ 3 & -1 & -1 & 0 \\ -23 & 7 & 3 & 1 \end{bmatrix}$

   f. $\begin{bmatrix} 5 & 8 & 6 & -13 \\ 3 & -1 & 0 & -9 \\ 0 & 1 & -1 & 0 \\ 2 & 4 & 3 & -5 \end{bmatrix}^{-1} = \begin{bmatrix} -68 & 21 & 9 & 139 \\ 3 & -1 & 0 & -6 \\ 3 & -1 & -1 & -6 \\ -23 & 7 & 3 & 47 \end{bmatrix}$

2. If A is a nonsingular square matrix, show that A^{-1} is invertible and $(A^{-1})^{-1} = A$.

3. If A, B, and X are conformable matrices, $XA = B - XB$, and $A + B$ is nonsingular, prove that $X = B(A + B)^{-1}$.

4. Let $A = \begin{bmatrix} 0 & -1 \\ 1 & -1 \end{bmatrix}$.
   a. Calculate A^2 and A^3.
   b. Show that A is invertible and that $A^{-1} = A^2$.

5. Define $M = \begin{bmatrix} k & 1 \\ 0 & 1 \end{bmatrix}$, where k is some nonzero constant.
   a. Find and prove a formula for M^n for any nonnegative integer n.
   b. Find and prove a formula for M^{-n} for any nonnegative integer n.
   c. Verify that $M^{-n} = (M^n)^{-1} = (M^{-1})^n$. (See Exercise B5.)

6. Using the formula derived in Example 3.5.4, find the inverses (if they exist) of the following matrices and check your answers:

   a. $\begin{bmatrix} 1 & 2 \\ 5 & 8 \end{bmatrix}$ b. $\begin{bmatrix} 2 & 3 \\ 4 & 6 \end{bmatrix}$ c. $\begin{bmatrix} -3 & 5 \\ 4 & 1 \end{bmatrix}$

   d. $\begin{bmatrix} -1 & -2 \\ -3 & -4 \end{bmatrix}$ e. $\begin{bmatrix} -2 & -1 \\ 6 & 3 \end{bmatrix}$

7.   Find the inverse of $\begin{bmatrix} a & \sqrt{1-a^2} \\ -\sqrt{1-a^2} & a \end{bmatrix}$, where $|\,a\,| < 1$.

8.   If the matrix $\begin{bmatrix} 1 & 1 \\ a & a^2 \end{bmatrix}$ has an inverse, what can you say about a?
     [*Hint*: What values can a *not* have?]

B.

1.   a. Show that a diagonal matrix is invertible if all its diagonal
        entries are nonzero. What is its inverse?
     b. Show that if a diagonal matrix is invertible, then all its
        diagonal entries are nonzero.

2.   Find all 3×3 diagonal matrices D such that $D^2 = I_3$.

3.   Prove the following generalization of property (3) of Theorem
     3.5.3: if each matrix $A_1, A_2, \ldots, A_k$ is $n \times n$ and invertible, then
     $A_1, A_2, \ldots, A_k$ is invertible and $(A_1, A_2, \ldots, A_{k-1}A_k)^{-1} =$
     $A_k^{-1}, A_{k-1}^{-1}, \ldots, A_2^{-1}A_1^{-1}$.

4.   Prove that if AB is invertible and A is square, then both A and B
     are invertible and $(AB)^{-1} = B^{-1}A^{-1}$. [*Hint*: Use Theorem 3.5.2.]

5.   Show that if A is nonsingular, so is A^m for any nonnegative
     integer m and $(A^m)^{-1} = (A^{-1})^m$.

6.   If A is nonsingular, prove that $A^{-m}A^{-n} = A^{-m-n}$ for any
     positive integers m and n. (Keep m fixed and use mathematical
     induction on n.)

7.   Are there any values of the scalar λ for which the matrix
     $\begin{bmatrix} 1 & 2 \\ 1 & 3 \end{bmatrix} + \lambda \begin{bmatrix} 2 & 0 \\ 1 & 1 \end{bmatrix}$ has no inverse?

8.   If C is invertible and A and B are the same size as C, show that
     a. $C^{-1}(A+B)C = C^{-1}AC + C^{-1}BC$.
     b. $C^{-1}(AB)C = (C^{-1}AC)(C^{-1}BC)$.
     c. $C^{-1}(A^m)C = (C^{-1}AC)^m$ for any nonnegative integer m.

9.   Show that multiplication of an $m \times n$ matrix A by a non-
     singular matrix does not change the rank of A. [*Hint*:
     Consider both BA and AC, where B and C are conformable
     nonsingular matrices, and use the result of Problem B30 of
     Exercises 3.3.]

C.

1. Show that if $A = \begin{bmatrix} A_{11} & A_{12} \\ O & A_{22} \end{bmatrix}$ is a partitioned matrix
    and A_{11}^{-1} and A_{22}^{-1} exist, then A is nonsingular and
    $A^{-1} = \begin{bmatrix} A_{11}^{-1} & -A_{11}^{-1}A_{12}A_{22}^{-1} \\ O & A_{22}^{-1} \end{bmatrix}$.

2. If $A = \begin{bmatrix} A_{11} & O \\ A_{21} & A_{22} \end{bmatrix}$, where A_{11} and A_{22} are invertible, find
    A^{-1}.

3. Let $P = \begin{bmatrix} P_{11} & P_{12} \\ P_{21} & P_{22} \end{bmatrix}$ be a $2n \times 2n$ matrix that is partitioned
    into $n \times n$ matrices. If P_{11}^{-1} and $N = (P_{22} - P_{21}P_{11}^{-1}P_{12})^{-1}$
    exist, show that

    $$P^{-1} = \begin{bmatrix} P_{11}^{-1} + P_{11}^{-1}P_{12}NP_{21}P_{11}^{-1} & -P_{11}^{-1}P_{12}N \\ -NP_{21}P_{11}^{-1} & N \end{bmatrix}.$$

3.6 Elementary Matrices

The essential idea in this section can be grasped quickly by examining a simple example. Suppose we define

$$M = \begin{bmatrix} 1 & 2 & 3 \\ 4 & 5 & 6 \\ 7 & 8 & 9 \end{bmatrix}$$

and then consider the matrix

$$E = \begin{bmatrix} 0 & 0 & 1 \\ 0 & 1 & 0 \\ 1 & 0 & 0 \end{bmatrix},$$

which we get by interchanging the first and third rows of I_3. Then,

$$EM = \begin{bmatrix} 7 & 8 & 9 \\ 4 & 5 & 6 \\ 1 & 2 & 3 \end{bmatrix},$$

the original matrix M with its first and third rows interchanged!

The significance of this last example is that we performed an elementary row operation (see Section 2.2) on M by multiplying M on the left (or *premultiplying M*) by E. Toward the end of Section 3.2, we saw that multiplying a vector by a matrix *transforms* the vector in a geometrical way, and now we begin to see how multiplying a matrix by a matrix can also have a transforming effect.

Definition 3.6.1

An $n \times n$ **elementary matrix** is a matrix obtained from I_n by applying a single elementary row operation to it.

We can give a similar definition of "elementary matrix" that involves a single operation on the *columns* of I_n.

Example 3.6.1: Premultiplying by an Elementary Matrix

Suppose

$$A = \begin{bmatrix} 1 & 2 & 3 & 4 \\ 5 & 6 & 7 & 8 \\ 9 & 10 & 11 & 12 \\ 13 & 14 & 15 & 16 \end{bmatrix}.$$

Now form the elementary matrix E you get from I_4 by adding twice row 3 to row 2:

$$E = \begin{bmatrix} 1 & 0 & 0 & 0 \\ 0 & 1 & 2 & 0 \\ 0 & 0 & 1 & 0 \\ 0 & 0 & 0 & 1 \end{bmatrix}.$$

Then, we see that

$$EA = \begin{bmatrix} 1 & 2 & 3 & 4 \\ 23 & 26 & 29 & 32 \\ 9 & 10 & 11 & 12 \\ 13 & 14 & 15 & 16 \end{bmatrix},$$

so the resulting product gives us the original matrix A with its second row increased by twice its third row.

It is not true that any matrix with a lot of 0s and 1s can be an elementary matrix. Reading carefully, we see that Definition 3.6.1 specifies that a *single* elementary operation is applied to the identity matrix.

Example 3.6.2: A Matrix That Is Not Elementary

The matrix $B = \begin{bmatrix} 2 & 0 & 1 \\ 0 & 1 & 0 \\ 0 & 0 & 1 \end{bmatrix}$ looks as if it could be elementary, but it requires *two* elementary operations to convert I_3 into B:

$$\begin{bmatrix} 1 & 0 & 0 \\ 0 & 1 & 0 \\ 0 & 0 & 1 \end{bmatrix} \xrightarrow{R_1 \to 2R_1} \begin{bmatrix} 2 & 0 & 0 \\ 0 & 1 & 0 \\ 0 & 0 & 1 \end{bmatrix} \xrightarrow{R_1 \to R_1 + R_3} \begin{bmatrix} 2 & 0 & 1 \\ 0 & 1 & 0 \\ 0 & 0 & 1 \end{bmatrix}.$$

Thus, B is not an elementary matrix.

So far in our discussion, all the matrices we have started with have been square. The next example shows that this restriction is not necessary and sets the stage for a general theorem.

Example 3.6.3: Premultiplying a Rectangular Matrix

Let $A = \begin{bmatrix} 1 & 2 & 3 \\ 4 & 5 & 6 \end{bmatrix}$ and observe that $E = \begin{bmatrix} 1 & 3 \\ 0 & 1 \end{bmatrix}$ is obtained from I_2 by adding three times row 2 to row 1. Then, $EA = \begin{bmatrix} 13 & 17 & 21 \\ 4 & 5 & 6 \end{bmatrix}$, which is A transformed by having three times its second row added to its first row.

The examples we have just seen are pretty convincing, but we need more. We need *proof* that this premultiplication process always works. Theorem 3.6.1 establishes that corresponding to the elementary row operations first seen in Section 2.2, we have three types of elementary matrices, each of which works its magic on a given matrix A by premultiplying A:

I. E_{I} is an elementary matrix that interchanges two rows of I_n.
II. E_{II} is an elementary matrix that multiplies a row of I_n by a nonzero scalar k.
III. E_{III} is an elementary matrix that adds k times a row of I_n to another row of I_n.

An important observation (whose proof is required in Exercise B1) is that **elementary matrices of types II and III are triangular matrices**.

In proving general statements about elementary matrices, it is necessary to consider three individual cases, corresponding to the three possible types of elementary matrices.

Theorem 3.6.1

Any elementary row operation on an $m \times n$ matrix A can be accomplished by multiplying A on the left (premultiplying) by an elementary matrix, the identity matrix I_m on which the same operation has been performed.

Proof (for type III elementary matrices): Suppose A is an $m \times n$ matrix and let $B = EA$, where E is an elementary matrix of type III. Suppose that row j of E is row j of I_m plus k times row i of I_m, where $i \neq j$ and $k \neq 0$. We will show that B is also the result of adding k times row i of A to row j of A ($i \neq j$, $k \neq 0$).

As in Section 3.3, we represent row r of a matrix M by $\text{row}_r(M)$. Then

$$
\begin{aligned}
\text{row}_j(B) = \text{row}_j(EA) &= [\text{row}_j(E)] \cdot A \quad \text{(by Theorem3.3.2)} \\
&= [\text{row}_j(I_m) + k \ \text{row}_i(I_m)] \cdot A \\
&= \text{row}_j(I_m) \cdot A + k \ \text{row}_i(I_m A) \cdot A \\
&= \text{row}_j(I_m A) + k \ \text{row}_i(I_m A) \\
&= \text{row}_j(A) + k \ \text{row}_i(A).
\end{aligned}
$$

In other words, we have shown that row j of $B = EA$ equals the row we get by adding k times row i of A to row j of A ($i \neq j$). It should be obvious that no other rows are affected.

The proofs for elementary matrices of types I and II are left as exercises (see Exercises B2 and B3).

Because an elementary matrix is derived from an identity matrix, we might expect it to have nice properties. The next result tells us that our expectations are justified.

Theorem 3.6.2

An elementary matrix is nonsingular, and its inverse is an elementary matrix of the same type.

Proof We prove this theorem for a type III elementary matrix, leaving the remaining parts of the proof as Exercises B4 and B5.

Suppose that E is a matrix obtained by adding k times row i of I_n to row j of I_n, where $i \neq j$ and k is a nonzero scalar. Notice that we can reverse the operation to see that I_n can be obtained from E via an elementary row operation of type III: add $(-k)$ times row i of E to row j of E.

By Theorem 3.6.1, there is an elementary matrix $\tilde{E}$ (of type III) such that $\tilde{E}E = I_n$. Therefore, by Theorem 3.5.2, E is invertible and $E^{-1} = \tilde{E}$.

Example 3.6.4: Inverse of an Elementary Matrix

The elementary matrix $E = \begin{bmatrix} 1 & 0 & 0 & 0 \\ 0 & 1 & 2 & 0 \\ 0 & 0 & 1 & 0 \\ 0 & 0 & 0 & 1 \end{bmatrix}$ in Example 3.6.1

is nonsingular and has its inverse equal to $\begin{bmatrix} 1 & 0 & 0 & 0 \\ 0 & 1 & -2 & 0 \\ 0 & 0 & 1 & 0 \\ 0 & 0 & 0 & 1 \end{bmatrix}$,

as we can confirm easily. If we premultiply I_n by E^{-1}, we get the matrix $\begin{bmatrix} 1 & 0 & 0 & 0 \\ 0 & 1 & -2 & 0 \\ 0 & 0 & 1 & 0 \\ 0 & 0 & 0 & 1 \end{bmatrix}$, which is I_4 changed by adding (-2) times row 3 to row 2—not the exact elementary row operation performed by E, but the same *type* of operation, which is all that is guaranteed by Theorem 3.6.2.

The equivalence of an elementary row operation and premultiplication by an elementary matrix can be exploited to provide a somewhat neater proof of Theorem 2.7.2: row equivalent matrices have identical row spaces.

To see this, suppose matrix B is the result of applying an elementary row operation to A. Then, by Theorem 3.6.1, there is an elementary matrix E such that $B = EA$. The row-times-column form of matrix multiplication shows that each row of B is a linear combination of the rows of A (see Exercises 3.3, B26). Hence, the row space of B is a subset of the row space of A. Because elementary matrices are invertible (Theorem 3.6.2), we can write $A = E^{-1}B$, so that we can argue as we did previously that the row space of A is a subset of the row space of B. The two statements (row space of B) $\subseteq$ (row space of A) and (row space of A) $\subseteq$ (row space of B) yield the conclusion that the row space of A is identical to the row space of B.

In a similar way, we can use Theorems 3.6.1 and 3.6.2 to show that the column space of a matrix A is unchanged when an elementary column operation is applied to A. The key fact here is that we can perform an elementary column operation on A by *postmultiplying* A by an appropriate elementary matrix (see Exercises 3.6, B7).

To continue our analysis of the transformation of one matrix into another via elementary operations, we must review and extend Definition 2.2.3: two matrices are **row [column] equivalent** if one matrix may be obtained from the other matrix by means of elementary row [column] operations. If A is row equivalent to B, we write $A \overset{R}{\sim} B$; and if A is column equivalent to B, we write $A \overset{C}{\sim} B$. (Elementary column operations are described briefly in Section 2.2.)

As we have already noted, just as elementary row operations can be accomplished by premultiplying by elementary matrices, the corresponding elementary column operations can be produced by *postmultiplying* a matrix (multiplying the matrix on the right) by appropriate elementary matrices.

Example 3.6.5: Row Equivalent and Column Equivalent Matrices

The matrices $A = \begin{bmatrix} 1 & -1 & 0 \\ 2 & 1 & 1 \end{bmatrix}$ and $B = \begin{bmatrix} -4 & -2 & -2 \\ 3 & 0 & 1 \end{bmatrix}$ are row equivalent. To see this, start with A and proceed as follows:

$$A = \begin{bmatrix} 1 & -1 & 0 \\ 2 & 1 & 1 \end{bmatrix} \xrightarrow{R_1 \to R_1 + R_2} \begin{bmatrix} 3 & 0 & 1 \\ 2 & 1 & 1 \end{bmatrix} \xrightarrow{R_2 \to (-2)R_2} \begin{bmatrix} 3 & 0 & 1 \\ -4 & -2 & -2 \end{bmatrix}$$

$$\xrightarrow{R_2 \leftrightarrow R_1} \begin{bmatrix} -4 & -2 & -2 \\ 3 & 0 & 1 \end{bmatrix} = B.$$

Because we get B after a finite sequence of elementary row operations, A and B are row equivalent. In terms of premultiplication by elementary matrices, we have

$$\begin{bmatrix} 0 & 1 \\ 1 & 0 \end{bmatrix} \begin{bmatrix} 1 & 0 \\ 0 & -2 \end{bmatrix} \begin{bmatrix} 1 & 1 \\ 0 & 1 \end{bmatrix} \begin{bmatrix} 1 & -1 & 0 \\ 2 & 1 & 1 \end{bmatrix} = \begin{bmatrix} -4 & -2 & -2 \\ 3 & 0 & 1 \end{bmatrix}.$$

The matrices $C = \begin{bmatrix} 1 & 2 & 3 \\ 0 & -1 & 4 \\ -2 & 5 & 0 \end{bmatrix}$ and $D = \begin{bmatrix} 2 & 7 & 10 \\ -1 & 8 & 12 \\ 5 & -2 & -2 \end{bmatrix}$

are column equivalent:

$$C = \begin{bmatrix} 1 & 2 & 3 \\ 0 & -1 & 4 \\ -2 & 5 & 0 \end{bmatrix} \xrightarrow{C_1 \to C_1 + 2C_3} \begin{bmatrix} 7 & 2 & 3 \\ 8 & -1 & 4 \\ -2 & 5 & 0 \end{bmatrix} \xrightarrow{C_1 \leftrightarrow C_2} \begin{bmatrix} 2 & 7 & 3 \\ -1 & 8 & 4 \\ 5 & -2 & 0 \end{bmatrix}$$

$$\xrightarrow{C_3 \to C_3 + C_2} \begin{bmatrix} 2 & 7 & 10 \\ -1 & 8 & 12 \\ 5 & -2 & -2 \end{bmatrix} = D$$

In terms of postmultiplication by elementary matrices, we have

$$\begin{bmatrix} 1 & 2 & 3 \\ 0 & -1 & 4 \\ -2 & 5 & 0 \end{bmatrix} \begin{bmatrix} 1 & 0 & 0 \\ 0 & 1 & 0 \\ 2 & 0 & 1 \end{bmatrix} \begin{bmatrix} 0 & 1 & 0 \\ 1 & 0 & 0 \\ 0 & 0 & 1 \end{bmatrix} \begin{bmatrix} 1 & 0 & 0 \\ 0 & 1 & 1 \\ 0 & 0 & 1 \end{bmatrix} = \begin{bmatrix} 2 & 7 & 10 \\ -1 & 8 & 12 \\ 5 & -2 & -2 \end{bmatrix}.$$

Note the order of the multiplication.

Key results in Sections 2.2 and 2.5 are expressed in terms of row equivalence. For instance, when we use Gauss–Jordan elimination we are showing that a matrix is row equivalent to a matrix in reduced row echelon form. Expressed in terms of row equivalence, we have the following important results.

Theorem 3.6.3

An $n \times n$ matrix A is invertible if and only if A is row equivalent to I_n.

Proof If we know that A is row equivalent to I_n, than we can find a finite sequence $\{E_i\}_{i=1}^{k}$ of elementary matrices such that $E_k, E_{k-1}, \ldots, E_2 E_1 A = I_n$. But, by the remark immediately after Theorem 3.5.2, we see that $B = E_k, E_{k-1}, \ldots, E_2 E_1 = A^{-1}$, so A is invertible.

On the other hand, if a matrix A is invertible, the system $A\mathbf{x} = \mathbf{b}$ has a unique solution, $\mathbf{x} = A^{-1}\mathbf{b}$, for every $\mathbf{b} \in \mathbb{R}^n$. Consequently, by Theorem 2.8.3, we can use Gauss–Jordan elimination, which involves a finite sequence of elementary row operations, to reduce A to a unique reduced row echelon form having n pivots. Because A is a square matrix, having n pivots means that $\text{rref}(A) = I_n$—that is, A is row equivalent to I_n.

Corollary 3.6.1(a)

An $n \times n$ matrix A is invertible if and only if the columns of A, regarded as elements of $\mathbb{R}^n$, are linearly independent.

Proof Suppose $A = [\mathbf{a}_1 \ \mathbf{a}_2 \ \ldots \ \mathbf{a}_n]$ is invertible and there are scalars $c_1, c_2, \ldots, c_n$ such that $c_1\mathbf{a}_1 + c_2\mathbf{a}_2 + \cdots + c_n\mathbf{a}_n = \mathbf{0}$, or $A\mathbf{c} = \mathbf{0}$, where $\mathbf{c} = [c_1 \ c_2 \ \ldots \ c_n]^T$. Because A is invertible, $A\mathbf{c} = \mathbf{0}$ has only the trivial solution $\mathbf{c} = A^{-1}\mathbf{0} = \mathbf{0}$. Thus, the columns of A are linearly independent.

(Alternatively, we could have used Theorem 2.8.1.)

On the other hand, if the columns of A are linearly independent, then rank $(A) = n$, indicating that the number of leading 1s (pivots) in rref (A) is n. Because A is square, this means that the pivots must appear on the main diagonal—that is, rref$(A) = I_n$. By Theorem 3.6.3, A is invertible.

We can restate Corollary 3.6.1(a) as follows: **an $n \times n$ matrix A is invertible if and only if rank$(A) = n$.** Alternatively, in view of Theorem 2.7.5, **an $n \times n$ matrix A is invertible if and only if nullity$(A) = 0$**—that is, A is invertible if and only if the only solution of $A\mathbf{x} = \mathbf{0}$ is the zero vector.

Another important consequence of Theorem 3.6.3 is a useful algorithm for computing the inverse of an invertible matrix.

Corollary 3.6.1(b)

If A is invertible, any sequence of elementary row operations reducing A to I_n also transforms I_n into A^{-1}.

Proof Suppose A is invertible and let $E_1, E_2, \ldots, E_k$ be a sequence of elementary row operations reducing A to I_n—that is, $E_k, E_{k-1}, \ldots, E_2E_1A = I_n$.

Then,

$$E_kE_{k-1}, \ldots, E_2E_1I_n = E_kE_{k-1}, \ldots, E_2E_1(AA^{-1})$$

$$= (E_k, E_{k-1}, \ldots, E_2E_1A)A^{-1} = I_nA^{-1} = A^{-1},$$

so that I_n has been transformed into A^{-1}.

Corollary 3.6.1(b) gives us a way to determine if an $n \times n$ matrix A is invertible and a systematic method of finding the inverse if it exists. We start by forming the augmented matrix $[A \vdots I_n]$. If we can find a sequence of elementary row operations that transforms A into I_n, then the same operations (in the same order) will change I_n into A^{-1}. The process can be expressed succinctly as

$$\left[A \vdots I_n \right] \xrightarrow{\text{Elementary row operations}} \left[I_n \vdots A^{-1} \right].$$

Example 3.6.6: A Matrix Inverse via Gauss–Jordan Elimination

Suppose we want to determine if $A = \begin{bmatrix} 1 & 4 & 3 \\ 2 & 5 & 4 \\ 1 & -3 & -2 \end{bmatrix}$ has an

inverse. We start with the augmented matrix $[A \vdots I_3]$ and try to reduce A to I_3 by Gauss–Jordan elimination:

$$\begin{bmatrix} 1 & 4 & 3 & \vdots & 1 & 0 & 0 \\ 2 & 5 & 4 & \vdots & 0 & 1 & 0 \\ 1 & -3 & -2 & \vdots & 0 & 0 & 1 \end{bmatrix} \xrightarrow[R_3 \to R_3 - R_1]{R_2 \to R_2 - 2R_1} \begin{bmatrix} 1 & 4 & 3 & \vdots & 1 & 0 & 0 \\ 0 & -3 & -2 & \vdots & -2 & 1 & 0 \\ 0 & -7 & -5 & \vdots & -1 & 0 & 1 \end{bmatrix}$$

$$\xrightarrow[\substack{R_2 \to -\frac{1}{3}R_2 \\ R_3 \to R_3 + 7R_2}]{R_1 \to R_1 + \frac{4}{3}R_2} \begin{bmatrix} 1 & 0 & \frac{1}{3} & \vdots & -\frac{5}{3} & \frac{4}{3} & 0 \\ 0 & 1 & \frac{2}{3} & \vdots & \frac{2}{3} & -\frac{1}{3} & 0 \\ 0 & 0 & -\frac{1}{3} & \vdots & \frac{11}{3} & -\frac{7}{3} & 1 \end{bmatrix}$$

$$\xrightarrow[\substack{R_2 \to R_2 + 2R_3 \\ R_3 \to -3R_3}]{R_1 \to R_1 + R_3} \begin{bmatrix} 1 & 0 & 0 & \vdots & 2 & -1 & 1 \\ 0 & 1 & 0 & \vdots & 8 & -5 & 2 \\ 0 & 0 & 1 & \vdots & -11 & 7 & -3 \end{bmatrix}$$

According to Corollary 3.6.1(b),

$$A^{-1} = \begin{bmatrix} 2 & -1 & 1 \\ 8 & -5 & 2 \\ -11 & 7 & -3 \end{bmatrix},$$

which can be checked easily.

Theorem 3.6.3 asserts that if our careful attempt to transform a square matrix A into the identity matrix via elementary row operations fails in some way, then A is not invertible.

Example 3.6.7: A Singular Matrix

Suppose we try to apply Theorem 3.6.3 to the matrix:

$$A = \begin{bmatrix} 4 & 0 & 8 \\ 0 & 1 & -6 \\ 2 & 0 & 4 \end{bmatrix}.$$

We start with the augmented matrix:

$$\begin{bmatrix} 4 & 0 & 8 & \vdots & 1 & 0 & 0 \\ 0 & 1 & -6 & \vdots & 0 & 1 & 0 \\ 2 & 0 & 4 & \vdots & 0 & 0 & 1 \end{bmatrix}.$$

A logical first move is to multiply row 1 by 1/4 to get

$$\begin{bmatrix} 1 & 0 & 2 & \vdots & \frac{1}{4} & 0 & 0 \\ 0 & 1 & -6 & \vdots & 0 & 1 & 0 \\ 2 & 0 & 4 & \vdots & 0 & 0 & 1 \end{bmatrix}$$

and then use row 1 to transform row 3:

$$\begin{bmatrix} 1 & 0 & 2 & \vdots & \frac{1}{4} & 0 & 0 \\ 0 & 1 & -6 & \vdots & 0 & 1 & 0 \\ 2 & 0 & 4 & \vdots & 0 & 0 & 1 \end{bmatrix} \xrightarrow{R_3 \to R_3 - 2R_1} \begin{bmatrix} 1 & 0 & 2 & \vdots & \frac{1}{4} & 0 & 0 \\ 0 & 1 & -6 & \vdots & 0 & 1 & 0 \\ 0 & 0 & 0 & \vdots & -\frac{1}{2} & 0 & 1 \end{bmatrix}.$$

Now we are stuck because the third row of the transformed version of A consists of zeros. There is no way we can continue the process and no other elementary operation we could have performed to avoid this situation. Therefore, we conclude that A *has no inverse*. (Think about the interpretation of this situation in terms of systems of equations.)

A common theme in many areas of mathematics is the idea of expressing some mathematical object in terms of its basic components—or at least simpler components. In arithmetic or number theory, for example, a fundamental fact is that every positive integer n can be expressed uniquely as the product of powers of primes: $n = p_1^{e_1}, p_2^{e_2}, \ldots, p_r^{e_r}$, where p_i is a prime for each i and e_i is a nonnegative integer for each i. For example, $60 = 2^2 \cdot 3 \cdot 5$. Then there is the factorization of a polynomial with complex coefficients into linear factors:

$$\begin{aligned} p(x) &= a_n x^n + a_{n-1} x^{n-1} + \cdots + a_1 x + a_0 \\ &= a_n(x - r_1)(x - r_2) \cdots (x - r_n), \end{aligned}$$

where r_i is a complex number (possibly with zero imaginary part) for $i = 1, 2, \ldots, n$. For instance, $x^5 + x^4 - 5x^3 + x^2 - 6x = x(x - 2)(x + 3)(x + i)(x - i)$. In Section 3.7, we will discuss the useful and important *LU factorization* for some matrices.

For now, let us show that the class of invertible matrices can be characterized in terms of such a decomposition. The next result is an important companion to Theorem 3.6.3.

Theorem 3.6.4

An $n \times n$ matrix A is invertible if and only if it can be written as a product of elementary matrices.

Proof We start with the assumption that an $n \times n$ matrix A can be written as $A = E_1 E_2 \cdots E_k$, where each E_i is an $n \times n$ elementary matrix. But we can also write this as $A = (E_1 E_2 \cdots E_{k-1} E_k)$ $I_n = (E_1 E_2 \cdots E_{k-1}) E_k I_n$, and so forth—that is, A can be obtained by successive elementary row operations on I_n. (We use the associativity of matrix multiplication and note that $E_k I_n$ represents a row operation on I_n, $E_{k-1}(E_k I_n)$ denotes a successive row operation, and so forth.) In this way, we see that A is row equivalent to I_n and so, by Theorem 3.6.3, is invertible.

Now suppose that A is an invertible $n \times n$ matrix. Then, by Theorem 3.6.3, A is row equivalent to I_n. Consequently, by Theorem 3.6.1, there must be a finite number of elementary matrices $E_1, E_2, \ldots, E_k$ such that

$$E_k E_{k-1} \cdots E_2 E_1 A = I_n. \qquad (*)$$

Because any elementary matrix is nonsingular and its inverse is also an elementary matrix (Theorem 3.6.2), we can solve for A by premultiplying both sides of (*) by the matrix product $E_1^{-1} E_2^{-1} \cdots E_{k-1}^{-1} E_k^{-1}$, so that

$$A = (E_1^{-1} E_2^{-1} \cdots E_{k-1}^{-1} E_k^{-1}) I_n = E_1^{-1} E_2^{-1} \cdots E_{k-1}^{-1} E_k^{-1},$$

a product of elementary matrices.

Example 3.6.8: Invertible Matrices as Products of Elementary Matrices

(a) Using the result of Example 3.5.4, we see that $A = \begin{bmatrix} -5 & 1 \\ -6 & 1 \end{bmatrix}$ is invertible. We can verify that $E_2 E_1 A = I_2$, where $E_1 = \begin{bmatrix} 1 & -1 \\ 0 & 1 \end{bmatrix}$ and $E_2 = \begin{bmatrix} 1 & 0 \\ 6 & 1 \end{bmatrix}$ are elementary matrices. Then, $A = E_1^{-1} E_2^{-1} = \begin{bmatrix} 1 & 1 \\ 0 & 1 \end{bmatrix} \begin{bmatrix} 1 & 0 \\ -6 & 1 \end{bmatrix}$, the product of elementary matrices.

(b) The matrix $B = \begin{bmatrix} 1 & 0 & 2 \\ -2 & 1 & 0 \\ 0 & 0 & 1 \end{bmatrix}$ is invertible: $E_3 E_2 E_1 B = I_3$, where $E_1 = \begin{bmatrix} 1 & 0 & 0 \\ 2 & 1 & 0 \\ 0 & 0 & 1 \end{bmatrix}$, $E_2 = \begin{bmatrix} 1 & 0 & -2 \\ 0 & 1 & 0 \\ 0 & 0 & 1 \end{bmatrix}$, and $E_3 = \begin{bmatrix} 1 & 0 & 0 \\ 0 & 1 & -4 \\ 0 & 0 & 1 \end{bmatrix}$ are elementary matrices.

Furthermore, $B = E_1^{-1} E_2^{-1} E_3^{-1} = \begin{bmatrix} 1 & 0 & 0 \\ -2 & 1 & 0 \\ 0 & 0 & 1 \end{bmatrix} \begin{bmatrix} 1 & 0 & 2 \\ 0 & 1 & 0 \\ 0 & 0 & 1 \end{bmatrix} \begin{bmatrix} 1 & 0 & 0 \\ 0 & 1 & 4 \\ 0 & 0 & 1 \end{bmatrix}$ and $B^{-1} = E_3 E_2 E_1 = \begin{bmatrix} 1 & 0 & 0 \\ 0 & 1 & -4 \\ 0 & 0 & 1 \end{bmatrix} \begin{bmatrix} 1 & 0 & -2 \\ 0 & 1 & 0 \\ 0 & 0 & 1 \end{bmatrix}$

$$\begin{bmatrix} 1 & 0 & 0 \\ 2 & 1 & 0 \\ 0 & 0 & 1 \end{bmatrix} = \begin{bmatrix} 1 & 0 & -2 \\ 2 & 1 & -4 \\ 0 & 0 & 1 \end{bmatrix}.$$

If we are allowed to use both elementary row operations and the similarly defined elementary column operations in transforming a matrix, we may carry out the transformation more efficiently. (As noted previously, in the definition of *elementary row operations*, in Section 2.2, just replace the word "row" by "column.")*

Definition 3.6.2

Two matrices are **equivalent** if one matrix may be obtained from the other matrix by means of a combination of finitely many elementary row and column operations. If A is equivalent to B, we write $A \sim B$.

Example 3.6.9: Equivalent Matrices

We show that $A = \begin{bmatrix} 1 & 5 & -3 \\ 0 & 1 & 2 \\ 1 & 6 & -1 \end{bmatrix}$ is equivalent to $B = \begin{bmatrix} 1 & 0 & 0 \\ 0 & 1 & 0 \\ 0 & 0 & 0 \end{bmatrix}$:

$$\begin{bmatrix} 1 & 5 & -3 \\ 0 & 1 & 2 \\ 1 & 6 & -1 \end{bmatrix} \xrightarrow{R_3 \rightarrow R_3 - R_1} \begin{bmatrix} 1 & 5 & -3 \\ 0 & 1 & 2 \\ 0 & 1 & 2 \end{bmatrix} \xrightarrow{R_3 \rightarrow R_3 - R_2} \begin{bmatrix} 1 & 5 & -3 \\ 0 & 1 & 2 \\ 0 & 0 & 0 \end{bmatrix}$$

$$\xrightarrow{R_1 \rightarrow R_1 - 5R_2} \begin{bmatrix} 1 & 0 & -13 \\ 0 & 1 & 2 \\ 0 & 0 & 0 \end{bmatrix} \xrightarrow{C_3 \rightarrow C_3 + 13C_1} \begin{bmatrix} 1 & 0 & 0 \\ 0 & 1 & 2 \\ 0 & 0 & 0 \end{bmatrix}$$

$$\xrightarrow{C_3 \rightarrow C_3 - 2C_2} \begin{bmatrix} 1 & 0 & 0 \\ 0 & 1 & 0 \\ 0 & 0 & 0 \end{bmatrix}.$$

Exercises 3.6

A.

1. Which of the following matrices are elementary matrices?

a. $\begin{bmatrix} 0 & -1 \\ 1 & 0 \end{bmatrix}$ b. $\begin{bmatrix} 1 & 0 \\ -2 & 1 \end{bmatrix}$ c. $\begin{bmatrix} 1 & -3 \\ 0 & 1 \end{bmatrix}$ d. $\begin{bmatrix} 1 & -2 & 0 \\ 0 & 0 & 1 \\ 0 & 0 & 1 \end{bmatrix}$

---

* For an example that uses both row and column operations to find the inverse of a matrix, see Section 6.4 of D.T. Finkbeiner II, *Introduction to Matrices and Linear Transformations*, 2nd edn. (San Francisco, CA: W.H. Freeman, 1966).

e. $\begin{bmatrix} 1 & 0 & 1 \\ 0 & 1 & 0 \\ -3 & 0 & -2 \end{bmatrix}$ f. $\begin{bmatrix} -1 & 0 & 0 \\ 0 & 1 & 0 \\ 0 & 0 & 1 \end{bmatrix}$ g. $\begin{bmatrix} 3 & 0 & 0 \\ 0 & 3 & 0 \\ 0 & 0 & 1 \end{bmatrix}$

h. $\begin{bmatrix} 0 & 0 & 4 \\ 0 & 1 & 0 \\ 1 & 0 & 1 \end{bmatrix}$

2. Suppose $A = \begin{bmatrix} 1 & -2 & 1 \\ 0 & 3 & -1 \\ -2 & 1 & -2 \\ 3 & 0 & -1 \end{bmatrix}$. For each part, find the

   elementary matrix E such that premultiplication by E performs the indicated row operation on A.

   a. Multiplies the third row by 3.
   b. Interchanges the first and third rows.
   c. Adds two times the first row to the fourth row.

3. a. Express B as the product of elementary matrices if

   $$B = \begin{bmatrix} 1 & 0 & 2 \\ 2 & 1 & 0 \\ 0 & 1 & 1 \end{bmatrix}.$$

   b. Use the result of part (a) to determine B^{-1}.

4. Use Gauss–Jordan elimination to determine the inverses of each of the following matrices, if possible.

   a. $\begin{bmatrix} 2 & 5 \\ 5 & 2 \end{bmatrix}$ b. $\begin{bmatrix} \frac{3}{2} & -2 \\ 2 & \frac{3}{2} \end{bmatrix}$ c. $\begin{bmatrix} 0 & 2 & 1 \\ 2 & 6 & 1 \\ 1 & 1 & 4 \end{bmatrix}$ d. $\begin{bmatrix} 1 & 2 & 3 \\ 4 & 5 & 6 \\ 7 & 8 & 9 \end{bmatrix}$

   e. $\begin{bmatrix} 1 & 2 & 3 \\ 3 & 5 & 5 \\ 2 & 1 & 2 \end{bmatrix}$ f. $\begin{bmatrix} 2 & 1 & 2 \\ 4 & 2 & 3 \\ 0 & -1 & 1 \end{bmatrix}$ g. $\begin{bmatrix} 2 & 1 & 0 & 1 \\ 0 & 0 & 1 & 3 \\ 1 & 0 & 0 & -1 \\ 0 & 0 & -2 & -5 \end{bmatrix}$

5. In Exercise A1(d) of Exercises 3.5, we claimed that $A = \begin{bmatrix} 1 & -2 & 3 \\ 2 & -5 & 10 \\ -1 & 2 & -2 \end{bmatrix}$ has inverse $A^{-1} = \begin{bmatrix} 10 & -2 & 5 \\ 6 & -1 & 4 \\ 1 & 0 & 1 \end{bmatrix}$. Express both A and A^{-1} as products of elementary matrices.

6. Example 3.6.6 calculated the inverse of $A = \begin{bmatrix} 1 & 4 & 3 \\ 2 & 5 & 4 \\ 1 & -3 & -2 \end{bmatrix}$ by Gauss–Jordan elimination. Express both A and A^{-1} as products of elementary matrices.

7. Show that $A = \begin{bmatrix} -1 & -3 & 1 \\ 0 & 2 & 1 \\ 1 & -1 & 0 \end{bmatrix}$ is row equivalent to

   $B = \begin{bmatrix} 1 & 0 & 0 \\ 0 & 1 & 0 \\ 0 & 0 & 1 \end{bmatrix}.$

8. Show that $A = \begin{bmatrix} 2 & 0 & 1 \\ 1 & 1 & 0 \\ 8 & 4 & 2 \\ 6 & 0 & 3 \end{bmatrix}$ is column equivalent to

$B = \begin{bmatrix} 1 & -3 & 0 \\ 0 & 1 & 2 \\ 2 & -2 & 8 \\ 3 & -9 & 0 \end{bmatrix}$.

9. How many different 4×4 elementary matrices of type I (row interchange) are there?

B.

1. Prove that any elementary matrix of type II or type III must be a triangular matrix.

2. Prove Theorem 3.6.1 for type I elementary matrices.

3. Prove Theorem 3.6.1 for type II elementary matrices.

4. Prove Theorem 3.6.2 for type I elementary matrices.

5. Prove Theorem 3.6.2 for type II elementary matrices.

6. Prove that E is an elementary matrix if and only if E^T is.

7. Show by 2×2 and 3×3 examples how each of the three types of elementary *column* operations can be performed on a matrix A by *postmultiplying* A by a suitable nonsingular matrix.

8. Suppose that A is an $m \times n$ matrix. Prove that if B can be obtained from A by an elementary row operation, then B^T can be obtained from A^T by the corresponding column operation. [You may use the result of Exercise B6.]

9. a. Do elementary matrices of type I commute with each other? Explain.
   b. Do elementary matrices of type II commute with each other? Explain.
   c. Do elementary matrices of type III commute with each other? Explain.

10. Prove that if E is an elementary matrix of type I, then $E^2 = I_n$.

11. Prove that B is equivalent to A if and only if $B = PAQ$ for suitable nonsingular matrices P and Q.

12. Prove that a square matrix is nonsingular if and only if it is equivalent to the identity matrix. [*Hint:* Use the result of the previous exercise.]

13. Show that matrices A and B are equivalent, where

$$A = \begin{bmatrix} 1 & 2 & 1 & 5 & 3 \\ -1 & 0 & 2 & -7 & -10 \\ 1 & 2 & 4 & -1 & -6 \end{bmatrix} \quad \text{and}$$

$$B = \begin{bmatrix} 0 & 0 & 1 & -2 & -3 \\ 0 & 1 & 0 & 2 & 1 \\ 1 & 0 & 0 & 3 & 4 \end{bmatrix}.$$

C.

1. Prove that row equivalence is an **equivalence relation** on the set of all $m \times n$ matrices, meaning that
    a. $A \overset{R}{\sim} A$ for all $m \times n$ matrices A.
    b. If $B \overset{R}{\sim} A$, then $A \overset{R}{\sim} B$.
    c. If $A \overset{R}{\sim} B$ and $B \overset{R}{\sim} C$, then $A \overset{R}{\sim} C$.

2. Define the square matrix E_{rs} to be the matrix with a 1 in row r, column s and 0s elsewhere. Expressed another way, $E_{rs} = [e_{ij}]$, where $e_{ij} = 0$ if $i \neq r$ or $j \neq s$ and $e_{rs} = 1$.
    a. If E is the matrix obtained from I by interchanging rows i and j, then show that $E = I - E_{ii} + E_{ji} - E_{jj} + E_{ij}$.
    b. If E is the matrix obtained from I by multiplying row i by the nonzero constant c, show that $E = I + (c - 1)\, E_{ii}$.
    c. If E is the matrix obtained from I by adding row i to row j, where $i \neq j$ show that $E = I + E_{ji}$.

3. If we define δ_{ij} to be 1 if $i = j$ and 0 otherwise, prove that $E_{ik}E_{hj} = \delta_{kh}E_{ij}$, where E_{ij} is defined in the previous exercise.

4. Prove that any $m \times n$ matrix can be transformed into a partitioned matrix of the form $\begin{bmatrix} I_r & \vdots & O \\ \cdots & \vdots & \cdots \\ O & \vdots & O \end{bmatrix}$ by using both elementary row and elementary column operations. (The nonnegative integer r is the rank of the matrix A. See Definition 2.7.2.)

5. If $A_1 \sim A_2$ and $B_1 \sim B_2$, show that $\begin{bmatrix} A_1 & O \\ O & B_1 \end{bmatrix} \sim \begin{bmatrix} A_2 & O \\ O & B_2 \end{bmatrix}$.

6. An applied linear algebra book* defines an *elementary matrix* as an $n \times n$ matrix having the form $I_n - \mathbf{u}\mathbf{v}^T$, where $\mathbf{u}$ and $\mathbf{v}$ are $n \times 1$ column vectors such that $\mathbf{v}^T\mathbf{u} \neq 1$. Define the column vector $\mathbf{e}_i$, $i = 1, 2, \ldots, n$, to be the $n \times 1$ column vector with 1 in the ith row and zeros everywhere else.

---

* C.D. Meyer, *Matrix Analysis and Applied Linear Algebra* (Philadelphia, PA: SIAM, 2000).

a. Show that by choosing $\mathbf{u} = \mathbf{e}_1 - \mathbf{e}_2$, the matrix $E = I_n - \mathbf{u}\mathbf{u}^{\mathrm{T}}$ is the matrix obtained from I_n by interchanging rows 1 and 2.

b. Show that the matrix $E = I_n - (1 - c)\mathbf{e}_2\mathbf{e}_2^{\mathrm{T}}$ derives from multiplying row 2 of I_n by the nonzero constant c.

c. Show that $E = I_n + c\mathbf{e}_3\mathbf{e}_1^{\mathrm{T}}$ is the matrix obtained by multiplying row 1 in I_n by c and adding the result to row 3.

3.7 The LU Factorization

As stated in the last section, just before Theorem 3.6.4, it is often useful to break down (factor, decompose) a mathematical object such as a natural number, a polynomial, or a matrix into simpler components. In particular, Theorem 3.6.4 described a very useful decomposition of an invertible matrix. In this section, we will develop an important matrix factorization that arises from Gaussian elimination and has particular usefulness in computer calculations.

To understand this factorization, let us reduce the matrix $A = \begin{bmatrix} 2 & 2 & -1 \\ 4 & 5 & 2 \\ -2 & 1 & 2 \end{bmatrix}$ to an echelon form that we will call U:

$$\begin{bmatrix} 2 & 2 & -1 \\ 4 & 5 & 2 \\ -2 & 1 & 2 \end{bmatrix} \xrightarrow{R_2 \rightarrow R_2 - 2R_1} \begin{bmatrix} 2 & 2 & -1 \\ 0 & 1 & 4 \\ -2 & 1 & 2 \end{bmatrix} \xrightarrow{R_3 \rightarrow R_3 + R_1} \begin{bmatrix} 2 & 2 & -1 \\ 0 & 1 & 4 \\ 0 & 3 & 1 \end{bmatrix}$$

$$\xrightarrow{R_3 \rightarrow R_3 - 3R_2} \begin{bmatrix} 2 & 2 & -1 \\ 0 & 1 & 4 \\ 0 & 0 & -11 \end{bmatrix} = U$$

As is the case for an echelon form of a square matrix, U is an upper triangular matrix.

Now let us express the transformation of A into U as a premultiplication of A by a finite sequence of elementary matrices, as we saw in Section 3.6 (see, especially, Theorem 3.6.1). Each elementary row operation applied to A must be applied to I_3 in the same order. This process yields the sequence of elementary matrices:

$$E_1 = \begin{bmatrix} 1 & 0 & 0 \\ -2 & 1 & 0 \\ 0 & 0 & 1 \end{bmatrix} \quad \langle R_2 \rightarrow R_2 - 2R_1 \rangle$$

$$E_2 = \begin{bmatrix} 1 & 0 & 0 \\ 0 & 1 & 0 \\ 1 & 0 & 1 \end{bmatrix} \quad \langle R_3 \rightarrow R_3 + 2R_1 \rangle$$

$$E_3 = \begin{bmatrix} 1 & 0 & 0 \\ 0 & 1 & 0 \\ 0 & -3 & 1 \end{bmatrix} \quad \langle R_3 \rightarrow R_3 - 3R_2 \rangle,$$

so $E_3 E_2 E_1 A = U$. We note that the product:

$$M = E_3 E_2 E_1 = \begin{bmatrix} 1 & 0 & 0 \\ -2 & 1 & 0 \\ 7 & -3 & 1 \end{bmatrix}$$

is a *lower triangular matrix*, which is not an accident. Finally, we can write

$$A = (E_3 E_2 E_1)^{-1} U = M^{-1} U = LU, \qquad (3.7.1)$$

where

$$L = M^{-1} = \begin{bmatrix} 1 & 0 & 0 \\ 2 & 1 & 0 \\ -1 & 3 & 1 \end{bmatrix},$$

another lower triangular matrix.

What we have just accomplished is to write A as the product of a **unit lower triangular matrix** (a lower triangular matrix with 1s on the main diagonal) L and an upper triangular matrix U:

$$A = \begin{bmatrix} 2 & 2 & -1 \\ 4 & 5 & 2 \\ -2 & 1 & 2 \end{bmatrix} = \overbrace{\begin{bmatrix} 1 & 0 & 0 \\ 2 & 1 & 0 \\ -1 & 3 & 1 \end{bmatrix}}^{L} \cdot \overbrace{\begin{bmatrix} 2 & 2 & -1 \\ 0 & 1 & 4 \\ 0 & 0 & -11 \end{bmatrix}}^{U}.$$

Definition 3.7.1

If A is a square matrix and $A = LU$, where U is upper triangular and L is a unit lower triangular matrix, this factorization is called an **LU factorization** (or **LU decomposition**) of A.*

First of all, as we shall see shortly, such an LU factorization may not exist. Then, because the echelon form of a matrix is not unique, we cannot, in general, expect a factorization involving lower triangular and upper triangular matrix factors to be unique.

One application for such a factorization is solving a system $Ax = b$, especially by computer. If (and we should be aware of the "if") $A = LU$, then the system can be written as $LUx = b$, or $L(Ux) = b$. Now the solution of $Ax = b$ is equivalent to the solution of two simpler problems:

(1) First let $Ux = y$ and solve the system $Ly = b$ for y.
(2) Then, solve the system $Ux = y$ for x.

---

* The formulation of the LU decomposition is generally credited to Alan Turing (1912–1954). See N.J. Higham, *Accuracy and Stability of Numerical Algorithms*, 2nd edn., p. 184 (Philadelphia, PA: SIAM, 2002).

Separating our original problem into two problems is a good move because *each subproblem involves triangular matrices*, which are easy to work with in the solution of systems.

Example 3.7.1: Solving a System via LU Factorization

We can express the system:

$$2a + 2b - c = 1$$
$$4a + 5b + 2c = 2$$
$$-2a + b + 2c = 3$$

as $A\mathbf{x} = \mathbf{b}$, where $A = \begin{bmatrix} 2 & 2 & -1 \\ 4 & 5 & 2 \\ -2 & 1 & 2 \end{bmatrix}$, $\mathbf{x} = \begin{bmatrix} a \\ b \\ c \end{bmatrix}$, and

$\mathbf{b} = \begin{bmatrix} 1 \\ 2 \\ 3 \end{bmatrix}$. Using the LU form of A found just before Definition 3.6.1, we write this system as $L(U\mathbf{x}) = \mathbf{b}$. Letting $U\mathbf{x} = \mathbf{y} = \begin{bmatrix} y_1 \\ y_2 \\ y_3 \end{bmatrix}$, we have

$$\begin{bmatrix} 1 & 0 & 0 \\ 2 & 1 & 0 \\ -1 & 3 & 1 \end{bmatrix} \begin{bmatrix} y_1 \\ y_2 \\ y_3 \end{bmatrix} = \begin{bmatrix} 1 \\ 2 \\ 3 \end{bmatrix},$$

which has the easily obtained solution (because we have a lower triangular matrix of coefficients) $y_1 = 1$, $y_2 = 0$, and $y_3 = 4$—the last two values obtained by **forward substitution**. That is, we started at the top by solving for y_1 and worked our way down, substituting to find y_2 and y_3.

Next, we consider the equation $U\mathbf{x} = \mathbf{y}$, or

$$\begin{bmatrix} 2 & 2 & -1 \\ 0 & 1 & 4 \\ 0 & 0 & -11 \end{bmatrix} \begin{bmatrix} a \\ b \\ c \end{bmatrix} = \begin{bmatrix} 1 \\ 0 \\ 4 \end{bmatrix},$$

which again has an easy solution because of the (upper) triangular matrix: $c = -\frac{4}{11}$, $b = \frac{16}{11}$, and $a = -\frac{25}{22}$ (the last two values obtained by back substitution). Thus,

$$\begin{bmatrix} -\frac{25}{22} \\ \frac{16}{11} \\ -\frac{4}{11} \end{bmatrix} = \frac{1}{22} \begin{bmatrix} -25 \\ 32 \\ -8 \end{bmatrix}$$ is the solution of our original system.

A potentially awkward part of getting the LU decomposition is finding the lower triangular matrix L by inverting $M = E_k, E_{k-1}, \ldots, E_2 E_1$ as indicated previously. However, it turns out we can find $M^{-1} = L = [l_{ij}]$ quite easily without calculating M first. To do this, we note the elementary operations used to convert A into U. Then, for $i \neq j$, l_{ij} is the number that multiplies the pivot row when it is subtracted from row i and produces a zero in position (i,j): $l_{ij} \leftrightarrow R_i - l_{ij} \underbrace{R_k}_{\text{pivot row}}$, $k < i$. Of course, $l_{ii} = 1$ for every i. To understand all this, we need a good example.

Example 3.7.2: Shortcut to the LU Factorization

Consider the factorization $A = \begin{bmatrix} 2 & 2 & -1 \\ 4 & 5 & 2 \\ -2 & 1 & 2 \end{bmatrix} = \begin{bmatrix} 1 & 0 & 0 \\ 2 & 1 & 0 \\ -1 & 3 & 1 \end{bmatrix}$

$\begin{bmatrix} 2 & 2 & -1 \\ 0 & 1 & 4 \\ 0 & 0 & -11 \end{bmatrix} = LU$ that we discussed in the previous example.

Let us focus on the numbers below the diagonal of the unit lower triangular matrix $L = [l_{ij}]$: $2, -1, 3$.

In going from A to $\begin{bmatrix} 2 & 2 & -1 \\ 0 & 1 & 4 \\ -2 & 1 & 2 \end{bmatrix}$, we subtracted twice row 1 from row 2. In other words, to produce a zero in position $(2, 1)$, we first multiplied row 1 by 2. Thus $l_{21} = 2$. Similarly, in going

from $\begin{bmatrix} 2 & 2 & -1 \\ 0 & 1 & 4 \\ -2 & 1 & 2 \end{bmatrix}$ to $\begin{bmatrix} 2 & 2 & -1 \\ 0 & 1 & 4 \\ 0 & 3 & 1 \end{bmatrix}$, thus zeroing out pos-

ition $(3, 1)$, we multiplied row 1 by 1 and *added* to row 3—that is, we *subtracted* (-1) times row 1: $R_3 + R_1 = R_3 - (-1)R_1$. Hence $l_{31} = -1$. Finally, in the last step we made the transformation

$\begin{bmatrix} 2 & 2 & -1 \\ 0 & 1 & 4 \\ 0 & 3 & 1 \end{bmatrix} \rightarrow \begin{bmatrix} 2 & 2 & -1 \\ 0 & 1 & 4 \\ 0 & 0 & -1 \end{bmatrix}$. Here, we zeroed out the $(3, 2)$

entry by subtracting three times row 2 from row 3. Thus, $l_{32} = 3$.

These three multipliers l_{21}, l_{31}, and l_{32} occupy positions $(2, 1)$, $(3, 1)$, and $(3, 2)$, respectively, below the diagonal of L, corresponding to the entries of A these multipliers helped zero out.

In the computer implementation of LU factorization, the entries of the original square matrix A are replaced gradually by the entries of L and U. There is a storage savings in ignoring the 1s on the main diagonal of L and just merging the lower portion of L with all of U. For example, the **L/U display** for the matrix A in Example 3.7.2 is

$$\begin{bmatrix} \boxed{2} & \boxed{2} & \boxed{-1} \\ 2 & \boxed{1} & \boxed{4} \\ -1 & 3 & \boxed{-11} \end{bmatrix},$$

where the entries of U are boxed and the entries of L (without diagonal 1s) appear in bold print.*

---

* See, for example, Section 1.7 of D.S. Watkins, *Fundamentals of Matrix Computations*, 2nd edn. (New York: Wiley, 2002).

Now is the time to acknowledge that we have put the cart before the horse by going through the mechanics of an LU decomposition and an example of its usefulness *without discussing whether such a decomposition always exists*. The bad news is that we may not be able to find an LU factorization in every situation.

Example 3.7.3: An LU Factorization Is Not Always Possible

The matrix $A = \begin{bmatrix} 0 & 1 \\ 1 & 0 \end{bmatrix}$ cannot be written in the form $\begin{bmatrix} 1 & 0 \\ a & 1 \end{bmatrix} \begin{bmatrix} b & c \\ 0 & d \end{bmatrix}$. Multiplying the triangular matrices, we get $\begin{bmatrix} b & c \\ ab & ac+d \end{bmatrix}$, which implies that $b=0$, forcing the $(2,1)$ entry of the product to be zero also. But this entry should be 1, to agree with A. The contradiction shows that there is no LU factorization of A.

The nature of the matrix A in the last example provides a clue about the essential condition for an LU decomposition. The next example provides additional evidence.

Example 3.7.4: An LU Factorization Is Not Always Possible

We can reduce the matrix:

$$A = \begin{bmatrix} 0 & 1 & 2 \\ -1 & 2 & 1 \\ 3 & 5 & -8 \end{bmatrix}$$

to echelon form U, for example, by using elementary matrices to get $E_3 E_2 E_1 A = U$, where

$$E_1 = \begin{bmatrix} 0 & 1 & 0 \\ 1 & 0 & 0 \\ 0 & 0 & 1 \end{bmatrix}, \quad E_2 = \begin{bmatrix} 1 & 0 & 0 \\ 0 & 1 & 0 \\ 3 & 0 & 1 \end{bmatrix},$$

$$E_3 = \begin{bmatrix} 1 & 0 & 0 \\ 0 & 1 & 0 \\ -11 & 0 & 1 \end{bmatrix}, \quad \text{and} \quad U = \begin{bmatrix} -1 & 2 & 1 \\ 0 & 1 & 2 \\ 0 & 0 & -27 \end{bmatrix}.$$

Then,

$$E_3 E_2 E_1 = \begin{bmatrix} 0 & 1 & 0 \\ 1 & 0 & 0 \\ 0 & -8 & 1 \end{bmatrix},$$

which is not a lower triangular matrix, unlike other examples we have seen. Also, $M^{-1} = (E_3 E_2 E_1)^{-1} = E_1^{-1} E_2^{-1} E_3^{-1} =$
$$\begin{bmatrix} 0 & 1 & 0 \\ 1 & 0 & 0 \\ 8 & 0 & 1 \end{bmatrix}$$ is not lower triangular. The procedure we have used previously to obtain an LU factorization does not work here!

The key element in Examples 3.7.3 and 3.7.4 is that *in using Gaussian elimination to reduce A to U, we had to interchange two rows of A* because of a zero in the first column. (In Example 3.7.3, we see that the matrix A there is just I_2 with its rows interchanged.)

Theorem 3.7.1

If A is an $n \times n$ matrix and we can reduce A to an upper triangular matrix U by Gaussian elimination *without any interchanges of rows*, then $A = LU$, where L is a unit lower triangular matrix.

Proof If there are only elementary operations of types II and III involved in the reduction of A to U, we have already given the algorithm for finding L in terms of the multipliers used in zeroing out entries below the diagonal of U.

Example 3.7.5: An LU Factorization of a Singular Matrix

The matrix $A = \begin{bmatrix} 1 & 1 & 1 \\ 2 & -2 & 2 \\ 3 & 5 & 3 \end{bmatrix}$ is singular and consequently not row equivalent to I_3. In particular, when we use Gaussian elimination to reduce A to an echelon form U, we should expect U to have at least one row of 0s. One possible LU factorization is

$$A = \begin{bmatrix} 1 & 1 & 1 \\ 2 & -2 & 2 \\ 3 & 5 & 3 \end{bmatrix} = \begin{bmatrix} 1 & 0 & 0 \\ 2 & 1 & 0 \\ 3 & -1/2 & 1 \end{bmatrix} \begin{bmatrix} 1 & 1 & 1 \\ 0 & -4 & 0 \\ 0 & 0 & 0 \end{bmatrix}.$$

Theorem 3.7.1 has an important consequence for invertible matrices.

Corollary 3.7.1

If an invertible $n \times n$ matrix A has an LU factorization, then the factorization is unique.

Proof Suppose A is invertible and $A = L_1 U_1 = L_2 U_2$, where L_1 and L_2 are unit lower triangular and U_1 and U_2 are upper triangular. Then all factor matrices are invertible (Problem B4, Exercises 3.5) and $L_2^{-1}L_1 = U_2U_1^{-1}$. But $L_2^{-1}L_1$ is a unit lower triangular matrix, whereas $U_2U_1^{-1}$ is an upper triangular matrix.* The only matrix that is both a unit lower triangular matrix and an upper triangular matrix is the identity matrix. Thus, $L_2^{-1}L_1 = I_n = U_2U_1^{-1}$, and so $L_1 = L_2$ and $U_1 = U_2$.

If we examine the LU decomposition carefully, we realize that the factorization $A = LU$ can be interpreted as the matrix formulation of Gaussian elimination, provided that no row interchanges are used. As one advanced treatise[†] puts it,

> The LU factorization is a "higher-level" algebraic description of Gaussian elimination. Expressing the outcome of a matrix algorithm in the "language" of matrix factorizations is a worthwhile activity. It facilitates generalization and highlights connections between algorithms that may appear very different at the scalar level.

Even if the reduction of a matrix A via Gaussian elimination uses an interchange of rows, there is a particular kind of factorization possible. First, we need a definition.

Definition 3.7.2

A **permutation matrix** P is any $n \times n$ matrix that results from rearranging (permuting) the order of the rows of I_n.

Alternatively, each row of a permutation matrix P contains exactly one nonzero entry, 1; and each column of P contains exactly one nonzero entry, 1. A permutation matrix is not necessarily an elementary matrix, but is clearly a product of elementary matrices of type I.

Example 3.7.6: Permutation Matrices

$$P_1 = \begin{bmatrix} 0 & 1 \\ 1 & 0 \end{bmatrix}, \quad P_2 = \begin{bmatrix} 0 & 1 & 0 \\ 0 & 0 & 1 \\ 1 & 0 & 0 \end{bmatrix}, \text{ and } P_3 = \begin{bmatrix} 0 & 0 & 1 & 0 \\ 0 & 0 & 0 & 1 \\ 0 & 1 & 0 & 0 \\ 1 & 0 & 0 & 0 \end{bmatrix}$$

are all permutation matrices.

Matrix P_1 is an elementary matrix, whereas P_2 and P_3 are products of two or more elementary matrices:

---

* We are assuming basic properties of triangular matrices. See Exercises B2, B3, B5, and B6.
[†] G.H. Golub and C.F. Van Loan, *Matrix Computations*, 3rd edn., p. 95 (Baltimore, MD: Johns Hopkins University Press, 1996).

$$P_2 = \begin{bmatrix} 1 & 0 & 0 \\ 0 & 0 & 1 \\ 0 & 1 & 0 \end{bmatrix} \begin{bmatrix} 0 & 1 & 0 \\ 1 & 0 & 0 \\ 0 & 0 & 1 \end{bmatrix},$$

$$P_3 = \begin{bmatrix} 0 & 1 & 0 & 0 \\ 1 & 0 & 0 & 0 \\ 0 & 0 & 1 & 0 \\ 0 & 0 & 0 & 1 \end{bmatrix} \begin{bmatrix} 0 & 0 & 0 & 1 \\ 0 & 1 & 0 & 0 \\ 0 & 0 & 1 & 0 \\ 1 & 0 & 0 & 0 \end{bmatrix} \begin{bmatrix} 1 & 0 & 0 & 0 \\ 0 & 0 & 1 & 0 \\ 0 & 1 & 0 & 0 \\ 0 & 0 & 0 & 1 \end{bmatrix}.$$

However, $M = \begin{bmatrix} 1 & 0 & 1 \\ 0 & 0 & 1 \\ 0 & 1 & 0 \end{bmatrix}$ is *not* a permutation matrix because row 1 is not the result of any rearrangement of the order of the rows of I_3. Also, we see that the first row and the third column of M contain more than one nonzero entry.

Before continuing, we need one simple result.

Theorem 3.7.2

The product of two permutation matrices is a permutation matrix.

Proof   If A and B are two permutation matrices, then $A = E_k E_{k-1} \cdots E_2 E_1$ and $B = \hat{E}_r \hat{E}_{r-1} \cdots \hat{E}_2 \hat{E}_1$, where each E_i and each $\hat{E}_i$ is a type I elementary matrix (representing a single interchange of rows). Thus, $AB = E_k E_{k-1} \cdots E_2 E_1 \hat{E}_r \hat{E}_{r-1} \cdots \hat{E}_2 \hat{E}_1$, a permutation matrix because successive permutations result in yet another permutation.

Now we can state a result that complements Theorem 3.7.1.

Theorem 3.7.3

If A is an $n \times n$ matrix, then there is a permutation matrix P such that $PA = LU$, where L is a unit lower triangular matrix and U is upper triangular. In the case that A is nonsingular, L and U are unique.

Proof   First, we go through the process of Gaussian elimination just to see what row interchanges there may be. Say that all these interchanges correspond to permutation matrices $P_1, P_2, \ldots, P_k$ in this order. Then, we form the new matrix $\tilde{A} = P_k, P_{k-1}, \ldots, P_2 P_1 A$ and apply the decomposition process to $\tilde{A}$ to get $\tilde{A} = LU$. Thus, $\tilde{A} = P_k, P_{k-1}, \ldots, P_2 P_1 A = LU$ or, letting $P = P_k, P_{k-1}, \ldots, P_2 P_1$, $PA = LU$. What we have done here is to perform all necessary row interchanges *first* and then apply Theorem 3.7.1 to matrix $\tilde{A}$, which requires no additional interchanges. The proof of uniqueness is left as Exercise B10.

It can be shown that if P is a permutation matrix, then $P^{-1} = P^{\mathrm{T}}$ (Exercise B8). Therefore, we can rewrite the result of Theorem 3.7.3 as $A = P^{\mathrm{T}} LU$.

Example 3.7.7: A $PA = LU$ Factorization

In Example 3.7.4, $A = \begin{bmatrix} 0 & 1 & 2 \\ -1 & 2 & 1 \\ 3 & 5 & -8 \end{bmatrix}$ is a nonsingular matrix,

and we can write $\overbrace{E_2}^{R_3-11R_2}\ \overbrace{E_1}^{R_3+3R_1}\ \overbrace{P}^{R_1\leftrightarrow R_2}\ A = \begin{bmatrix} -1 & 2 & 1 \\ 0 & 1 & 2 \\ 0 & 0 & -27 \end{bmatrix} = U.$

Then $\quad PA = (E_2 E_1)^{-1} U = (E_1^{-1} E_2^{-1}) U = LU.$ Here,

$P = \begin{bmatrix} 0 & 1 & 0 \\ 1 & 0 & 0 \\ 0 & 0 & 1 \end{bmatrix},\ E_1 = \begin{bmatrix} 1 & 0 & 0 \\ 0 & 1 & 0 \\ 3 & 0 & 1 \end{bmatrix},\ E_2 = \begin{bmatrix} 1 & 0 & 0 \\ 0 & 1 & 0 \\ -11 & 0 & 1 \end{bmatrix},$

$L = \begin{bmatrix} 1 & 0 & 0 \\ 0 & 1 & 0 \\ -3 & 11 & 1 \end{bmatrix},\quad$ and $\quad$ we $\quad$ can $\quad$ verify $\quad$ that

$PA = \begin{bmatrix} -1 & 2 & 1 \\ 0 & 1 & 2 \\ 3 & 5 & -8 \end{bmatrix} = LU.$

Example 3.7.8: A $PA = LU$ Factorization
for a Singular Matrix

Reducing the singular matrix $A = \begin{bmatrix} 0 & 1 & 0 \\ 2 & -2 & 2 \\ 3 & 5 & 3 \end{bmatrix}$ to an echelon
form requires one interchange of rows because of the zero in the $(1, 1)$ position of A. By swapping rows 1 and 2 of A, for example, the new matrix will need no further interchanges and we can write

$\begin{bmatrix} 0 & 1 & 0 \\ 1 & 0 & 0 \\ 0 & 0 & 1 \end{bmatrix} \begin{bmatrix} 0 & 1 & 0 \\ 2 & -2 & 2 \\ 3 & 5 & 3 \end{bmatrix} = \begin{bmatrix} 1 & 0 & 0 \\ 0 & 1 & 0 \\ 3/2 & 8 & 1 \end{bmatrix} \begin{bmatrix} 2 & -2 & 2 \\ 0 & 1 & 0 \\ 0 & 0 & 0 \end{bmatrix}.$

Note that U has a zero row because A is singular.

If we have the factorization $PA = LU$, we can solve a matrix equation $A\mathbf{x} = \mathbf{b}$ by first multiplying on the left by P to get $PA\mathbf{x} = P\mathbf{b}$. Letting $P\mathbf{b} = \tilde{\mathbf{b}}$ and substituting $PA = LU$, we obtain $LU\mathbf{x} = \tilde{\mathbf{b}}$. Now, as we did before, we let $\mathbf{y} = U\mathbf{x}$ and

(1) Solve $L\mathbf{y} = \tilde{\mathbf{b}}$ for $\mathbf{y}$ by forward substitution
and then
(2) Solve $U\mathbf{x} = \mathbf{y}$ for $\mathbf{x}$ by back substitution

Example 3.7.9: Solving a System via a $PA = LU$ Factorization

Let us solve the system:

$$b + 2c = 1$$
$$-a + 2b + c = 2$$
$$3a + 5b - 8c = 3$$

We write this as

$$\begin{bmatrix} 0 & 1 & 2 \\ -1 & 2 & 1 \\ 3 & 5 & -8 \end{bmatrix} \begin{bmatrix} a \\ b \\ c \end{bmatrix} = \begin{bmatrix} 1 \\ 2 \\ 3 \end{bmatrix}.$$

Conveniently, from Example 3.7.7 we know that we have

$$\overbrace{\begin{bmatrix} 0 & 1 & 0 \\ 1 & 0 & 0 \\ 0 & 0 & 1 \end{bmatrix}}^{P} \overbrace{\begin{bmatrix} 0 & 1 & 2 \\ -1 & 2 & 1 \\ 3 & 5 & -8 \end{bmatrix}}^{A} = \overbrace{\begin{bmatrix} 1 & 0 & 0 \\ 0 & 1 & 0 \\ -3 & 11 & 1 \end{bmatrix}}^{L} \overbrace{\begin{bmatrix} -1 & 2 & 1 \\ 0 & 1 & 2 \\ 0 & 0 & -27 \end{bmatrix}}^{U},$$

so that $PA\mathbf{x} = P\mathbf{b}$ is really

$$\overbrace{\begin{bmatrix} 1 & 0 & 0 \\ 0 & 1 & 0 \\ -3 & 11 & 1 \end{bmatrix}}^{L} \overbrace{\begin{bmatrix} -1 & 2 & 1 \\ 0 & 1 & 2 \\ 0 & 0 & -27 \end{bmatrix}}^{U} \overbrace{\begin{bmatrix} a \\ b \\ c \end{bmatrix}}^{\mathbf{x}} = \overbrace{\begin{bmatrix} 0 & 1 & 0 \\ 1 & 0 & 0 \\ 0 & 0 & 1 \end{bmatrix}}^{P} \overbrace{\begin{bmatrix} 1 \\ 2 \\ 3 \end{bmatrix}}^{\mathbf{b}}$$

$$= \begin{bmatrix} 2 \\ 1 \\ 3 \end{bmatrix} = \tilde{\mathbf{b}}.$$

If we let $\mathbf{y} = \begin{bmatrix} y_1 \\ y_2 \\ y_3 \end{bmatrix} = U\mathbf{x}$, then $L\mathbf{y} = \tilde{\mathbf{b}}$ is

$$\begin{bmatrix} 1 & 0 & 0 \\ 0 & 1 & 0 \\ -3 & 11 & 1 \end{bmatrix} \begin{bmatrix} y_1 \\ y_2 \\ y_3 \end{bmatrix} = \begin{bmatrix} 2 \\ 1 \\ 3 \end{bmatrix},$$

so that by forward substitution (top-down) $y_1 = 2$, $y_2 = 1$, and $y_3 = -2$.

Now we solve $U\mathbf{x} = \mathbf{y}$, or

$$\begin{bmatrix} -1 & 2 & 1 \\ 0 & 1 & 2 \\ 0 & 0 & -27 \end{bmatrix} \begin{bmatrix} a \\ b \\ c \end{bmatrix} = \begin{bmatrix} 2 \\ 1 \\ -2 \end{bmatrix},$$

by back substitution (bottom-up) to obtain $c = 2/27$, $b = 23/27$, and $a = -6/27$. Therefore,

$$\mathbf{x} = \begin{bmatrix} -\frac{6}{27} \\ \frac{23}{27} \\ \frac{2}{27} \end{bmatrix} = \frac{1}{27} \begin{bmatrix} -6 \\ 23 \\ 2 \end{bmatrix}.$$

So far, the work in this section has been about factoring *square* matrices, but the same decomposition techniques can be applied to rectangular matrices as well.

Example 3.7.10: An LU Factorization of a Rectangular Matrix*

Consider the matrix $A = \begin{bmatrix} 1 & 2 \\ 3 & 4 \\ 5 & 6 \end{bmatrix}$. We reduce A to echelon form as follows:

$$A = \begin{bmatrix} 1 & 2 \\ 3 & 4 \\ 5 & 6 \end{bmatrix} \xrightarrow[R_3 \to R_3 - 5R_1]{R_2 \to R_2 - 3R_1} \begin{bmatrix} 1 & 2 \\ 0 & -2 \\ 0 & -4 \end{bmatrix} \xrightarrow{R_3 \to R_3 - 2R_2} \begin{bmatrix} 1 & 2 \\ 0 & -2 \\ 0 & 0 \end{bmatrix} = U.$$

In this situation, U is called *upper trapezoidal* rather than upper triangular.

Using the shortcut illustrated in Example 3.7.2, we note the multipliers l_{ij} that are used to produce zeros in position (i, j) of U: $l_{21} = 3$, $l_{31} = 5$,

$l_{32} = 2$. Therefore, $L = \begin{bmatrix} 1 & 0 & 0 \\ 3 & 1 & 0 \\ 5 & 2 & 1 \end{bmatrix}$ and

$$LU = \begin{bmatrix} 1 & 0 & 0 \\ 3 & 1 & 0 \\ 5 & 2 & 1 \end{bmatrix} \begin{bmatrix} 1 & 2 \\ 0 & -2 \\ 0 & 0 \end{bmatrix} = \begin{bmatrix} 1 & 2 \\ 3 & 4 \\ 5 & 6 \end{bmatrix} = A.$$

As we have seen, an LU form of a matrix (when it exists) is very useful in solving systems of linear equations; so algorithms for LU factorization exist in many calculators, computer algebra systems for personal computers, and on mainframe computers. However, because of nonuniqueness and a desire to preserve numerical accuracy, **some computer algorithms may not return the same result that you would get by a hand calculation.**[†]

For example, at the beginning of this section we saw that we can write

$$A = \begin{bmatrix} 2 & 2 & -1 \\ 4 & 5 & 2 \\ -2 & 1 & 2 \end{bmatrix} = \overbrace{\begin{bmatrix} 1 & 0 & 0 \\ 2 & 1 & 0 \\ -1 & 3 & 1 \end{bmatrix}}^{L} \overbrace{\begin{bmatrix} 2 & 2 & -1 \\ 0 & 1 & 4 \\ 0 & 0 & -11 \end{bmatrix}}^{U}.$$

---

[*] This is based on an example of B.N. Datta, *Numerical Linear Algebra and Applications*, pp. 118–119 (Pacific Grove, CA: Brooks/Cole, 1995).

[†] For information on MATLAB®, see D.J. Higham and N.J. Higham, *MATLAB Guide*, 2nd edn. (Philadelphia, PA: SIAM, 2005) or D.R. Hill, *Experiments in Computational Matrix Algebra* (New York: Random House, 1988).

However, the MATLAB command $[LU] = lu(A)$ returns:

$$L = \begin{bmatrix} 0.5000 & -0.1429 & 1.0000 \\ 1.0000 & 0 & 0 \\ -0.5000 & 1.0000 & 0 \end{bmatrix} \text{ and}$$

$$U = \begin{bmatrix} 4.0000 & 5.0000 & 2.0000 \\ 0 & 3.5000 & 3.0000 \\ 0 & 0 & -1.5714 \end{bmatrix}.$$

The product of this L (a *permuted* lower triangular matrix) and U is

$$\begin{bmatrix} 2 & 1.9998 & -1.0001 \\ 4 & 5 & 2.0000 \\ -2.0000 & 1.0000 & 2.0000 \end{bmatrix}.$$

The MATLAB command **lu** uses *partial pivoting* (see Section 2.4) to calculate $PA = LU$, and entering $[LU] = lu(A)$ yields $L = P^T L$, where P is a permutation matrix. The command $[LUP] = lu(A)$ produces the permutation matrix P as well as the triangular factors. In our example, this expanded command gives us

$$L = \begin{bmatrix} 1.0000 & 0 & 0 \\ -0.5000 & 1.0000 & 0 \\ 0.5000 & -0.1429 & 1.0000 \end{bmatrix},$$

$$U = \begin{bmatrix} 4.0000 & 5.0000 & 2.0000 \\ 0 & 3.5000 & 3.0000 \\ 0 & 0 & -1.5714 \end{bmatrix}, \quad P = \begin{bmatrix} 0 & 1 & 0 \\ 0 & 0 & 1 \\ 1 & 0 & 0 \end{bmatrix}.$$

The *Maple*® command $(\mathbf{p, l, u}) := \mathbf{LUDecomposition(A)}$ produces a more familiar permutation matrix, a permuted unit lower triangular matrix, and an upper triangular matrix:

$$p = \begin{bmatrix} 1 & 0 & 0 \\ 0 & 1 & 0 \\ 0 & 0 & 1 \end{bmatrix}, \quad l = \begin{bmatrix} 1 & 0 & 0 \\ 2 & 1 & 0 \\ -1 & 3 & 1 \end{bmatrix}, \quad u = \begin{bmatrix} 2 & 2 & -1 \\ 0 & 1 & 4 \\ 0 & 0 & -11 \end{bmatrix}.$$

(It should be noted that the MATLAB Symbolic Math Toolbox is based on the *Maple*® kernel.)

Exercises 3.7

A.

1. Find the LU factorization, if possible, of each of the following matrices:

a. $\begin{bmatrix} 1 & 2 \\ -3 & -1 \end{bmatrix}$ b. $\begin{bmatrix} 2 & -4 \\ 3 & 1 \end{bmatrix}$ c. $\begin{bmatrix} 1 & 2 & 3 \\ 2 & 6 & 7 \\ 2 & 2 & 4 \end{bmatrix}$

d. $\begin{bmatrix} 2 & 2 & 2 \\ 4 & 7 & 7 \\ 6 & 18 & 22 \end{bmatrix}$ e. $\begin{bmatrix} 1 & 4 & 5 \\ 4 & 18 & 26 \\ 3 & 16 & 30 \end{bmatrix}$ f. $\begin{bmatrix} 2 & 2 & -1 \\ 4 & 0 & 4 \\ 3 & 4 & 4 \end{bmatrix}$

g. $\begin{bmatrix} 1 & 2 & -1 & 4 \\ -3 & -5 & 6 & -5 \\ 1 & 4 & 6 & 20 \\ -1 & 6 & 20 & 43 \end{bmatrix}$

2. If your Social Security number is abc-de-fghi, find the LU decomposition, if possible, of

$$\begin{bmatrix} a & b & c \\ d & e & f \\ g & h & i \end{bmatrix}.$$

3. Find the LU decomposition, if possible, of

$$\begin{bmatrix} -1 & -2 & 4 & 2 & 3 & 1 \\ 7 & 8 & 3 & -1 & 0 & -9 \\ 1 & 1 & -1 & 3 & 1 & 3 \\ -2 & 5 & 8 & 1 & -1 & 3 \\ -5 & 2 & 0 & 2 & -5 & 6 \\ 0 & 1 & -1 & -3 & 6 & 2 \end{bmatrix}.$$

4. Give two distinct LU factorizations L_1U_1 and L_2U_2 of the 3×3 zero matrix and explain why this does not contradict Corollary 3.7.1

5. Using the LU factorization found in Example 3.7.2, solve the system:

$$2a + 2b - c = 0$$
$$4a + 5b + 2c = -1.$$
$$-2a + b + 2c = 2$$

In Exercises 6–11, solve each system $Ax = b$ using the given LU factorizations and the given vector b.

6. $A = \begin{bmatrix} 2 & -1 \\ 2 & 5 \end{bmatrix} = \begin{bmatrix} -1 & 0 \\ -1 & 1 \end{bmatrix} \begin{bmatrix} -2 & 1 \\ 0 & 6 \end{bmatrix}$, $b = \begin{bmatrix} 2 \\ 3 \end{bmatrix}$

7. $A = \begin{bmatrix} 1 & 2 \\ 3 & 4 \end{bmatrix} = \begin{bmatrix} 1 & 0 \\ 3 & 1 \end{bmatrix} \begin{bmatrix} 1 & 2 \\ 0 & -2 \end{bmatrix}$, $b = \begin{bmatrix} -1 \\ 2 \end{bmatrix}$

8. $A = \begin{bmatrix} 2 & 1 & -2 \\ -2 & 3 & -4 \\ 4 & -3 & 0 \end{bmatrix} = \begin{bmatrix} 1 & 0 & 0 \\ -1 & 1 & 0 \\ 2 & -\frac{5}{4} & 1 \end{bmatrix} \begin{bmatrix} 2 & 1 & -2 \\ 0 & 4 & -6 \\ 0 & 0 & -\frac{7}{2} \end{bmatrix}$,

$b = \begin{bmatrix} -3 \\ 1 \\ 0 \end{bmatrix}$

9. $A = \begin{bmatrix} 2 & 1 & 3 \\ 4 & -1 & 3 \\ -2 & 5 & 5 \end{bmatrix} = \begin{bmatrix} 1 & 0 & 0 \\ 2 & 1 & 0 \\ -1 & -2 & 1 \end{bmatrix} \begin{bmatrix} 2 & 1 & 3 \\ 0 & -3 & -3 \\ 0 & 0 & 2 \end{bmatrix}$,

$\mathbf{b} = \begin{bmatrix} 1 \\ 2 \\ 3 \end{bmatrix}$

10. $A = \begin{bmatrix} 4 & -2 & 1 \\ 20 & -7 & 12 \\ -8 & 13 & 17 \end{bmatrix} = \begin{bmatrix} 1 & 0 & 0 \\ 5 & 1 & 0 \\ -2 & 3 & 1 \end{bmatrix} \begin{bmatrix} 4 & -2 & 1 \\ 0 & 3 & 7 \\ 0 & 0 & -2 \end{bmatrix}$,

$\mathbf{b} = \begin{bmatrix} 11 \\ 70 \\ 17 \end{bmatrix}$

11. $A = \begin{bmatrix} 3 & 2 & 1 & 4 \\ 6 & 5 & 8 & 7 \\ 9 & 10 & 11 & 12 \\ 13 & 14 & 15 & 16 \end{bmatrix} = \begin{bmatrix} 1 & 0 & 0 & 0 \\ 2 & 1 & 0 & 0 \\ 3 & 4 & 1 & 0 \\ \frac{13}{3} & \frac{16}{3} & \frac{4}{3} & 1 \end{bmatrix} \begin{bmatrix} 3 & 2 & 1 & 4 \\ 0 & 1 & 6 & -1 \\ 0 & 0 & -16 & 4 \\ 0 & 0 & 0 & -\frac{4}{3} \end{bmatrix}$,

$\mathbf{b} = \begin{bmatrix} 1 \\ 2 \\ 3 \\ 4 \end{bmatrix}$

12.   If $A = \begin{bmatrix} 0 & 0 & 4 \\ 1 & 2 & 3 \\ 1 & 4 & 1 \end{bmatrix}$, find a permutation matrix P, a unit lower triangular matrix L, and an upper triangular matrix U such that $PA = LU$.

13.   Use the $PA = LU$ factorization in Examples 3.7.8 and 3.7.9 to solve the system:

$$-2b + 2c = 0$$
$$a + 2b - c = -1 .$$
$$3a + 5b - 8c = 2$$

B.

1.   Prove that the product of two lower triangular matrices is lower triangular.

2.   Show that the product of two *unit* lower triangular matrices is unit lower triangular.

3.   Show that the product of two upper triangular matrices is upper triangular.

4.   a. Prove that if a lower triangular matrix is invertible, then all its diagonal entries are nonzero.
     b. Prove that if all the diagonal entries of a lower triangular matrix are nonzero, then the matrix is invertible.

5. Show that the inverse of an invertible (unit) lower triangular matrix is (unit) lower triangular.

6. Show that the inverse of an invertible upper triangular matrix is upper triangular.

7. Suppose that an $n \times n$ matrix A can be written $A = LU$, where L is an $n \times n$ lower triangular matrix and U is an $n \times n$ upper triangular matrix. Prove that for any partition

$$A = \begin{bmatrix} A_{11} & A_{12} \\ A_{21} & A_{22} \end{bmatrix}, \quad L = \begin{bmatrix} L_{11} & O \\ L_{21} & L_{22} \end{bmatrix}, \quad U = \begin{bmatrix} U_{11} & U_{12} \\ O & U_{22} \end{bmatrix}$$

    with A_{11}, L_{11}, and U_{11} $k \times k$ matrices, $k \leq n$, we have
    a. $L_{11}U_{11} = A_{11}$ b. $L_{11}U_{12} = A_{12}$ c. $L_{21}U_{11} = A_{21}$
    d. $L_{21}U_{12} + L_{22}U_{22} = A_{22}$

8. Prove that if P is a permutation matrix, then $P^{-1} = P^{T}$.

9. Show that the matrix

$$\begin{bmatrix} 1 & 2 & 3 \\ 2 & 4 & 7 \\ 3 & 5 & 3 \end{bmatrix}$$

    does not have an LU decomposition.

10. In Theorem 3.7.3, if A is nonsingular, prove that the matrices L and U are unique for a fixed matrix P.

11. Find an LU factorization of

$$A = \begin{bmatrix} 2 & 3 & -1 \\ -6 & -6 & 5 \\ 4 & 18 & 6 \\ -2 & -9 & -3 \end{bmatrix}.$$

    [*Hint*: L will be 4×4 and U will be 4×3.]

C.

1. Determine all values of x for which

$$A = \begin{bmatrix} x & 2 & 0 \\ 1 & x & 1 \\ 0 & 1 & x \end{bmatrix}$$

    fails to have an LU factorization.

2. If A is a nonsingular matrix that contains only integer entries and all of its pivots are 1, explain why A^{-1} must also have only integer entries. [If the row echelon form of A is obtained without any multiplication of rows by nonzero scalars, then a *pivot* of A is defined to be any pivot of the echelon form.]

3. An LU factorization can be converted into an **LDU factorization**, where L is unit lower triangular, D is diagonal, and U is an upper triangular matrix with 1s on the diagonal (i.e., a *unit* upper triangular matrix). Find the LDU factorization of the matrix A in Example 3.7.2.

4. Prove that if an invertible matrix has an LDU factorization (see Exercise C3), then the LDU factors are uniquely determined.

5. A **generalized inverse** of an $m \times n$ matrix A is any $n \times m$ matrix G such that $AGA = A$. (Note that if A is nonsingular, its only generalized inverse is A^{-1}; otherwise, it has infinitely many generalized inverses.)

    Suppose that A is an $n \times n$ matrix that has an LDU decomposition (see Exercise C3), say $A = LDU$, and define $G = U^{-1}D^{-1}L^{-1}$.
    a. Show that G is a generalized inverse of A.
    b. Show that $G = D^{-1}L^{-1} + (I_n - U)G = U^{-1}D^{-1} + G(I_n - L)$.

6. Prove that an $n \times n$ matrix M commutes with every $n \times n$ permutation matrix if and only if M has the form $aI_n + bE$, where E is an $n \times n$ matrix all of whose entries are 1s and a and b are scalars.

3.8 Summary

The **addition** and **subtraction** of matrices having the same size is carried out in an entry-by-entry fashion. Matrix addition is **commutative** and **associative**. For the set of all $m \times n$ matrices, there is an additive **identity element**, the **zero matrix** O_{mn}, all of whose entries are zero. Each matrix A has an **additive inverse**, $-A = (-1)A$, such that $A + (-A) = O_{mn}$. There is also **scalar–matrix multiplication** and Theorem 3.1.2 provides the algebraic properties of scalar multiplication. The **transpose** of an $m \times n$ matrix A is the $n \times m$ matrix A^{T} obtained by interchanging rows and columns in A. This definition leads to the consideration of **symmetric matrices** and **skew-symmetric matrices**.

If A is an $m \times n$ matrix whose jth column is denoted by $\mathbf{a}_j$ and $\mathbf{x} = [x_1 \quad x_2 \quad \ldots \quad x_n]^{\mathrm{T}}$, then we define the **matrix–vector product** $A\mathbf{x}$ as $A\mathbf{x} = x_1\mathbf{a}_1 + x_2\mathbf{a}_2 + \cdots + x_n\mathbf{a}_n$. For any positive integer n, there is an **identity matrix**, denoted by I_n, the unique $n \times n$ matrix such that $I_n\mathbf{x} = \mathbf{x}$ for every vector $\mathbf{x} \in \mathbb{R}^n$. Other important types of matrices are **scalar matrices**, **diagonal matrices**, **upper triangular matrices**, and **lower triangular matrices**. An equivalent way of defining the product of any $m \times n$ matrix and any vector in $\mathbb{R}^n$ is as a row-by-column dot product. The multiplication of vectors in $\mathbb{R}^n$ by a fixed $m \times n$ matrix A defines a **linear transformation**, $T_A : \mathbb{R}^n \to \mathbb{R}^m$. This transformation T_A is defined by $T_A(\mathbf{v}) = A\mathbf{v}$ for all $\mathbf{v} \in \mathbb{R}^n$. A linear transformation is defined by the properties (1) $T_A(\mathbf{v}_1 + \mathbf{v}_2) = T_A(\mathbf{v}_1) + T_A(\mathbf{v}_2)$ for all vectors $\mathbf{v}_1$ and $\mathbf{v}_2$ in and (2) $T_A(k\mathbf{v}) = kT_A(\mathbf{v})$ for any vector $\mathbf{v}$ in $\mathbb{R}^n$ and any real number k.

Geometrically, multiplying a vector by a matrix *transforms* the vector—for example, by changing the vector's position and/or altering its length within its Euclidean space.

Matrices A and B are said to be **conformable** for multiplication in the order AB if A has exactly as many columns as B has rows—that is, if A is $m \times p$ and B is $p \times n$ for some positive integers m, p, and n. If $A = [a_{ij}]$ is an $m \times p$ matrix and $B = [b_{ij}]$ is a $p \times n$ matrix, then we can define the **product** AB is an $m \times n$ matrix whose (i, j) entry is the dot product of $[\text{row}_i(A)]^T$ and $\text{col}_j(B)$. Also, column j of the product AB is given by $\text{col}_j(AB) = A \cdot \text{col}_j(B)$, so $AB = A\ [\text{col}_1(B)\ \text{col}_2(B) \ldots \text{col}_n(B)] = [A\ \text{col}_1(B)\ A\ \text{col}_2(B) \ldots A\ \text{col}_n(B)]$. Furthermore, row i of the product AB is given by $\text{row}_i(AB) = \text{row}_i(A) \cdot B$. In general, **matrix multiplication is not commutative**. The basic algebraic properties of matrix multiplication are given in Theorem 3.3.3.

Powers of an $n \times n$ matrix are defined recursively. Such powers obey the law of exponents: $A^m A^n = A^{m+n}$ and $(A^m)^n = A^{mn}$ for all nonnegative integers m and n.

An $m \times n$ matrix A is said to be **partitioned into submatrices** when it is written in the form:

$$A = \begin{bmatrix} A_{11} & A_{12} & \ldots & A_{1s} \\ A_{21} & A_{22} & \ldots & A_{2s} \\ \vdots & \vdots & \ldots & \vdots \\ A_{r1} & A_{r2} & \ldots & A_{rs} \end{bmatrix},$$

where each A_{ij} is an $m_i \times n_j$ submatrix of A with $m_1 + m_2 + \cdots + m_r = m$ and $n_1 + n_2 + \cdots + n_s = n$. These submatrices A_{ij} are also called **blocks**, and the partitioned form given prior is called a **block matrix**. If $A(m \times n)$ and $B(n \times p)$ are partitioned into blocks as indicated:

$$A = \begin{bmatrix} A_{11} & A_{12} & \cdots & A_{1n} \\ A_{21} & A_{22} & \cdots & A_{2n} \\ \vdots & \vdots & \vdots & \vdots \\ A_{m1} & A_{m2} & \cdots & A_{mn} \end{bmatrix}, \quad B = \begin{bmatrix} B_{11} & B_{12} & \cdots & B_{1p} \\ B_{21} & B_{22} & \cdots & B_{2p} \\ \vdots & \vdots & \vdots & \vdots \\ B_{n1} & B_{n2} & \cdots & B_{np} \end{bmatrix},$$

and if each pair (A_{ik}, B_{kj}) is conformable for multiplication, then A and B are said to be **conformably partitioned**. If matrices A and B are conformably partitioned, we can calculate the product $AB = C$ in any of the usual ways, regarding blocks as entries. Partitioning is used to reduce computer storage requirements when multiplying large matrices.

If A is an $n \times n$ matrix, and if there is an $n \times n$ matrix B such that $AB = I_n$ and $BA = I_n$, then A is called **nonsingular** (or **invertible**) and B is called an **inverse** of A, denoted by $B = A^{-1}$. A matrix A that has no inverse is called a **singular** (or **noninvertible**) matrix. If a square matrix A has an inverse, then that inverse is unique. A number of important properties of nonsingular matrices are proved in Sections 3.5 and 3.6. Negative integer powers of square matrices can be defined, and the laws of exponents for square matrices hold for *all* integers.

An **elementary matrix** is a matrix obtained from I_n by applying a single elementary row operation to it. **Any elementary row operation on** an $m \times n$ matrix A can be accomplished by multiplying A on the left (premultiplying) by an elementary matrix, the identity matrix I_m on which the same operation has been performed. There are three types of elementary matrices. An elementary matrix is nonsingular, and its inverse is an elementary matrix of the same type. Two matrices are row [column] equivalent if one matrix may be obtained from the other matrix by means of elementary row [column] operations. A key result is that an $n \times n$ matrix A is invertible if and only if A is row equivalent to I_n. If A is invertible, any sequence of elementary row operations reducing A to I_n also transforms I_n into A^{-1}: $\left[A \vdots I_n \right] \xrightarrow{\text{elementary row operations}} \left[I_n \vdots A^{-1} \right].$

Another useful result is that an $n \times n$ matrix A is invertible if and only if it can be written as a product of elementary matrices.

If A is a square matrix and $A = LU$, where U is upper triangular and L is a unit lower triangular matrix (diagonal entries are all ones), this factorization, which is not necessarily unique, is called an LU factorization (or LU decomposition) of A. Not all square matrices can be factored in this way. However, if A is an $n \times n$ matrix and we can reduce A to an upper triangular matrix U by Gaussian elimination *without any interchanges of rows*, then $A = LU$, where L is a unit lower triangular matrix. The factorization $A = LU$ is the matrix formulation of Gaussian elimination, provided that no row interchanges are used. If an *invertible* $n \times n$ matrix, A has an LU factorization, then the factorization is unique. When it exists, an LU factorization can be used to solve a system $A\mathbf{x} = \mathbf{b}$. A permutation matrix P is any $n \times n$ matrix that results from rearranging (permuting) the order of the rows of I_n. If A is any $n \times n$ matrix, then there is a permutation matrix P such that $PA = LU$, where L is a unit lower triangular matrix and U is upper triangular. This kind of factorization can also be used to solve systems of linear equations.

4

Eigenvalues, Eigenvectors, and Diagonalization

There are various important scalars associated with a square matrix, numbers which are somehow characteristic of the nature of the matrix and which simplify calculations with the matrix. One of the most useful of these scalars is the *determinant*. The concept of a determinant has an ancient and honorable history strewn with the names of many prominent mathematicians and scientists. Its "modern" development begins in the late seventeenth century with the independent work of the Japanese mathematician Seki Kowa (1642–1708) and of Gottfried Wilhelm Leibniz (1646–1716), the coinventor of calculus. At one time, determinants were the most important tool used to analyze and solve systems of linear equations, while matrix theory played only a supporting role. Currently, there are various schools of thought on the importance of determinants in modern mathematical and scientific work. As one author* has described the situation:

> ...mathematics, like a river, is everchanging in its course, and major branches can dry up to become minor tributaries while small trickling brooks can develop into raging torrents. This is precisely what occurred with determinants and matrices. The study and use of determinants eventually gave way to Cayley's matrix algebra, and today matrix and linear algebra are in the main stream of applied mathematics, while the role of determinants has been relegated to a minor backwater position.

NEVERTHELESS, the determinant remains important in the theory of linear algebra and in many disciplines using matrices as a useful tool, even if its calculation and more obscure properties no longer deserve as much attention as in the past.[†] It is through the determinant that we will understand the important concepts of *eigenvalues* and *eigenvectors*, leading to the technique of *diagonalizing* a matrix.

---

* C.D. Meyer, *Matrix Analysis and Applied Linear Algebra*, pp. 459–460 (Philadelphia, PA: SIAM, 2000).
[†] See S. Axler, Down with determinants!, *Amer. Math. Monthly* **102** (1995): 139–154 for a development of important concepts in linear algebra without using determinants. Also see his book, *Linear Algebra Done Right*, 2nd edn. (New York: Springer-Verlag, 1997).

4.1 Determinants

Determinants are often introduced in the context of solving systems of linear equations. For example, given the system:

$$ax + by = e$$
$$cx + dy = f$$

where we assume that $a \neq 0$, we can row reduce the augmented matrix of this system and obtain

$$\begin{bmatrix} a & b & e \\ 0 & \frac{ad-bc}{a} & \frac{af-ce}{a} \end{bmatrix},$$

leading to the solution:

$$x = \frac{de - bf}{ad - bc} \quad \text{and} \quad y = \frac{af - ce}{ad - bc},$$

provided that $ad - bc \neq 0$. (Note that in the row reduction, we have avoided the operation of simply multiplying a row by a scalar.) The number $ad - bc$ is called the **determinant** of the system because it *determines* whether the system has a unique solution. Alternatively, the determinant indicates whether the square matrix of coefficients $A = \begin{bmatrix} a & b \\ c & d \end{bmatrix}$ has an inverse (see Example 3.5.4). The dependence of the determinant on the entries of the matrix of coefficients is shown by the notation det(A) or $\begin{vmatrix} a & b \\ c & d \end{vmatrix}$ for the determinant. In our example, we note that det(A) is the product of the diagonal elements of the echelon form of A: $\det(A) = a \cdot ((ad - bc)/a) = ad - bc$.

Looking at this situation another way, if $a \neq 0$, we can find an LU factorization of A: $\begin{bmatrix} a & b \\ c & d \end{bmatrix} = \begin{bmatrix} 1 & 0 \\ c/a & 1 \end{bmatrix} \begin{bmatrix} a & b \\ 0 & (ad - bc)/a \end{bmatrix}$. We note that the product of the main diagonal entries of the upper triangular matrix U is $ad - bc$, the determinant of A.

Computationally, for a 2×2 matrix A we have $\det(A) = \begin{vmatrix} a & b \\ c & d \end{vmatrix} =$ (the product of the main diagonal elements a and d) $-$ (the product of the secondary diagonal elements b and c). For future reference, we note that if $\hat{A} = \begin{bmatrix} c & d \\ a & b \end{bmatrix}$, the matrix obtained from A by interchanging its rows, then $\det(\hat{A}) = cb - ad = -(ad - bc) = -\det(A)$. (For example, in trying to reduce A, we might have to perform this interchange if $a = 0$.)

Similarly, if we row reduce the matrix of coefficients of a third-order system:

$$a_{11}x + a_{12}y + a_{13}z = b_1$$
$$a_{21}x + a_{22}y + a_{23}z = b_2$$
$$a_{31}x + a_{32}y + a_{33}z = b_3$$

assuming $a_{11} \neq 0$ and avoiding a simple scalar multiplication of a row, we get the upper triangular matrix:

$$\begin{bmatrix} a_{11} & a_{12} & a_{13} \\ 0 & \dfrac{a_{22}a_{11} - a_{21}a_{12}}{a_{11}} & \dfrac{a_{23}a_{11} - a_{21}a_{13}}{a_{11}} \\ 0 & 0 & \dfrac{a_{33}a_{11}a_{22} - a_{33}a_{21}a_{12} - a_{31}a_{13}a_{22} - a_{32}a_{11}a_{23} + a_{32}a_{21}a_{13} + a_{31}a_{12}a_{23}}{a_{22}a_{11} - a_{21}a_{12}} \end{bmatrix}.$$

Once again, the existence of a unique solution depends on whether a particular number, the determinant, is nonzero. In this 3×3 case, the determinant is

$$a_{33}a_{11}a_{22} - a_{33}a_{21}a_{12} - a_{31}a_{13}a_{22} - a_{32}a_{11}a_{23} + a_{32}a_{21}a_{13} + a_{31}a_{12}a_{23},$$

the numerator of the (3,3) entry, the product of the diagonal entries of the upper triangular matrix we end up with. (In trying to solve the system, we wind up *dividing* by this number.) It is important to recall that an echelon form is not unique, a point we will discuss further in a little while.

In general, a determinant is a particular scalar associated with a square matrix, a scalar that arises naturally from the algebra required to determine the solution or solutions of a linear system. As we have already noted, in this limited context of solving such a system all that matters is whether the determinant is zero or nonzero. We shall see other uses of the determinant later, for example, in Section 4.2.

Definition 4.1.1

Given an $n \times n$ matrix A, the **determinant** of A, denoted by $\det(A)$ or $|A|$, is defined as follows.

If A has the factorization $PA = LU$, where P is a permutation matrix and $U = [u_{ij}]$, then $\det(A) = (-1)^k u_{11} \; u_{22} \; \cdots \; u_{nn}$, where k is the number of row interchanges represented by P:

$$\det(A) = (-1)^k \prod_{i=1}^{n} u_{ii} \quad \text{if } PA = LU,$$

where k is the number of row interchanges represented by P.

Of course, if we can write $A = LU$, then there are no row interchanges in obtaining U (i.e., $k = 0$) and the determinant of A is just the product of the diagonal elements of U.

In Theorem 4.1.1, we will see how an elementary row operation of type II (multiplying a row by a nonzero scalar) affects the basic calculation of a determinant. Such an operation is usually applied for mathematical neatness—for example, to remove fractions from a row or to rationalize a denominator—and it is usually not essential to the reduction to upper triangular form.

This is not one of the "classic" definitions of the determinant (see the exposition in Section 4.3 for one); but, relying as it does on the Gaussian elimination, it provides a relatively inexpensive way of actually calculating this important number. For someone beginning the study of linear

algebra, it is more important to understand the significance of the determinant in various theoretical and applied situations than to worry about
using methods of computation. These days any graphing calculator can
compute determinants of fairly large matrices. In most cases, a calculator
or computer algebra system (CAS) will provide accurate *approximations*.

The prime difficulty with Definition 4.1.1 is that the LU and PA = LU
factorizations, when they exist, are not necessarily unique. For example, if
$A = LU$ and D is a nonsingular diagonal matrix, then $\tilde{L} = LD$ is lower
triangular   and $\tilde{U} = D^{-1}U$ is   upper   triangular.   Hence
$A = LU = LDD^{-1}U = \tilde{L}\tilde{U}$, and $\tilde{L}\tilde{U}$ is also an LU factorization. However,
**if A is nonsingular and $A = LU$ or $PA = LU$, then these factorizations
are unique** (Corollary 3.7.1 and Theorem 3.7.3), **and the determinant is
uniquely defined. Furthermore, the value of det(A) calculated by
Definition 4.1.1 will agree with the value calculated in a more traditional way. In addition, it can be shown that even when the factorizations are not unique, the product of the diagonal entries of U does not
change**.

The determinant of a singular matrix is particularly significant. If A is
singular, then A can be row reduced to an upper triangular matrix U with
at least one zero row. Otherwise, we could continue reducing U until the
reduced row echelon form of A is I_n. But arriving at I_n would say (by
Theorem 3.6.3) that A is invertible—a contradiction. If row i of U consists
of zeros, then $u_{ii} = 0$ and $\det(A) = \pm \prod_{k=1}^{n} u_{kk} = 0$. Thus **if A is singular, $\det(A) = 0$**. We will prove the converse of this last statement as part
of Theorem 4.1.4.

Example 4.1.1: Some Determinants, Using
Definition 4.1.1

(a) Consider   the   matrix $A = \begin{bmatrix} 1 & 2 & 3 \\ -1 & -2 & -3 \\ 2 & 4 & 6 \end{bmatrix} =$

$$\begin{bmatrix} 1 & 0 & 0 \\ -1 & 1 & 0 \\ 2 & 0 & 1 \end{bmatrix} \begin{bmatrix} 1 & 2 & 3 \\ 0 & 0 & 0 \\ 0 & 0 & 0 \end{bmatrix} = LU.$$

Using   Definition   4.1.1,   we   calculate   det(A) =
$(1)(0)(0) = 0$. However, we can provide another LU factorization:

$$A = \begin{bmatrix} 1 & 2 & 3 \\ -1 & -2 & -3 \\ 2 & 4 & 6 \end{bmatrix} = \begin{bmatrix} 1 & 0 & 0 \\ -1 & 1 & 0 \\ 2 & 5 & 1 \end{bmatrix} \begin{bmatrix} 1 & 2 & 3 \\ 0 & 0 & 0 \\ 0 & 0 & 0 \end{bmatrix} = \hat{L}U,$$

where we have changed the (3, 2) entry of L. The definition still gives us det(A) = $(1)(0)(0) = 0$. The fact that
the decomposition of A is not unique indicates that A is a
singular matrix, and we expect its determinant to be zero.

The matrix $B = \begin{bmatrix} 0 & 2 & 3 \\ 2 & -4 & 7 \\ 1 & -2 & 5 \end{bmatrix}$ has the following $PB = LU$ factorization, where premultiplication by P causes two row interchanges:

$$\begin{bmatrix} 0 & 0 & 1 \\ 1 & 0 & 0 \\ 0 & 1 & 0 \end{bmatrix} \begin{bmatrix} 0 & 2 & 3 \\ 2 & -4 & 7 \\ 1 & -2 & 5 \end{bmatrix} = \begin{bmatrix} 1 & 0 & 0 \\ 0 & 1 & 0 \\ 2 & 0 & 1 \end{bmatrix} \begin{bmatrix} 1 & -2 & 5 \\ 0 & 2 & 3 \\ 0 & 0 & -3 \end{bmatrix},$$

from which we calculate $\det(B) = (-1)^2(1)(2)(-3) = -6$. Changing P provides a different decomposition:

$$\begin{bmatrix} 0 & 1 & 0 \\ 1 & 0 & 0 \\ 0 & 0 & 1 \end{bmatrix} \begin{bmatrix} 0 & 2 & 3 \\ 2 & -4 & 7 \\ 1 & -2 & 5 \end{bmatrix} = \begin{bmatrix} 1 & 0 & 0 \\ 0 & 1 & 0 \\ 1/2 & 0 & 1 \end{bmatrix} \begin{bmatrix} 2 & -4 & 7 \\ 0 & 2 & 3 \\ 0 & 0 & 3/2 \end{bmatrix}.$$

Again we find that $\det(B) = (-1)^1(2)(2)(3/2) = -6$.

(b) If $A = \begin{bmatrix} 3 & 4 & 5 \\ 0 & 1 & 2 \\ 0 & 0 & 6 \end{bmatrix}$, then $\det(A) = (3)(1)(6) = 18$. In Sec-

tion 3.7, we saw that $B = \begin{bmatrix} 2 & 2 & -1 \\ 4 & 5 & 2 \\ -2 & 1 & 2 \end{bmatrix} =$

$$\overbrace{\begin{bmatrix} 1 & 0 & 0 \\ 2 & 1 & 0 \\ -1 & 3 & 1 \end{bmatrix}}^{L} \overbrace{\begin{bmatrix} 2 & 2 & -1 \\ 0 & 1 & 4 \\ 0 & 0 & -11 \end{bmatrix}}^{U}.$$ Therefore, $\det(B) = \det(U) =$

$(2)(1)(-11) = -22$. Finally, in Example 3.7.7 we saw that

if $C = \begin{bmatrix} 0 & 1 & 2 \\ -1 & 2 & 1 \\ 3 & 5 & -8 \end{bmatrix}$, we can write $PC = \begin{bmatrix} -1 & 2 & 1 \\ 0 & 1 & 2 \\ 3 & 5 & -8 \end{bmatrix} =$

$$\begin{bmatrix} 1 & 0 & 0 \\ 0 & 1 & 0 \\ -3 & 11 & 1 \end{bmatrix} \begin{bmatrix} -1 & 2 & 1 \\ 0 & 1 & 2 \\ 0 & 0 & -27 \end{bmatrix} = LU,$$ where the permuta-

tion matrix P involves *one* interchange of rows. Therefore,

$$\det(C) = (-1)^1(-1)(1)(-27) = -27.$$

A simple but very useful consequence of Definition 4.1.1 is that **the determinant of any triangular matrix is the product of its main diagonal elements**. Of course, this statement includes all diagonal matrices. In particular, **det $(I_n) = 1$ for any positive integer n**.

An important property to notice is that if A and B are both upper triangular (or lower triangular) $n \times n$ matrices, then the product AB is also upper triangular (or lower triangular)—see Problems B1 and B3 of Exercise 3.7—and $\det(AB) = \det(A) \det(B)$: (The main diagonal of AB) = (the main diagonal of A)·(the main diagonal of B). The proof of this result is left as Exercise B1. The fact that $\det(AB) = \det(A) \det(B)$ for *any* $n \times n$ matrices A and B is fundamental to further work with determinants and will be proved shortly.

Now we prove the basic properties of determinant that will simplify calculation and provide information about the determinant of a product. Because the LU and PA = LU decompositions can be accomplished by premultiplying by elementary matrices, we will begin by focusing on the determinants of elementary matrices. Recall from the last chapter that there are three types of elementary matrix:

I. E_I is a matrix that interchanges two rows of I_n.
II. E_{II} is a matrix that multiplies a row of I_n by a nonzero scalar k.
III. E_{III} is a matrix that adds k times a row of I_n to another row of I_n, where k is a nonzero scalar.

Furthermore, any elementary matrix is invertible, and its inverse is an elementary matrix of the same type (Theorem 3.6.2).

Theorem 4.1.1

If E_I, E_{II}, and E_{III} are $n \times n$ elementary matrices of types I, II, and III, respectively, then

(i) $\det(E_I) = -1$.
(ii) $\det(E_{II}) = k$, where premultiplication by E_{II} multiplies row i of a matrix by the nonzero scalar k.
(iii) $\det(E_{III}) = 1$.

Proof (1) Suppose that E is an $n \times n$ matrix that is just I_n with rows i and j interchanged for some i and j. Then $E^2 = EE = I_n$. We can write this as $PE = I_n I_n$, where $P = E$ is a permutation matrix and I_n is both a unit lower triangular matrix and an upper triangular matrix. By Definition 4.1.1,

$$\det(E) = (-1)^1 \cdot \overbrace{(1 \cdot 1 \cdots 1)}^{n \text{ factors}} = -1.$$

(2) If E is an $n \times n$ elementary matrix of type II, then E is just I_n with one row, say i, multiplied by a nonzero scalar k. But this means that E is a diagonal matrix, $(n-1)$ of whose diagonal elements are 1, with the remaining diagonal element equal to k. Thus, by the comments immediately after Definition 4.1.1,

$$\det(E) = \overbrace{1 \cdot 1 \cdots 1}^{n-1 \text{ factors}} \cdot k = k.$$

(3) If E is an $n \times n$ elementary matrix of type III, then E is just I_n with one row increased by a scalar times another row. In particular, E must be triangular. If $\text{row}_i(E) = \text{row}_i(I_n) + k \ \text{row}_j(I_n)$ for some $j > i$, then E is an *upper* triangular matrix whose main diagonal elements are still all ones. Similarly, if $j < i$ and $\text{row}_i(E) = \text{row}_i(I_n) + k \ \text{row}_j(I_n)$, then E is a *lower* triangular matrix, each of whose main diagonal entries is 1. Therefore, $\det(E) = 1$.

As an aid to proving the next theorem, we first establish an auxiliary result that has some interest in its own right.

Lemma 4.1.1

If A is a nonsingular $n \times n$ matrix and B is a singular $n \times n$ matrix, then AB is singular. (Similarly, BA is singular.)

Proof   Suppose that AB is nonsingular. If there exists a matrix C such that $C(AB) = I_n$, then $(CA)B = I_n$, implying (by Theorem 3.5.2) that CA is the inverse of B—a contradiction.

Theorem 4.1.2

If E is an $n \times n$ elementary matrix and A is any $n \times n$ matrix, then $\det(EA) = \det(E) \det(A)$. (Similarly, $\det(AE) = \det(A) \det(E)$.)

Proof   (*For a type* I *matrix*)   Suppose that E is a type I elementary matrix and hence nonsingular.

If A is nonsingular, then EA is also nonsingular (Theorem 3.5.3(3)) and there exists a permutation matrix P such that $P(EA) = LU$. Thus $\det(EA) = (-1)^k \prod_{i=1}^{n} u_{ii}$, where k is the number of interchanges induced by P. But $(PE)A = LU$ also, and PE is a permutation matrix with one more interchange than P. Therefore, $\det(A) = (-1)^{k+1} \prod_{i=1}^{n} u_{ii}$. It follows that $\det(E) \det(A) = (-1)$ $(-1)^{k+1} \prod_{i=1}^{n} u_{ii} = (-1)^{k+2} \prod_{i=1}^{n} u_{ii} = (-1)^2 (-1)^k \prod_{i=1}^{n} u_{ii} = \det(EA)$.

On the other hand, if A is singular, then EA is also singular, by Lemma 4.1.1. Thus $\det(EA) = 0 = (-1)(0) = \det(E) \det(A)$.

The proofs for elementary matrices of types II and III are left as Exercises B2 and B3.

Corollary 4.1.1

If A is an $n \times n$ matrix with two identical rows, then $\det(A) = 0$.

Proof   Let E be the elementary matrix corresponding to the interchange of the identical rows of A. Then $EA = A$ and $\det(A) = \det(EA) = \det(E) \det(A) = -\det(A)$, where we have used Theorems 4.1.1(i) and 4.1.2. But the only number that equals its own negative is zero.

Theorem 4.1.3

(i) If a matrix B is obtained by interchanging two rows of matrix A, then $\det(B) = -\det(A)$.
(ii) If a matrix B is obtained by multiplying any row of A by a scalar k, then $\det(B) = k \det(A)$.
(iii) If a matrix B is obtained by adding a scalar multiple of a row of A to another row of A, then $\det(B) = \det(A)$.

Proof Each of the procedures described in parts (i)–(iii) to convert A into B can be carried out via the premultiplication of A by an appropriate elementary matrix.

To prove parts (i), (ii), and (iii) of Theorem 4.1.3, we can use parts (i), (ii), and (iii), respectively, of Theorem 4.1.1, together with Theorem 4.1.2.

For example, to prove (iii), write $B = EA$, where E is a type III elementary matrix. Then $\det(B) = \det(EA) = \det(E) \det(A) = \det(A)$, using Theorems 4.1.2 and 4.1.1(iii) in that order.

Corollary 4.1.2

If any row of A consists entirely of zeros, then $\det(A) = 0$.

Proof The matrix A can be considered as resulting from another matrix $\hat{A}$ by multiplying one of its rows by the scalar 0. Now use Theorem 4.1.3(ii) to conclude that $\det(A) = 0 \cdot \det(\hat{A}) = 0$.

Example 4.1.2: Properties of Determinants

We will use the matrices from Example 4.1.1 to illustrate some basic properties of determinants.

$$\text{Consider} \quad A = \begin{bmatrix} 3 & 4 & 5 \\ 0 & 1 & 2 \\ 0 & 0 & 6 \end{bmatrix}, \quad B = \begin{bmatrix} 2 & 2 & -1 \\ 4 & 5 & 2 \\ -2 & 1 & 2 \end{bmatrix},$$

$$C = \begin{bmatrix} 0 & 1 & 2 \\ -1 & 2 & 1 \\ 3 & 5 & -8 \end{bmatrix},$$

where $\det(A) = 18, \det(B) = -22$, and $\det(C) = -27$.
First let us interchange rows 1 and 3 of A to get

$$\tilde{A} = \begin{bmatrix} 0 & 0 & 6 \\ 0 & 1 & 2 \\ 3 & 4 & 5 \end{bmatrix}.$$

Then

$$\overbrace{\begin{bmatrix} 0 & 0 & 1 \\ 0 & 1 & 0 \\ 1 & 0 & 0 \end{bmatrix}}^{P} \overbrace{\begin{bmatrix} 0 & 0 & 6 \\ 0 & 1 & 2 \\ 3 & 4 & 5 \end{bmatrix}}^{\tilde{A}} = \overbrace{\begin{bmatrix} 1 & 0 & 0 \\ 0 & 1 & 0 \\ 0 & 0 & 1 \end{bmatrix}}^{L} \overbrace{\begin{bmatrix} 3 & 4 & 5 \\ 0 & 1 & 2 \\ 0 & 0 & 6 \end{bmatrix}}^{U},$$

giving us $\det(\tilde{A}) = (-1)^1(3)(1)(6) = -18 = -\det(A)$.

Now suppose that we multiply the second row of B by -2. Then

$$\tilde{B} = \begin{bmatrix} 2 & 2 & -1 \\ -8 & -10 & -4 \\ -2 & 1 & 2 \end{bmatrix}$$

$$= \begin{bmatrix} 1 & 0 & 0 \\ -4 & 1 & 0 \\ -1 & -3/2 & 1 \end{bmatrix} \begin{bmatrix} 2 & 2 & -1 \\ 0 & -2 & -8 \\ 0 & 0 & -11 \end{bmatrix},$$

and $\det(\tilde{B}) = 44 = (-2)\det(B)$.

If we change C by adding three times row 3 to row 2, we get

$$\tilde{C} = \begin{bmatrix} 0 & 1 & 2 \\ 8 & 17 & -23 \\ 3 & 5 & -8 \end{bmatrix}.$$ Then

$$\begin{bmatrix} 0 & 1 & 0 \\ 1 & 0 & 0 \\ 0 & 0 & 1 \end{bmatrix} \begin{bmatrix} 0 & 1 & 2 \\ 8 & 17 & -23 \\ 3 & 5 & -8 \end{bmatrix} = \begin{bmatrix} 1 & 0 & 0 \\ 0 & 1 & 0 \\ 3/8 & -11/8 & 1 \end{bmatrix} \begin{bmatrix} 8 & 17 & -23 \\ 0 & 1 & 2 \\ 0 & 0 & 27/8 \end{bmatrix},$$

so $\det(\tilde{C}) = (8)(1)(27/8) = 27 = \det(C)$.

An important consequence of Theorem 4.1.3 is that we can use elementary row operations (and/or elementary column operations) to simplify calculation of a determinant. The idea is to reduce a determinant to the determinant of a triangular matrix while keeping track of the elementary operations used.

Example 4.1.3: Determinants, Using Elementary Row Operations

(a) Suppose that we want to evaluate $\det(A)$, where

$$A = \begin{bmatrix} 4 & 3 & 1 \\ 0 & 2 & 5 \\ 1 & 0 & -3 \end{bmatrix}.$$ Because it's convenient to have 1 as

a pivot, we will interchange rows 1 and 3. But Theorem 4.1.3 tells us that the new determinant will be the negative of the original determinant: $\det \begin{bmatrix} 1 & 0 & -3 \\ 0 & 2 & 5 \\ 4 & 3 & 1 \end{bmatrix} =$

$-\det \begin{bmatrix} 4 & 3 & 1 \\ 0 & 2 & 5 \\ 1 & 0 & -3 \end{bmatrix} = -\det(A)$. Also, we can interchange

rows 2 and 3, so that the initial zero is in the last

row. Then $\det \begin{bmatrix} 1 & 0 & -3 \\ 4 & 3 & 1 \\ 0 & 2 & 5 \end{bmatrix} = -\det \begin{bmatrix} 1 & 0 & -3 \\ 0 & 2 & 5 \\ 4 & 3 & 1 \end{bmatrix}$

$= -\left\{ -\det \begin{bmatrix} 4 & 3 & 1 \\ 0 & 2 & 5 \\ 1 & 0 & -3 \end{bmatrix} \right\} = \det \begin{bmatrix} 4 & 3 & 1 \\ 0 & 2 & 5 \\ 1 & 0 & -3 \end{bmatrix} = \det(A).$

So after two row interchanges, we have a determinant equal to the determinant of the original matrix A. The remaining elementary row operations we will use to simplify our determinant will all involve adding a scalar multiple of a row to another row and so will not change the value of the successive determinants (Theorem 4.1.3(iii)):

$\det(A) = \cdots$

$= \det \begin{bmatrix} 1 & 0 & -3 \\ 4 & 3 & 1 \\ 0 & 2 & 5 \end{bmatrix} \to \det \begin{bmatrix} 1 & 0 & -3 \\ 0 & 3 & 13 \\ 0 & 2 & 5 \end{bmatrix} \to \det \begin{bmatrix} 1 & 0 & -3 \\ 0 & 3 & 13 \\ 0 & 0 & -11/3 \end{bmatrix}.$

Finally, we see that the determinant of A equals the determinant of an upper triangular matrix:

$\det(A) = \det \begin{bmatrix} 1 & 0 & -3 \\ 0 & 3 & 13 \\ 0 & 0 & -11/3 \end{bmatrix} = (1)(3)(-11/3) = -11.$

(b) Let $B = \begin{bmatrix} 1 & 3 & -1 \\ 2 & 0 & 1 \\ 1 & 1 & 4 \end{bmatrix}$. We see that

$\det(B) = \det \begin{bmatrix} 1 & 3 & -1 \\ 2 & 0 & 1 \\ 1 & 1 & 4 \end{bmatrix} \overset{R_2 \to R_2 - 2R_1}{\underset{R_3 \to R_3 - R_1}{=}} \det \begin{bmatrix} 1 & 3 & -1 \\ 0 & -6 & 3 \\ 0 & -2 & 5 \end{bmatrix}$

because the elementary row operations we use do not change the value of a determinant. However, if we want to have 1 as the pivot in the second row of this last determinant (for the sake of convenience), we must multiply row 2 by $-1/6$—an operation that *does* change the value of the determinant, by part (ii) of Theorem 4.1.3:

$\det \begin{bmatrix} 1 & 3 & -1 \\ 0 & 1 & -1/2 \\ 0 & -2 & 5 \end{bmatrix} = -\frac{1}{6} \det \begin{bmatrix} 1 & 3 & -1 \\ 0 & -6 & 3 \\ 0 & -2 & 5 \end{bmatrix},$ or

$\det \begin{bmatrix} 1 & 3 & -1 \\ 0 & -6 & 3 \\ 0 & -2 & 5 \end{bmatrix} = (-6) \det \begin{bmatrix} 1 & 3 & -1 \\ 0 & 1 & -1/2 \\ 0 & -2 & 5 \end{bmatrix}.$

Thus, we conclude that

$$\det(B) = \det \begin{bmatrix} 1 & 3 & -1 \\ 2 & 0 & 1 \\ 1 & 1 & 4 \end{bmatrix} = \det \begin{bmatrix} 1 & 3 & -1 \\ 0 & -6 & 3 \\ 0 & -2 & 5 \end{bmatrix}$$

$$= (-6)\det \begin{bmatrix} 1 & 3 & -1 \\ 0 & 1 & -1/2 \\ 0 & -2 & 5 \end{bmatrix}$$

$$= (-6)\det \begin{bmatrix} 1 & 3 & -1 \\ 0 & 1 & -1/2 \\ 0 & 0 & 4 \end{bmatrix} = (-6)(1)(1)(4) = -24.$$

We could have used a different sequence of elementary row operations to reach a (possibly different) triangular form, but so long as we kept track of the effect of these operations via Theorem 4.1.3, we would have arrived at the same determinant value.

(c) We can also use row operations to evaluate determinants containing parameters.

For example, suppose that $A = \begin{bmatrix} 1 & \lambda & \lambda^2 \\ 1 & 1 & 1 \\ 1 & 2 & 4 \end{bmatrix}$ and we want to determine the values of λ for which $\det(A) = 0$. We can achieve a triangular form as follows:

$$\det(A) = \begin{bmatrix} 1 & \lambda & \lambda^2 \\ 1 & 1 & 1 \\ 1 & 2 & 4 \end{bmatrix} \overset{\substack{R_2 \to R_2 - R_1 \\ R_3 \to R_3 - R_1}}{=} \det \begin{bmatrix} 1 & \lambda & \lambda^2 \\ 0 & 1-\lambda & 1-\lambda^2 \\ 0 & 2-\lambda & 4-\lambda^2 \end{bmatrix}$$

$$\overset{\text{Theorem } 4.1.3(ii)}{=} (1-\lambda)(2-\lambda)\det \begin{bmatrix} 1 & \lambda & \lambda^2 \\ 0 & 1 & 1+\lambda \\ 0 & 1 & 2+\lambda \end{bmatrix}$$

$$\overset{R_3 \to R_3 - R_2}{=} (1-\lambda)(2-\lambda)\det \begin{bmatrix} 1 & \lambda & \lambda^2 \\ 0 & 1 & 1+\lambda \\ 0 & 0 & 1 \end{bmatrix}$$

$$= (1-\lambda)(2-\lambda)(1)(1)(1) = (1-\lambda)(2-\lambda).$$

Thus $\det(A) = 0$ if and only if $\lambda = 1$ or $\lambda = 2$.

In various areas of applied mathematics, the problem of *polynomial interpolation* arises. This involves "threading" a polynomial graph through given data points. More precisely, we are given n ordered pairs $(x_1, y_1), (x_2, y_2), \ldots, (x_n, y_n)$, where the values x_i are distinct, and we want to find a unique polynomial $p(x) = a_0 + a_1 x + a_2 x^2 + \cdots + a_{n-1}x^{n-1}$

such that $p(x_1) = y_1, p(x_2) = y_2, \ldots, p(x_n) = y_n$. This polynomial is called the *interpolating polynomial*.

The problem is equivalent to finding the unique solution of the $n \times n$ system:

$$a_0 + a_1 x_1 + a_2 x_2^2 + \cdots + a_{n-1} x_1^{n-1} = y_1$$
$$a_0 + a_1 x_2 + a_2 x_2^2 + \cdots + a_{n-1} x_2^{n-1} = y_2$$
$$\vdots \qquad\qquad \vdots \qquad\qquad \vdots$$
$$a_0 + a_1 x_n + a_2 x_n^2 + \cdots + a_{n-1} x_n^{n-1} = y_n$$

In matrix form, we have $A\mathbf{x} = \mathbf{b}$, where

$$A = \begin{bmatrix} 1 & x_1 & x_1^2 & \cdots & x_1^{n-1} \\ 1 & x_2 & x_2^2 & \cdots & x_2^{n-1} \\ \vdots & \vdots & \vdots & \cdots & \vdots \\ 1 & x_n & x_n^2 & \cdots & x_n^{n-1} \end{bmatrix}, \quad \mathbf{x} = \begin{bmatrix} a_0 \\ a_1 \\ \vdots \\ a_{n-1} \end{bmatrix}, \quad \mathbf{b} = \begin{bmatrix} y_1 \\ y_2 \\ \vdots \\ y_n \end{bmatrix}.$$

If A is nonsingular (i.e., if $\det(A) \neq 0$), then $\mathbf{x} = A^{-1}\mathbf{b}$ is the unique vector of coefficients we want.

Matrices (or their transposes) with the pattern shown by A have important applications, so we investigate the determinants of such matrices in more detail.*

Example 4.1.4: The Vandermonde Determinant

The $n \times n$ *Vandermonde matrix V*, named for the French mathematician Alexandre-Théophile Vandermonde (1735–1796), is a matrix with a geometric progression in each row: $V = [v_{ij}]$, where $v_{ij} = x_i^{j-1}$ for $i, j = 1, 2, \ldots, n$. This looks like

$$V = \begin{bmatrix} 1 & x_1 & x_1^2 & \cdots & x_1^{n-1} \\ 1 & x_2 & x_2^2 & \cdots & x_2^{n-1} \\ \vdots & \vdots & \vdots & \cdots & \vdots \\ 1 & x_n & x_n^2 & \cdots & x_n^{n-1} \end{bmatrix},$$

and we want to calculate $\det(V)$, the *Vandermonde determinant*. To illustrate the method, we will consider $n = 4$. A key algebraic fact that we use repeatedly is that $x - y$ is a factor of $x^n - y^n$ for any positive integer $n : x^n - y^n = (x - y)(x^{n-1} + x^{n-2}y + x^{n-3}y^2 + \cdots + x^2 y^{n-3} + xy^{n-2} + y^{n-1})$.

If we start by using the operation $R_k \rightarrow R_k - R_1$ for $k = 2, 3, 4$, we see that

---

* For more on polynomial interpolation, see any numerical analysis book. C.D. Meyer, *Matrix Analysis and Applied Linear Algebra* (Philadelphia, PA: SIAM, 2000) contains useful information on Vandermonde matrices and their applications.

$$\det(V) = \begin{vmatrix} 1 & x_1 & x_1^2 & x_1^3 \\ 1 & x_2 & x_2^2 & x_2^3 \\ 1 & x_3 & x_3^2 & x_3^3 \\ 1 & x_4 & x_4^2 & x_4^3 \end{vmatrix} = \begin{vmatrix} 1 & x_1 & x_1^2 & x_1^3 \\ 0 & x_2 - x_1 & x_2^2 - x_1^2 & x_2^3 - x_1^3 \\ 0 & x_3 - x_1 & x_3^2 - x_1^2 & x_3^3 - x_1^3 \\ 0 & x_4 - x_1 & x_4^2 - x_1^2 & x_4^3 - x_1^3 \end{vmatrix}$$

$$= (x_2 - x_1)(x_3 - x_1)(x_4 - x_1) \begin{vmatrix} 1 & x_1 & x_1^2 & x_1^3 \\ 0 & 1 & x_2 + x_1 & x_2^2 + x_1 x_2 + x_1^2 \\ 0 & 1 & x_3 + x_1 & x_3^2 + x_1 x_3 + x_1^2 \\ 0 & 1 & x_4 + x_1 & x_4^2 + x_1 x_4 + x_1^2 \end{vmatrix},$$

where we have factored each row appropriately, not changing the original determinant by Theorem 4.1.3(iii). Now use the operation $R_k \rightarrow R_k - R_2$ for $k = 3, 4$ to obtain

$\det(V)$

$$= (x_2 - x_1)(x_3 - x_1)(x_4 - x_1) \begin{vmatrix} 1 & x_1 & x_1^2 & x_1^3 \\ 0 & 1 & x_2 + x_1 & x_2^2 + x_1 x_2 + x_1^2 \\ 0 & 1 & x_3 + x_1 & x_3^2 + x_1 x_3 + x_1^2 \\ 0 & 1 & x_4 + x_1 & x_4^2 + x_1 x_4 + x_1^2 \end{vmatrix}$$

$$= (x_2 - x_1)(x_3 - x_1)(x_4 - x_1) \begin{vmatrix} 1 & x_1 & x_1^2 & x_1^3 \\ 0 & 1 & x_2 + x_1 & x_2^2 + x_1 x_2 + x_1^2 \\ 0 & 0 & x_3 - x_2 & (x_3^2 - x_2^2) + x_1 (x_3 - x_2) \\ 0 & 0 & x_4 - x_2 & (x_4^2 - x_2^2) + x_1 (x_4 - x_2) \end{vmatrix}$$

$$= (x_2 - x_1)(x_3 - x_1)(x_4 - x_1)(x_3 - x_2)(x_4 - x_2) \begin{vmatrix} 1 & x_1 & x_1^2 & x_1^3 \\ 0 & 1 & x_2 + x_1 & x_2^2 + x_1 x_2 + x_1^2 \\ 0 & 0 & 1 & (x_3 + x_2) + x_1 \\ 0 & 0 & 1 & (x_4 + x_2) + x_1 \end{vmatrix}.$$

Finally, we subtract row 2 from row 3 to conclude

$\det(V)$

$$= (x_2 - x_1)(x_3 - x_1)(x_4 - x_1)(x_3 - x_2)(x_4 - x_2) \begin{vmatrix} 1 & x_1 & x_1^2 & x_1^3 \\ 0 & 1 & x_2 + x_1 & x_2^2 + x_1 x_2 + x_1^2 \\ 0 & 0 & 1 & (x_3 + x_2) + x_1 \\ 0 & 0 & 1 & (x_4 + x_2) + x_1 \end{vmatrix}$$

$$= (x_2 - x_1)(x_3 - x_1)(x_4 - x_1)(x_3 - x_2)(x_4 - x_2) \begin{vmatrix} 1 & x_1 & x_1^2 & x_1^3 \\ 0 & 1 & x_2 + x_1 & x_2^2 + x_1 x_2 + x_1^2 \\ 0 & 0 & 1 & (x_3 + x_2) + x_1 \\ 0 & 0 & 0 & x_4 - x_3 \end{vmatrix}$$

$$= (x_2 - x_1)(x_3 - x_1)(x_4 - x_1)(x_3 - x_2)(x_4 - x_2)(x_4 - x_3)$$

$$= \prod_{1 \le j < i \le 4} (x_i - x_j),$$

where $\prod_{1 \le j < i \le 4} (x_i - x_j)$ denotes the product of all factors $x_i - x_j$ with $j < i$ and with both i and j ranging from 1 to 4. Evaluating $\det(V)$ for any positive integer greater than or equal to 2 is left as Problem B3 of Exercise 4.3.

> This solution tells us that $\det(V) = 0$ if and only if $x_i = x_j$ for some i and j, $i \neq j$. In terms of the polynomial interpolation problem that motivated our discussion of the Vandermonde determinant, we see that there is a unique third degree polynomial whose graph passes through three points (x_1, y_1), (x_2, y_2), (x_3, y_3) if and only if the values x_1, x_2, and x_3 are distinct.

Using previously established results, we can state and prove several important properties of determinants.

Theorem 4.1.4

If A and B are $n \times n$ matrices, then

(1) $\det(AB) = \det(A)\det(B)$.
(2) A is invertible if and only if $\det(A)$ is nonzero.
(3) $\det(A^{\mathrm{T}}) = \det(A)$.
(4) If A is invertible, $\det(A^{-1}) = \frac{1}{\det(A)}$.

Proof (1) First suppose that A is nonsingular. Then A is the product of elementary matrices (Theorem 3.6.4): $A = E_1 E_2 \cdots E_k$. Unfortunately, the product of elementary matrices is not an elementary matrix. (Such a product is equivalent to using more than one elementary operation.) However, repeated application of Theorem 4.1.2 gives us

$$
\begin{aligned}
\det(AB) &= \det(E_1 E_2 \cdots E_k B) \\
&= \det(E_1)\det(E_2 \cdots E_{k-1}E_k B) \\
&= \det(E_1)\det(E_2)\det(E_3 \cdots E_{k-1}E_k B) \\
&= \cdots = \det(E_1)\det(E_2)\det(E_3)\cdots \det(E_{k-1})\det(E_k)\det(B) \\
&= \det(E_1)\det(E_2)\det(E_3)\cdots \det(E_{k-1}E_k)\det(B) \\
&= \cdots = \det(E_1)\det(E_2, E_3 \cdots E_{k-1}E_k)\det(B) \\
&= \det(E_1 E_2 E_3 \cdots E_{k-1}E_k)\det(B) \\
&= \det(A)\det(B).
\end{aligned}
$$

On the other hand, by Lemma 4.1.1, if A is singular, then so is AB. Thus, $\det(AB) = 0 = 0 \cdot \det(B) = \det(A)\det(B)$.

(2) If A is invertible, then A is the product of elementary matrices: $A = E_1 E_2 \cdots E_k$. By repeated use of property (1), $\det(A) = \det(E_1) \cdot \det(E_2) \cdots \det(E_k) \neq 0$ because the determinant of any elementary matrix is nonzero (Theorem 4.1.1).

Conversely, let us suppose that $\det(A) \neq 0$. If A were singular, we would have $\det(A) = 0$, a contradiction. Therefore, A must be nonsingular (invertible).

(3) If A is invertible, then $A = E_1 E_2 \cdots E_k$ the product of elementary matrices. Then $A^{\mathrm{T}} = (E_1 E_2 \cdots E_{k-1}E_k)^{\mathrm{T}} = E_k^{\mathrm{T}} E_{k-1}^{\mathrm{T}} \cdots E_2^{\mathrm{T}} E_1^{\mathrm{T}}$ [Theorem 3.3.3(f) generalized], and

$$
\begin{aligned}
\det(A^{\mathrm{T}}) &= \det(E_k^{\mathrm{T}})\det(E_{k-1}^{\mathrm{T}})\cdots \det(E_1^{\mathrm{T}}) = \det(E_k)\det(E_{k-1})\cdots \det(E_1) \\
&= \det(E_1 E_2 \cdots E_{k-1}E_k) = \det(A)
\end{aligned}
$$

because the transpose of an elementary matrix is also elementary and has the same determinant (Exercise B5). On the other hand, if A is not invertible, the relation $(A^T)^{-1} = (A^{-1})^T$ shows that A^T cannot be invertible either—that is, $\det(A) = 0 = \det(A^T)$.

(4) If A is invertible, $\det(A) \neq 0$ by Property (2). Then we have $1 = \det(I_n) = \det(AA^{-1}) = \det(A)\det(A^{-1})$, so $\det(A^{-1}) = \frac{1}{\det(A)}$.

It is an easy exercise in using mathematical induction to show that Property (1) can be extended to the product of any finite number of $n \times n$ matrices:

$$\det(A_1 A_2 \cdots A_k) = \det(A_1)\det(A_2) \cdots \det(A_k).$$

Property (3) has a useful consequence, as the next result shows.

Corollary 4.1.3

In every theorem about determinants, it is legitimate to interchange the words "row" and "column" throughout.

For example, **if A is an $n \times n$ matrix with two identical columns, then det $(A) = 0$.** (If A has two equal columns, then A^T must have two equal rows. Thus $\det(A) = \det(A^T) = 0$ by Property (3) of Theorem 4.1.4 and Corollary 4.1.1.)

Example 4.1.5: Properties of Determinants

Suppose that $A = \begin{bmatrix} 2 & 2 & -1 \\ 4 & 5 & 2 \\ -2 & 1 & 2 \end{bmatrix} = \begin{bmatrix} 1 & 0 & 0 \\ 2 & 1 & 0 \\ -1 & 3 & 1 \end{bmatrix}\begin{bmatrix} 2 & 2 & -1 \\ 0 & 1 & 4 \\ 0 & 0 & -11 \end{bmatrix}$

and $B = \begin{bmatrix} 0 & 1 & 2 \\ -1 & 2 & 1 \\ 3 & 5 & -8 \end{bmatrix}$, where $\begin{bmatrix} 0 & 1 & 0 \\ 1 & 0 & 0 \\ 0 & 0 & 1 \end{bmatrix}\begin{bmatrix} 0 & 1 & 2 \\ -1 & 2 & 1 \\ 3 & 5 & -8 \end{bmatrix} =$

$\begin{bmatrix} 1 & 0 & 0 \\ 0 & 1 & 0 \\ -3 & 11 & 1 \end{bmatrix}\begin{bmatrix} -1 & 2 & 1 \\ 0 & 1 & 2 \\ 0 & 0 & -27 \end{bmatrix}$.   Then $\det(A) = -22$ and $\det(B) = -27$.

Now

$AB = \begin{bmatrix} -5 & 1 & 14 \\ 1 & 24 & -3 \\ 5 & 10 & -19 \end{bmatrix} = \begin{bmatrix} 1 & 0 & 0 \\ -1/5 & 1 & 0 \\ -1 & 5/11 & 1 \end{bmatrix}\begin{bmatrix} -5 & 1 & 14 \\ 0 & 121/5 & -1/5 \\ 0 & 0 & -54/11 \end{bmatrix}$,

so $\det(AB) = (-5)(121/5)(-54/11) = 594 = (-22)(-27) = \det(A)\det(B)$.

Because $\det(A)$ and $\det(B)$ are nonzero, we know that A and B are invertible. In fact,

$$A^{-1} = \frac{1}{11} \begin{bmatrix} -4 & 5/2 & -9/2 \\ 6 & -1 & 4 \\ -7 & 3 & -1 \end{bmatrix} \text{ and } B^{-1} = \frac{1}{9} \begin{bmatrix} 7 & -6 & 1 \\ 5/3 & 2 & 2/3 \\ 11/3 & -1 & -1/3 \end{bmatrix}.$$

Furthermore, with the aid of a CAS, we find that $\det(A^{-1}) = -1/22 = 1/\det(A)$ and $\det(B^{-1}) = -1/27 = 1/\det(B)$. Also, $\det(A^{T}) = \det \begin{bmatrix} 2 & 4 & -2 \\ 2 & 5 & 1 \\ -1 & 2 & 2 \end{bmatrix} = -22 = \det(A)$

and $\det(B^{T}) = \det \begin{bmatrix} 0 & -1 & 3 \\ 1 & 2 & 5 \\ 2 & 1 & -8 \end{bmatrix} = -27 = \det(B)$.

(See Exercise A22.)

4.1.1 Cramer's Rule

In 1750, at a time when determinants dominated the landscape of linear systems (100 years or so before matrix theory began), the Swiss mathematician Gabriel Cramer (1704–1752) published his treatise *Introduction à l'Analyse des Lignes Courbes Algébriques*. An appendix to this work contains the famous method for solving a nonsingular system of n linear equations in n unknowns.* (An interesting version of Cramer's Rule for nonsquare matrices appears as problem 10618—proposed by S. Lakshminarayanan, S.L. Shah, and K. Nandakumar—in the November 1999 issue of *The American Mathematical Monthly*.)

Theorem 4.1.5: Cramer's Rule

If $A\mathbf{x} = \mathbf{b}$ represents a system of n linear equations in n unknowns, with nonsingular matrix of coefficients A (so $\det(A) \neq 0$), then x_i, the ith unknown, is given uniquely by the formula:

$$x_i = \frac{\det([\,\mathbf{a}_1 \quad \cdots \quad \mathbf{a}_{i-1} \quad \mathbf{b} \quad \mathbf{a}_{i+1} \quad \cdots \quad \mathbf{a}_n\,])}{\det(A)},$$

where $[\,\mathbf{a}_1 \quad \cdots \quad \mathbf{a}_{i-1} \quad \mathbf{b} \quad \mathbf{a}_{i+1} \quad \cdots \quad \mathbf{a}_n\,]$ denotes matrix A with column i replaced by vector $\mathbf{b}$.

---

* It is generally acknowledged that this formula was known (to Leibniz, for example) long before Cramer rediscovered and published it. See C.D. Meyer, *Matrix Analysis and Applied Linear Algebra*, p. 459 (Philadelphia, PA: SIAM, 2000) or V.J. Katz, *A History of Mathematics: An Introduction*, 2nd edn., p. 612 (Reading, MA: Addison-Wesley, 1998). However, also see A.A. Kosinski, Cramer's rule is due to Cramer, *Math. Mag.* **74** (2001): 310–312.

Proof* Let $I(\mathbf{x};i) = [\, \mathbf{e}_1 \quad \ldots \quad \mathbf{e}_{i-1} \quad \mathbf{x} \quad \mathbf{e}_{i+1} \quad \ldots \quad \mathbf{e}_n\,]$ denote the identity matrix I_n with column $i\ (=\mathbf{e}_i)$ replaced by the column vector $\mathbf{x}$. Then

$$AI(\mathbf{x};i) = A[\, \mathbf{e}_1 \quad \ldots \quad \mathbf{e}_{i-1} \quad \mathbf{x} \quad \mathbf{e}_{i+1} \quad \ldots \quad \mathbf{e}_n\,]$$
$$= [\, A\,\mathbf{e}_1 \quad \ldots \quad A\,\mathbf{e}_{i-1} \quad A\,\mathbf{x} \quad A\,\mathbf{e}_{i+1} \quad \ldots \quad A\,\mathbf{e}_n\,]$$
$$= [\, \mathbf{a}_1 \quad \ldots \quad \mathbf{a}_{i-1} \quad \mathbf{b} \quad \mathbf{a}_{i+1} \quad \ldots \quad \mathbf{a}_n\,].$$

Therefore,

$$\det(AI(\mathbf{x};i)) = \det(A)\det(I(\mathbf{x};i))$$
$$= \det([\, \mathbf{a}_1 \quad \ldots \quad \mathbf{a}_{i-1} \quad \mathbf{b} \quad \mathbf{a}_{i+1} \quad \ldots \quad \mathbf{a}_n\,]),$$

and

$$\det(I(\mathbf{x};i)) = \frac{\det([\, \mathbf{a}_1 \quad \ldots \quad \mathbf{a}_{i-1} \quad \mathbf{b} \quad \mathbf{a}_{i+1} \quad \ldots \quad \mathbf{a}_n\,])}{\det(A)}.$$

Now let us pause to analyze $I(\mathbf{x};i)$. If $\mathbf{x}$ replaces $\mathbf{e}_i$ as column i of I_n, then x_i replaces the 1 in diagonal position (i, i). The remaining $n-1$ diagonal positions are occupied by 1's. Thus if $\mathbf{x}$ replaces $\mathbf{e}_1$, $I(\mathbf{x};1)$ is a lower triangular matrix whose determinant—the product of its diagonal elements—is x_1. Similarly, $I(\mathbf{x};n)$ is an upper triangular matrix whose determinant is x_n. For any other value of i ($i \neq 1$ or n), we can use x_i as a pivot to reduce all elements in column i below x_i to zero, producing an upper triangular matrix with determinant x_i. Therefore, $\det(I(\mathbf{x};i)) = x_i$ for every i, $i = 1, 2, \ldots, n$, and so

$$\det(I(\mathbf{x};i)) = x_i = \frac{\det([\, \mathbf{a}_1 \quad \ldots \quad \mathbf{a}_{i-1} \quad \mathbf{b} \quad \mathbf{a}_{i+1} \quad \ldots \quad \mathbf{a}_n\,])}{\det(A)}.$$

Example 4.1.6: Cramer's Rule

The system

$$x + y - 2z = -3$$
$$2x - y - z = 0$$
$$x + 2y + 3z = 13$$

can be expressed as

$$\begin{bmatrix} 1 & 1 & -2 \\ 2 & -1 & -1 \\ 1 & 2 & 3 \end{bmatrix} \begin{bmatrix} x \\ y \\ z \end{bmatrix} = \begin{bmatrix} -3 \\ 0 \\ 13 \end{bmatrix},$$

where the determinant of the matrix of coefficients is -18.

---

* Based on S.H. Friedberg, An alternate proof of Cramer's rule, *Coll. Math. J.* **19** (1988): 171. Also see R. Ehrenborg, A conceptual proof of Cramer's rule, *Math. Mag.* **77** (2004): 308.

Then Cramer's formula yields

$$x = \frac{\det\begin{bmatrix} -3 & 1 & -2 \\ 0 & -1 & -1 \\ 13 & 2 & 3 \end{bmatrix}}{\det\begin{bmatrix} 1 & 1 & -2 \\ 2 & -1 & -1 \\ 1 & 2 & 3 \end{bmatrix}} = \frac{-36}{-18} = 2,$$

$$y = \frac{\det\begin{bmatrix} 1 & -3 & -2 \\ 2 & 0 & -1 \\ 1 & 13 & 3 \end{bmatrix}}{\det\begin{bmatrix} 1 & 1 & -2 \\ 2 & -1 & -1 \\ 1 & 2 & 3 \end{bmatrix}} = \frac{-18}{-18} = 1, \quad \text{and}$$

$$z = \frac{\det\begin{bmatrix} 1 & 1 & -3 \\ 2 & -1 & 0 \\ 1 & 2 & 13 \end{bmatrix}}{\det\begin{bmatrix} 1 & 1 & -2 \\ 2 & -1 & -1 \\ 1 & 2 & 3 \end{bmatrix}} = \frac{-54}{-18} = 3.$$

Cramer's Rule provides a quick proof of the fact that if $A\mathbf{x} = \mathbf{0}$ is a system of n equations in n unknowns and if $\det(A) \neq 0$, then the only solution is $\mathbf{x} = \mathbf{0}$: If $\mathbf{x} = [x_1 \quad x_2 \quad \ldots \quad x_n]^{\mathrm{T}}$, then $x_i = \det([\mathbf{a}_1 \quad \ldots \quad \mathbf{a}_{i-1} \quad \mathbf{0} \quad \mathbf{a}_{i+1} \quad \ldots \quad \mathbf{a}_n])/\det(A)$, where column i of A has been replaced by the zero vector, $i = 1, 2, \ldots, n$; but the determinant of a matrix with a zero column must equal zero (via the transposed version of Corollary 4.1.2), and this proves the result we want.

Cramer's Rule is more important in theory than in practice. If the system has a square coefficient matrix and has a unique solution, Cramer's formula does illustrate clearly the dependence of the solution on the entries of $\mathbf{b}$ and A and gives the closed form solution in terms of determinants; but the practical solution of large systems by computer relies on variants of the direct Gaussian elimination (which, after all, applies to *any* linear system).* In fact, even the determinants of large matrices are most efficiently evaluated by some form of Gaussian elimination.

Section 4.3 will introduce some new methods of computing determinants that are especially useful for hand calculation.

---

* See N.J. Higham, *Accuracy and Stability of Numerical Algorithms*, 2nd edn. (Philadelphia, PA: SIAM, 2002), Section 1.10.1, for additional comments and supporting data.

Exercises 4.1

A.

1. Find the determinant of each of the following matrices:

   a. $\begin{bmatrix} 2 & 1 \\ 1 & 1 \end{bmatrix}$ b. $\begin{bmatrix} -4 & 2 \\ 6 & -3 \end{bmatrix}$ c. $\begin{bmatrix} 2 & 1 \\ 1 & 1 \end{bmatrix}$

   d. $\begin{bmatrix} \cos\theta & \sin\theta \\ -\sin\theta & \cos\theta \end{bmatrix}$ e. $\begin{bmatrix} 1 & 2 & 0 \\ 0 & 2 & 0 \\ 1 & 0 & 1 \end{bmatrix}$

   f. $\begin{bmatrix} 2 & -1 & 3 \\ 1 & 4 & 4 \\ 1 & 0 & 2 \end{bmatrix}$ g. $\begin{bmatrix} 10 & 20 & 30 \\ 40 & 10 & -10 \\ 0 & 0 & 50 \end{bmatrix}$

   h. $\begin{bmatrix} 100 & -237 & 4 \\ 8 & 9 & 14 \\ 100 & -237 & 4 \end{bmatrix}$

2. Show the details of evaluating all determinants in Example 4.1.6.

3. If A is an $n \times n$ matrix such that $\det(A) = -2$, calculate the values of
   a. $\det(3A)$, b. $\det(5A^{-1})$, c. $\det(A^3)$, d. $\det(A - 4A)$.

4. If A is an $n \times n$ matrix, is it possible that $\det(AA^{\mathsf{T}}) = -5$? *Explain.*

5. Suppose that A and B are $n \times n$ matrices such that $AB = I_n$. Use determinants to show that both A and B are invertible.

6. Evaluate $\begin{vmatrix} 1 & a & a^2 \\ a & a^2 & 1 \\ a^2 & 1 & a \end{vmatrix}$. [*Hint*: Consider the cases $a = 1, -1$ separately.]

7. Evaluate $\begin{vmatrix} 3 & 1 & -2 & 4 \\ 6 & 0 & -11 & 1 \\ 1 & -1 & 2 & 6 \\ -2 & 3 & -2 & 3 \end{vmatrix}$.

8. Evaluate $\begin{vmatrix} 1 & 1 & 0 & 0 & 0 \\ -1 & 1 & 1 & 0 & 0 \\ 0 & -1 & 1 & 1 & 0 \\ 0 & 0 & -1 & 1 & 1 \\ 0 & 0 & 0 & -1 & 1 \end{vmatrix}$.

9. Evaluate $\begin{vmatrix} 5 & 2 & 7 & -1 & 2 \\ 0 & 1 & 2 & 6 & 1 \\ 0 & 0 & 0 & -3 & -1 \\ 0 & 2 & 7 & 0 & 3 \\ 0 & 3 & 2 & 1 & 0 \end{vmatrix}$.

10. Evaluate $\begin{vmatrix} 1 & 0 & 0 & 1 & -1 & 1 \\ 0 & 0 & 1 & 0 & 2 & 0 \\ 1 & 0 & 0 & 0 & 0 & 1 \\ 0 & 0 & 0 & 1 & 1 & 0 \\ 1 & 0 & 2 & 0 & 0 & 1 \\ 0 & 0 & 1 & 1 & 0 & 0 \end{vmatrix}$.

11. Show that $\begin{vmatrix} a & a & a \\ b & a & a \\ 0 & b & a \end{vmatrix} = a(a-b)^2$.

12. Find $\begin{bmatrix} x+a & a & a \\ a & x+a & a \\ a & a & x+a \end{bmatrix}$.

13. Find $\begin{bmatrix} a & a & a \\ a & 0 & a \\ a & a & b \end{bmatrix}$.

14. Evaluate $\begin{bmatrix} 1 & 2 & 3 & 4 & 5 \\ 6 & 7 & 8 & 9 & 10 \\ 11 & 12 & 13 & 14 & 15 \\ 16 & 17 & 18 & 19 & 20 \\ 21 & 22 & 23 & 24 & 25 \end{bmatrix}$.

15. Find $\det(AB)$ and $\det(BA)$ for

$$A = \begin{bmatrix} a_{11} & a_{12} \\ a_{21} & a_{22} \\ a_{31} & a_{32} \end{bmatrix} \quad \text{and} \quad B = \begin{bmatrix} b_{11} & b_{12} & b_{13} \\ b_{21} & b_{22} & b_{23} \end{bmatrix}.$$

[Note that $\det(A)$ and $\det(B)$ are not defined.]

16. If A is a 3×3 matrix, $\det(A) = 3$, B is obtained from A by multiplying the second row by 4, C is obtained from B by adding five times row 2 to row 3, and D is obtained from C by interchanging rows 1 and 3, what is $\det(D)$?

17. Suppose that

$$A = \begin{bmatrix} \mathbf{x} \\ \mathbf{y} \\ \mathbf{z} \\ \mathbf{w} \end{bmatrix},$$

where $\mathbf{x}, \mathbf{y}, \mathbf{z}$, and $\mathbf{w}$ are row vectors, the transposes of vectors in $\mathbb{R}^n$, and $\det(A) = -3$.

a. Find $\det(B)$ if

$$B = \begin{bmatrix} x - y \\ y \\ x + y - 3w \\ z + x \end{bmatrix},$$

b. Find $\det(C)$ if

$$C = \begin{bmatrix} x + y \\ x - 2y - w \\ x + 2y - 4z + 2w \\ x - y + z - w \end{bmatrix}.$$

Exercises 18–21 are for students who have studied calculus. If f is a differentiable function, we denote its derivative at any point x_0 of its domain by $f'(x_0)$. Suppose that x_0 is a fixed number in the common domains of the functions $f_1, f_2, \ldots, f_n$ (a finite number of functions). We define the **Wronskian** $w(f_1, f_2, \ldots, f_n; x_0)$ *at x_0 to be the value of the determinant of the square matrix as given in the following:*

$$w(f_1, f_2, \ldots, f_n; x_0) = \det \begin{bmatrix} f_1(x_0) & f_2(x_0) & \cdots & f_n(x_0) \\ f_1'(x_0) & f_2'(x_0) & \cdots & f_n'(x_0) \\ \vdots & \vdots & \vdots & \vdots \\ f_1^{(n-1)}(x_0) & f_2^{(n-1)}(x_0) & \cdots & f_n^{(n-1)}(x_0) \end{bmatrix}.$$

The Wronskian is used to determine whether a set of functions (regarded as "vectors"—see Chapter 6) is linearly independent.

18. Evaluate $w(e^x, e^{2x}; 0)$.

19. Evaluate $w(\sin x, \sin 2x; \frac{\pi}{4})$.

20. Evaluate $w(1, \sin 2x, \cos 2x; 0)$.

21. Evaluate $w(\sin x, \sin^2 x, \sin^3 x, \sin^4 x; \frac{\pi}{4})$.

22. Using the matrices A and B in Example 4.1.5:
     a. verify A^{-1} and B^{-1}.
     b. calculate $\det(A^{-1})$ and $\det(B^{-1})$.
     c. calculate $\det(A^T)$ and $\det(B^T)$.

23. For any positive integer $n \geq 2$, define the $n \times n$ matrix $A = [a_{ij}]$, where $a_{ij} = (-1)^{i+j}$. Compute $\det(A)$.

24. a. Show that $x_1 = x_2 = x_3 = x_4 = 1$ is a solution of the system:

$$2x_1 - 3x_2 + 4x_3 - 3x_4 = 0$$
$$3x_1 - x_2 + 11x_3 - 13x_4 = 0$$
$$4x_1 + 5x_2 - 7x_3 - 2x_4 = 0$$
$$13x_1 - 25x_2 + x_3 + 11x_4 = 0$$

b. What is the determinant of the matrix of coefficients of this system? (This is a *thought* problem, not a computational one.)

25. Evaluate $\begin{vmatrix} 1 & 1 & 1 & 1 \\ 2 & 3 & 4 & 5 \\ 4 & 9 & 16 & 25 \\ 8 & 27 & 64 & 125 \end{vmatrix}$. [*Hint:* Compare Example 4.1.4.]

26. Given two vectors $\mathbf{u} = \begin{bmatrix} u_1 \\ u_2 \\ u_3 \end{bmatrix}$ and $\mathbf{v} = \begin{bmatrix} v_1 \\ v_2 \\ v_3 \end{bmatrix}$ in $\mathbb{R}^3$, the **cross-product** (also called the **vector product**) $\mathbf{u} \times \mathbf{v}$ is the vector defined as $\mathbf{u} \times \mathbf{v} = \begin{bmatrix} u_2 v_3 - u_3 v_2 \\ u_3 v_1 - u_1 v_3 \\ u_1 v_2 - u_2 v_1 \end{bmatrix}$. (See Exercise 1.2, Problems A10 and A11, B19–B22.)

    a. Form the matrix $[\mathbf{u} \quad \mathbf{v}] = \begin{bmatrix} u_1 & v_1 \\ u_2 & v_2 \\ u_3 & v_3 \end{bmatrix}$ and verify that

$$\mathbf{u} \times \mathbf{v} = \begin{bmatrix} \det\begin{bmatrix} u_2 & v_2 \\ u_3 & v_3 \end{bmatrix} \\ -\det\begin{bmatrix} u_1 & v_1 \\ u_3 & v_3 \end{bmatrix} \\ \det\begin{bmatrix} u_1 & v_1 \\ u_2 & v_2 \end{bmatrix} \end{bmatrix}.$$

    b. If $\mathbf{u}$, $\mathbf{v}$, and $\mathbf{w}$ are vectors in $\mathbb{R}^3$, show that $\det[\mathbf{u} \ \mathbf{v} \ \mathbf{w}] = \mathbf{u} \cdot (\mathbf{v} \times \mathbf{w})$.

27. Use Cramer's Rule (Theorem 4.1.5) to solve for x in the system:

$$2x + y - z = 4$$
$$x + 7y - z = 1$$
$$3x - y - z = 1$$

B.

1. Without using Theorem 4.1.4(1), prove that if A and B are $n \times n$ upper triangular matrices, then $\det(AB) = \det(A)\det(B)$.

2. Prove that if E is an $n \times n$ elementary matrix of type II and A is any $n \times n$ matrix, then $\det(EA) = \det(E)\det(A)$. (Do not use Theorem 4.1.4(1).)

3. Prove that if E is an $n \times n$ elementary matrix of type III and A is any $n \times n$ matrix, then $\det(EA) = \det(E)\det(A)$. (Do not use Theorem 4.1.4(1).)

4. If A is an $n \times n$ matrix and k is a constant, show that $\det(kA) = k^n \det(A)$.

5. Prove that the transpose of an elementary matrix E is also elementary and has the same determinant as E. (Do not use Theorem 4.1.4(3).)

6. a. If B is invertible, show that $\det(B^{-1}AB) = \det(A)$.
   b. If A, B, and C are invertible, show that $\det(AB^{-1}CA^{-1}BC^{-1}) = 1$.

7. An $n \times n$ matrix A with real entries is **orthogonal** if $A^{\mathsf{T}} = A^{-1}$. If A is an orthogonal matrix, prove that $\det(A) = \pm 1$. (Problem B8, Exercise 3.7 indicates that any permutation matrix is orthogonal.)

8. If A is an $n \times n$ matrix such that $A^3 = A$, what are the possible values for $\det(A)$?

9. Suppose that a square matrix A has all entries 0's or 1's. Must $\det(A) = 0$ or ± 1? [*Hint*: Look at some 3×3 matrices.]

10. Suppose that A is a 3×3 matrix, all entries are either 0 or 1, and there are exactly four 1's. What are the possible values of $\det(A)$?

11. If A and B are two $n \times n$ *orthogonal* matrices (see Exercise B7):
    a. Prove that $A + B = A(A^{\mathsf{T}} + B^{\mathsf{T}})B$.
    b. Using the result of part (a), show that if $\det(A) + \det(B) = 0$, then $A + B$ is singular.

12. Let A be an $n \times n$ matrix:
    a. If $A^{\mathsf{T}} = -A$ and n is odd, show that $|A| = 0$.
    b. Show that if $A^2 + I_n = O_n$, then n must be even.*

13. Let D_n be the $n \times n$ determinant whose (i, j) entry is $i + j$. Show that $D_n = 0$ if $n > 2$.

14. Determine a necessary and sufficient condition for $\det(A + B) = \det(A) + \det(B)$, where A and B are 2×2 matrices. (Equality does not hold in general.)

15. Use Cramer's Rule (Theorem 4.1.5) to solve the following system of equations:

$$4x + y + 4z = 1$$
$$x + 3y - z = 2$$
$$3x + 2z = -1$$

16. Use Cramer's Rule (Theorem 4.1.5) to prove that if the augmented matrix of an $n \times n$ system of linear equations has rational entries, then the solution consists of rational numbers.

---

* Part (b) is not true if A has complex number entries.

C.

1. Suppose that $A = [a_{ij}]$ is an $n \times n$ nonsingular matrix and $A^{-1} = [c_{ij}] = [\mathbf{c}_1 \quad \mathbf{c}_2 \quad \ldots \quad \mathbf{c}_n]$, where $\mathbf{c}_i \in \mathbb{R}^n$ for $i = 1, 2, \ldots, n$. Let $A(\mathbf{x};i)$ denote the matrix obtained by replacing column i of A by the vector $\mathbf{x}$.

   a. Show that $A\mathbf{c}_i = \mathbf{e}_i$, where $\mathbf{e}_i$ is column i of I_n.

   b. Use Cramer's Rule to show that the (i, j) entry of A^{-1} is given by

   $$c_{ij} = \frac{\det[A(\mathbf{e}_j;i)]}{\det(A)}.$$

   c. If A_{ij} is the submatrix of A obtained by deleting row i and column j of A, prove that $\det[A(\mathbf{e}_j;i)] = (-1)^{i+j} \det(A_{ji})$. (The determinant $\det(A_{ij})$ is called the *minor* of a_{ij} and the expression $(-1)^{i+j}\det(A_{ji})$ is called the *cofactor* of a_{ij}.)

   d. The *adjugate* of A (often called the *classical adjoint* of A) is the matrix adj (A) whose (i, j) entry is $(-1)^{i+j}\det(A_{ji})$. Show that

   $$A^{-1} = \frac{\text{adj}(A)}{\det(A)}.$$

   e. Use the formula found in part (d) to find the inverse of
   $$A = \begin{bmatrix} 1 & 0 & 2 & 3 \\ 0 & 1 & 1 & 2 \\ 0 & 0 & 1 & 3 \\ 0 & 0 & 0 & 1 \end{bmatrix}.$$

2. It is known that the numbers 14,529, 15,197, 20,541, 38,911, 59,619 are multiples of 167. Without actually calculating, prove that the determinant of the 5×5 matrix A is also a multiple of 167, where

   $$A = \begin{pmatrix} 1 & 4 & 5 & 2 & 9 \\ 1 & 5 & 1 & 9 & 7 \\ 2 & 0 & 5 & 4 & 1 \\ 3 & 8 & 9 & 1 & 1 \\ 5 & 9 & 6 & 1 & 9 \end{pmatrix}. \quad *$$

   [*Hint*: Use elementary column operations and the fact that if an integer a divides the product of integers bc and a has no factors in common with b, then a must divide c.]

---

* Z. Mustafaev, V. Dontsov, and E. Maevaki, Problem 1175, *Pi Mu Epsilon J.* **12** (Spring 2008).

*4.2 Determinants and Geometry

In addition to the algebraic significance of determinants, there is a long and fruitful tradition of interpreting determinants geometrically, starting in $\mathbb{R}^2$ and $\mathbb{R}^3$, and expanding to any Euclidean space $\mathbb{R}^n$.

We begin with the problem of calculating the area of a triangle formed by two vectors $\mathbf{x} = [x_1 \quad x_2]^T$ and $\mathbf{y} = [y_1 \quad y_2]^T$ (Figure 4.1).

The area of triangle OPQ is (area of triangle OAQ) + (area of trapezoid ABPQ) − (area of triangle OBP):

$$= \tfrac{1}{2}y_1y_2 + \tfrac{1}{2}(x_1 - y_1)(x_2 + y_2) - \tfrac{1}{2}x_1x_2$$

$$\tfrac{1}{2}(x_1y_2 - y_1x_2) = \tfrac{1}{2}\begin{vmatrix} x_1 & y_1 \\ x_2 & y_2 \end{vmatrix},$$

one-half the determinant of the matrix whose columns are the vectors $\mathbf{x}$ and $\mathbf{y}$. We orient ourselves so that the positive direction is counterclockwise—that is, the columns of the determinant are selected according to the order in which we encounter the vectors $\mathbf{x}$ and $\mathbf{y}$ as we move counterclockwise from the positive x-axis. Note that if we interchange the columns of the determinant, obtaining $\begin{vmatrix} y_1 & x_1 \\ y_2 & x_2 \end{vmatrix}$, the final result is the

negative of what we found before. Because negative area does not make sense, we agree to define the *signed area* of the triangle OPQ to be the area of OPQ when we turn *counterclockwise* from $\mathbf{x}$ to $\mathbf{y}$ and the *negative* of the area when we turn *clockwise* from $\mathbf{y}$ to $\mathbf{x}$.

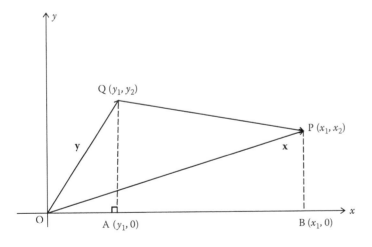

Figure 4.1 Area of a triangle.

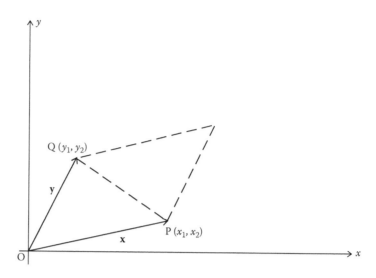

Figure 4.2 Area of a parallelogram.

Similarly, we can see that the signed area of the *parallelogram* spanned by **x** and **y** (Figure 4.2) is twice the area of the triangle in Figure 4.1, or
$$\begin{vmatrix} x_1 & y_1 \\ x_2 & y_2 \end{vmatrix}.$$

Technically, in terms of our previous use of the term *span*, the expression "parallelogram spanned by **x** and **y**" refers to the set $\{a\mathbf{x} + b\mathbf{y} | 0 \le a, b \le 1|\}$, which is a proper subset of *span*$\{\mathbf{x}, \mathbf{y}\}$.

Example 4.2.1: The Determinant as an Area

Let us look at the parallelogram OABC spanned by the vectors $\mathbf{x} = \begin{bmatrix} 7 \\ 5 \end{bmatrix}$ and $\mathbf{y} = \begin{bmatrix} 1 \\ 6 \end{bmatrix}$ (Figure 4.3).

The area of the parallelogram is $\det([\mathbf{x} \quad \mathbf{y}]) = \begin{vmatrix} 7 & 1 \\ 5 & 6 \end{vmatrix} = 7(6) - (1)(5) = 37$ square units, and the area of triangle OAC is $\frac{1}{2} \begin{vmatrix} 7 & 1 \\ 5 & 6 \end{vmatrix} = 18.5$ square units.

We can use the notation $\det(\mathbf{x}, \mathbf{y}) = \begin{vmatrix} x_1 & y_1 \\ x_2 & y_2 \end{vmatrix}$ to emphasize that in this geometric context the determinant is a function of the (ordered) pair of vectors $\mathbf{x}, \mathbf{y}$. Some interesting properties follow from this geometric interpretation. If $\mathbf{x}, \mathbf{y}$, and $\mathbf{z}$ are vectors in $\mathbb{R}^2$, then

(1) $\det(\mathbf{x}, \mathbf{y}) = -\det(\mathbf{y}, \mathbf{x})$.
(2) $\det(\mathbf{x}, c\mathbf{y}) = c \det(\mathbf{x}, \mathbf{y})$ and $\det(c\mathbf{x}, \mathbf{y}) = c \det(\mathbf{x}, \mathbf{y})$.
(3) $\det(\mathbf{x}, \mathbf{y} + \mathbf{z}) = \det(\mathbf{x}, \mathbf{y}) + \det(\mathbf{x}, \mathbf{z})$.

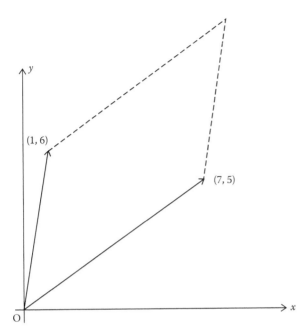

Figure 4.3 Area of a parallelogram.

The proofs of these properties are left as exercises (see B1).

Before we consider the geometry of $\mathbb{R}^3$, one more example is useful. Suppose that we consider the parallelogram spanned by $\mathbf{e}_1 = [\,1 \quad 0\,]^T$ and $\mathbf{e}_2 = [\,0 \quad 1\,]^T$, a subspace of $\mathbb{R}^2$. This is called the **standard unit square** (Figure 4.4).

Now suppose, for example, that we multiply each vector in the span of $\mathbf{e}_1$ and $\mathbf{e}_2$ (i.e., each point in the square) by the matrix $M = \begin{bmatrix} 1 & 1 \\ -3 & 1 \end{bmatrix}$.

What happens is that $\mathbf{e}_1$ and $\mathbf{e}_2$ get transformed into vectors $\hat{\mathbf{e}}_1 = \begin{bmatrix} 1 \\ -3 \end{bmatrix}$

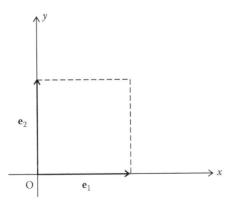

Figure 4.4 The standard unit square.

and $\hat{\mathbf{e}}_2 = \begin{bmatrix} 1 \\ 1 \end{bmatrix}$, respectively, and *every vector in the original square gets transformed into a vector in the span of $\hat{\mathbf{e}}_1$ and $\hat{\mathbf{e}}_2$*: if $M\mathbf{e}_1 = \hat{\mathbf{e}}_1$ and $M\mathbf{e}_2 = \hat{\mathbf{e}}_2$, then $M(\overbrace{c_1\ \mathbf{e}_1 + c_2\ \mathbf{e}_2}^{\in\ span\{\mathbf{e}_1,\mathbf{e}_2\}}) = c_1(M\mathbf{e}_1) + c_2(M\mathbf{e}_2) = \overbrace{c_1\hat{\mathbf{e}}_1 + c_2\hat{\mathbf{e}}_2}^{\in\ span\{\hat{\mathbf{e}}_1,\hat{\mathbf{e}}_2\}}$.

Figure 4.5 illustrates how the standard unit square is distorted by the matrix multiplication.

The (signed) area of the transformed square in Figure 4.5 is just

$$\det(\hat{\mathbf{e}}_1, \hat{\mathbf{e}}_2) = \begin{vmatrix} 1 & 1 \\ -3 & 1 \end{vmatrix} = 4.$$

Symbolically, we can write

$$\text{Area}\,(M(\square)) = |\det(M)| \cdot \text{Area}(\square) = 4 \cdot 1 = 4.$$

In other words, the area of the transformed square equals the absolute value of the determinant of the transforming matrix times the area of the original square. Thus the determinant of M measures the change in area when every vector in the square is multiplied by M. The absolute value of the determinant is a **scaling factor**, or a **magnification factor**.

In $\mathbb{R}^3$, determinants can be used to represent volumes. If we have three linearly independent vectors $\mathbf{x} = [x_1\ \ x_2\ \ x_3]^T, \mathbf{y} = [y_1\ \ y_2\ \ y_3]^T$, and $\mathbf{z} = [z_1\ \ z_2\ \ z_3]^T$, they determine (span) a *parallelepiped** in $\mathbb{R}^3$ (Figure 4.6).

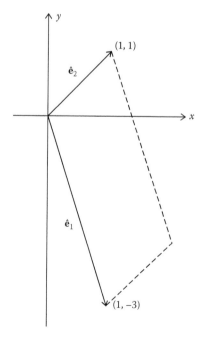

Figure 4.5 Distortion of the standard unit square.

---

* In $\mathbb{R}^3$, a **parallelepiped** is a solid with six faces, each a parallelogram and each being parallel to the opposite face. Such an object can be defined using three basis vectors. In $\mathbb{R}^n$, a parallelepiped (or n-parallelepiped) is a special set spanned by n vectors.

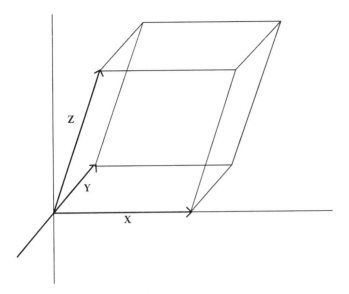

Figure 4.6 A parallelepiped in $\mathbb{R}^3$.

The determinant

$$\det(\mathbf{x}, \mathbf{y}, \mathbf{z}) = \begin{vmatrix} x_1 & y_1 & z_1 \\ x_2 & y_2 & z_2 \\ x_3 & y_3 & z_3 \end{vmatrix}$$

provides the **signed volume** of the parallelepiped spanned by $\mathbf{x}, \mathbf{y},$ and $\mathbf{z}$. We shall not prove this.* The absolute value of this determinant gives the volume.

Example 4.2.2: The Determinant as a Volume

The volume of the parallelepiped spanned by the vectors $\begin{bmatrix} 1 \\ 1 \\ 1 \end{bmatrix}, \begin{bmatrix} 2 \\ 3 \\ 1 \end{bmatrix}$, and $\begin{bmatrix} 0 \\ 1 \\ 1 \end{bmatrix}$ in $\mathbb{R}^3$ is given by the determinant $\begin{bmatrix} 1 & 2 & 0 \\ 1 & 3 & 1 \\ 1 & 1 & 1 \end{bmatrix}$, which has the value 2. Therefore, the volume is 2 cubic units.

---

* For further information on the geometric meaning of 3×3 determinants, see T. Banchoff and J. Wermer, *Linear Algebra Through Geometry*, 2nd edn., pp. 158–162 (New York: Springer-Verlag, 1992).

Finally, if $\mathbf{x}_1, \mathbf{x}_2, \ldots, \mathbf{x}_n$ are vectors in $\mathbb{R}^n$, then the signed volume of the parallelepiped spanned by the ordered n-tuple $[\,\mathbf{x}_1 \quad \mathbf{x}_2 \quad \ldots \quad \mathbf{x}_n\,]$ is given by

$$\det(\mathbf{x}_1, \mathbf{x}_2, \ldots, \mathbf{x}_n) = \begin{vmatrix} x_1^1 & x_1^2 & \cdots & x_1^n \\ x_2^1 & x_2^2 & \cdots & x_2^n \\ \vdots & \vdots & \vdots & \vdots \\ x_n^1 & x_n^2 & \cdots & x_n^n \end{vmatrix},$$

where x_i^k denotes entry i of vector $\mathbf{x}_k$.

The geometric view of determinants leads to a rather abstract definition in terms of what is called an **alternating multilinear function.***

Definition 4.2.1

Suppose that $A = [\,\mathbf{a}_1 \quad \mathbf{a}_2 \quad \ldots \quad \mathbf{a}_n\,]$ is any $n \times n$ matrix written in column form.

The **determinant** of A, written $D(A)$ or $D(\mathbf{a}_1, \mathbf{a}_2, \ldots, \mathbf{a}_n)$, is the unique number assigned to A that satisfies the following three properties:

(1) $D(I_n) = D(\mathbf{e}_1, \mathbf{e}_2, \ldots, \mathbf{e}_n) = 1$, where $\{\mathbf{e}_1, \mathbf{e}_2, \ldots, \mathbf{e}_n\}$ is the standard basis for $\mathbb{R}^n$,
(2) $D(\mathbf{a}_1, \ldots, \mathbf{a}_j, \ldots, \mathbf{a}_k, \ldots, \mathbf{a}_n) = -D(\mathbf{a}_1, \ldots, \mathbf{a}_k, \ldots, \mathbf{a}_j, \ldots, \mathbf{a}_n)$,
(3) $D(\mathbf{a}_1, \ldots, c\,\mathbf{a}_j + \mathbf{x}, \ldots, \mathbf{a}_n)$
$= cD(\mathbf{a}_1, \ldots, \mathbf{a}_j, \ldots, \mathbf{a}_n) + D(\mathbf{a}_1, \ldots, \mathbf{x}, \ldots, \mathbf{a}_n)$,

where c is a scalar and $\mathbf{x} \in \mathbb{R}^n$.

This definition defines the determinant as a function from the Cartesian product (see Appendix A) $\mathbb{R}^n \times \mathbb{R}^n \times \cdots \times \mathbb{R}^n$ into $\mathbb{R}$, $D : \mathbb{R}^n \times \mathbb{R}^n \times \cdots \times \mathbb{R}^n \to \mathbb{R}$. Property (2) states that D is an **alternating function**: if you interchange two columns of A, the resulting determinant is the negative of $D(A)$. Property (3) states that D is a linear transformation if $n - 1$ of the n vectors are kept fixed. This can be split into two separate statements:

(a) $D(\mathbf{a}_1, \ldots, c\,\mathbf{a}_j, \ldots, \mathbf{a}_n) = cD(\mathbf{a}_1, \ldots, \mathbf{a}_j, \ldots, \mathbf{a}_n)$,

and

(b) $D(\mathbf{a}_1, \ldots, \mathbf{a}_j + \mathbf{x}, \ldots, \mathbf{a}_n) = D(\mathbf{a}_1, \ldots, \mathbf{a}_j, \ldots, \mathbf{a}_n) + D(\mathbf{a}_1, \ldots, \mathbf{x}, \ldots, \mathbf{a}_n)$.

In this case, D is said to be a **multilinear function**. We will not prove that these properties define the determinant uniquely. Some of the following exercises explore consequences of this definition.

---

* Some definitions of such a function differ from the one assumed here, and consequently the definition of the determinant will be slightly different. The axiomatic definition of the determinant was formulated by the great German mathematician Weierstrass in the 1880s.

Exercises 4.2

A.

1. Use determinants to calculate the area of each of the following triangles spanned by the given vectors:

   a. $\begin{bmatrix} 1 \\ 1 \end{bmatrix}, \begin{bmatrix} 2 \\ -1 \end{bmatrix}$ b. $\begin{bmatrix} -5 \\ 7 \end{bmatrix}, \begin{bmatrix} 4 \\ 8 \end{bmatrix}$ c. $\begin{bmatrix} 5 \\ -2 \end{bmatrix}, \begin{bmatrix} 8 \\ -1 \end{bmatrix}$

2. Use determinants to calculate the volume of each of the following parallelepipeds spanned by the given vectors:

   a. $\begin{bmatrix} 1 \\ -2 \\ 1 \end{bmatrix}, \begin{bmatrix} 1 \\ 0 \\ -1 \end{bmatrix}, \begin{bmatrix} 1 \\ 1 \\ 1 \end{bmatrix}$ b. $\begin{bmatrix} 1 \\ 1 \\ 0 \end{bmatrix}, \begin{bmatrix} 1 \\ -1 \\ 0 \end{bmatrix}, \begin{bmatrix} 0 \\ 0 \\ 1 \end{bmatrix}$

   c. $\begin{bmatrix} 1 \\ 2 \\ 0 \end{bmatrix}, \begin{bmatrix} -2 \\ 1 \\ 0 \end{bmatrix}, \begin{bmatrix} 0 \\ 0 \\ 1 \end{bmatrix}$ d. $\begin{bmatrix} 2 \\ 1 \\ 2 \\ 3 \end{bmatrix}, \begin{bmatrix} 1 \\ 1 \\ 0 \\ 6 \end{bmatrix}, \begin{bmatrix} 1 \\ -4 \\ -3 \\ 1 \end{bmatrix}, \begin{bmatrix} 5 \\ -1 \\ 1 \\ 2 \end{bmatrix}$

   e. $\begin{bmatrix} 1 \\ -3 \\ 0 \\ 2 \end{bmatrix}, \begin{bmatrix} 0 \\ 1 \\ 4 \\ 3 \end{bmatrix}, \begin{bmatrix} -2 \\ 1 \\ -1 \\ 0 \end{bmatrix}, \begin{bmatrix} 3 \\ 2 \\ 1 \\ 1 \end{bmatrix}$

3. If $A = [\mathbf{a}_1 \quad \mathbf{a}_2 \quad \ldots \quad \mathbf{a}_n]$, use Definition 4.2.1 to find $D(A)$ in each case:

   a. $\mathbf{a}_1 = \mathbf{e}_1 + 4\mathbf{e}_2, \mathbf{a}_2 = 3\mathbf{e}_1 + 5\mathbf{e}_2$.
   b. $\mathbf{a}_1 = 2\mathbf{e}_1 + 4\mathbf{e}_2, \mathbf{a}_2 = 3\mathbf{e}_1 + 6\mathbf{e}_2$.
   c. $\mathbf{a}_1 = \mathbf{e}_1 + \mathbf{e}_2 + \mathbf{e}_3, \mathbf{a}_2 = \mathbf{e}_1 + 2\mathbf{e}_2 + 3\mathbf{e}_3, \mathbf{a}_3$
      $= \mathbf{e}_1 + 3\mathbf{e}_2 + 6\mathbf{e}_3$.
   d. $\mathbf{a}_1 = \mathbf{e}_1 + \mathbf{e}_2 + \mathbf{e}_3, \mathbf{a}_2 = 3\mathbf{e}_3 - \mathbf{e}_1, \mathbf{a}_3 = \mathbf{e}_1 + \mathbf{e}_2$.
   e. $\mathbf{a}_1 = 3\mathbf{e}_1 + 8\mathbf{e}_2 + 2\mathbf{e}_3, \mathbf{a}_2 = 4\mathbf{e}_1 + 7\mathbf{e}_2 - \mathbf{e}_3, \mathbf{a}_3$
      $= -5\mathbf{e}_1 - 2\mathbf{e}_2 + 8\mathbf{e}_3$.

B.

1. If $\mathbf{x}, \mathbf{y},$ and $\mathbf{z}$ are vectors in $\mathbb{R}^2$, show that
   a. $\det(\mathbf{x}, \mathbf{y}) = -\det(\mathbf{y}, \mathbf{x})$.
   b. $\det(\mathbf{x}, c\,\mathbf{y}) = c\det(\mathbf{x}, \mathbf{y})$ and $\det(c\,\mathbf{x}, \mathbf{y}) = c\det(\mathbf{x}, \mathbf{y})$.
   c. $\det(\mathbf{x}, \mathbf{y} + \mathbf{z}) = \det(\mathbf{x}, \mathbf{y}) + \det(\mathbf{x}, \mathbf{z})$.

2. a. Suppose that the vertices of a triangle have coordinates $(x_1, y_1), (x_2, y_2),$ and (x_3, y_3). Prove that the area of the triangle is the absolute value of $\frac{1}{2}\begin{vmatrix} x_1 & y_1 & 1 \\ x_2 & y_2 & 1 \\ x_3 & y_3 & 1 \end{vmatrix}$.

   b. Use the formula from part (a) to calculate the area of the triangle having vertices $(-2, 1), (4, 3),$ and $(0, 7)$.

3. Using Definition 4.2.1, prove that if $A = [\mathbf{a}_1 \quad \mathbf{a}_2 \quad \ldots \quad \mathbf{a}_n]$ and if $\mathbf{a}_i = \mathbf{0}$ for some i, then $D(A) = 0$.

4. Using Definition 4.2.1, prove that if $\mathbf{a}_j = \mathbf{a}_k$ for $j \neq k$, then $D(A) = 0$.

5. Using Definition 4.2.1, prove that if $A = [\mathbf{a}_1 \quad \mathbf{a}_2 \quad \ldots \quad \mathbf{a}_n]$ and $\{\mathbf{a}_1, \mathbf{a}_2, \ldots, \mathbf{a}_n\}$ is a linearly dependent set, then $D(A) = 0$.

6. If B is the matrix obtained from A by adding a scalar multiple of one column to another, then $D(B) = D(A)$. (Use Definition 4.2.1.)

C.

1. Suppose that B is a fixed nonsingular $n \times n$ matrix. Define the function Δ on the set of all $n \times n$ matrices as

$$\Delta(\mathbf{a}_1, \mathbf{a}_2, \ldots, \mathbf{a}_n) = \frac{\det(BA)}{\det(B)},$$

where $A = [\mathbf{a}_1 \quad \mathbf{a}_2 \quad \ldots \quad \mathbf{a}_n]$.
a. Show that Δ is an alternating multilinear function.
b. Show that $\Delta(I_n) = 1$.
c. Show how the conclusions of parts (a) and (b) together with Definition 4.2.1 allow you to prove $\det(AB) = \det(A)\det(B)$.

4.3 The Manual Calculation of Determinants

The definition of the determinant in Section 4.1 in terms of the $PA = LU$ factorization of a matrix places a great deal of emphasis on row and column operations. Although variants of Gaussian elimination are used in the computer solution of large linear systems and in the computation of the related large determinants, this algorithm is not the most appropriate for calculating smaller determinants by hand, especially those determinants containing parameters (variables).

It is easy to deal with 2×2 determinants manually:

$$\det \begin{bmatrix} a_{11} & a_{12} \\ a_{21} & a_{22} \end{bmatrix} = a_{11}a_{22} - a_{12}a_{21}.$$

For 3×3 determinants, there is a neat rule usually attributed to the French mathematician P.F. Sarrus (1798–1861) but probably known to Seki Kowa (see the first paragraph of this chapter) and others before its appearance in an 1846 algebra text. This scheme, which we will denote as the *Sarrus rule*, requires that we repeat the first two columns of the 3×3 matrix right after the third column, giving us five columns in a row. Then we add the products of the diagonals going from left to right and subtract the products of the diagonals going from right to left:

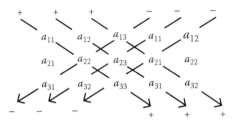

Thus

$$
\det \begin{bmatrix} a_{11} & a_{12} & a_{13} \\ a_{21} & a_{22} & a_{23} \\ a_{31} & a_{32} & a_{33} \end{bmatrix} = a_{11}a_{22}a_{33} + a_{12}a_{23}a_{31} + a_{13}a_{21}a_{32} - a_{12}a_{21}a_{33}
$$

$$
- a_{11}a_{23}a_{32} - a_{13}a_{22}a_{31}.
$$

Anticipating the last method in this section, we note that we can write the formula for the determinant of a 3×3 matrix as follows:

$$
\det \begin{bmatrix} a_{11} & a_{12} & a_{13} \\ a_{21} & a_{22} & a_{23} \\ a_{31} & a_{32} & a_{33} \end{bmatrix} = a_{11}a_{22}a_{33} + a_{12}a_{23}a_{31} + a_{13}a_{21}a_{32}
$$

$$
- a_{12}a_{21}a_{33} - a_{11}a_{23}a_{32} - a_{13}a_{22}a_{31}
$$

$$
= a_{11}(a_{22}a_{33} - a_{23}a_{32}) - a_{12}(a_{21}a_{33} - a_{23}a_{31})
$$

$$
+ a_{13}(a_{21}a_{32} - a_{22}a_{31})
$$

$$
= a_{11} \det \begin{bmatrix} a_{22} & a_{23} \\ a_{32} & a_{33} \end{bmatrix} - a_{12} \det \begin{bmatrix} a_{21} & a_{23} \\ a_{31} & a_{33} \end{bmatrix}
$$

$$
+ a_{13} \det \begin{bmatrix} a_{21} & a_{22} \\ a_{31} & a_{32} \end{bmatrix}.
$$

Example 4.3.1: The Sarrus Rule

Let $A = \begin{bmatrix} 1 & 2 & 6 \\ 3 & -5 & 7 \\ 0 & 8 & 9 \end{bmatrix}$. We repeat the first two columns of A to form a 3×5 array:

$$
\begin{array}{ccccc} 1 & 2 & 6 & 1 & 2 \\ 3 & -5 & 7 & 3 & -5 \\ 0 & 8 & 9 & 0 & 8 \end{array}
$$

Then

$$
\det(A) = (1)(-5)(9) + (2)(7)(0) + (6)(3)(8)
$$

$$
- (2)(3)(9) - (1)(7)(8) - (6)(-5)(0)
$$

$$
= -45 + 144 - 54 - 56 = -11.
$$

Unfortunately, this easily remembered scheme does not generalize to larger determinants, and so we turn to a standard method called the *Laplace expansion*,* which is often given as the *definition* of the determinant, however unmotivated it may be. This algorithm proceeds inductively, reducing the evaluation of an $n \times n$ determinant to a linear combination of $(n-1) \times (n-1)$ determinants, each of which can be written as a sum of $(n-2) \times (n-2)$ determinants, and so forth, until we reduce the original problem to one of evaluating 2×2 determinants. The method uses the entries in any row or any column to structure the calculation.

To understand this procedure, we need to define some pieces of the process. In the definitions that follow, we assume that A is an $n \times n$ matrix.

Definition 4.3.1

(a) The (i, j)-**minor** of A (or the minor of the entry a_{ij}) is the determinant of the $(n-1) \times (n-1)$ submatrix formed by removing row i and column j from A. The minor is denoted by $\left| A_{ij} \right|$ $(= \det(A_{ij}))$.

(b) The (i, j)-**cofactor** of A (or the cofactor of the entry a_{ij}) is the number $(-1)^{i+j} \left| A_{ij} \right|$ and is denoted by C_{ij}.

Example 4.3.2: Minors and Cofactors

Let $\quad A = \begin{bmatrix} 1 & 0 & -2 & 4 \\ 5 & 6 & 0 & 1 \\ -2 & 0 & 6 & 3 \\ 4 & 3 & 2 & 1 \end{bmatrix}.$ Then,   for   example,

row 2, col. 3
of A omitted

$A_{23} = \begin{bmatrix} 1 & 0 & 4 \\ -2 & 0 & 3 \\ 4 & 3 & 1 \end{bmatrix}$ and   the   (2,3)-minor   of A is

$\left| A_{23} \right| = \det \begin{bmatrix} 1 & 0 & 4 \\ -2 & 0 & 3 \\ 4 & 3 & 1 \end{bmatrix}.$ The   (2,3)-cofactor   of A is

$C_{23} = (-1)^{2+3} \left| A_{23} \right| = -\det \begin{bmatrix} 1 & 0 & 4 \\ -2 & 0 & 3 \\ 4 & 3 & 1 \end{bmatrix}.$

---

* Named for Pierre Simon Laplace (1749–1827), who made important contributions to analysis, probability, and astronomy.

row 4, col. 2
of A omitted

$$\text{Similarly, } A_{42} = \begin{bmatrix} 1 & -2 & 4 \\ 5 & 0 & 1 \\ -2 & 6 & 3 \end{bmatrix}, \; |A_{42}| = \det \begin{bmatrix} 1 & -2 & 4 \\ 5 & 0 & 1 \\ -2 & 6 & 3 \end{bmatrix},$$

$$\text{and } C_{42} = (-1)^{4+2}|A_{42}| = \det \begin{bmatrix} 1 & -2 & 4 \\ 5 & 0 & 1 \\ -2 & 6 & 3 \end{bmatrix}.$$

Laplace's expansion expresses the determinant of an $n \times n$ matrix as a linear combination of the cofactors of A. For this reason, this algorithm is sometimes referred to as the *cofactor expansion*.

The Laplace Expansion

Suppose that $A = [a_{ij}]$ is an $n \times n$ matrix.

(a) Choose any row, say row i. Then

$$\det(A) = \sum_{k=1}^{n} a_{ik} C_{ik} = \sum_{k=1}^{n} (-1)^{i+k} a_{ik} |A_{ik}|$$
$$= (-1)^{i+1} a_{i1} |A_{i1}| + (-1)^{i+2} a_{i2} |A_{i2}| + \cdots + (-1)^{i+n} a_{in} |A_{in}|.$$

[Note that the row index stays equal to i, while the column index varies.]

or

(b) Choose any column, say column j. Then

$$\det(A) = \sum_{k=1}^{n} a_{kj} C_{kj} = \sum_{k=1}^{n} (-1)^{k+j} a_{kj} |A_{kj}|$$
$$= (-1)^{1+j} a_{1j} |A_{1j}| + (-1)^{2+j} a_{2j} |A_{2j}| + \cdots + (-1)^{n+j} a_{nj} |A_{nj}|.$$

[Note that the column index stays equal to j, while the row index varies.]

This algorithm looks daunting in its mathematically compact form. A few examples should help establish a comfort level. In using this procedure, it is important to choose a particular row or column that simplifies the arithmetic—often the row or column that contains the greatest number of zeros.

The correct placement of the plus and minus signs associated with the cofactors—from the factors $(-1)^{i+j}$ in the Laplace expansion—can be aided by thinking of the following $n \times n$ checkerboard pattern superimposed on the matrix:

$$
\begin{array}{ccccc}
+ & - & + & - & \cdots \\
- & + & - & + & \cdots \\
+ & - & + & - & \cdots \\
- & + & - & + & \cdots \\
\vdots & \vdots & \vdots & \vdots & \ddots
\end{array}
$$

Start with a $+$ in the upper left-hand corner and then alternate signs. For example, from the diagram we see that if we expand along the second column, the pattern of signs attached to the terms $a_{k2}|A_{k2}|$ is $- + - + - + \cdots, k = 1, 2, \ldots, n$.

Example 4.3.3: The Laplace Expansion of a 3 × 3 Determinant

Let us calculate the determinant of $A = \begin{bmatrix} 4 & 2 & 1 \\ -2 & -6 & 3 \\ -7 & 5 & 0 \end{bmatrix}$ (a) by using the first row; (b) by using the third column.

(a) Using the first row, Laplace's expansion takes the form:

$$
\begin{aligned}
\det(A) &= a_{11}C_{11} + a_{12}C_{12} + a_{13}C_{13} \\
&= (-1)^{1+1}a_{11}|A_{11}| + (-1)^{1+2}a_{12}|A_{12}| + (-1)^{1+3}a_{13}|A_{13}|
\end{aligned}
$$

$$
\begin{array}{ccc}
\text{row 1, col. 1} & \text{row 1, col. 2} & \text{row 1, col. 3} \\
\text{of } A \text{ omitted} & \text{of } A \text{ omitted} & \text{of } A \text{ omitted}
\end{array}
$$

$$
\begin{aligned}
&= 4 \cdot \det \begin{bmatrix} -6 & 3 \\ 5 & 0 \end{bmatrix} - 2 \cdot \begin{bmatrix} -2 & 3 \\ -7 & 0 \end{bmatrix} + 1 \cdot \begin{bmatrix} -6 & 3 \\ 5 & 0 \end{bmatrix} \\
&= 4(-15) - 2(21) + (-52) = -154.
\end{aligned}
$$

(b) Expanding along the third column gives us

$$
\begin{aligned}
\det(A) &= a_{13}C_{13} + a_{23}C_{23} + a_{33}C_{33} \\
&= (-1)^{1+3}a_{13}|A_{13}| + (-1)^{2+3}a_{23}|A_{23}| + (-1)^{3+3}a_{33}|A_{33}|
\end{aligned}
$$

$$
\begin{array}{ccc}
\text{row 1, col. 3} & \text{row 2, col. 3} & \text{row 3, col. 3} \\
\text{of } A \text{ omitted} & \text{of } A \text{ omitted} & \text{of } A \text{ omitted}
\end{array}
$$

$$
\begin{aligned}
&= 1 \cdot \det \begin{bmatrix} -2 & -6 \\ -7 & 5 \end{bmatrix} - 3 \cdot \begin{bmatrix} 4 & 2 \\ -7 & 5 \end{bmatrix} + 0 \cdot \begin{bmatrix} 4 & 2 \\ -2 & -6 \end{bmatrix} \\
&= (-52) - 3(34) = -154.
\end{aligned}
$$

Notice how calculation (b) in Example 4.4.3 was simplified by choosing a column with a zero in it. We could have used the third row for the same reason.

Example 4.3.4: The Laplace Expansion of a 4 × 4 Determinant

Now let us compute det(A), where

$$A = \begin{bmatrix} 1 & 0 & -2 & 4 \\ 5 & 6 & 0 & 1 \\ -2 & 0 & 6 & 3 \\ 4 & 3 & 2 & 1 \end{bmatrix}.$$

Noting that column 2 contains two zeros, we choose to expand along this column:

$$\det(A) = (-1)^{1+2} \cdot 0 \cdot |A_{12}| + (-1)^{2+2} \cdot 6 \cdot |A_{22}|$$
$$+ (-1)^{3+2} \cdot 0 \cdot |A_{32}| + (-1)^{4+2} \cdot 3 \cdot |A_{42}|$$

$$= 6|A_{22}| + 3|A_{42}| = 6 \overbrace{\begin{bmatrix} 1 & -2 & 4 \\ -2 & 6 & 3 \\ 4 & 2 & 1 \end{bmatrix}}^{B} + 3 \overbrace{\begin{bmatrix} 1 & -2 & 4 \\ 5 & 0 & 1 \\ -2 & 6 & 3 \end{bmatrix}}^{C}.$$

To complete the calculation of det (A), we have to compute the determinants of the 3 × 3 matrices B and C. Because there are no zeros in either matrix, we arbitrarily choose to expand using the first row of each matrix:

$$\det(B) = 1 \cdot \det \begin{bmatrix} 6 & 3 \\ 2 & 1 \end{bmatrix} - (-2) \det \begin{bmatrix} -2 & 3 \\ 4 & 1 \end{bmatrix} + 4 \ \det \begin{bmatrix} -2 & 6 \\ 4 & 2 \end{bmatrix}$$
$$= 0 + 2(-14) + 4(-28) = -140.$$
$$\det(C) = 1 \cdot \det \begin{bmatrix} 0 & 1 \\ 6 & 3 \end{bmatrix} - (-2) \det \begin{bmatrix} 5 & 1 \\ -2 & 3 \end{bmatrix} + 4 \ \det \begin{bmatrix} 5 & 0 \\ -2 & 6 \end{bmatrix}$$
$$= (-6) + 2(17) + 4(30) = 148.$$

Then det(A) = 6 det(B) + 3 det(C) = 6(−140) + 3(148) = −396.

Beginning in Section 4.4, we will have to compute the determinants of matrices containing parameters.

Example 4.3.5: Determinants of Matrices with Parameters

(a) In Example 4.1.3(c) we let $A = \begin{bmatrix} 1 & \lambda & \lambda^2 \\ 1 & 1 & 1 \\ 1 & 2 & 4 \end{bmatrix}$ and asked

for the values of λ that make this matrix singular (non-invertible). Using Laplace's method along the first row, we find that

$$\det(A) = 1 \cdot \det\begin{bmatrix} 1 & 1 \\ 2 & 4 \end{bmatrix} - \lambda \det\begin{bmatrix} 1 & 1 \\ 1 & 4 \end{bmatrix} + \lambda^2 \det\begin{bmatrix} 1 & 1 \\ 1 & 2 \end{bmatrix}.$$

$$= 2 - \lambda(3) + \lambda^2(1) = \lambda^2 - 3\lambda + 2 = (\lambda - 2)(\lambda - 1)$$

The matrix A is clearly singular if $\lambda = 1$ or $\lambda = 2$.

(b) Given $A = \begin{bmatrix} 3 & 2 & 4 \\ 2 & 0 & 2 \\ 4 & 2 & 3 \end{bmatrix}$, we want to find the values of λ

for which $\det(A - \lambda I) = 0$. (The rationale for this computation will be made clear in Section 4.4.)

We have $\lambda I = \begin{bmatrix} \lambda & 0 & 0 \\ 0 & \lambda & 0 \\ 0 & 0 & \lambda \end{bmatrix}$ and $A - \lambda I = \begin{bmatrix} 3 - \lambda & 2 & 4 \\ 2 & -\lambda & 2 \\ 4 & 2 & 3 - \lambda \end{bmatrix}$.

Using the first row, we calculate

$\det(A - \lambda I)$

$$= (3 - \lambda) \det\begin{bmatrix} -\lambda & 2 \\ 2 & 3 - \lambda \end{bmatrix} - 2 \det\begin{bmatrix} 2 & 2 \\ 4 & 3 - \lambda \end{bmatrix}$$

$$+ 4 \det\begin{bmatrix} 2 & -\lambda \\ 4 & 2 \end{bmatrix}$$

$$= (3 - \lambda)[-\lambda(3 - \lambda) - 4] - 2[2(3 - \lambda) - 8] + 4[4 + 4\lambda]$$

$$= (3 - \lambda)(\lambda - 4)(\lambda + 1) + 20(\lambda + 1) = (\lambda + 1)(-\lambda^2 + 7\lambda + 8)$$

$$= -(\lambda + 1)^2(\lambda - 8).$$

Hence $\det(A - \lambda I) = 0$ if and only if $\lambda = -1$ or $\lambda = 8$.

Laplace's expansion is useful for finding the determinants of small matrices. Computationally, however, this method of evaluating an $n \times n$ determinant is terribly inefficient for large values of n because it requires more than $n!$ additions, subtractions, and multiplications* compared to about n^3 operations for Gauss–Jordan reduction and $\frac{2}{3}n^3$ for Gaussian elimination. For comparison purposes, when $n = 100$, the value of n^3 is 1000, whereas $n! \approx 10^{158}$.

For calculating determinants manually, a combination of row/column manipulations and Laplace's expansion will usually do the trick. For example, row and/or column operations can be used to produce zeros, thereby preparing the determinant for a more efficient use of Laplace's method.

---

* See C. Dubbs and D. Siegel, Computing determinants, *Coll. Math. J.* **18** (1987): 48–50.

Exercises 4.3

A.

1. Using the Laplace expansion, find the determinant of the matrix in Problem A1(g), Exercise 4.1.

2. Use the Laplace expansion to find the determinant of the matrix in Problem A6, Exercise 4.1.

3. Use the Laplace expansion to find the determinant of the matrix in Problem A9, Exercise 4.1.

4. Consider the matrix $A = \begin{bmatrix} 1 & 4 & 7 \\ 2 & 5 & 8 \\ 3 & 6 & 9 \end{bmatrix}$. Compute $\det(A)$ by using:

   (a) Row and/or column operations
   (b) The Laplace expansion
   (c) A mixture of the methods used in parts (a) and (b)

5. a. Use the Laplace expansion to evaluate the determinant of

   $$\begin{bmatrix} 1 & 2 & 3 \\ 4 & 5 & 6 \\ 7 & 8 & 9 \end{bmatrix}.$$

   b. Use the Laplace expansion to evaluate the determinant of

   $$\begin{bmatrix} 1 & -2 & 3 & -4 \\ 5 & 1 & 2 & 3 \\ -4 & 2 & 3 & 0 \\ 4 & 3 & 2 & 1 \end{bmatrix}.$$

6. If

   $$\begin{bmatrix} 1 & a & -1 \\ 2 & 4 & 0 \\ -1 & b & 1 \end{bmatrix} = 120,$$

   what are all possible values for a and b?

7. If $A = \begin{bmatrix} 0 & 4 & 1 \\ 1 & 3 & 4 \\ 3 & 2 & -1 \end{bmatrix}$, evaluate $\det(A)$ by using the Sarrus rule.

8. By using a combination of row/column operations and the Laplace expansion, show that the 3×3 *Vandermonde matrix* (see Example 4.1.4)

   $$\begin{bmatrix} 1 & a & a^2 \\ 1 & b & b^2 \\ 1 & c & c^2 \end{bmatrix}$$

   has determinant $-(a - b)(a - c)(b - c)$.

9. Let $A = \begin{bmatrix} 1 & 3 & -1 \\ 0 & x & 5 \\ -2 & -4 & 4 \end{bmatrix}$, where x is a real variable.

(a) Find $\det(A)$.

(b) For what values of x will A have an inverse?

(c) For those values of x such that A^{-1} exists, find $\det(A^{-1})$.

B.

1. Suppose that $A = \begin{bmatrix} 0 & 1 & 1 & 1 \\ 1 & 1 & 1 & 1 \\ 1 & 1 & 3 & 1 \\ 2 & 1 & 3 & 4 \end{bmatrix}$.

(a) Find $\det(A)$ by using row reduction techniques.

(b) Find $\det(A)$ by using the Laplace expansion.

2. Use the cofactor expansion to evaluate $\det(A)$, where

$$A = \begin{bmatrix} 3 & 0 & 0 & -2 & 4 \\ 0 & 2 & 0 & 0 & 0 \\ 0 & -1 & 0 & 5 & -3 \\ -4 & 0 & 1 & 0 & 6 \\ 0 & -1 & 0 & 3 & 2 \end{bmatrix}.$$

3. Prove that the $n \times n$ *Vandermonde determinant* of Example 4.1.4 is given by

$$\prod_{1 \le i < j \le n} (x_j - x_i).$$

4. Let

$$T_5 = \begin{bmatrix} a & b & b & b & b \\ b & a & b & b & b \\ b & b & a & b & b \\ b & b & b & a & b \\ b & b & b & b & a \end{bmatrix}.$$

Show that $\det(T_5) = (a - b)^4(a + 4b)$. [This is a type of matrix called a **circulant** because of the way successive columns (or rows) are shifted one position downward (or across) and wrapped around at the top (or at the left end).]

5. Generalize the result of the previous exercise to any $n \times n$ matrix $T_n = [t_{ij}]$, where

$$t_{ij} = \begin{cases} a & \text{if } i = j \\ b & \text{otherwise.} \end{cases}$$

That is, show that $\det(T_n) = (a - b)^{n-1}[a + (n - 1)b]$.

6. If J_n is the $n \times n$ matrix all of whose entries equal 1, show that $\det(J_n - nI_n) = 0$.

7. Show that $\det \begin{bmatrix} 1 + a_1 & a_1 & \cdots & a_1 \\ a_2 & 1 + a_2 & \cdots & a_2 \\ \vdots & \vdots & \vdots & \vdots \\ a_n & a_n & \cdots & 1 + a_n \end{bmatrix} = 1 + a_1 + a_2 + \cdots + a_n.$

C.

1. Suppose that A is an $n \times n$ matrix with more than $n^2 - n$ entries that are 0. Show that $\det(A) = 0$.

2. If A and B are $n \times n$ matrices, show that

$$\det \begin{bmatrix} A & O_n \\ O_n & B \end{bmatrix} = (\det A)(\det B).$$

3. Let A be a square matrix that can be partitioned as

$$A = \begin{bmatrix} P & Q \\ O & S \end{bmatrix},$$

where P and S are square matrices. Prove that $\det(A) = \det(P)\det(S)$.

4.4 Eigenvalues and Eigenvectors

In Section 3.2, we saw matrix–vector multiplication interpreted as a function or transformation that takes as input vectors in some Euclidean space $\mathbb{R}^n$ and returns vectors in some space $\mathbb{R}^m$ as output. For a fixed $n \times n$ matrix A, we can define the function symbolically as $f_A : \mathbb{R}^n \to \mathbb{R}^m$, where $f_A(\mathbf{x}) = A\mathbf{x}$ for every $\mathbf{x} \in \mathbb{R}^n$.

Loosely speaking, such a transformation scrambles the vectors in $\mathbb{R}^n$ processing input vectors by some combination of rotation, stretching, and shrinking. In this way, matrix–vector multiplication affects both the magnitude and direction of geometric vectors. Under such a function, it sometimes happens that a nonzero vector $\mathbf{x}$ may be transformed into a scalar multiple of itself—a vector in the same or opposite direction as itself, depending on whether the scalar is positive or negative. We first encountered this idea in Section 3.2. (Also review the discussion between Problems A26 and A27 in Exercise 3.2.)

Definition 4.4.1

Let A be an $n \times n$ matrix. If $A\mathbf{v} = \lambda\mathbf{v}$ holds for some scalar λ and a nonzero vector $\mathbf{v}$, then λ is called an **eigenvalue*** of A and $\mathbf{v}$ is called an **eigenvector** of A. (We also say that λ is the eigenvalue of A associated with $\mathbf{v}$ or $\mathbf{v}$ is an eigenvector associated with λ.) The set of all eigenvalues of A is called the **spectrum** of A and is denoted by $\sigma(A)$.

---

* The terms **eigenvector** and **eigenvalue** are German–English hybrids, with the word *eigen* meaning "characteristic of" or "peculiar to." The terms **characteristic vector** (**value**), **proper vector** (**value**), and **latent vector** (**value**) are also used.

Example 4.4.1: Eigenvectors and Eigenvalues

Consider the matrix $A = \begin{bmatrix} 1 & 1 \\ 2 & 0 \end{bmatrix}$ and the vectors $\mathbf{v}_1 = \begin{bmatrix} -1 \\ 2 \end{bmatrix}$ and $\mathbf{v}_2 = \begin{bmatrix} 1 \\ 1 \end{bmatrix}$. Then

$$A\mathbf{v}_1 = \begin{bmatrix} 1 & 1 \\ 2 & 0 \end{bmatrix} \begin{bmatrix} -1 \\ 2 \end{bmatrix} = \begin{bmatrix} 1 \\ -2 \end{bmatrix} = (-1)\begin{bmatrix} -1 \\ 2 \end{bmatrix} = (-1)\mathbf{v}_1$$

and

$$A\mathbf{v}_2 = \begin{bmatrix} 1 & 1 \\ 2 & 0 \end{bmatrix} \begin{bmatrix} 1 \\ 1 \end{bmatrix} = \begin{bmatrix} 2 \\ 2 \end{bmatrix} = (2)\begin{bmatrix} 1 \\ 1 \end{bmatrix} = (2)\mathbf{v}_2,$$

so $\mathbf{v}_1$ is an eigenvector of A with associated eigenvalue -1 and $\mathbf{v}_2$ is an eigenvector of A with associated eigenvalue 2.

The matrix $B = \begin{bmatrix} 3 & 0 & 1 \\ 2 & 5 & 5 \\ 0 & 0 & 3 \end{bmatrix}$ has eigenvectors $\mathbf{v}_1 = \begin{bmatrix} 0 \\ 1 \\ 0 \end{bmatrix}$

and $\mathbf{v}_2 = \begin{bmatrix} -1 \\ 1 \\ 0 \end{bmatrix}$, with associated eigenvalues $\lambda_1 = 5$ and $\lambda_2 = 3$, respectively:

$$B\mathbf{v}_1 = \begin{bmatrix} 3 & 0 & 1 \\ 2 & 5 & 5 \\ 0 & 0 & 3 \end{bmatrix} \begin{bmatrix} 0 \\ 1 \\ 0 \end{bmatrix} = \begin{bmatrix} 0 \\ 5 \\ 0 \end{bmatrix} = 5\begin{bmatrix} 0 \\ 1 \\ 0 \end{bmatrix} = 5\mathbf{v}_1,$$

$$B\mathbf{v}_2 = \begin{bmatrix} 3 & 0 & 1 \\ 2 & 5 & 5 \\ 0 & 0 & 3 \end{bmatrix} \begin{bmatrix} -1 \\ 1 \\ 0 \end{bmatrix} = \begin{bmatrix} -3 \\ 3 \\ 0 \end{bmatrix} = 3\begin{bmatrix} -1 \\ 1 \\ 0 \end{bmatrix} = 3\mathbf{v}_2.$$

Vectors that behave this way, preserving or reversing their direction under a linear transformation, turn out to be fundamental in understanding the matrix B used in defining the transformation. There are also important applications that require eigenvalues and eigenvectors, as we will see later.

Definition 4.4.2

Given a square matrix A, an ordered pair $(\lambda, \mathbf{v})$, with λ an eigenvalue of A and $\mathbf{v}$ its corresponding eigenvector, is called an **eigenpair** for A.

The $n \times 1$ zero vector $\mathbf{0}$ is never allowed to be an eigenvector. (Clearly $A\mathbf{0} = \lambda\mathbf{0}$ for every scalar λ, so *every* scalar would be an eigenvalue of A associated with $\mathbf{0}$.) On the other hand, the scalar 0 can be an eigenvalue: If $\mathbf{v}$ is an eigenvector of A, $A\mathbf{v} = 0(\mathbf{v}) = \mathbf{0}$ simply means that $\mathbf{v}$ is a nonzero vector in the *null space* of A. (See Example 1.6.4.)

A crucial observation is that $A\mathbf{v} = \lambda\mathbf{v}$ is equivalent to $A\mathbf{v} - \lambda\mathbf{v} = \mathbf{0}$, or

$$(A - \lambda I_n)\mathbf{v} = \mathbf{0}.* \qquad\qquad (*)$$

We see that $\mathbf{v}$ is an eigenvector of A with a corresponding eigenvalue λ if and only if $\mathbf{v}$ is a nonzero vector belonging to the null space of $A - \lambda I_n$. From (*), if the matrix $A - \lambda I_n$ were invertible, we would have $\mathbf{v} = (A - \lambda I_n)^{-1}\mathbf{0} = \mathbf{0}$, so there is no eigenvector corresponding to the scalar λ. Thus **the only way for a scalar λ to be an eigenvalue is for the matrix $A - \lambda I_n$ to be singular.** We summarize and formalize this discussion in the next theorem.

Theorem 4.4.1

If A is an $n \times n$ matrix and λ is a scalar, then the following statements are equivalent:

(i) λ is an eigenvalue of A.
(ii) The matrix $A - \lambda I_n$ is singular (noninvertible).
(iii) $\det(A - \lambda I_n) = 0$.

Proof The formal proof of this result consists of the circle of implications:

$$(i) \Rightarrow (ii) \Rightarrow (iii) \Rightarrow (i).$$

(i) $\Rightarrow$ (ii): Let λ be an eigenvalue of A. Then, by definition, there is a nonzero vector $\mathbf{v}$ such that $A\mathbf{v} = \lambda\mathbf{v}$, or $A\mathbf{v} - \lambda\mathbf{v} = (A - \lambda I_n)\mathbf{v} = \mathbf{0}$. If $A - \lambda I_n$ were nonsingular, then we could conclude that $\mathbf{v} = (A - \lambda I_n)^{-1}\mathbf{0} = \mathbf{0}$—a contradiction. Thus $A - \lambda I$ must be singular.

(ii) $\Rightarrow$ (iii): Suppose that $A - \lambda I_n$ is singular. Then Theorem 4.1.4(2) implies that $\det(A - \lambda I_n) = 0$.

(iii) $\Rightarrow$ (i): If $\det(A - \lambda I_n) = 0$, then $A - \lambda I_n$ is singular. Thus the homogeneous system $(A - \lambda I)\mathbf{x} = \mathbf{0}$ has a nonzero solution $\mathbf{x}$—that is, $A\mathbf{x} = \lambda\mathbf{x}$ for some nonzero vector $\mathbf{x}$. By definition, this makes $\mathbf{x}$ an eigenvector of A with associated eigenvalue λ.

One important consequence of Theorem 4.4.1 is that **a matrix A is nonsingular if and only if 0 is not an eigenvalue of A (if and only if $\det(A) \neq 0$).**

Example 4.4.2: An Application of Theorem 4.4.1

As an interesting consequence of Theorem 4.4.1, we prove that *any square matrix is the sum of two nonsingular matrices.*

---

* Note that the factorization $A\mathbf{v} - \lambda\mathbf{v} = (A - \lambda)\mathbf{v}$ does not make sense because A is a matrix and λ is a scalar. We cannot subtract a scalar from a matrix; however, λI_n (and hence $A - \lambda I_n$) is a matrix.

The proof is neat. Given an $n \times n$ matrix A, choose any scalar λ that is not an eigenvalue of either A or $-A$. Then

$$A = \frac{1}{2}[(A + \lambda I_n) + (A - \lambda I_n)] = \frac{1}{2}(A + \lambda I_n) + \frac{1}{2}(A - \lambda I_n).$$

Because λ is not an eigenvalue of $\pm A$, $\det[(1/2)(A - \lambda I_n)] = (1/2)^n \det(A - \lambda I_n) \neq 0$, and $\det[(1/2)(A + \lambda I_n)] = (1/2)^n (-1)^n$ $\det(-A - \lambda I_n) \neq 0$, which shows that $(1/2)(A + \lambda I_n)$ and $(1/2)(A - \lambda I_n)$ are nonsingular. (In simplifying the determinants, we used the fact that $\det(kA) = k^n \det(A)$—Problem B4 of Exercise 4.1, a natural extension of Theorem 4.1.3(ii).) Clearly this decomposition is not unique.

Before we continue proving results about eigenvalues and eigenvectors, we should establish the *existence* of what we are discussing. Note that the following theorem is determinant-free.

Theorem 4.4.2

Every $n \times n$ matrix A has an eigenvalue (possibly complex).

Proof * Suppose that $\mathbf{x}$ is any nonzero vector in $\mathbb{R}^n$. The vectors $\mathbf{x}, A\mathbf{x}, A^2\mathbf{x}, \ldots, A^n\mathbf{x}$ cannot be linearly independent because they are $n + 1$ vectors in a space having dimension n. Therefore, there exist real numbers $a_0, a_1, \ldots, a_n$, not all 0 such that $a_0\mathbf{x} + a_1 A\mathbf{x} + \cdots + a_n A^n\mathbf{x} = \mathbf{0}$. Let m be the largest subscript such that $a_m \neq 0$. Because $\mathbf{x} \neq \mathbf{0}$, the coefficients $a_1, a_2, \ldots, a_m$ cannot all be 0, so $0 < m \leq n$. Now make the a_i's the coefficients of an ordinary nth degree polynomial, which can then be written in factored form as

$$a_0 + a_1 z + \cdots + a_n z^n = a_m(z - r_1)(z - r_2) \cdots (z - r_m),$$

where a_m is a nonzero complex number, each r_j is complex, and the equation holds for all complex numbers z. Then we can replace z by A and write

$$\mathbf{0} = (a_0 I_n + a_1 A + \cdots + a_n A^n)\mathbf{x}$$
$$= a_m(A - r_1 I_n)(A - r_2 I_n) \cdots (A - r_m I_n)\mathbf{x}.$$

This means that at least one of the matrices, say $A - r_j I_n$, is noninvertible; otherwise $M = (A - r_1 I_n)(A - r_2 I_n) \cdots (A - r_m I_n)$, as the product of invertible matrices, would be invertible and so $\mathbf{x} = M^{-1}\mathbf{0} = \mathbf{0}$, a contradiction. But Theorem 4.4.1 says that r_j is an eigenvalue of A.

Even a simple matrix such as $A = \begin{bmatrix} 0 & -1 \\ 1 & 0 \end{bmatrix}$ fails to have a *real* eigenvalue: Theorem 4.4.1 says that λ is an eigenvalue of A if and only

---

* This proof is based on one by Sheldon Axler.

if $\det(A - \lambda I_n) = \begin{vmatrix} -\lambda & -1 \\ 1 & -\lambda \end{vmatrix} = \lambda^2 + 1 = 0$, which is impossible for a real number λ. Even though we are not dealing with complex eigenvalues now, we will expand our horizons appropriately in Chapter 7.

The determinant criterion

$$\boxed{\det(A - \lambda I_n) = 0}$$

is called the **characteristic equation** of A. It is important because it tells us how to determine the eigenvalues of A. It is customary to solve the characteristic equation for the eigenvalues and then calculate the corresponding eigenvectors for each eigenvalue. It will make our algebraic life easier further on in this section to note that the equation $\det(A - \lambda I_n) = 0$ is equivalent to the equation $\det(\lambda I_n - A) = 0$ (see Exercise A8).

Example 4.4.3: Eigenvalues and Eigenvectors

Let us revisit Example 4.1.3, where $A = \begin{bmatrix} 1 & 1 \\ 2 & 0 \end{bmatrix}$. Form the matrix $A - \lambda I_2 = \begin{bmatrix} 1 & 1 \\ 2 & 0 \end{bmatrix} - \lambda \begin{bmatrix} 1 & 0 \\ 0 & 1 \end{bmatrix} = \begin{bmatrix} 1 - \lambda & 1 \\ 2 & -\lambda \end{bmatrix}$. Then the characteristic equation $\det(A - \lambda I_2) = (1 - \lambda)(-\lambda) - (1)(2) = \lambda^2 - \lambda - 2 = 0$ implies that there are two eigenvalues: $\lambda_1 = -1$ and $\lambda_2 = 2$.

Now let $\mathbf{v}_1 = \begin{bmatrix} x_1 \\ y_1 \end{bmatrix}$ be an eigenvector corresponding to the eigenvalue -1. The matrix–vector equation $A\mathbf{v}_1 = \lambda_1 \mathbf{v}_1$ is equivalent to the system:

$$\begin{array}{c} x_1 + y_1 = -x_1 \\ 2x_1 = y_1 \end{array} \quad \text{or} \quad \begin{array}{c} 2x_1 + y_1 = 0 \\ 2x_1 + y_1 = 0 \end{array}$$

whose solution is any vector of the form $\begin{bmatrix} x_1 \\ -2x_1 \end{bmatrix} = x_1 \begin{bmatrix} 1 \\ -2 \end{bmatrix}$ — that is, any vector lying along the line $y = -2x$ is an eigenvector. For convenience, we choose $x_1 = 1$, so our *representative* eigenvector corresponding to -1 is $\mathbf{v}_1 = \begin{bmatrix} 1 \\ -2 \end{bmatrix}$.

Similarly, $A\mathbf{v}_2 = \lambda_2 \mathbf{v}_2$ yields the system:

$$\begin{array}{c} x_2 + y_2 = 2x_2 \\ 2x_2 = 2y_2 \end{array} \quad \text{or} \quad \begin{array}{c} -x_2 + y_2 = 0 \\ x_2 - y_2 = 0 \end{array}$$

with solution vector $\begin{bmatrix} x_2 \\ x_2 \end{bmatrix} = x_2 \begin{bmatrix} 1 \\ 1 \end{bmatrix}$. Letting $x_2 = 1$, we get our simplified eigenvector $\mathbf{v}_2 = \begin{bmatrix} 1 \\ 1 \end{bmatrix}$.

Generalizing what we saw in Example 4.4.3, let us accept the fact that *if A is an $n \times n$ matrix, then the left side of the equivalent characteristic equation $\det(\lambda I_n - A) = 0$ is a polynomial of degree n in λ, with leading* *coefficient 1.* (The *Laplace expansion* introduced in Section 4.3 makes it easier to see this fact.) This polynomial is called the **characteristic polynomial** of A. Theorem 4.4.1(iii) states that **a scalar is an eigenvalue of** A **if and only if it is a zero of the characteristic polynomial of** A. Such zeros may be complex numbers, but for the sake of focusing on the basics, we will postpone a discussion of complex eigenvalues (and complex eigenvectors!) until Chapter 7. We note that there is no general formula for the zeros of a polynomial of degree greater than or equal to five.* A CAS or graphing calculator uses various algorithms to *approximate* such zeros.

Example 4.4.4: Approximate Eigenvalues

The matrix

$$A = \begin{bmatrix} 0 & 0 & 2 & 2 & 0 \\ 1 & 1 & 1 & 0 & 2 \\ 2 & 0 & 1 & 0 & 1 \\ 2 & 1 & 2 & 0 & 1 \\ 0 & 1 & 2 & 2 & 2 \end{bmatrix}$$

has the characteristic equation $\lambda^5 - 4\lambda^4 - 9\lambda^3 + 15\lambda^2 + 14\lambda + 2 = 0$.

A CAS (*Maple*® in this case) approximates the eigenvalues:

$$\lambda_1 = 1.803591727, \quad \lambda_2 = 5.080576287,$$
$$\lambda_3 = -0.1818101748, \quad \lambda_4 = -5.604901079,$$
$$\lambda_5 = -2.141867731.$$

The problem of calculating (estimating) eigenvalues and eigenvectors accurately is an important part of numerical analysis. For a good discussion of some of the algorithms used to solve eigenvalue problems, see Part V of *Numerical Linear Algebra* by L.N. Trefethen and D. Bau, III (Philadelphia, PA: SIAM, 1997).

Now suppose that $A = \begin{bmatrix} a & b \\ c & d \end{bmatrix}$ is any 2×2 matrix. The characteristic polynomial of A is $\det(\lambda I_2 - A) = \det\begin{bmatrix} \lambda - a & b \\ c & \lambda - d \end{bmatrix} = (\lambda - a)(\lambda - d) - bc = \lambda^2 - (a+d)\lambda + (ad - bc)$. The number $a + d$, the sum of the diagonal entries of A, is called the **trace** of A and is denoted by trace(A). The constant term of the characteristic polynomial is the determinant of A. In general, the **trace** of an $n \times n$ matrix $A = [a_{ij}]$ is the sum of its diagonal entries:

$$\text{trace}(A) = a_{11} + a_{22} + \cdots + a_{nn}.$$

(Exercise B11 lists several important properties of the trace of a matrix.)

Let us look at the situation for a 3×3 matrix.

---

* See Section 15.2.3 of V.J. Katz, *A History of Mathematics*, 2nd edn. (Reading, MA: Addison-Wesley, 1998).

Example 4.4.5: The Characteristic Polynomial of a 3 × 3 Matrix

Let $A = \begin{bmatrix} 2 & 3 & 1 \\ 3 & 2 & 4 \\ 0 & 0 & -1 \end{bmatrix}$. Then $\lambda I_3 - A = \begin{bmatrix} \lambda - 2 & 3 & 1 \\ 3 & \lambda - 2 & 4 \\ 0 & 0 & \lambda + 1 \end{bmatrix}$.

Expanding by the third row, we find that

$$c_A(\lambda) = \det(\lambda I_3 - A) = 0 \cdot |A_{31}| - 0 \cdot |A_{32}| + (\lambda + 1)|A_{33}|$$

$$= (\lambda + 1) \det \begin{bmatrix} \lambda - 2 & 3 \\ 3 & \lambda - 2 \end{bmatrix}$$

$$= (\lambda + 1)[(\lambda - 2)^2 - 9] = (\lambda + 1)^2(\lambda - 5).$$

The coefficients of the characteristic polynomial $\det(\lambda I_n - A)$ can always be expressed in terms of the individual entries of A. If n is large, some of these expressions are very cumbersome, but we can use induction to prove that the coefficient of λ^n in the characteristic polynomial is 1 and the coefficient of λ^{n-1} is $-(a_{11} + a_{22} + \cdots + a_{nn})$, which is $-\text{trace}(A)$. If we substitute $\lambda = 0$ in the characteristic polynomial $\det(\lambda I_n - A)$, we conclude that **the constant term in the characteristic polynomial is $\det(-A) = (-1)^n \det(A)$.** We summarize the discussion in this paragraph by saying that the characteristic polynomial $c_A(\lambda)$ is given by

$$\boxed{\begin{aligned} c_A(\lambda) = \det(\lambda I_n - A) &= \lambda^n - \text{trace}(A)\lambda^{n-1} + \cdots + c_2\lambda^2 + c_1\lambda \\ &\quad + (-1)^n \det(A). \end{aligned}} \tag{4.4.1}$$

In factored form, we have

$$\boxed{c_A(\lambda) = (\lambda - \lambda_1)(\lambda - \lambda_2)\ldots(\lambda - \lambda_n),} \tag{4.4.2}$$

where some of the scalars $\lambda_1, \lambda_2, \ldots, \lambda_n$ may be repeated.

Theorem 4.4.3

If A is an $n \times n$ matrix with eigenvalues $\lambda_1, \lambda_2, \ldots, \lambda_n$, counting multiplicity (i.e., counting repetitions), then

(1) $\det(A) = \lambda_1 \lambda_2 \cdots \lambda_n$
(2) $\text{trace}(A) = \lambda_1 + \lambda_2 + \cdots + \lambda_n.$

Proof   (1) Equation 4.4.2 states that $c_A(\lambda) = \det(\lambda I_n - A) = (\lambda - \lambda_1)(\lambda - \lambda_2) \cdots (\lambda - \lambda_n)$. Therefore, letting $\lambda = 0$, we have $\det(-A) = (-\lambda_1)(-\lambda_2) \cdots (-\lambda_n)$, or $(-1)^n \det(A) = (-1)^n \lambda_1 \lambda_2 \cdots \lambda_n$, which implies $\det(A) = \lambda_1 \lambda_2 \cdots \lambda_n$.

(2) From a previous discussion, we know that the negative of the trace is the coefficient of λ^{n-1} in $c_A(\lambda)$. Now let us examine the factored form of the characteristic polynomial, $(\lambda - \lambda_1)(\lambda - \lambda_2) \cdots (\lambda - \lambda_n)$, to see how we can obtain terms that look like scalar

multiples of λ^{n-1}. When we multiply the factors, every term multiplies every other term. Thus we can take $-\lambda_1$ from the first set of parentheses and multiply it by a λ from each of the remaining $n-1$ sets of parentheses, giving us the term $-\lambda_1 \cdot \overbrace{\lambda \cdot \lambda, \ldots, \lambda}^{n-1 \text{ factors}} = -\lambda_1 \lambda^{n-1}$. Similarly, we can do this for each remaining λ_i, obtaining the terms $-\lambda_i \cdot \overbrace{\lambda \cdot \lambda, \ldots, \lambda}^{n-1 \text{ factors}} = -\lambda_i \lambda^{n-1}$, $i = 2, 3, \ldots, n$. Collecting all these like terms, we see that the total number of occurrences of λ^{n-1} is $-\lambda_1 - \lambda_2 - \cdots - \lambda_n = -(\lambda_1 + \lambda_2 + \cdots + \lambda_n)$, so trace$(A) = \lambda_1 + \lambda_2 + \cdots + \lambda_n$. This property can also be proved by mathematical induction.

Property (2) of Theorem 4.4.3 can be used as a quick check of our eigenvalue calculations: if the sum of our calculated eigenvalues does not equal the easily calculated trace of the matrix, we know that we have made a mistake.

Example 4.4.6

The matrix $B = \begin{bmatrix} 3 & 0 & 1 \\ 2 & 5 & 5 \\ 0 & 0 & 3 \end{bmatrix}$ from Example 4.4.1 has eigenvalues $\lambda_1 = 5, \lambda_2 = 3$, and $\lambda_3 = 3$. (The fact that 3 is a repeated root of the characteristic equation of B was not significant in the context of Example 4.4.1, but it *is* important here.) Clearly, trace$(B) = 11 = \lambda_1 + \lambda_2 + \lambda_3$.

Expanding by the third row of B, we see that

$$\det(B) = 3 \det \begin{bmatrix} 3 & 0 \\ 2 & 5 \end{bmatrix} = 45 = \lambda_1 \lambda_2 \lambda_3.$$

The eigenvectors $\mathbf{v}_1$ and $\mathbf{v}_2$ in Example 4.4.1 and in Example 4.4.3 are linearly independent vectors in $\mathbb{R}^2$. This is a consequence of the next theorem.

Theorem 4.4.4

Let A be an $n \times n$ matrix. Then any set of eigenvectors corresponding to distinct eigenvalues of A is linearly independent.

Proof We give a proof by contradiction. Suppose that $S = \{\mathbf{v}_1, \mathbf{v}_2, \ldots, \mathbf{v}_k\}$ is a set of eigenvectors corresponding to distinct eigenvalues $\lambda_1, \lambda_2, \ldots, \lambda_k$ and that S is linearly dependent. Let $\mathbf{v}_i$ be the first vector that can be written as a linear combination of the preceding ones:

$$\mathbf{v}_i = a_1 \mathbf{v}_1 + \cdots + a_{i-1} \mathbf{v}_{i-1}, \tag{*}$$

where $\mathbf{v}_1, \mathbf{v}_2, \ldots, \mathbf{v}_{i-1}$ are linearly independent and $a_1, \ldots, a_{i-1}$ are not all zero.

Then

$$A(\mathbf{v}_i) = A(a_1\mathbf{v}_1 + \cdots + a_{i-1}\mathbf{v}_{i-1}) = a_1(A\mathbf{v}_1) + \cdots + a_{i-1}(A\mathbf{v}_{i-1}),$$

or

$$\lambda_i\mathbf{v}_i = a_1\lambda_1\mathbf{v}_1 + \cdots + a_{i-1}\lambda_{i-1}\mathbf{v}_{i-1}. \qquad (**)$$

If we multiply $(*)$ by $-\lambda_i$ and add it to $(**)$, we get

$$\mathbf{0} = a_1(\lambda_1 - \lambda_i)\mathbf{v}_1 + \cdots + a_{i-1}(\lambda_{i-1} - \lambda_i)\mathbf{v}_{i-1}.$$

Because we are assuming $\mathbf{v}_1, \mathbf{v}_2, \ldots, \mathbf{v}_{i-1}$ are linearly independent, we see that

$$a_1(\lambda_1 - \lambda_i) = 0 = \cdots = a_{i-1}(\lambda_{i-1} - \lambda_i).$$

Furthermore, by assumption, (at least) one of the a's, say a_j, is nonzero, so $a_j(\lambda_j - \lambda_i) = 0$ implies $\lambda_j - \lambda_i = 0$, or $\lambda_j = \lambda_i$—a contradiction of the assumption that the λ's are distinct. Therefore, none of the eigenvectors can be written as a linear combination of the preceding eigenvectors, and so they are linearly independent.

As we will see shortly (Example 4.4.8), even if some eigenvalues are repeated we may still be able to find enough linearly independent eigenvectors to form a basis for $\mathbb{R}^n$.

Now suppose that A is a 2×2 matrix with linearly independent eigenvectors $\mathbf{v}_1$ and $\mathbf{v}_2$ in $\mathbb{R}^2$. Then these eigenvectors constitute a basis (proved at the end of Section 1.4) and any vector $\mathbf{x} \in \mathbb{R}^2$ can be written uniquely as $c_1\mathbf{v}_1 + c_2\mathbf{v}_2$ for some scalars c_1 and c_2. Now we can apply A repeatedly as follows:

$$A\mathbf{x} = A(c_1\mathbf{v}_1 + c_2\mathbf{v}_2) = c_1(A\mathbf{v}_1) + c_2(A\mathbf{v}_2) = c_1(\lambda_1\mathbf{v}_1) + c_2(\lambda_2\mathbf{v}_2)$$
$$= (c_1\lambda_1)\mathbf{v}_1 + (c_2\lambda_2)\mathbf{v}_2,$$
$$A^2\mathbf{x} = A(A\mathbf{x}) = A[(c_1\lambda_1)\mathbf{v}_1 + (c_2\lambda_2)\mathbf{v}_2] = (c_1\lambda_1)A\mathbf{v}_1 + (c_2\lambda_2)A\mathbf{v}_2$$
$$= (c_1\lambda_1^2)\mathbf{v}_1 + (c_2\lambda_2^2)\mathbf{v}_2,$$
$$\vdots$$
$$A^k\mathbf{x} = A(A^{k-1}\mathbf{x}) = A[(c_1\lambda_1^{k-1})\mathbf{v}_1 + (c_2\lambda_2^{k-1})\mathbf{v}_2]$$
$$= (c_1\lambda_1^{k-1})A\mathbf{v}_1 + (c_2\lambda_2^{k-1})A\mathbf{v}_2$$
$$= (c_1\lambda_1^k)\mathbf{v}_1 + (c_2\lambda_2^k)\mathbf{v}_2.$$

Therefore, we can see the effect of multiplying the vector $\mathbf{x}$ by positive integer powers of A *without calculating the powers of the matrix*. The next example shows a practical use of this technique, the solution of a *linear discrete dynamical system*.

Example 4.4.7: An Application of Eigenvalues and Eigenvectors

A car rental company has offices in Manhattan and the Bronx* (Figure 4.7). Relying on its records, the company knows that on a monthly basis 40% of rentals from the Manhattan office are returned there and 60% are one-way rentals that are dropped off in the Bronx. Similarly, 70% of rentals from the Bronx office are returned there, whereas 30% are dropped off in Manhattan.

Let m_k and b_k denote the number of cars at the depots in Manhattan and the Bronx, respectively, at the beginning of month k ($k=0,1,2\ldots$). One month later, the cars at the Manhattan location consist of those returned there during the previous month (i.e., 40% of m_k), together with those dropped off on a one-way rental from the Bronx office (i.e., 30% of b_k). Similarly, the cars at the Bronx location consist of those returned there during the previous month (i.e., 70% of b_k), together with those dropped off on a one-way rental from the Manhattan office (i.e., 60% of m_k). We can express this information in terms of *difference equations* (which are analogous to differential equations):

$$\left\{ \begin{array}{l} m_{k+1} = 0.4m_k + 0.3b_k \\ b_{k+1} = 0.6m_k + 0.7b_k \end{array} \right\} \quad k = 0, 1, 2, \ldots.$$

The matrix–vector form of this system is

$$\begin{bmatrix} m_{k+1} \\ b_{k+1} \end{bmatrix} = \begin{bmatrix} 0.4 & 0.3 \\ 0.6 & 0.7 \end{bmatrix} \begin{bmatrix} m_k \\ b_k \end{bmatrix} \quad k = 0, 1, 2 \ldots,$$

which we can write compactly as $\mathbf{X}_{k+1} = A\mathbf{X}_k$. We notice that

$$\mathbf{X}_1 = A\mathbf{X}_0,$$
$$\mathbf{X}_2 = A\mathbf{X}_1 = A^2\mathbf{X}_0$$
$$\mathbf{X}_3 = A\mathbf{X}_2 = A^3\mathbf{X}_0$$
$$\vdots \qquad \vdots$$
$$\mathbf{X}_k = A^k\mathbf{X}_0$$

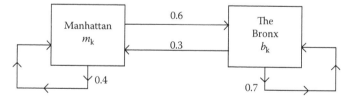

Figure 4.7 Percentage of one- and two-way rentals.

---

* We say "the" Bronx because the area was first settled by Jonas Bronck (1600–1643), a Swede in the service of the Dutch West India Company. People would speak of visiting *the Broncks*, who farmed this area for eight generations. This region became part of Greater New York in 1898.

With a little effort, we find the eigenpairs for A, $(\lambda_1, \mathbf{v}_1) = \left(1, \begin{bmatrix} 1 \\ 2 \end{bmatrix}\right)$, and $(\lambda_2, \mathbf{v}_2) = \left(0.1, \begin{bmatrix} 1 \\ -1 \end{bmatrix}\right)$. Noting that the eigenvectors form a basis for $\mathbb{R}^2$, we can write any initial vector $\mathbf{X}_0$ uniquely as $\mathbf{X}_0 = c_1 \mathbf{v}_1 + c_2 \mathbf{v}_2$. Using the result of the discussion preceding this example, we see that

$$A^k \mathbf{X}_0 = (c_1 \lambda_1^k)\mathbf{v}_1 + (c_2 \lambda_2^k)\mathbf{v}_2 = c_1 (1)^k \begin{bmatrix} 1 \\ 2 \end{bmatrix} + c_2 (0.1)^k \begin{bmatrix} 1 \\ -1 \end{bmatrix}$$

$$= \begin{bmatrix} c_1 + c_2 (0.1)^k \\ 2c_1 - c_2 (0.1)^k \end{bmatrix},$$

so $m_{k+1} = c_1 + c_2 (0.1)^k$ and $b_{k+1} = 2c_1 - c_2 (0.1)^k$.

Notice that as time passes (mathematically, as $k \to \infty$), $m_k \to c_1$, and $b_k \to 2c_1$, which we can write as $\begin{bmatrix} m_\infty \\ b_\infty \end{bmatrix} = \begin{bmatrix} c_1 \\ 2c_1 \end{bmatrix}$. Thus, over time, the number of cars in the Manhattan depot tends to a value that's half the number of cars at the Bronx office.

Theorem 4.4.4 assumes that we have distinct eigenvalues and draws a conclusion about the corresponding eigenvectors. The next result starts with an eigenvector and states something about the associated eigenvalue(s).

Theorem 4.4.5

The eigenvalue associated with an eigenvector is uniquely determined.

Proof We give a proof by contradiction. Suppose that $\mathbf{v}$ is an eigenvector of A and we have $A\mathbf{v} = \lambda\mathbf{v}$ and $A\mathbf{v} = \mu\mathbf{v}$ for distinct scalars λ and μ. Then $\mathbf{0} = A\mathbf{v} - A\mathbf{v} = \lambda\mathbf{v} - \mu\mathbf{v} = (\lambda - \mu)\mathbf{v}$. But $\mathbf{v} \neq \mathbf{0}$ implies that $\lambda - \mu = 0$—that is, $\lambda = \mu$.

Although an eigenvector can have only one eigenvalue associated with it, an eigenvalue may have more than one corresponding eigenvector.

Example 4.4.8: An Eigenvalue with More Than One Eigenvector

The matrix $C = \begin{bmatrix} 4 & 1 & -1 \\ 2 & 5 & -2 \\ 1 & 1 & 2 \end{bmatrix}$ has characteristic polynomial $\lambda^3 - 11\lambda^2 - 39\lambda - 45 = (\lambda - 3)^2 (\lambda - 5)$, so the eigenvalues are 3, 3, and 5. The eigenvalue $\lambda = 3$ has two linearly independent eigenvectors, $\begin{bmatrix} 1 \\ -1 \\ 0 \end{bmatrix}$ and $\begin{bmatrix} 1 \\ 0 \\ 1 \end{bmatrix}$. (The eigenvalue 5 has only one independent eigenvector, $\begin{bmatrix} 1 \\ 2 \\ 1 \end{bmatrix}$.) Also see

Example 4.4.3, where each eigenvalue gives rise to an infinite family of eigenvectors.

If $\mathbf{v}$ is an eigenvector of A with associated eigenvalue λ, then any scalar multiple of $\mathbf{v}$ is also an eigenvector of A with associated eigenvalue λ:

$$A(k\mathbf{v}) = k(A\mathbf{v}) = k(\lambda\mathbf{v}) = \lambda(k\mathbf{v}) \quad \text{for any scalar } k.$$

The following more general result holds true.

Theorem 4.4.6

If $\mathbf{v}_1, \mathbf{v}_2 \in \mathbb{R}^n$ are eigenvectors with the same eigenvalue λ, then for every pair of scalars c_1, c_2 the vector $c_1\mathbf{v}_1 + c_2\mathbf{v}_2$, *if it is nonzero*, is also an eigenvector with eigenvalue λ.

Proof $A(c_1\mathbf{v}_1 + c_2\mathbf{v}_2) = c_1(A\mathbf{v}_1) + c_2(A\mathbf{v}_2) = c_1(\lambda\mathbf{v}_1) + c_2(\lambda\mathbf{v}_2) = \lambda(c_1\mathbf{v}_1 + c_2\mathbf{v}_2)$.

An important consequence of the last theorem is that **given an $n \times n$ matrix A and a scalar λ, the set of all eigenvectors of A with associated eigenvalue λ, together with the zero vector, is a *subspace* of $\mathbb{R}^n$ called the *eigenspace of A associated with* (or *corresponding to*) λ.** We will denote this subspace by $S_\lambda(A)$.

Example 4.4.9: Some Eigenspaces

Let us consider the diagonal matrix $D = \begin{bmatrix} 1 & 0 & 0 \\ 0 & 2 & 0 \\ 0 & 0 & 3 \end{bmatrix}$. The

representative eigenvectors for D are $\mathbf{e}_1 = \begin{bmatrix} 1 \\ 0 \\ 0 \end{bmatrix}, \mathbf{e}_2 = \begin{bmatrix} 0 \\ 1 \\ 0 \end{bmatrix}$,

and $\mathbf{e}_3 = \begin{bmatrix} 0 \\ 0 \\ 1 \end{bmatrix}$, with associated eigenvalues $\lambda_1 = 1, \lambda_2 = 2$,

and $\lambda_3 = 3$, respectively. Thus, D has three eigenspaces:

$$S_1(D) = \left\{ c\begin{bmatrix} 1 \\ 0 \\ 0 \end{bmatrix} : c \in \mathbb{R} \right\} = \left\{ \begin{bmatrix} c \\ 0 \\ 0 \end{bmatrix} : c \in \mathbb{R} \right\}$$

$$S_2(D) = \left\{ k\begin{bmatrix} 0 \\ 1 \\ 0 \end{bmatrix} : k \in \mathbb{R} \right\} = \left\{ \begin{bmatrix} 0 \\ k \\ 0 \end{bmatrix} : k \in \mathbb{R} \right\}$$

$$S_3(D) = \left\{ m\begin{bmatrix} 0 \\ 0 \\ 1 \end{bmatrix} : m \in \mathbb{R} \right\} = \left\{ \begin{bmatrix} 0 \\ 0 \\ m \end{bmatrix} : m \in \mathbb{R} \right\}$$

Each eigenspace is a straight line through the origin in $\mathbb{R}^3$.

As a comment on Theorem 4.4.6, we note that the sum of two eigenvectors with *different* eigenvalues is *not* an eigenvector.

In this example, take $\mathbf{v} = \mathbf{e}_1 + \mathbf{e}_2 = \begin{bmatrix} 1 \\ 1 \\ 0 \end{bmatrix}$. Then

$$D\mathbf{v} = \begin{bmatrix} 1 & 0 & 0 \\ 0 & 2 & 0 \\ 0 & 0 & 3 \end{bmatrix} \begin{bmatrix} 1 \\ 1 \\ 0 \end{bmatrix} = \begin{bmatrix} 1 \\ 2 \\ 0 \end{bmatrix}, \text{ which is not a scalar multiple}$$

of $\mathbf{v}$ (see Exercise B22).

This last example illustrates a general principle that enables us to determine eigenvalues easily in some important cases.

Theorem 4.4.7

The eigenvalues of a triangular matrix are the entries on its main diagonal.

Proof If A is a triangular matrix, then so is $\lambda I_n - A$. Therefore, $\det(\lambda I_n - A)$ equals the product of the diagonal elements of $\lambda I_n - A$—that is,

$$\det(\lambda I_n - A) = (\lambda - a_{11})(\lambda - a_{22}) \cdots (\lambda - a_{nn}).$$

(see Section 4.1). Clearly $\det(\lambda I_n - A) = 0$ implies that the eigenvalues of A are the scalars $a_{11}, a_{22}, \dots, a_{nn}$.

Exercises 4.4

A.

1. For each of the following matrices, determine which of the given vectors are eigenvectors. For each eigenvector, find the corresponding eigenvalue.

a. $\begin{bmatrix} 3 & 1 \\ 2 & 2 \end{bmatrix}$; $\mathbf{x} = \begin{bmatrix} 1 \\ 2 \end{bmatrix}$, $\mathbf{y} = \begin{bmatrix} -1 \\ 2 \end{bmatrix}$, $\mathbf{u} = \begin{bmatrix} 1 \\ 1 \end{bmatrix}$, $\mathbf{v} = \begin{bmatrix} 1 \\ -1 \end{bmatrix}$.

b. $\begin{bmatrix} 6 & 0 \\ 1 & 9 \end{bmatrix}$; $\mathbf{x} = \begin{bmatrix} -3 \\ 1 \end{bmatrix}$, $\mathbf{y} = \begin{bmatrix} 0 \\ 1 \end{bmatrix}$, $\mathbf{u} = \begin{bmatrix} 0 \\ 0 \end{bmatrix}$, $\mathbf{v} = \begin{bmatrix} -6 \\ 2 \end{bmatrix}$.

c. $\begin{bmatrix} -3 & -4 & 7 \\ 12 & 8 & -20 \\ 3 & -4 & 1 \end{bmatrix}$; $\mathbf{x} = \begin{bmatrix} -1 \\ 8 \\ 5 \end{bmatrix}$, $\mathbf{y} = \begin{bmatrix} 11 \\ 3 \\ 3 \end{bmatrix}$, $\mathbf{u} = \begin{bmatrix} 1 \\ -2 \\ 1 \end{bmatrix}$, $\mathbf{v} = \begin{bmatrix} 1 \\ -8 \\ -5 \end{bmatrix}$.

d. $\begin{bmatrix} -3 & 1 & -3 \\ 20 & 3 & 10 \\ 2 & -2 & 4 \end{bmatrix}$; $\mathbf{x} = \begin{bmatrix} 2 \\ -1 \\ 4 \end{bmatrix}$, $\mathbf{y} = \begin{bmatrix} -1 \\ 0 \\ 2 \end{bmatrix}$, $\mathbf{u} = \begin{bmatrix} 0 \\ -3 \\ 2 \end{bmatrix}$,

$\mathbf{v} = \begin{bmatrix} -1 \\ 2 \\ 1 \end{bmatrix}$.

e. $\begin{bmatrix} -3 & 6 & 6 & 4 \\ 0 & 0 & 0 & -2 \\ 0 & 0 & 0 & 2 \\ 0 & 0 & 0 & 3 \end{bmatrix}$; $\mathbf{x} = \begin{bmatrix} 1 \\ -2 \\ 2 \\ 3 \end{bmatrix}$, $\mathbf{y} = \begin{bmatrix} 2 \\ 0 \\ 1 \\ 0 \end{bmatrix}$, $\mathbf{u} = \begin{bmatrix} 2 \\ -2 \\ 2 \\ 3 \end{bmatrix}$,

$\mathbf{v} = \begin{bmatrix} 2 \\ 1 \\ 0 \\ 0 \end{bmatrix}$.

2. For each of the following matrices, all eigenvalues are given. Find the corresponding eigenvectors.

a. $\begin{bmatrix} 4 & -2 \\ 1 & 1 \end{bmatrix}$; $\lambda_1 = 2, \lambda_2 = 3$.

b. $\begin{bmatrix} 1 & -8 \\ 1 & -5 \end{bmatrix}$; $\lambda_1 = -3, \lambda_2 = -1$.

c. $\begin{bmatrix} 1 & 3 & -2 \\ 3 & 1 & 2 \\ -4 & 4 & -2 \end{bmatrix}$; $\lambda_1 = -6, \lambda_2 = 2, \lambda_3 = 4$.

d. $\begin{bmatrix} -3 & 3 & 1 \\ 3 & -3 & -1 \\ 2 & -2 & -6 \end{bmatrix}$; $\lambda_1 = -8, \lambda_2 = -4, \lambda_3 = 0$.

3. Find all eigenvalues and corresponding eigenvectors for each of the following matrices:

a. $\begin{bmatrix} 2 & 3 \\ 3 & 2 \end{bmatrix}$ b. $\begin{bmatrix} 0 & 1 \\ 1 & 0 \end{bmatrix}$ c. $\begin{bmatrix} 4 & -2 \\ 3 & -1 \end{bmatrix}$

d. $\begin{bmatrix} 1 & 3 & -2 \\ 3 & 1 & 2 \\ -4 & 4 & -2 \end{bmatrix}$ e. $\begin{bmatrix} -3 & 3 & 1 \\ 3 & -3 & -1 \\ 2 & -2 & -6 \end{bmatrix}$

f. $\begin{bmatrix} -1 & -1 & 1 \\ 2 & -4 & 2 \\ -1 & 1 & -3 \end{bmatrix}$ g. $\begin{bmatrix} 3 & 2 & 2 \\ 1 & 2 & 2 \\ -1 & -1 & 0 \end{bmatrix}$

h. $\begin{bmatrix} 1 & 3 & 2 & -2 \\ 0 & 2 & 3 & -4 \\ 0 & 0 & 3 & -1 \\ 0 & 0 & 0 & 4 \end{bmatrix}$ i. $\begin{bmatrix} 1 & 2 & 0 & 0 & 0 \\ 0 & 2 & 0 & 0 & 0 \\ 0 & 0 & 3 & 2 & 0 \\ 0 & 0 & 0 & 1 & 0 \\ 0 & 0 & 0 & 0 & 5 \end{bmatrix}$.

4. For each matrix A described in the following, find all eigenvalues and their corresponding eigenspaces.

a. $A\begin{bmatrix} x \\ y \end{bmatrix} = \begin{bmatrix} x+y \\ y \end{bmatrix}$ b. $A\begin{bmatrix} x \\ y \end{bmatrix} = \begin{bmatrix} x+y \\ x-y \end{bmatrix}$

5. Use Example 4.4.7 to answer the following questions:
    a. If $m_0 = 20$ and $b_0 = 10$, what are the values of m_k and b_k for $k = 1, 2, 3, 4, 5$? [Round to the nearest integer.]
    b. If $m_0 = 10$ and $b_0 = 20$, what are the values of m_k and b_k for $k = 1, 2, 3, 4, 5$? [Round to the nearest integer.]

6. Suppose that there is some biological organism that reproduces itself every month, and suppose that the new offspring must wait 2 months before it begins to reproduce. If none of the organisms dies, this situation can be represented by the difference equation $x_n = x_{n-1} + x_{n-2}$ $(n = 2, 3 \ldots)$, where x_n represents the number of organisms after n months. Assume that $x_0 = 1$ and $x_1 = 1$. We wish to find a general expression for x_n.*
    a. Letting $y_n = x_{n-1}$, $n = 1, 2 \ldots$, obtain the system:

    $$\begin{aligned} x_n &= x_{n-1} + x_{n-2} \\ y_n &= x_{n-1} \end{aligned} \quad \text{or} \quad \begin{bmatrix} x_n \\ y_n \end{bmatrix} = \begin{bmatrix} 1 & 1 \\ 1 & 0 \end{bmatrix} \begin{bmatrix} x_{n-1} \\ y_{n-1} \end{bmatrix}.$$

    Show that

    $$\begin{bmatrix} x_n \\ y_n \end{bmatrix} = \begin{bmatrix} 1 & 1 \\ 1 & 0 \end{bmatrix}^{n-1} \begin{bmatrix} x_1 \\ y_1 \end{bmatrix}$$

    for $n = 2, 3 \ldots$.
    b. Find the eigenvalues of $A = \begin{bmatrix} 1 & 1 \\ 1 & 0 \end{bmatrix}$.
    c. Find the eigenvectors corresponding to the eigenvalues found in part (b).
    d. As in Example 4.4.7, use the eigenvalues and eigenvectors of A to find a formula for x_n.

7. Solve the system of difference equations:

    $$\begin{aligned} x_{n+1} &= x_n + 2y_n \\ y_{n+1} &= 2x_n + y_n \end{aligned} \quad \text{where} \quad \begin{aligned} x_1 &= 0 \\ y_1 &= 1 \end{aligned}.$$

8. Prove that $\det(A - \lambda I_n) = 0$ if and only if $\det(\lambda I_n - A) = 0$.

9. Let $A = \begin{bmatrix} 51 & -12 & -21 \\ 60 & -40 & -28 \\ 57 & -68 & 1 \end{bmatrix}$. Suppose someone tells you (correctly) that -48 and 24 are eigenvalues of A. Without using a calculator or computer—and without writing anything down—find the third eigenvalue of A.†

---
* The terms of this sequence—1, 1, 2, 3, 5, 8, 13...—are known as **Fibonacci numbers** (named for Leonardo of Pisa (ca. 1170–1240), also called Fibonacci) and their study began in the thirteenth century.
† This exercise is found in S. Axler, *Linear Algebra Done Right*, 2nd edn., Chap. 10 (New York: Springer, 1997).

B.

1. Suppose that $A = \begin{bmatrix} a & b \\ c & d \end{bmatrix}$, where a, b, c, and d are positive.
    Prove that
    a. The eigenvalues of A are real and distinct.
    b. A has at least one positive eigenvalue.
    c. Corresponding to the larger eigenvalue of A there are infinitely many eigenvectors with both components positive.

2. Show that the slopes m of the eigenvectors of $A = \begin{bmatrix} a & b \\ c & d \end{bmatrix}$
    satisfy the quadratic equation $cm^2 + (a - d)m - b = 0$.

3. Show that if A is a 2×2 matrix such that $A^2 = I_2$, and if $\mathbf{x}$ is any vector in $\mathbb{R}^2$, then $\mathbf{y} = \mathbf{x} + A\mathbf{x}$ and $\mathbf{z} = \mathbf{x} - A\mathbf{x}$ are eigenvectors of A. Find their corresponding eigenvalues.

4. If c is a scalar, find a relationship between the eigenvalues and eigenvectors of a matrix A and those of each of the matrices:
    a. cA b. $A + cI$.

5. Prove that if $\mathbf{v}$ is an eigenvector of a matrix A, then it is also an eigenvector of the matrix A^k for every positive integer k. How are the associated eigenvalues related? [*Hint*: Use mathematical induction.]

6. Suppose that λ is an eigenvalue of a matrix A. Show that $p(\lambda)$ is an eigenvalue of $p(A)$ for any polynomial p. [*Hint*: Use the answer to the previous exercise.]

7. Prove that if $\mathbf{v}$ is an eigenvector of an invertible matrix A and its associated eigenvalue is not zero, then it is also an eigenvector of the matrix A^{-1}. How are the associated eigenvalues related?

8. Suppose that A and S are $n \times n$ matrices with S nonsingular. Prove that if the first column of the matrix $S^{-1}AS$ is $[\lambda \quad 0 \quad 0 \quad \cdots \quad 0]^T$, then the first column of S is an eigenvector of A corresponding to the eigenvalue λ.

9. Prove that the (representative) eigenvectors of an $n \times n$ diagonal matrix can be the standard basis vectors $\mathbf{e}_1, \mathbf{e}_2, \ldots, \mathbf{e}_n$.

10. Prove that the characteristic polynomial of A^T is the same as the characteristic polynomial of A.

11.   Suppose that A and B are $n \times n$ matrices. Prove the following properties:

    a. trace(kA) = k trace(A) for any scalar k.

    b. trace($A \pm B$) = trace(A) $\pm$ trace(B).

    c. trace($\alpha A + \beta B$) = α trace(A) + β trace(B), where α and β are scalars.

    d. trace(AB) = trace(BA).

    e. trace(TAT^{-1}) = trace(A), where T is any $n \times n$ nonsingular matrix.

12.   If A and B are $n \times n$ matrices and $A^2 = A$, show that trace(AB) = trace(ABA). (You may use a property from Exercise 11.)

13.   Prove that if $n \geq 2$, there are no $n \times n$ matrices A and B such that $AB - BA = I_n$. [*Hint*: Use a property from Exercise 11.]

14.   Is it true that if trace(A) = 0, then det(A) = 0? If so, prove it. If not, give a counterexample.

15.   If n eigenvalues $\lambda_1, \lambda_2, \ldots, \lambda_n$ of an $(n+1) \times (n+1)$ matrix A are given, how can you find another eigenvalue?

16.   a. Prove that if A is nilpotent of index k for some integer $k \geq 2$, then the only eigenvalue of A is 0 [see Definition 3.3.3].

    b. Prove that if A is nilpotent of index k for some integer $k \geq 2$, then trace(A) = 0.

17.   If the sum of entries in each row of an $n \times n$ matrix A is α, show that α is an eigenvalue of A. What is the corresponding eigenvector?

18.   Using just the basic relation $A\mathbf{x} = \lambda\mathbf{x}$ rather than the characteristic polynomial, find all eigenvalues of the matrix:

$$A = \begin{bmatrix} 2 & 5 & 5 & 5 \\ 5 & 2 & 5 & 5 \\ 5 & 5 & 2 & 5 \\ 5 & 5 & 5 & 2 \end{bmatrix}.$$

Explain why you know you have found *all* the eigenvalues. [*Hint*: The result of Exercise B17 may help.]

19.   A *semimagic square* is an $n \times n$ matrix with real entries in which every row and every column has the same sum s (see Exercise 3.3, B23). Prove that an $n \times n$ matrix A is a semimagic square if and only if the vector $\mathbf{v} = [1 \quad 1 \quad \ldots \quad 1]^T$ consisting of n 1s as entries is an eigenvector of both A and A^T with the same eigenvalue.

20. Let A be an $n \times n$ matrix and E a row echelon form of A. Are the eigenvalues of A necessarily on the main diagonal of E? If not, find an example.

21. Let A and B be $n \times n$ matrices, each with n distinct eigenvalues. Prove that A and B have the same eigenvectors if and only if $AB = BA$.

22. Suppose that $(\lambda_1, \mathbf{v}_1), (\lambda_2, \mathbf{v}_2), \ldots, (\lambda_k, \mathbf{v}_k)$ are eigenpairs of A with $\lambda_i \neq \lambda_j$ for $i \neq j$. Show that $a_1 \mathbf{v}_1 + a_2 \mathbf{v}_2 + \cdots + a_k \mathbf{v}_k$ is never an eigenvector of A.

23. Suppose that A is an $n \times n$ matrix and that $S_\lambda(A)$ is the eigenspace of A associated with the scalar λ. Show that $S_\lambda(A)$ is an **invariant subspace** of $\mathbb{R}^n$ under A—that is, show that $A\mathbf{x} \in S_\lambda(A)$ for every $\mathbf{x} \in S_\lambda(A)$.

C.

1. Find the eigenvalues and eigenvectors of the $n \times n$ matrix:

$$M_n = \begin{bmatrix} 1 & 1 & \cdots & 1 \\ 1 & 1 & \cdots & 1 \\ \vdots & \vdots & \cdots & \vdots \\ 1 & 1 & \cdots & 1 \end{bmatrix}.$$

2. If A and B are $n \times n$ matrices, prove that AB and BA have the same characteristic polynomial.

3. Let A be an $m \times n$ matrix and B an $n \times m$ matrix. Prove that the $m \times m$ matrix AB and the $n \times n$ matrix BA have the same nonzero eigenvalues.

$$\left[Hint: \begin{bmatrix} AB & O \\ B & O \end{bmatrix} \begin{bmatrix} I & A \\ O & I \end{bmatrix} = \begin{bmatrix} I & A \\ O & I \end{bmatrix} \begin{bmatrix} O & O \\ B & BA \end{bmatrix}. \text{ Also note} \right.$$

$$\left. \text{that } \begin{bmatrix} I & A \\ O & I \end{bmatrix} \text{ has the inverse } \begin{bmatrix} I & -A \\ O & I \end{bmatrix}. \right]$$

4. Given the partitioned matrix:

$$M = \begin{bmatrix} A & O \\ B & C \end{bmatrix},$$

where A and C are square matrices, and the fact that $\det(M) = \det A \det C$ (see Problem C3, Exercise 4.3, and use the transpose), how can you find the eigenvalues of M in terms of A and C?

5. Show that if the square matrix A can be partitioned as

$$A = \begin{bmatrix} P & Q \\ O & S \end{bmatrix},$$

where P and S are square matrices, then the characteristic polynomial of A, $c_A(\lambda)$, is given by $c_A(\lambda) = c_P(\lambda) c_Q(\lambda)$.

4.5 Similar Matrices and Diagonalization

Diagonal matrices are important in the theory of linear algebra, are particularly easy to work with from the point of view of matrix algebra, and often appear in applications. The identity matrix I_n is a diagonal matrix for any positive integer n. If $D = \text{diag}\,(d_1, d_2, \ldots, d_m) = [d_{ij}]$, where $d_{ij} = 0$ if $i \neq j$, is an $m \times m$ diagonal matrix and $A = [a_{ij}]$ is any $m \times n$ matrix, then $C = DA$ is the $m \times n$ matrix whose kth row is $d_{kk} \cdot (\text{row } k \text{ of } A)$, $k = 1, 2, \ldots, m$: for any $k, 1 \leq k \leq m$, $c_{kj} = \sum_{i=1}^{m} d_{ki} a_{ij} = d_{kk} a_{kj}, j = 1, 2, \ldots, n$. Also, if A is $m \times n$ and D is $n \times n$, then AD is the $m \times n$ matrix whose kth column is $d_{kk}(\text{column } k \text{ of } A)$, $k = 1, 2, \ldots, n$: $AD = A[\,d_1\mathbf{e}_1 \quad d_2\mathbf{e}_2 \quad \cdots \quad d_n\mathbf{e}_n\,] = [\,d_1(A\mathbf{e}_1) \quad d_2(A\mathbf{e}_2) \quad \cdots \quad d_n(A\mathbf{e}_n)\,] = [\,d_1 \text{col}_1(A) \quad d_2 \text{col}_2(A) \quad \cdots \quad d_n \text{col}_n(A)\,]$.

Furthermore, *provided that no diagonal element of D is zero*, the inverse of D is just the diagonal matrix whose elements are the reciprocals of the diagonal elements of D:

$$D = \begin{bmatrix} d_{11} & 0 & 0 & \cdots & 0 \\ 0 & d_{22} & 0 & \cdots & 0 \\ 0 & 0 & d_{33} & \cdots & 0 \\ \vdots & \vdots & \vdots & \vdots & \vdots \\ 0 & 0 & 0 & \cdots & d_{nn} \end{bmatrix} \Leftrightarrow D^{-1}$$

$$= \begin{bmatrix} 1/d_{11} & 0 & 0 & \cdots & 0 \\ 0 & 1/d_{22} & 0 & \cdots & 0 \\ 0 & 0 & 1/d_{33} & \cdots & 0 \\ \vdots & \vdots & \vdots & \vdots & \vdots \\ 0 & 0 & 0 & \cdots & 1/d_{nn} \end{bmatrix}.$$

In applications such as Example 4.4.7 that require us to calculate positive or negative integer powers of a matrix A, computations are much easier when A is a diagonal matrix:

$$A = \begin{bmatrix} a_{11} & 0 & \cdots & 0 \\ 0 & a_{22} & \cdots & 0 \\ \vdots & \vdots & \vdots & \vdots \\ 0 & 0 & \cdots & a_{nn} \end{bmatrix} \Rightarrow A^k = \begin{bmatrix} a_{11}^k & 0 & \cdots & 0 \\ 0 & a_{22}^k & \cdots & 0 \\ \vdots & \vdots & \vdots & \vdots \\ 0 & 0 & \cdots & a_{nn}^k \end{bmatrix}.$$

Finally, any diagonal matrix D is a triangular matrix (both upper and lower at the same time), and thus the determinant of D is the product of its diagonal entries. Theorem 4.4.7 tells us that the eigenvalues of D are just the diagonal entries of D.

Example 4.5.1: Calculations with a Diagonal Matrix

Let $\quad D = \begin{bmatrix} -1 & 0 & 0 \\ 0 & 2 & 0 \\ 0 & 0 & -3 \end{bmatrix}$, $A = \begin{bmatrix} 1 & 2 & 3 & 4 \\ 5 & 6 & 7 & 8 \\ 9 & 10 & 11 & 12 \end{bmatrix}$, and

$\hat{D} = \begin{bmatrix} 2 & 0 & 0 & 0 \\ 0 & -1 & 0 & 0 \\ 0 & 0 & 3 & 0 \\ 0 & 0 & 0 & 4 \end{bmatrix}$. Then $D^{-1} = \begin{bmatrix} -1 & 0 & 0 \\ 0 & 1/2 & 0 \\ 0 & 0 & -1/3 \end{bmatrix}$,

$$D^k = \begin{bmatrix} (-1)^k & 0 & 0 \\ 0 & 2^k & 0 \\ 0 & 0 & (-3)^k \end{bmatrix}, \det(D) = (-1)(2)(-3) = 6, \text{ and}$$

the eigenvalues of D are -1, 2, and -3, with corresponding

eigenvectors $\begin{bmatrix} 1 \\ 0 \\ 0 \end{bmatrix}, \begin{bmatrix} 0 \\ 1 \\ 0 \end{bmatrix}$, and $\begin{bmatrix} 0 \\ 0 \\ 1 \end{bmatrix}$. Furthermore,

$$DA = \begin{bmatrix} -1 & -2 & -3 & -4 \\ 10 & 12 & 14 & 16 \\ -27 & -30 & -33 & -36 \end{bmatrix}, \text{ the result of multiplying}$$

row k of A by d_{kk}; and $A\hat{D} = \begin{bmatrix} 2 & -2 & 9 & 16 \\ 10 & -6 & 21 & 32 \\ 18 & -10 & 33 & 48 \end{bmatrix}$, the result

of multiplying column j of A by entry $\hat{d}_{jj}$ of $\hat{D}$ ($k = 1, 2, 3$; $j = 1, 2, 3, 4$).

Shortly, we are going to show that under certain conditions, a matrix A may share important properties of a related diagonal matrix. An example gives the basic idea.

Example 4.5.2: "Diagonalizing" a Matrix

Let $A = \begin{bmatrix} 1 & 1 \\ 2 & 0 \end{bmatrix}$. Example 4.4.3 showed that A has the

eigenpairs $(\lambda_1, \mathbf{v}_1) = \left(-1, \begin{bmatrix} 1 \\ -2 \end{bmatrix}\right)$ and $(\lambda_2, \mathbf{v}_2) = \left(2, \begin{bmatrix} 1 \\ 1 \end{bmatrix}\right)$.

We know (by Theorem 4.4.4) that the eigenvectors $\mathbf{v}_1$ and $\mathbf{v}_2$ are linearly independent. Now form the matrices $P = [\mathbf{v}_1 \ \ \mathbf{v}_2] = \begin{bmatrix} 1 & 1 \\ -2 & 1 \end{bmatrix}$ and $D = \begin{bmatrix} \lambda_1 & 0 \\ 0 & \lambda_2 \end{bmatrix} = \begin{bmatrix} -1 & 0 \\ 0 & 2 \end{bmatrix}$.

Then, for $j = 1$, 2, column j of AP is just $A\operatorname{col}_j(P) = A\mathbf{v}_j = \lambda_j \mathbf{v}_j$ and column j of PD is $P\operatorname{col}_j(D) = \lambda_j \operatorname{col}_j(P) = \lambda_j \mathbf{v}_j$. Thus, $AP = PD$, or $A = PDP^{-1}$, where we know P has an inverse because $\det(P) \neq 0$.

Alternatively, we can write $P^{-1}AP = D$ and say that we have **diagonalized** A.

The result of the manipulative magic demonstrated in this last example is important enough to justify special terms.

Definition 4.5.1

Two $n \times n$ matrices A and B are **similar** if there exists an invertible matrix P such that $A = PBP^{-1}$. (If we let $P^{-1} = Q$, we can write this condition in the form $A = Q^{-1}BQ$.) We also say "A is similar to B" and write $A \sim B$.

The symbol "$\sim$" was introduced in Section 2.2 in the context of row/column equivalence and is used in general to denote *equivalence relations*. Similarity of matrices is such a relation (Theorem 4.5.1). From now on we will use this symbol to denote only the similarity of matrices as described in Definition 4.5.1.

Example 4.5.3: Similar Matrices

The matrices $A = \begin{bmatrix} 0 & 0 & 2 \\ 1 & 1 & 0 \\ 1 & 1 & 0 \end{bmatrix}$ and $B = \begin{bmatrix} 1 & 0 & 1 \\ 1 & 0 & 0 \\ 2 & 0 & 0 \end{bmatrix}$ are simi-

lar because $A = \begin{bmatrix} 1 & -1 & 0 \\ 0 & 1 & 0 \\ 0 & 0 & \frac{1}{2} \end{bmatrix} \begin{bmatrix} 1 & 0 & 1 \\ 1 & 0 & 0 \\ 2 & 0 & 0 \end{bmatrix} \begin{bmatrix} 1 & 1 & 0 \\ 0 & 1 & 0 \\ 0 & 0 & 2 \end{bmatrix} =$

$\begin{bmatrix} 0 & 0 & 2 \\ 1 & 1 & 0 \\ 1 & 1 & 0 \end{bmatrix}$ —that is, $A = PBP^{-1}$, where $P = \begin{bmatrix} 1 & -1 & 0 \\ 0 & 1 & 0 \\ 0 & 0 & \frac{1}{2} \end{bmatrix}$

and $P^{-1} = \begin{bmatrix} 1 & 1 & 0 \\ 0 & 1 & 0 \\ 0 & 0 & 2 \end{bmatrix}$, as can be verified easily.

The determination of the invertible matrix P in the last example is not a feat of mathematical guesswork. Although we will explore the connection between similar matrices in greater depth in Chapter 6, we can give some insight now, using the idea of *change of basis* described in Section 1.5.

If we use the standard basis $\mathcal{E}_3 = \{\mathbf{e}_1, \mathbf{e}_2, \mathbf{e}_3\}$ for $\mathbb{R}^3$, then the coordinates of a vector $\mathbf{x} = \begin{bmatrix} x \\ y \\ z \end{bmatrix}$ with respect to the basis $\mathcal{E}_3$ can be denoted by $[\mathbf{x}]_{\mathcal{E}_3}$. In Example 4.5.3, if we use the basis

$\mathcal{B} = \left\{ \mathbf{v}_1 = \begin{bmatrix} 1 \\ 0 \\ 0 \end{bmatrix}, \mathbf{v}_2 = \begin{bmatrix} 1 \\ 1 \\ 0 \end{bmatrix}, \mathbf{v}_3 = \begin{bmatrix} 0 \\ 0 \\ 2 \end{bmatrix} \right\}$—the column vectors of

P^{-1}—we can find the coordinates of $\begin{bmatrix} x \\ y \\ z \end{bmatrix}$ with respect to this new ordered basis:

$$\begin{bmatrix} x \\ y \\ z \end{bmatrix} = (x - y)\mathbf{v}_1 + y\,\mathbf{v}_2 + \frac{z}{2}\mathbf{v}_3,$$

so $[\mathbf{x}]_{\mathcal{B}} = \begin{bmatrix} x - y \\ y \\ z/2 \end{bmatrix}$.

Notice that $P \begin{bmatrix} x \\ y \\ z \end{bmatrix} = \begin{bmatrix} 1 & -1 & 0 \\ 0 & 1 & 0 \\ 0 & 0 & 1/2 \end{bmatrix} \begin{bmatrix} x \\ y \\ z \end{bmatrix} = \begin{bmatrix} x - y \\ y \\ z/2 \end{bmatrix}$. In other

words, premultiplying by P converts from coordinates with respect to the standard basis $\mathcal{E}_3$ to coordinates with respect to the basis $\mathcal{B}$. We can call P a **change-of-basis matrix**, a **transition matrix**, or a **change-of-coordinates matrix**: $P[\mathbf{x}]_{\mathcal{E}_3} = [\mathbf{x}]_{\mathcal{B}}$. Premultiplying by P^{-1} changes the basis from $\mathcal{B}$ back to $\mathcal{E}_3$:

$$P^{-1}\begin{bmatrix} x - y \\ y \\ z/2 \end{bmatrix} = \begin{bmatrix} 1 & 1 & 0 \\ 0 & 1 & 0 \\ 0 & 0 & 2 \end{bmatrix}\begin{bmatrix} x - y \\ y \\ z/2 \end{bmatrix} = \begin{bmatrix} x \\ y \\ z \end{bmatrix}.$$

In Example 4.5.2, we used linearly independent eigenvectors of A as the columns of matrix P.

Theorem 4.5.1

If A, B, and C are $n \times n$ matrices, then

(1) $A \sim A$.
(2) $A \sim B$ implies $B \sim A$.
(3) $A \sim B$ and $B \sim C$ imply that $A \sim C$.

The properties stated in Theorem 4.5.1 indicate that similarity is an **equivalence relation**. The proof of Theorem 4.5.1 is left as Exercise B6. (See Problem C1 of Exercise 3.6 for a similar exercise—no pun intended.)

Definition 4.5.2

A square matrix A is **diagonalizable** if it is similar to a diagonal matrix D–that is, if there is an invertible matrix P such that $P^{-1}AP = D$.

In Example 4.5.4, we show that A was similar to a diagonal matrix D. It is important to realize that not every square matrix is diagonalizable.

Example 4.5.4: A Matrix That Is Not Diagonalizable

Suppose that $A = \begin{bmatrix} 1 & 1 \\ 0 & 1 \end{bmatrix}$ and we can find an invertible matrix $P = \begin{bmatrix} a & b \\ c & d \end{bmatrix}$ and a matrix $D = \begin{bmatrix} \delta_1 & 0 \\ 0 & \delta_2 \end{bmatrix}$ such that $PAP^{-1} = D$, or $PA = DP$. This equality is equivalent to $\begin{bmatrix} a & a+b \\ c & c+d \end{bmatrix} = \begin{bmatrix} \delta_1 a & \delta_1 b \\ \delta_2 c & \delta_2 d \end{bmatrix}$, which implies $\delta_1 = \delta_2 = 1$, leading to the conclusion that $a = 0$ and $c = 0$. But $P = \begin{bmatrix} 0 & b \\ 0 & d \end{bmatrix}$ is not invertible because $\det(P) = 0$, and this contradiction shows that A is not diagonalizable.

What does "similarity" get us in general? We shall see that there are important consequences of having one matrix similar to another. For example, the next theorem states that the characteristic equation, determinant, and eigenvalues are **similarity invariants**—that is, they remain unchanged under the similarity transformation $(\ ) \to P(\ )P^{-1}$.

Theorem 4.5.2

If A and B are similar $n \times n$ matrices, then

(1) the characteristic equation of A equals the characteristic equation of B.
(2) $\det(A) = \det(B)$.
(3) the eigenvalues of B are the same as those of A.

Proof   (1) Suppose that $B = PAP^{-1}$. Then the characteristic equation of B is

$$0 = \det(B - \lambda I_n) = \det(PAP^{-1} - \lambda I_n)$$
$$= \det\left[P(A - \lambda I_n)P^{-1}\right] = \det(P)\det(A - \lambda I_n)\det(P^{-1})$$
$$= \det(P)\det(A - \lambda I_n)\frac{1}{\det(P)} = \det(A - \lambda I_n).$$

Thus, A and B have the same characteristic equation.

(2)

$$\det(B) = \det(PAP^{-1}) = \det(P)\det(A)\det(P^{-1})$$
$$= \det(P)\det(A)\frac{1}{\det(P)} = \det(A).$$

(3) The eigenvalues of A and B are the roots of the characteristic equations of A and B, respectively. But Property (1) states that these characteristic equations are identical, so their roots must be the same.

Each of the three conditions of the last theorem is a *necessary* condition for similarity but is not a *sufficient* condition. For example, there are matrices with the same eigenvalues that are not similar (see Example 4.5.5). Exercise B8 asks about the converse of Property (2). Also, even if A is similar to B, the eigenvectors of A and B are not necessarily the same (Exercise B9).

We can generalize the procedure shown in Example 4.5.2 to state and prove an important characterization of a diagonalizable matrix.

Theorem 4.5.3

An $n \times n$ matrix A is diagonalizable if and only if A has n linearly independent eigenvectors $\mathbf{v}_1, \mathbf{v}_2, \ldots, \mathbf{v}_n$. Furthermore, if A is diagonalizable, then A is similar to a diagonal matrix whose diagonal elements are the corresponding eigenvalues $\lambda_1, \lambda_2, \ldots, \lambda_n$, possibly repeated.

Proof First assume that A has n linearly independent eigenvectors $\mathbf{v}_1, \mathbf{v}_2, \ldots, \mathbf{v}_n$. Then matrix $M = \begin{bmatrix} \mathbf{v}_1 & \mathbf{v}_2 & \cdots & \mathbf{v}_n \end{bmatrix}$ is nonsingular (Corollary 3.6.1(a)) and

$$
\begin{aligned}
M^{-1}AM = M^{-1}A\begin{bmatrix} \mathbf{v}_1 & \mathbf{v}_2 & \cdots & \mathbf{v}_n \end{bmatrix} &= M^{-1}\begin{bmatrix} A\mathbf{v}_1 & A\mathbf{v}_2 & \cdots & A\mathbf{v}_n \end{bmatrix} \\
&= M^{-1}\begin{bmatrix} \lambda_1\mathbf{v}_1 & \lambda_2\mathbf{v}_2 & \cdots & \lambda_n\mathbf{v}_n \end{bmatrix} \\
&= \begin{bmatrix} \lambda_1 M^{-1}\mathbf{v}_1 & \lambda_2 M^{-1}\mathbf{v}_2 & \cdots & \lambda_n M^{-1}\mathbf{v}_n \end{bmatrix}.
\end{aligned} \tag{$*$}
$$

We also have

$$
\begin{aligned}
I_n = M^{-1}M = M^{-1}\begin{bmatrix} \mathbf{v}_1 & \mathbf{v}_2 & \cdots & \mathbf{v}_n \end{bmatrix} \\
= \begin{bmatrix} M^{-1}\mathbf{v}_1 & M^{-1}\mathbf{v}_2 & \cdots & M^{-1}\mathbf{v}_n \end{bmatrix}.
\end{aligned} \tag{$**$}
$$

Comparing $(*)$ and $(**)$, we see that the ith column of $(*)$ is just λ_i times the ith column of $(**)$, which is the ith column of I_n. Therefore, we can write $(*)$ as $M^{-1}AM = \text{diag}(\lambda_1, \lambda_2, \ldots, \lambda_n) = D$.

Now suppose that A is similar to a diagonal matrix D with diagonal entries $d_1, d_2, \ldots, d_n$. Then the standard basis vectors $\mathbf{e}_1, \mathbf{e}_2, \ldots, \mathbf{e}_n$ are eigenvectors for D because $D\mathbf{e}_1 = d_1\mathbf{e}_1, D\mathbf{e}_2 = d_2\mathbf{e}_2, \ldots, D\mathbf{e}_n = d_n\mathbf{e}_n$ (see B8 in Exercise 4.4). Also, the diagonal entries $d_1, d_2, \ldots, d_n$ are the corresponding eigenvalues of D (Theorem 4.4.7) and hence of A (Theorem 4.5.2(3)). Furthermore, these diagonal entries are the *only* eigenvalues. If $\mathbf{x} = \begin{bmatrix} x_1 \\ x_2 \\ \vdots \\ x_n \end{bmatrix} \neq \mathbf{0}$ is any eigenvector of D, so that $D\mathbf{x} = \lambda\mathbf{x}$ for an appropriate eigenvalue λ, then $D\mathbf{x} = \begin{bmatrix} d_1 x_1 \\ d_2 x_2 \\ \vdots \\ d_n x_n \end{bmatrix} = \begin{bmatrix} \lambda x_1 \\ \lambda x_2 \\ \vdots \\ \lambda x_n \end{bmatrix}$, or $d_i x_i = \lambda x_i$ for all i.

Because $x_i \neq 0$ for some i, this proves that $d_i = \lambda$ for this i, and the eigenvalue λ is indeed some diagonal entry d_i.

Corollary 4.5.1

An $n \times n$ matrix A with n distinct eigenvalues is diagonalizable.

Proof If A has n distinct eigenvalues, it follows from Theorem 4.4.4 that A has n linearly independent eigenvectors. Then Theorem 4.5.3 implies that A is diagonalizable.

Example 4.5.5: Nonsimilar Matrices with the Same Eigenvalues

Both $A = \begin{bmatrix} 1 & -3 & 3 \\ 3 & -5 & 3 \\ 6 & -6 & 4 \end{bmatrix}$ and $B = \begin{bmatrix} -3 & 1 & -1 \\ -7 & 5 & -1 \\ -6 & 6 & -2 \end{bmatrix}$ have eigen-

values -2, -2, and 4. Matrix A has corresponding (linearly

independent) eigenvectors $\begin{bmatrix} 0 \\ 1 \\ 1 \end{bmatrix}$, $\begin{bmatrix} 1 \\ 1 \\ 0 \end{bmatrix}$, and $\begin{bmatrix} 1 \\ 1 \\ 2 \end{bmatrix}$; but B has

only the single eigenvector $\begin{bmatrix} 1 \\ 1 \\ 0 \end{bmatrix}$ corresponding to the double

eigenvalue -2 and the eigenvector $\begin{bmatrix} 0 \\ 1 \\ 1 \end{bmatrix}$ corresponding to the

eigenvalue 4. By Theorem 4.5.3, A is similar to the diagonal

matrix $D = \begin{bmatrix} -2 & 0 & 0 \\ 0 & -2 & 0 \\ 0 & 0 & 4 \end{bmatrix}$, but B cannot be diagonalized

because Theorem 4.5.3 would imply the existence of a second (linearly independent) eigenvector corresponding to $\lambda = -2$.

Example 4.5.6: Example 4.5.4 Revisited

Viewed from the perspective of Theorem 4.5.3, $A = \begin{bmatrix} 1 & 1 \\ 0 & 1 \end{bmatrix}$ cannot be diagonalizable because it has only one independent eigenvector, $\begin{bmatrix} 1 \\ 0 \end{bmatrix}$, corresponding to the sole eigenvalue 1.

Exercises 4.5

A.

1. Show that if $D = \text{diag}(d_1, d_2, \ldots, d_n)$, with $d_i \neq 0$ for all i, then

$$D^{-1} = \text{diag}(d_1^{-1}, d_2^{-1}, \ldots, d_n^{-1}).$$

2. Show that if $D = \text{diag}(d_1, d_2, \ldots, d_n)$, then

$$D^k = \text{diag}(d_1^k, d_2^k, \ldots, d_n^k) \text{ for every positive integer } k.$$

3. If $D = \text{diag}(d_1, d_2, \ldots, d_n)$, with $d_i \geq 0$, what should the matrix $D^{1/2}$ look like? Show that your answer is correct.

4. In Example 4.5.5, show that the eigenvalues and eigenvectors for A and B are as described.

5. Suppose that $A = \begin{bmatrix} 1 & 1 & -2 \\ -1 & 2 & 1 \\ 0 & 1 & -1 \end{bmatrix}$.

    a. Find the eigenvalues and eigenvectors of A.
    b. Find an invertible matrix P such that $P^{-1}AP$ is a diagonal matrix.
    c. Use the result of part (b) to find A^{-1}.

6. If $k \geq 2$ is a positive integer, evaluate each of the following powers:

    a. $\begin{bmatrix} 2 & 1 \\ 2 & 3 \end{bmatrix}^k$ b. $\begin{bmatrix} a & 1 \\ 0 & a \end{bmatrix}^k$ c. $\begin{bmatrix} 0 & 1 & 0 \\ 0 & 0 & 1 \\ 0 & 0 & 0 \end{bmatrix}^k$

    d. $\begin{bmatrix} 0 & 1 & 0 \\ 0 & 0 & 1 \\ 1 & 0 & 0 \end{bmatrix}^k$ e. $\begin{bmatrix} 2 & 2 & 0 \\ 1 & 2 & 1 \\ 1 & 2 & 1 \end{bmatrix}^k$

7. Solve the problem in Example 4.4.7 by diagonalizing matrix A.

8. Solve Exercise A6 in Exercise 4.4 by diagonalizing matrix A.

9. Solve the system of equations:

    $$x_{n+1} = 2x_n + 6y_n$$
    $$y_{n+1} = 6x_n - 3y_n$$

    given that $x_1 = 0$ and $y_1 = -1$.

10. If $A = \begin{bmatrix} 4 & -1 & 0 \\ -1 & 5 & -1 \\ 0 & -1 & 4 \end{bmatrix}$, find a nonsingular matrix P and a diagonal matrix D such that $A = PDP^{-1}$.

11. Find an invertible matrix P such that $P^{-1}AP = D$, where matrices A and D are given in Example 4.5.5.

B.

1. If A is a diagonalizable matrix, prove that A^T is diagonalizable.

2. If A is a nonsingular diagonalizable matrix, prove that A^{-1} is diagonalizable.

3. Suppose that A is an $n \times n$ upper triangular matrix with all main diagonal entries distinct. Prove that A is diagonalizable.

4. Describe those $n \times n$ matrices that are similar only to themselves.

5. If A and B are similar $n \times n$ matrices, prove that trace(A) = trace(B).

6. Prove Theorem 4.5.1.

7. Suppose that A is a square matrix.
   a. Prove that if A is diagonalizable, then so is A^2.
   b. Find a matrix A that is *not* diagonalizable, but is such that A^2 *is* diagonalizable.

8. Examine the converse of Theorem 4.5.2(2) by considering the matrices:

$$A = \begin{bmatrix} -1 & 0 & 0 \\ 0 & -1 & 0 \\ 0 & 0 & 1 \end{bmatrix} \quad \text{and} \quad I_3 = \begin{bmatrix} 1 & 0 & 0 \\ 0 & 1 & 0 \\ 0 & 0 & 1 \end{bmatrix}.$$

That is, is it true that if $\det(A) = \det(B)$ then A and B are similar?

9. Let $A = \begin{bmatrix} 1 & 0 \\ 0 & 2 \end{bmatrix}$ and $B = \begin{bmatrix} -4 & -15 \\ 2 & 7 \end{bmatrix}$.
   a. Show that $B = PAP^{-1}$, where $P = \begin{bmatrix} 3 & -5 \\ -1 & 2 \end{bmatrix}$, so that A and B are similar.
   b. Determine the eigenvectors of A and B.
   c. Is the statement "Similar matrices have equal eigenvectors" true?

10. a. Find 2×2 matrices A and B such that $AB = O_2$ and $BA \neq O_2$.
    b. Explain why the example found in part (a) shows that AB and BA are not necessarily similar.

11. For any $n \times n$ matrices A and B, show that AB and BA are similar if A or B is nonsingular. (Compare this result with the conclusion of Exercise B10.)

12. If A and B are similar $n \times n$ matrices and k is a positive integer, prove that $A^k \sim B^k$.

13. A square matrix A is called **idempotent** if $A^2 = A$. Show that if an $n \times n$ matrix B is similar to an $n \times n$ idempotent matrix, then B is idempotent.

14. If $A = P^{-1}DP$, where P is an invertible matrix and D is diagonal, use mathematical induction to show that $A^k = P^{-1}D^kP$ for any integer k.

15. Suppose that $A = PBP^{-1}$, where P is an invertible matrix. If $\mathbf{x}$ is an eigenvector of B corresponding to eigenvalue λ, then $P\mathbf{x}$ is an eigenvector of A.

16. Suppose A is an $n \times n$ idempotent matrix (see Exercise B13).
    a. Show that each eigenvalue of A is either 1 or 0.
    b. Show that A is diagonalizable.

17. Let $A = \begin{bmatrix} 0 & 1 \\ 1 & 1 \end{bmatrix}$. Show that $\text{trace}(A^k) = \text{trace}(A^{k-1}) + \text{trace}(A^{k-2})$ for $k \geq 2$.

18. For any scalars a, b, and c, show that
    a. $A = \begin{bmatrix} b & c & a \\ c & a & b \\ a & b & c \end{bmatrix}, B = \begin{bmatrix} c & a & b \\ a & b & c \\ b & c & a \end{bmatrix}, C = \begin{bmatrix} a & b & c \\ b & c & a \\ c & a & b \end{bmatrix}$ are similar.
    b. if $BC = CB$, then A has two zero eigenvalues.

19. Prove that if P is a matrix whose columns are n linearly independent eigenvectors of the $n \times n$ matrix A, then P is nonsingular and $P^{-1}AP$ is diagonal.

C.

1. Diagonalize $A = \begin{bmatrix} \cos t & \sin t \\ \sin t & -\cos t \end{bmatrix}$. [*Hint*: Use the identity $\tan(t/2) = ((1 - \cos t)/\sin t) = (\sin t/(1 + \cos t))$.]

2. Consider the sequence described by $(1/1), (3/2), (7/5), \ldots,$ $(a_n/b_n) \ldots,$ where $a_{n+1} = a_n + 2b_n$ and $b_{n+1} = a_n + b_n$. Find a matrix A such that
   $$\begin{bmatrix} a_{n+1} \\ b_{n+1} \end{bmatrix} = A \begin{bmatrix} a_n \\ b_n \end{bmatrix}$$
   by diagonalizing A, find explicit formulas for a_n and b_n and then show that $\lim_{n \to \infty} (a_n/b_n) = \sqrt{2}$.

3. Suppose that A is an $n \times n$ matrix with n distinct eigenvalues. Prove that there exists an $n \times n$ nonsingular matrix X such that $A^T X - XA = O_n$.

4. Suppose that A is an $n \times n$ matrix with n distinct eigenvalues. Prove that any matrix that commutes with A is a polynomial in A.

5. Show that any two eigenvectors of a symmetric matrix that correspond to different eigenvalues are orthogonal. [*Hint*: The transpose of a scalar, regarded as a 1×1 matrix, is itself.]

6. Prove that if **u** and **v** are eigenvectors, belonging to different eigenvalues, of a matrix A and its transpose A^{T}, respectively, then they are orthogonal to each other.

7. Suppose that A is a symmetric $n \times n$ matrix with distinct eigenvalues $\lambda_1, \lambda_2, \ldots, \lambda_n$ and corresponding eigenvectors $\mathbf{x}_1, \mathbf{x}_2, \ldots, \mathbf{x}_n$. Assume $\mathbf{x}_i^{\mathrm{T}} \mathbf{x}_i = 1$ for $1 \leq i \leq n$.
   Show that $A = \lambda_1 \mathbf{x}_1 \mathbf{x}_1^{\mathrm{T}} + \lambda_2 \mathbf{x}_2 \mathbf{x}_2^{\mathrm{T}} + \cdots + \lambda_n \mathbf{x}_n \mathbf{x}_n^{\mathrm{T}}.$* [*Hint*: You may use the result of Exercise 5.]

4.6 Algebraic and Geometric Multiplicities of Eigenvalues

Although the Fundamental Theorem of Algebra[†] guarantees that any polynomial of degree n with complex coefficients has exactly n complex zeros (counting multiplicity), so that any $n \times n$ matrix has exactly n eigenvalues, there is no guarantee that all the eigenvalues will be *distinct*.

The number of times λ appears as the eigenvalue of a particular $n \times n$ matrix A is called the **algebraic multiplicity** of λ. More precisely, when the characteristic polynomial of an $n \times n$ matrix A is written in the form:

$$\det(\lambda I_n - A) = (\lambda - \lambda_1)^{m_1}(\lambda - \lambda_2)^{m_2} \cdots (\lambda - \lambda_k)^{m_k},$$

with $\lambda_i \neq \lambda_j$ for $1 \leq i \neq j \leq k$ and $m_1 + m_2 + \cdots + m_k = n$, the positive integer m_i is called the **algebraic multiplicity** of the eigenvalue λ_i. For example, the characteristic polynomial of A in Example 4.4.8 can be written as $(\lambda - 5)(\lambda - (-1))^2$, so the eigenvalue 5 has algebraic multiplicity 1 and the eigenvalue -1 has algebraic multiplicity 2.

The number of linearly independent eigenvectors corresponding to an eigenvalue λ is called the **geometric multiplicity** of λ. Put another way, the geometric multiplicity of an eigenvalue λ is the dimension of the eigenspace $S_\lambda(A)$ corresponding to λ. Thus, in Example 4.5.5, the eigenvalue -2 of B has algebraic multiplicity 2, but geometric multiplicity 1.[‡] Considered as an eigenvalue of matrix A, -2 has both algebraic and geometric multiplicity 2.

---

* This result has been described as a **spectral decomposition of A** or an **eigenvalue decomposition of A**. See the paper by D. Kalman, A singularly valuable decomposition: The SVD of a matrix, *Coll. Math. J.* (27): 2–23. We will return to this concept in Chapter 8.

[†] For a comprehensive discussion, including a variety of proofs, see B. Fine and G. Rosenberger, *The Fundamental Theorem of Algebra* (New York: Springer, 1997).

[‡] If a matrix is such that some eigenvalue does *not* have associated with it the same number of linearly independent eigenvectors as its algebraic multiplicity—that is, the geometric multiplicity of the eigenvalue is less than its algebraic multiplicity—the matrix is called **defective**. Thus matrix B in Example 4.5.5 is defective.

Example 4.6.1: Algebraic and Geometric Multiplicities

Let us consider the matrix:

$$A = \begin{bmatrix} 3 & 4 & -1 \\ -1 & -2 & 1 \\ 3 & 9 & 0 \end{bmatrix}.$$

The characteristic polynomial of A is $\lambda^3 - \lambda^2 - 8\lambda + 12 = (\lambda + 3)(\lambda - 2)^2$, so the eigenvalues of A are -3, 2, and 2. The algebraic multiplicity of -3 is 1, while the eigenvalue 2 has algebraic multiplicity 2. The eigenvalue -3 has only one independent eigenvector, $\begin{bmatrix} -1 \\ 1 \\ -2 \end{bmatrix}$, so its geometric multiplicity is 1, equal to its algebraic multiplicity. In examining the situation for the repeated eigenvalue 2, we find that $\begin{bmatrix} -1 \\ 1 \\ 3 \end{bmatrix}$ is the only linearly independent eigenvector, so the eigenvalue 2 has geometric multiplicity 1, which is less than its algebraic multiplicity.

Although examples show that the two kinds of multiplicity are not necessarily equal, there *is* a relationship between them.

Theorem 4.6.1

If λ is an eigenvalue of an $n \times n$ matrix A, the geometric multiplicity of λ is less than or equal to the algebraic multiplicity of λ.

Proof   Suppose that λ^* is an eigenvalue of A of geometric multiplicity k. Let $\{v_1, v_2, \ldots, v_k\}$ be the set of k linearly independent eigenvectors of A corresponding to λ^*. Then $\{v_1, v_2, \ldots, v_k\}$ is a basis for the eigenspace $S_\lambda(A)$. Now extend this basis to a basis $\{v_1, v_2, \ldots, v_k, v_{k+1}, \ldots, v_n\}$ for $\mathbb{R}^n$. Then the matrix $P = \begin{bmatrix} v_1 & v_2 & \cdots & v_k & v_{k+1} & \cdots & v_n \end{bmatrix}$ is a nonsingular $n \times n$ matrix, which we can partition in a different way:

$P = \begin{bmatrix} V \vdots W \end{bmatrix}$, where $V = \begin{bmatrix} v_1 & v_2 & \cdots & v_k \end{bmatrix}$ and $W = \begin{bmatrix} v_{k+1} & v_{k+2} & \cdots & v_n \end{bmatrix}$. Now partition P^{-1} as $\begin{bmatrix} C \\ D \end{bmatrix}$, where C is a $k \times n$ matrix and D is an $(n - k) \times n$ matrix. Because the columns of V are eigenvectors corresponding to λ^*, $AV = \lambda^* V$. We also have

$$\begin{bmatrix} I_k & \vdots & O_{k,n-k} \\ \cdots & \vdots & \cdots \\ O_{n-k,k} & \vdots & I_{n-k} \end{bmatrix} = I_n = P^{-1}P = \begin{bmatrix} C \\ D \end{bmatrix} \begin{bmatrix} V & W \end{bmatrix} = \begin{bmatrix} CV & \vdots & CW \\ \cdots & \vdots & \cdots \\ DV & \vdots & DW \end{bmatrix},$$

which implies that $CV = I_k, CW = O_{k,n-k}, DV = O_{n-k,k}$, and $DW = I_{n-k}$.

Therefore,

$$P^{-1}AP = \begin{bmatrix} C \\ D \end{bmatrix} A [V \quad W] = \begin{bmatrix} CAV & \vdots & CAW \\ \ldots & \vdots & \ldots \\ DAV & \vdots & DAW \end{bmatrix} = \begin{bmatrix} \lambda^*CV & \vdots & CAW \\ \ldots & \vdots & \ldots \\ \lambda^*DV & \vdots & DAW \end{bmatrix}$$

$$= \begin{bmatrix} \lambda^*I_k & \vdots & CAW \\ \ldots & \vdots & \ldots \\ O_{n-k,k} & \vdots & DAW \end{bmatrix}.$$

By Problem B4 of Exercise 4.1 and Problem C2, Exercises 4.3, we know that

$$\det(P^{-1}AP - \lambda I_n) = \det \begin{bmatrix} (\lambda^* - \lambda)I_k & \vdots & CAW \\ \ldots & \vdots & \ldots \\ O_{n-k,k} & \vdots & DAW - \lambda I_{n-k} \end{bmatrix}$$

$$= \det(\lambda^* - \lambda)I_k \cdot \det(DAW - \lambda I_{n-k})$$

$$= (\lambda^* - \lambda)^k \det(DAW - \lambda I_{n-k}). \qquad (*)$$

But $\det(P^{-1}AP - \lambda I_n)$ is the characteristic polynomial of $P^{-1}AP$, which is the same as the characteristic polynomial of A by Theorem 4.5.2(1). Therefore, statement $(*)$ implies that the algebraic multiplicity of the eigenvalue λ^* is at least k, its geometric multiplicity. (There may be additional factors of the form $(\lambda^* - \lambda)$ in $\det(DAW - \lambda I_{n-k})$.)

Theorem 4.6.2

For an $n \times n$ matrix A, the following statements are equivalent:

(1) Matrix A is diagonalizable.
(2) For every eigenvalue λ of A, the geometric multiplicity of λ equals the algebraic multiplicity of λ.

Proof First suppose that the $n \times n$ matrix A is diagonalizable. Then A has n linearly independent eigenvectors (Theorem 4.5.3). Let the eigenvalues of A be $\lambda_1, \lambda_2, \ldots, \lambda_k$, each λ_i having algebraic multiplicity m_i and geometric multiplicity γ_i. Theorem 4.6.1 implies

$$\gamma_1 + \gamma_2 + \cdots + \gamma_k \leq m_1 + m_2 + \cdots + m_k = n.$$

If there is an eigenvalue $\hat{\lambda}$ whose geometric multiplicity $\hat{\gamma}$ is strictly less than its algebraic multiplicity $\hat{m}$, then $\gamma_1 + \gamma_2 + \cdots + \hat{\gamma} + \cdots + \gamma_k < m_1 + m_2 + \cdots + \hat{m} + \cdots + m_k = n$, implying that A has fewer than n linearly independent eigenvectors. This contradiction tells us that the geometric multiplicity of each eigenvalue must *equal* its algebraic multiplicity.

Conversely, if the geometric multiplicity equals the algebraic multiplicity for each eigenvalue, then the union of the bases for the eigenspaces is a set of n linearly independent eigenvectors—that is, A is diagonalizable by Theorem 4.5.3.

Example 4.6.2: Example 4.5.5 Revisited

In Example 4.5.5, the eigenvalue -2 of $B = \begin{bmatrix} -3 & 1 & -1 \\ -7 & 5 & -1 \\ -6 & 6 & -2 \end{bmatrix}$ has algebraic multiplicity 2, but geometric multiplicity 1. Thus, by Theorem 4.6.2, B is not diagonalizable. However, each eigenvalue of matrix $A = \begin{bmatrix} 1 & -3 & 3 \\ 3 & -5 & 3 \\ 6 & -6 & 4 \end{bmatrix}$ has its geometric multiplicity equal to its algebraic multiplicity and so is diagonalizable.

Exercises 4.6

A.

1. Given $A = \begin{bmatrix} 2 & -1 \\ -1 & 2 \end{bmatrix}$, which has $\lambda = 3$ as an eigenvalue,
    a. Find a basis for the eigenspace $S_3(A)$.
    b. Determine the algebraic and geometric multiplicities of the eigenvalue $\lambda = 3$.

2. Given $A = \begin{bmatrix} -6 & -1 & 2 \\ 3 & 2 & 0 \\ -14 & -2 & 5 \end{bmatrix}$, which has $\lambda = 1$ as an eigenvalue.
    a. Find a basis for the eigenspace $S_1(A)$.
    b. Determine the algebraic and geometric multiplicities of the eigenvalue $\lambda = 1$.

3. For each of the following matrices, determine (i) the eigenvalues; (ii) the geometric multiplicity of each eigenvalue; and (iii) whether the matrix is diagonalizable. For those matrices A that are diagonalizable, find a nonsingular matrix P such that PAP^{-1} is diagonal.
    a. $\begin{bmatrix} 3 & -1 \\ -1 & 3 \end{bmatrix}$ b. $\begin{bmatrix} 3 & -1 & 1 \\ -1 & 5 & -1 \\ 1 & -1 & 3 \end{bmatrix}$ c. $\begin{bmatrix} 7 & -1 & 2 \\ -1 & 7 & 2 \\ -2 & 2 & 10 \end{bmatrix}$

d. $\begin{bmatrix} 2 & 1 & -1 \\ 0 & 2 & 1 \\ 0 & 0 & 1 \end{bmatrix}$ e. $\begin{bmatrix} 1 & 0 & 1 \\ 0 & 2 & 1 \\ -1 & 0 & 3 \end{bmatrix}$ f. $\begin{bmatrix} 1 & -1 & 1 \\ 0 & -2 & 1 \\ 0 & 0 & -1 \end{bmatrix}$

4. Consider the matrix:

$$A = \begin{bmatrix} 1 & -4 & -4 \\ 8 & -11 & -8 \\ -8 & 8 & 5 \end{bmatrix}.$$

a. Find the eigenvalues of A.
b. What are the algebraic multiplicities of the eigenvalues found in part (a)?
c. What are the geometric multiplicities of the eigenvalues of A?
d. Is A diagonalizable?

5. Suppose that $M = \begin{bmatrix} 1 & -1 & -1 & -1 \\ -1 & 1 & -1 & -1 \\ -1 & -1 & 1 & -1 \\ -1 & -1 & -1 & 1 \end{bmatrix}$, which has $\lambda = 2$ as one of its eigenvalues.
a. Find the characteristic polynomial of M.
b. Find a basis for the eigenspace $S_2(M)$.
c. Determine the algebraic and geometric multiplicities of the eigenvalue $\lambda = 2$.

B.

1. Let $A = \begin{bmatrix} 0 & 2 & 2 & 0 & 0 \\ 2 & 0 & 2 & 0 & 0 \\ 2 & 2 & 0 & 0 & 0 \\ 0 & 0 & 0 & 2 & 0 \\ 2 & 2 & 2 & 2 & 2 \end{bmatrix}$.
a. Determine the eigenvalues of A.
b. Determine the algebraic multiplicity of each eigenvalue of A.
c. Determine a basis for each eigenspace $S_\lambda(A)$.
d. Determine the geometric multiplicity of each eigenvalue of A.
e. Is A diagonalizable? If so, diagonalize A.

2. Let $A = \begin{bmatrix} 5 & 8 & 0 & 2 & 6 & -6 \\ 0 & 1 & 0 & 0 & 0 & 0 \\ 6 & 18 & -1 & 1 & 13 & -9 \\ 3 & 6 & 0 & 4 & 6 & -6 \\ 4 & 14 & -2 & 0 & 11 & -6 \\ 6 & 18 & -2 & 1 & 13 & -8 \end{bmatrix}$.
a. Determine the eigenvalues of A.
b. Determine the algebraic multiplicity of each eigenvalue of A.
c. Determine a basis for each eigenspace $S_\lambda(A)$.

d. Determine the geometric multiplicity of each eigenvalue of A.

e. Is A diagonalizable? If so, diagonalize A.

C.

1. Given the matrix:

$$A = \begin{bmatrix} 6.5 & -2.5 & 2.5 \\ -2.5 & 6.5 & -2.5 \\ 0 & 0 & 4 \end{bmatrix}$$

Find a matrix B such that $B^2 = A$. [*Hint*: First diagonalize A.]

*4.7 The Diagonalization of Real Symmetric Matrices

In Sections 4.5 and 4.6, we discussed the similarity of matrices and the idea of diagonalizing a matrix. As we saw, not every matrix is similar to a diagonal matrix, but when this relationship exists, various calculations with the matrix are simplified.

It is a fact, which we will prove shortly, that **every real symmetric matrix is diagonalizable**—a result that will be generalized and proved in the context of complex matrices (matrices with complex number entries) in Section 8.2. To prepare the way for the proof of this remarkable theorem, we will use an example to highlight the important properties that a symmetric matrix has and outline the steps leading to diagonalization (see also Example 4.5.2).

Example 4.7.1: Diagonalizing a Real Symmetric Matrix

Consider the symmetric matrix $A = \begin{bmatrix} 4 & -2 & -2 \\ -2 & 1 & 1 \\ -2 & 1 & 1 \end{bmatrix}$. The characteristic equation of A is $\lambda^2(\lambda - 6) = 0$, so the eigenvalues are the (*real*) numbers 0 and 6. We note that the eigenvalue 0 is repeated, so we cannot use Corollary 4.5.1 to conclude that A is diagonalizable. However, corresponding to the eigenvalue 0 we have the linearly independent eigenvectors $\begin{bmatrix} 1 \\ 0 \\ 2 \end{bmatrix}$ and $\begin{bmatrix} 1 \\ 2 \\ 0 \end{bmatrix}$; and corresponding to the eigenvalue 6 there is the eigenvector $\begin{bmatrix} -2 \\ 1 \\ 1 \end{bmatrix}$. We can determine that *the algebraic multiplicity of each eigenvalue equals its geometric multiplicity*, so Theorem 4.6.2 guarantees that A is similar to a diagonal matrix.

Now we form the matrix P with these eigenvectors as columns:

$$P = \begin{bmatrix} -2 & 1 & 1 \\ 1 & 0 & 2 \\ 1 & 2 & 0 \end{bmatrix}$$

(There is no significance to the order in which we used the eigenvectors to form P.) It is easy to see that *these eigenvectors of A are linearly independent*, again implying (Theorem 4.5.3) that A is diagonalizable. The columns of P make up a basis for $\mathbb{R}^3$, so the matrix P has an inverse (Corollary 3.6.1(a)). With a little effort (using Corollary 3.6.1(b)) we calculate

$$P^{-1} = \begin{bmatrix} -\frac{1}{3} & \frac{1}{6} & \frac{1}{6} \\ \frac{1}{6} & -\frac{1}{12} & \frac{5}{12} \\ \frac{1}{6} & \frac{5}{12} & -\frac{1}{12} \end{bmatrix}$$

and $P^{-1}AP = \begin{bmatrix} 6 & 0 & 0 \\ 0 & 0 & 0 \\ 0 & 0 & 0 \end{bmatrix}$, a diagonal matrix with the eigen-

values of A down the main diagonal. (If we had taken the columns of P in a different order, we would have obtained a diagonal matrix in which the entries appeared in the corresponding order.) One final comment: *The eigenvectors of A corresponding to distinct eigenvalues are orthogonal.*

The properties highlighted in the previous example are characteristic of real symmetric matrices. Before we state and prove the main result of this section, we will need some auxiliary results, the first of which requires familiarity with the complex number system (see Appendix D, if necessary). We will also need to preview some facts from Section 7.1. In what follows, a bar denotes the *complex conjugate*: If $\mathbf{z} = [\, z_1 \quad z_2 \quad \ldots \quad z_n \,]^T$, where some or all the z_i's are complex numbers, then $\bar{\mathbf{z}} = [\, \bar{z}_1 \quad \bar{z}_2 \quad \ldots \quad \bar{z}_n \,]^T$; if $C = [c_{ij}]$ is a matrix with complex entries, then $\overline{C} = [\bar{c}_{ij}]$; $\overline{(\alpha C)} = \bar{\alpha}\overline{C}$, where C is a complex matrix and α is a complex number. Note that C is a real matrix if and only if $\overline{C} = [\bar{c}_{ij}] = [c_{ij}] = C$.

Lemma 4.7.1

If A is a real symmetric matrix, then all the eigenvalues of A are real and A has at least one real eigenvector (i.e., an eigenvector in $\mathbb{R}^n$).

Proof   Theorem 4.4.2 established the existence of eigenvalues for any square matrix, but indicated that some of these eigenvalues might be complex numbers.

To carry out a proof by contradiction, let us assume that the symmetric matrix A has a complex eigenvalue λ and a corresponding eigenvector $\mathbf{v}$, which may have one or more complex components. Thus

$$A\mathbf{v} = \lambda\mathbf{v}. \qquad (*)$$

Now we take the complex conjugate of each side of (∗) to get the equation:

$$A\bar{\mathbf{v}} = \bar{\lambda}\bar{\mathbf{v}}. \qquad (**)$$

Multiply (∗) on the left by $\bar{\mathbf{v}}^T$ and (∗∗) on the left by $\mathbf{v}^T$, then subtract the resulting equations to get

$$\bar{\mathbf{v}}^T A \mathbf{v} - \mathbf{v}^T A \bar{\mathbf{v}} = \lambda(\bar{\mathbf{v}}^T \mathbf{v}) - \bar{\lambda}(\bar{\mathbf{v}}^T \mathbf{v})^T$$
$$= \lambda(\bar{\mathbf{v}}^T \mathbf{v}) - \bar{\lambda}(\bar{\mathbf{v}}^T \mathbf{v}) = (\lambda - \bar{\lambda})\bar{\mathbf{v}}^T \mathbf{v}. \qquad (***)$$

Each term on the far left-hand side of (∗∗∗) is a scalar (and so equal to its own transpose) and because A is symmetric, the expression on the far left is in fact equal to zero: $\bar{\mathbf{v}}^T A \mathbf{v} = (\mathbf{v}^T A \bar{\mathbf{v}})^T = \mathbf{v}^T A^T (\bar{\mathbf{v}}^T)^T = \mathbf{v}^T A \bar{\mathbf{v}}$. But on the far right-hand side of (∗∗∗), $\bar{\mathbf{v}}^T \mathbf{v}$ is the sum of products of complex numbers times their conjugates, which can never equal zero unless all the numbers themselves are zero. Therefore, $\lambda - \bar{\lambda} = 0$, or $\bar{\lambda} = \lambda$—which means that λ is a real number.

A similar argument shows that a real symmetric matrix has at least one real eigenvector.

Lemma 4.7.2

If the columns of a real $n \times n$ matrix M form an orthonormal basis for $\mathbb{R}^n$, then $M^{-1} = M^T$.

Proof First of all, the hypothesis that the columns of M are linearly independent vectors in $\mathbb{R}^n$ tells us that M has an inverse (Corollary 3.6.1(a)). We will prove that $M^{-1} = M^T$ by showing that $M^T M = I_n$.

Suppose that $M = [m_{ij}]$, so $M^T = [m_{ij}^T]$, and let $M^T M = C = [c_{ij}]$. Then

$$c_{ij} = \sum_{k=1}^n m_{ik}^T m_{kj} = \sum_{k=1}^n m_{ki} m_{kj} = \text{col}_i(M) \bullet \text{col}_j(M) = \begin{cases} 0 & \text{if } i \neq j \\ 1 & \text{if } i = j \end{cases}.$$

But this last equation describes I_n. Thus, $M^T M = I_n$ and so $M^{-1} = M^T$.

Now we are ready for the main event.

Theorem 4.7.1

If A is an $n \times n$ symmetric matrix with real entries, then there exists an invertible matrix P such that $P^{-1}AP = D$, where D is a diagonal matrix.

Proof Let A be an $n \times n$ symmetric matrix. We will prove the theorem by mathematical induction on n. If $n = 1$, we can take $P = I$ and $D = A$.

Now assume that the result is true for an $(n - 1) \times (n - 1)$ symmetric matrix with real entries, and let $\mathbf{x}$ be a real eigenvector of the $n \times n$ symmetric matrix A corresponding to the real eigenvalue λ. So $\mathbf{x} \neq \mathbf{0}$ and $A\mathbf{x} = \lambda\mathbf{x}$. Moreover, we can assume that $\| \mathbf{x} \| = 1$. (If $\| \mathbf{x} \| \neq 1$, divide $\mathbf{x}$ by $\| \mathbf{x} \|$. The new vector, of unit length, is an eigenvector of A: $A(\mathbf{x}/\| \mathbf{x} \|) = (1/\| \mathbf{x} \|)A(\mathbf{x}) = \lambda(\mathbf{x}/\| \mathbf{x} \|)$.)

Corollary 1.5.2 guarantees that we can extend $\{\mathbf{x}\}$ to a basis $\{\mathbf{x}, \mathbf{x}_2, \dots, \mathbf{x}_n\}$ for $\mathbb{R}^n$. The algorithm illustrated in Example 1.5.2 can be generalized to guarantee that the basis $\{\mathbf{x}, \mathbf{x}_2, \dots, \mathbf{x}_n\}$ is an *orthonormal* basis. Let P_1 be the matrix whose columns are the orthonormal vectors $\mathbf{x}, \mathbf{x}_2, \dots, \mathbf{x}_n: P_1 = [\mathbf{x} \ \ \mathbf{x}_2 \ \ \dots \ \ \mathbf{x}_n]$. Then P_1 is invertible (Corollary 3.6.1(a)) and we note that the first column of $P_1^{-1}AP_1$ is $[\lambda \ \ 0 \ \ \dots \ \ 0]^T$: If $\mathbf{e}_1 = [1 \ \ 0 \ \ \dots \ \ 0]^T$, then $\text{col}_1(P_1^{-1}AP_1) = (P_1^{-1}AP_1)\mathbf{e}_1 = (P_1^{-1}A)(P_1\mathbf{e}_1) = P_1^{-1}A\mathbf{x} = P_1^{-1}(\lambda\mathbf{x}) = \lambda(P_1^{-1}\mathbf{x}) = \lambda\mathbf{e}_1$. Now we use Lemma 4.7.2 to see that $P_1^{-1}AP_1$ is symmetric: $(P_1^{-1}AP_1)^T = P_1^T A^T (P_1^{-1})^T = P_1^{-1}AP_1$. This symmetry means that $P_1^{-1}AP_1$ can be written in the block form:

$$\begin{bmatrix} \lambda & \mathbf{0}^T \\ \mathbf{0} & B \end{bmatrix},$$

where $\mathbf{0} \in \mathbb{R}^{n-1}$ and B is an $(n - 1) \times (n - 1)$ symmetric matrix.

By our inductive hypothesis, there exists an invertible matrix Q such that $Q^{-1}BQ$ is diagonal—that is, $BQ = QD$, where D is a diagonal *matrix*. Finally, set $P = P_1 \begin{bmatrix} 1 & \mathbf{0}^T \\ \mathbf{0} & Q \end{bmatrix}$ and note that $\begin{bmatrix} 1 & \mathbf{0}^T \\ \mathbf{0} & Q \end{bmatrix}^{-1} = \begin{bmatrix} 1 & \mathbf{0}^T \\ \mathbf{0} & Q^{-1} \end{bmatrix}$. Then P, as the product of two invertible matrices, is invertible and

$$
\begin{aligned}
P^{-1}AP &= \left(P_1 \begin{bmatrix} 1 & \mathbf{0}^T \\ \mathbf{0} & Q \end{bmatrix} \right)^{-1} A \left(P_1 \begin{bmatrix} 1 & \mathbf{0}^T \\ \mathbf{0} & Q \end{bmatrix} \right) \\
&= \begin{bmatrix} 1 & \mathbf{0}^T \\ \mathbf{0} & Q \end{bmatrix}^{-1} P_1^{-1}AP_1 \begin{bmatrix} 1 & \mathbf{0}^T \\ \mathbf{0} & Q \end{bmatrix} \\
&= \begin{bmatrix} 1 & \mathbf{0}^T \\ \mathbf{0} & Q^{-1} \end{bmatrix} \begin{bmatrix} \lambda & \mathbf{0}^T \\ \mathbf{0} & B \end{bmatrix} \begin{bmatrix} 1 & \mathbf{0}^T \\ \mathbf{0} & Q \end{bmatrix} = \begin{bmatrix} 1 & \mathbf{0}^T \\ \mathbf{0} & Q^{-1} \end{bmatrix} \begin{bmatrix} \lambda & \mathbf{0}^T \\ \mathbf{0} & BQ \end{bmatrix} \\
&= \begin{bmatrix} 1 & \mathbf{0}^T \\ \mathbf{0} & Q^{-1} \end{bmatrix} \begin{bmatrix} \lambda & \mathbf{0}^T \\ \mathbf{0} & QD \end{bmatrix} = \begin{bmatrix} \lambda & \mathbf{0}^T \\ \mathbf{0} & D \end{bmatrix}.
\end{aligned}
$$

Because D is an $(n - 1) \times (n - 1)$ diagonal matrix, $P^{-1}AP = \begin{bmatrix} \lambda & \mathbf{0}^T \\ \mathbf{0} & D \end{bmatrix}$ is an $n \times n$ diagonal matrix. By mathematical induction, then, any $n \times n$ symmetric matrix with real entries can be diagonalized.

What is not obvious from the proof of Theorem 4.7.1 is that the eigenvalue λ together with the main diagonal elements of D constitute all the eigenvalues of the symmetric matrix A, repeated according to their multiplicity. Also, in the proof, the matrix P_1 is an **orthogonal matrix**, a matrix whose columns (and rows) are mutually orthogonal and of unit length. It is a fact that *a real $n \times n$ matrix A is symmetric if and only if it is orthogonally diagonalizable—that is, if and only if there is an orthogonal matrix P such that $P^{-1}AP$ is a diagonal matrix*. Thus, the conclusion of Theorem 4.7.1 can be written as $P^TAP = D$ by Lemma 4.7.2. We did not use an orthogonal matrix in Example 4.7.1.

The practical advantage of using an *orthogonal* matrix P to produce a diagonal matrix is that because $P^{-1} = P^T$ we avoid potential difficulties in calculating an inverse. We will explore these ideas further in Chapter 7.

Because every real symmetric $n \times n$ matrix A is diagonalizable, we can invoke Theorem 4.5.3 to conclude that such a matrix A must have n linearly independent eigenvectors. This information allows us to use the general algorithm illustrated in Example 4.5.2 to diagonalize any real $n \times n$ symmetric matrix A:

1. Determine n linearly independent eigenvectors $\mathbf{v}_1, \mathbf{v}_2, \ldots, \mathbf{v}_n$ of A.
2. Form the matrix $P = [\, \mathbf{v}_1 \quad \mathbf{v}_2 \quad \ldots \quad \mathbf{v}_n]$.
3. Calculate $P^{-1}AP$, which will be a diagonal matrix having the eigenvalues as its entries (repeated according to multiplicity).

Section 7.4 shows how to find an *orthogonal* matrix P, so that $P^{-1} = P^T$.

Example 4.7.2: Diagonalizing a Real Symmetric Matrix

Let $A = \begin{bmatrix} 1 & -2 & 4 \\ -2 & 4 & 2 \\ 4 & 2 & 1 \end{bmatrix}$. The determinant of $\lambda I - A$ is $(\lambda - 5)^2(\lambda + 4)$, giving us the eigenvalues 5 and -4.

Corresponding to $\lambda = 5$, we have the independent eigenvectors $\begin{bmatrix} 1 \\ 0 \\ 1 \end{bmatrix}$ and $\begin{bmatrix} -1 \\ 2 \\ 0 \end{bmatrix}$. The eigenvalue $\lambda = -4$ has the associated eigenvector $\begin{bmatrix} -2 \\ -1 \\ 2 \end{bmatrix}$. These three eigenvectors form a basis for $\mathbb{R}^3$. Using these vectors as columns, we form the matrix:

$$P = \begin{bmatrix} 1 & -1 & -2 \\ 0 & 2 & -1 \\ 1 & 0 & 2 \end{bmatrix}.$$

Then

$$P^{-1} = \frac{1}{9} \begin{bmatrix} 4 & 2 & 5 \\ -1 & 4 & 1 \\ -2 & -1 & 2 \end{bmatrix}$$

and

$$P^{-1}AP = \frac{1}{9}\begin{bmatrix} 4 & 2 & 5 \\ -1 & 4 & 1 \\ -2 & -1 & 2 \end{bmatrix}\begin{bmatrix} 1 & -2 & 4 \\ -2 & 4 & 2 \\ 4 & 2 & 1 \end{bmatrix}\begin{bmatrix} 1 & -1 & -2 \\ 0 & 2 & -1 \\ 1 & 0 & 2 \end{bmatrix}$$

$$= \begin{bmatrix} 5 & 0 & 0 \\ 0 & 5 & 0 \\ 0 & 0 & -4 \end{bmatrix}.$$

$$\left(\text{If we had taken } P \text{ to be } \begin{bmatrix} 1 & -2 & -1 \\ 0 & -1 & 2 \\ 1 & 2 & 0 \end{bmatrix}, \quad \text{for} \quad \text{example,}\right.$$

switching the original second and third columns, then $P^{-1}AP$

$$\left.\text{would have been } \begin{bmatrix} 5 & 0 & 0 \\ 0 & -4 & 0 \\ 0 & 0 & 5 \end{bmatrix}.\right)$$

If we had transformed the three linearly independent eigenvectors into orthonormal vectors by means of the algorithm hinted at in Example 1.5.2, the matrix P (an orthogonal matrix) would be

$$\begin{bmatrix} \sqrt{2/2} & -\sqrt{2/6} & -2/3 \\ 0 & 2\sqrt{2/3} & -1/3 \\ \sqrt{2/2} & \sqrt{2/6} & 2/3 \end{bmatrix}.$$

Then

$$P^{-1}AP = P^{\mathsf{T}}AP = \begin{bmatrix} \sqrt{2/2} & 0 & \sqrt{2/2} \\ -\sqrt{2/6} & 2\sqrt{2/3} & \sqrt{2/6} \\ -2/3 & -1/3 & 2/3 \end{bmatrix}\begin{bmatrix} 1 & -2 & 4 \\ -2 & 4 & 2 \\ 4 & 2 & 1 \end{bmatrix}$$

$$\times \begin{bmatrix} \sqrt{2/2} & -\sqrt{2/6} & -2/3 \\ 0 & 2\sqrt{2/3} & -1/3 \\ \sqrt{2/2} & \sqrt{2/6} & 2/3 \end{bmatrix} = \begin{bmatrix} 5 & 0 & 0 \\ 0 & 5 & 0 \\ 0 & 0 & -4 \end{bmatrix}.$$

Exercises 4.7

A.

1. Diagonalize the matrix $A = \begin{bmatrix} 2 & -1 \\ -1 & 2 \end{bmatrix}$ as follows:

    (a) Determine the eigenvalues of A and find two linearly independent eigenvectors of A.

    (b) Show that the vectors found in part (a) are orthogonal to each other.

    (c) Normalize each vector by dividing by its length.

(d) Form a matrix P using the normalized vectors found in part (c) as columns.

(e) Show that $P^{-1}AP$ is a diagonal matrix, whose entries are the eigenvalues of A.

2. Consider $A = \begin{bmatrix} 1 & 2 \\ 2 & 4 \end{bmatrix}$.

   (a) Diagonalize A using all the steps given in the previous exercise.

   (b) Diagonalize A without using step (c) of the previous exercise.

3. Use diagonalization to compute A^5, where $A = \begin{bmatrix} 1 & -3 \\ -3 & 1 \end{bmatrix}$.

4. Given that the eigenvalues of a 2×2 matrix A are 2 and -1, with corresponding eigenvectors $\begin{bmatrix} 1 \\ 2 \end{bmatrix}$ and $\begin{bmatrix} 1 \\ 3 \end{bmatrix}$, respectively, diagonalize A to determine A^4.

5. Diagonalize the matrix $\begin{bmatrix} 0 & 1 & 0 \\ 1 & 0 & 0 \\ 0 & 0 & 1 \end{bmatrix}$ without using an orthogonal matrix.

6. Diagonalize the matrix $\begin{bmatrix} 6 & 2 & 0 \\ 2 & 6 & 0 \\ 0 & 0 & -4 \end{bmatrix}$ without using an orthogonal matrix.

7. Diagonalize the matrix $\begin{bmatrix} 1 & 2 & 2 \\ 2 & 1 & -2 \\ 2 & -2 & 1 \end{bmatrix}$ without using an orthogonal matrix.

8. If A is a square matrix, show that AA^T and A^TA are orthogonally diagonalizable.

9. (a) If $\mathbf{v}$ is an $n \times 1$ matrix, show that $I_n - \mathbf{v}\mathbf{v}^T$ is orthogonally diagonalizable.

   (b) Find a matrix P that orthogonally diagonalizes $I_n - \mathbf{v}\mathbf{v}^T$ if $\mathbf{v} = \begin{bmatrix} 2 \\ 3 \end{bmatrix}$.

B.

1. Suppose that A is an $n \times n$ real matrix. Prove that A is symmetric if and only if

$$(A\mathbf{x}) \cdot \mathbf{y} = \mathbf{x} \cdot (A\mathbf{y}) \quad \text{for all } \mathbf{x}, \mathbf{y} \in \mathbb{R}^n.$$

2. Suppose that A is a real symmetric matrix. Let $\mathbf{x}$ and $\mathbf{y}$ be eigenvectors of A corresponding to different eigenvalues. Prove that $\mathbf{x}$ and $\mathbf{y}$ are orthogonal.

3. If A is an orthogonal matrix, prove that $\det(A) = \pm 1$.

4. Suppose that A is an $n \times n$ real symmetric matrix. Let $\mathbf{x}$ be an eigenvector of A and let S be the $(n - 1)$-dimensional subspace of $\mathbb{R}^n$ consisting of all vectors orthogonal to $\mathbf{x}$: $S = \{\mathbf{y} \in \mathbb{R}^n : \mathbf{x} \cdot \mathbf{y} = 0\}$. Prove that S is invariant with respect to A—that is, $\mathbf{y} \in S$ implies $A\mathbf{y} \in S$.

5. Assuming that $b \neq 0$, find an orthogonal matrix P such that $P^{-1} \begin{bmatrix} a & b \\ b & a \end{bmatrix} P$ is a diagonal matrix.

C.

1. Consider the symmetric matrix:

$$A = \begin{bmatrix} 1 & 1 & 1 & 1 & 1 \\ 1 & 1 & 1 & 1 & 1 \\ 1 & 1 & 1 & 1 & 1 \\ 1 & 1 & 1 & 1 & 1 \\ 1 & 1 & 1 & 1 & 1 \end{bmatrix}.$$

   (a) Find five linearly independent eigenvectors of A.
   (b) Form the matrix P using the vectors found in part (a) as its columns.
   (c) Calculate P^{-1}.
   (d) Calculate $P^{-1}AP$.

4.8 The Cayley–Hamilton Theorem (a First Look)/ the Minimal Polynomial

So far, the characteristic equation introduced in Section 4.4 has been used to find the eigenvalues of a matrix. There are, however, other applications of this concept in both the theory and applications of linear algebra. One of the neatest is the following theorem. We will prove only a useful special case now, saving the general proof for Chapter 8.

Theorem 4.8.1:   (The Cayley–Hamilton Theorem)*

Every square matrix satisfies its own characteristic equation—that is, if an $n \times n$ matrix A has the characteristic polynomial $c_A(\lambda)$, then $c_A(A) = O_n$.

Proof (*for A diagonalizable*) Suppose that A is a diagonalizable matrix with characteristic polynomial $c_A(\lambda) = \lambda^n + c_{n-1}\lambda^{n-1} + \cdots + c_1\lambda + c_0$. Let $\lambda_1, \lambda_2, \ldots, \lambda_n$ be the eigenvalues of A (possibly with repetitions), and let P be a nonsingular matrix such that $PAP^{-1} = D = \text{diag}(\lambda_1, \lambda_2, \ldots, \lambda_n)$. Because each λ_j is a zero of the characteristic polynomial $c_A(\lambda)$, we have $c_A(\lambda_j) = 0$ for $j = 1, 2, \ldots, n$, and it follows that

$$c_A(D) = D^n + c_{n-1}D^{n-1} + \cdots + c_1D + c_0I_n$$

$$= \begin{bmatrix} \lambda_1^n & \cdots & 0 \\ \vdots & \ddots & \vdots \\ 0 & \cdots & \lambda_n^n \end{bmatrix} + c_{n-1}\begin{bmatrix} \lambda_1^{n-1} & \cdots & 0 \\ \vdots & \ddots & \vdots \\ 0 & \cdots & \lambda_n^{n-1} \end{bmatrix}$$

$$+ \cdots + c_1\begin{bmatrix} \lambda_1 & \cdots & 0 \\ \vdots & \ddots & \vdots \\ 0 & \cdots & \lambda_1 \end{bmatrix} + c_0\begin{bmatrix} 1 & \cdots & 0 \\ \vdots & \ddots & \vdots \\ 0 & \cdots & 1 \end{bmatrix}$$

$$= \begin{bmatrix} \lambda_1^n + c_{n-1}\lambda_1^{n-1} + \cdots + c_1\lambda_1 + c_0 & \cdots & 0 \\ \vdots & \ddots & \vdots \\ 0 & \cdots & \lambda_n^n + c_{n-1}\lambda_n^{n-1} + \cdots + c_1\lambda_n + c_0 \end{bmatrix}$$

$$= \text{diag}[c_A(\lambda_1), c_A(\lambda_2), \ldots, c_A(\lambda_n)] = O_n.$$

(see Problem A2, Exercises 4.5). Therefore, because $A = P^{-1}DP$, we can see that $c_A(A) = c_A(P^{-1}DP) = P^{-1}[c_A(D)]P = O_n$—that is, $c_A(A) = A^n + c_{n-1}A^{n-1} + \cdots + c_1A + c_0I_n = O_n$.

Example 4.8.1: The Cayley–Hamilton Theorem

If $A = \begin{bmatrix} 1 & 2 \\ 3 & 4 \end{bmatrix}$, the characteristic polynomial for A is

$c_A(\lambda) = \det(A - \lambda I_2) = \det\begin{bmatrix} 1-\lambda & 2 \\ 3 & 4-\lambda \end{bmatrix} = \lambda^2 - 5\lambda - 2$. Now

$A^2 = \begin{bmatrix} 7 & 10 \\ 15 & 22 \end{bmatrix}$,   so   that $c_A(A) = A^2 - 5A - 2I_2 =$

$\begin{bmatrix} 7 & 10 \\ 15 & 22 \end{bmatrix} - \begin{bmatrix} 5 & 10 \\ 15 & 20 \end{bmatrix} - \begin{bmatrix} 2 & 0 \\ 0 & 2 \end{bmatrix} = \begin{bmatrix} 0 & 0 \\ 0 & 0 \end{bmatrix} = O_2.$

---

* The result was established by William Rowan Hamilton (1805–1865) for a special class of matrices (**quaternions**) in 1853 and stated generally (without proof except for $n = 2$ and 3) 5 years later by Arthur Cayley (1821–1895).

Similarly, $B = \begin{bmatrix} 1 & 0 & 2 \\ 0 & -3 & 4 \\ 5 & 6 & 7 \end{bmatrix}$ has the characteristic polyno-

mial $c_B(\lambda) = \lambda^3 - 5\lambda^2 - 51\lambda + 15$. We calculate

$B^2 = \begin{bmatrix} 11 & 12 & 16 \\ 20 & 33 & 16 \\ 40 & 24 & 83 \end{bmatrix}$ and $B^3 = \begin{bmatrix} 91 & 60 & 182 \\ 100 & -3 & 284 \\ 455 & 426 & 757 \end{bmatrix}$ and

see that $c_B(B) = B^3 - 5B^2 - 51B + 15I_3 = \begin{bmatrix} 0 & 0 & 0 \\ 0 & 0 & 0 \\ 0 & 0 & 0 \end{bmatrix} = O_3$.

If A is an $n \times n$ matrix, the Cayley–Hamilton Theorem allows us to express A^m for $m \geq n$ ($m = n + k, k \geq 0$) in terms of lower powers of A. If $c_A(\lambda) = \lambda^n + c_{n-1}\lambda^{n-1} + \cdots + c_1\lambda + c_0$ is the characteristic polynomial of A, then

$$c_A(A) = A^n + c_{n-1}A^{n-1} + \cdots + c_1 A + c_0 I_n = O_n, \text{ so}$$
$$A^n = -c_{n-1}A^{n-1} - \cdots - c_1 A - c_0 I_n,$$

and

$$A^m = A^k A^n = A^k \left(-c_{n-1}A^{n-1} - \cdots - c_1 A - c_0 I_n \right)$$
$$= -c_{n-1}A^{k+n-1} - \cdots - c_1 A^{k+1} - c_0 A^k,$$

where $k + n - 1 < m$. In fact, we can reduce the computation of a high power of a square matrix quite a bit, as Example 4.8.2 below shows. Furthermore, if A is a nonsingular $n \times n$ matrix with characteristic polynomial $c_A(\lambda) = \lambda^n + c_{n-1}\lambda^{n-1} + \cdots + c_1\lambda + c_0$, the Cayley–Hamilton Theorem gives us an expression for the inverse of A:

$A^n + c_{n-1}A^{n-1} + \cdots + c_1 A + c_0 I_n = O_n$, so $I_n = \left(-\frac{1}{c_0} \right)(A^n + c_{n-1}A^{n-1} + \cdots +$
$c_1 A)$; and multiplication by A^{-1} yields:

$$A^{-1} = \left(-\frac{1}{c_0} \right)(A^{n-1} + c_{n-1}A^{n-2} + \cdots + c_1 I_n).$$

(The coefficient c_0 is nonzero because it equals $\pm \det(A)$—see Section 4.4— and we are assuming that A is nonsingular.)

Example 4.8.2: Powers of a Matrix via Cayley–Hamilton

Suppose that $A = \begin{bmatrix} 11 & 17 \\ -6 & -10 \end{bmatrix}$ and we want to find A^5 and A^{-1}. The first thing we do is calculate the characteristic polynomial of A: $c_A(\lambda) = \lambda^2 - \lambda - 8$. The Cayley–Hamilton Theorem gives us $A^2 - A - 8I_2 = O_2$, or $A^2 = A + 8I_2$. Then

$$A^3 = A(A^2) = A(A + 8I_2) = A^2 + 8A = (A + 8I_2) + 8A = 9A + 8I_2,$$
$$A^4 = A(A^3) = A(9A + 8I_2) = 9A^2 + 8A = 9(A + 8I_2) + 8A$$
$$= 17A + 72I_2,$$
$$A^5 = A(A^4) = A(17A + 72I_2) = 17A^2 + 72A$$
$$= 17(A + 8I_2) + 72A = 89A + 136I_2.$$

In this way, we have shown that the fifth power of A equals a linear expression in A. Some simple matrix algebra gives us

$$A^5 = 89 \begin{bmatrix} 11 & 17 \\ -6 & -10 \end{bmatrix} + 136 \begin{bmatrix} 1 & 0 \\ 0 & 1 \end{bmatrix} = \begin{bmatrix} 1115 & 1513 \\ -534 & -754 \end{bmatrix}.$$

Another way to calculate A^5 is to use algebraic long division (the Division Algorithm) as follows:

$$\lambda^5 = (\lambda^3 + \lambda^2 + 9\lambda + 17) \overbrace{(\lambda^2 - \lambda - 8)}^{c_A(\lambda)} + (89\lambda + 136).$$

Replacing λ by A (and 1 by I_2) gives us

$$A^5 = (A^3 + A^2 + 9A + 17) \overbrace{(A^2 - A - 8I_2)}^{c_A(A) = O_2} + (89A + 136I_2)$$
$$= 89A + 136I_2,$$

as before.

Finally, $A^2 - A - 8I_2 = 0$ implies $I_2 = \frac{1}{8}(A^2 - A)$, so that

$$A^{-1} = A^{-1} \cdot I_2 = A^{-1} \cdot \frac{1}{8}(A^2 - A) = \frac{1}{8}(A - I_2) = \begin{bmatrix} \frac{10}{8} & \frac{17}{8} \\ -\frac{6}{8} & -\frac{11}{8} \end{bmatrix}.$$

(Of course, we could have used the formula derived in Example 3.5.4.)

Sometimes the Cayley–Hamilton Theorem is stated picturesquely (perhaps violently) by saying that the characteristic polynomial of a matrix A **annihilates** A or is an **annihilating polynomial** of A: $c_A(A) = O_n$. Trivially, $kc_A(\lambda)$, where k is a scalar, also annihilates A, as does $q(\lambda)c_A(\lambda)$ for any polynomial q. The next example suggests there are many less trivial polynomials that annihilate a given $n \times n$ matrix A.

Example 4.8.3: A Nontrivial Annihilator

The matrix $A = \begin{bmatrix} 1 & 0 & 0 \\ 0 & 0 & 0 \\ 0 & 0 & 0 \end{bmatrix}$ has characteristic polynomial $c_A(\lambda) = \lambda^3 - \lambda^2 = \lambda(\lambda^2 - \lambda)$.

But $p(\lambda) = \lambda^2 - \lambda$ also annihilates A: $p(A) = A^2 - A = O_3$.

If we consider the set of all polynomials that annihilate a given $n \times n$ matrix A, we can find a polynomial $f(x)$ of smallest degree* such that $f(A) = O_n$. In fact, if $f(x) = a_k x^k + a_{k-1} x^{k-1} + \cdots + a_1 x + a_0$, where $a_k \neq 0$, is such a polynomial of smallest degree, we can divide through by a_k to get $g(x) = \frac{1}{a_k} f(x) = x^k + \left(\frac{a_{k-1}}{a_k}\right) x^{k-1} + \cdots + \left(\frac{a_1}{a_k}\right) x + \left(\frac{a_0}{a_k}\right)$, a **monic** polynomial (one whose highest power has coefficient 1) that annihilates A.

Furthermore, this annihilating monic polynomial of lowest degree must be *unique*. To see this, suppose k is the smallest degree of any nonzero polynomial that annihilates A and let f and g be two monic polynomials of degree k annihilating A. Then $h = f - g$ also annihilates A, and h has degree less than k because the terms x^k in f and g cancel each other in forming h. This contradiction of the minimal nature of k forces us to conclude that $h \equiv 0$ and $f \equiv g$. The uniqueness we have just established allows us to give the following definition.

Definition 4.8.1

If A is an $n \times n$ matrix, then the monic polynomial f of smallest degree such that $f(A) = O_n$ is called the **minimal polynomial** of A and is denoted by $m_A(x)$.

In Example 4.8.3, the second-degree polynomial p is the minimal polynomial of A. There is no monic linear polynomial that annihilates A.

As we will see later in this chapter and in subsequent chapters, the minimal polynomial is a valuable tool in investigating matrices and their behavior. The Cayley–Hamilton Theorem indicates that the degree of the minimal polynomial of an $n \times n$ matrix A is at most n, but this information provides only a partial insight. An important step in determining the minimal polynomial is proving that the minimal polynomial actually *divides* any nonzero polynomial annihilating A.

Theorem 4.8.2

The minimal polynomial of A divides every polynomial that annihilates A.[†]

Proof Let $f(x)$ be the minimal polynomial of A and suppose that g is a polynomial such that $g(A) = O_n$. Because f is nonzero, there exist (by the Division Algorithm) polynomials $q(x)$ and $r(x)$ such that $g(x) = f(x)q(x) + r(x)$, where the degree of $r(x)$ is less than that of

---

* This follows from the **Well-Ordering Principle** applied to the degrees of such annihilating polynomials: If S is a nonempty set of natural numbers, then S has a smallest element. See, for example, J. D'Angelo and D. West, *Mathematical Thinking: Problem-Solving and Proofs*, 2nd edn., chap. 3 (Upper Saddle River, NJ: Prentice-Hall, 2000).

† In more abstract algebraic terms, we can say that the minimal polynomial of A is the *unique monic generator of the ideal of polynomials which annihilate A*. See K. Hoffman, *Linear Algebra*, 2nd edn., p. 191 (Englewood Cliffs, NJ: Prentice-Hall, 1971).

$f(x)$. Then $O_n = g(A) = f(A)q(A) + r(A) = r(A)$. In other words, $r(x)$ annihilates A. By the minimality of f, it follows that $r(x) \equiv 0$ and so $g(x) = f(x)q(x)$—that is, f divides g.

A significant consequence of Theorem 4.8.2 is that **the minimal polynomial of A divides the characteristic polynomial of A**.

Example 4.8.4: The Minimal Polynomial of a Matrix

Let $A = \begin{bmatrix} 2 & 1 & 1 \\ 0 & 2 & 0 \\ 0 & 0 & 2 \end{bmatrix}$. The characteristic polynomial of A is

$(\lambda - 2)^3$. The minimal polynomial of A must divide $(\lambda - 2)^3$, so we have only three possibilities: (1) $\lambda - 2$; (2) $(\lambda - 2)^2$; and (3) $(\lambda - 2)^3$. First, $\lambda - 2$ is not the minimal polynomial because $A - 2I_3 \neq O_3$. Next, we find that $(A - 2I_3)^2 = \begin{bmatrix} 0 & 1 & 1 \\ 0 & 0 & 0 \\ 0 & 0 & 0 \end{bmatrix}^2 = O_3$, so the minimal polynomial of A is $m_A(\lambda) = (\lambda - 2)^2$.

The next two theorems describe important properties of the minimal polynomial.

Theorem 4.8.3

If A is an $n \times n$ matrix, then every eigenvalue of A is a zero of $m_A(x)$, the minimal polynomial of A. Furthermore, any zero of $m_A(x)$ is an eigenvalue of A.

Proof Because the minimal polynomial of A divides the characteristic polynomial of A—that is, $c_A(x) = q(x)m_A(x)$ for some polynomial $q(x)$—it is clear that any zero of $m_A(x)$ is a zero of $c_A(x)$—that is, an eigenvalue of A.

Now suppose that λ is an eigenvalue of A, $\mathbf{v}$ is an eigenvector corresponding to λ, and $f(x)$ is a polynomial. Then $A\mathbf{v} = \lambda\mathbf{v}$. Premultiplying by A, we get $A^2\mathbf{v} = \lambda\,A\mathbf{v} = \lambda^2\mathbf{v}$. Continuing this premultiplication by A, we see that $A^k\mathbf{v} = \lambda^k\mathbf{v}$ for all $k \geq 0$. Therefore, $f(A)\mathbf{v} = f(\lambda)\mathbf{v}$, which is easily proved. Because $\mathbf{v} \neq \mathbf{0}$, it follows that $f(\lambda)$ is an eigenvalue of $f(A)$. Thus, if $f(A) = O_n$, then $f(\lambda) = 0$. Replacing $f(x)$ by $m_A(x)$, the minimal polynomial of A, we see that any eigenvalue of A must be a zero of $m_A(x)$.

An equivalent way of stating Theorem 4.8.3 is that **the set of distinct zeros of the minimal polynomial of A coincides with the set of distinct eigenvalues of A**.

Theorem 4.8.4

Similar matrices have the same minimal polynomial.

Proof   Suppose that A and B are similar matrices and let $B = PAP^{-1}$. Then $f(B) = Pf(A)P^{-1}$ for any polynomial f.* Thus, $f(B) = O_n$ if $f(A) = O_n$. On the other hand, $f(A) = P^{-1}f(B)P$, so $f(A) = O_n$ if $f(B) = O_n$. Thus $f(A) = O_n$ if and only if $f(B) = O_n$, so A and B have the same minimal polynomial.

At this point—perhaps even earlier—the astute reader may be wondering (may have wondered) why we are paying so much attention to the minimal polynomial, when the characteristic polynomial is more easily calculated. The simple answer is that the minimal polynomial tells us things about a matrix that may not be obvious from the characteristic polynomial—for example, diagonalizability, as we will see in Theorem 4.8.5.

Example 4.8.5

The matrices $I_2 = \begin{bmatrix} 1 & 0 \\ 0 & 1 \end{bmatrix}$ and $A = \begin{bmatrix} 1 & 1 \\ 0 & 1 \end{bmatrix}$ both have the characteristic polynomial $(\lambda - 1)^2$; but I_2 is clearly diagonalizable while A is not (Example 4.5.4). Thus the characteristic equation alone cannot tell us if a matrix is diagonalizable. We note that the minimal polynomial of I_2 is $\lambda - 1$, while A has minimal polynomial $(\lambda - 1)^2$.

The last example provides a clue about the relationship between the minimal polynomial and diagonalizability. The next theorem supplies the answer by giving necessary and sufficient conditions for diagonalizability in terms of the factors of the minimal polynomial. Because one part of the proof is a bit intricate, we first establish a lemma that has some interest in its own right. The representation of a matrix A in hypothesis (1) of the lemma is called a **spectral decomposition** of A. Any theorem that asserts the existence of such a spectral decomposition is called a **spectral theorem** (see Problem C7 of Exercise 4.5).

Lemma 4.8.1[†]

Let A be an $n \times n$ matrix. If there exist an integer k, distinct scalars $d_1, d_2, \ldots, d_k$, and nonzero $n \times n$ matrices $E_1, E_2, \ldots, E_k$ such that (1) $\sum_{j=1}^{k} d_j E_j = A$; (2) $\sum_{j=1}^{k} E_j = I_n$; and (3) $E_i E_j = O_n$ if $i \neq j$, then

---

* If $f(x) = \sum_{k=0}^{n} c_k x^k$, then $f(B) = \sum_{k=0}^{n} c_k B^k = \sum_{k=0}^{n} c_k (PA^k P^{-1}) = P\left(\sum_{k=0}^{n} c_k A^k\right)P^{-1} = Pf(A)P^{-1}$. See problem B12 of Exercises 4.5.

[†] This lemma resembles *Cochran's Theorem*, a well-known statistical result used in the analysis of variance. See M.J.R. Healy, *Matrices for Statistics*, 2nd edn. (New York: Oxford University Press, 2000), Section 8.5 or W.C. Waterhouse, Cochran's theorem for rings, *Amer. Math. Monthly* **108** (2001): 58–59.

(a) The matrices E_i are *idempotent* (*i.e.*, $E_i^2 = E_i$ for every i).

(b) For an arbitrary vector $\mathbf{v} \in \mathbb{R}^n$, $E_i\mathbf{v}$ is either an eigenvector of A or the zero vector, $i = 1, 2, \ldots, k$.

Proof (a) For any i, $E_i = E_iI_n = E_i \sum_{j=1}^{k} E_j = \sum_{j=1}^{k} E_iE_j = E_i^2$ by hypotheses (2) and (3). Thus, E_i is idempotent for any i, $1 \leq i \leq k$.

(b) For any $i, i = 1, 2, \ldots, k$, $A(E_i \mathbf{v}) = \left(\sum_{j=1}^{k} d_jE_j\right)E_i\mathbf{v} = \left(\sum_{j=1}^{k} d_jE_jE_i\right)\mathbf{v} = d_iE_i^2\mathbf{v} = d_i(E_i\mathbf{v})$, where the last equality follows from conclusion (a). Clearly, if $E_i \mathbf{v} = \mathbf{0}$ we have $A(\mathbf{0}) = d_i(\mathbf{0})$. Otherwise, the equation $A(E_i\mathbf{v}) = d_i(E_i\mathbf{v})$ states that $E_i \mathbf{v}$ is an eigenvector of A with corresponding eigenvalue d_i.

Theorem 4.8.5

An $n \times n$ matrix A is diagonalizable if and only if its minimal polynomial is a product of distinct linear factors.

Proof If A is diagonalizable, then A is similar to a diagonal matrix D. The minimal polynomial of a diagonal matrix D is easily seen to be $(\lambda - d_1)(\lambda - d_2) \cdots (\lambda - d_k)$, where $d_1, d_2, \ldots, d_k$ are the distinct diagonal entries of D (see Exercise B3 in the following text). But Theorem 4.8.4 says that A and D, as similar matrices, have the same minimal polynomial. Thus, A has a minimal polynomial that is a product of distinct linear factors.

To prove the converse, we will construct a family of matrices $\{E_i\}$ satisfying the hypotheses of Lemma 4.8.1. Now suppose that A has a minimal polynomial:

$$f(\lambda) = (\lambda - d_1)(\lambda - d_2) \cdots (\lambda - d_k),$$

where $d_1, d_2, \ldots, d_k$ are distinct (possibly complex) numbers. Define k other polynomials p_i by the relations:

$$f(\lambda) = (\lambda - d_i)p_i(\lambda), \quad i = 1, 2, \ldots, k.$$

The degree of each polynomial p_i is $k - 1$ and $p_i(d_j) = 0$ if and only if $i \neq j$: clearly $p_i(d_j) = 0$ if $i \neq j$; but if $p_i(d_j) = 0$ for *all* i ($1 \leq i \leq k$), we have a polynomial of degree $k - 1$ that vanishes at k distinct values, implying that p_i is the zero polynomial.*

Now define the polynomial g by

$$g(\lambda) = 1 - \sum_{i=1}^{k} [p_i(d_i)]^{-1}p_i(\lambda), \tag{*}$$

---

* A polynomial of degree n cannot have more than n zeros (counting multiplicity) unless it is the zero polynomial. In our situation, the degree of p_i is less than k, but p_i has k zeros.

and notice that g is a polynomial of degree less than k such that $g(d_j) = 0$ for $j = 1, 2, \ldots, k$. Hence $g(x) = 0$ for *every* x, and each coefficient of g is 0. Hence $g(B) = O_n$ for every $n \times n$ matrix B.

Now consider the k matrices:

$$E_i = [p_i(d_i)]^{-1} p_i(A), \quad i = 1, 2, \ldots, k.$$

Because k is the least degree of any annihilating polynomial of A, and we have noted that the degree of each p_i is less than k, we see that $E_i \neq O_n$.

Furthermore, replacing λ by A in formula $(*)$ we have $\sum_{i=1}^{k} E_i = I_n$, hypothesis (2) of Lemma 4.8.1. Also,

$$f(A) = (A - d_i I_n) p_i(A) = O_n,$$

so $d_i p_i(A) = A p_i(A)$. Thus

$$\sum_{i=1}^{k} d_i E_i = \sum_{i=1}^{k} [p_i(d_i)]^{-1} A p_i(A) = A \sum_{i=1}^{k} [p_i(d_i)]^{-1} p_i(A) = A \sum_{i=1}^{k} E_i$$

$$= A \cdot I_n = A,$$

hypothesis (1) of Lemma 4.8.1. Finally, we establish hypothesis (3) of the lemma. If $i \neq j$, we have

$$E_i E_j = \left[p_i(d_i) p_j(d_j) \right]^{-1} p_i(A) p_j(A) = c f(A) h(A),$$

where c is a scalar and $h(A)$ is the product of the $k - 2$ matrices $A - d_r I_n$, $r \neq i$ and $i \neq j$. Therefore, because $f(A) = O_n, E_i E_j = O_n$ if $i \neq j$.

Because we have already established that $E_i \neq O_n$ for all i, conclusion (b) of the lemma implies that if $\mathbf{v} \in \mathbb{R}^n$, then

$$\mathbf{v} = I_n \mathbf{v} = \left(\sum_{i=1}^{k} E_i \right) \mathbf{v} = \sum_{i=1}^{k} (E_i \mathbf{v}).$$

Thus any vector $\mathbf{v}$ is a linear combination of eigenvectors of A. This means that if we combine all the bases of the eigenspaces of A, we get n linearly independent eigenvectors. By Theorem 4.5.3, A is diagonalizable.

Example 4.8.5 is more meaningful in light of this last theorem. The matrix I_2 has minimal polynomial $\lambda - 1$ and is therefore diagonalizable, whereas $A = \begin{bmatrix} 1 & 1 \\ 0 & 1 \end{bmatrix}$ cannot be diagonalized because its minimal polynomial has repeated linear factors. Exercise C3 in the following text asks for the application of the various ideas in the proof of Theorem 4.8.5 to a concrete example.

Exercises 4.8

A.

1. For each matrix, find its characteristic polynomial and verify the Cayley–Hamilton Theorem.

   a. $\begin{bmatrix} 1 & 1 \\ 0 & 1 \end{bmatrix}$ b. $\begin{bmatrix} 1 & 2 \\ 3 & -4 \end{bmatrix}$ c. $\begin{bmatrix} -2 & 5 & 7 \\ 1 & 0 & -1 \\ -1 & 1 & 2 \end{bmatrix}$

   d. $\begin{bmatrix} 1 & 0 & 1 \\ 0 & 1 & 0 \\ 1 & 0 & 1 \end{bmatrix}$ e. $\begin{bmatrix} -3 & -7 & 19 \\ -2 & -1 & 8 \\ -2 & -3 & 10 \end{bmatrix}$ f. $\begin{bmatrix} 5 & 0 & 13 \\ 1 & 3 & 14 \\ -2 & 0 & -5 \end{bmatrix}$

2. Show directly that any 2×2 matrix $\begin{bmatrix} a & b \\ c & d \end{bmatrix}$ satisfies its characteristic equation.

3. Given $A = \begin{bmatrix} -3 & 2 & 2 \\ -12 & 7 & 6 \\ 0 & 0 & 1 \end{bmatrix}$, find $f(A) = A^4 - 4A^3 + 4A^2 - 4A + 3I_3$.

4. Show directly that the matrix $A = \begin{bmatrix} a & b & 0 \\ c & d & 0 \\ 0 & 0 & k \end{bmatrix}$, where $a, b, c, d, k \in \mathbb{R}$, satisfies its characteristic equation.

5. Find A^{-1}, where A is the matrix in Exercise 3.

6. Suppose $A = \begin{bmatrix} 3 & 4 \\ 1 & 3 \end{bmatrix}$.
   a. By diagonalizing A, find A^3 and A^{-1}.
   b. Use the Cayley–Hamilton Theorem to evaluate A^3 and A^{-1} in terms of I_2 and powers of A. (Make sure your answers agree with those given in part (a).)

7. If $A = \begin{bmatrix} -3 & -4 \\ 2 & 3 \end{bmatrix}$, find A^{593}. [*Hint*: Use algebraic long division and note the pattern of the quotient terms.]

8. Let $A = \begin{bmatrix} 1 & 0 \\ 0 & 1 \end{bmatrix}$ and $B = \begin{bmatrix} 1 & 1 \\ 0 & 1 \end{bmatrix}$. Show that B has the same rank, determinant, trace, and characteristic polynomial as A, but that, nevertheless, B is not similar to A.

9. Solve the problem in Example 4.4.7 by using the Cayley–Hamilton Theorem.

10. Find the minimal polynomial of each of the following matrices:

a. $\begin{bmatrix} 3 & 2 \\ 1 & 2 \end{bmatrix}$ b. $\begin{bmatrix} 1 & 1 \\ -2 & 4 \end{bmatrix}$ c. $\begin{bmatrix} 3 & 1 & 0 \\ 1 & 3 & 0 \\ 0 & 0 & 2 \end{bmatrix}$ d. $\begin{bmatrix} 2 & 1 & 0 \\ 0 & 2 & 1 \\ 0 & 0 & 2 \end{bmatrix}$

11. Use Theorem 4.8.5 to show that the matrix $\begin{bmatrix} 1 & c \\ 0 & 1 \end{bmatrix}$, where $c \neq 0$, is not similar to a diagonal matrix. (See Exercise 8 earlier for a concrete example.)

12. Find a 3×3 matrix for which the minimal polynomial is x^2.

B.

1. Criticize the following alleged proof of the Cayley–Hamilton Theorem: The characteristic polynomial of A is $p_A(\lambda) = \det(A - \lambda I)$. Substituting $\lambda = A$ on both sides, we get $p_A(A) = \det(A - AI) = 0$. Therefore, the matrix A satisfies its own characteristic equation.

2. Suppose that A is a singular $n \times n$ matrix.
   a. Prove that if $n = 2$, then A^2 is proportional to A.
   b. Show that A^2 need not be proportional to A whenever $n > 2$.

3. Prove that the minimal polynomial of a diagonal matrix A is $(\lambda - d_1)(\lambda - d_2) \cdots (\lambda - d_k)$, where $d_1, d_2, \ldots, d_k$ are the distinct diagonal entries of A.

4. Consider $A = \begin{bmatrix} 0 & 1 \\ 0 & 0 \end{bmatrix}$ and $B = \begin{bmatrix} 0 & 0 \\ 0 & 1 \end{bmatrix}$ to show that the minimal polynomials of AB and BA need not be the same. The characteristic polynomials of AB and BA are the same, however (see Problem C2 in Exercise 4.4). Explain why there is this difference between the characteristic and minimal polynomials.

5. Let A and B be 3×3 nilpotent matrices of index k (see Definition 3.3.3). Show that A and B are similar if and only if A and B have the same minimal polynomial. Is this true for 4×4 nilpotent matrices?

C.

1. Suppose that $A = \begin{bmatrix} a & b \\ c & d \end{bmatrix}$ has two unequal eigenvalues.
   a. Show that $A^n = \beta_1 I + \beta_2 A$, where β_1 and β_2 are constants.
   b. Find formulas for β_1 and β_2 involving n and the unequal eigenvalues λ_1 and λ_2.

2. If $M = \begin{bmatrix} 1 - a + a^2 & 1 - a \\ a - a^2 & a \end{bmatrix}$, where a is a real number, determine M^n for any positive integer n.* [*Hint*: Consider the case $a = -1$ separately.]

3. Suppose that $A = \begin{bmatrix} 2 & 2 & -5 \\ 3 & 7 & -15 \\ 1 & 2 & -4 \end{bmatrix}$. Using the proof of Theorem 4.6.5 as a model, go through each of the following steps:
   a. Determine the characteristic polynomial of A.
   b. Determine the minimal polynomial of A.
   c. Find the polynomials $p_i(\lambda)$.
   d. Determine the matrices E_i.
   e. Show that $\sum_{j=1}^{k} E_j = I_3$ for the appropriate value of k.
   f. Show that $E_i E_j = O_3$ for $i \neq j$.
   g. Show that for any nonzero $\mathbf{v} \in \mathbb{R}^3$, $E_i \mathbf{v}$ is an eigenvector of A $(i = 1, 2, \ldots, k)$.
   h. Show that for any nonzero vector $\mathbf{v} \in \mathbb{R}^3$, the set $\{E_i \mathbf{v}\}$ spans $\mathbb{R}^3$.
   i. Find a basis for each eigenspace of A.
   j. Diagonalize A.

4. Let the minimal polynomial of an $n \times n$ matrix A be $\prod (x - \lambda_i)^{n_i}$. Prove that the minimal polynomial of $M = \begin{bmatrix} A & I_n \\ O_n & A \end{bmatrix}$ is $\prod (x - \lambda_i)^{n_i + 1}$ as follows:
   a. If p is any polynomial, show that $p\left(\begin{bmatrix} A & I_n \\ O_n & A \end{bmatrix} \right) = \begin{bmatrix} p(A) & p'(A) \\ O_n & p(A) \end{bmatrix}$, where p' denotes the derivative of p.
   b. Show that if $q(x) = \prod (x - \lambda_i)^{n_i}$ is the minimal polynomial of A and p is an annihilating polynomial of M, then p and p' are divisible by q.
   c. Using the result of part (b), show that among all annihilating polynomials p, the polynomial $\prod (x - \lambda_i)^{n_i + 1}$ has the lowest degree.

4.9 Summary

Associated with any square matrix A is a unique number called the **determinant** of A, $\det(A)$. If we let $U = [u_{ij}]$, then $\det(A) = (-1)^k \prod_{i=1}^{n} u_{ii}$ if $PA = LU$, where k is the number of row interchanges represented by P. If A is nonsingular, then the LU or $PA = LU$ decomposition of a matrix A is unique, so $\det(A)$ is well defined. Some properties

---

* This is Problem 64-19 by J.F. Foley, *N*th power of a matrix, *SIAM Rev.* **6** (1964): 456. The problem arose in determining the result of repeated partial mixing of two completely miscible fluids, and it has also appeared in electrical circuit analysis.

of the determinant follow almost immediately from the definition: (1) If A is singular, $\det(A) = 0$; and (2) the determinant of any triangular matrix is the product of its main diagonal elements. In particular, $\det(I_n) = 1$.

Some other properties of determinants are (1) If A has two identical rows, $\det(A) = 0$. (2) If B is obtained from A by interchanging two rows of A, $\det(B) = -\det(A)$; (3) If B is obtained from A by multiplying any row of A by a scalar k, $\det(B) = k\det(A)$. (3) If B is obtained from A by adding a scalar multiple of a row of A to another row of A, $\det(B) = \det(A)$. (4) $\det(AB) = \det(A)\det(B)$. (5) A is invertible if and only if $\det(A) \neq 0$. (6) $\det(A^T) = \det(A)$. (7) If A is invertible, $\det(A^{-1}) = \frac{1}{\det(A)}$. Property 6 implies that we can interchange the words "row" and "column" in any theorem about determinants.

A result that is more useful theoretically than computationally is **Cramer's Rule,** which gives the unique solution (if there is one) of an $n \times n$ system of linear equations as the quotient of two $n \times n$ determinants.

[Geometrically, the determinant of a 2×2 matrix A represents the **signed area** of the parallelogram spanned by the column vectors of A. In $\mathbb{R}^2$ the absolute value of the determinant of A is a **scaling** or **magnification factor**, measuring the change in area when every vector in a plane figure is multiplied by A. In $\mathbb{R}^3$, a determinant provides the **signed volume** of the parallelepiped spanned by the columns of a matrix. More abstractly, the determinant of an $n \times n$ matrix is the unique **alternating multilinear function** satisfying the three conditions given in Definition 4.2.1.]

If A is an $n \times n$ matrix, an **eigenvector** of A is a nonzero vector $\mathbf{v}$ such that $A\mathbf{v} = \lambda\mathbf{v}$ for some scalar λ. The scalar λ is called an **eigenvalue** of A, or the **eigenvalue of A associated with $\mathbf{v}$**. An ordered pair $(\lambda, \mathbf{v})$, with λ an eigenvalue of A and $\mathbf{v}$ its corresponding eigenvector, is called an **eigenpair** for A. Every $n \times n$ matrix has an eigenvalue (possibly complex). For an $n \times n$ matrix A and a scalar λ, the following statements are equivalent: (i) λ is an eigenvalue of A; (ii) the matrix $A - \lambda I_n$ is singular; and (iii) $\det(A - \lambda I_n) = 0$. Thus, a matrix is nonsingular if and only if 0 is not an eigenvalue of A.

The determinant of the matrix $\lambda I_n - A$ is an nth degree polynomial in λ called the **characteristic polynomial** $c_A(\lambda)$ of A. The zeros of the characteristic polynomial are the eigenvalues of A, from which the eigenvectors are found. In general, $\det(\lambda I_n - A) = c_A(\lambda) = \lambda^n - \text{trace}(A)\lambda^{n-1} + \cdots (-1)^n \det(A)$. Det (A) is the product of its eigenvalues, counting multiplicity. The **trace** of a square matrix—the sum of its main diagonal entries—is the sum of its eigenvalues (counting multiplicity). The eigenvalues of a triangular matrix are the entries on its main diagonal.

Any set of eigenvectors corresponding to distinct eigenvalues of a matrix is linearly independent. The eigenvalue associated with an eigenvector is unique. However, an eigenvalue may have more than one corresponding eigenvector. Given an $n \times n$ matrix A and a scalar λ, the set of all eigenvectors of A with associated eigenvalue λ, together with the zero vector, is a subspace of $\mathbb{R}^n$ called the **eigenspace of A associated with** (or **corresponding to**) λ. We denote this subspace by $S_\lambda(A)$.

Two $n \times n$ matrices A and B are **similar** if there exists an invertible matrix P such that $A = PBP^{-1}$. If A and B are similar $n \times n$ matrices, then

(1) the characteristic equations of A and B are the same; (2) $\det(A) = \det(B)$; and (3) the eigenvalues of B are the same as those of A.

A square matrix is **diagonalizable** if it is similar to a diagonal matrix. An $n \times n$ matrix A is diagonalizable if and only if A has n linearly independent eigenvectors. Furthermore, if A is diagonalizable, then A is similar to a diagonal matrix whose elements are the corresponding eigenvalues $\lambda_1, \lambda_2, \ldots, \lambda_n$, possibly repeated. An $n \times n$ matrix with n distinct eigenvalues is diagonalizable.

When the characteristic polynomial of an $n \times n$ matrix A is written in the form $\det(\lambda I_n - A) = (\lambda - \lambda_1)^{m_1}(\lambda - \lambda_2)^{m_2} \cdots (\lambda - \lambda_k)^{m_k}$, with $\lambda_i \neq \lambda_j$ for $1 \leq i \neq j \leq k$ and $m_1 + m_2 + \cdots + m_k = n$, the positive integer m_i is called the **algebraic multiplicity** of the eigenvalue λ_i. The number of linearly independent eigenvectors corresponding to an eigenvalue λ is called the **geometric multiplicity** of λ. The geometric multiplicity of an eigenvalue λ is the dimension of the eigenspace $S_\lambda(A)$ corresponding to λ. If λ is an eigenvalue of an $n \times n$ matrix A, the geometric multiplicity of λ is less than or equal to the algebraic multiplicity of λ. The following statements are equivalent: (1) Matrix A is diagonalizable; and (2) for every eigenvalue λ of A, the geometric multiplicity of λ equals the algebraic multiplicity of λ.

The **Cayley–Hamilton Theorem** states that **every square matrix satisfies its own characteristic equation**—that is, if an $n \times n$ matrix A has the characteristic polynomial $p_A(\lambda)$, then $p_A(A) = O_n$. The monic polynomial f of smallest degree such that $f(A) = O_n$ is called the **minimal polynomial** of A, denoted by $m_A(\lambda)$. The minimal polynomial of A divides every polynomial that annihilates A, in particular the characteristic polynomial of A. If A is an $n \times n$ matrix, then every eigenvalue of A is a zero of $m_A(\lambda)$. Furthermore, any zero of $m_A(\lambda)$ is an eigenvalue of A. Similar matrices have the same minimal polynomial; and a matrix is diagonalizable if and only if its minimal polynomial is a product of distinct linear factors.

5

Vector Spaces

In previous chapters, we have worked with vectors and matrices algebraically, combining like mathematical objects, solving equations, and so forth. If we look carefully at the basic properties of vector operations (Theorem 1.1.1) and at the characteristics of matrix algebra (Theorems 3.1.1 and 3.1.2), we notice some striking similarities. We have defined operations of addition and scalar multiplication for both vectors and matrices, and these operations follow rules that are identical, if you ignore the obvious differences between vectors (rows or columns) and matrices (generally rectangular). In Chapter 1, some proofs of results about $\mathbb{R}^n$ (e.g., see Theorem 1.4.1 or Theorem 1.5.1) depend on general algebraic properties and do not make explicit use of the fact that the vectors are n-tuples of real numbers.

These operations and algebraic properties (for example, commutativity and associativity) are familiar to us from arithmetic as well as from more advanced areas of mathematics. For example, in studying the concept of a *function*, we define the sum and difference of two functions f and g as

$$(f + g)(x) = f(x) + g(x)$$
$$(f - g)(x) = f(x) - g(x)$$

for all x in the intersection of the domain of f and the domain of g. We also describe the multiplication of a function by a real number: $(c \cdot f)(x) = c \cdot f(x)$, for any real number c and all x in the domain of f. If we take these definitions and explore their consequences, we find that functions possess many of the same algebraic properties as vectors and matrices.

In this chapter, we examine several kinds of familiar (and unfamiliar) mathematical objects to focus on their common algebraic structure, in particular to marvel at their resemblance to vectors as we have seen them so far, as elements of $\mathbb{R}^n$.

5.1 Vector Spaces

The investigation of the common algebraic behavior of different sets of mathematical objects led, in the late 1880s, to formal statements of the important common properties of such mathematical systems. The system we will study now is called a *vector space* or a *linear space*.*

---

* The axiomatic definition of a vector space is developed from the work of Hermann Grassmann (1808–1887), Giuseppe Peano (1858–1932), and Hermann Weyl (1885–1955).

Definition 5.1.1

Let V be a nonempty set on which two operations, *addition* and *scalar multiplication*, are defined. This means that if $\mathbf{u}$ and $\mathbf{v}$ are elements of V, then $\mathbf{u} + \mathbf{v}$ is a unique element in V and $c \cdot \mathbf{u}$ is a unique element in V for every real number c [*Closure*]. The set V together with the operations of addition and scalar multiplication is called a **vector space**—and the elements of V are called **vectors**—if the following rules (axioms) are satisfied for every $\mathbf{u}, \mathbf{v}$, and $\mathbf{w}$ in V and all scalars (real numbers) c and d.

(1) $\mathbf{u} + \mathbf{v} = \mathbf{v} + \mathbf{u}$.   [*Commutativity*]
(2) $\mathbf{u} + (\mathbf{v} + \mathbf{w}) = (\mathbf{u} + \mathbf{v}) + \mathbf{w}$. [*Associativity*]
(3) There exists an element $\mathbf{0}$ in V such that $\mathbf{u} + \mathbf{0} = \mathbf{u} = \mathbf{0} + \mathbf{u}$ for each $\mathbf{u} \in V$.   [*Zero element*]
(4) For each $\mathbf{u} \in V$, there is an element $-\mathbf{u}$ in V such that $\mathbf{u} + (-\mathbf{u}) = \mathbf{0}$.   [*Negative of an element*]
(5) $c \cdot (\mathbf{u} + \mathbf{v}) = c \cdot \mathbf{u} + c \cdot \mathbf{v}$.   [*Distributivity*]
(6) $(c + d) \cdot \mathbf{u} = c \cdot \mathbf{u} + d \cdot \mathbf{u}$.   [*Distributivity*]
(7) $c \cdot (d \cdot \mathbf{u}) = (cd) \cdot \mathbf{u}$.   [*Associativity*]
(8) $1 \cdot \mathbf{u} = \mathbf{u}$ for every $\mathbf{u} \in V$.   [*Identity*]

Note that we can define the *subtraction* of vectors in any vector space by writing $\mathbf{u} - \mathbf{v} = \mathbf{u} + (-\mathbf{v})$, just as we did in $\mathbb{R}^n$.

To emphasize the fact that a vector space consists of a set together with two operations, we will sometimes use the notation $\{V, +, \cdot\}$, where the first operation, $+$, represents vector addition and the second symbol, $\cdot$, denotes the multiplication of a vector by a scalar. Most of the time, we will simply use V or some other capital letter to denote a vector space.

Technically, we have just defined a **real vector space**, or a **vector space over** $\mathbb{R}$, because we are using the term *scalar* to mean *real number*. In later chapters, we will allow the scalars to be complex numbers. In general, abstract vector space theory assumes that the scalars come from a mathematical structure called a **field**, of which the real and complex number systems are specific examples.*

To understand the power and generality of the vector space concept, we examine some important examples.

Example 5.1.1: Some Vector Spaces

(a) Looking back at Theorem 1.1.1, it is obvious that $\{\mathbb{R}^n, +, \cdot\}$, where $+$ denotes the addition of n-tuples and $\cdot$ signifies multiplication of an n-tuple by a real number, is a vector space for any positive integer n.

---

* See, for example, the definition of **field** given at the very beginning of P.R. Halmos, *Finite-Dimensional Vector Spaces* (New York: Springer-Verlag, 1987).

(b) The **complex number system** $\mathbb{C} = \left\{ x + iy \mid x, y \in \mathbb{R}; i = \sqrt{-1} \right\}$ is a vector space over $\mathbb{R}$. (See Appendix D for algebraic properties of complex numbers.)

(c) For positive integers m and n, the set of all $m \times n$ matrices with real entries is a vector space. Theorems 3.1.1 and 3.1.2 confirm that the axioms of Definition 5.1.1 are satisfied. *From now on, we will denote this vector space by M_{mn}.* Thus any $m \times n$ matrix will be called a *vector* in this new, general sense. If $m = n$, we will just write M_n; context and spacing will determine if, for example, M_{42} refers to 42×42 matrices or to 4×2 matrices: M_{42} vs. M_{42}. Also, $M_{42,3}$ represents the set of all 42×3 real matrices together with the operations of addition and scalar multiplication of matrices, whereas $M_{4,23}$ and M_{423} represent spaces of 4×23 and 423×423 matrices, respectively.

(d) For any positive integer n, define $P_n[x] = \{a_n x^n + a_{n-1} x^{n-1} + \cdots + a_1 x + a_0 \mid a_i \in \mathbb{R} \text{ for } i = 0, 1, 2, \ldots, n\}$, the set of all polynomials (in the variable x) of degree less than or equal to n with real coefficients. We define the sum of two elements of $P_n[x]$ in the usual way, by combining "like terms." For example, if $-2x^3 + 3x^2 - x + 4$ and $4x^2 - 3x - 5$ are two elements of $P_3[x]$, then we can add them to get the element $-2x^3 + 7x^2 - 4x - 1$ in $P_3[x]$. Multiplying a polynomial in $P_n[x]$ by a real number is done in the obvious way:

$$c \cdot \left(a_n x^n + a_{n-1} x^{n-1} + \cdots + a_1 x + a_0 \right)$$
$$= (ca_n)x^n + (ca_{n-1})x^{n-1} + \cdots + (ca_1)x + (ca_0).$$

It is easy to show that $P_n[x]$ is a vector space with these operations.

(e) $C[a, b]$, the set of all real-valued functions that are continuous on the interval $[a, b]$, is a vector space if we define the addition of functions and the scalar multiple of a function in the usual ways. It is a basic calculus fact that the sum of a finite number of continuous functions is again a continuous function. Clearly the *zero function*, the function that is identically equal to zero on the interval $[a, b]$, is continuous and has the appropriate algebraic property. It is easy to verify all the axioms of a vector space for $\{C[a, b], +, \cdot\}$ (Exercise A1).

It is important to realize that not every set with an addition and a scalar multiplication defined is a vector space.

Example 5.1.2: Not a Vector Space

For a particular positive integer n, suppose we modify part (d) of Example 5.1.1 and consider the set S of all nth degree polynomials with real coefficients—that is, the subset of $P_n[x]$ consisting of polynomials whose degree is exactly n. It is easy to show that $\{S,+,\cdot\}$ is not a vector space. For example, the set S is not closed under addition. To illustrate this, we can take $-3x^n + 2x^2 + 7x - 2$ and $3x^n - 4x^2 + 5$, both elements of S. The sum of these elements is $-2x^2 + 7x + 3$, which is *not* an element of S. (Exercise A2 asks for a check of the other axioms for this example.)

The last example highlights two important aspects of vector spaces. The first is that *closure* is an important part of the definition of a vector space that is often overlooked. In fact, we can highlight closure by adding two properties to the numbered properties in Definition 5.1.1: (I) If $\mathbf{u}$ and $\mathbf{v}$ are elements of V, then $\mathbf{u} + \mathbf{v} \in V$; and (II) if $\mathbf{u}$ is in V, then $c \cdot \mathbf{u} \in V$ for every real number c. To be a vector space, $\{V, +, \cdot\}$ must satisfy all *ten* axioms.

The second observation, illustrated by Example 5.1.2, is that *a subset of a vector space is not necessarily a vector space in its own right*. We will say more about this phenomenon in Section 5.2.

With little effort, we can establish some simple properties of a vector space—useful algebraic properties that we will use often, usually without thinking. They are very easy to prove in $\mathbb{R}^n$, and almost as easy to demonstrate in a more general setting.

Theorem 5.1.1

If $\mathbf{u}$ and $\mathbf{v}$ are vectors in a vector space V, it follows that

(a) $0 \cdot \mathbf{u} = \mathbf{0}$ and $c \cdot \mathbf{0} = \mathbf{0}$, where 0 is the scalar zero, $\mathbf{0}$ is the zero vector of V, and c is any scalar.
(b) if $\mathbf{u} + \mathbf{v} = \mathbf{0}$, then $\mathbf{u} = -\mathbf{v}$.
(c) $(-1) \cdot \mathbf{u} = -\mathbf{u}$.
(d) if $c \cdot \mathbf{u} = \mathbf{0}$, then either $c = 0$ or $\mathbf{u} = \mathbf{0}$ (or both).

Proof of (a)   Using property (6) in Definition 5.1.1, we can write

$$0 \cdot \mathbf{u} = (0 + 0) \cdot \mathbf{u} = 0 \cdot \mathbf{u} + 0 \cdot \mathbf{u}.$$

Add the negative of $0 \cdot \mathbf{u}$—that is, $-(0 \cdot \mathbf{u})$—to the far left and far right sides of the last equation to get

$$0 \cdot \mathbf{u} + [-(0 \cdot \mathbf{u})] = (0 \cdot \mathbf{u} + 0 \cdot \mathbf{u}) + [-(0 \cdot \mathbf{u})],$$

or

$$\mathbf{0} = 0 \cdot \mathbf{u} + ((0 \cdot \mathbf{u}) + -(0 \cdot \mathbf{u})) = 0 \cdot \mathbf{u} + \mathbf{0} = 0 \cdot \mathbf{u},$$

using axioms (2), (3), and (4). This proves the first part of (a). The second part is left as Exercise A3.

Proof of (b)   If $\mathbf{u} + \mathbf{v} = \mathbf{0}$, then $(\mathbf{u} + \mathbf{v}) + (-\mathbf{v}) = \mathbf{0} + (-\mathbf{v})$ or, using associativity, $\mathbf{u} + (\mathbf{v} + (-\mathbf{v})) = -\mathbf{v}$. This gives $\mathbf{u} + \mathbf{0} = -\mathbf{v}$, or $\mathbf{u} = -\mathbf{v}$.

Proof of (c)   Using axioms (6), (8), and part (a), we can write

$$\mathbf{u} + (-1) \cdot \mathbf{u} = 1 \cdot \mathbf{u} + (-1) \cdot \mathbf{u} = (1 + (-1)) \cdot \mathbf{u} = 0 \cdot \mathbf{u} = \mathbf{0}.$$

Hence $(-1) \cdot \mathbf{u} = -\mathbf{u}$ by part (b).

Proof of (d)   Suppose that $c \cdot \mathbf{u} = \mathbf{0}$ and $c \neq 0$. Then we have

$$\mathbf{u} = 1 \cdot \mathbf{u} = \left(\frac{1}{c}c\right) \cdot \mathbf{u} = \frac{1}{c}(c \cdot \mathbf{u}) = \frac{1}{c} \cdot \mathbf{0} = \mathbf{0}.$$

Notice that if $c \cdot \mathbf{u} = \mathbf{0}$ and $\mathbf{u} \neq \mathbf{0}$, this forces $c = 0$. (Otherwise, if $c \neq 0$, we proceed as in the first part of this proof and conclude that $\mathbf{u} = \mathbf{0}$—a contradiction to our assumption.)

Exercises 5.1

A.

1. Verify that $\{C[a, b], +, \cdot\}$, as defined in part (e) of Example 5.1.1, is a vector space.

2. Determine which axioms of a vector space are satisfied by $\{S, +, \cdot\}$ in Example 5.1.2.

3. Prove that if $\mathbf{0}$ is the zero vector of a vector space V, then $c \cdot \mathbf{0} = \mathbf{0}$, where c is any scalar. (See the proof of Theorem 5.1.1, part (a).)

4. Suppose $S = \mathbb{R}^3$ with the usual definition of vector addition, but with scalar multiplication defined as follows:
   $$c \cdot \begin{bmatrix} x \\ y \\ z \end{bmatrix} = \begin{bmatrix} cx \\ cy \\ 0 \end{bmatrix}.$$ Is $\{S, +, \cdot\}$ a vector space? Explain your answer.

5. Suppose $S = \left\{ \begin{bmatrix} a \\ b \end{bmatrix} : a, b \in \mathbb{R}; a > 0, b > 0 \right\}$. Define the addition of elements of S by $\begin{bmatrix} a \\ b \end{bmatrix} \oplus \begin{bmatrix} c \\ d \end{bmatrix} = \begin{bmatrix} ac \\ bd \end{bmatrix}$ and define scalar multiplication by $k \odot \begin{bmatrix} a \\ b \end{bmatrix} = \begin{bmatrix} a^k \\ b^k \end{bmatrix}$, where k is any real number. Show that $\{S, \oplus, \odot\}$ is a vector space.

6. Let $\mathbb{R}^+$ denote the set of all positive real numbers.

    a. If we define the "addition" $\alpha \boxplus \beta = \alpha\beta$, $\alpha, \beta \in \mathbb{R}^+$, and the "scalar multiplication" $a \boxdot \alpha = \alpha^a$, $\alpha \in \mathbb{R}^+$, $a \in \mathbb{R}$, is $\mathbb{R}^+$ a vector space over $\mathbb{R}$?

    b. If we keep the vector addition in part (a), but change the definition of scalar multiplication to $a \boxtimes \alpha = a^\alpha$, $\alpha \in \mathbb{R}^+$, $a \in \mathbb{R}$, is $\mathbb{R}^+$ a vector space over $\mathbb{R}$?

7. Show that the set of all 3×3 diagonal matrices with the usual operations of addition and scalar multiplication is a vector space.

8. Show that the set of all real-valued functions f that have at least two derivatives is a vector space V if we define the addition and scalar multiplication of functions in the usual way.

9. Show that the set of all $n \times n$ semimagic squares with sum s (see Exercises 4.4, B19) is a vector space.

10. A *magic square* is an $n \times n$ matrix with real entries in which each row, each column, and each diagonal [entries $(1, 1)$ to (n, n) and $(1, n)$ to $(n, 1)$] have the same sum s. If T is the set of all $n \times n$ magic squares with $s \neq 0$, is T a vector space with the usual matrix addition and scalar multiplication? Does your answer change if $s = 0$?

B.

1. Why does a real vector space have either one vector or infinitely many vectors?

2. Prove that the commutative law (property (1) of Definition 5.1.1) follows from the other axioms of a vector space.

3. Let S be the set of all elements $\mathbf{u}$ of $\mathbb{R}^3$ such that $\|\mathbf{u}\| = 1$, where $\|\cdot\|$ denotes the Euclidean norm (Section 1.2). Is $\{S, +, \cdot\}$ a vector space with the usual definition of n-tuple addition and scalar multiplication?

4. Define $P[x]$ to be the set of all polynomials (of all degrees) in the variable x, and define the addition of polynomials and multiplication of a polynomial by a scalar in the usual ways. Determine which of the following sets are vector spaces:

    a. $\{p \in P[x] \mid p(0) = 0\}$

    b. $\{p \in P[x] \mid 2p(0) - 3p(1) = 0\}$

    c. $\{p \in P[x] \mid p(0) = 1\}$

    d. $\{p \in P[x] \mid p(1) + p(2) + \cdots + p(k) = 0; \ k$ a fixed real number$\}$

5. Show that the set of functions $\{a\cos(t + b) \mid a, b \in \mathbb{R}\}$ is a vector space under the usual operations of function addition and scalar multiplication.

6. Let $\{C_{\frac{2}{3}}[0, 1], +, \cdot\}$ denote the set of all continuous functions on the interval $[0, 1]$ that vanish at $x = \frac{2}{3}$, together with the usual operations of function addition and scalar multiplication. Is $\{C_{\frac{2}{3}}[0, 1], +, \cdot\}$ a vector space?

7. A real function defined on an interval $[a, b]$ is called a *simple function* if it takes on (i.e., equals) only finitely many values on $[a, b]$.* Is the set of all simple functions on $[a, b]$, with the usual addition of functions and multiplication by a real number, a vector space?

8. Suppose that S is the set of all functions that satisfy the differential equation

$$f'' + f = 0.$$

(For example, $f(x) = \sin x$ and $f(x) = \cos x$ are in S.) Show that $\{S, +, \cdot\}$ is a vector space with the usual operations of function addition and scalar multiplication.

9. Prove that the set of all 2×2 matrices that commute with $\begin{bmatrix} 1 & 1 \\ 0 & 1 \end{bmatrix}$ (with respect to matrix multiplication) is a vector space.

C.

1. Is the set of all $n \times n$ nonsymmetric matrices, with the usual matrix operations, a vector space? (A *nonsymmetric matrix A* is a square matrix such that $A \neq A^{\mathsf{T}}$.)

2. Let S be the set of all infinite sequences of real numbers $\mathbf{x} = (a_1, a_2, \ldots, a_n, \ldots)$, where operations on S are defined as follows: (1) If $\mathbf{x} = (a_1, a_2, \ldots, a_n, \ldots)$ and $\mathbf{y} = (b_1, b_2, \ldots, b_n, \ldots)$, then $\mathbf{x} + \mathbf{y} = (a_1 + b_1, a_2 + b_2, \ldots, a_n + b_n, \ldots)$; (2) for any $\lambda \in \mathbb{R}$, $\lambda \mathbf{x} = (\lambda a_1, \lambda a_2, \ldots, \lambda a_n, \ldots)$. Is S a vector space?

3. Let $\mathfrak{F}$ be the set of all infinite sequences $(a_1, a_2, \ldots, a_n, \ldots)$ of real numbers whose entries satisfy the relation $a_k = a_{k-1} + a_{k-2}$ for $k = 3, 4, \ldots$. The operations on elements of $\mathfrak{F}$ are defined as in the previous problem. Is $\mathfrak{F}$ a vector space?

5.2 Subspaces

As Example 5.1.2 showed, a subset of a vector space is not necessarily a vector space on its own. Given a vector space $\{V, +, \cdot\}$, it is significant when a subset of vectors $S \subseteq V$ forms a vector space with respect to the

---

* More precisely, given subsets $A_1, A_2, \ldots, A_n$ of a set A, a **simple function** defined on A is a finite sum $\sum_{k=1}^{n} a_k \chi_{A_k}$, where $a_k \in \mathbb{R}$ and $\chi_{A_k}(x) = 1$ if $x \in A_k$ and $= 0$ if $x \in A \backslash A_k$.

same operations of addition and scalar multiplication as its "parent." For example, it happens (see Problem A7 of Exercises 5.1) that the set

$$S = \left\{ \begin{bmatrix} a & 0 & 0 \\ 0 & b & 0 \\ 0 & 0 & c \end{bmatrix} : a, b, c \in \mathbb{R} \right\}$$

of all 3×3 diagonal matrices is a subset of M_3 that satisfies all the properties of a vector space—that is, $\{S, +, \cdot\}$ is a vector space contained within the larger vector space M_3.

Definition 5.2.1

Let S be a nonempty subset of a vector space $\{V, +, \cdot\}$. If S is also a vector space using the same addition and scalar multiplication, then S is called a **subspace** of $\{V, +, \cdot\}$.

The subset $S = \{\mathbf{0}\}$, consisting of the zero element alone, and V itself are called **trivial subspaces**. Otherwise, a subspace is called **nontrivial**.

If S is a subspace of V, then S is a miniature version of V that has the same algebraic structure as V. Before considering additional examples, we should realize that it is not necessary to verify all eight properties given in Definition 5.1.1 to determine if a subset of a vector space is a subspace. We are saved the tedium of all this checking by the following fact (see Definition 1.6.1).

Theorem 5.2.1

A nonempty subset S of a vector space V is a subspace of V if and only if, for all vectors $\mathbf{u}$ and $\mathbf{v}$ in S and every scalar c,

   (1) $\mathbf{u} + \mathbf{v} \in S$,

     and

   (2) $c \cdot \mathbf{u} \in S$.

Proof   First suppose that S is a subspace of V. This means that S is a vector space and that all the properties given in Definition 5.1.1 hold. In particular, the closure of vector addition and of scalar multiplication is guaranteed.

Conversely, assume that S is a nonempty subset of V satisfying properties (1) and (2). As a subset of V, S automatically shares all the vector space properties of V except possibly the closure of vector addition and scalar multiplication and properties (3) and (4) of Definition 5.1.1. But we are given closure as our hypothesis, so we just have to verify that S contains a zero element and a negative for every one of its elements.

To see that all this is true, we note that hypothesis (2) of our theorem implies that $0 \cdot \mathbf{u} \in S$ for $\mathbf{u} \in S$. But Theorem 5.1.1(a) states that $0 \cdot \mathbf{u} = \mathbf{0}$, the zero element of V, so S contains the zero element. Similarly, Theorem 5.1.1(c) tells us that $(-1) \cdot \mathbf{u} = -\mathbf{u}$ for $\mathbf{u} \in S$. Then assumption (2) of Theorem 5.2.1 yields the conclusion that the negative of any element of S must be in S. Thus $\{S, +, \cdot\}$ is a subspace of V.

The criteria in Theorem 5.2.1 are often combined: A nonempty subset S of a vector space V is a subspace of V if and only if $a\mathbf{u} + b\mathbf{v} \in S$ for every $\mathbf{u}, \mathbf{v} \in S$ and $a, b \in \mathbb{R}$.

Example 5.2.1: A Subspace of Functions

The set of all real-valued functions f that have at least two derivatives is a vector space V if we define the addition and scalar multiplication of functions in the usual way (see Exercises 5.1, A8). We can use Theorem 5.2.1 to show that the set $S = \{f \in V : f'' + f = 0\}$ is a subspace of V (see Exercises 5.1, B8). First of all, we see that $f(x) = \sin x$ is a vector in V, so V is nonempty.

Now suppose f_1 and f_2 are two functions in S. If we let $f = f_1 + f_2$, we have $f'' + f = (f_1 + f_2)'' + (f_1 + f_2) = f_1'' + f_2'' + f_1 + f_2 = (f_1'' + f_1) + (f_2'' + f_2) = 0 + 0 = 0$ by hypothesis. Next, suppose $f \in S$ and c is any real number. Then $(cf)'' + (cf) = c(f'') + c(f) = c(f'' + f) = c(0) = 0$.

The conditions of Theorem 5.2.1 are satisfied, and we have proved that S is a subspace of V.

The equation $f'' + f = 0$ is a particular example of a *homogeneous second-order linear differential equation with constant coefficients*. Such equations play an important role in many applications of mathematics.* Example 5.2.1 shows that the set of solutions of this differential equation is a subspace of a vector space of functions.

Let us look at a few more examples for good measure.

Example 5.2.2: A Subspace of $P_3[x]$

Recall from Example 5.1.1(d) that $P_3[x]$, the set of all polynomials of degree less than or equal to three with real coefficients, is a vector space. Now consider $P_2[x] = \{a_k x^k + a_{k-1} x^{k-1} + \cdots + a_1 x + a_0 \mid a_i \in \mathbb{R}$ for $i = 0, 1, 2, \ldots, k; k \leq 2\}$. Clearly, $P_2[x]$ is a nonempty subset of $P_3[x]$. If we add two polynomials in $P_2[x]$, the resulting polynomial is also of degree less than or equal to two; and if we multiply an element of $P_2[x]$ by a real number, the new polynomial is clearly in $P_2[x]$. Thus $P_2[x]$ is a subspace of $P_3[x]$.

---

* Euler's investigation of the differential equation $f'' + f = 0$ around 1740 led to the definition of the complex exponential, $\exp(z)$, where z is any complex number. See Section 6.2 of P.J. Nahin, *An Imaginary Tale: The Story of* $\sqrt{-1}$ (Princeton, NJ: Princeton University Press, 1998).

Example 5.2.3: A Matrix Subspace

We show that $S = \left\{ \begin{bmatrix} a & b \\ c & d \end{bmatrix} : a - d = 0,\ b + c = 0 \right\}$ is a subspace of M_2. First, the set is not empty—for example, the vector $\begin{bmatrix} 2 & 5 \\ -5 & 2 \end{bmatrix}$ is in S. Now suppose that $U = \begin{bmatrix} a & b \\ c & d \end{bmatrix}$ and $V = \begin{bmatrix} x & y \\ z & w \end{bmatrix}$ are vectors in S. (This implies that $d = a$, $c = -b$, $w = x$, and $z = -y$.) Then $U + V = \begin{bmatrix} a & b \\ c & d \end{bmatrix} + \begin{bmatrix} x & y \\ z & w \end{bmatrix} = \begin{bmatrix} a+x & b+y \\ c+z & d+w \end{bmatrix} = \begin{bmatrix} a+x & b+y \\ -b-y & a+x \end{bmatrix} = \begin{bmatrix} a+x & b+y \\ -(b+y) & a+x \end{bmatrix} \in S$.

Also, if k is a scalar, we have $kU = k \begin{bmatrix} a & b \\ c & d \end{bmatrix} = \begin{bmatrix} ka & kb \\ kc & kd \end{bmatrix} = \begin{bmatrix} ka & kb \\ k(-b) & ka \end{bmatrix} = \begin{bmatrix} ka & kb \\ -kb & ka \end{bmatrix} \in S$. Thus S is a subspace of M_2.

Example 1.6.4 and Section 2.7 provided introductory discussions of the *null space* associated with a system of equations or a matrix. The next example provides a broader view of the null space as well as a look at its faithful companion, the *range*. Example 5.2.4 will be referred to and its content developed more fully in Chapter 6.

Example 5.2.4: The Null Space and Range of a Linear Transformation

Suppose that $\{V, +, \cdot\}$ and $\{W, \boxplus, \odot\}$ are two vector spaces with the same set of scalars and T is a *linear transformation* between them, a function $T : V \rightarrow W$ such that $T(a_1 \cdot \mathbf{v}_1 + a_2 \cdot \mathbf{v}_2) = [a_1 \odot T(\mathbf{v}_1)] \boxplus [a_2 \odot T(\mathbf{v}_2)]$ for all vectors $\mathbf{v}_1, \mathbf{v}_2 \in V$ and all scalars a_1 and a_2. (This concept was introduced less generally in Section 3.2.) We are being careful here by using different symbols for operations because the vectors in V and W may be different mathematical objects—for example, V may consist of matrices and W may be made up of polynomials—so the definitions of vector addition and scalar multiplication may be different in the two spaces.

The **null space** (or **kernel**) of T is defined as the set of all vectors in V that are transformed into the zero vector of W:

$$\text{null space of } T = \{\mathbf{v} \in V : T(\mathbf{v}) = \mathbf{0}_W\}$$

We show that the null space is a subspace of V. First suppose that $\mathbf{v}_1$ and $\mathbf{v}_2$ are vectors in the null space of T and c is a scalar. Then $T(\mathbf{v}_1 + \mathbf{v}_2) = T(\mathbf{v}_1) \boxplus T(\mathbf{v}_2) = \mathbf{0}_W \boxplus \mathbf{0}_W = \mathbf{0}_W$, so $\mathbf{v}_1 + \mathbf{v}_2$ is in the null space. Similarly, $T(c \cdot \mathbf{v}_1) = c \odot T(\mathbf{v}_1) = c \odot \mathbf{0}_W = \mathbf{0}_W$, showing that $c \cdot \mathbf{v}_1$ is an element of the null space. Hence the null space of T is a subspace of V.

The **range** of the linear transformation T is defined as the set of images of V under the transformation T:

range of $T = \{\mathbf{w} \in W : T(\mathbf{v}) = \mathbf{w} \text{ for some } \mathbf{v} \in V\}$.

This set of vectors is a subspace of W. To see this, let $\mathbf{w}_1$ and $\mathbf{w}_2$ be any vectors in the range of T and let c be any scalar. Then $\mathbf{w}_1 = T(\mathbf{v}_1)$ for some $\mathbf{v}_1 \in V$ and $\mathbf{w}_2 = T(\mathbf{v}_2)$ for $\mathbf{v}_2 \in V$. This gives us $\mathbf{w}_1 \boxplus \mathbf{w}_2 = T(\mathbf{v}_1) \boxplus T(\mathbf{v}_2) = T(\mathbf{v}_1 + \mathbf{v}_2)$. Because $\mathbf{v}_1 + \mathbf{v}_2 \in V$ and $\mathbf{w}_1 \boxplus \mathbf{w}_2 \in W$ by closure, we see that $\mathbf{w}_1 \boxplus \mathbf{w}_2$ is in the range of T. Also, $c \odot \mathbf{w}_1 = c \odot T(\mathbf{v}_1) = T(c \cdot \mathbf{v}_1)$, so $c \odot \mathbf{w}_1$ is in the range of T, and the range is a subspace of W.

5.2.1 The Sum and Intersection of Subspaces

There are various ways to combine subspaces of a vector space V to form new subspaces of V. If we regard subspaces as subsets of a vector space V, it is natural to consider the union and intersection of subspaces. We will prove that the intersection of two subspaces is a subspace, but it turns out that the union of two subspaces may not be a subspace. For example, in $\mathbb{R}^2$, the x-axis ($=\mathrm{span}\{\mathbf{e}_1\}$) and the y-axis ($=\mathrm{span}\{\mathbf{e}_2\}$) are each subspaces; but the union of these subspaces is not closed under vector addition—e.g., $\mathbf{e}_1 + \mathbf{e}_2 = \begin{bmatrix} 1 \\ 1 \end{bmatrix}$ is not in the union. (See Exercises B4 and B5 for basic facts about unions of subspaces.)

However, in addition to using the operations of union and intersection, there is another way to produce a new subspace from old subspaces.

Definition 5.2.2

If U and W are subspaces of a vector space V, then we can define the **sum** of U and W as follows:

$$U + W = \{\mathbf{u} + \mathbf{w} \mid \mathbf{u} \in U \text{ and } \mathbf{w} \in W\}.$$

Letting $\mathbf{u} = \mathbf{0}$ or $\mathbf{w} = \mathbf{0}$, we see that $U \subseteq U + W$, $W \subseteq U + W$, and so $U \cap W \subseteq U + W$. We can define the sum of a finite number of subspaces in the obvious way:

$$U_1 + U_2 + \cdots + U_m = \{\mathbf{u}_1 + \mathbf{u}_2 + \cdots + \mathbf{u}_m \mid \mathbf{u}_i \in U_i, \, i = 1, 2, \ldots, m\}.$$

Theorem 5.2.2

If U and W are subspaces of a vector space V, then $U \cap W$ and $U + W$ are subspaces of V.

Proof Let us first consider the intersection of subspaces. Assume that U and W are nontrivial subspaces and that $\mathbf{v}_1$ and $\mathbf{v}_2$ are in the intersection of U and W. Then $\mathbf{v}_1 \in U$, $\mathbf{v}_1 \in W$, $\mathbf{v}_2 \in U$, and $\mathbf{v}_2 \in W$. Because U and W are subspaces, $\mathbf{v}_1 + \mathbf{v}_2 \in U$, and $\mathbf{v}_1 + \mathbf{v}_2 \in W$, so $\mathbf{v}_1 + \mathbf{v}_2 \in U \cap W$. Also, for any scalar c, $c \cdot \mathbf{v}_1 \in U$ and $c \cdot \mathbf{v}_1 \in W$ imply that $c \cdot \mathbf{v}_1 \in U \cap W$. By Theorem 5.2.1, $U \cap W$ is a subspace of V.

To prove that the sum of subspaces is a subspace, suppose that $\mathbf{u}_1$, $\mathbf{u}_2$ and $\mathbf{w}_1, \mathbf{w}_2$ are vectors in U and W, respectively, and c is a scalar. Then $\mathbf{u}_i + \mathbf{w}_i \in U + W$ $(i = 1, 2)$ and

$$(\mathbf{u}_1 + \mathbf{w}_1) + (\mathbf{u}_2 + \mathbf{w}_2) = (\mathbf{u}_1 + \mathbf{u}_2) + (\mathbf{w}_1 + \mathbf{w}_2) \in U + W$$

and

$$c(\mathbf{u}_1 + \mathbf{w}_1) = c\mathbf{u}_1 + c\mathbf{w}_1 \in U + W.$$

Thus $U + W$ is closed with respect to addition and scalar multiplication, and so $U + W$ is a subspace.

Corollary 5.2.1

If $U_1, U_2, \ldots, U_n$ are subspaces of a vector space V, then $\bigcap_{i=1}^{n} U_i$ is a subspace of V.

The proof is requested in Exercise B3.

Corollary 5.2.2

If $U_1, U_2, \ldots, U_n$ are subspaces of a vector space V, then $U_1 + U_2 + \cdots + U_n$ is a subspace of V.

The proof is requested in Exercise B7.

Example 5.2.5: The Sum and Intersection of Subspaces in $\mathbb{R}^4$

In $\mathbb{R}^4$, consider the subspaces $U = \left\{ [x_1 \quad x_2 \quad x_3 \quad x_4]^\mathsf{T} : x_1 + x_2 = 0, x_3 + x_4 = 0 \right\}$ and $W = \left\{ [x_1 \quad x_2 \quad x_3 \quad x_4]^\mathsf{T} : x_1 + x_3 = 0, x_2 + x_4 = 0 \right\}$. First we will prove that

$$U + W = \left\{ [x_1 \quad x_2 \quad x_3 \quad x_4]^\mathsf{T} : x_1 + x_2 + x_3 + x_4 = 0 \right\}.$$

Let S denote the subspace $\left\{ [x_1 \quad x_2 \quad x_3 \quad x_4]^\mathsf{T} : x_1 + x_2 + x_3 + x_4 = 0 \right\}$. (Problem A3 asks for the proof that S is a subspace.) We will show that $U + W \subseteq S$ and also $S \subseteq U + W$, which implies $U + W = S$. Now, if $\mathbf{u} = [u_1 \quad u_2 \quad u_3 \quad u_4]^\mathsf{T} \in U$ and

$\mathbf{w} = [\,w_1 \quad w_2 \quad w_3 \quad w_4\,]^{\mathsf{T}} \in W$, then $\mathbf{u} + \mathbf{w} = [\,u_1 + w_1 \quad u_2 + w_2$
$u_3 + w_3 \quad u_4 + w_4\,]^{\mathsf{T}} \in S$ because $(u_1 + w_1) + (u_2 + w_2) +$
$(u_3 + w_3) + (u_4 + w_4) = (u_1 + u_2) + (u_3 + u_4) + (w_1 + w_3) +$
$(w_2 + w_4) = 0 + 0 + 0 + 0 = 0$. Thus $U + W \subseteq S$.

To prove that $S \subseteq U + W$, let $\mathbf{x} = [\,x_1 \quad x_2 \quad x_3 \quad x_4\,]^{\mathsf{T}}$ be any vector in S. (Note that $x_1 + x_2 + x_3 + x_4 = 0$ implies $x_4 = -x_1 - x_2 - x_3$.) We can express $\mathbf{x}$ as the sum of a vector in U and a vector in W:

$$\mathbf{x} = [\,0 \quad 0 \quad x_1 + x_3 \quad -x_1 - x_3\,]^{\mathsf{T}} + [\,x_1 \quad x_2 \quad -x_1 \quad -x_2\,]^{\mathsf{T}}.$$

Hence $S \subseteq U + W$ and it follows that $S = U + W$.

A vector in $U \cap W$ must satisfy the conditions of both U and W:

$$U \cap W = \left\{ [\,x_1 \quad x_2 \quad x_3 \quad x_4\,]^{\mathsf{T}} : x_1 + x_2 = 0, \; x_3 + x_4 = 0, \right.$$

$$\left. x_1 + x_3 = 0, \; x_2 + x_4 = 0 \right\}.$$

We solve the homogeneous system of four linear equations to find that $U \cap W = \left\{ [\,x \quad -x \quad -x \quad x\,]^{\mathsf{T}} : x \in \mathbb{R} \right\} = \left\{ x[\,1 \quad -1 \quad -1 \quad 1\,]^{\mathsf{T}} : x \in \mathbb{R} \right\}$, a one-dimensional subspace of $\mathbb{R}^4$.

Example 5.2.6: The Sum and Intersection of Subspaces in M_2

Consider the vector space M_2 and two of its subspaces (see Exercise A5),

$$U = \left\{ \begin{bmatrix} a & b \\ 0 & c \end{bmatrix} : a, b, c \in \mathbb{R} \right\} \quad \text{and} \quad W = \left\{ \begin{bmatrix} x & 0 \\ y & z \end{bmatrix} : x, y, z \in \mathbb{R} \right\}.$$

We want to determine $U + W$ and $U \cap W$. First, we notice that $M_2 \subseteq U + W$: Any matrix $A = \begin{bmatrix} a_{11} & a_{12} \\ a_{21} & a_{22} \end{bmatrix} \in M_2$ can be expressed as $\begin{bmatrix} a_{11} + 1 & a_{12} \\ 0 & a_{22} - 1 \end{bmatrix} + \begin{bmatrix} -1 & 0 \\ a_{21} & 1 \end{bmatrix}$, for example. Because $U + W \subseteq M_2$, we see that, in fact, $U + W = M_2$.

On the other hand, a matrix A is in $U \cap W$ if and only if A is both upper triangular and lower triangular. A moment's thought yields the insight that A must be a diagonal matrix. Thus

$$U \cap W = \left\{ \begin{bmatrix} d_1 & 0 \\ 0 & d_2 \end{bmatrix} : d_1, d_2 \in \mathbb{R} \right\}.$$

We can think of the subspaces $U \cap W$ and $U + W$ in other ways. First, $U \cap W$ is the **largest subspace of V contained in both U and W.** Next, $U + W$ is the **smallest subspace of V containing both U and W.** (See Figure 5.1) The proofs of these statements are solicited in Exercise C2.

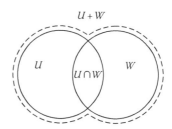

Figure 5.1 The sum and intersection of subspaces.

In Example 5.2.6, we saw that the sum of the subspaces U and W was the entire space M_2—that is, any vector in M_2 could be expressed as a sum $\mathbf{u} + \mathbf{w}$, where $\mathbf{u} \in U$ and $\mathbf{v} \in W$. However, this representation of a vector in M_2 is not unique: For example, $\begin{bmatrix} 1 & 2 \\ 3 & 4 \end{bmatrix} = \begin{bmatrix} 2 & 2 \\ 0 & 2 \end{bmatrix} + \begin{bmatrix} -1 & 0 \\ 3 & 2 \end{bmatrix} = \begin{bmatrix} -2 & 2 \\ 0 & -3 \end{bmatrix} + \begin{bmatrix} 3 & 0 \\ 3 & 7 \end{bmatrix}$.

Now suppose we take

$$U = \left\{ \begin{bmatrix} 0 & b \\ 0 & 0 \end{bmatrix} : b \in \mathbb{R} \right\} \quad \text{and} \quad W = \left\{ \begin{bmatrix} x & 0 \\ y & z \end{bmatrix} : x, y, z \in \mathbb{R} \right\}.$$

First, it should be clear that $U + W = M_2$. Now suppose that $A \in M_2$ has two distinct representations as a vector in $U + W$:

$$A = \begin{bmatrix} 0 & b_1 \\ 0 & 0 \end{bmatrix} + \begin{bmatrix} x_1 & 0 \\ y_1 & z_1 \end{bmatrix} = \begin{bmatrix} 0 & b_2 \\ 0 & 0 \end{bmatrix} + \begin{bmatrix} x_2 & 0 \\ y_2 & z_2 \end{bmatrix},$$

or

$$\begin{bmatrix} x_1 & b_1 \\ y_1 & z_1 \end{bmatrix} = \begin{bmatrix} x_2 & b_2 \\ y_2 & z_2 \end{bmatrix}.$$

This implies that $x_1 = x_2, b_1 = b_2, y_1 = y_2$, and $z_1 = z_2$—that is, the representation of A as a vector in $U + W$ is unique. This leads to the definition of a special kind of sum of subspaces.

Definition 5.2.3

If U and W are subspaces of V, we say that V is the **direct sum** of U and W, denoted as $V = U \oplus W$, if every $\mathbf{v} \in V$ has a *unique* representation $\mathbf{v} = \mathbf{u} + \mathbf{w}$, where $\mathbf{u} \in U$ and $\mathbf{w} \in W$.

More generally, if U and W are subspaces of V, then the subspace $U + W$ is called a **direct sum of U and W**, denoted by $U \oplus W$, if every element of $U + W$ can be written uniquely as $\mathbf{u} + \mathbf{w}$, where $\mathbf{u} \in U$ and $\mathbf{w} \in W$. If U is a subspace of a vector space V, then a subspace W of V is said to be a **complement** of U if $U \oplus W = V$. Of course, in this case, U is also a complement of W, and we say that U and W are **complementary subspaces**. Trivially, the subspaces $\{\mathbf{0}\}$ and V are always complements of each other.

We can define the direct sum of a finite number of subspaces in the obvious way: $W = U_1 \oplus U_2 \oplus \cdots \oplus U_m$ if each element $\mathbf{w} \in W$ can be expressed *uniquely* as $\mathbf{u}_1 + \mathbf{u}_2 + \cdots + \mathbf{u}_m$, where $\mathbf{u}_i \in U_i, i = 1, 2, \ldots, m$.

Theorem 5.2.3

If U and W are subspaces of V, then $V = U \oplus W$ if and only if $V = U + W$ and $U \cap W = \{\mathbf{0}\}$.

Proof Suppose that $V = U \oplus W$. Then every $\mathbf{v} \in V$ can be expressed as $\mathbf{v} = \mathbf{u} + \mathbf{w}$ for a unique $\mathbf{u} \in U$ and a unique $\mathbf{w} \in W$. Clearly, $V = U + W$. Now assume $\mathbf{z}$ belongs to $U \cap W$. Then $\mathbf{z} = \mathbf{z} + \mathbf{0}$, where $\mathbf{z} \in U$ and $\mathbf{0} \in W$; and $\mathbf{z} = \mathbf{0} + \mathbf{z}$, where $\mathbf{0} \in U$ and $\mathbf{z} \in W$. Because any such representation of $\mathbf{z}$ must be unique, we must have $\mathbf{z} = \mathbf{0}$—that is, $U \cap W = \{\mathbf{0}\}$.

Conversely, suppose that $V = U + W$ and $U \cap W = \{\mathbf{0}\}$. Then every $\mathbf{v} \in V$ can be expressed as $\mathbf{v} = \mathbf{u} + \mathbf{w}$ for some $\mathbf{u} \in U$ and some $\mathbf{w} \in W$, and we need only to prove uniqueness. Now suppose that we also have $\mathbf{v} = \hat{\mathbf{u}} + \hat{\mathbf{w}}$. Then $\mathbf{u} + \mathbf{w} = \hat{\mathbf{u}} + \hat{\mathbf{w}}$, so $\mathbf{u} - \hat{\mathbf{u}} = \hat{\mathbf{w}} - \mathbf{w}$. But $\mathbf{u} - \hat{\mathbf{u}} \in U$ and $\hat{\mathbf{w}} - \mathbf{w} \in W$, and by hypothesis U and W have only the zero vector in common. Hence $\mathbf{u} - \hat{\mathbf{u}} = \mathbf{0} = \hat{\mathbf{w}} - \mathbf{w}$, or $\mathbf{u} = \hat{\mathbf{u}}$ and $\hat{\mathbf{w}} = \mathbf{w}$, so the representation is unique. Therefore, $V = U \oplus W$.

For a generalization of Theorem 5.2.3, see Exercise C7.

Exercises 5.2

A.

1. Is the set $S = \left\{ \begin{bmatrix} x \\ y \end{bmatrix} : x^2 + y^2 \geq 0 \right\}$ a subspace of $\mathbb{R}^2$? Explain your answer and interpret S geometrically.

2. Prove that the set $W = \left\{ \begin{bmatrix} x \\ y \\ z \end{bmatrix} : x + y - z = 0 \right\}$ is a subspace of $\mathbb{R}^3$. Interpret W geometrically.

3. Prove that the set $S = \left\{ [x_1 \ \ x_2 \ \ x_3 \ \ x_4]^T : x_1 + x_2 + x_3 + x_4 = 0 \right\}$ used in Example 5.2.5 is a subspace of $\mathbb{R}^4$. (Do not assume $S = U + W$.)

4. With respect to Example 5.2.5, show that a vector $\mathbf{x}$ in S can be expressed as $\mathbf{u} + \mathbf{w}$ with $\mathbf{u} \in U$ and $\mathbf{w} \in W$ in more than one way—that is, find an expression for $\mathbf{x}$ that differs from the one in the example.

5. Verify that the subsets U and W in Example 5.2.6 are subspaces of M_2.

6. Let S be the subset of all vectors $[\,x \quad y \quad z\,]^T$ in $\mathbb{R}^3$ such that
    $$\frac{x}{3} = \frac{y}{4} = \frac{z}{2}.$$
    (For example, $[\,12 \quad 16 \quad 8\,]^T \in S$.) Is S a subspace of $\mathbb{R}^3$? What is the geometric interpretation of S?

7. Let U be the line $y = x$ in $\mathbb{R}^2$. Write $\mathbb{R}^2$ in two ways as a direct sum of U and another subspace.

8. Let $U = \left\{ \begin{bmatrix} a \\ b \\ 0 \end{bmatrix} : a, b \in \mathbb{R} \right\}$ and $W = \left\{ \begin{bmatrix} 0 \\ 0 \\ c \end{bmatrix} : c \in \mathbb{R} \right\}$.
    a. Show that U and W are subspaces of $\mathbb{R}^3$.
    b. Show that $\mathbb{R}^3 = U \oplus W$.

9. Is the set of all 2×2 matrices $\begin{bmatrix} a & b \\ c & d \end{bmatrix}$ such that $ad = 0$, a subspace of M_2? Explain your answer.

10. Suppose $P[x]$ is the set of all polynomials (i.e., of any degree) in the variable x. Let $S = \{p \in P[x]:$ all exponents of x in p are even$\}$. Is S a subspace of $P[x]$? Explain your answer.

11. Which of the following subsets of M_2 are subspaces of M_2? Explain your answers.
    a. All 2×2 diagonal matrices
    b. All 2×2 matrices such that the second row consists of zeros and the sum of entries in the first row is 4
    c. The set of all 2×2 matrices with determinant 4
    d. The set of all 2×2 matrices of rank 2

12. Suppose $P[x]$ is the set of all polynomials (i.e., of any degree) in the variable x. Determine if each of the following subsets of $P[x]$ is a subspace. Explain your answers.
    a. $\{p(x) \mid p(x)$ is a third degree polynomial$\}$
    b. $\{p(x) \mid p(0) = 0\}$
    c. $\{p(x) \mid 2p(0) = p(1)\}$
    d. $\{p(x) \mid p(x) \geq 0$ for all values of $x\}$

13. Use Theorem 5.2.1 to prove that the set of all 3×3 semimagic squares (see Exercises 4.4, B19) is a subspace of M_3.

14. Show that the set M of all $n \times n$ magic squares (see Exercises 5.1, A10) is a subspace of the vector space of $n \times n$ semimagic squares.

15. Show that the set of all 2×2 skew-symmetric matrices is a subspace of M_2.

16.   If $C(-1, 1)$ denotes the vector space of all functions continuous on the open interval $(-1, 1)$ with the usual operations of function addition and scalar multiplication, show that each of the following sets is a subspace of $C(-1, 1)$.
   a.   All functions differentiable on $(-1, 1)$.
   b.   All functions f of $C(-1, 1)$ with $f'(0) = 0$.

17.   Suppose U is a subspace of V. What is $U + U$?

18.   Let S be the set of all functions $f \in C[1, 2]$ which satisfy $\int_1^2 f(x)dx = 0$. Prove that S is a subspace of $C[1, 2]$.

B.

1.   In $\mathbb{R}^2$, show that any straight line that does not pass through the origin cannot be a subspace. Furthermore, show that the *only* nontrivial subspaces of $\mathbb{R}^2$ are straight lines through the origin.

2.   Let S be the subset of M_n that consists of those matrices A such that $A\mathbf{x} = \mathbf{0}$, where $\mathbf{x}$ is a fixed nonzero vector in $\mathbb{R}^n$. Show that S is a subspace of M_n.

3.   If $U_1, U_2, \ldots, U_n$ are subspaces of a vector space V, show that $\bigcap_{i=1}^n U_i$ is a subspace of V.* (This generalizes the first part of Theorem 5.2.2.)

4.   Let $U_1, U_2, \ldots, U_n$ be subspaces of a vector space V such that $V = \bigcup_{i=1}^n U_i$. Show that $V = U_j$ for some j.

5.   If U_1 and U_2 are subspaces of V, prove that $U_1 \cup U_2$ is a subspace of V if and only if $U_1 \subseteq U_2$ or $U_2 \subseteq U_1$.

6.   If S, U, and W are subspaces of a vector space V,
   a.   Show that $(S + U) \cap W \supseteq (S \cap W) + (U \cap W)$.
   b.   Show by an example in $\mathbb{R}^2$ that equality need not hold.

7.   If $U_1, U_2, \ldots, U_n$ are subspaces of a vector space V, show that $U_1 + U_2 + \cdots + U_n$ is a subspace of V. (This generalizes the second part of Theorem 5.2.2.)

8.   Let U, V, and W be subspaces of some vector space, and suppose that U is a subset of W. Prove that $(U + V) \cap W = U + (V \cap W)$.

---

* In fact, the intersection of *any* collection of subspaces is a subspace: If $\{U_\alpha\}_{\alpha \in \Gamma}$ is a collection of subspaces of V, where Γ is an arbitrary index set, then $\bigcap_{\alpha \in \Gamma} U_\alpha$ is a subspace of V.

9. Let U_1, U_2, and U_3 be subspaces of $\mathbb{R}^5$ which consist of all vectors of the forms $[0 \ \ 0 \ \ a \ \ 0 \ \ 0]^T$, $[0 \ \ b \ \ 0 \ \ c \ \ 0]^T$, and $[d \ \ 0 \ \ 0 \ \ 0 \ \ e]^T$, respectively, where a, b, c, d, and e are arbitrary real numbers. Show that $\mathbb{R}^5 = U_1 \oplus U_2 \oplus U_3$.

10. Let U and W denote the sets of all $n \times n$ real symmetric and skew-symmetric matrices, respectively.
    a. Show that U and W are subspaces of M_n.
    b. Show that $M_n = U \oplus W$. [*Hint:* See Problem C1 of Exercises 3.1.]

11. Show that $M_2 = W_1 \oplus W_2$, where

$$W_1 = \left\{ \begin{bmatrix} a & b \\ -b & a \end{bmatrix} : a, b \in \mathbb{R} \right\} \quad \text{and}$$

$$W_2 = \left\{ \begin{bmatrix} c & d \\ d & -c \end{bmatrix} : c, d \in \mathbb{R} \right\}.$$

12. Find three distinct subspaces U, V, W of $\mathbb{R}^2$ such that

$$\mathbb{R}^2 = U \oplus V = V \oplus W = W \oplus U.$$

13. Show by example that $U_1 + U_2 + U_3$ is not necessarily a direct sum even though

$$U_1 \cap U_2 = U_1 \cap U_3 = U_2 \cap U_3 = \{\mathbf{0}\}.$$

C.

1. Suppose $V = C[0,1]$, $W_{\frac{1}{2}} = \{f \in V | f(\frac{1}{2}) = 0\}$, and $W_{\frac{1}{3}} = \{f \in V | f(\frac{1}{3}) = 0\}$.
    a. Prove that $V = W_{\frac{1}{2}} + W_{\frac{1}{3}}$.
    b. Is $W_{\frac{1}{2}} + W_{\frac{1}{3}}$ a direct sum?

2. Prove that
    a. $U \cap W$ is the largest subspace of V contained in both U and W. [*Hint:* Assume there is a subspace S such that $U \cap W \subseteq S \subseteq U$ and $U \cap W \subseteq S \subseteq W$. Then show $S \subseteq U \cap W$.]
    b. $U + W$ is the smallest subspace of V containing both U and W. [*Hint:* Assume there is a subspace S such that $U \subseteq S \subseteq U + W$ and $W \subseteq S \subseteq U + W$. Then show $U + W \subseteq S$.]

3. Prove that if S, U, and W are subspaces of a vector space V, then

$$S \cap (U + (S \cap W)) = (S \cap U) + (S \cap W).$$

4. If S, U, and W are subspaces of a vector space V, prove that

$$S \cap (U + W) = (S \cap U) + (S \cap W)$$

if and only if every vector in V is a scalar multiple of a fixed vector $\mathbf{x}$.

5. Show that if U is a subspace of a vector space V, and if there is a *unique* subspace W such that $V = U \oplus W$, then $U = V$.

6. Let $U_1, U_2, \ldots, U_n$ be subspaces of a vector space V. Prove that the following statements are equivalent:
   (1) $V = U_1 \oplus U_2 \oplus \cdots \oplus U_n$
   (2) $\mathbf{u}_i \in U_i$ and $\mathbf{u}_1 + \mathbf{u}_2 + \cdots + \mathbf{u}_n = \mathbf{0}$ imply $\mathbf{u}_i = \mathbf{0}$, $i = 1, 2, \ldots, n$.

7. Let $U_1, U_2, \ldots, U_n$ be subspaces of a vector space V. Prove that $V = U_1 \oplus U_2 \oplus \cdots \oplus U_n$ if and only if $V = U_1 + U_2 + \cdots + U_n$ and $U_j \cap \sum_{i \neq j} U_i = \{\mathbf{0}\}$ for each j $(1 \leq j \leq n)$.

8. Let $V = U_1 \oplus U_2$, $U_1 = V_1 \oplus W_1$, and $U_2 = V_2 \oplus W_2$. Using the result of Exercise C6, prove that $V = V_1 \oplus W_1 \oplus V_2 \oplus W_2$.

5.3 Linear Independence and the Span

The closure property of vector spaces and subspaces makes it clear that adding finitely many vectors in such spaces produces a vector in the space. For example, if $\mathbf{v}_1$, $\mathbf{v}_2$, and $\mathbf{v}_3$ are elements of a vector space V, we can define the sum $\mathbf{v}_1 + \mathbf{v}_2 + \mathbf{v}_3$ as either $\mathbf{v}_1 + (\mathbf{v}_2 + \mathbf{v}_3)$ or $(\mathbf{v}_1 + \mathbf{v}_2) + \mathbf{v}_3$. In both cases, we know that the expression in parentheses is some vector in V, so finally we are dealing with the sum of *two* vectors—a result that is in V because of closure. As seen in Section 1.3, such sums of vectors play an important role in the theory and applications of vector spaces.

Definition 5.3.1

Given a nonempty finite set of vectors $S = \{\mathbf{v}_1, \mathbf{v}_2, \ldots, \mathbf{v}_k\}$ in a vector space V, a **linear combination** of these vectors is any vector of the form $a_1 \mathbf{v}_1 + a_2 \mathbf{v}_2 + \cdots + a_k \mathbf{v}_k$, where $a_1, a_2, \ldots, a_k$ are scalars.

Definition 5.3.1 is identical to Definition 1.3.1 except that $\mathbb{R}^n$ has been replaced by a general vector space V.

Example 5.3.1: Linear Combinations

The functions x^3, x, and 1 are elements of the vector space $P_3[x]$, the set of all polynomials of degree less than or equal to three with real coefficients. We could, for example, form the linear combination $(-3) \cdot x^3 + (\pi) \cdot x + (-7) \cdot 1 = -3x^3 + \pi x - 7$, resulting in another vector in $P_3[x]$.

In the vector space M_2, we can form linear combinations of the following kind:

$$2 \cdot \begin{bmatrix} 1 & -3 \\ 0 & 4 \end{bmatrix} + (-3) \cdot \begin{bmatrix} 0 & 1 \\ -5 & 2 \end{bmatrix} + \sqrt{2} \cdot \begin{bmatrix} -2 & 0 \\ 3 & 4 \end{bmatrix}$$

$$= \begin{bmatrix} 2 - 2\sqrt{2} & -9 \\ 15 + 3\sqrt{2} & 2 + 4\sqrt{2} \end{bmatrix}$$

or

$$(1) \cdot \begin{bmatrix} 0 & 1 \\ 3 & -2 \end{bmatrix} + (-1) \cdot \begin{bmatrix} 1 & -1 \\ 2 & 1 \end{bmatrix} + (1) \cdot \begin{bmatrix} 1 & -2 \\ -1 & 3 \end{bmatrix} = \begin{bmatrix} 0 & 0 \\ 0 & 0 \end{bmatrix}.$$

Now that we have generalized the concept of *linear combination*, we can extend other important ideas we first discussed in the context of $\mathbb{R}^n$.

Definition 5.3.2

A nonempty finite set of vectors $S = \{\mathbf{v}_1, \mathbf{v}_2, \ldots, \mathbf{v}_k\}$ in a vector space V is called **linearly independent** if the only way that $a_1\mathbf{v}_1 + a_2\mathbf{v}_2 + \cdots + a_k\mathbf{v}_k = \mathbf{0}$, where $a_1, a_2, \ldots, a_k$ are scalars, is if $a_1 = a_2 = \cdots = a_k = 0$. Otherwise S is called **linearly dependent**.

Example 5.3.2: Linear Independence and Linear Dependence

In Example 5.3.1, the vectors $A_1 = \begin{bmatrix} 0 & 1 \\ 3 & -2 \end{bmatrix}$, $A_2 = \begin{bmatrix} 1 & -1 \\ 2 & 1 \end{bmatrix}$, and $A_3 = \begin{bmatrix} 1 & -2 \\ -1 & 3 \end{bmatrix}$ form a linearly dependent set of vectors in M_2 because a linear combination of the three vectors turns out to be the zero element (the zero matrix) without having all the scalar multipliers equal to zero: $A_1 - A_2 + A_3 = O_2$.

On the other hand, the vectors $1, x, x^2$, and x^3 are linearly independent in $P_3[x]$ because the only way for $a \cdot 1 + b \cdot x + c \cdot x^2 + d \cdot x^3$ to equal the zero polynomial—that is, the polynomial equal to 0 for all values of x—is to have $a = b = c = d = 0$. (See Exercise B1.)

Just as we did in Section 1.3 for $\mathbb{R}^n$, we can characterize linear dependence (and hence independence) in any vector space in an alternative way.

Theorem 5.3.1

A set $S = \{\mathbf{v}_1, \mathbf{v}_2, \ldots, \mathbf{v}_k\}$ in a vector space V that contains at least two vectors is linearly dependent if and only if some vector $\mathbf{v}_j$ is a linear combination of the remaining vectors in the set.

The proof of this theorem is identical to that of Theorem 1.3.1. The earlier proof was purely algebraic and made no use of the fact that we were dealing with n-tuples of real numbers.

In $\mathbb{R}^n$, we made good use of special vectors that played an important role in understanding the nature of that vector space. We saw, for example, that any vector $\mathbf{v} = \begin{bmatrix} x_1 \\ x_2 \\ \vdots \\ x_n \end{bmatrix}$ in $\mathbb{R}^n$ could be written as a (unique) linear combination of the n vectors

$$\mathbf{e}_1 = \begin{bmatrix} 1 \\ 0 \\ \vdots \\ 0 \end{bmatrix}, \mathbf{e}_2 = \begin{bmatrix} 0 \\ 1 \\ \vdots \\ 0 \end{bmatrix}, \ldots, \mathbf{e}_n = \begin{bmatrix} 0 \\ 0 \\ \vdots \\ 1 \end{bmatrix}.$$

In a general vector space, we are interested in vectors that reach across (span) the entire space when all linear combinations of vectors in the set are formed.

Definition 5.3.3

Given a nonempty set of vectors $S = \{\mathbf{v}_1, \mathbf{v}_2, \ldots, \mathbf{v}_k\}$ in a vector space V, the **span** of S, denoted by $span(S)$, is the set of all linear combinations of vectors from S:

$$span(S) = \{a_1\mathbf{v}_1 + a_2\mathbf{v}_2 + \cdots + a_k\mathbf{v}_k \,|\, a_i \in \mathbb{R}, \mathbf{v}_i \in S, i = 1, 2, \ldots, k\}.$$

(We can also say that S is a **generating set** of $span(S)$.)

Example 5.3.3: Span of a Set of Vectors in M_2

Suppose the subset S of M_2 consists of the vectors $A_1 = \begin{bmatrix} 1 & 0 \\ 0 & 1 \end{bmatrix}$, $A_2 = \begin{bmatrix} 0 & 1 \\ 1 & 0 \end{bmatrix}$, and $A_3 = \begin{bmatrix} 1 & 0 \\ 0 & -1 \end{bmatrix}$. Then $span(S) = \{aA_1 + bA_2 + cA_3 \,|\, a, b, c \in \mathbb{R}\} = \left\{ \begin{bmatrix} a+c & b \\ b & a-c \end{bmatrix} : a, b, c \in \mathbb{R} \right\}$.

We see that the span of S is not the whole vector space M_2—for example, the vector $\begin{bmatrix} 0 & 1 \\ -1 & 0 \end{bmatrix}$ is not in $span(S)$.

Example 5.3.4: Span of Vectors in $P_2[x]$

The functions $1 + x^2$, $x^2 - x$, and $3 - 2x$ are elements of $P_2[x]$, the vector space of all polynomials of degree less than or equal to two with real coefficients. Then

$$span(S) = \{a(1 + x^2) + b(x^2 - x) + c(3 - 2x) | a, b, c \in \mathbb{R}\}$$
$$= \{(a + b)x^2 + (-b - 2c)x + (a + 3c) | a, b, c \in \mathbb{R}\}.$$

For example, letting $a = b = c = 1$, we see that the quadratic polynomial $2x^2 - 3x + 4$ is in $span(S)$. The interesting fact is that *every vector of $P_2[x]$ is in $span(S)$*—that is, $span(S) = P_2[x]$. One way to see this is to select an arbitrary polynomial in $P_2[x]$, $Ax^2 + Bx + C$, and consider the system

$$a + b = A$$
$$-b - 2c = B$$
$$a + 3c = C$$

We can solve this system to find that $a = \frac{1}{5}(3A + 3B + 2C)$, $b = \frac{1}{5}(2A - 3B - 2C)$, and $c = \frac{1}{5}(-A - B + C)$.

The next theorem is important, but should not be surprising. Its proof, for $\mathbb{R}^n$, was left as Problem B1 of Exercises 1.6, but we will prove it here in the context of a general vector space.

Theorem 5.3.2

Given a set of vectors $S = \{\mathbf{v}_1, \mathbf{v}_2, \ldots, \mathbf{v}_k\}$ in a vector space V, $span(S)$ is a subspace of V.

Proof We show that $span(S)$ is closed under vector addition and scalar multiplication and then use Theorem 5.2.1. First, suppose that $\mathbf{u}$ and $\mathbf{v}$ are in $span(S)$. Then $\mathbf{u} = a_1\mathbf{v}_1 + a_2\mathbf{v}_2 + \cdots + a_k\mathbf{v}_k$ and $\mathbf{v} = b_1\mathbf{v}_1 + b_2\mathbf{v}_2 + \cdots + b_k\mathbf{v}_k$ for scalars a_i and b_i.
Now

$$\mathbf{u} + \mathbf{v} = (a_1\mathbf{v}_1 + a_2\mathbf{v}_2 + \cdots + a_k\mathbf{v}_k) + (b_1\mathbf{v}_1 + b_2\mathbf{v}_2 + \cdots + b_k\mathbf{v}_k)$$
$$= (a_1 + b_1)\mathbf{v}_1 + (a_2 + b_2)\mathbf{v}_2 + \cdots + (a_k + b_k)\mathbf{v}_k \in span(S)$$

and

$$c \cdot \mathbf{u} = c \cdot (a_1\mathbf{v}_1 + a_2\mathbf{v}_2 + \cdots + a_k\mathbf{v}_k)$$
$$= (ca_1)\mathbf{v}_1 + (ca_2)\mathbf{v}_2 + \cdots + (ca_k)\mathbf{v}_k \in span(S).$$

Thus $span(S)$ is a subspace of V.

Example 5.3.4 leads us to the following definition for a general vector space.

Definition 5.3.4

A nonempty set S of vectors in a vector space V **spans** V (or is a **spanning set** for the space V) if every vector in V is an element of $span(S)$—that is, if every vector in V is a linear combination of vectors in S.

By closure, we know that *span*(*S*) ⊆ *V* for any set *S* of vectors in *V*. If *S* is a spanning set for *V*, we can also write *V* ⊆ *span*(*S*). Thus, if *S* is a spanning set, we can simply write *span*(*S*) = *V*.

With Definition 5.3.4 in hand, we see that the set of polynomials $\{1 + x^2, x^2 - x, 3 - 2x\}$ in Example 5.3.4 is a spanning set for $P_2[x]$. Examples 1.3.3 and 1.3.4 represent spanning and nonspanning sets in $\mathbb{R}^3$. One more example should help us understand this concept.

Example 5.3.5: A Spanning Set for M_2

The vectors

$$A_1 = \begin{bmatrix} 1 & 1 \\ 0 & 0 \end{bmatrix}, \quad A_2 = \begin{bmatrix} 0 & 0 \\ 1 & 1 \end{bmatrix}, \quad A_3 = \begin{bmatrix} 1 & 0 \\ 0 & 1 \end{bmatrix}, \quad \text{and} \quad A_4 = \begin{bmatrix} 0 & 1 \\ 1 & 1 \end{bmatrix}$$

span the vector space M_2. To see this, let $\begin{bmatrix} x & y \\ z & w \end{bmatrix}$ be any element of M_2 and consider

$$\begin{bmatrix} x & y \\ z & w \end{bmatrix} = aA_1 + bA_2 + cA_3 + dA_4,$$

which is equivalent to the system of equations

$$a + c = x$$
$$a + d = y$$
$$b + d = z$$
$$b + c + d = w$$

The coefficient matrix of this system of equations is

$$\begin{bmatrix} 1 & 0 & 1 & 0 \\ 1 & 0 & 0 & 1 \\ 0 & 1 & 0 & 1 \\ 0 & 1 & 1 & 1 \end{bmatrix},$$

which has rank equal to 4 and hence is invertible by Corollary 3.6.1(a). Therefore the system has a unique solution, implying that the set $\{A_1, A_2, A_3, A_4\}$ spans M_2.

We have seen the following result in the setting of $\mathbb{R}^n$ in Section 1.4. It shows a relationship between the sizes of spanning sets and linearly independent sets in any vector space *V*. The proof is the same as the proof of Theorem 1.4.2 because that earlier demonstration did not assume any properties peculiar to $\mathbb{R}^n$.

Theorem 5.3.3

Suppose that the vectors $\mathbf{v}_1, \mathbf{v}_2, \ldots, \mathbf{v}_m$ span a vector space *V* and that the nonzero vectors $\mathbf{w}_1, \mathbf{w}_2, \ldots, \mathbf{w}_k$ in *V* are linearly independent. Then $k \leq m$. (That is, in a vector space the size of any linearly independent set is always less than or equal to the size of any spanning set.)

Exercises 5.3

A.

1. Is the set of vectors $\{1 + x + x^2 + 2x^3, 6 + 7x + 8x^2 + 10x^3,$ $-3 - x + x^2 - 10x^3\}$ a linearly independent subset of $P_3[x]$? Explain your answer.

2. Determine which (if any) of the following sets of matrices is linearly independent in M_2.

   a. $\left\{ \begin{bmatrix} 1 & 0 \\ 0 & 1 \end{bmatrix}, \begin{bmatrix} 1 & 1 \\ 0 & 0 \end{bmatrix}, \begin{bmatrix} 0 & 1 \\ 1 & 1 \end{bmatrix} \right\}$

   b. $\left\{ \begin{bmatrix} 1 & 0 \\ 1 & 2 \end{bmatrix}, \begin{bmatrix} 1 & 1 \\ -1 & 1 \end{bmatrix}, \begin{bmatrix} 0 & -1 \\ 2 & 1 \end{bmatrix} \right\}$

3. a. Show that $\{1, x, x^2, (x-1)^2\}$ is a linearly dependent subset of $P_2[x]$.

   b. Find two distinct linear combinations for which $x^2 + 1 = a + bx + cx^2 + d(x-1)^2$.

4. Show that $P_2[x] = span\{1, 1+x, 1+x+x^2\}$.

5. What is the subspace spanned by the polynomials $1, x, x^2$ in $P_5[x]$?

6. What is the subspace spanned by the polynomials 1 and $x^2 - 1$ in $P_3[x]$?

7. Show that the matrices $\begin{bmatrix} 1 & 1 \\ 0 & 1 \end{bmatrix}, \begin{bmatrix} -1 & 1 \\ 0 & -1 \end{bmatrix}, \begin{bmatrix} 0 & 1 \\ 0 & 0 \end{bmatrix}$ do not span M_2.

8. Show that the matrix $\begin{bmatrix} 0 & 1 \\ -1 & 0 \end{bmatrix}$ spans the vector space of 2×2 skew-symmetric matrices.

9. Show that the matrices

$$\begin{bmatrix} 1 & 0 & 0 \\ 0 & 1 & 0 \\ 0 & 0 & 1 \end{bmatrix}, \begin{bmatrix} 1 & 0 & 0 \\ 0 & -1 & 0 \\ 0 & 0 & 0 \end{bmatrix}, \begin{bmatrix} 1 & 0 & 0 \\ 0 & 1 & 0 \\ 0 & 0 & -1 \end{bmatrix}$$

span the vector space of all 3×3 diagonal matrices.

10. For what values of c is the set of vectors $\{x + 3, 2x + c^2 + 2\}$ in $P_1[x]$ linearly independent?

11. Given that $\mathbf{p}_1 = 1 + x$, $\mathbf{p}_2 = x + x^2$, and $\mathbf{p}_3 = 1 + x^2$,
   a. Write $\pi x^2 + 2x - 1$ as a linear combination of $\mathbf{p}_1$, $\mathbf{p}_2$, and $\mathbf{p}_3$.
   b. Show that $\mathbf{p}_1$, $\mathbf{p}_2$, and $\mathbf{p}_3$ span $P_2[x]$.

12. Determine whether the set $S = \{1 + x^3,\ 1 + x + x^2,\ x + x^3,\ 1 + x + x^2 + x^3\}$ of vectors in $P_3[x]$ is linearly independent or linearly dependent. If dependent, write one vector as a linear combination of the others.

13. In the vector space $P_n[x]$ $(n \geq 2)$, show that
   a. $x^2 - x - 1 \in span\{x + 2, x^2 - x + 1, 3x^2 - 2x\}$.
   b. $x^2 - x - 1 \notin span\{x + 2, 3x^2 - 2x\}$.

14. Consider the following 3×3 semimagic squares (see Exercises 5.2, A13):

$$A_1 = \begin{bmatrix} 1 & 0 & 0 \\ 0 & 1 & 0 \\ 0 & 0 & 1 \end{bmatrix}, \quad A_2 = \begin{bmatrix} 1 & 0 & 0 \\ 0 & 0 & 1 \\ 0 & 1 & 0 \end{bmatrix}, \quad A_3 = \begin{bmatrix} 0 & 0 & 1 \\ 1 & 0 & 0 \\ 0 & 1 & 0 \end{bmatrix},$$

$$A_4 = \begin{bmatrix} 0 & 1 & 0 \\ 1 & 0 & 0 \\ 0 & 0 & 1 \end{bmatrix}, \quad A_5 = \begin{bmatrix} 0 & 1 & 0 \\ 0 & 0 & 1 \\ 1 & 0 & 0 \end{bmatrix}.$$

   a. Show that the set $\{A_i\}_{i=1}^5$ spans the vector space of all 3×3 semimagic squares.
   b. Show that $\{A_i\}_{i=1}^5$ is a linearly independent set.

B.

1. Suppose a, b, c, and d are real numbers such that $a + bx + cx^2 + dx^3 = 0$ for all values of the variable x. Show that $a = b = c = d = 0$. (See Example 5.3.2.)

2. Consider the functions $f(x) = x$, $g(x) = \cos x$, and $h(x) = e^x$ as vectors in $C(-\infty, \infty)$, the space of continuous functions over $\mathbb{R}$. Is this collection of vectors linearly independent? [*Hint*: Use differentiation.]

3. Prove that $\cos(nt) \in span\{\cos^k(t): k = 0, 1, 2, \ldots, n\}$.

4. Prove that $\sin(nt) \in span\{\sin(t)\cos^k(t): k = 0, 1, 2, \ldots, n-1\}$.

5. Prove that $1 \in span\{x^k (1 - x)^{n-k}: k = 0, 1, \ldots, n\}$.

6. Suppose f is the *identity function*, $f(x) \equiv x$ for all $x \in (-\infty, \infty)$. Show that $f \notin span\{e^{-kx}: k = 1, 2, \ldots, n\}$.
   [*Hint*: If f were in the span, then $x = \sum_{k=1}^{n} a_k e^{-kx}$. Now let $x \to \infty$.]

7. If $S = \{\mathbf{u}, \mathbf{v}, \mathbf{w}\}$ is a linearly independent subset of a vector space V, show that $\{\mathbf{u} + \mathbf{v} + \mathbf{w}, \mathbf{u} - \mathbf{w}, 2\mathbf{v} + 3\mathbf{w}\}$ is also a linearly independent set.

8. Suppose V is a real vector space and suppose that $S = \{\mathbf{v}_1, \mathbf{v}_2, \ldots, \mathbf{v}_k\}$ is a linearly independent subset of V. If $\mathbf{v} = \sum_{i=1}^{k} a_i \mathbf{v}_i$, where each $a_i \in \mathbb{R}$ prove that the set $\hat{S} = \{\mathbf{v} - \mathbf{v}_1, \mathbf{v} - \mathbf{v}_2, \ldots, \mathbf{v} - \mathbf{v}_k\}$ is linearly independent if and only if $\sum_{i=1}^{k} a_i \neq 1$.

9. Suppose $A_1, A_2, \ldots, A_k$ are matrices in M_{mn}. Let X and Y be invertible $m \times m$ and $n \times n$ matrices, respectively, and let $B \in M_{np}$. Prove that
   a. $\{A_1, A_2, \ldots, A_k\}$ is a linearly independent subset of M_{mn} if and only if $\{XA_1Y, XA_2Y, \ldots, XA_kY\}$ is a linearly independent subset of M_{mn}.
   b. If $\{A_1B, A_2B, \ldots, A_kB\}$ is a linearly independent subset of M_{mp}, then $\{A_1, A_2, \ldots, A_k\}$ is a linearly independent subset of M_{mn}.

10. In $\mathbb{R}^3$, suppose that $\mathbf{u} = \begin{bmatrix} 2 \\ 1 \\ 0 \end{bmatrix}$, $\mathbf{v} = \begin{bmatrix} 1 \\ 0 \\ -1 \end{bmatrix}$, $\mathbf{w} = \begin{bmatrix} -1 \\ 1 \\ 1 \end{bmatrix}$, and $\mathbf{z} = \begin{bmatrix} 0 \\ -1 \\ 1 \end{bmatrix}$. Let $U = span\{\mathbf{u}, \mathbf{v}\}$ and $W = span\{\mathbf{w}, \mathbf{z}\}$.
    a. Describe $U \cap W$.
    b. Describe $U + W$.

11. Suppose that $U = span\left\{ \begin{bmatrix} 0 \\ 1 \\ -1 \end{bmatrix}, \begin{bmatrix} 0 \\ 1 \\ 1 \end{bmatrix} \right\}$ and $W = span\left\{ \begin{bmatrix} 1 \\ 1 \\ 0 \end{bmatrix} \right\}$. Show that $\mathbb{R}^3 = U \oplus W$.

12. In $\mathbb{R}^3$, let $\mathbf{u} = \begin{bmatrix} 1 \\ 1 \\ 0 \end{bmatrix}$, $\mathbf{v} = \begin{bmatrix} 0 \\ 0 \\ 1 \end{bmatrix}$, $\mathbf{w} = \begin{bmatrix} 1 \\ 0 \\ 0 \end{bmatrix}$, $\mathbf{z} = \begin{bmatrix} 0 \\ 2 \\ 1 \end{bmatrix}$, $U = span\{\mathbf{u}, \mathbf{v}\}$, and $W = span\{\mathbf{w}, \mathbf{z}\}$.
    a. Show that $\mathbb{R}^3 = U + W$.
    b. Determine $U \cap W$.
    c. Explain why $\mathbb{R}^3$ is not the direct sum of U and W.
    d. If we redefine W as $span\{\mathbf{w}\}$, show that now $\mathbb{R}^3 = U \oplus W$.

13. Let $V = \mathbb{R}^4$ and let $W = span\left\{ [1 \ 1 \ 0 \ 0]^T, [1 \ 0 \ 1 \ 0]^T \right\}$. Find subspaces W_1 and W_2 of V such that $V = W \oplus W_1$ and $V = W \oplus W_2$ but $W_1 \neq W_2$.

14. Prove that a subset S of a vector space V is linearly dependent if and only if there exists an $\mathbf{x} \in S$ such that $\mathbf{x} \in span(S - \{\mathbf{x}\})$.

C.

1. If U and W are subspaces, prove that if S_U and S_W span U and W, respectively, then $S_U \cup S_W$ spans $U + W$. [Alternatively, $span(U \cup W) = U + W$.]

2. For each positive integer k, let $f_k : \mathbb{R} \to \mathbb{R}$ be given by $f_k(x) = e^{r_k x}$, where $r_k \in \mathbb{R}$. Prove that the set of functions $\{f_1, f_2, \ldots, f_n\}$ is linearly independent if and only if $r_1, r_2, \ldots, r_n$ are distinct. [*Hint*: Think mathematical induction. Some calculus might help.]

5.4 Bases and Dimension

As we saw in Chapter 1, the combination of linear independence and spanning is a powerful concept. We extend Definition 1.5.1 to any vector space.

Definition 5.4.1

A nonempty set B of vectors in a vector space V is called a **basis for** (or **of**) V if both the following conditions hold:

(1) The set B is linearly independent.

(2) The set B spans V.

Furthermore, the vector space V is **finite-dimensional** if it has a finite basis—that is, if V is spanned by a finite linearly independent set of vectors $\{v_1, v_2, \ldots, v_k\}$. Otherwise, the vector space is called **infinite-dimensional**.

For any positive integer n, we have seen the *standard basis* for $\mathbb{R}^n$:

$$\mathcal{E}_n = \{e_1, e_2, \ldots, e_n\}, \quad \text{where} \quad e_i = [0 \ \ 0 \ldots 0 \ \ \overset{\overset{position}{\frown}}{1} \ \ 0 \ldots 0 \ldots, 0]^T.$$

We can easily determine that the set $\{1, x, x^2, \ldots, x^n\}$ is a basis (called the **standard basis**) for $P_n[x]$ and that the spanning set of matrices in Example 5.3.5 is a linearly independent set. Thus the matrices A_1, A_2, A_3, and A_4 constitute a basis for M_2, which is therefore finite-dimensional. We note that each of the vector spaces $\mathbb{R}^n$, $P_n[x]$, and M_2 contains infinitely many elements, yet each space is generated (spanned) by a finite set of vectors.

However, not all vector spaces are finite-dimensional.

Example 5.4.1: An Infinite-Dimensional Vector Space

The vector space $P[x]$ consisting of all polynomials in the variable x with real coefficients is infinite-dimensional. It is easy to see (Exercise B4) that no finite set of polynomials can generate $P[x]$. However, the infinite set of polynomials $\{1, x, x^2, \ldots, x^n, \ldots\}$ is linearly independent* (Exercise B11) and spans $P[x]$. Thus the set is a basis for $P[x]$.

By its very definition, a finite-dimensional vector space has a basis—in fact, as we have seen for $\mathbb{R}^n$, it may have many bases. It can be proved that **every vector space V has a basis**. However, the justification of this claim for an infinite-dimensional vector space requires heavy theoretical artillery and goes beyond the scope of this course.[†]

An important characterization of any finite basis is given by the following theorem, an extension of Theorem 1.5.3 and Problem B1 of Exercises 1.5. The "only if" part of this proof is the same as the proof of Theorem 1.5.3, whereas the "if" part was left as an exercise. For completeness, we provide the entire proof in the context of a general vector space.

Theorem 5.4.1

A nonempty subset $S = \{\mathbf{v}_1, \mathbf{v}_2, \ldots, \mathbf{v}_n\}$ of a vector space V is a basis if and only if every vector $\mathbf{v} \in V$ can be written as a linear combination

$$\mathbf{v} = a_1\mathbf{v}_1 + a_2\mathbf{v}_2 + \cdots + a_n\mathbf{v}_n$$

in one and only one way.

Proof   Suppose that S is a basis. Then $span(S) = V$ and so every vector $\mathbf{v} \in V$ is a linear combination of elements of S. Now suppose that a vector $\mathbf{v}$ in V has *two* representations as a linear combination of basis vectors:

$$\mathbf{v} = a_1\mathbf{v}_1 + a_2\mathbf{v}_2 + \cdots + a_n\mathbf{v}_n,$$

and

$$\mathbf{v} = b_1\mathbf{v}_1 + b_2\mathbf{v}_2 + \cdots + b_n\mathbf{v}_n.$$

---

* Up to now, we have defined linear independence/dependence in terms of *finite* sets of vectors. However, we can extend these concepts in an obvious way. For example, any subset X (not necessarily finite) of a vector space V is said to be **linearly independent** if every finite subset of X is linearly independent. Otherwise, X is called **linearly dependent**.

[†] See N. Jacobson, *Lectures in Abstract Algebra*, vol. II, Chap. IX (New York: Van Nostrand, 1953). You can also consult the discussion of a **Hamel basis** in C.G. Small, *Functional Equations and How to Solve Them*, Chap. 6 (New York: Springer, 2007).

Then

$$0 = \mathbf{v} - \mathbf{v} = a_1\mathbf{v}_1 + a_2\mathbf{v}_2 + \cdots + a_n\mathbf{v}_n - (b_1\mathbf{v}_1 + b_2\mathbf{v}_2 + \cdots + b_n\mathbf{v}_n)$$
$$= (a_1 - b_1)\mathbf{v}_1 + (a_2 - b_2)\mathbf{v}_2 + \cdots + (a_n - b_n)\mathbf{v}_n.$$

But the linear independence of the vectors $\mathbf{v}_i$ implies that $(a_1 - b_1) = (a_2 - b_2) = \cdots = (a_n - b_n) = 0$—that is, $a_1 = b_1, a_2 = b_2, \ldots, a_n = b_n$, so $\mathbf{v}$ has a unique representation as a linear combination of basis vectors.

Conversely, suppose every $\mathbf{v} \in V$ can be written uniquely as a linear combination of elements of S. Firstly, this says that $V = span(S)$—that is, S is a spanning set for V. Then the hypothesis implies, in particular, that the zero element of V can be expressed in only one way as a linear combination of the vectors in S: $\mathbf{0} = a_1\mathbf{v}_1 + a_2\mathbf{v}_2 + \cdots + a_n\mathbf{v}_n$. Because we can also write $\mathbf{0} = 0 \cdot \mathbf{v}_1 + 0 \cdot \mathbf{v}_2 + \cdots + 0 \cdot \mathbf{v}_n$, we must have $a_i = 0$ for $i = 1, 2, \ldots, n$. Therefore, S must be a linearly independent set. As a linearly independent spanning set, S must be a basis for V.

As we saw for $\mathbb{R}^n$ in Section 1.5, if $S = \{\mathbf{v}_1, \mathbf{v}_2, \ldots, \mathbf{v}_n\}$ is a basis for a vector space V and vector $\mathbf{v}$ in V is written as a linear combination of these basis vectors,

$$\mathbf{v} = a_1\mathbf{v}_1 + a_2\mathbf{v}_2 + \cdots + a_n\mathbf{v}_n,$$

then the uniquely defined scalars a_i are called the **coordinates** of the vector $\mathbf{v}$ **relative to** (or **with respect to**) **the basis** S. Thus whenever we use a particular basis in V, we are establishing a coordinate system according to which any vector in V is identified uniquely by the vector of its coordinates $\begin{bmatrix} a_1 & a_2 & \cdots & a_n \end{bmatrix}^{\mathrm{T}}$ with respect to the **coordinate axes** $\mathbf{v}_1, \mathbf{v}_2, \ldots, \mathbf{v}_n$. To be precise, we must have an *ordered* basis, so that the vector of coordinates with respect to this basis is unambiguous.

Example 1.5.3 provided two coordinate systems for $\mathbb{R}^3$ corresponding to two different bases for $\mathbb{R}^3$. Even without the geometric interpretation of a coordinate system, we can still use such a description in discussing a basis for any finite-dimensional vector space V. In fact, the key point here is that

> **Every finite-dimensional real vector space is algebraically identical to a vector space of n-tuples with real components**.

If V is a finite-dimensional vector space and $S = \{\mathbf{v}_1, \mathbf{v}_2, \ldots, \mathbf{v}_n\}$ is any basis of V, then the correspondence (*mapping*) $\mathbf{v} \leftrightarrow \begin{bmatrix} a_1 & a_2 & \cdots & a_n \end{bmatrix}^{\mathrm{T}}$, which associates each vector $\mathbf{v} \in V$ with its unique coordinate vector in $\mathbb{R}^n$, is a special type of mapping called an **isomorphism**[*] and establishes that V **and** $\mathbb{R}^n$ **behave the same way with respect to the algebraic rules required for a vector space**. Algebraic behavior on one side of the double arrow $\leftrightarrow$ is mirrored by the algebraic behavior on the other side. We have to realize that one side of the double arrow refers to $\{V, \oplus, \mathbb{R}\}$, a set of mathematical objects that have an "addition" indicated by $\oplus$ and a "scalar multiplication" indicated by $\odot$, and the other side refers to

---

[*] For an extensive treatment of this important algebraic concept, see J.A. Gallian, *Contemporary Abstract Algebra*, 7th edn. (Belmont, CA: Brooks/Cole, 2009).

Euclidean n-space $\{\mathbb{R}^n, +, \cdot\}$. We will discuss this concept further in Chapter 6. (See Exercises B7 and B16 of this section for other examples.) For now, let us consider an illustration of the usefulness of this fact.

Example 5.4.2: A Basis for a Subspace of $P_3[x]$

Suppose we want to find a basis for the subspace of $P_3[x]$ spanned by the vectors $1 - 2x - x^3, 3x - x^2, 1 + x + x^2 + x^3$, and $4 + 7x + x^2 + 2x^3$. We can first determine the coordinate vectors of these polynomials with respect to the standard ordered basis vectors for $P_3[x]$, $1, x, x^2$, and x^3, thereby establishing an algebraic identification (isomorphism) between the

vector spaces $P_3[x]$ and $\mathbb{R}^4 : a_0 + a_1 x + a_2 x^2 + a_3 x^3 \leftrightarrow \begin{bmatrix} a_0 \\ a_1 \\ a_2 \\ a_3 \end{bmatrix}$.

In our example, we have the correspondences

$$1 - 2x - x^3, \quad 3x - x^2, \quad 1 + x + x^2 + x^3, \quad 4 + 7x + x^2 + 2x^3$$
$$\updownarrow \qquad\qquad \updownarrow \qquad\qquad \updownarrow \qquad\qquad\qquad \updownarrow$$
$$\begin{bmatrix} 1 \\ -2 \\ 0 \\ -1 \end{bmatrix} \quad \begin{bmatrix} 0 \\ 3 \\ -1 \\ 0 \end{bmatrix} \quad \begin{bmatrix} 1 \\ 1 \\ 1 \\ 1 \end{bmatrix} \quad \begin{bmatrix} 4 \\ 7 \\ 1 \\ 2 \end{bmatrix}.$$

Next we form the matrix A having these vectors as columns:

$$A = \begin{bmatrix} 1 & 0 & 1 & 4 \\ -2 & 3 & 1 & 7 \\ 0 & -1 & 1 & 1 \\ -1 & 0 & 1 & 2 \end{bmatrix}.$$

The column space of A is algebraically equivalent to (isomorphic to) the span of the four polynomials we started with. Now we determine the basis of the column space by computing rref(A) and noting the pivot columns (see Theorem 2.7.3):

$$\text{rref}(A) = \begin{bmatrix} 1 & 0 & 0 & 1 \\ 0 & 1 & 0 & 2 \\ 0 & 0 & 1 & 3 \\ 0 & 0 & 0 & 0 \end{bmatrix}.$$

The pivot columns, columns 1, 2 and 3 of matrix A form a basis for the column space:

$$\begin{bmatrix} 1 \\ -2 \\ 0 \\ -1 \end{bmatrix}, \begin{bmatrix} 0 \\ 3 \\ -1 \\ 0 \end{bmatrix}, \begin{bmatrix} 1 \\ 1 \\ 1 \\ 1 \end{bmatrix}.$$

Translating back into the language of $P_3[x]$, we conclude that a basis for the three-dimensional subspace spanned by the four given polynomials is $\{1 - 2x - x^3, 3x - x^2, 1 + x + x^2 + x^3\}$.

Notice that the fourth original polynomial is a linear combination of the first three:

$$4 + 7x + x^2 + 2x^3 = 1 \cdot (1 - 2x - x^3) + 2 \cdot (3x - x^2)$$
$$+ 3 \cdot (1 + x + x^2 + x^3).$$

We could also have found a basis by determining the reduced column echelon form of A,

$$\begin{bmatrix} 1 & 0 & 0 & 0 \\ 0 & 1 & 0 & 0 \\ 0 & 0 & 1 & 0 \\ -1/3 & 1/3 & 1 & 0 \end{bmatrix},$$

and then transforming the nonzero columns directly into basis vectors for the polynomial subspace: $1 - \frac{1}{3}x^3, x + \frac{1}{3}x^3, x^2 + x^3$. (If $B = \text{rref}(A^T)$, then B^T is A in reduced column echelon form.)

As we have seen, a vector space can have many bases. For a Euclidean space $\mathbb{R}^n$, we have shown that any basis must have exactly n elements. This assertion remains true for an arbitrary finite-dimensional vector space V. The proof is identical to the proof of Theorem 1.5.1 (using Theorem 5.3.2 instead of Theorem 1.4.2).

Theorem 5.4.2

If $S = \{\mathbf{u}_1, \mathbf{u}_2, \ldots, \mathbf{u}_k\}$ and $T = \{\mathbf{v}_1, \mathbf{v}_2, \ldots, \mathbf{v}_r\}$ are two bases for a finite-dimensional vector space V, then $k = r$.

This last theorem establishes that the dimension of a finite-dimensional vector space is now *well-defined*—we can speak of *the* dimension, rather than *a* dimension.

Definition 5.4.2

The unique number of elements in any basis of a finite-dimensional vector space V is called the **dimension** of V and is denoted by $\dim(V)$. If $\dim(V) = n$ for some nonnegative integer n, we say that V is an **n-dimensional vector space** (or is **n-dimensional**).

The vector space consisting of the zero element alone is defined to have dimension 0: If $V = \{\mathbf{0}\}$, then $\dim(V) = 0$. Examples we have seen suggest that $\dim(M_n) = n^2$ and $\dim(P_n[x]) = n + 1$.

We can even assign a dimension, a *cardinal number*, to an infinite-dimensional vector space. Furthermore, **any two bases for an infinite-dimensional vector space must have the same cardinality.**[*]

---

[*] See N. Jacobson, *Lectures in Abstract Algebra*, vol. II, Chap. IX (New York: Van Nostrand, 1953). You can also consult the discussion of a **Hamel basis** in C.G. Small, *Functional Equations and How to Solve Them*, Chap. 6 (New York: Springer, 2007).

We can use the unambiguous nature of the term *dimension* in a finite-dimensional vector space to state and prove a fundamental result about linearly independent/dependent sets of vectors and a basic statement about subspaces.

Theorem 5.4.3

Suppose that V is a vector space of dimension n and $A = \{v_1, v_2, \ldots, v_k\}$ is a set of elements in V. If $k > n$—that is, if the number of vectors in A exceeds the dimension of the space—then A is linearly dependent.

Proof Because $\dim(V) = n$, V has a spanning set consisting of n vectors. Theorem 5.3.2 implies if A is a set of k linearly independent vectors in V, then $k \leq n$. Therefore any set of $k > n$ vectors must be dependent.

Another way of stating the result of Theorem 5.4.3 is to say that **if $A = \{v_1, v_2, \ldots, v_k\}$ is a linearly independent subset of an n-dimensional space, then $k \leq n$.**

Theorem 5.4.4

If W is a subspace of a finite-dimensional vector space V, then (a) $\dim(W) \leq \dim(V)$ and (b) $\dim(W) = \dim(V)$ if and only if $W = V$.

Proof Suppose W is a subspace of the vector space V with $\dim(V) = n$ and $\dim(W) = m$. Any basis of W is a linearly independent set of m elements of V and any basis of V is a spanning set having n elements. By Theorem 5.3.3, $m \leq n$, or $\dim(W) \leq \dim(V)$. Thus part (a) is proved.

Now suppose that $\dim(W) = \dim(V)$ and $W \neq V$. Then we can find a vector v such that $v \in V$ but $v \notin W$. If S is a basis for W, then $v \notin span(S)$ and the set $T = S \cup \{v\}$ must be a linearly independent subset of V (Exercises 1.4, B1, which is true for any vector space). But then T contains $m + 1 = n + 1$ vectors—impossible because $\dim(V) = n$ and any set of $n + 1$ vectors is linearly dependent according to Theorem 5.4.3. This contradiction establishes that the hypothesis $W \neq V$ must be false and thus we have proved part (b).

In Theorem 5.4.4, the statement that $\dim(W) = \dim(V)$ implies $W = V$ is not valid if V is not finite-dimensional. For example, if $V = C[0, 1]$ and $W = \{f \mid f' \text{ exists on } [0, 1]\}$, then both vector spaces contain the linearly independent functions $1, x, x^2, \ldots, x^n, \ldots$ and so are infinite-dimensional. But $W \neq V$ because there are continuous functions on $[0, 1]$ that are not differentiable on the interval—for example, $f(x) = |x - 0.5|$. In this case, W is a proper subspace of V: $W \subset V$. (Recall that a function that is differentiable at a point $x = a$ is continuous at $x = a$.)

Theorem 5.3.3 established a useful relationship between the sizes of spanning sets and linearly independent sets in a vector space V. The next result, the general vector space version of Theorem 1.5.2, provides another important connection between spanning sets and linearly independent sets. The proofs of Theorem 5.4.5 and its corollaries are identical to the proofs of the related results in Section 1.5.

Theorem 5.4.5

Suppose that V is a finite-dimensional vector space. If S is a finite spanning set for V and if I is a linearly independent subset of V such that $I \subseteq S$, then there is a basis B of V such that $I \subseteq B \subseteq S$.

The next results are very useful when working with sets of linearly independent vectors and will be used often in later sections.

Corollary 5.4.1

Every linearly independent subset I of a finite-dimensional vector space V can be extended to form a basis.

Corollary 5.4.2

If $\dim(V) = n$, then any set of n linearly independent vectors is a basis of V.

Note that Theorem 5.4.5 also implies that **any spanning set can be reduced (by discarding vectors) to a linearly independent set**.

To deepen our understanding of the relationships among linearly independent sets, spanning sets, and bases in a general vector space, we need some terminology. A linearly independent subset I of a vector space V is called a **maximal linearly independent set** if no larger set containing I is linearly independent. Equivalently, if I is a maximal linearly independent set and $I \subseteq \hat{I}$, where $\hat{I}$ is a linearly independent set, then $I = \hat{I}$. A subset S of a vector space V is called a **minimal spanning set** if S spans V and no subset of S spans V. Equivalently, if S is a minimal spanning set and $\hat{S}$ spans V with $\hat{S} \subseteq S$, then $\hat{S} = S$.

Theorem 5.4.6

If S is a subset of a finite-dimensional vector space V, then the following statements are equivalent:

(1) S is a basis.
(2) S is a minimal spanning set.
(3) S is a maximal linearly independent subset

Proof We will prove the implications (1) $\Rightarrow$ (2) $\Rightarrow$ (3) $\Rightarrow$ (1). (1) $\Rightarrow$ (2): Suppose S is a basis and $\check{S}$ is a subset of S such that $V = span(\check{S})$.

Note that since $\check{S} \subseteq S$, the number of elements in $\check{S}$ must be less than or equal to the number of elements in S. Because S is a linearly independent set, Theorem 5.3.3 tells us that the number of elements in S is less than or equal to the number of elements in $\check{S}$. Hence $S = \check{S}$ and S is a minimal spanning set. (2) $\Rightarrow$ (3): Suppose that $S = \{v_1, v_2, \ldots, v_k\}$ is a minimal spanning set. First we show that S is a linearly independent set. Suppose S is *not* linearly independent. Then, by Theorem 5.3.1, there exists a vector $v_j \in S$ which is a linear combination of the remaining vectors in S:

$$v_j = c_1 v_1 + \cdots + c_{j-1} v_{j-1} + c_{j+1} v_{j+1} + \cdots + c_k v_k.$$

Now because S spans V, any vector $v \in V$ can be written as a linear combination of vectors in S:

$$
\begin{aligned}
v &= c_1 v_1 + \cdots + c_{j-1} v_{j-1} + c_j v_j + c_{j+1} v_{j+1} + \cdots + c_k v_k \\
&= c_j v_j + \left(c_1 v_1 + \cdots + c_{j-1} v_{j-1} + c_{j+1} v_{j+1} + \cdots + c_k v_k \right) \\
&= c_j \left(c_1 v_1 + \cdots + c_{j-1} v_{j-1} + c_{j+1} v_{j+1} + \cdots + c_k v_k \right) \\
&\quad + \left(c_1 v_1 + \cdots + c_{j-1} v_{j-1} + c_{j+1} v_{j+1} + \cdots + c_k v_k \right) = \sum_{\substack{1 \le i \le k \\ i \ne j}} b_i v_i.
\end{aligned}
$$

This indicates that $S \backslash \{v_j\}$, a subset of S, is also a spanning set for V, contradicting the minimality of S. Thus S must be linearly independent. To show that S is *maximally* linearly independent, suppose that $S \subseteq \hat{S}$, where $\hat{S}$ is linearly independent. If $v \in \hat{S} \backslash S$, then $v = c_1 v_1 + c_2 v_2 + \cdots + c_k v_k$ for some scalars c_i because S is a spanning set. Consequently, $c_1 v_1 + c_2 v_2 + \cdots + c_k v_k - v = 0$, showing that $\{v_1, v_2, \ldots, v_k, v\} \subseteq \hat{S}$ is linearly dependent. But then $\hat{S}$ would have to be linearly dependent—a contradiction. Therefore, there can be no vector $v \in \hat{S} \backslash S$, implying $S = \hat{S}$; and S is thus a maximally linearly independent set. (3) $\Rightarrow$ (1): Suppose v is any element of V and let $S = \{v_1, v_2, \ldots, v_k\}$ be a maximal linearly independent subset of V. We will show that S spans V. If $v \notin S$, then $S \cup \{v\}$ is a linearly *dependent* set because $S \subset S \cup \{v\}$ and S is a maximal linearly independent set. Now $v \in S \cup \{v\}$ and there are scalars $c, c_1, c_2, \ldots, c_k$, not all of which are zero, such that $cv + c_1 v_1 + c_2 v_2 + \cdots + c_k v_k = 0$. If $c = 0$, the linear independence of S implies that $c_1 = c_2 = \cdots = c_k = 0$, which contradicts our hypothesis of linear dependence. Thus $c \ne 0$ and we can write $v = \left(-\frac{c_1}{c} \right) v_1 + \left(-\frac{c_2}{c} \right) v_2 + \cdots + \left(-\frac{c_k}{c} \right) v_k$. Hence $v \in span(S)$ and S is a basis for V.

We end this section with an important relation connecting the dimensions of various subspaces of a vector space. It is analogous to the formula for the number of elements in the union of two subsets of a finite set (and consequently related to the formula for the probability of the union of two events).*

---

* See, for example, Sections 5.1 and 6.1 of K.H. Rosen, *Discrete Mathematics and its Applications*, 6th edn. (New York: McGraw-Hill, 2007).

Theorem 5.4.7

If U and W are subspaces of a finite-dimensional vector space V, then

$$\dim(U + W) = \dim(U) + \dim(W) - \dim(U \cap W).$$

Proof Let $\{\mathbf{v}_1, \mathbf{v}_2, \ldots, \mathbf{v}_k\}$ be a basis of the subspace $U \cap W$, so $\dim(U \cap W) = k$. Because $\{\mathbf{v}_1, \mathbf{v}_2, \ldots, \mathbf{v}_k\}$ is a basis of $U \cap W$, it is a linearly independent subset of U and can be extended (by Corollary 5.4.1) to a basis $\{\mathbf{v}_1, \mathbf{v}_2, \ldots, \mathbf{v}_k, \mathbf{y}_1, \mathbf{y}_2, \ldots, \mathbf{y}_r\}$ of U. Thus $\dim(U) = k + r$. Similarly, we can extend $\{\mathbf{v}_1, \mathbf{v}_2, \ldots, \mathbf{v}_k\}$ to a basis $\{\mathbf{v}_1, \mathbf{v}_2, \ldots, \mathbf{v}_k, \mathbf{z}_1, \mathbf{z}_2, \ldots, \mathbf{z}_s\}$ for W, so $\dim(W) = k + s$.

Now we show that $S = \{\mathbf{v}_1, \mathbf{v}_2, \ldots, \mathbf{v}_k, \mathbf{y}_1, \mathbf{y}_2, \ldots, \mathbf{y}_r, \mathbf{z}_1, \mathbf{z}_2, \ldots, \mathbf{z}_s\}$ is a basis for $U + W$. This would complete the proof because then we will have $\dim(U + W) = k + r + s = (k + r) + (k + s) - k = \dim(U) + \dim(W) - \dim(U \cap W)$.

Clearly $span(S) = span\{\mathbf{v}_1, \mathbf{v}_2, \ldots, \mathbf{v}_k, \mathbf{y}_1, \mathbf{y}_2, \ldots, \mathbf{y}_r, \mathbf{z}_1, \mathbf{z}_2, \ldots, \mathbf{z}_s\}$ is a subspace containing U and W and therefore contains $U + W$ (see the comments right after Example 5.2.6).

To show that S is linearly independent, suppose there are scalars $a_1, \ldots, a_k, b_1, \ldots, b_r, c_1, \ldots, c_s$ such that

$$a_1\mathbf{v}_1 + \cdots + a_k\mathbf{v}_k + b_1\mathbf{y}_1 + \cdots + b_r\mathbf{y}_r + c_1\mathbf{z}_1 + \cdots + c_s\mathbf{z}_s = \mathbf{0}. \ (*)$$

Then $c_1\mathbf{z}_1 + c_2\mathbf{z}_2 + \cdots + c_s\mathbf{z}_s = -a_1\mathbf{v}_1 - a_2\mathbf{v}_2 - \cdots - a_k\mathbf{v}_k - b_1\mathbf{y}_1 - b_2\mathbf{y}_2 - \cdots - b_r\mathbf{y}_r \in span\{\mathbf{v}_1, \mathbf{v}_2, \ldots, \mathbf{v}_k, \mathbf{y}_1, \mathbf{y}_2, \ldots, \mathbf{y}_r\} = U$, which shows that $c_1\mathbf{z}_1 + c_2\mathbf{z}_2 + \cdots + c_s\mathbf{z}_s \in U$. But all the $\mathbf{z}_i$'s are in W, so we have $c_1\mathbf{z}_1 + c_2\mathbf{z}_2 + \cdots + c_s\mathbf{z}_s \in U \cap W$. Because $\{\mathbf{v}_1, \mathbf{v}_2, \ldots, \mathbf{v}_k\}$ is a basis of $U \cap W$, we can write

$$c_1\mathbf{z}_1 + c_2\mathbf{z}_2 + \cdots + c_s\mathbf{z}_s = d_1\mathbf{v}_1 + d_2\mathbf{v}_2 + \cdots + d_k\mathbf{v}_k,$$

or

$$c_1\mathbf{z}_1 + c_2\mathbf{z}_2 + \cdots + c_s\mathbf{z}_s - d_1\mathbf{v}_1 - d_2\mathbf{v}_2 - \cdots - d_k\mathbf{v}_k = \mathbf{0}.$$

But, as a basis for W, $\{\mathbf{v}_1, \mathbf{v}_2, \ldots, \mathbf{v}_k, \mathbf{z}_1, \mathbf{z}_2, \ldots, \mathbf{z}_s\}$ is a linearly independent set, so the last equation implies that all the c_i's and d_i's equal 0. Then equation (*) becomes

$$a_1\mathbf{v}_1 + \cdots + a_k\mathbf{v}_k + b_1\mathbf{y}_1 + \cdots + b_r\mathbf{y}_r = \mathbf{0},$$

which implies that all the a_i's and b_i's are 0 because the set $\{\mathbf{v}_1, \mathbf{v}_2, \ldots, \mathbf{v}_k, \mathbf{y}_1, \mathbf{y}_2, \ldots, \mathbf{y}_r\}$ is a basis for U and thus is linearly independent. Therefore all the coefficients in (*) are 0 and the set S must be linearly independent.

Corollary 5.4.3

If U and W are subspaces of a finite-dimensional vector space, then

$$\dim(U \oplus W) = \dim(U) + \dim(W).$$

Proof Theorem 5.2.3 implies $U \cap W = \{\mathbf{0}\}$ and Theorem 5.4.7 yields $\dim(U \oplus W) = \dim(U + W) = \dim(U) + \dim(W) - \dim(\{\mathbf{0}\}) = \dim(U) + \dim(W)$.

Exercises 5.4

A.

1. Show that the matrices $E_1 = \begin{bmatrix} 1 & 1 \\ 0 & 0 \end{bmatrix}$, $E_2 = \begin{bmatrix} 0 & 0 \\ 1 & 1 \end{bmatrix}$, $E_3 = \begin{bmatrix} 1 & 0 \\ 0 & 1 \end{bmatrix}$, and $E_4 = \begin{bmatrix} 0 & 1 \\ 1 & 1 \end{bmatrix}$ (see Example 5.3.5) constitute a basis for the space M_2.

2. If $\mathbb{R}^3 = U \oplus W$ for nontrivial subspaces U and W, show that the geometric representation of one of the subspaces is a plane and the other is a line. [*Hint*: Use a dimension argument.]

3. Extend the linearly independent set $\{[\, 1 \quad -1 \quad 1 \quad -1 \,]^{\mathrm{T}}, [\, 1 \quad 1 \quad -1 \quad 1 \,]^{\mathrm{T}}\}$ to a basis for $\mathbb{R}^4$.

4. Suppose that U and W are subspaces of $\mathbb{R}^{10}$ with dimensions 6 and 8, respectively. What is the smallest possible dimension for $U \cap W$?

5. Suppose that U and W are subspaces of $\mathbb{R}^8$ such that $\dim(U) = 3$, $\dim(W) = 5$, and $U + W = \mathbb{R}^8$. Explain why $U \cap W = \{\mathbf{0}\}$.

6. Suppose that U and W are both four-dimensional subspaces of $\mathbb{R}^7$. Explain why $U \cap W \neq \{\mathbf{0}\}$.

7. Show that if U and W are subspaces of $\mathbb{R}^3$ such that $\dim(U) = \dim(W) = 2$, then $U \cap W \neq \{\mathbf{0}\}$. What is the geometric interpretation of this?

8. Suppose U and W are subspaces of $\mathbb{R}^8$ with dimensions 4 and 6, respectively, and that U is not contained in W. What can be said about $\dim(U \cap W)$?

9. Suppose V is a vector space of dimension 55, and let U, V be two subspaces of dimensions 36 and 28, respectively. What is the *least* possible value and the *greatest* possible value of $\dim(U + V)$?

B.

1. Find a basis for the subspace of $P_3[x]$ spanned by the polynomials

    $$1 - x - 2x^3, \quad 1 + x^3, \quad 1 + x + 4x^3, \quad x^2.$$

2. Find a basis for the subspace spanned by the vectors

    $$\begin{bmatrix} 3 & 4 \\ 1 & 2 \end{bmatrix}, \quad \begin{bmatrix} 2 & 5 \\ 1 & 1 \end{bmatrix}, \quad \begin{bmatrix} 0 & -7 \\ -1 & 1 \end{bmatrix}$$

    in M_2.

3. Find a basis for M_2 that consists of matrices A with $A^2 = A$.

4. Show that no finite set of polynomials can generate $P[x]$, the vector space consisting of all polynomials in the variable x with real coefficients. (See Example 5.4.1.)

5. Show that $B = \left\{ \begin{bmatrix} 1 & 1 & 1 \\ 1 & 1 & 1 \\ 1 & 1 & 1 \end{bmatrix}, \begin{bmatrix} 0 & 1 & -1 \\ -1 & 0 & 1 \\ 1 & -1 & 0 \end{bmatrix}, \begin{bmatrix} -1 & 1 & 0 \\ 1 & 0 & -1 \\ 0 & -1 & 1 \end{bmatrix} \right\}$ is a basis for the space M of all 3×3 magic squares. (Note that row, column, and diagonal sums will be equal for any one matrix in M; but these sums may differ from matrix to matrix.)

6. Find a basis for the range of the linear transformation $T_A : \mathbb{R}^4 \to \mathbb{R}^4$ defined by $T_A(\mathbf{v}) = A\mathbf{v}$, where

    $$A = \begin{bmatrix} 1 & 1 & -1 & 0 \\ 0 & 1 & 1 & 1 \\ 0 & -1 & -1 & -1 \\ -1 & -1 & 1 & 0 \end{bmatrix}. \text{ (See Example 5.2.4.)}$$

7. Consider the isomorphism $\begin{bmatrix} a & b \\ c & d \end{bmatrix} \leftrightarrow \begin{bmatrix} a \\ b \\ c \\ d \end{bmatrix}$ between M_2 and $\mathbb{R}^4$. Match each basis matrix E_i in Exercise A1 to a vector in $\mathbb{R}^4$ and show that these vectors form a basis for $\mathbb{R}^4$.

8. Consider the isomorphism

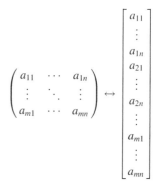

between M_{mn} and $\mathbb{R}^{m \cdot n}$. Generalize the previous exercise by showing that any basis for M_{mn} determines a corresponding basis for $\mathbb{R}^{m \cdot n}$.

9. a. Show that the infinite sequences $\mathbf{f}_1 = (2, 3, 5, 8, 13, \ldots .)$ and $\mathbf{f}_2 = (1, 2, 3, 5, 8, \ldots .)$ are basis vectors for the vector space $\mathfrak{F}$ described in Problem C3 of Exercises 5.1.

    b. Express the sequence $\alpha = (1, 1, 2, 3, 5, 8, \ldots)$ as a linear combination of $\mathbf{f}_1$ and $\mathbf{f}_2$.

10. Let U and W be unequal subspaces, each of dimension $n - 1$, in an n-dimensional vector space V.
    a. Prove that $U + W = V$.
    b. Prove that $\dim(U \cap W) = n - 2$.

11. Show that the infinite set of polynomials $\{1, x, x^2, \ldots, x^n, \ldots\}$ is linearly independent and spans $P[x]$. (See Example 5.4.1.) [Recall that any subset X (not necessarily finite) of a vector space V is said to be *linearly independent* if every finite subset of X is linearly independent.]

12. Problem B2, Exercises 5.2, asserts that the subset of M_n consisting of those matrices A such that $A\mathbf{x} = \mathbf{0}$, where $\mathbf{x}$ is a fixed nonzero vector in $\mathbb{R}^n$, is a subspace of M_n. Prove that $\dim(S) = n^2 - n$.

13. Suppose that U and W are subspaces of a finite-dimensional space and that $\dim(U \cap W) = \dim(U + W)$. Prove that $U = W$.

14. Let U and W be subspaces of the finite-dimensional space V. Show that $V = U \oplus W$ if any two of the following conditions hold:
    (1) $V = U + W$.
    (2) $U \cap W = \{\mathbf{0}\}$.
    (3) $\dim(V) = \dim(U) + \dim(W)$.
    (Note that we have three pairs of conditions to consider.)

15. Define the $m \times n$ matrices E_{ij} as follows: For $i = 1, 2, \ldots, m$ and $j = 1, 2, \ldots, n$, the entry in row i and column j of E_{ij} is 1, whereas all other entries are 0. Show that $\{E_{ij}\}$ is a basis for M_{mn}. What does this say about the dimension of M_{mn}?

16. Let $\mathfrak{C}$ denote the set of all 2×2 matrices of the form $\begin{bmatrix} a & b \\ -b & a \end{bmatrix}$, where $a, b \in \mathbb{R}$, and consider the isomorphism
$\begin{bmatrix} a & b \\ -b & a \end{bmatrix} \leftrightarrow a + bi$, where $i = \sqrt{-1}$.

    a. Prove that $\mathfrak{C}$ is a subspace of M_2.

    b. What matrix in $\mathfrak{C}$ corresponds to the number 1? What matrix in $\mathfrak{C}$ corresponds to the complex number i?

    c. Show that

    $$\begin{bmatrix} a & b \\ -b & a \end{bmatrix} + \begin{bmatrix} c & d \\ -d & c \end{bmatrix} \leftrightarrow (a + bi) + (c + di),$$

    and

    $$\begin{bmatrix} a & b \\ -b & a \end{bmatrix} \cdot \begin{bmatrix} c & d \\ -d & c \end{bmatrix} \leftrightarrow (a + bi) \cdot (c + di).$$

    d. Show that

    $$\begin{bmatrix} a & b \\ -b & a \end{bmatrix}^{-1} \leftrightarrow \frac{1}{a + bi},$$

    where not both a and b are zero.

    e. What is the dimension of $\mathfrak{C}$? What does your answer suggest about the dimension of $\mathbb{C}$, the vector space of complex numbers over the reals? (See Example 5.1.1(b).)

17. a. Show that the set $\{\sin x, \cos x\}$ is a linearly independent subset of the vector space $S = \{f \in V : f'' + f = 0\}$ described in Example 5.2.1.

    b. Show that the set $\{\sin x, \cos x\}$ spans the space S and hence is a basis for S.

18. a. If A and B are $m \times n$ matrices, show that

    $$\text{rank}(A + B) \leq \text{rank}(A) + \text{rank}(B).$$

    [*Hint*: First show that the column space of $A + B$ is a subset of (the column space of A) + (the column space of B).]

    b. Show that $\text{rank}(A + B) = \text{rank}(A) + \text{rank}(B)$ implies (the column space of A) ∩ (the column space of B) = $\{\mathbf{0}\}$.

19. Let $A, B \in M_n$. Show that if $AB = O_n$, then $\text{rank}(A) + \text{rank}(B) \leq n$.

20. If A and B are conformable matrices, prove that $\text{nullity}(AB) \leq \text{nullity}(A) + \text{nullity}(B)$.

C.

1. Let V be a finite-dimensional vector space with a basis B, and let $B_1, B_2, \ldots, B_k$ be a *partition* of B—that is, $B_1, B_2, \ldots, B_k$ are subsets of B such that $B = B_1 \cup B_2 \cup \cdots \cup B_k$ and $B_i \cap B_j = \emptyset$ if $i \neq j$. Prove that $V = span(B_1) \oplus span(B_2) \oplus \cdots \oplus span(B_k)$.

2. The set $\mathcal{M}(\mathbb{R}, \mathbb{R})$ of all functions from $\mathbb{R}$ to $\mathbb{R}$ is a vector space under addition and scalar multiplication. Let n be a positive integer and let $\mathcal{F}_n$ be the set of all functions $f: \mathbb{R} \to \mathbb{R}$ of the form $f(x) = a_0 + \sum_{k=1}^{n} (a_k \cos\ kx + b_k \sin\ kx)$, where $a_k, b_k \in \mathbb{R}$ for every k.
    a. Show that $\mathcal{F}_n$ is a subspace of $\mathcal{M}(\mathbb{R}, \mathbb{R})$.
    b. If $f \in \mathcal{F}_n$ is the zero function, prove that all the coefficients a_k, b_k must be zero. [*Hint*: Find a formula for $f'' + n^2 f$ and use induction.]
    c. Use the result of part (b) to prove that the $2n + 1$ functions

    $$f(x) \equiv 1, \ f(x) = \cos kx, \ f(x) = \sin kx \quad (k = 1, 2, \ldots, n)$$

    constitute a basis for $\mathcal{F}_n$.

3. If $V = U \oplus W_1$ and $V = U \oplus W_2$, then $\dim(V) - 2 \dim(U) \leq \dim(W_1 \cap W_2) \leq \dim(V) - \dim(U)$.

4. Suppose that $\dim(V) = n$. Prove that there exist one-dimensional subspaces $U_1, U_2, \ldots, U_n$ of V such that $V = U_1 \oplus U_2 \oplus \cdots \oplus U_n$.

5. Comparison of Theorem 5.4.7 with various counting formulas in combinatorics and discrete probability suggests the generalization

$$\begin{aligned}
\dim(U_1 + U_2 + U_3) = {}&\dim(U_1) + \dim(U_2) + \dim(U_3) \\
&- \dim(U_1 \cap U_2) - \dim(U_1 \cap U_3) \\
&- \dim(U_2 \cap U_3) + \dim(U_1 \cap U_2 \cap U_3),
\end{aligned}$$

    where U_1, U_2, and U_3 are subspaces of a finite-dimensional space. Prove this result or give a counterexample.

6. Suppose that $U_1, U_2, \ldots, U_m$ are subspaces of V such that $V = U_1 + U_2 + \cdots + U_m$. Prove that $V = U_1 \oplus U_2 \oplus \cdots \oplus U_m$ if and only if $\dim(V) = \sum_{i=1}^{m} \dim(U_i)$.

5.5 Summary

A nonempty set V on which two operations, *addition* and *scalar multiplication*, are defined is called a **vector space**—and the elements of V are called **vectors**—if the axioms of Definition 5.1.1 are satisfied for every **u**, **v**, and **w** in V and all scalars c and d. These axioms are general versions of the properties of $\mathbb{R}^n$.

If a nonempty subset S of a vector space $\{V, +, \cdot\}$ is also a vector space using the same addition and scalar multiplication, then S is called a **subspace** of $\{V, +, \cdot\}$. A nonempty subset S of a vector space V is a subspace of V if and only if for all vectors $\mathbf{u}$ and $\mathbf{v}$ in S and every scalar c, (1) $\mathbf{u} + \mathbf{v} \in S$ and (2) $c \cdot \mathbf{u} \in S$.

If U and W are subspaces of a vector space V, then we can define the **sum** of U and W as follows: $U + W = \{\mathbf{u} + \mathbf{w} | \mathbf{u} \in U \text{ and } \mathbf{w} \in W\}$. Similarly, $U_1 + U_2 + \cdots + U_m = \{\mathbf{u}_1 + \mathbf{u}_2 + \cdots + \mathbf{u}_m | \mathbf{u}_i \in U_i, \ i = 1, 2, \ldots, m\}$. If U and W are subspaces of a vector space V, then $U \cap W$ and $U + W$ are also subspaces of V. We can prove that $U \cap W$ **is the largest subspace of** V **contained in both** U **and** W; whereas $U + W$ **is the smallest subspace of** V **containing both** U **and** W.

If U and W are subspaces of V, we say that V is the **direct sum** of U and W, denoted as $V = U \oplus W$, if every $\mathbf{v} \in V$ has a *unique* representation $\mathbf{v} = \mathbf{u} + \mathbf{w}$, where $\mathbf{u} \in U$ and $\mathbf{w} \in W$. More generally, if U and W are subspaces of V, then the subspace $U + W$ is called a **direct** sum if every element of $U + W$ can be written uniquely as $\mathbf{u} + \mathbf{w}$, where $\mathbf{u} \in U$ and $\mathbf{w} \in W$. If U is a subspace of a vector space V, then a subspace W of V is said to be a **complement** of U if $U \oplus W = V$. **If** U **and** W **are subspaces of** V**, then** $V = U \oplus W$ **if and only if** $V = U + W$ **and** $U \cap W = \{\mathbf{0}\}$.

As in Chapter 1, we can define a **linear combination** of a nonempty finite set of vectors and discuss the important related concept of **linearly independent** (**linearly dependent**) sets. A set $S = \{\mathbf{v}_1, \mathbf{v}_2, \ldots, \mathbf{v}_k\}$ in a vector space V that contains at least two vectors is linearly dependent if and only if some vector $\mathbf{v}_j$ is a linear combination of the remaining vectors in the set.

A nonempty set S of vectors in V **spans** V (or is a **spanning set** for V) if every vector in V is an element of *span*(S). In a vector space, **the size of any linearly independent set is always less than or equal to the size of any spanning set**.

A set B of vectors in a vector space V is called a **basis** for (or of) V if B is both a linearly independent set and a spanning set for V. The vector space V is **finite-dimensional** if it has a finite basis. Otherwise, the vector space is called **infinite-dimensional. A nonempty subset** $S = \{\mathbf{v}_1, \mathbf{v}_2, \ldots, \mathbf{v}_n\}$ **of a vector space** V **is a basis if and only if every vector** $\mathbf{v} \in V$ **can be written as a linear combination** $\mathbf{v} = a_1\mathbf{v}_1 + a_2\mathbf{v}_2 + \cdots + a_n\mathbf{v}_n$ **in one and only one way.**

If $S = \{\mathbf{v}_1, \mathbf{v}_2, \ldots, \mathbf{v}_n\}$ is a basis for a vector space V and vector $\mathbf{v}$ in V is written as a linear combination of these basis vectors, $\mathbf{v} = a_1\mathbf{v}_1 + a_2\mathbf{v}_2 + \cdots + a_n\mathbf{v}_n$, then the uniquely defined scalars a_i are called the **coordinates** of the vector $\mathbf{v}$ **relative to** (or **with respect to**) **the basis** S. Whenever we use a particular basis in V, we are establishing a coordinate system according to which any vector in V is identified uniquely by the vector of its coordinates $[a_1 \quad a_2 \quad \cdots \quad a_n]^{\mathrm{T}}$ with respect to the **coordinate axes** $\mathbf{v}_1, \mathbf{v}_2, \ldots, \mathbf{v}_n$. We must have an *ordered* basis, so that the vector of coordinates with respect to this basis is unambiguous. It is a fundamental fact that **every finite-dimensional vector space is algebraically identical to a vector space of** n**-tuples with real components**. A consequence of this algebraic connection is that **any two bases for a finite-dimensional vector space** V **must have the same number of elements**. Thus we can define the **dimension** of a finite-dimensional vector space as the unique number of elements in any basis for V. If V is a vector space of dimension n and A is a set of elements in V such that the number of vectors in A exceeds n, then A is linearly dependent. Alternatively, if $A = \{\mathbf{v}_1, \mathbf{v}_2, \ldots, \mathbf{v}_k\}$ is a linearly independent subset of an n-dimensional space, then $k \leq n$. If W is a subspace of a finite-dimensional vector space V, then $\dim(W) \leq \dim(V)$ and $\dim(W) = \dim(V)$ if and only if $W = V$.

A powerful result is that **any linearly independent subset of a finite-dimensional vector space V can be extended to form a basis for V.** Also, **if $\dim(V) = n$, then any set of n linearly independent vectors is a basis for V.** If S is a subset of a finite-dimensional vector space V, then the following statements are equivalent: (1) S is a basis; (2) S is a minimal spanning set for V; and (3) S is a maximal linearly independent subset of V.

An important relation is that if U and W are subspaces of a finite-dimensional vector space V, then $\mathbf{\dim(U + W) = \dim(U) + \dim(W) - \dim(U \cap W)}$.

6

Linear Transformations

In Section 3.2, we saw that an $m \times n$ matrix A could be used to define a function (*transformation or mapping*) from $\mathbb{R}^n$ to $\mathbb{R}^m$: For any $\mathbf{v} \in \mathbb{R}^n$, define $T_A(\mathbf{v}) = A\mathbf{v}$. This function has the following important properties:

(1) $T_A(\mathbf{v}_1 + \mathbf{v}_2) = T_A(\mathbf{v}_1) + T_A(\mathbf{v}_2)$ [*Additivity*] for any vectors $\mathbf{v}_1$ and $\mathbf{v}_2$ in $\mathbb{R}^n$.

(2) $T_A(k\mathbf{v}) = kT_A(\mathbf{v})$ [*Homogeneity*] for any vector $\mathbf{v}$ in $\mathbb{R}^n$ and any real number k.

Any function with these two properties is called a **linear transformation** from $\mathbb{R}^n$ to $\mathbb{R}^n$. (Example 5.2.4 extended this idea slightly.)

In this chapter, we will discuss such transformations in the context of general vector spaces. Furthermore, instead of starting with a matrix and using it to define a function, we will focus on functions from one vector space to another and show how to derive an appropriate matrix from this situation. A reasonable definition of linear algebra is **the study of linear transformations on vector spaces.**

6.1 Linear Transformations

We start with basic terminology, working with vector spaces instead of arbitrary sets.

(See Appendix A for a background in set theory.)

Definition 6.1.1

If V and W are vector spaces, a **transformation** (**mapping** or **function**) T from V to W is a rule that assigns to each vector $\mathbf{v} \in V$ one and only one element $T(\mathbf{v}) \in W$, called the **image of v under T**. We will denote such a mapping by $T : V \to W$.

The vector spaces V and W are called the **domain** and the **codomain** of T, respectively. The set of all images of vectors of V under the transformation T is called the **range** of T (or the **image** of T), written range T: range

$T = T(V) = \{T(\mathbf{v}) : \mathbf{v} \in V\}$. Equivalently, range $T = \{\mathbf{w} \in W : \mathbf{w} = T(\mathbf{v})$ for some $\mathbf{v} \in V\}$. Note that range $T \subseteq W$.

> From this point on, we will assume that two vector spaces linked by a transformation have the same set (field) of scalars, generally $\mathbb{R}$.

Example 6.1.1: Transformations between Vector Spaces

(a) $T : \mathbb{R}^2 \to \mathbb{R}^3$, defined by $T\left(\begin{bmatrix} x \\ y \end{bmatrix}\right) = \begin{bmatrix} x + 1 \\ x + y \\ y - 2 \end{bmatrix}$.

(b) $T : M_n \to \mathbb{R}$, defined by $T(A) = \text{trace}(A)$ for any $A \in M_n$.

(c) $T : P_n[x] \to P_n[x]$, defined by $T(p(x)) = xp'(x)$ for any $p \in P_n[x]$.

(d) If V is a vector space, define $I : V \to V$ by $I(\mathbf{v}) = \mathbf{v}$ for all $\mathbf{v} \in V$. This is called the **identity transformation** on V. In this case, V is the domain, codomain, and range of I.

(e) If V and W are vector spaces, define $T : V \to W$ by $T(\mathbf{v}) = \mathbf{0}_W$ for all $\mathbf{v} \in V$. This is called the **null** (or **zero**) **transformation** on V. Here, the codomain is W, but the range of T is $\{\mathbf{0}_W\}$, a proper subset of W.

In Example 6.1.1(b), we realize that two distinct $n \times n$ matrices may have the same image (the trace). Similarly, in part (c) the transformation may send distinct vectors into the same image vector. In part (e) of this example, we notice that all vectors in V have the same image, $\mathbf{0}_W$, under T. However, in parts (a) and (d) the transformations send distinct vectors into distinct vectors.

A transformation $T : V \to W$ is said to be **one-to-one** (**1-1** or **injective**) if T maps distinct vectors in V to distinct vectors in W. Thus the transformations in parts (a) and (d) of Example 6.1.1 are one-to-one, whereas those in parts (b), (c), and (e) are not.

In Example 6.1.1(a), we can determine that range $T \neq \mathbb{R}^3$, that is, range T is a proper subset of the codomain. For example, the zero vector in $\mathbb{R}^3$ is not the image of any vector in $\mathbb{R}^2$ under T. In part (e), unless W is the trivial vector space, $\{\mathbf{0}_W\}$, there are vectors in W that are not the images of vectors in V. In part (c), no nonzero constant polynomial $p(x) \equiv k \in P_n[x]$, where $k \in \mathbb{R}$, is the image of a vector in $P_n[x]$ under T. However, in parts (b) and (d), we can see that the range equals the codomain in each case—that is, every vector in the codomain of the transformation is the image of at least one vector in the domain of the transformation.

A transformation $T : V \to W$ is called **onto** (or **surjective**, from the French word *sur*, meaning "on" or "onto") if given any $\mathbf{w} \in W$, there is at least one vector $\mathbf{v} \in V$ such that $T(\mathbf{v}) = \mathbf{w}$. Thus, the transformations in parts (b) and (d) of Example 6.1.1 are onto, whereas those in parts (a), (c), and (e) are not.

A transformation $T: V \rightarrow W$ is called **one-to-one onto** (or **bijective**) if T is both one-to-one and onto. The concepts "one-to-one" and "onto" are independent of each other. In Example 6.1.1, only transformation (d) is bijective.

Definition 6.1.2

A transformation T from a vector space V to a vector space W is called a **linear transformation** if $T(a\mathbf{v}_1 + b\mathbf{v}_2) = aT(\mathbf{v}_1) + bT(\mathbf{v}_2)$ for all $\mathbf{v}_1, \mathbf{v}_2 \in V$ and $a, b \in \mathbb{R}$. A linear transformation from V to itself is often called an **operator on V**.

This definition is equivalent to the two-part description given in the "Introduction" to this chapter: In Definition 6.1.2, take $a = b = 1$ to get $T(\mathbf{v}_1 + \mathbf{v}_2) = T(\mathbf{v}_1) + T(\mathbf{v}_2)$ [*Additivity*], and let $b = 0$ to see that $T(a\mathbf{v}_1) = aT(\mathbf{v}_1)$ [*Homogeneity*]; conversely, if we start with the two properties in the "Introduction," it follows that $T(a\mathbf{v}_1 + b\mathbf{v}_2) = T(a\mathbf{v}_1) + T(b\mathbf{v}_2) = aT(\mathbf{v}_1) + bT(\mathbf{v}_2)$. Another way to express the relation $T(a\mathbf{v}_1 + b\mathbf{v}_2) = aT(\mathbf{v}_1) + bT(\mathbf{v}_2)$ is to say that **the image of a linear combination of vectors is a linear combination of the images of these vectors**. More concisely, **a linear transformation T preserves vector sums and scalar multiples**—that is, **T preserves linear combinations**.

In Example 5.2.4, we were careful to acknowledge that the definitions of vector addition and scalar multiplication may be different in two vector spaces connected by a linear transformation. Mathematical accuracy demands that we indicate the domain and codomain vector spaces as $\{V, +, \cdot\}$ and $\{W, \boxplus, \odot\}$, for example, so that the property of linearity looks like $T(a \cdot \mathbf{v}_1 + b \cdot \mathbf{v}_2) = a \odot T(\mathbf{v}_1) \boxplus b \odot T(\mathbf{v}_2)$. (Note that axiom (6) of Definition 5.1.1, $(c + d) \cdot \mathbf{u} = c \cdot \mathbf{u} + d \cdot \mathbf{u}$, abuses the symbol $+$, using it to indicate both vector addition and the addition of two scalars.) However, because any two finite-dimensional vector spaces are isomorphic to Euclidean spaces (Section 5.4), from now on we will use the same symbols $+$ and $\cdot$ to denote the (possibly different) operations in both vector spaces. The context, the mathematical environment we find ourselves in, will usually make it clear what operations we are using.

Two linear transformations $T_1, T_2 : V \rightarrow W$ are **equal** if $T_1(\mathbf{v}) = T_2(\mathbf{v})$ for all $\mathbf{v} \in V$. Equivalently, as we will see (Theorem 6.1.2), two linear transformations from V to W are equal if and only if they agree on a basis for V.

Example 6.1.2: Some Linear Transformations

(a) The function $T: \mathbb{R}^3 \rightarrow \mathbb{R}^2$ defined by the rule

$$T\left(\begin{bmatrix} x \\ y \\ z \end{bmatrix}\right) = \begin{bmatrix} x \\ y \end{bmatrix}$$

is linear:

$$T\left(a\begin{bmatrix}x_1\\y_1\\z_1\end{bmatrix}+b\begin{bmatrix}x_2\\y_2\\z_2\end{bmatrix}\right)=T\left(\begin{bmatrix}ax_1+bx_2\\ay_1+by_2\\az_1+bz_2\end{bmatrix}\right)=\begin{bmatrix}ax_1+bx_2\\ay_1+by_2\end{bmatrix}$$

$$=a\begin{bmatrix}x_1\\y_1\end{bmatrix}+b\begin{bmatrix}x_2\\y_2\end{bmatrix}=aT\left(\begin{bmatrix}x_1\\y_1\\z_1\end{bmatrix}\right)$$

$$+bT\left(\begin{bmatrix}x_2\\y_2\\z_2\end{bmatrix}\right).$$

Furthermore, T is not 1-1 but *is* onto.

(b) For each positive integer i, $1\leq i\leq n$, the mapping $pr_i\colon\mathbb{R}^n\to\mathbb{R}$ described by

$$pr_i\left(\begin{bmatrix}x_1\\x_2\\\vdots\\x_n\end{bmatrix}\right)=x_i$$

(i.e., the mapping that selects the ith component of a vector in $\mathbb{R}^n$) is called the ith *projection* of $\mathbb{R}^n$ onto $\mathbb{R}$. This transformation is easily shown to be linear and onto, but not 1-1 (Exercise A5). A linear transformation from a vector space to its set of scalars is called a *linear functional*.

(c) For a fixed angle θ, the transformation T_θ from $\mathbb{R}^2$ to itself defined by

$$T_\theta\left(\begin{bmatrix}x\\y\end{bmatrix}\right)=\begin{bmatrix}x\cos\theta-y\sin\theta\\x\sin\theta+y\cos\theta\end{bmatrix}$$

is a linear transformation, which rotates each vector in $\mathbb{R}^2$ (regarded as an arrow emanating from the origin) about the origin and through the angle θ in a counterclockwise direction (Figure 6.1).

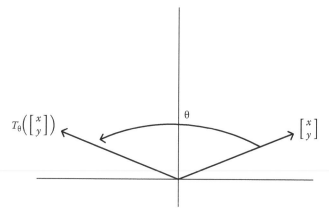

Figure 6.1 A linear transformation that rotates vectors.

Furthermore, this transformation preserves the length of any vector, $\|T_\theta(\mathbf{v})\| = \|\mathbf{v}\|$ for any $\mathbf{v} \in \mathbb{R}^2$, where $\|\cdot\|$ denotes the usual Euclidean norm (Definition 1.2.2). The mapping T is bijective. (Exercise A6 asks for the verification of this mapping's linearity and its length-preserving nature.)

(d) If $C^\infty[a, b]$ is the infinite-dimensional vector space of functions having derivatives of all orders on the closed interval $[a, b]$, then the *derivative transformation* (or *derivative operator*) $D: C^\infty[a, b] \to C^\infty[a, b]$ defined by $D(f)(x) = f'(x)$, is linear by the well-known properties of differentiation. This operator is onto, but not 1-1. [Sometimes it is not obvious that a function is the image of another function under D. Note, for example, that $F(x) = \int_0^x e^{t^2} dt \in C^\infty[0, b]$ for any $b > 0$ by the Fundamental Theorem of Calculus; therefore, $e^{x^2} \in C^\infty[0, b]$ is the image of $F(x)$ under D, that is, $F(x)$ is an *anti-derivative* of e^{x^2}.]

(e) If $A = [a_{ij}]$ is an $m \times n$ matrix, $n \geq 2$, define $T: M_{mn} \to M_{m2}$

by $T(A) = \begin{bmatrix} a_{11} & a_{1n} \\ a_{21} & a_{2n} \\ \vdots & \vdots \\ a_{m1} & a_{mn} \end{bmatrix}$, that is, T deletes all but the first

and last columns of A.

This transformation is easily shown to be linear and onto, but not 1-1 (Exercise A7).

However, not all transformations from one vector space to another are linear.

Example 6.1.3: Some Nonlinear Transformations

(a) If we define $T: \mathbb{R}^2 \to \mathbb{R}^2$ by $T\left(\begin{bmatrix} x \\ y \end{bmatrix}\right) = \begin{bmatrix} x + 1 \\ y \end{bmatrix}$, then T is not a linear transformation. For example,

$$T\left(\begin{bmatrix} x_1 \\ y_1 \end{bmatrix} + \begin{bmatrix} x_2 \\ y_2 \end{bmatrix}\right) = T\left(\begin{bmatrix} x_1 + x_2 \\ y_1 + y_2 \end{bmatrix}\right) = \begin{bmatrix} x_1 + x_2 + 1 \\ y_1 + y_2 \end{bmatrix}$$

$$\neq \begin{bmatrix} x_1 + 1 \\ y_1 \end{bmatrix} + \begin{bmatrix} x_2 + 1 \\ y_2 \end{bmatrix}$$

$$= T\left(\begin{bmatrix} x_1 \\ y_1 \end{bmatrix}\right) + T\left(\begin{bmatrix} x_2 \\ y_2 \end{bmatrix}\right).$$

(b) As another example of a nonlinear transformation, consider the mapping $\mathbf{D}$ from M_n, the vector space of all real $n \times n$ matrices ($n \geq 2$) into the vector space $\mathbb{R}$, defined by

> **D**$(A) = \det(A)$ for $A \in M_n$. In general, $\det(A+B) \neq \det(A) + \det(B)$ and $\det(kA) \neq k \det(A)$ unless $k = 0$ or 1. (See Exercises 4.1, Problems B4 and B14.)

Before we get deeper into theory and applications, let us establish some basic but important properties of linear transformations.

Theorem 6.1.1

If T is a linear transformation from a vector space V to a vector space W, then

(1) $T(\mathbf{0}_V) = \mathbf{0}_W$, where $\mathbf{0}_V$ denotes the zero vector of V and $\mathbf{0}_W$ denotes the zero vector of W.

(2) $T(\mathbf{v}_1 - \mathbf{v}_2) = T(\mathbf{v}_1) - T(\mathbf{v}_2)$ for $\mathbf{v}_1, \mathbf{v}_2 \in V$.

(3) $T(c_1\mathbf{v}_1 + c_2\mathbf{v}_2 + \cdots + c_k\mathbf{v}_k) = c_1T(\mathbf{v}_1) + c_2T(\mathbf{v}_2) + \cdots + c_kT(\mathbf{v}_k)$ for $\mathbf{v}_1, \mathbf{v}_2, \ldots, \mathbf{v}_k \in V$ and any scalars $c_1, c_2, \ldots, c_k$.

Proof (1) $T(\mathbf{0}_V) = T(\mathbf{0}_V + \mathbf{0}_V) = T(\mathbf{0}_V) + T(\mathbf{0}_V) = 2T(\mathbf{0}_V)$. Therefore, adding $-T(\mathbf{0}_V)$ to both sides of $T(\mathbf{0}_V) = 2T(\mathbf{0}_V)$, we see that $T(\mathbf{0}_V) = \mathbf{0}_W$. (Alternatively, $T(\mathbf{0}_V) = T(0 \cdot \mathbf{0}_V) = 0 \cdot T(\mathbf{0}_V) = \mathbf{0}_W$.)

(2) $T(\mathbf{v}_1 - \mathbf{v}_2) = T(\mathbf{v}_1 + (-1)\mathbf{v}_2) = T(\mathbf{v}_1) + T((-1)\mathbf{v}_2) = T(\mathbf{v}_1) + (-1)T(\mathbf{v}_2) = T(\mathbf{v}_1) - T(\mathbf{v}_2)$, where we have used property (c) of Theorem 5.1.1.

(3) This proof is inductive, just a matter of successively grouping vectors, two at a time:

$$
\begin{aligned}
T(c_1\mathbf{v}_1 + c_2\mathbf{v}_2 + \cdots + c_k\mathbf{v}_k) &= T(c_1\mathbf{v}_1 + [c_2\mathbf{v}_2 + \cdots + c_k\mathbf{v}_k]) \\
&= c_1T(\mathbf{v}_1) + T(c_2\mathbf{v}_2 + \cdots + c_k\mathbf{v}_k) \\
&= c_1T(\mathbf{v}_1) + T(c_2\mathbf{v}_2 + [c_3\mathbf{v}_3 + \cdots + c_k\mathbf{v}_k]) \\
&= c_1T(\mathbf{v}_1) + T(c_2\mathbf{v}_2) + T(c_3\mathbf{v}_3 + \cdots + c_k\mathbf{v}_k) \\
&= \ldots \\
&= c_1T(\mathbf{v}_1) + c_2T(\mathbf{v}_2) + \cdots + c_kT(\mathbf{v}_k).
\end{aligned}
$$

Corollary 6.1.1

If T is a one-to-one linear transformation from V to W, then $T(\mathbf{v}) = \mathbf{0}_W$ if and only if $\mathbf{v} = \mathbf{0}_V$.

Proof First, suppose that $T(\mathbf{v}) = \mathbf{0}_W$. Since $T(\mathbf{0}_V) = \mathbf{0}_W$ by Theorem 6.1.1 and T is 1-1, we must have $\mathbf{v} = \mathbf{0}_V$. Conversely, if $\mathbf{v} = \mathbf{0}_V$, we have $T(\mathbf{v}) = T(\mathbf{0}_V) = \mathbf{0}_W$ by Theorem 6.1.1.

Let T be a mapping from a vector space V to a vector space W. The next theorem states that **if T is linear, then T is completely determined by its effect on any basis for V**.

Theorem 6.1.2

Let $\{\mathbf{v}_1, \mathbf{v}_2, \ldots, \mathbf{v}_n\}$ be a basis for V and let $\{\mathbf{w}_1, \mathbf{w}_2, \ldots, \mathbf{w}_n\}$ be a set of any n vectors in W. There is exactly one linear transformation $T : V \to W$ for which $T(\mathbf{v}_i) = \mathbf{w}_i, i = 1, 2, \ldots, n$.

Proof   We can write any vector $\mathbf{v}$ of V uniquely in terms of its basis vectors: $\mathbf{v} = \sum_{i=1}^{n} c_i \mathbf{v}_i$. Now define $T(\mathbf{v}) = \sum_{i=1}^{n} c_i \mathbf{w}_i$, where we reuse the unique coefficients c_i from the representation of $\mathbf{v}$. We note that $T(\mathbf{v}_i) = 1 \cdot \mathbf{w}_i = \mathbf{w}_i$.

Then T is linear because for all $\mathbf{v} = \sum_{i=1}^{n} c_i \mathbf{v}_i$, $\tilde{\mathbf{v}} = \sum_{i=1}^{n} \tilde{c}_i \mathbf{v}_i \in V$ and all scalars a, b,

$$
T(a\mathbf{v} + b\tilde{\mathbf{v}}) = T\left(a \sum_{i=1}^{n} c_i \mathbf{v}_i + b \sum_{i=1}^{n} \tilde{c}_i \mathbf{v}_i \right) = T\left(\sum_{i=1}^{n} (ac_i + b\tilde{c}_i)\mathbf{v}_i \right)
$$

$$
= \sum_{i=1}^{n} (ac_i + b\tilde{c}_i)\mathbf{w}_i = a \sum_{i=1}^{n} c_i \mathbf{w}_i + b \sum_{i=1}^{n} \tilde{c}_i \mathbf{w}_i
$$

$$
= aT(\mathbf{v}) + bT(\tilde{\mathbf{v}}).
$$

Now we prove the uniqueness by showing that any two linear transformations that agree on a basis must agree everywhere. Let S and T be two linear transformations such that $S(\mathbf{v}_i) = T(\mathbf{v}_i)(= \mathbf{w}_i)$ for $i = 1, 2, \ldots, n$. Then, for any $\mathbf{v} \in V$, $\mathbf{v} = \sum_{i=1}^{n} c_i \mathbf{v}_i$, and by linearity we have $S(\mathbf{v}) = S\left(\sum_{i=1}^{n} c_i \mathbf{v}_i \right) = \sum_{i=1}^{n} c_i S(\mathbf{v}_i) = \sum_{i=1}^{n} c_i T(\mathbf{v}_i) = T\left(\sum_{i=1}^{n} c_i \mathbf{v}_i \right) = T(\mathbf{v})$. Therefore, S and T are equal on all of V.

We can express Theorem 6.1.2 in another way: **Two linear transformations $T_1, T_2 : V \to W$ are equal if and only if $T_1(\mathbf{v}_i) = T_2(\mathbf{v}_i)$ for $i = 1, 2, \ldots, n$, where $\{\mathbf{v}_1, \mathbf{v}_2, \ldots, \mathbf{v}_n\}$ is a basis for V.**

Example 6.1.4: A Linear Transformation on a Basis

Let $T : \mathbb{R}^3 \to \mathbb{R}^2$ be a linear transformation such that

$$
T\left(\begin{bmatrix} 1 \\ 0 \\ 0 \end{bmatrix} \right) = \begin{bmatrix} 1 \\ 2 \end{bmatrix}, \quad T\left(\begin{bmatrix} 0 \\ 1 \\ 0 \end{bmatrix} \right) = \begin{bmatrix} 3 \\ -2 \end{bmatrix}, \quad \text{and} \quad T\left(\begin{bmatrix} 0 \\ 0 \\ 1 \end{bmatrix} \right) = \begin{bmatrix} 1 \\ -1 \end{bmatrix}.
$$

Because we know the image of each vector in a basis for $\mathbb{R}^3$, the linear transformation is completely determined. This means we can easily calculate the image of a general vector in $\mathbb{R}^3$.

$$
\text{We have } \begin{bmatrix} x \\ y \\ z \end{bmatrix} = x \begin{bmatrix} 1 \\ 0 \\ 0 \end{bmatrix} + y \begin{bmatrix} 0 \\ 1 \\ 0 \end{bmatrix} + z \begin{bmatrix} 0 \\ 0 \\ 1 \end{bmatrix}. \text{ Therefore,}
$$

$$T\left(\begin{bmatrix} x \\ y \\ z \end{bmatrix}\right) = xT\left(\begin{bmatrix} 1 \\ 0 \\ 0 \end{bmatrix}\right) + yT\left(\begin{bmatrix} 0 \\ 1 \\ 0 \end{bmatrix}\right) + zT\left(\begin{bmatrix} 0 \\ 0 \\ 1 \end{bmatrix}\right)$$

$$= x\begin{bmatrix} 1 \\ 2 \end{bmatrix} + y\begin{bmatrix} 3 \\ -2 \end{bmatrix} + z\begin{bmatrix} 1 \\ -1 \end{bmatrix} = \begin{bmatrix} x + 3y + z \\ 2x - 2y - z \end{bmatrix}.$$

For example, $T\left(\begin{bmatrix} 1 \\ -2 \\ 3 \end{bmatrix}\right) = \begin{bmatrix} 1 + 3(-2) + 3 \\ 2(1) - 2(-2) - 3 \end{bmatrix} = \begin{bmatrix} -2 \\ 3 \end{bmatrix}.$

If we do not have enough information, it may not be possible to determine the effect of a linear transformation on all vectors of a vector space.

Example 6.1.5: A Mystery Linear Transformation

Suppose we know that $T: \mathbb{R}^3 \to \mathbb{R}^3$ is linear and is such that

$$T\left(\begin{bmatrix} 1 \\ 1 \\ 1 \end{bmatrix}\right) = \begin{bmatrix} 1 \\ 1 \\ 1 \end{bmatrix}, \ T\left(\begin{bmatrix} 2 \\ 2 \\ 3 \end{bmatrix}\right) = \begin{bmatrix} 3 \\ 3 \\ 5 \end{bmatrix}, \text{ and } T\left(\begin{bmatrix} 1 \\ 1 \\ 2 \end{bmatrix}\right) = \begin{bmatrix} 2 \\ 2 \\ 4 \end{bmatrix}.$$

Can we find $T\left(\begin{bmatrix} x \\ y \\ z \end{bmatrix}\right)$ for all $\begin{bmatrix} x \\ y \\ z \end{bmatrix} \in \mathbb{R}^3$?

The answer, unfortunately, is *no*. The reason is that the set

$\left\{\begin{bmatrix} 1 \\ 1 \\ 1 \end{bmatrix}, \begin{bmatrix} 2 \\ 2 \\ 3 \end{bmatrix}, \begin{bmatrix} 1 \\ 1 \\ 2 \end{bmatrix}\right\}$ does not span $\mathbb{R}^3$. For example, $\begin{bmatrix} 1 \\ 0 \\ 0 \end{bmatrix}$ is

not a linear combination of these vectors, so $T\left(\begin{bmatrix} 1 \\ 0 \\ 0 \end{bmatrix}\right)$ cannot

be determined from the given information.

Exercises 6.1

A.

1.   Determine which of the following mappings $T: \mathbb{R}^3 \to \mathbb{R}^3$ are linear. (For typographical convenience, vectors are written in transposed form.)

(a) $T([x \quad y \quad z]^\mathrm{T}) = [x - 1 \quad x \quad y]^\mathrm{T}$.

(b) $T([x \quad y \quad z]^\mathrm{T}) = [z \quad -y \quad x]^\mathrm{T}$.

(c) $T([x \quad y \quad z]^\mathrm{T}) = [2x \quad y - z \quad 3y]^\mathrm{T}$.

(d) $T\big([\,x\quad y\quad z\,]^{\mathrm{T}}\big) = [\,|x|\quad 0\quad -y\,]^{\mathrm{T}}.$

(e) $T\big([\,x\quad y\quad z\,]^{\mathrm{T}}\big) = [\,y\quad x-y\quad z+x\,]^{\mathrm{T}}.$

(f) $T\big([\,x\quad y\quad z\,]^{\mathrm{T}}\big) = [\,\sin x\quad y\quad z\,]^{\mathrm{T}}.$

(g) Example 6.1.1(a).

(h) Example 6.1.1(b).

2. Determine which of the following mappings $T: \mathbb{R}^3 \to \mathbb{R}^2$ or $T: \mathbb{R}^2 \to \mathbb{R}^3$ are linear. (For typographical convenience, vectors are written in transposed form.)

(a) $T\big([\,x\quad y\quad z\,]^{\mathrm{T}}\big) = [\,3x+y\quad x-2y+z\,]^{\mathrm{T}}.$

(b) $T\big([\,x\quad y\,]^{\mathrm{T}}\big) = [\,0\quad 0\quad x\,]^{\mathrm{T}}.$

(c) $T\big([\,x\quad y\,]^{\mathrm{T}}\big) = [\,x\quad -x\quad x\,]^{\mathrm{T}}.$

(d) $T\big([\,x\quad y\quad z\,]^{\mathrm{T}}\big) = [\,x\quad x+y+z+1\,]^{\mathrm{T}}.$

(e) $T\big([\,x\quad y\quad z\,]^{\mathrm{T}}\big) = [\,0\quad 0\,]^{\mathrm{T}}.$

(f) $T\big([\,x\quad y\,]^{\mathrm{T}}\big) = [\,0\quad 0\quad 1\,]^{\mathrm{T}}.$

3. (a) Show that the *identity transformation* I_V from a vector space V to itself, defined by $I_V(\mathbf{v}) = \mathbf{v}$ for all $\mathbf{v} \in V$, is linear.

(b) Show that the *zero transformation* O_{VW} from V to W, defined by $O_{VW}(\mathbf{v}) = \mathbf{0}_W$ for all $\mathbf{v} \in V$, is linear.

4. Is the transformation $T: V \to V$ defined by $T(\mathbf{v}) = a\mathbf{v} + \boldsymbol{\beta}$, where $a \in \mathbb{R}$ and $\boldsymbol{\beta}$ is a fixed vector in V, a linear transformation?

5. Show that the transformation $pr_i: \mathbb{R}^n \to \mathbb{R}$, described in Example 6.1.2(b), is linear and onto but not 1-1.

6. (a) Show that the transformation T_θ from $\mathbb{R}^2$ into itself defined by

$$T_\theta\left(\begin{bmatrix} x \\ y \end{bmatrix}\right) = \begin{bmatrix} x\cos\theta - y\sin\theta \\ x\sin\theta + y\cos\theta \end{bmatrix}$$

is linear. (See Example 6.1.2(c).)

(b) Show that $\|T(\mathbf{v})\| = \|\mathbf{v}\|$, where $\|\cdot\|$ denotes the Euclidean norm (Definition 1.2.2).

7. Show that the transformation $T: M_{mn} \to M_{m2}$, described in Example 6.1.2(e), is linear and onto but not 1-1.

8. Suppose a linear transformation $T: \mathbb{R}^3 \to \mathbb{R}^4$ satisfies

$$T\left(\begin{bmatrix} 0 \\ 0 \\ 1 \end{bmatrix}\right) = \mathbf{0}_{\mathbb{R}^4}, \quad T\left(\begin{bmatrix} 0 \\ 1 \\ 1 \end{bmatrix}\right) = \mathbf{0}_{\mathbb{R}^4}, \quad T\left(\begin{bmatrix} 1 \\ 1 \\ 1 \end{bmatrix}\right) = \mathbf{0}_{\mathbb{R}^4}.$$

Show that $T(\mathbf{v}) = \mathbf{0}_{\mathbb{R}^4}$ for all vectors $\mathbf{v} \in \mathbb{R}^3$.

9. Let B be a fixed nonzero matrix in M_n. Is the mapping $T: M_n \to M_n$ defined by $T(A) = AB - BA$ linear?

10. Consider the *transpose operator* $T: M_2 \to M_2$ defined by

$$T\left(\begin{bmatrix} a & b \\ c & d \end{bmatrix} \right) = \begin{bmatrix} a & c \\ b & d \end{bmatrix}.$$

Is T linear?

11. Show that the operator $T: M_2 \to M_2$ defined by $T(A) = \frac{A+A^\mathsf{T}}{2}$ is linear.

12. Consider the vector space $C[a,b]$ of all functions that are continuous on the closed interval $[a,b]$. Define $T: C[a,b] \to C[a,b]$ as $T(f)(x) = \int_a^x f(t)\,dt$ for all $f \in C[a,b]$ and $x \in [a,b]$. Show that T is a linear transformation.

13. Is the mapping $T: \mathbb{R} \to \mathbb{R}$ defined by $T(x) = \int_0^x \frac{e^t}{t}\,dt$ linear?

14. If $T: \mathbb{R}^2 \to \mathbb{R}^3$ is a linear transformation such that $T\left(\begin{bmatrix} -1 \\ 2 \end{bmatrix} \right) = \begin{bmatrix} 1 \\ 1 \\ -2 \end{bmatrix}$ and $T\left(\begin{bmatrix} 3 \\ -5 \end{bmatrix} \right) = \begin{bmatrix} 0 \\ 4 \\ 7 \end{bmatrix}$, what is $T\left(\begin{bmatrix} 9 \\ -17 \end{bmatrix} \right)$?

15. Is there a linear transformation $T: \mathbb{R}^3 \to \mathbb{R}^2$ such that $T\left(\begin{bmatrix} 3 \\ 1 \\ 0 \end{bmatrix} \right) = \begin{bmatrix} 1 \\ 1 \end{bmatrix}$ and $T\left(\begin{bmatrix} -6 \\ -2 \\ 0 \end{bmatrix} \right) = \begin{bmatrix} 2 \\ 1 \end{bmatrix}$? If yes, can you find a formula for $T(\mathbf{x})$, where $\mathbf{x}$ is any vector in $\mathbb{R}^3$?

16. Consider the basis $\left\{ \begin{bmatrix} 1 \\ -1 \end{bmatrix}, \begin{bmatrix} 2 \\ 0 \end{bmatrix} \right\}$ for $\mathbb{R}^2$. Find a linear transformation $T: \mathbb{R}^2 \to \mathbb{R}^4$ such that

$$T\left(\begin{bmatrix} 1 \\ -1 \end{bmatrix} \right) = \begin{bmatrix} -1 \\ 0 \\ 1 \\ 0 \end{bmatrix} \quad \text{and} \quad T\left(\begin{bmatrix} 2 \\ 0 \end{bmatrix} \right) = \begin{bmatrix} 1 \\ 1 \\ 0 \\ -1 \end{bmatrix}.$$

B.

1. Show that the mapping $T: M_2 \to M_2$ that transforms a 2×2 matrix to its reduced row echelon form is not linear.

2. Let C be a fixed invertible matrix in M_2. Define the mapping $T: M_2 \to M_2$ by $T(A) = C^{-1}AC$ for all $A \in M_2$. Is T linear?

3. Give an example of a mapping $T: \mathbb{R}^2 \to \mathbb{R}$ such that $T(c\mathbf{v}) = cT(\mathbf{v})$ for all $c \in \mathbb{R}$ and all $\mathbf{v} \in \mathbb{R}^2$, but T is not linear.

4. (a) Suppose that $\mathbf{v}_1, \mathbf{v}_2, \ldots, \mathbf{v}_k$ are vectors in a vector space V. Prove that if T is a linear transformation from V to a vector space W and $\{T(\mathbf{v}_1), T(\mathbf{v}_2), \ldots, T(\mathbf{v}_k)\}$ is a linearly independent set in W, then $\{\mathbf{v}_1, \mathbf{v}_2, \ldots, \mathbf{v}_k\}$ is a linearly independent set in V. (See Problem C2 of Exercises 3.2.)

    (b) Show by example that the converse of the result in part (a) is not true.

5. If U_1 and U_2 are *invariant subspaces* of an operator $T: V \to V$—that is, if U_1 and U_2 are subspaces of V such that $T(U_1) \subseteq U_1$ and $T(U_2) \subseteq U_2$—prove that $U_1 + U_2$ and $U_1 \cap U_2$ are also invariant subspaces of T.

6. What can be said about an operator $T: V \to V$ such that *every* subspace of V is an invariant subspace of T? (See the previous exercise for a definition.)

C.

1. Prove that if $T: \mathbb{R}^n \to \mathbb{R}^n$ is a *distance-preserving transformation*—that is, $\|T(\mathbf{v}_1 - \mathbf{v}_2)\| = \|\mathbf{v}_1 - \mathbf{v}_2\|$ for every $\mathbf{v}_1, \mathbf{v}_2 \in \mathbb{R}^n$—with the property that $T(\mathbf{0}) = \mathbf{0}$, then T is linear.

6.2 The Range and Null Space of a Linear Transformation

In this section, we focus on two subspaces that play an important role in the study of linear transformations: the range and the null space of a linear transformation. These were first defined and discussed in Example 5.2.4.

Example 6.2.1: Range and Null Space

Let T be the linear transformation from $\mathbb{R}^3$ to $\mathbb{R}^2$ defined by

$$T\left(\begin{bmatrix} x \\ y \\ z \end{bmatrix}\right) = \begin{bmatrix} x - y + 2z \\ y - x - 3z \end{bmatrix}.$$

The null space of T consists of all $\mathbf{v} = \begin{bmatrix} x \\ y \\ z \end{bmatrix} \in \mathbb{R}^3$ such that

$T(\mathbf{v}) = \begin{bmatrix} x - y + 2z \\ y - x - 3z \end{bmatrix} = \begin{bmatrix} 0 \\ 0 \end{bmatrix}$. We can solve the system

$$x - y + 2z = 0$$
$$y - x - 3z = 0$$

by finding the reduced row echelon form of the coefficient matrix:

$$\begin{bmatrix} 1 & -1 & 2 \\ -1 & 1 & -3 \end{bmatrix} \rightarrow \begin{bmatrix} 1 & -1 & 2 \\ 0 & 0 & -1 \end{bmatrix} \rightarrow \begin{bmatrix} 1 & -1 & 0 \\ 0 & 0 & 1 \end{bmatrix}.$$

Therefore $z = 0$ and $x - y = 0$, so $\mathbf{v} = \begin{bmatrix} x \\ x \\ 0 \end{bmatrix} = x \begin{bmatrix} 1 \\ 1 \\ 0 \end{bmatrix}$ and the

null space of T is the subspace $\left\{ x \begin{bmatrix} 1 \\ 1 \\ 0 \end{bmatrix} : x \in \mathbb{R} \right\}$.

Also, we see that the range of T is $T(\mathbb{R}^3) = \left\{ \begin{bmatrix} x - y + 2z \\ y - x - 3z \end{bmatrix} : \right.$

$\left. x, y, z \in \mathbb{R} \right\} = \left\{ (x - y + 2z) \begin{bmatrix} 1 \\ 0 \end{bmatrix} + (y - x - 3z) \begin{bmatrix} 0 \\ 1 \end{bmatrix} : x, y, z \in \mathbb{R} \right\} =$

$\left\{ r \begin{bmatrix} 1 \\ 0 \end{bmatrix} + s \begin{bmatrix} 0 \\ 1 \end{bmatrix} : r, s \in \mathbb{R} \right\} = \mathbb{R}^2$ since the vectors $\begin{bmatrix} 1 \\ 0 \end{bmatrix}$ and $\begin{bmatrix} 0 \\ 1 \end{bmatrix}$

constitute a basis for $\mathbb{R}^2$.

Example 6.2.2: The Null Space of a Differential Operator

The (*second-order homogeneous linear*) *differential equation*, $3y'' + 2y' + y = 0$, can be expressed in terms of an operator T acting on the vector space V of all functions having at least two derivatives on an interval $[a,b]$.

Define $T: V \rightarrow V$ by $(T(f))(x) = 3f''(x) + 2f'(x) + f(x)$. Then the null space of T is precisely the set of all solutions (the *solution space*) of the differential equation. In this case, the null space is $\left\{ e^{-x/3} \left(a \sin\left(\frac{\sqrt{2}}{3} x \right) + b \cos\left(\frac{\sqrt{2}}{3} x \right) \right) : a, b \in \mathbb{R} \right\}$. (It is easy to check that any function of the form $e^{-x/3} \left(a \sin\left(\frac{\sqrt{2}}{3} x \right) + b \cos\left(\frac{\sqrt{2}}{3} x \right) \right)$ is in the null space. A further knowledge of differential equations would tell us that *only* functions of this form are elements of the null space.)

In discussing the null space and the range, it is important to recognize that these are *subspaces*, not just subsets. As subspaces, the range and the null space of a linear transformation have bases and, therefore, *dimensions*.

Definition 6.2.1

(a) The **rank** of a linear transformation T, denoted by rank T, is the dimension of its range.

(b) The **nullity** of a linear transformation T, denoted by nullity T, is the dimension of its null space.

In Example 6.2.1, the nullity of T is 1 and the rank of T is 2. The form of the null space in Example 6.2.2 suggests that the nullity of the differential operator is 2, although this is difficult to see without knowing something about the theory of differential equations.

There are important relationships between the range and null space of a linear transformation whose domain is finite-dimensional, as the next three results show.

Lemma 6.2.1

Suppose V is a finite-dimensional vector space and $T: V \rightarrow W$ is a linear transformation. Let $\{\mathbf{v}_1, \mathbf{v}_2, \ldots, \mathbf{v}_r\}$ be a basis for the null space of T. Extend this basis to any basis $\{\mathbf{v}_1, \mathbf{v}_2, \ldots, \mathbf{v}_r, \mathbf{v}_{r+1}, \ldots, \mathbf{v}_n\}$ for V [by Corollary 5.4.1]. Then $\{T(\mathbf{v}_{r+1}), T(\mathbf{v}_{r+2}), \ldots, T(\mathbf{v}_n)\}$ is a basis for the range of T.

Proof   Let $\{\mathbf{v}_1, \mathbf{v}_2, \ldots, \mathbf{v}_n\}$ be a basis for V, chosen as in the statement of the theorem. Any vector $\mathbf{w}$ in the range of T is of the form $T(\mathbf{v})$ for some $\mathbf{v} \in V$. By letting $\mathbf{v} = \sum_{i=1}^{n} c_i \mathbf{v}_i$, we have

$$T(\mathbf{v}) = \sum_{i=1}^{n} c_i T(\mathbf{v}_i) = \sum_{i=r+1}^{n} c_i T(\mathbf{v}_i)$$

because $T(\mathbf{v}_i) = \mathbf{0}_W$ for $i = 1, 2, \ldots, r$. Hence $\{T(\mathbf{v}_{r+1}), T(\mathbf{v}_{r+2}), \ldots, T(\mathbf{v}_n)\}$ spans the range of T. Because we do not know the dimension of the range, we must also show linear independence.

Suppose that we can find scalars a_i, not all zero, such that

$$\mathbf{0}_W = \sum_{i=r+1}^{n} a_i T(\mathbf{v}_i) = T\left(\sum_{i=r+1}^{n} a_i \mathbf{v}_i \right).$$

Then $\sum_{i=r+1}^{n} a_i \mathbf{v}_i$ is a vector in the null space. But $\{\mathbf{v}_1, \mathbf{v}_2, \ldots, \mathbf{v}_r\}$ spans the null space, so, for suitable scalars b_i,

$$\sum_{i=r+1}^{n} a_i \mathbf{v}_i = \sum_{i=1}^{r} b_i \mathbf{v}_i, \quad \text{or} \quad \mathbf{0}_W = \sum_{i=1}^{r} b_i \mathbf{v}_i + \sum_{i=r+1}^{n} (-a_i) \mathbf{v}_i.$$

This contradicts the linear independence of $\{v_1, v_2, \ldots, v_n\}$, so the vectors $T(v_{r+1}), T(v_{r+2}), \ldots, T(v_n)$ are linearly independent and, therefore, form a basis for the range of T.

The next theorem expresses an important relationship between the range and the null space of a linear transformation.

Theorem 6.2.1

If T is a linear transformation from V to W, where the dimension of V is n, then

$$\boxed{\text{rank } T + \text{nullity } T = n.}$$

Proof This is an immediate consequence of the proof of Lemma 6.2.1, which shows that rank $T =$ the dimension of the range of $T = n - r = n -$ nullity T.

Corollary 6.2.1

If $T : V \to W$ is a linear transformation, where $\dim(V) = n$ and $\dim(W) = m$, then

$$\boxed{\text{rank } T \leq \min(m, n).}$$

[Min (m, n) denotes the smaller of the numbers m and n. If $m = n$, $\min(m, n) = m = n$.]

Proof The inclusion (range of T) $\subseteq W$ implies (by Theorem 5.4.4) that rank $T = \dim$(range of T) $\leq \dim(W) = m$. But we also have, by Theorem 6.2.1, that rank $T = n -$ nullity $T \leq n$.

Example 6.2.3: Rank + Nullity = n

If V denotes the vector space of real-valued functions continuous on $(-\infty, \infty)$, then $W = span\{e^x, e^{-x}, x\} \subset V$ is a three-dimensional subspace whose vectors have derivatives of all orders. Let $T : W \to W$ be the linear transformation defined as $(T(f))(x) = \frac{d^2}{dx^2} f(x) - f(x)$. If $v = c_1 e^x + c_2 e^{-x} + c_3 x$, where c_1, $c_2, c_3 \in \mathbb{R}$, then $T(v) = -c_3 x$, so the range of $T = \{rx \mid r \in \mathbb{R}\}$ and rank $T = 1$.

The fact that $T(c_1 e^x + c_2 e^{-x}) = 0$, the function identically equal to zero, tells us that $span\{e^x, e^{-x}\} \subseteq$ (null space T). Now we will show that (null space T) $\subseteq span\{e^x, e^{-x}\}$. Suppose that $v = c_1 e^x + c_2 e^{-x} + c_3 x$ is in the null space. Then $0 = T(v) = -c_3 x$ for all $x \in \mathbb{R}$ implies that $c_3 = 0$ (let $x = 1$, for example), that is, $v = c_1 e^x + c_2 e^{-x}$. Thus, (null space T) $\subseteq span\{e^x, e^{-x}\}$, and so the null space equals $span\{e^x, e^{-x}\}$. (For example, the function $-2e^x + \pi e^{-x}$ is in the null space.) We see that nullity $T = 2$ and rank $T +$ nullity $T = 1 + 2 = 3 = \dim(W)$.

Example 6.2.4: Rank $+$ Nullity $= n$

Suppose U and V are finite-dimensional spaces, $T: U \to V$ is linear, and $\dim(U) > \dim(V)$. We will use Theorem 6.2.1 to show that there exists a nonzero vector $\mathbf{u} \in U$ such that $T(\mathbf{u}) = \mathbf{0}_V$, that is, the null space of T is not equal to $\{\mathbf{0}_U\}$. To see this, we start with the fact that rank $T +$ nullity $T = \dim(U)$. Then nullity $T = \dim(U) -$ rank $T > \dim(V) -$ rank T.

Now the fact that the range of T is a subspace of V implies that rank $T \le \dim(V)$. It follows that nullity $T \ge 1$, and hence the null space contains a nonzero element $\mathbf{u}$, for which $T(\mathbf{u}) = \mathbf{0}_V$. (Recall that $\dim \{\mathbf{0}_U\} = 0$.)

Consider the matrix transformation $T_A : \mathbb{R}^n \to \mathbb{R}^m$ defined by $T_A(\mathbf{v}) = A\mathbf{v}$ for a fixed $m \times n$ matrix A and every $\mathbf{v} \in \mathbb{R}^n$. In this case, a matrix is given and leads to a linear transformation. This circumstantial evidence suggests a possible connection between linear transformations and matrices. As we will see in Section 6.4, it is possible to start with a linear transformation between finite-dimensional vector spaces and define a matrix corresponding to the transformation. To continue on the road to understanding the connection between linear transformations and matrices, we prove that the rank of a particular linear transformation equals the rank of the matrix inducing that transformation.

Theorem 6.2.2

Let A be an $m \times n$ matrix and let $T_A : \mathbb{R}^n \to \mathbb{R}^m$ be the linear transformation defined by $T_A(\mathbf{v}) = A\mathbf{v}$ for every $\mathbf{v} \in \mathbb{R}^n$. Then, rank $T_A = \text{rank}(A)$.

Proof   First, we see that the null space of T_A is the solution space S of the homogeneous $m \times n$ linear system $A\mathbf{v} = \mathbf{0}$, where $\mathbf{0} \in \mathbb{R}^m$, so the dimension of S is the nullity of T_A. But Theorem 2.7.5 gives us $\dim(S) = n - \text{rank}(A)$, and Theorem 6.2.1 tells us that the nullity of T_A equals $n -$ rank T_A. It follows that $n - \text{rank}(A) = \dim(S) =$ nullity $T_A = n -$ rank T_A, and we conclude that $\text{rank}(A) =$ rank T_A.

Example 6.2.5: Rank $T_A = \text{rank}(A)$

Suppose $A = \begin{bmatrix} 1 & 1 & -1 \\ 1 & -1 & -7 \\ 2 & 1 & -5 \end{bmatrix}$ and $T_A : \mathbb{R}^3 \to \mathbb{R}^3$ is the linear transformation defined by

$$T_A\left(\begin{bmatrix} x \\ y \\ z \end{bmatrix}\right) = \begin{bmatrix} 1 & 1 & -1 \\ 1 & -1 & -7 \\ 2 & 1 & -5 \end{bmatrix}\begin{bmatrix} x \\ y \\ z \end{bmatrix} = \begin{bmatrix} x+y-z \\ x-y-7z \\ 2x+y-5z \end{bmatrix}.$$

The reduced row echelon form of A is

$$\begin{bmatrix} 1 & 0 & -4 \\ 0 & 1 & 3 \\ 0 & 0 & 0 \end{bmatrix},$$

so rank$(A) = 2$ and Corollary 6.2.2 tells us that T_A also has rank 2. We can determine the rank of T_A directly by examining its range. We see that $T_A\left(\begin{bmatrix} x \\ y \\ z \end{bmatrix}\right) = \begin{bmatrix} 1 & 1 & -1 \\ 1 & -1 & -7 \\ 2 & 1 & -5 \end{bmatrix}\begin{bmatrix} x \\ y \\ z \end{bmatrix} =$

$\begin{bmatrix} x+y-z \\ x-y-7z \\ 2x+y-5z \end{bmatrix} = (x-4z)\begin{bmatrix} 1 \\ 1 \\ 2 \end{bmatrix} + (y+3z)\begin{bmatrix} 1 \\ -1 \\ 1 \end{bmatrix}$, where we

have used the pivot columns of A as basis vectors for the range (recall Theorem 2.7.3). Thus rank $T_A = 2$.

Exercises 6.2

A.

1. If $pr_i : \mathbb{R}^n \to \mathbb{R}$ is the transformation described in Example 6.1.2(b), find the range of T and the null space of T.

2. For each of the following linear transformations T, find the null space of T and nullity T, and describe the null space geometrically.

    (a) $T\left(\begin{bmatrix} x \\ y \end{bmatrix}\right) = x - 2y$ (b) $T\left(\begin{bmatrix} x \\ y \\ z \end{bmatrix}\right) = \begin{bmatrix} y-x \\ 2x+y+z \end{bmatrix}$

    (c) $T\left(\begin{bmatrix} x \\ y \end{bmatrix}\right) = \begin{bmatrix} x-y \\ 2y \end{bmatrix}$ (d) $T(x) = \begin{bmatrix} 2x \\ 0 \\ -x \end{bmatrix}$

3. For each linear transformation in the preceding exercise, find the range of T, rank T, and describe the range geometrically.

4. Define $T: \mathbb{R}^3 \to \mathbb{R}^3$ by $T\left(\begin{bmatrix} x \\ y \\ z \end{bmatrix}\right) = \begin{bmatrix} x+y \\ x+z \\ y+z \end{bmatrix}$. Find the range of T and the null space of T.

5. Let V be the vector space of all functions from $\mathbb{R}$ to itself and consider the subspace of V that is defined as $W = span\{\sin x, \cos x\}$. Define the linear transformation $T: W \to \mathbb{R}$ by $T(f) = \int_0^\pi f(x)dx$. Determine the range of T and the null space of T.

6. For each of the following linear transformations T, find the range, null space, rank of T, and the nullity of T.

(a) $T: \mathbb{R}^3 \to \mathbb{R}^2$, $T\left(\begin{bmatrix} x \\ y \\ z \end{bmatrix}\right) = \begin{bmatrix} x - 2y \\ 2x + z \end{bmatrix}$.

(b) $T: \mathbb{R}^3 \to \mathbb{R}^2$, $T\left(\begin{bmatrix} x \\ y \\ z \end{bmatrix}\right) = \begin{bmatrix} x - y + 2z \\ y - x - 3z \end{bmatrix}$.

(c) $T: \mathbb{R}^3 \to \mathbb{R}^3$, $T\left(\begin{bmatrix} x \\ y \\ z \end{bmatrix}\right) = \begin{bmatrix} x \\ x + y + z \\ y + z \end{bmatrix}$.

(d) Let V be the space of all polynomials with real coefficients. Let T be the linear transformation from V to V defined by $T(p) = p''$ (the second derivative of p) for $p \in V$.

(e) Let V be the vector space of all real functions continuous on $(-\infty, \infty)$, and let W be the space of all real-valued functions. Define $T: V \to W$ by $(Tf)(x) = \int_a^x f(t)dt$, where a is a fixed real number.

7. Find bases for the null space and the range of the linear transformation $T: P_2[x] \to P_2[x]$ defined by

$$T(ax^2 + bx + c) = (4a - b + 3c)x^2 + (a - 2c)x + (11a - 2b).$$

8. Find a basis for the null space of the linear transformation $T: \mathbb{R}^3 \to \mathbb{R}^3$ defined by $T\left(\begin{bmatrix} x \\ y \\ z \end{bmatrix}\right) = \begin{bmatrix} 2x - y + z \\ x - 4z \end{bmatrix}$. Interpret the null space geometrically.

In the exercises that follow, assume the following information:

*Given vector spaces U, V, and W, and linear transformations T_1 from U to V and T_2 from V to W, we define the transformation T_2T_1 as follows: $(T_2T_1)(\mathbf{u}) = T_2(T_1(\mathbf{u}))$ for $\mathbf{u} \in U$. The transformation T_2T_1 is linear. The **zero transformation** O_{VW} from V to W is defined by $O_{VW}(\mathbf{v}) = \mathbf{0}_W$ for all $\mathbf{v} \in V$. If T is a transformation from V to itself, we define **powers of** T inductively: $T^2(\mathbf{x}) = T(T(\mathbf{x})), T^3(\mathbf{x}) = T(T^2(\mathbf{x})), \ldots, T^k(\mathbf{x}) = T(T^{k-1}(\mathbf{x}))$ for all $\mathbf{x} \in V$. We define $T^0 = I_V$, the **identity transformation** on V, given by $I_V(\mathbf{v}) = \mathbf{v}$ for all $\mathbf{v} \in V$. The usual laws of exponents are valid for operators: $T^m T^n = T^{m+n}$, $(T^m)^n = T^{mn}$ for nonnegative integers m, n.*

B.

1. If $T: V \to V$ is a linear transformation of rank 1, prove that $T^2 = kT$ for some scalar k.

2. Consider the subspace V of $\mathbb{R}^4$ given by

$$V = \left\{ \begin{bmatrix} x \\ 0 \\ z \\ 0 \end{bmatrix} : x, z \in \mathbb{R} \right\}.$$

   Determine linear transformations $T, S: \mathbb{R}^4 \to \mathbb{R}^4$ such that
   (a) the range of T is V and (b) the null space of S is V.

3. Let T be a linear transformation from V to W.
   (a) Show that if two distinct elements of V have the same image, then their difference is in the null space of T.
   (b) Suppose $T(\mathbf{v}_1) = T(\mathbf{v}_2) = T(\mathbf{v}_3) = \mathbf{w}$. How would you choose c_1, c_2, and c_3 in order that $T(c_1\mathbf{v}_1 + c_2\mathbf{v}_2 + c_3\mathbf{v}_3) = \mathbf{w}$?

4. If U and V are finite-dimensional vector spaces, and $S: U \to V$, $T: V \to W$ are linear transformations, prove that nullity $TS \leq$ nullity $S +$ nullity T.

5. If S is a linear transformation from V to W and if T is a linear transformation from W to Y, prove that
   (a) (range TS) $\subseteq$ (range of T) and rank $TS \leq$ rank T.
   (b) (null space of TS) $\supseteq$ (null space of S) and nullity $TS \geq$ nullity S.

6. If T is a linear transformation from V to itself, prove that
   (a) $V \supseteq$ range of $T \supseteq$ range of $T^2 \supseteq \cdots \supseteq$ range of $T^k \supseteq \cdots$.
   (b) $\{\mathbf{0}_V\} \subseteq$ null space of $T \subseteq$ null space of $T^2 \subseteq \cdots \subseteq$ null space of $T^k \subseteq \cdots$.

7. If $T, S: V \to V$ are linear transformations, show that $TS = O_V$ if and only if (range of S) $\subset$ (null space of T).

8. If T is an operator on a finite-dimensional space V, show that rank $T =$ rank T^2 if and only if (null space of T) $\cap$ (range of T) $= \{\mathbf{0}_V\}$.

9. Suppose T is an operator on $\mathbb{R}^3$ such that $T \neq O_{\mathbb{R}^3}$ but $T^2 = O_{\mathbb{R}^3}$. Show that rank $T = 1$.

10. If $T: \mathbb{R}^n \to \mathbb{R}^n$ is a linear transformation such that $T^{n-1} \neq O_{\mathbb{R}^3}$ but $T^n = O_{\mathbb{R}^3}$, show that rank $T = n - 1$. Give an example of such a T.

11. If $T: V \to V$ is linear, where $\dim(V) = n$, and $T^2 = O_V$, show that rank $T \leq \frac{1}{2}n$.

12. Recall from Section 6.1 that a mapping f from U to V is called *onto* (or *surjective*) if for every element $v \in V$ there is an element $u \in U$ such that $f(u) = v$. Prove that if $T: \mathbb{R}^4 \to \mathbb{R}^2$ is a linear transformation such that the null space of T is the

$$\text{subspace } \left\{ \begin{bmatrix} x_1 \\ x_2 \\ x_3 \\ x_4 \end{bmatrix} : x_1 = 3x_2 \text{ and } x_3 = 5x_4 \right\}, \text{ then } T \text{ is surjec-}$$

tive. (See Appendix A if necessary.)

13. Prove that there does not exist a linear transformation from $\mathbb{R}^5$ to $\mathbb{R}^2$ whose null space equals

$$\left\{ \begin{bmatrix} x_1 \\ x_2 \\ x_3 \\ x_4 \\ x_5 \end{bmatrix} : x_1 = 4x_2 \text{ and } x_3 = x_4 = x_5 \right\}.$$

14. If T is an operator on an n-dimensional space V and T is idempotent ($T^2 = T$), show that
    (a) $(I_V - T)$ is idempotent,
    (b) The null space of $T = \{ \mathbf{v} - T(\mathbf{v}) | \mathbf{v} \in V \} = $ the range of $I_V - T$,
    (c) $V = $ (the range of T) $\oplus$ (the null space of T), where $\oplus$ denotes the direct sum.

15. Let $T: V \to W$ be a linear transformation and suppose $\mathbf{y}$ is in the range of T. Then there is a vector $\mathbf{x}^* \in V$ such that $T(\mathbf{x}^*) = \mathbf{y}$. Show that *every* vector that T maps into $\mathbf{y}$ can be written as a sum $\mathbf{x}^* + \mathbf{v}$, where $\mathbf{v}$ is in the null space of T. Conversely, show that every such sum $\mathbf{x}^* + \mathbf{v}$ maps into $\mathbf{y}$. (Compare this result to Theorem 2.8.5.)

C.

1. Suppose the vector space V is either $\mathbb{R}^2$ or $\mathbb{R}^3$ and that $T: V \to V$ is a linear transformation such that $T \neq O_V, T \neq I_V$, and (range of T) $\cap$ (null space of T) $\neq \{\mathbf{0}_V\}$.
    (a) Is it possible to have the range of T equal to the null space of T?
    (b) What about (range of T) $\subset$ (null space of T)?
    (c) What about (null space of T) $\subset$ (range of T)?

2. Suppose that V is a vector space of dimension n. If $T: V \to V$ is linear, prove that the following statements are equivalent.
    (a) (range of T) $=$ (null space of T).
    (b) $T^2 = O_V, T \neq O_V, n$ is even, and rank $T = \frac{1}{2}n$. (See Exercise B11.)

3. Prove that if there exists a linear transformation on V whose null space and range are both finite-dimensional, then V is finite-dimensional.

4. A diagram of finite-dimensional vector spaces and linear transformations of the form

$$V_1 \xrightarrow{T_1} V_2 \xrightarrow{T_2} V_3 \xrightarrow{T_3} \cdots \xrightarrow{T_{n-2}} V_{n-1} \xrightarrow{T_{n-1}} V_n \xrightarrow{T_n} V_{n+1}$$

is called an *exact sequence* if (a) T_1 is one-to-one (i.e., $T_1(\mathbf{x}_1) = T_1(\mathbf{x}_2)$ implies $\mathbf{x}_1 = \mathbf{x}_2$), (b) T_n is onto (i.e., the range of $T_n = V_{n+1}$), and (c) the range of $T_i =$ the null space of T_{i+1} for $i = 1, 2, \ldots, n-1$. Prove that, for an exact sequence, $\sum_{i=1}^{n+1} (-1)^i \dim(V_i) = 0$.

6.3 The Algebra of Linear Transformations

Given linear transformations from a vector space V to a vector space W, where V and W have the same set of scalars, we can combine them algebraically to obtain new transformations. This follows a pattern that is similar to the study of functions.

Definition 6.3.1

For two linear transformations T and S from V to W, the **sum** $T + S$, **difference** $T - S$, and **scalar multiple** cT are the transformations from V to W defined as follows:

1. $(T + S)(\mathbf{v}) = T(\mathbf{v}) + S(\mathbf{v})$ for every $\mathbf{v} \in V$.
2. $(T - S)(\mathbf{v}) = T(\mathbf{v}) - S(\mathbf{v})$ for every $\mathbf{v} \in V$.
3. $(cT)(\mathbf{v}) = c(T(\mathbf{v}))$ for every $\mathbf{v} \in V$ and every scalar c.

It is easy to show that $T + S, T - S$, and cT are linear transformations (Exercise A1). Parts 1 and 2 of Definition 6.3.1 can be extended to any finite sum or difference of linear transformations in an obvious way.

Given two vector spaces V and W, the **zero transformation** O_{VW} is defined as the mapping from V to W that sends every vector $\mathbf{v}$ in V to the zero vector of $W: O_{VW}(\mathbf{v}) = \mathbf{0}_W$ for every $\mathbf{v} \in V$. (Instead of O_{VV}, we will write only O_V.) Both the zero transformation and the identity transformation are linear (Exercises 6.1, A3). Clearly, $T + O_{VW} = T = O_{VW} + T$ for every linear transformation T from a vector space V to a vector space W.

With respect to the operations of addition and scalar multiplication just introduced, **the set $L(V,W)$ of all linear transformations from a vector space V to a vector space W is itself a vector space** (Exercise B11). We assume that vector spaces V, W, and $L(V, W)$ use the same scalars. If $V = W$, we write $L(V)$ instead of $L(V, V)$ and we usually refer to the vectors in $L(V)$ as **operators** on V.

Example 6.3.1: Algebraic Combinations of Linear Transformations

Consider $S: \mathbb{R}^3 \to \mathbb{R}^3$ and $T: \mathbb{R}^3 \to \mathbb{R}^3$ defined by

$$T\left(\begin{bmatrix} x \\ y \\ z \end{bmatrix}\right) = \begin{bmatrix} z \\ y \\ x \end{bmatrix} \quad \text{and} \quad S\left(\begin{bmatrix} x \\ y \\ z \end{bmatrix}\right) = \begin{bmatrix} x - y \\ 0 \\ y - z \end{bmatrix}.$$

Letting $\mathbf{v} = \begin{bmatrix} x \\ y \\ z \end{bmatrix}$, suppose that we want to calculate $-2T$,
$3S + T$, and $S - 2T$.

First, $(-2T)(\mathbf{v}) = -2(T(\mathbf{v})) = -2 \begin{bmatrix} z \\ y \\ x \end{bmatrix} = \begin{bmatrix} -2z \\ -2y \\ -2x \end{bmatrix}$. Then,

$$(3S + T)(\mathbf{v}) = (3S)(\mathbf{v}) + T(\mathbf{v}) = 3(S(\mathbf{v})) + T(\mathbf{v}) = 3\begin{bmatrix} x - y \\ 0 \\ y - z \end{bmatrix} + \begin{bmatrix} z \\ y \\ x \end{bmatrix} =$$

$$\begin{bmatrix} 3x - 3y \\ 0 \\ 3y - 3z \end{bmatrix} + \begin{bmatrix} z \\ y \\ x \end{bmatrix} = \begin{bmatrix} 3x - 3y + z \\ y \\ x + 3y - 3z \end{bmatrix}. \quad \text{Finally,} \quad (S - 2T)(\mathbf{v}) =$$

$$S(\mathbf{v}) - (2T)(\mathbf{v}) = S(\mathbf{v}) - 2(T(\mathbf{v})) = \begin{bmatrix} x - y \\ 0 \\ y - z \end{bmatrix} - 2\begin{bmatrix} z \\ y \\ x \end{bmatrix} = \begin{bmatrix} x - y - 2z \\ -2y \\ -2x + y - z \end{bmatrix}.$$

We can also define a *composition* of linear transformations under proper circumstances. To emphasize the algebraic connections between linear transformations and matrices, this is often described as a *product* of linear transformations.

Definition 6.3.2

Given vector spaces U, V, W and linear transformations T_1 from U to V and T_2 from V to W, we can define the **composite transformation** $T_2 \circ T_1$ as follows: $(T_2 \circ T_1)(\mathbf{u}) = T_2(T_1(\mathbf{u}))$ for $\mathbf{u} \in U$.

In the language of functions, we can say that the range of T_1 becomes the domain of T_2 and the output of T_1 becomes the input of T_2. We will usually suppress the composition symbol $\circ$ and write $T_2 \circ T_1$ as $T_2 T_1$. Note that $T_2 T_1$ is a mapping from U to W and that $T_1 T_2$ is not necessarily defined.

It is easy to show that the composition of two linear transformations is again a linear transformation (Exercise B3). Similarly, a composition $T = T_k T_{k-1} \cdots T_2 T_1$ of finitely many linear transformations between appropriate vector spaces $V_1, V_2, \ldots, V_{k+1}$ is defined by the rule

$$T(\mathbf{x}) = T_k(T_{k-1}(\cdots(T_2(T_1(\mathbf{x})))\cdots)) \quad \text{for all } \mathbf{x} \in V_1 \quad \text{(Figure 6.2)}.$$

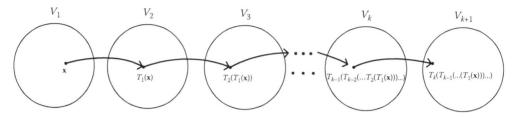

Figure 6.2 A composition of finitely many linear transformations.

If $T \in L(V)$, then we can define **powers** of the transformation T:

$$T^k(\mathbf{x}) = T_k(T_{k-1}(\cdots(T_2(T_1(\mathbf{x})))\cdots)) \quad \text{for all } \mathbf{x} \in V,$$

where $T = T_1 = T_2 = \cdots = T_k$, or, as we defined powers of a matrix inductively in Section 3.3,

$$T^2(\mathbf{x}) = T(T(\mathbf{x})), T^3(\mathbf{x}) = T(T^2(\mathbf{x})), \ldots, T^k(\mathbf{x})$$
$$= T(T^{k-1}(\mathbf{x})) \quad \text{for all } \mathbf{x} \in V.$$

We define $T^0 = I_V$. The usual laws of exponents are valid for operators: $T^m T^n = T^{m+n}$, $(T^m)^n = T^{mn}$ for nonnegative integers m, n.

Example 6.3.2: Compositions of Linear Transformations

(a) Suppose we take $T: M_2 \to \mathbb{R}^2$ defined by $T\left(\begin{bmatrix} a & b \\ c & d \end{bmatrix}\right) = \begin{bmatrix} a \\ d \end{bmatrix}$, $S: \mathbb{R}^2 \to \mathbb{R}^3$ given by $S\left(\begin{bmatrix} x \\ y \end{bmatrix}\right) = \begin{bmatrix} x \\ 0 \\ y \end{bmatrix}$, and $U: \mathbb{R}^3 \to \mathbb{R}$ defined by $U\left(\begin{bmatrix} x \\ y \\ z \end{bmatrix}\right) = x + z$.

Then the linear transformation UST is from M_2 to $\mathbb{R}$ and is defined by $(UST)\left(\begin{bmatrix} a & b \\ c & d \end{bmatrix}\right) = U\left(S\left(T\left(\begin{bmatrix} a & b \\ c & d \end{bmatrix}\right)\right)\right) = US\left(\begin{bmatrix} a \\ d \end{bmatrix}\right) = U\left(\begin{bmatrix} a \\ 0 \\ d \end{bmatrix}\right) = a + d$. This says that UST is a

linear functional on M_2, a linear transformation from a vector space to its set of scalars. [Note that $(UST)\left(\begin{bmatrix} a & b \\ c & d \end{bmatrix}\right) = \text{trace}\left(\begin{bmatrix} a & b \\ c & d \end{bmatrix}\right)$.]

(b) Define $T: \mathbb{R}^4 \to \mathbb{R}^4$ by $T\left(\begin{bmatrix} x \\ y \\ z \\ w \end{bmatrix}\right) = \begin{bmatrix} 0 \\ y \\ 2z \\ 3w \end{bmatrix}$. Then, for

example, $T^2 + 6T$ is the operator defined by

$$(T^2 + 6T)\left(\begin{bmatrix} x \\ y \\ z \\ w \end{bmatrix}\right) = T\left(T\left(\begin{bmatrix} x \\ y \\ z \\ w \end{bmatrix}\right)\right) + 6T\left(\begin{bmatrix} x \\ y \\ z \\ w \end{bmatrix}\right)$$

$$= T\left(\begin{bmatrix} 0 \\ y \\ 2z \\ 3w \end{bmatrix}\right) + 6\begin{bmatrix} 0 \\ y \\ 2z \\ 3w \end{bmatrix}$$

$$= \begin{bmatrix} 0 \\ y \\ 4z \\ 9w \end{bmatrix} + \begin{bmatrix} 0 \\ 6y \\ 12z \\ 18w \end{bmatrix} = \begin{bmatrix} 0 \\ 7y \\ 16z \\ 27w \end{bmatrix}.$$

(c) Define $S, T: C^\infty(-\infty, \infty) \to C^\infty(-\infty, \infty)$ by $(T(f))(x) = f'(x)$ and $(S(f))(x) = \int_0^x f(t)dt$. Then, by the Fundamental Theorem of Calculus, $((TS)(f))(x) = T(S(f)(x)) = T(\int_0^x f(t)dt) = f(x)$ and $(ST)(f(x)) = S(T(f)(x)) = S(f'(x)) = f(x) - f(0)$. Thus $TS \neq ST$ in general, even when both products are defined. (We note that $TS = ST$ if $f(0) = 0$.)

The observation made at the end of the last example deserves emphasis: Even when both compositions TS and ST exist, $TS \neq ST$ in general. Similarly, $ST = SU$ with $S \neq O$ does not imply that $T = U$. (See Example 3.3.8 for the matrix versions of this kind of algebraic (mis)behavior.)

Recalling that $L(V,W)$ denotes the vector space of all linear mappings from a vector space V to a vector space W, we can state some useful algebraic properties of linear transformations.

Theorem 6.3.1

Suppose that U, V, W, and X are vector spaces. Then

(1) $RI_V = I_W R = R$ for $R \in L(V, W)$.
(2) $RO_V = O_W R = O_W$ for $R \in L(V, W)$.
(3) $(RS)T = R(ST)$ for $R \in L(W, X)$, $S \in L(V, W)$, and $T \in L(U, V)$.
(4) $(R + S)T = RT + ST$ for $R, S \in L(V, W)$ and $T \in L(U, V)$.
(5) $T(R + S) = TR + TS$ for $T \in L(V, W)$ and $R, S \in L(U, V)$.
(6) $c(RS) = (cR)S = R(cS)$ for $R \in L(W, X)$ and $S \in L(V, W)$.

Proof We prove property (4) and leave the rest as exercises (see B12 through B15).

Let **u** be any vector in U. Then $T(\mathbf{u}) = \mathbf{v} \in V$ and $[(R+S)T]](\mathbf{u}) = (R+S)(T(\mathbf{u})) = (R+S)(\mathbf{v}) = R(\mathbf{v}) + S(\mathbf{v}) = R(T(\mathbf{u})) + S(T(\mathbf{u})) = (RT)(\mathbf{u}) + (ST)(\mathbf{u}) = (RT + ST)(\mathbf{u})$.   Since $(R+S)T$ and $RT + ST$ agree on every vector **u** in U, they must be equal transformations.

The vector space $L(V)$, with the additional operation of the composition of transformations, forms an algebraic structure called a **linear algebra with identity.***

Exercises 6.3

A.

1. Prove that transformations $T + S$, $T - S$, and cT described in Definition 6.3.1 are linear transformations.

2. If T is the operator defined in Example 6.3.2(b), what is $T^4 - 6T^3 + 11T^2 - 6T$?

3. Let S and T be linear operators on $\mathbb{R}^2$ defined by
    $$S\left(\begin{bmatrix} x \\ y \end{bmatrix}\right) = \begin{bmatrix} y \\ x \end{bmatrix} \text{ and } T\left(\begin{bmatrix} x \\ y \end{bmatrix}\right) = \begin{bmatrix} x \\ 0 \end{bmatrix}.$$
    (a) Describe S and T geometrically.
    (b) Find $S + T, TS, ST, S^2$, and T^2.

4. If $T_\theta\left(\begin{bmatrix} x \\ y \end{bmatrix}\right) = \begin{bmatrix} (\cos\theta)x - (\sin\theta)y \\ (\sin\theta)x + (\cos\theta)y \end{bmatrix}$, describe the composition $T_{-\theta}T_\theta$. Explain your result geometrically.

5. Let S, T, and U be operators on $\mathbb{R}^2$ defined by
    $$S\left(\begin{bmatrix} x \\ y \end{bmatrix}\right) = \begin{bmatrix} x+y \\ y \end{bmatrix}, T\left(\begin{bmatrix} x \\ y \end{bmatrix}\right) = \begin{bmatrix} x \\ -y \end{bmatrix}, U\left(\begin{bmatrix} x \\ y \end{bmatrix}\right) = \begin{bmatrix} y \\ x \end{bmatrix}.$$
    Calculate each of the following compositions.
    (a) ST (b) TS (c) S^2 (d) S^3 (e) $(ST)U$ (f) $S(TU)$
    (g) $ST - TS$ (h) $TU - UT$ (i) $U^2 - I_{\mathbb{R}^2}$ (j) $S^2 - 2S + I_{\mathbb{R}^2}$
    (k) $SU - US$

6. If S and T are the operators defined in the previous problem,
    (a) what is S^n for a positive integer n?
    (b) what is T^n for any positive integer n?

---

* See, for example, Section 4.1 of K. Hoffman, *Linear Algebra*, 2nd edn. (Englewood Cliffs, NJ: Prentice-Hall, 1971).

7. Let S, T, and U be operators on $P_3[x]$ defined by $S(f) = f + f'$, $T(f) = f + xf' + x^2f''$, and $U(f) = f''$, where the primes denote derivatives.

Calculate each of the following operators.

(a) SU (b) US (c) UT (d) TU (e) $TU - UT$ (f) U^2
(g) S^2 (h) S^3 (i) TS (j) ST

8. If S is the operator defined in the previous problem, what is S^n if n is any positive integer?

9. If $A \in M_n$, define $T_1(A) = \frac{1}{2}(A + A^T)$ and $T_2(A) = \frac{1}{2}(A - A^T)$.

(a) Show that $T_1^2 = T_1$ (i.e., T_1 is *idempotent*).
(b) Show that $T_2^2 = T_2$ (i.e., T_2 is *idempotent*).
(c) Show that $T_1 T_2 = T_2 T_1 = O_{M_n}$.

10. For any $f \in P_n[x]$, define $(T(f))(x) = f(-x)$. Show that $T^2 = I_{P_n[x]}$.

11. If D is the differential operator on $P_n[x]$, $D(f) = f'$ for $f \in P_n[x]$, what is D^{n+1}?

B.

1. Define a transformation $S: \mathbb{R}^n \to \mathbb{R}^n$ by

$$ S\left(\begin{bmatrix} x_1 \\ x_2 \\ \vdots \\ x_n \end{bmatrix}\right) = \begin{bmatrix} 0 \\ x_1 \\ x_2 \\ \vdots \\ x_{n-1} \end{bmatrix}. $$

(a) Show that S is a linear transformation. (S is called the *right shift operator*.)
(b) Show that $S^n = O_{\mathbb{R}^n}$.

2. Let B be a fixed matrix in M_n. Show that the mapping $T: M_n \to M_n$ defined by $T(A) = (A + B)^2 - (A + 2B)(A - 3B)$ for all $A \in M_n$ is linear if and only if $B^2 = O_n$.

3. Given vector spaces U, V, W and linear transformations T from U to V and S from V to W, show that the composite transformation ST is a linear transformation from U to W.

4. Suppose $S, T \in L(V)$ and $TS = O_V$. Does it follow that $ST = O_V$? Explain.

5. Give an example of two operators $S \neq O_V$ and $T \neq O_V$ on a vector space V such that $ST = TS = O_V$.

6. If $T, S \in L(V)$ and $(T+S)^2 = T^2 + 2ST + S^2$, show that S and T commute (i.e., $ST = TS$ on V.)

7. Prove that if p and q are polynomials and S and T are commuting operators on V, then $p(S)$ and $q(T)$ commute.

8. Let $U = span\{t, t^2, \ldots, t^n\} \subset P_n[t]$ and let $W = P_{n-1}[t]$. Consider the derivative operator $D : U \to W$ and consider $T(p)(t) = \int_0^t p(x)dx$, a mapping from W into U. Show that $TD = I_U$ and $DT = I_W$.

9. Let T be an operator on a finite-dimensional vector space V. Show that T commutes with all other operators on V if and only if $T = \alpha I_V$ for some scalar α.

10. If $\{\mathbf{v}_1, \mathbf{v}_2, \ldots, \mathbf{v}_n\}$ is a basis for the vector space V, show that the operator defined by $T(\mathbf{v}_1) = \mathbf{v}_2, T(\mathbf{v}_2) = \mathbf{v}_3, \ldots, T(\mathbf{v}_{n-1}) = \mathbf{v}_n, T(\mathbf{v}_n) = \alpha\mathbf{v}_1$, where α is a scalar, satisfies the relation $T^n = \alpha I_V$.

11. Prove that the set $L(V, W)$ of all linear transformations from a vector space V to a vector space W is itself a vector space.

12. Prove properties (1) and (2) of Theorem 6.3.1.

13. Prove property (3) of Theorem 6.3.1.

14. Prove property (5) of Theorem 6.3.1.

15. Prove property (6) of Theorem 6.3.1.

C.

1. Suppose that V is a finite-dimensional vector space. Prove that any linear transformation on a subspace of V can be extended to a linear transformation on V.

    In other words, show that if U is a subspace of V and $S \in L(U, W)$, then there exists $T \in L(V, W)$ such that $T(\mathbf{u}) = S(\mathbf{u})$ for all $\mathbf{u} \in U$.

6.4 Matrix Representation of a Linear Transformation

If V and W are finite-dimensional vector spaces with dimensions n and m, respectively, then each linear transformation T from V to W can be represented by an $m \times n$ matrix. This is the development promised at the beginning of this chapter and it is one of the high points in any linear algebra course. Before we justify this claim, let us consider a concrete example.

Example 6.4.1: Matrix Representation of a Linear Transformation

Let us consider the particular case of Example 6.1.2(c) and let $T_{\pi/2}$ be the linear transformation from $\mathbb{R}^2$ to itself that rotates each vector through an angle of 90° ($\pi/2$ rad) in a counter-clockwise direction. If we choose the standard basis for $\mathbb{R}^2$ (for both the domain and the codomain of $T_{\pi/2}$), we get

$$T_{\pi/2}(\mathbf{e}_1) = T_{\pi/2}\left(\begin{bmatrix} 1 \\ 0 \end{bmatrix}\right) = \begin{bmatrix} 0 \\ 1 \end{bmatrix} = 0\mathbf{e}_1 + 1\mathbf{e}_2,$$

$$T_{\pi/2}(\mathbf{e}_2) = T_{\pi/2}\left(\begin{bmatrix} 0 \\ 1 \end{bmatrix}\right) = \begin{bmatrix} -1 \\ 0 \end{bmatrix} = -1\mathbf{e}_1 + 0\mathbf{e}_2.$$

Now we can represent $T_{\pi/2}$ by the matrix $A_{T_{\pi/2}} = \begin{bmatrix} 0 & -1 \\ 1 & 0 \end{bmatrix}$, where we have used the coordinate vectors of $T_{\pi/2}(\mathbf{e}_1)$ and $T_{\pi/2}(\mathbf{e}_2)$ (see Sections 1.5 and 4.5) as columns:

$$A_{T_{\pi/2}} = [\mathbf{a}_1 \quad \mathbf{a}_2] = \left[[T_{\pi/2}(\mathbf{e}_1)]_{\mathcal{E}_2} \ \ [T_{\pi/2}(\mathbf{e}_2)]_{\mathcal{E}_2} \right] = \begin{bmatrix} 0 & -1 \\ 1 & 0 \end{bmatrix}.$$

Saying that this matrix represents the transformation means that the effect of the transformation on a vector can be obtained by multiplying the vector by the matrix:

$$A_{T_{\pi/2}} \begin{bmatrix} x \\ y \end{bmatrix} = \begin{bmatrix} 0 & -1 \\ 1 & 0 \end{bmatrix} \begin{bmatrix} x \\ y \end{bmatrix} = \begin{bmatrix} -y \\ x \end{bmatrix} = T_{\pi/2}\left(\begin{bmatrix} x \\ y \end{bmatrix}\right).$$ To see
that this outcome makes sense, let $\theta = 90°$ or $\pi/2$ radians in the formula provided in Example 6.1.2(c): $\begin{bmatrix} x\cos\frac{\pi}{2} - y\sin\frac{\pi}{2} \\ x\sin\frac{\pi}{2} + y\cos\frac{\pi}{2} \end{bmatrix} = \begin{bmatrix} -y \\ x \end{bmatrix}$ (Figure 6.3).

If we choose another basis for $\mathbb{R}^2$, say $B = \{\mathbf{v}_1, \mathbf{v}_2\}$, where $\mathbf{v}_1 = \begin{bmatrix} 1 \\ 1 \end{bmatrix}$ and $\mathbf{v}_2 = \begin{bmatrix} -1 \\ 0 \end{bmatrix}$, then

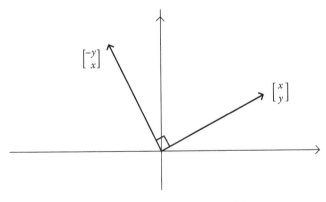

Figure 6.3 Rotation of a vector through an angle of 90°.

$$T_{\pi/2}(\mathbf{v}_1) = T_{\pi/2}\left(\begin{bmatrix} 1 \\ 1 \end{bmatrix}\right) = \begin{bmatrix} -1 \\ 1 \end{bmatrix} = 1\mathbf{v}_1 + 2\mathbf{v}_2, \text{ so } [T_{\pi/2}(\mathbf{v}_1)]_B =$$

$$\begin{bmatrix} 1 \\ 2 \end{bmatrix} T_{\pi/2}(\mathbf{v}_2) = T_{\pi/2}\left(\begin{bmatrix} -1 \\ 0 \end{bmatrix}\right) = \begin{bmatrix} 0 \\ -1 \end{bmatrix} = -1\mathbf{v}_1 + (-1)\mathbf{v}_2,$$

so $[T_{\pi/2}(\mathbf{v}_2)]_B = \begin{bmatrix} -1 \\ -1 \end{bmatrix}$, and the matrix representing the trans-

formation is $A_{T_{\pi/2}} = \left[[T_{\pi/2}(\mathbf{v}_1)]_B \quad [T_{\pi/2}(\mathbf{v}_2)]_B \right] = \begin{bmatrix} 1 & -1 \\ 2 & -1 \end{bmatrix}$.

For example, if we choose $\mathbf{v} = \begin{bmatrix} 0 \\ 1 \end{bmatrix} = \mathbf{v}_1 + \mathbf{v}_2$, then (using

the change-of-basis notation from Sections 1.5 and 4.5)

$[\mathbf{v}]_B = \begin{bmatrix} 1 \\ 1 \end{bmatrix}$. Also, $T_{\pi/2}(\mathbf{v}) = \begin{bmatrix} -1 \\ 0 \end{bmatrix} = 0\mathbf{v}_1 + 1\mathbf{v}_2$, so $[T_{\pi/2}(\mathbf{v})]_B =$

$\begin{bmatrix} 0 \\ 1 \end{bmatrix}$ and $A_{T_{\pi/2}}[\mathbf{v}]_B = \begin{bmatrix} 1 & -1 \\ 2 & -1 \end{bmatrix}\begin{bmatrix} 1 \\ 1 \end{bmatrix} = \begin{bmatrix} 0 \\ 1 \end{bmatrix} = [T_{\pi/2}(\mathbf{v})]_B$. Geometrically, this looks like Figure 6.4.

Physically (geometrically), the vector $\mathbf{v}$ has been rotated 90°
and its coordinates in the new axis system induced by B have
changed accordingly.

Generalizing the procedure we followed in Example 6.4.1, let us
consider a linear transformation T from $\mathbb{R}^n$ to $\mathbb{R}^m$ and assume that the
standard bases $\mathcal{E}_n$ and $\mathcal{E}_m$ are used for $\mathbb{R}^n$ and $\mathbb{R}^m$, respectively. We show
that there is an $m \times n$ matrix $[T]_{\mathcal{E}_n}^{\mathcal{E}_m}$ that represents the transformation T in
the sense that $[T]_{\mathcal{E}_n}^{\mathcal{E}_m}[\mathbf{v}]_{\mathcal{E}_n} = [T(\mathbf{v})]_{\mathcal{E}_m}$ for $\mathbf{v} \in \mathbb{R}^n$. The notation may seem
awkward, but *it emphasizes the dependence of the matrix not only on the
transformation but also on the bases of the domain and the codomain* (see
Figure 6.5).

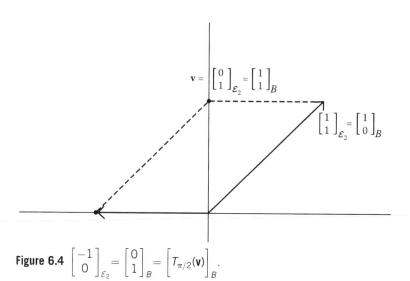

Figure 6.4 $\begin{bmatrix} -1 \\ 0 \end{bmatrix}_{\mathcal{E}_2} = \begin{bmatrix} 0 \\ 1 \end{bmatrix}_B = [T_{\pi/2}(\mathbf{v})]_B$.

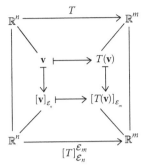

Figure 6.5 $[T]_{\mathcal{E}_n}^{\mathcal{E}_m}[\mathbf{v}]_{\mathcal{E}_n} = [T(\mathbf{v})]_{\mathcal{E}_m}$.

We form this matrix by taking column j of $[T]_{\mathcal{E}_n}^{\mathcal{E}_m}$ to be $[T(\mathbf{e}_j)]_{\mathcal{E}_m}$, $j = 1, 2, \ldots, n$, where $\mathcal{E}_n = \{\mathbf{e}_1, \mathbf{e}_2, \ldots, \mathbf{e}_n\}$ is the standard basis for $\mathbb{R}^n$:

$$[T]_{\mathcal{E}_n}^{\mathcal{E}_m} = [\mathbf{t}_1 \quad \mathbf{t}_2 \quad \cdots \quad \mathbf{t}_n] = [[T(\mathbf{e}_1)]_{\mathcal{E}_m} \quad [T(\mathbf{e}_2)]_{\mathcal{E}_m} \quad \cdots \quad [T(\mathbf{e}_n)]_{\mathcal{E}_m}].$$

To see the effect of this matrix, first express an arbitrary vector $\mathbf{v}$ in $\mathbb{R}^n$ as a unique linear combination of its basis vectors: $\mathbf{v} = v_1\mathbf{e}_1 + v_2\mathbf{e}_2 + \cdots + v_n\mathbf{e}_n$. Then $T(\mathbf{v}) = v_1 T(\mathbf{e}_1) + v_2 T(\mathbf{e}_2) + \cdots + v_n T(\mathbf{e}_n)$

and

$$[T(\mathbf{v})]_{\mathcal{E}_m} = v_1[T(\mathbf{e}_1)]_{\mathcal{E}_m} + v_2[T(\mathbf{e}_2)]_{\mathcal{E}_m} + \cdots + v_n[T(\mathbf{e}_n)]_{\mathcal{E}_m}$$

$$= v_1\mathbf{t}_1 + v_2\mathbf{t}_2 + \cdots + v_n\mathbf{t}_n = [\mathbf{t}_1 \quad \mathbf{t}_2 \quad \cdots \quad \mathbf{t}_n]\begin{bmatrix} v_1 \\ v_2 \\ \vdots \\ v_n \end{bmatrix} = [T]_{\mathcal{E}_n}^{\mathcal{E}_m}[\mathbf{v}]_{\mathcal{E}_n}.$$

The matrix $[T]_{\mathcal{E}_n}^{\mathcal{E}_m}$ we have constructed is called the **standard matrix representation of T**. Theorem 6.1.2 guarantees that this matrix is unique.

Example 6.4.2: Standard Matrix Representation of a Linear Transformation

Let us revisit the linear transformation $T: \mathbb{R}^3 \to \mathbb{R}^2$ in Example 6.1.2: $T\left(\begin{bmatrix} x \\ y \\ z \end{bmatrix}\right) = \begin{bmatrix} x \\ y \end{bmatrix}$. Using the standard bases for $\mathbb{R}^3$ and $\mathbb{R}^2$, we construct the matrix $[T]_{\mathcal{E}_3}^{\mathcal{E}_2}$ by taking column j of $[T]_{\mathcal{E}_3}^{\mathcal{E}_2}$ to be $[T(\mathbf{e}_j)]_{\mathcal{E}_2}$, $j = 1, 2, 3$. Then, $[T]_{\mathcal{E}_3}^{\mathcal{E}_2} = [\mathbf{t}_1 \quad \mathbf{t}_2 \quad \mathbf{t}_3]$, where

$$\mathbf{t}_1 = [T(\mathbf{e}_1)]_{\mathcal{E}_2} = \left[T\left([1 \quad 0 \quad 0]^T\right)\right]_{\mathcal{E}_2} = \begin{bmatrix} 1 \\ 0 \end{bmatrix},$$

$$\mathbf{t}_2 = [T(\mathbf{e}_2)]_{\mathcal{E}_2} = \left[T\left([0 \quad 1 \quad 0]^T\right)\right]_{\mathcal{E}_2} = \begin{bmatrix} 0 \\ 1 \end{bmatrix},$$

$$\mathbf{t}_3 = [T(\mathbf{e}_3)]_{\mathcal{E}_2} = \left[T\left([0 \quad 0 \quad 1]^T\right)\right]_{\mathcal{E}_2} = \begin{bmatrix} 0 \\ 0 \end{bmatrix}.$$

Thus $[T]_{\mathcal{E}_3}^{\mathcal{E}_2} = \begin{bmatrix} 1 & 0 & 0 \\ 0 & 1 & 0 \end{bmatrix}$ is the standard matrix representation

of T. We see that $[T]_{\mathcal{E}_3}^{\mathcal{E}_2}\begin{bmatrix} x \\ y \\ z \end{bmatrix}_{\mathcal{E}_3} = \begin{bmatrix} 1 & 0 & 0 \\ 0 & 1 & 0 \end{bmatrix}\begin{bmatrix} x \\ y \\ z \end{bmatrix}_{\mathcal{E}_3} = \begin{bmatrix} x \\ y \end{bmatrix}_{\mathcal{E}_2}$,

as expected.

The second part of Example 6.4.1 illustrates how we should proceed if we do not use the standard bases in finding a linear transformation from $\mathbb{R}^n$ to $\mathbb{R}^m$. If we consider $T: \mathbb{R}^n \to \mathbb{R}^m$, with bases $B_n = \{\mathbf{v}_1, \mathbf{v}_2, \ldots, \mathbf{v}_n\}$ for $\mathbb{R}^n$ and B_m for $\mathbb{R}^m$, we can modify our algorithm easily. We construct $[T]_{B_n}^{B_m}$ by letting column j of the matrix equal $\left[T(\mathbf{v}_j)\right]_{B_m}$, $j = 1, 2, \ldots, n$. Then it can be verified that $[T]_{B_n}^{B_m}[\mathbf{v}]_{B_n} = [T(\mathbf{v})]_{B_m}$ for all $\mathbf{v} \in \mathbb{R}^n$.

If we think about the process of finding a matrix representation of a linear transformation, we realize that there is nothing special about using Euclidean spaces and standard bases. Our previous analysis allows us to state the following important result.

Theorem 6.4.1

Suppose B_V and B_W are the ordered bases for the finite-dimensional vector spaces V and W, respectively. Then corresponding to each linear transformation $T: V \to W$, there is a unique $m \times n$ matrix $[T]_{B_V}^{B_W}$ such that $[T]_{B_V}^{B_W}[\mathbf{v}]_{B_V} = [T(\mathbf{v})]_{B_W}$ for every $\mathbf{v} \in V$.

We say that $[T]_{B_V}^{B_W}$ is the **matrix representing T relative to the ordered bases B_V and B_W**.

To prove this theorem, we set column j of $[T]_{B_V}^{B_W}$ equal to $[T(\mathbf{v}_j)]_{B_W}$ for $j = 1, 2, \ldots, n$, and then verify that $[T]_{B_V}^{B_W}[\mathbf{v}]_{B_V} = [T(\mathbf{v})]_{B_W}$. Uniqueness of the matrix follows from Theorem 6.1.2.

The relationship in Theorem 6.4.1 deserves to be highlighted:

$$\boxed{[T]_{B_V}^{B_W}[\mathbf{v}]_{B_V} = [T(\mathbf{v})]_{B_W}.} \tag{6.4.1}$$

Assuming that V and W are finite-dimensional with bases B_V and B_W, respectively, and using Equation 6.4.1, we can see that $[O_{VW}(\mathbf{v})]_{B_W} = O_{mn}[\mathbf{v}]_{B_V} = \mathbf{0}_W$, or $[O_{VW}]_{B_V}^{B_W} = O_{mn}$, where O_{VW} is a transformation and O_{mn} is a matrix. Similarly, if I_V is the identity transformation on V, then $[I_V(\mathbf{v})]_{B_V} = I_n[\mathbf{v}]_{B_V} = [\mathbf{v}]_{B_V}$, so $[I_V]_{B_V} = I_n$, the $n \times n$ identity matrix. We can see these relationships in Figure 6.6.

Corollary 6.4.1

If $[T]_{B_V}^{B_W}$ is the matrix representation of the linear transformation $T: V \to W$ with respect to the bases B_V and B_W, where $\dim(V) = n$ and $\dim(W) = m$, then the rank of $[T]_{B_V}^{B_W}$ equals the rank of T.

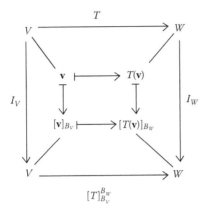

Figure 6.6 $[T]_{B_V}^{B_W}[\mathbf{v}]_{B_V} = [T(\mathbf{v})]_{B_W}$.

Proof We know that $[T(\mathbf{v})]_{B_W} = [T]_{B_V}^{B_W}[\mathbf{v}]_{B_V}$ for $\mathbf{v} \in V$ and we can apply Theorem 6.2.2 to conclude that $\text{rank}([T]_{B_V}^{B_W}) = \text{rank } T$.

Example 6.4.3: Matrix Representation of a Linear Transformation

Suppose that we have the vector spaces M_2 and $P_2[x]$, with the ordered bases $M = \{E_1, E_2, E_3, E_4\} = \left\{ \begin{bmatrix} 1 & 1 \\ 0 & 0 \end{bmatrix}, \begin{bmatrix} 0 & 0 \\ 1 & 1 \end{bmatrix}, \right.$

$\left. \begin{bmatrix} 1 & 0 \\ 0 & 1 \end{bmatrix}, \begin{bmatrix} 0 & 1 \\ 1 & 1 \end{bmatrix} \right\}$ and $F = \{\mathbf{f}_1, \mathbf{f}_2, \mathbf{f}_3\} = \{1, -x, 1+x+x^2\}$, respectively. Now consider the linear transformation $T : M_2 \to P_2[x]$ defined by $T\left(\begin{bmatrix} a & b \\ c & d \end{bmatrix} \right) = a + bx + cx^2$. To find $[T]_M^F$, the matrix of T relative to the two bases, we must calculate $[T(E_j)]_F$ for $j = 1, 2, 3, 4$. We have

$$T(E_1) = T\left(\begin{bmatrix} 1 & 1 \\ 0 & 0 \end{bmatrix} \right) = 1 + x = 1[1] + (-1)[-x] + 0[1+x+x^2],$$

so $[T(E_1)]_F = \begin{bmatrix} 1 \\ -1 \\ 0 \end{bmatrix}$.

$$T(E_2) = T\left(\begin{bmatrix} 0 & 0 \\ 1 & 1 \end{bmatrix} \right) = x^2 = (-1)[1] + 1[-x] + 1[1+x+x^2],$$

so $[T(E_2)]_F = \begin{bmatrix} -1 \\ 1 \\ 1 \end{bmatrix}$.

$$T(E_3) = T\left(\begin{bmatrix} 1 & 0 \\ 0 & 1 \end{bmatrix}\right) = 1 = 1[1] + 0[-x] + 0[1 + x + x^2],$$

so $[T(E_3)]_F = \begin{bmatrix} 1 \\ 0 \\ 0 \end{bmatrix}$.

$$T(E_4) = T\left(\begin{bmatrix} 0 & 1 \\ 1 & 1 \end{bmatrix}\right) = x + x^2 = (-1)[1] + 0[-x] + 1[1 + x + x^2],$$

so $[T(E_4)]_F = \begin{bmatrix} -1 \\ 0 \\ 1 \end{bmatrix}$.

Thus $[T]_M^F = \begin{bmatrix} 1 & -1 & 1 & -1 \\ -1 & 1 & 0 & 0 \\ 0 & 1 & 0 & 1 \end{bmatrix}$.

To see how this matrix represents T, let us examine what happens to the vector $\mathbf{v} = \begin{bmatrix} 1 & 2 \\ 3 & 4 \end{bmatrix} = 0 \cdot E_1 + 1 \cdot E_2 +$

$1 \cdot E_3 + 2 \cdot E_4$ under T. We have $[\mathbf{v}]_M = \begin{bmatrix} 0 \\ 1 \\ 1 \\ 2 \end{bmatrix}$ and

$$T(\mathbf{v}) = 1 + 2x + 3x^2 = -2 \cdot [1] + 1 \cdot [-x] + 3 \cdot [1 + x + x^2],$$

so $[T(\mathbf{v})]_F = \begin{bmatrix} -2 \\ 1 \\ 3 \end{bmatrix}$. Finally, we see that $[T]_M^F[\mathbf{v}]_M =$

$$\begin{bmatrix} 1 & -1 & 1 & -1 \\ -1 & 1 & 0 & 0 \\ 0 & 1 & 0 & 1 \end{bmatrix} \begin{bmatrix} 0 \\ 1 \\ 1 \\ 2 \end{bmatrix} = \begin{bmatrix} -2 \\ 1 \\ 3 \end{bmatrix} = [T(\mathbf{v})]_F,$$ as guaranteed by

Theorem 6.4.1.

It is important to realize that the formula $[T]_{B_V}^{B_W}[\mathbf{v}]_{B_V} = [T(\mathbf{v})]_{B_W}$ enables us to treat any linear transformation $T : V \to W$ between finite-dimensional vector spaces as if it were the matrix transformation $T_A : \mathbb{R}^n \to \mathbb{R}^m$ defined by $T_A(\mathbf{v}) = A\mathbf{v}$ for a suitable $m \times n$ matrix A, as in the proof of Corollary 6.4.1. Thus, in many cases, questions about a linear transformation T can be answered by analyzing an appropriate $m \times n$ matrix. The next theorem establishes a basic connection between the algebra of linear transformations and the algebra of matrices.

Theorem 6.4.2

Suppose T_1 and T_2 are two linear transformations from V to W, with the ordered bases B_V and B_W, respectively. Then,

1. $[T_1 + T_2]_{B_V}^{B_W} = [T_1]_{B_V}^{B_W} + [T_2]_{B_V}^{B_W}$.
2. $[T_1 - T_2]_{B_V}^{B_W} = [T_1]_{B_V}^{B_W} - [T_2]_{B_V}^{B_W}$.
3. $[cT_i]_{B_V}^{B_W} = c[T_i]_{B_V}^{B_W}$, $i = 1, 2$.

Proof Exercise B2.

In words: If we add (or subtract) linear transformations T_1 and T_2, the resulting linear mapping is represented by a matrix that is the sum (or difference) of the matrices that represent T_1 and T_2. The scalar multiple cT_i is represented by a matrix that is just c times the matrix representing T_i. Using Equation 6.4.1, we can write

$$\left[(T_1 + T_2)(\mathbf{v})\right]_{B_W} = [T_1 + T_2]_{B_V}^{B_W}[\mathbf{v}]_{B_V} = \left([T_1]_{B_V}^{B_W} + [T_2]_{B_V}^{B_W}\right)[\mathbf{v}]_{B_V}$$
$$= [T_1]_{B_V}^{B_W}[\mathbf{v}]_{B_V} + [T_2]_{B_V}^{B_W}[\mathbf{v}]_{B_V},$$

$$\left[(T_1 - T_2)(\mathbf{v})\right]_{B_W} = [T_1 - T_2]_{B_V}^{B_W}[\mathbf{v}]_{B_V} = \left([T_1]_{B_V}^{B_W} - [T_2]_{B_V}^{B_W}\right)[\mathbf{v}]_{B_V}$$
$$= [T_1]_{B_V}^{B_W}[\mathbf{v}]_{B_V} - [T_2]_{B_V}^{B_W}[\mathbf{v}]_{B_V},$$

and

$$\left[(cT_i)(\mathbf{v})\right]_{B_W} = \left(c[T_i]_{B_V}^{B_W}\right)[\mathbf{v}]_{B_V} = c\left([T_i]_{B_V}^{B_W}[\mathbf{v}]_{B_V}\right).$$

Example 6.4.4: The Algebra of Linear Transformations/Matrices

Let us consider the vector spaces M_2 and $\mathbb{R}^2$, with standard bases $\mathcal{E} = \left\{ \begin{bmatrix} 1 & 0 \\ 0 & 0 \end{bmatrix}, \begin{bmatrix} 0 & 1 \\ 0 & 0 \end{bmatrix}, \begin{bmatrix} 0 & 0 \\ 1 & 0 \end{bmatrix}, \begin{bmatrix} 0 & 0 \\ 0 & 1 \end{bmatrix} \right\}$ and $\mathcal{E}_2 = \left\{ \begin{bmatrix} 1 \\ 0 \end{bmatrix}, \begin{bmatrix} 0 \\ 1 \end{bmatrix} \right\}$, respectively.

Now define two linear transformations from M_2 to $\mathbb{R}^2$:

$$T_1\left(\begin{bmatrix} a & b \\ c & d \end{bmatrix}\right) = \begin{bmatrix} a \\ d \end{bmatrix} \quad \text{and} \quad T_2\left(\begin{bmatrix} a & b \\ c & d \end{bmatrix}\right) = \begin{bmatrix} a + b \\ c + d \end{bmatrix}.$$

It is easy to calculate

$$[T_1]_{\mathcal{E}}^{\mathcal{E}_2} = \begin{bmatrix} 1 & 0 & 0 & 0 \\ 0 & 0 & 0 & 1 \end{bmatrix} \quad \text{and} \quad [T_2]_{\mathcal{E}}^{\mathcal{E}_2} = \begin{bmatrix} 1 & 1 & 0 & 0 \\ 0 & 0 & 1 & 1 \end{bmatrix}.$$

Furthermore, $(T + S)\left(\begin{bmatrix} a & b \\ c & d \end{bmatrix}\right) = \begin{bmatrix} 2a + b \\ c + 2d \end{bmatrix}$, so that

$$[T_1 + T_2]_{\mathcal{E}}^{\mathcal{E}_2} = \begin{bmatrix} 2 & 1 & 0 & 0 \\ 0 & 0 & 1 & 2 \end{bmatrix} = \begin{bmatrix} 1 & 0 & 0 & 0 \\ 0 & 0 & 0 & 1 \end{bmatrix}$$
$$+ \begin{bmatrix} 1 & 1 & 0 & 0 \\ 0 & 0 & 1 & 1 \end{bmatrix} = [T_1]_{\mathcal{E}}^{\mathcal{E}_2} + [T_2]_{\mathcal{E}}^{\mathcal{E}_2}.$$

Similarly, we can verify that $[T_1 - T_2]_{\mathcal{E}}^{\mathcal{E}_2} = [T_1]_{\mathcal{E}}^{\mathcal{E}_2} - [T_2]_{\mathcal{E}}^{\mathcal{E}_2}$ and

$$[cT_i]_{\mathcal{E}}^{\mathcal{E}_2} = c[T_i]_{\mathcal{E}}^{\mathcal{E}_2} \quad \text{for } i = 1, 2.$$

It should not be surprising that we can establish a connection between the composition of linear transformations and the product of matrices, just as Theorem 6.4.2 linked the addition and scalar multiplication of linear transformations to the addition and scalar multiplication of matrices. In other words, the next theorem states that *the matrix representation of a composition of two linear transformations is the product of the matrix representations of the individual transformations.*

Theorem 6.4.3

Suppose U, V, and W are finite-dimensional vector spaces, with the ordered bases $B_U = \{\mathbf{u}_i\}_{i=1}^n$, B_V, and B_W, respectively, and we have linear transformations $T_1 : U \to V$ and $T_2 : V \to W$. Then $[T_2 T_1]_{B_U}^{B_W} = [T_2]_{B_V}^{B_W} \cdot [T_1]_{B_U}^{B_V}$.

Proof We will demonstrate that $[T_2 T_1]_{B_U}^{B_W} = [T_2]_{B_V}^{B_W} \cdot [T_1]_{B_U}^{B_V}$ by showing that column j of $[T_2 T_1]_{B_U}^{B_W}$ equals column j of the product $[T_2]_{B_V}^{B_W} \cdot [T_1]_{B_U}^{B_V}$ for $j = 1, 2, \ldots, n$.

From the proof of Theorem 6.4.1, we see that for $j = 1, 2, \ldots, n$, column j of $[T_2 T_1]_{B_U}^{B_W} = \left[(T_2 T_1)(\mathbf{u}_j) \right]_{B_W} = \left[T_2 (T_1(\mathbf{u}_j)) \right]_{B_W}$. This last expression, in turn, equals $[T_2]_{B_V}^{B_W} \left[T_1(\mathbf{u}_j) \right]_{B_V}$ by Equation 6.4.1.

Now by Theorem 3.3.1 (the "column view" of matrix–matrix multiplication), column j of $[T_2]_{B_V}^{B_W} \cdot [T_1]_{B_U}^{B_V} = [T_2]_{B_V}^{B_W} \cdot$ column j of $[T_1]_{B_U}^{B_V} = [T_2]_{B_V}^{B_W} \cdot \left[T_1(\mathbf{u}_j) \right]_{B_V}$.

We have just shown that column j of $[T_2 T_1]_{B_U}^{B_W}$ equals column j of $[T_2]_{B_V}^{B_W} \cdot [T_1]_{B_U}^{B_V}$ for $j = 1, 2, \ldots, n$, that is, $[T_2 T_1]_{B_U}^{B_W} = [T_2]_{B_V}^{B_W} \cdot [T_1]_{B_U}^{B_V}$ (Figure 6.7).

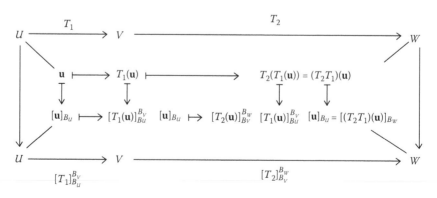

Figure 6.7 $[(T_2 T_1)(\mathbf{u})]_{B_W} = [T_2(\mathbf{u})]_{B_V}^{B_W} [T_1(\mathbf{u})]_{B_U}^{B_V} [\mathbf{u}]_{B_U}$.

Example 6.4.5: The Algebra of Linear Transformations/Matrices

Let us define linear transformations $T: \mathbb{R}^2 \to \mathbb{R}^2$ and $S: \mathbb{R}^2 \to \mathbb{R}^3$ as follows:

$$T\left(\begin{bmatrix} x \\ y \end{bmatrix}\right) = \begin{bmatrix} 2x - 3y \\ x + 5y \end{bmatrix} \quad \text{and} \quad S\left(\begin{bmatrix} x \\ y \end{bmatrix}\right) = \begin{bmatrix} x - 2y \\ x + 3y \\ 7x - 5y \end{bmatrix}.$$

Then, with respect to the standard bases for $\mathbb{R}^2$ and $\mathbb{R}^3$, it is easy to calculate $[T]_{\mathcal{E}_2} = \begin{bmatrix} 2 & -3 \\ 1 & 5 \end{bmatrix}$ and $[S]_{\mathcal{E}_2}^{\mathcal{E}_3} = \begin{bmatrix} 1 & -2 \\ 1 & 3 \\ 7 & -5 \end{bmatrix}$. Furthermore,

$$(ST)\left(\begin{bmatrix} x \\ y \end{bmatrix}\right) = \begin{bmatrix} (2x - 3y) - 2(x + 5y) \\ (2x - 3y) + 3(x + 5y) \\ 7(2x - 3y) - 5(x + 5y) \end{bmatrix} = \begin{bmatrix} -13y \\ 5x + 12y \\ 9x - 46y \end{bmatrix}$$

$$= \begin{bmatrix} 0 & -13 \\ 5 & 12 \\ 9 & -46 \end{bmatrix} \begin{bmatrix} x \\ y \end{bmatrix} \quad \text{and}$$

$$[S]_{\mathcal{E}_2}^{\mathcal{E}_3} \cdot [T]_{\mathcal{E}_2} = \begin{bmatrix} 1 & -2 \\ 1 & 3 \\ 7 & -5 \end{bmatrix} \begin{bmatrix} 2 & -3 \\ 1 & 5 \end{bmatrix} = \begin{bmatrix} 0 & -13 \\ 5 & 12 \\ 9 & -46 \end{bmatrix} = [ST]_{\mathcal{E}_2}^{\mathcal{E}_3}.$$

Exercises 6.4

A.

1. Let $T: \mathbb{R}^2 \to \mathbb{R}^2$ be a linear transformation defined by $T(\mathbf{v}) = \begin{bmatrix} 2 & 4 \\ 2 & 6 \end{bmatrix} \mathbf{v}$ for every $\mathbf{v} \in \mathbb{R}^2$. Find the matrix of T with respect to the following (ordered) bases for $\mathbb{R}^2$.

    (a) $\begin{bmatrix} 1 \\ -1 \end{bmatrix}, \begin{bmatrix} 1 \\ 1 \end{bmatrix}$ (b) $\begin{bmatrix} 1 \\ 1 \end{bmatrix}, \begin{bmatrix} 1 \\ -1 \end{bmatrix}$ (c) $\begin{bmatrix} 1 \\ 0 \end{bmatrix}, \begin{bmatrix} 3 \\ 1 \end{bmatrix}$ (d) $\begin{bmatrix} 3 \\ 1 \end{bmatrix}, \begin{bmatrix} 1 \\ 0 \end{bmatrix}$

2. Suppose $T: \mathbb{R}^3 \to \mathbb{R}^3$ is linear such that $T\left(\begin{bmatrix} 1 \\ 0 \\ 0 \end{bmatrix}\right) = \begin{bmatrix} 2 \\ 3 \\ -2 \end{bmatrix}$,

    $T\left(\begin{bmatrix} 1 \\ 1 \\ 0 \end{bmatrix}\right) = \begin{bmatrix} 4 \\ 1 \\ 4 \end{bmatrix}$, and $T\left(\begin{bmatrix} 1 \\ 1 \\ 1 \end{bmatrix}\right) = \begin{bmatrix} 5 \\ -1 \\ 7 \end{bmatrix}$. Find the matrix of T with respect to the standard basis of $\mathbb{R}^3$.

3. Find the matrix that represents the linear transformation $T: \mathbb{R}^3 \rightarrow \mathbb{R}^3$ given by $T\left(\begin{bmatrix} x \\ y \\ z \end{bmatrix}\right) = \begin{bmatrix} 2x + z \\ y - x + z \\ 3z \end{bmatrix}$ with respect to the basis $\left\{ \begin{bmatrix} 1 \\ -1 \\ 0 \end{bmatrix}, \begin{bmatrix} 1 \\ 0 \\ -1 \end{bmatrix}, \begin{bmatrix} 1 \\ 0 \\ 0 \end{bmatrix} \right\}$.

4. For $n \geq 1$, let $P_n[x]$ be the vector space of the polynomials of a degree less than or equal to n with real coefficients. Determine the matrix of the differentiation mapping $D: P_n[x] \rightarrow P_{n-1}[x]$ $(D(p) = p'$ for $p \in P_n[x])$ with respect to each of the following ordered bases, using the same basis for the domain and the codomain.
   (a) $\{1, x, x^2, \ldots, x^n\}$ (b) $\{x^n, x^{n-1}, \ldots, x, 1\}$ (c) $\{1, 1+x, 1+x^2, \ldots, 1+x^n\}$.

   If $n = 4$, use the matrix found in part (a) and the polynomial $2 - 3x + 4x^2 + x^3 - 5x^4$ to illustrate the relation in Equation 6.4.1.

5. Let $T: \mathbb{R}^2 \rightarrow \mathbb{R}^2$ be the linear transformation defined by $T(\mathbf{v}) = \mathbf{v} - \frac{1}{2}(\mathbf{v} \bullet \mathbf{x})\mathbf{x}$, where $\mathbf{x} = \begin{bmatrix} 1 \\ 1 \end{bmatrix}$. Find the matrix which represents T relative to the standard basis for $\mathbb{R}^2$.

6. Define the linear transformation $T: M_2 \rightarrow M_2$ by $T(A) = \frac{A + A^T}{2}$. Find the matrix which represents T relative to the ordered basis $\left\{ \begin{bmatrix} 1 & 1 \\ 0 & 0 \end{bmatrix}, \begin{bmatrix} 0 & 0 \\ 1 & 1 \end{bmatrix}, \begin{bmatrix} 1 & 0 \\ 0 & 1 \end{bmatrix}, \begin{bmatrix} 0 & 1 \\ 1 & 1 \end{bmatrix} \right\}$ for M_2.

B.

1. Let $T: \mathbb{R}^3 \rightarrow \mathbb{R}^3$ be a linear transformation such that $T\left(\begin{bmatrix} 5 \\ 8 \\ 8 \end{bmatrix}\right) = \begin{bmatrix} 1 \\ 5 \\ 3 \end{bmatrix}, T\left(\begin{bmatrix} 3 \\ 3 \\ 5 \end{bmatrix}\right) = \begin{bmatrix} 1 \\ 2 \\ 2 \end{bmatrix}$, and $T\left(\begin{bmatrix} a \\ b \\ c \end{bmatrix}\right) = \begin{bmatrix} 9 \\ 9 \\ 9 \end{bmatrix}$.
   (a) Find $T\left(\begin{bmatrix} 1 \\ 2 \\ 3 \end{bmatrix}\right)$, given that $a = b = c = 9$.
   (b) In part (a), what values must $a, b,$ and c have in order that $T\left(\begin{bmatrix} 1 \\ 2 \\ 3 \end{bmatrix}\right) = \begin{bmatrix} -1 \\ -2 \\ -3 \end{bmatrix}$?
   (c) Write down the matrices corresponding to T in parts (a) and (b).

2. Prove Theorem 6.4.2.

C.

1. Let T be an operator on an n-dimensional space V. If rank $(T) = 1$, show that the matrix representation of T relative to any basis of V is of the form

$$\begin{bmatrix} a_1b_1 & a_2b_1 & \dots & a_nb_1 \\ a_1b_2 & a_2b_2 & \dots & a_nb_2 \\ \vdots & \vdots & \dots & \vdots \\ a_1b_n & a_2b_n & \dots & a_nb_n \end{bmatrix}$$

where $a_i, b_i \in \mathbb{R}$, $1 \le i \le n$.

6.5 Invertible Linear Transformations

In both the theory and applications of mathematics, it is frequently necessary to try to reverse a transformation, so that we go from output back to input. We want to undo the effect of a transformation and get back to where we started. Algebraically, we try to solve a matrix–vector equation $A\mathbf{x} = \mathbf{b}$ by finding the inverse of the matrix A. If we can find such an inverse, then $\mathbf{x} = A^{-1}\mathbf{b}$. Similarly, there are equations $T(\mathbf{v}) = \mathbf{w}$ that can be solved, provided that we can somehow reverse the effect of the transformation T. Problem A4 of Exercises 6.3 asks the reader to consider the transformation from $\mathbb{R}^2$ to itself defined by $T_\theta\left(\begin{bmatrix} x \\ y \end{bmatrix}\right) = \begin{bmatrix} (\cos \theta)x - (\sin \theta)y \\ (\sin \theta)x + (\cos \theta)y \end{bmatrix}$ and describe the composite mapping $T_{-\theta}T_\theta$. It turns out that $T_{-\theta}$ reverses the effect of T_θ, so that $T_{-\theta}T_\theta = I_{\mathbb{R}^2}$, the identity transformation on $\mathbb{R}^2$.

To understand this situation, we need some careful preliminary definitions.

Definition 6.5.1

A linear transformation T from V to W is said to be **invertible** if there exists a mapping $\hat{T}$ from W to V such that $\hat{T}T = I_V$ and $T\hat{T} = I_W$, where I_V is the identity transformation on V and I_W is the identity transformation on W. Otherwise, the linear transformation is called **noninvertible**. Such a mapping $\hat{T}$ is called an **inverse** of T.

Even though we define "an" inverse, it turns out that if T is invertible, then T has a *unique* inverse, which we denote by T^{-1} (Figure 6.8). This means we can speak of "the" inverse of T.

To verify this uniqueness, suppose that $T: V \rightarrow W$ is a linear transformation and $\tilde{T}$ and $\hat{T}$ are inverses of T. Then, for any $\mathbf{w} \in W$, $\tilde{T}(\mathbf{w}) = \tilde{T}I_W(\mathbf{w}) = \tilde{T}(T\hat{T}(\mathbf{w})) = (\tilde{T}T)\hat{T}(\mathbf{w}) = I_V\hat{T}(\mathbf{w}) = \hat{T}(\mathbf{w})$.

Because $\tilde{T}$ and $\hat{T}$ agree on every vector in W, $\tilde{T} = \hat{T}$.

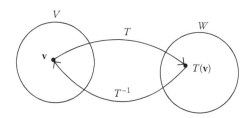

Figure 6.8 An inverse transformation.

Definition 6.5.1 does not say anything about the linearity or nonlinearity of T^{-1} if such a mapping exists. As we will see in Theorem 6.5.2, **if it exists, T^{-1} must be linear.**

Example 6.5.1: Some Invertible Linear Transformations

(a) The linear transformation $T: M_2 \rightarrow P_3[x]$ defined by $T\left(\begin{bmatrix} a & b \\ c & d \end{bmatrix}\right) = a + bx + cx^2 + dx^3$ is invertible. If we define $T^{-1}: P_3[x] \rightarrow M_2$ by $T^{-1}(c_0 + c_1 x + c_2 x^2 + c_3 x^3) = \begin{bmatrix} c_0 & c_1 \\ c_2 & c_3 \end{bmatrix}$, then $T^{-1}T\left(\begin{bmatrix} a & b \\ c & d \end{bmatrix}\right) = T^{-1}(a + bx + cx^2 + dx^3) = \begin{bmatrix} a & b \\ c & d \end{bmatrix}$, so $T^{-1}T = I_{M_2}$ and $TT^{-1}(c_0 + c_1 x + c_2 x^2 + c_3 x^3) = T\left(\begin{bmatrix} c_0 & c_1 \\ c_2 & c_3 \end{bmatrix}\right) = c_0 + c_1 x + c_2 x^2 + c_3 x^3$, showing that $TT^{-1} = I_{P_3[x]}$.

(b) Let A be an invertible $n \times n$ matrix and let $T_A: \mathbb{R}^n \rightarrow \mathbb{R}^n$ be defined by $T_A(\mathbf{v}) = A\mathbf{v}$. Now consider the operator $T_{A^{-1}}: \mathbb{R}^n \rightarrow \mathbb{R}^n$ defined by $T_{A^{-1}}(\mathbf{v}) = A^{-1}\mathbf{v}$. Then, for any $\mathbf{v} \in \mathbb{R}^n$, $(T_{A^{-1}} T_A)(\mathbf{v}) = T_{A^{-1}}(T_A(\mathbf{v})) = T_{A^{-1}}(A\mathbf{v}) = A^{-1}(A\mathbf{v}) = \mathbf{v}$, so $T_{A^{-1}} T_A = I_{\mathbb{R}^n}$. Similarly, we can show that $T_A T_{A^{-1}} = I_{\mathbb{R}^n}$.

We can prove that a transformation T is invertible fairly easily, although the actual construction of an inverse for T may be difficult.

Theorem 6.5.1

A linear transformation $T: V \rightarrow W$ is invertible if and only if T is one-to-one and onto.

Proof Suppose T is invertible. To show that T is one-to-one, assume that $\mathbf{v}_1$ and $\mathbf{v}_2$ are vectors in V such that $T(\mathbf{v}_1) = T(\mathbf{v}_2)$. Because T is invertible, there is a unique mapping T^{-1} such that $\mathbf{v}_1 = T^{-1}(T(\mathbf{v}_1)) = T^{-1}(T(\mathbf{v}_2)) = \mathbf{v}_2$, so T is 1-1.

To show that T is onto, we take any $\mathbf{w} \in W$. Then $\mathbf{w} = T\left(\overbrace{T^{-1}(\mathbf{w})}^{\in V}\right)$, which shows that $\mathbf{w}$ is in the range of T, that is, W *is* the range of T, and so T is onto.

Conversely, now we suppose that T is 1-1 and onto. Then, for every $\mathbf{w} \in W$, we can define $\hat{T}(\mathbf{w})$ to be the unique vector in V such that $T(\hat{T}(\mathbf{w})) = \mathbf{w}$. The existence of such a vector in V follows because T is onto; the uniqueness of $\hat{T}(\mathbf{w})$ follows from the 1-1 nature of T. Clearly $T\hat{T}$ is the identity transformation on W. To show that $\hat{T}T$ is the identity mapping on V, choose any $\mathbf{v} \in V$. Then, $T(\hat{T}(T(\mathbf{v}))) = (T\hat{T})(T(\mathbf{v})) = I_W(T(\mathbf{v})) = T(\mathbf{v})$. Since T is 1-1, the fact that $T\left(\overbrace{\hat{T}(T(\mathbf{v}))}^{\mathbf{v}_2 \in V}\right) = T(\mathbf{v})$, or $T(\mathbf{v}_2) = T(\mathbf{v})$, implies that $\mathbf{v}_2 = \mathbf{v}$, that is, $\hat{T}(T(\mathbf{v})) = \mathbf{v}$, so $\hat{T}T$ is the identity mapping on V. Thus T is invertible.

Example 6.5.2: An Invertible Transformation

Suppose that T is a linear operator on vector space V of dimension n, and $\sum_{i=0}^{k} c_i T^i = O_V$, with $c_0 \neq 0$. Then T is invertible. We prove this by showing that T is 1-1 and onto.

If T were *not* 1-1, we could find distinct vectors $\mathbf{v}_1$ and $\mathbf{v}_2$ in V such that $T(\mathbf{v}_1) = T(\mathbf{v}_2)$—that is, $\mathbf{v}^* = \mathbf{v}_1 - \mathbf{v}_2 \neq \mathbf{0}_V$ such that $T(\mathbf{v}^*) = \mathbf{0}_V$. This implies $T^i(\mathbf{v}^*) = \mathbf{0}_V$ for all positive integers i; and so, starting with $i = 1$, we have $\sum_{i=1}^{k} c_i T^i(\mathbf{v}^*) = \left(\sum_{i=1}^{k} c_i T^i\right)(\mathbf{v}^*) = \mathbf{0}_V$. Therefore, because $c_0 \neq 0$ and $\mathbf{v}^* \neq \mathbf{0}_V$,

$$\left(\sum_{i=0}^{k} c_i T^i\right)(\mathbf{v}^*) = c_0 I_V(\mathbf{v}^*) + \left(\sum_{i=1}^{k} c_i T^i\right)(\mathbf{v}^*) = c_0 \mathbf{v}^* \neq \mathbf{0}_V,$$

contradicting our hypothesis that $\sum_{i=0}^{k} c_i T^i$ is the zero operator on V. Thus, T is 1-1.

We have just demonstrated that the null space of T is $\{\mathbf{0}_V\}$. Now Theorem 6.2.1 tells us that rank $T +$ nullity $T = n$. But nullity $T = 0$; and so rank $T = n = \dim(V)$, and Theorem 5.4.4 allows us to conclude that the range of T is all of V, that is, T is onto. (Exercise B2 contains the explicit representation of T^{-1}.)

Recall that Definition 6.4.1 just describes the inverse of a linear transformation as a *mapping*, not necessarily linear. Now it is time for the other shoe to drop.

Theorem 6.5.2

Let V and W be vector spaces. If $T: V \to W$ is an invertible linear transformation, then its inverse $T^{-1}: W \to V$ is a linear transformation.

Proof Suppose that $\mathbf{w}_1, \mathbf{w}_2 \in W$, and let a and b be any scalars. Since T is invertible, it is onto. Therefore, there exist vectors $\mathbf{v}_1, \mathbf{v}_2 \in V$ such that $T(\mathbf{v}_1) = \mathbf{w}_1$ and $T(\mathbf{v}_2) = \mathbf{w}_2$. (Note that $\mathbf{v}_1 = T^{-1}(\mathbf{w}_1)$ and $\mathbf{v}_2 = T^{-1}(\mathbf{w}_2)$.) Then,

$$T^{-1}(a\mathbf{w}_1 + b\mathbf{w}_2) = T^{-1}(aT(\mathbf{v}_1) + bT(\mathbf{v}_2)) = T^{-1}(T(a\mathbf{v}_1 + b\mathbf{v}_2))$$
$$= (T^{-1}T)(a\mathbf{v}_1 + b\mathbf{v}_2) = I_V(a\mathbf{v}_1 + b\mathbf{v}_2) = a\mathbf{v}_1 + b\mathbf{v}_2$$
$$= aT^{-1}(\mathbf{w}_1) + bT^{-1}(\mathbf{w}_2),$$

so T^{-1} is linear.

As we might expect, an invertible linear transformation has properties identical to those stated for invertible matrices in Theorem 3.5.3.

Theorem 6.5.3:   *Properties of the Inverse of a Linear Transformation*

If $T : V \to W$ is invertible and c is a nonzero scalar, then

(1) $T^{-1} : W \to V$ is invertible.
(2) $(T^{-1})^{-1} = T$.
(3) $(cT)^{-1} = c^{-1}T^{-1}$.

Proof The proof is left as Exercise B3.

Theorem 6.5.4

If $S : U \to V$ and $T : V \to W$ are invertible linear transformations, then $TS : U \to W$ is invertible and $(TS)^{-1} = S^{-1}T^{-1}$.

Proof We start by showing that TS is 1-1 and onto. First suppose that TS is *not* 1-1. Then, $(TS)(\mathbf{u}_1) = (TS)(\mathbf{u}_2)$ for some vectors $\mathbf{u}_1, \mathbf{u}_2 \in U$, $\mathbf{u}_1 \neq \mathbf{u}_2$. This means that $T(S(\mathbf{u}_1)) = T(S(\mathbf{u}_2))$, or $T(\mathbf{v}_1) = T(\mathbf{v}_2)$ with $\mathbf{v}_1 = S(\mathbf{u}_1)$ and $\mathbf{v}_2 = S(\mathbf{u}_2)$ in V. Because S is 1-1, $\mathbf{v}_1 \neq \mathbf{v}_2$. This, in turn, implies that $T(\mathbf{v}_1) \neq T(\mathbf{v}_2)$ because T is 1-1. This contradiction shows that TS must be 1-1.

To prove that TS is onto, choose any vector $\mathbf{w}$ in W. Because T is onto, there is a vector $\mathbf{v} \in V$ such that $T(\mathbf{v}) = \mathbf{w}$. But S is also onto, so there is a vector $\mathbf{u} \in U$ such that $S(\mathbf{u}) = \mathbf{v}$. Then $(TS)(\mathbf{u}) = T(S(\mathbf{u})) = T(\mathbf{v}) = \mathbf{w}$, and so TS is onto. As a 1-1 onto transformation, TS is invertible.

To demonstrate that $(TS)^{-1} = S^{-1}T^{-1}$, we let $P = S^{-1}T^{-1} : W \to U$ and show that $P(TS) = I_U$ and $(TS)P = I_W$. Now we see that $P(TS) = (S^{-1}T^{-1})(TS) = S^{-1}((T^{-1}T)S) = S^{-1}(I_VS) = S^{-1}S = I_U$ and $(TS)P = (TS)S^{-1}T^{-1} = T(SS^{-1})T^{-1} = (TI_V)T^{-1} = TT^{-1} = I_W$, which indicates that $P = (TS)^{-1}$, or $S^{-1}T^{-1} = (TS)^{-1}$. Figure 6.9 shows the relationships among the various transformations.

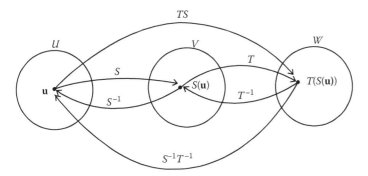

Figure 6.9 The inverse of a composite linear transformation.

We have stated that the set $L(V)$ of all operators on a finite-dimensional vector space V is itself a vector space. The set of all *invertible* operators in $L(V)$, together with the zero operator, is a subspace of $L(V)$ (Exercise B4).

Example 6.5.3: An Invertible Operator on an Infinite-Dimensional Space

Let $V = C(-\infty,\infty)$, the infinite-dimensional vector space of all functions that are continuous on the whole real number line. (It is an interesting problem in advanced mathematics to prove the existence of a basis for such a space!)

Now for any real number c, define $T_c : V \rightarrow V$ by $(T_c f)(x) = f(x+c)$. Then, T_c is invertible with $T_c^{-1} = \hat{T}_c$ defined on V by $(\hat{T}_c f)(x) = f(x-c)$:

$$(\hat{T}_c T_c f)(x) = \hat{T}_c((T_c f)(x)) = \hat{T}_c(f(x+c)) = f(x+c-c) = f(x)$$

and $(T_c \hat{T}_c f)(x) = T_c((\hat{T}_c f)(x)) = T_c(f(x-c)) = f(x-c+c) = f(x)$.

Example 6.5.4: A Noninvertible Operator on an Infinite-Dimensional Space

Let $P[x]$ denote the vector space of all polynomials in x with real coefficients:

$P[x] = \{\sum_{k=0}^{n} a_k x^k : n$ is any nonnegative integer and $a_k \in \mathbb{R}\}$.
(In set-theoretical terms, $P[x] = \bigcup_{n=0}^{\infty} P_n[x]$.)

Now define $T : P[x] \rightarrow P[x]$ by $T(p(x)) = xp(x)$ for all $p(x) \in P[x]$. Then, T is not invertible because T is not *onto*: No polynomial in $P[x]$ with a nonzero constant term is the image of a polynomial in $P[x]$. For example, $p(x) = 1 + 2x$ cannot be the image of any polynomial $q(x) = \sum_{k=0}^{n} a_k x^k$ because $T(q(x)) = xq(x) = x\sum_{k=0}^{n} a_k x^k = \sum_{k=0}^{n} a_k x^{k+1} = a_0 x + a_1 x^2 + \cdots + a_n x^{n+1} = 1 + 2x$ implies $n=0$, $a_0=2$, and $0=1$ (because we have $2x = 1 + 2x$).

The definition of the inverse of a linear transformation involves a two-sided composition. The next theorem simplifies the work of showing that the inverse of a linear transformation exists and makes it easier to verify that a particular transformation is an inverse. In particular, part (d) is analogous to Theorem 3.5.2 for matrices.

Theorem 6.5.5

Suppose that $T: V \to W$ and $S: W \to V$ are linear transformations, and let $\dim(V) = \dim(W) = n$. Then,

   (a) If T is 1-1, then T is invertible.
   (b) If T is onto, then T is invertible.
   (c) If rank $T = n$, then T is invertible.
   (d) If $ST = I_V$, then $TS = I_W$, that is, $S = T^{-1}$ and $S^{-1} = T$.

Proof   (a) If we assume that T is 1-1, it follows that null space $T = \{\mathbf{0}_V\}$ (Corollary 6.1.1). Then, rank $T = n -$ nullity $T = n$. But range T is a subspace of W and both spaces have the same dimension. Hence, Theorem 5.4.4 implies that range $T = W$. Thus, T is 1-1 and onto and so must have an inverse.

   (b) The fact that T is onto says that range $T = W$. Then, nullity $T = n -$ rank $T = 0$. This implies that null space $T = \{\mathbf{0}_V\}$, so T is 1-1 as well as onto, which says that T is invertible.

   (c) Rank $T = n$ implies (by Theorem 5.4.4) that range $T = W$. This means that T is onto and hence invertible by part (c).

   (d) The condition $ST = I_V$ implies that T is 1-1: If $\mathbf{v}_1, \mathbf{v}_2 \in V$ with $\mathbf{v}_1 \neq \mathbf{v}_2$ and $T(\mathbf{v}_1) = T(\mathbf{v}_2)$, then

$$\mathbf{v}_1 = I_V(\mathbf{v}_1) = (ST)(\mathbf{v}_1) = S(T(\mathbf{v}_1)) = S(T(\mathbf{v}_2)) = (ST)(\mathbf{v}_2)$$

$$= I_V(\mathbf{v}_2) = \mathbf{v}_2,$$

a contradiction. By part (a), T has an inverse T^{-1}. Now, $T^{-1} = I_V T^{-1} = (ST)T^{-1} = S(TT^{-1}) = SI_W = S$. Theorem 6.5.4 (2) gives us $S^{-1} = T$.

It follows that the operator version of Theorem 3.5.2 concerning one-sided inverses holds.

Corollary 6.5.1

If V is finite-dimensional and $S, T \in L(V)$, then $ST = I_V$ if and only if $TS = I_V$.

Proof   First we suppose that $ST = I_V$. It follows from part (d) of Theorem 6.5.5 that $TS = I_V$ also. Reversing the positions of S and T in the first part of this proof gives us the fact that $TS = I_V$ implies $ST = I_V$.

Theorem 6.4.3 established a link between the composition of linear transformations and the product of their matrix representations. The next result concerns the matrix corresponding to an invertible transformation.

Theorem 6.5.6

Suppose T is a linear transformation from V to W, B_V and B_W are the bases of V and W, respectively, and $\dim(V) = \dim(W) = n$. If T is invertible, then the matrix representing T is also invertible and $[T^{-1}]_{B_W}^{B_V} = ([T]_{B_V}^{B_W})^{-1}$.

Proof If T^{-1} exists, then Theorem 6.4.3 gives us $[T]_{B_V}^{B_W} [T^{-1}]_{B_W}^{B_V} = [TT^{-1}]_{B_W} = [I_W]_{B_W} = I_n$. This says that the matrix $[T^{-1}]_{B_W}^{B_V}$ is a *right-hand* inverse of $[T]_{B_V}^{B_W}$. But Theorem 3.5.2 lets us conclude that $[T^{-1}]_{B_W}^{B_V}$ is in fact a two-sided inverse, so $[T^{-1}]_{B_W}^{B_V} = ([T]_{B_V}^{B_W})^{-1}$.

Example 6.5.5:* Integration via an Inverse Operator

We are going to integrate $f(x) = 2\sin x$ using linear algebra. This is an easy calculus problem, but we want to illustrate the use of operators.

We take $V = span\{f_1, f_2\} = \{af_1 + bf_2 | a, b \in \mathbb{R}\}$, where $f_1(x) = \sin x$ and $f_2(x) = \cos x$. Using algebra and basic calculus, the differentiation mapping D can be shown to be a 1-1 operator on V, so D^{-1} exists by Theorem 6.5.5(a).

It is easy to see that f_1 and f_2 are linearly independent, and so form a basis B for V. Also, $f(x) = 2\sin x = 2 \cdot f_1 + 0 \cdot f_2$, which gives us $[f]_B = \begin{bmatrix} 2 \\ 0 \end{bmatrix}$.

We have

$$(D(f_1))(x) = \cos x = 0 \cdot f_1(x) + 1 \cdot f_2(x),$$
$$(D(f_2))(x) = -\sin x = (-1) \cdot f_1(x) + 0 \cdot f_2(x),$$

so $[D]_B = \begin{bmatrix} 0 & -1 \\ 1 & 0 \end{bmatrix}$ and $[D]_B^{-1} = \begin{bmatrix} 0 & 1 \\ -1 & 0 \end{bmatrix}$. Theorem 6.5.6 tells us $[D^{-1}(f)]_B = [D]_B^{-1}[f]_B$, so $[D]_B^{-1}$ yields an antiderivative for f:

$$[D]_B^{-1}[f]_B = \begin{bmatrix} 0 & 1 \\ -1 & 0 \end{bmatrix} \begin{bmatrix} 2 \\ 0 \end{bmatrix} = \begin{bmatrix} 0 \\ -2 \end{bmatrix},$$

or an antiderivative of $2\sin x$ is $(0) \cdot f_1 + (-2)f_2 = -2\cos x$.

(To get the most general antiderivative, $\int 2\sin x \, dx$, just add an arbitrary constant C to the answer just found.)

---

* Based on W. Swartz, Integration by matrix inversion, *Amer. Math. Monthly* **65** (1958): 282–283; reprinted in *Selected Papers on Calculus*, T.M. Apostol et al., eds. (Washington, DC: MAA, 1969).

Exercises 6.5

A.

1. Consider the linear transformation $T_A(\mathbf{v}) = A\mathbf{v}$ on $\mathbb{R}^n$ induced by the $n \times n$ matrix A. Which of the following matrices induce *invertible* linear operators on the appropriate space $\mathbb{R}^n$? Determine the inverse transformation if it exists.

   (a) $\begin{bmatrix} -3 & 2 \\ 1 & 4 \end{bmatrix}$ (b) $\begin{bmatrix} -4 & 6 \\ 2 & -3 \end{bmatrix}$ (c) $\begin{bmatrix} \cos\theta & -\sin\theta \\ \sin\theta & \cos\theta \end{bmatrix}$

   (d) $\begin{bmatrix} 1 & -3 & 2 \\ 4 & 1 & 0 \\ 5 & -2 & 2 \end{bmatrix}$ (e) $\begin{bmatrix} 1 & 1 & -2 \\ -7 & 3 & 0 \\ 0 & 1 & 1 \end{bmatrix}$

2. For each of the following linear transformations, find a formula for the inverse of the transformation or explain the difficulties you encounter.

   (a) $T:\mathbb{R}^3 \to \mathbb{R}^3$ defined by $T\left(\begin{bmatrix} x \\ y \\ z \end{bmatrix}\right) = \begin{bmatrix} 2x + y - z \\ -x + 2z \\ x + y + z \end{bmatrix}$.

   (b) $S:P_2[x] \to P_2[x]$ defined by $(S(p))(x) = 2p(x+1)$.

   (c) $R:M_2 \to P_2[x]$ defined by $R\left(\begin{bmatrix} a & b \\ c & d \end{bmatrix}\right) = a + 2bx + (c+d)x^2$.

   (d) $T:M_2 \to M_2$ defined by $T\left(\begin{bmatrix} a & b \\ c & d \end{bmatrix}\right) = \begin{bmatrix} a+b & a \\ c & c+d \end{bmatrix}$.

3. If a and b are real numbers, let $T_{a,b}$ be the linear operator on $P_n[x]$ defined by $(T_{a,b}(f))(x) = f(ax+b)$.
   (a) Show that $T_{a,b} \circ T_{c,d} = T_{ac,bc+d}$.
   (b) If $a \neq 0$, show that $T_{a,b}$ is invertible and find its inverse.

4. Let S be the linear transformation defined in Problem A2(b).
   (a) Find the matrix of S relative to the basis $B = \{1, x, x^2\}$.
   (b) Find the matrix of S^{-1} relative to B *without using Theorem 6.5.6*.
   (c) Confirm that $[S^{-1}]_B = ([S]_B)^{-1}$.

5. Following the method of Example 6.5.5, find a formula for an antiderivative of $f(x) = 2x^3 e^x$. [*Hint:* Let $V = span\{e^x, xe^x, x^2 e^x, x^3 e^x\}$.]

B.

1. Let $T:\mathbb{R}^3 \to \mathbb{R}^2$ be a linear transformation and let U be a linear transformation from $\mathbb{R}^2$ into $\mathbb{R}^3$.
   (a) Prove that the transformation UT from $\mathbb{R}^3$ to itself is not invertible. [You may use the result of Problem B5(a), Exercises 6.2.]

(b) Generalize the result of part (a) to the case of two mappings, one in $L(\mathbb{R}^m, \mathbb{R}^n)$ and one in $L(\mathbb{R}^n, \mathbb{R}^m)$, where $m > n$.

2. In Example 6.3.2, show that $T^{-1} = -\frac{1}{c_0}\left(\sum_{i=1}^{k} c_i T^{i-1}\right)$.

3. Prove Theorem 6.5.3.

4. Prove that the set of all invertible operators in $L(V)$, together with the zero operator, is a subspace of $L(V)$.

5. Prove that for any invertible operator T on a vector space V and an arbitrary operator S on V, $(T + S)T^{-1}(T - S) = (T - S)T^{-1}(T + S)$.

6. An operator T on a vector space V is said to be *nilpotent of index q* if $T^q = O_V$ but $T^r \neq O_V$ for any $r < q$. If T is nilpotent of index q, prove that the operator $I_V - T$ is invertible and that $(I_V - T)^{-1} = I_V + T + T^2 + \cdots + T^{q-1}$.

7. Let $T: V \to W$ be a linear transformation, and let $n = \dim(V)$, $m = \dim(W)$. Show that
   (a) If $n < m$, then T is not onto.
   (b) If $n > m$, then T is not one-to-one.

8. Suppose $B = \{\mathbf{v}_1, \mathbf{v}_2, \ldots, \mathbf{v}_n\}$ is a basis for V. Prove that the mapping $T: V \to \mathbb{R}^n$ defined by $T(\mathbf{v}) = [\mathbf{v}]_B$ is
   (a) A linear transformation
   (b) Invertible

9. Suppose B and C are matrices in the vector space M_n. Define the linear transformation T on M_n by $T(A) = BAC$ for all $A \in M_n$. Show that T has an inverse if and only if both B and C are invertible matrices.

10. Let W be an invariant subspace of a linear transformation T on a finite-dimensional vector space V—that is, $T(W) \subseteq W$, or $T(\mathbf{w}) \in W$ for every $\mathbf{w} \in W$. Show that if T is invertible, then W is also invariant under T^{-1} (i.e., $T^{-1}(W) \subseteq W$).

C.

1. Let $P[x] = \{a_k x^k + a_{k-1} x^{k-1} + \cdots + a_1 x + a_0 | a_i \in \mathbb{R} \text{ for all } i\}$, the vector space of all polynomials in the variable x with real coefficients, and let $P_0[x]$ be the subspace of all real polynomials p for which $p(0) = 0$. Consider the linear transformations defined as follows:

$$T(p)(x) = \int_0^x p(t)dt \quad \text{for all } p(x) \in P[x],$$

$$D(p)(x) = \frac{d}{dx}p(x) \quad \text{for all } p(x) \in P[x],$$

$$D_0(p)(x) = \frac{d}{dx}p(x) \quad \text{for all } p(x) \in P_0[x].$$

(a) Determine the domain and range of each of the transformations: T, D, D_0, DT, TD, D_0T, and TD_0.

(b) Which of the four product transformations in part (a) is (are) the identity transformation(s) on its (their) domain(s)?

(c) Which of the seven transformations in part (a) is (are) invertible?

2. Prove that if V is a finite-dimensional vector space with $\dim(V) > 1$, then the set of noninvertible operators on V is not a subspace of $L(V)$.

6.6 Isomorphisms

The existence of an invertible linear transformation between two vector spaces is a powerful fact with useful consequences, both theoretical and applied.

Relying on our discussions of invertible linear transformations, we can formalize this important bond that may exist between certain vector spaces, a connection we first mentioned in Section 5.4.

Definition 6.6.1

An invertible linear transformation $T: V \to W$ is called an **isomorphism**.* In this case, we say that V and W are **isomorphic** (to each other), and denote this by writing $V \cong W$.

Expressed another way, a vector space V is isomorphic to another vector space W if there exists a transformation T from V to W that is 1-1 and onto and preserves linear combinations: $T(c_1\mathbf{v}_1 + c_2\mathbf{v}_2) = c_1 T(\mathbf{v}_1) + c_2 T(\mathbf{v}_2)$ for every $\mathbf{v}_1, \mathbf{v}_2 \in V$ and all scalars, c_1, c_2. (Here we are ignoring possible differences in the notations and terminologies between V and W.)

To say that V and W are isomorphic is to declare that they are algebraically indistinguishable as vector spaces. Their elements (vectors) and operations may look different, but isomorphic vector spaces behave the same way algebraically with respect to vector addition and scalar multiplication.

---

* The word **isomorphism** is derived from Greek roots meaning "equal" and "form" (or "shape" or "structure"). For a broader algebraic treatment of isomorphisms, see J.A. Gallian, *Contemporary Abstract Algebra*, 7th edn. (Belmont, CA: Brooks/Cole, 2009).

An important example of isomorphic vector spaces consists of the two vector spaces $\mathbb{R}^2$ and $\mathbb{C}$, the complex number system, where we take the scalars to be the set of real numbers in each space. There is a "natural" isomorphism $T : \mathbb{C} \to \mathbb{R}^2$ defined by $T(a + bi) = \begin{bmatrix} a \\ b \end{bmatrix}$, where $i = \sqrt{-1}$ and $a, b \in \mathbb{R}$. Similarly, the correspondence $\begin{bmatrix} a & b \\ -b & a \end{bmatrix} \overset{\hat{T}}{\longleftrightarrow} a + bi$ defines an isomorphism between a subspace of M_2 and $\mathbb{C}$ (see Problem B16 in Exercises 5.4).

However, there may be some operations and properties that are specific to a given vector space but that do not translate to a vector space isomorphic to it. For example, we can multiply two vectors in M_2, but this operation has no natural counterpart in $\mathbb{R}^4$, even though the spaces are isomorphic: $\begin{bmatrix} a & b \\ c & d \end{bmatrix} \longleftrightarrow [a \quad b \quad c \quad d]^{\mathsf{T}}$. Similarly, quantities such as angle measure and distance may not be preserved by an isomorphism from $\mathbb{R}^n$ to itself.

Example 6.6.1: Some Isomorphisms

(a) If $\mathbb{C}^n = \left\{ \begin{bmatrix} c_1 \\ c_2 \\ \vdots \\ c_n \end{bmatrix} : c_i \in \mathbb{R} \text{ for all } i \right\}$ is taken as a *real* vector

space—that is, with scalars from $\mathbb{R}$, then $\mathbb{C}^n$ is

isomorphic to $\mathbb{R}^{2n}$. Noting that $\begin{bmatrix} c_1 \\ c_2 \\ \vdots \\ c_n \end{bmatrix} = \begin{bmatrix} a_1 + b_1 i \\ a_2 + b_2 i \\ \vdots \\ a_n + b_n i \end{bmatrix} =$

$\begin{bmatrix} a_1 \\ a_2 \\ \vdots \\ a_n \end{bmatrix} + i \begin{bmatrix} b_1 \\ b_2 \\ \vdots \\ b_n \end{bmatrix}$, where $a_i, b_i \in \mathbb{R}$ for $1 \leq i \leq n$ we can

define the isomorphism: $T \left(\begin{bmatrix} c_1 \\ c_2 \\ \vdots \\ c_n \end{bmatrix} \right) =$

$[a_1 \quad a_2 \quad \dots \quad a_n \quad b_1 \quad b_2 \quad \dots \quad b_n]^{\mathsf{T}} \in \mathbb{R}^{2n}$.

(b) The vector spaces M_2 and $P_3[x]$ are isomorphic because

(for example) $T : M_2 \to P_3[x]$ defined by $T \left(\begin{bmatrix} a & b \\ c & d \end{bmatrix} \right) =$

$a + bx + cx^2 + dx^3$ is an invertible linear transformation:

$T^{-1}(a + bx + cx^2 + dx^3) = \begin{bmatrix} a & b \\ c & d \end{bmatrix}$.

(c) Consider the vector space $V = \{(x_1, x_2, \dots) | x_i \in \mathbb{R} \text{ for all } i\}$ of all infinite sequences of real numbers with the expected component-by-component addition and scalar multiplication. The set S consisting of all those elements of V that have a first component equal to zero is a proper subspace of V that is isomorphic to V!* Just define the *(right shift)* operator $T: V \to S$ by $T(x_1, x_2, \dots, x_n, \dots) = (0, x_1, x_2, \dots, x_{n-1}, \dots)$. Then T is linear and T^{-1} is defined by $T^{-1}((0, x_1, x_2, \dots, x_{n-1}, \dots)) = (x_1, x_2, \dots, x_n, \dots)$.

(d) The operator $\alpha: \mathbb{R}^n \to \mathbb{R}^n$ defined by $\alpha(\mathbf{x}) = \alpha\left(\begin{bmatrix} x_1 & x_2 & \dots & x_n \end{bmatrix}^T\right) = \begin{bmatrix} 2x_1 & 2x_2 & \dots & 2x_n \end{bmatrix}^T$ is an isomorphism, but $\|T(\mathbf{x})\| = 2\sqrt{x_1^2 + x_2^2 + \dots + x_n^2} = 2\|\mathbf{x}\|$, so lengths are not preserved under this isomorphism.

At first glance, it may seem difficult to tell when an isomorphism exists. However, a simple criterion for determining if two vector spaces are isomorphic is contained in the following theorem.

Theorem 6.6.1

Two finite-dimensional vector spaces V and W with the same set of scalars are isomorphic if and only if $\dim(V) = \dim(W)$.

Proof First suppose that V and W are isomorphic vector spaces with $\dim(V) = m$ and $\dim(W) = n$. Then there exists an invertible linear transformation T from V to W. This implies that the null space of T is $\{\mathbf{0}_V\}$ and the range of T is all of W. Thus nullity $T = 0$, rank $T = \dim(W) = n$, and the relation $m = \dim(V) = \text{rank } T + \text{nullity } T$ becomes $m = n$.

Conversely, suppose that V and W are vector spaces with $\dim(V) = n = \dim(W)$. Let $\{\mathbf{v}_1, \mathbf{v}_2, \dots, \mathbf{v}_n\}$ be a basis for V and $\{\mathbf{w}_1, \mathbf{w}_2, \dots, \mathbf{w}_n\}$ be a basis for W. Define the linear transformation $T: V \to W$ by

$$T(a_1\mathbf{v}_1 + a_2\mathbf{v}_2 + \dots + a_n\mathbf{v}_n) = a_1\mathbf{w}_1 + a_2\mathbf{w}_2 + \dots + a_n\mathbf{w}_n.$$

Then T is *onto* because $\{\mathbf{w}_1, \mathbf{w}_2, \dots, \mathbf{w}_n\}$ spans W, and every vector $\mathbf{w} = c_1\mathbf{w}_1 + c_2\mathbf{w}_2 + \dots + c_n\mathbf{w}_n \in W$ is the image of the vector

---

* This example points out the sometimes paradoxical nature of infinite sets. For amusement and enlightenment, read N.Ya. Vilenkin, The extraordinary hotel, or the thousand and first journey of ion the quiet, in *Stories About Sets* (New York: Academic Press, 1968).

$\mathbf{v} = c_1\mathbf{v}_1 + c_2\mathbf{v}_2 + \cdots + c_n\mathbf{v}_n \in V$. Furthermore, T is 1-1 because $\{\mathbf{w}_1,$ $\mathbf{w}_2, \ldots, \mathbf{w}_n\}$ is a linearly independent set: If $\mathbf{u}_1 = a_1\mathbf{v}_1 + a_2\mathbf{v}_2 + \cdots + a_n\mathbf{v}_n, \mathbf{u}_2 = b_1\mathbf{v}_1 + b_2\mathbf{v}_2 + \cdots + b_n\mathbf{v}_n$, and $T(\mathbf{u}_1) = T(\mathbf{u}_2)$, then $\mathbf{0}_W = T(\mathbf{u}_1 - \mathbf{u}_2) = T((a_1 - b_1)\mathbf{v}_1 + (a_2 - b_2)\mathbf{v}_2 + \cdots + (a_n - b_n)\mathbf{v}_n) = (a_1 - b_1)\mathbf{w}_1 + (a_2 - b_2)\mathbf{w}_2 + \cdots + (a_n - b_n)\mathbf{w}_n$—impossible unless $a_i = b_i$ for $1 \leq i \leq n$, which implies that $\mathbf{u}_1 = \mathbf{u}_2$.

Because T is both 1-1 and onto, T is invertible, and is therefore an isomorphism from V to W.

This theorem has a profound consequence (see Section 5.4).

Corollary 6.6.1

Any n-dimensional vector space V (with scalars $\mathbb{R}$) is isomorphic to the Euclidean space $\mathbb{R}^n$.

Example 6.6.2

If V and W are isomorphic, write $V \cong W$. From previous examples, we can deduce, for instance, that

$$M_{34} \cong L(\mathbb{R}^4, \mathbb{R}^3) \cong P_{11}[x] \cong \mathbb{C}^6 \cong \mathbb{R}^{12}.$$

The property of being isomorphic is an *equivalence relation* (see Section 3.6, Exercise C1, and Exercise B7 following this section).

Corollary 6.6.1 states that any finite-dimensional linear algebra problem can be handled as a problem in n-tuples or geometric vectors. It is not always advantageous to take the "coordinate" approach because of the dependence on the basis chosen. However, there are occasions when being able to replace an abstract operator problem with matrix calculations is invaluable. This idea is captured in an oft-quoted remark by Irving Kaplansky about Paul Halmos:*

We share a philosophy about linear algebra: we think basis-free, but when the chips are down, we close the office door and compute with matrices like fury.

---

* Irving Kaplansky (1917–2006) was a prominent algebraist and Paul Halmos (1916–2006) was noted for his work in operator theory and Hilbert space, among other areas. The remark can be found in *Paul Halmos: Celebrating 50 Years of Mathematics*, J. Ewing and F.W. Gehring, eds. (New York: Springer, 1991).

Exercises 6.6

A.

1. Show that $T: \mathbb{R}^n \to \mathbb{R}^n$ defined by $T\left(\begin{bmatrix} x_1 \\ x_2 \\ \vdots \\ x_n \end{bmatrix}\right) = \begin{bmatrix} \alpha x_1 \\ x_2 \\ \vdots \\ x_n \end{bmatrix}$,

    where $\alpha \neq 0$ is a fixed scalar, is an isomorphism.

2. Show that $T: \mathbb{R}^n \to \mathbb{R}^n$ defined by $T\left(\begin{bmatrix} x_1 \\ x_2 \\ \vdots \\ x_n \end{bmatrix}\right) =$

    $\begin{bmatrix} x_1 + \beta x_2 \\ x_2 \\ \vdots \\ x_n \end{bmatrix}$, where β is a fixed scalar, is an isomorphism.

3. Which of the following linear transformations are isomorphisms? Justify your answers.

    (a) $T: \mathbb{R}^3 \to \mathbb{R}^2$ defined by $T\left(\begin{bmatrix} x \\ y \\ z \end{bmatrix}\right) = \begin{bmatrix} x + 2y \\ z \end{bmatrix}$.

    (b) $S: \mathbb{R}^3 \to \mathbb{R}^3$ defined by $S\left(\begin{bmatrix} x \\ y \\ z \end{bmatrix}\right) = \begin{bmatrix} z \\ -y \\ 0 \end{bmatrix}$.

    (c) $T: \mathbb{R}^3 \to \mathbb{R}^3$ defined by $T\left(\begin{bmatrix} x \\ y \\ z \end{bmatrix}\right) = \begin{bmatrix} x - y \\ y \\ x + z \end{bmatrix}$.

4. Show that $T: M_{mn} \to M_{nm}$ defined by $T(A) = A^{\mathrm{T}}$ is an isomorphism.

5. Determine an isomorphism between M_{32} and $\mathbb{R}^6$. In general, describe why M_{mn} and $\mathbb{R}^{mn}$ are isomorphic.

6. Find an isomorphism between the subspace of $n \times n$ diagonal matrices and $\mathbb{R}^n$

7. Let $V = P_4[x]$ and $W = \{p \in P_5[x] : p(0) = 0\}$. Show that V and W are isomorphic.

8. Define $T: P_n[x] \to P_n[x]$ by $T(p) = p + p'$, where p' is the derivative of p. Show that T is an isomorphism.

9. Define $T: C[0, 1] \to C[3, 4]$ by $T(f(x)) = f(x - 3)$. Show that T is an isomorphism.

10. Show that the transformation $T(p(x)) = xp'(x)$ is *not* an isomorphism from $P_n[x]$ into $P_n[x]$.

11. If $f : \mathbb{R} \to \mathbb{R}$ is a linear transformation and $T : \mathbb{R}^2 \to \mathbb{R}^2$ is given by $T\left(\begin{bmatrix} x \\ y \end{bmatrix}\right) = \begin{bmatrix} x \\ y - f(x) \end{bmatrix}$, prove that T is an isomorphism.

12. Let $V = \left\{ \begin{bmatrix} a & a+b \\ 0 & c \end{bmatrix} : a, b, c \in \mathbb{R} \right\}$. Construct an isomorphism from V to $\mathbb{R}^3$.

13. Which pairs of the following vector spaces are isomorphic?
$$\mathbb{R}^7 \quad \mathbb{R}^{15} \quad M_3 \quad M_{3,5} \quad P_6[x] \quad P_8[x] \quad P_{14}[x]$$

14. Determine which of the following pairs of vector spaces are isomorphic. Justify your answers.
(a) $\mathbb{R}^3$ and $P_3[x]$ (b) $\mathbb{R}^4$ and $P_3[x]$ (c) M_2 and $P_3[x]$
(d) $V = \{A \in M_2 : \text{trace}(A) = 0\}$ and $\mathbb{R}^4$

B.

1. For what values of λ is the linear transformation on $P_n[x]$ defined by $T(f) = \lambda f - xf'$, an isomorphism?

2. Which of the following pairs of vector spaces are isomorphic? Justify your answers.
(a) 3×3 skew-symmetric matrices and $\mathbb{R}^3$
(b) $n \times n$ diagonal matrices and $P_n[x]$
(c) $(n-1) \times (n-1)$ symmetric matrices and $n \times n$ skew-symmetric matrices

3. Construct an isomorphism between the vector space of 3×4 matrices and that of 2×6 matrices. When, in general, is it possible to find an isomorphism between the space of $k \times l$ and that of $m \times n$ matrices?

4. For what value of m is the vector space of $n \times n$ symmetric matrices isomorphic to $\mathbb{R}^m$?

5. For what value of n is $\mathbb{R}^n$ isomorphic to the subspace of all vectors in $\mathbb{R}^4$ that are orthogonal to $[\,1 \quad -3 \quad 0 \quad 4\,]^T$?

6. Let $T : P_n[x] \to P_n[x]$ be the linear transformation defined by $T(p) = p + a_1 p' + a_2 p'' + \cdots + a_n p^{(n)}$, where $a_1, a_2, \ldots, a_n$ are real numbers and $p^{(k)}$ denotes the kth derivative of p. Show that T is an isomorphism.

7. Prove that the property of being isomorphic is an *equivalence relation* on the set V of all vector spaces over $\mathbb{R}$—that is, show (1) $V \cong V$ for all $V \in \mathsf{V}$; (2) if $V_1 \cong V_2$, then $V_2 \cong V_1$ for

$V_1, V_2 \in \mathsf{V}$; (3) if $V_1 \cong V_2$ and $V_2 \cong V_3$, then $V_1 \cong V_3$ for $V_1, V_2, V_3 \in \mathsf{V}$.

8. For each subspace U below, determine the dimension d of U and find an isomorphism from U to $\mathbb{R}^d$.

(a) $U = span\left\{ \begin{bmatrix} 1 \\ -1 \\ 1 \end{bmatrix}, \begin{bmatrix} 2 \\ -2 \\ 2 \end{bmatrix}, \begin{bmatrix} 0 \\ 1 \\ 0 \end{bmatrix}, \begin{bmatrix} 1 \\ 0 \\ 1 \end{bmatrix} \right\}$ in $\mathbb{R}^3$.

(b) $U = span\{p_1(x), p_2(x), p_3(x), p_4(x)\}$ in $P_3[x]$, where $p_1(x) = x^3 + x^2 + 1$, $p_2(x) = x^2 + x + 1$, $p_3(x) = 2x^3 + 3x^2 + x + 3$, and $p_4(x) = x^3 - x$.

(c) $U = span\left\{ \begin{bmatrix} 1 & 0 \\ -1 & 1 \end{bmatrix}, \begin{bmatrix} -2 & 0 \\ 2 & -2 \end{bmatrix}, \begin{bmatrix} 1 & 1 \\ 1 & 1 \end{bmatrix}, \begin{bmatrix} 2 & 1 \\ 0 & 2 \end{bmatrix}, \begin{bmatrix} 0 & 1 \\ 2 & 0 \end{bmatrix} \right\}$ in M_2.

9. Consider the linear transformation $T : \mathbb{R}^2 \to \mathbb{R}^2$ defined by $T\left(\begin{bmatrix} x \\ y \end{bmatrix} \right) = \begin{bmatrix} 3x + y \\ x - 2y \end{bmatrix}$.

(a) Show that T is an isomorphism.

(b) What is the distance between $T\left(\begin{bmatrix} 0 \\ 0 \end{bmatrix} \right)$ and $T\left(\begin{bmatrix} 1 \\ 0 \end{bmatrix} \right)$?

(c) Does T take points on every straight line to points on another straight line?

(Consider lines passing through the origin as well as those not passing through the origin.)

(d) Show that T takes a parallelogram to a parallelogram.

10. Let $\{V, +, \cdot\}$ be a vector space and $\mathbf{t}$ be a fixed vector in V. Define new operations on the set of vectors V by

$$\mathbf{x} \boxplus \mathbf{y} = \mathbf{x} + \mathbf{y} + \mathbf{t},$$
$$c * \mathbf{x} = c \cdot \mathbf{x} + (1 - c)\mathbf{t}.$$

(a) Show that $\{V, \boxplus, *\}$ is a vector space.

(b) Show that the mapping $T : \{V, +, \cdot\} \to \{V, \boxplus, *\}$ defined by $T(\mathbf{x}) = \mathbf{x} + \mathbf{t}$ for all $\mathbf{x} \in V$ is an isomorphism.

C.

1. Here are the rules of a game that we can call the Fifteen Game:* Two players take turns selecting integers from 1 to 9. No number can be selected more than once. Each player tries to select numbers so that three of them add up to 15. The first player to succeed wins the game. Try playing this game a few times. Show that the Fifteen Game is isomorphic (in a general sense) to another familiar game.

---

* Adapted from R. Messer, *Linear Algebra: Gateway to Mathematics*, p. 258 (New York: HarperCollins College Publishers, 1994).

2. Suppose U and W are subspaces of V. Define $\psi: U \oplus W \to V$ by $\psi(\mathbf{u}, \mathbf{w}) = \mathbf{u} + \mathbf{w}$, where $\mathbf{u} \in U$ and $\mathbf{w} \in W$. Show that the null space of ψ is isomorphic to $U \cap W$.

3. Let V and W be finite-dimensional vector spaces with the ordered bases $B_V = \{\mathbf{v}_1, \mathbf{v}_2, \ldots, \mathbf{v}_n\}$ and $B_W = \{\mathbf{w}_1, \mathbf{w}_2, \ldots, \mathbf{w}_m\}$, respectively. By Theorem 6.1.2, there exist linear transformations $T_{ij}: V \to W$ such that

$$T_{ij}(\mathbf{v}_k) = \begin{cases} \mathbf{w}_i & \text{if } i = j \\ \mathbf{0}_W & \text{if } i \neq j \end{cases}.$$

(a) Prove that $\{T_{ij} : 1 \leq i \leq m, 1 \leq j \leq n\}$ is a basis for $L(V, W)$.
(b) Now let $M(i,j)$ be the $m \times n$ matrix with 1 in row i, and column j and 0 elsewhere, and prove that $[T_{ij}]_{B_V}^{B_W} = M(i,j)$.
(c) Again, by Theorem 6.1.2, there exists a linear transformation $\varphi: L(V, W) \to M_{mn}$ such that $\varphi(T_{ij}) = M(i,j)$. Prove that φ is an isomorphism. (This shows, by Theorem 6.6.1, that $\dim(L(V, W)) = \dim(M_{mn}) = mn = \dim(W) \cdot \dim(V)$.)

6.7 Similarity

One of the difficulties in using a matrix representation of a linear transformation $T: V \to W$ is the dependence of the matrix on the bases chosen for V and W. The convenience of working (calculating) with a matrix depends very much on the form of the matrix and, therefore, on the choice of bases. For example, if the matrix associated with a linear transformation is diagonal, calculating products (compositions) of the transformation is equivalent to elementary calculations with diagonal matrices.

The first step in understanding the sensitivity of a matrix to the choice of bases is to see the effect a change of basis has on the coordinate representation of a vector. (Recall that coordinate change has been discussed in Section 1.3 and in Section 4.5.) A fundamental idea, discussed briefly in Section 4.5 in the context of diagonalization of a matrix, is that we can effect a change of basis (coordinates) by premultiplying by a unique **transition matrix** (change-of-basis matrix or change-of-coordinates matrix), $P: P[\mathbf{x}]_{B_1} = [\mathbf{x}]_{B_2}$. Now we can view this idea of a change-of-basis matrix in light of the fundamental relationship $[T]_{B_V}^{B_W}[\mathbf{v}]_{B_V} = [T(\mathbf{v})]_{B_W}$, which describes the matrix of a linear transformation relative to two bases.

Example 6.7.1: Change of Coordinates

Let us consider the space $\mathbb{R}^2$ with the standard basis $\mathcal{E}_2 = \{\mathbf{e}_1, \mathbf{e}_2\}$. If $\mathbf{v} \in \mathbb{R}^2$ and $\mathbf{v} = x\mathbf{e}_1 + y\mathbf{e}_2$, we write $[\mathbf{v}]_{\mathcal{E}_2} = \begin{bmatrix} x \\ y \end{bmatrix}$.

Now suppose we want to change the basis for $\mathbb{R}^2$

to $B = \left\{ \begin{bmatrix} 1 \\ -1 \end{bmatrix}, \begin{bmatrix} 1 \\ 1 \end{bmatrix} \right\}$. What are the coordinates of $\mathbf{v} \in \mathbb{R}^2$ with respect to this new basis and what is the relationship between $[\mathbf{v}]_{\mathcal{E}_2}$ and $[\mathbf{v}]_B$?

To answer this question, notice what happens in the relation $[T]_{B_V}^{B_W}[\mathbf{v}]_{B_V} = [T(\mathbf{v})]_{B_W}$ if we replace T by $I_{\mathbb{R}^2}$, B_V by $\mathcal{E}_2$, and B_W by B:

$$\left[I_{\mathbb{R}^2} \right]_{\mathcal{E}_2}^{B} [\mathbf{v}]_{\mathcal{E}_2} = \left[I_{\mathbb{R}^2}(\mathbf{v}) \right]_B = [\mathbf{v}]_B. \qquad (*)$$

Simply put, this last line says that to change the coordinates of a vector $\mathbf{v}$ with respect to the standard basis to coordinates with respect to a basis B, *just premultiply the original coordinate vector by the matrix representation of the identity transformation with respect to the bases $\mathcal{E}_2$ and B.*

From our work in Section 6.4, we see that

$$\left[I_{\mathbb{R}^2} \right]_{\mathcal{E}_2}^{B} = \left[[I_{\mathbb{R}^2}(\mathbf{e}_1)]_B \quad [I_{\mathbb{R}^2}(\mathbf{e}_2)]_B \right] = \left[[\mathbf{e}_1]_B \quad [\mathbf{e}_2]_B \right]. \text{ Now}$$

$$\mathbf{e}_1 = \begin{bmatrix} 1 \\ 0 \end{bmatrix} = \frac{1}{2} \begin{bmatrix} 1 \\ -1 \end{bmatrix} + \frac{1}{2} \begin{bmatrix} 1 \\ 1 \end{bmatrix},$$

$$\mathbf{e}_2 = \begin{bmatrix} 0 \\ 1 \end{bmatrix} = -\frac{1}{2} \begin{bmatrix} 1 \\ -1 \end{bmatrix} + \frac{1}{2} \begin{bmatrix} 1 \\ 1 \end{bmatrix},$$

so $\left[I_{\mathbb{R}^2} \right]_{\mathcal{E}_2}^{B} = \begin{bmatrix} \frac{1}{2} & -\frac{1}{2} \\ \frac{1}{2} & \frac{1}{2} \end{bmatrix}$. Thus, for example, the vector $\mathbf{v} = \begin{bmatrix} -3 \\ 4 \end{bmatrix}$ with respect to the standard basis of $\mathbb{R}^2$ becomes $\left[I_{\mathbb{R}^2} \right]_{\mathcal{E}_2}^{B} \mathbf{v} = \begin{bmatrix} \frac{1}{2} & -\frac{1}{2} \\ \frac{1}{2} & \frac{1}{2} \end{bmatrix} \begin{bmatrix} -3 \\ 4 \end{bmatrix} = \begin{bmatrix} -\frac{7}{2} \\ \frac{1}{2} \end{bmatrix}$. We can verify this by noting that $\begin{bmatrix} -3 \\ 4 \end{bmatrix} = -\frac{7}{2} \begin{bmatrix} 1 \\ -1 \end{bmatrix} + \frac{1}{2} \begin{bmatrix} 1 \\ 1 \end{bmatrix}$.

Finally, we note that $\left[I_{\mathbb{R}^2} \right]_{\mathcal{E}_2}^{B}$ is an invertible matrix, so we can reverse the direction of the relationship $(*)$ to write

$$[\mathbf{v}]_{\mathcal{E}_2} = \left(\left[I_{\mathbb{R}^2} \right]_{\mathcal{E}_2}^{B} \right)^{-1} [\mathbf{v}]_B.$$

Now let us generalize a bit by considering the identity mapping $I : V \to V$ with respect to B_V and $\hat{B}_V$, any two ordered bases for a vector space V. If we take B_V to be the basis for the first copy of V and $\hat{B}_V$ to be the basis for the second copy of V, we can transform the coordinates of a vector $\mathbf{v} \in V$ with respect to the basis B_V to its coordinates with respect to the basis $\hat{B}_V$, as we did in Example 6.7.1. In the equation $[T]_{B_V}^{B_W}[\mathbf{v}]_{B_V} = [T(\mathbf{v})]_{B_W}$, we can replace T by I and B_W by $\hat{B}_V$, and write

$$[I]_{B_V}^{\hat{B}_V}[\mathbf{v}]_{B_V} = [\mathbf{v}]_{\hat{B}_V}.$$

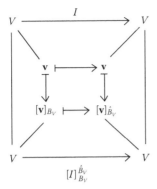

Figure 6.10 $[I]^{\hat{B}_V}_{B_V}[v]_{B_V} = [v]_{\hat{B}_V}$.

This invertible matrix, $[I]^{\hat{B}_V}_{B_V}$, is called the **transition matrix (change-of-basis matrix** or **change-of-coordinates matrix**) from B_V to $\hat{B}_V$ (see Figure 6.10). (Also see Section 4.5.)

We can wrap all these calculations up in a theorem.

Theorem 6.7.1

Let B_1 and B_2 be ordered bases of a finite-dimensional vector space, V. Then $[I]^{B_2}_{B_1}$ is the invertible transition matrix from B_1 to B_2:

$$[I]^{B_2}_{B_1}[\mathbf{v}]_{B_1} = [\mathbf{v}]_{B_2} \quad \text{and} \quad [\mathbf{v}]_{B_1} = \left([I]^{B_2}_{B_1}\right)^{-1}[\mathbf{v}]_{B_2} \quad \text{for every } \mathbf{v} \in V.$$

Now that we have seen how a change of basis for a vector space V affects the coordinates of a vector $\mathbf{v} \in V$, we can move on to determine the relationship between two matrices, each of which represents the same linear transformation, $T: V \to W$, with respect to independent choices of bases for V and W. The next example sets the stage for the general situation.

Example 6.7.2: Change of Basis, Change of Matrix Representation

Consider the linear transformation $T: \mathbb{R}^2 \to \mathbb{R}^3$ defined by $T\left(\begin{bmatrix} x \\ y \end{bmatrix}\right) = \begin{bmatrix} x+y \\ x-y \\ x+2y \end{bmatrix}$. With respect to the standard bases for $\mathbb{R}^2$ and $\mathbb{R}^3$, the matrix representation of T is $[T]^{\mathcal{E}_3}_{\mathcal{E}_2} = \begin{bmatrix} 1 & 1 \\ 1 & -1 \\ 1 & 2 \end{bmatrix}$.

Now suppose that we change the bases to $B_1 = \left\{ \begin{bmatrix} 1 \\ -1 \end{bmatrix}, \begin{bmatrix} 1 \\ 1 \end{bmatrix} \right\}$ for $\mathbb{R}^2$ and $B_2 = \left\{ \begin{bmatrix} 2 \\ 4 \\ 2 \end{bmatrix}, \begin{bmatrix} 3 \\ 2 \\ 0 \end{bmatrix}, \begin{bmatrix} 1 \\ -2 \\ 2 \end{bmatrix} \right\}$ for $\mathbb{R}^3$.

This change of bases results in a new matrix representation $[T]_{B_1}^{B_2}$ for T. We want to understand the relationship between $[T]_{\mathcal{E}_2}^{\mathcal{E}_3}$ and $[T]_{B_1}^{B_2}$.

Proceeding in the standard way—representing the images of the basis vectors in B_1 as linear combinations of the basis vectors in B_2—we see that the matrix representation of T with respect to B_1 and B_2 is $[T]_{B_1}^{B_2} = \begin{bmatrix} \frac{1}{8} & \frac{1}{2} \\ \frac{1}{8} & 0 \\ -\frac{5}{8} & 1 \end{bmatrix}$. The relationship between $[T]_{\mathcal{E}_2}^{\mathcal{E}_3}$ and $[T]_{B_1}^{B_2}$ is not obvious. To see farther into the mist, we have to examine the change of coordinates from $\mathcal{E}_2 = \{\mathbf{e}_1, \mathbf{e}_2\}$ to B_1 and from $\mathcal{E}_3 = \{\mathbf{e}_1, \mathbf{e}_2, \mathbf{e}_3\}$ to B_2.

From Example 6.7.1, we see that the invertible matrix $M_1 = [I_{\mathbb{R}^2}]_{\mathcal{E}_2}^{B_1} = \begin{bmatrix} \frac{1}{2} & -\frac{1}{2} \\ \frac{1}{2} & \frac{1}{2} \end{bmatrix}$ transforms coordinates with respect to the standard basis for $\mathbb{R}^2$ into coordinates with respect to the basis B_1. Similarly, we can determine that the matrix $M_2 = [I_{\mathbb{R}^2}]_{\mathcal{E}_3}^{B_2} = \begin{bmatrix} -\frac{1}{8} & \frac{3}{16} & \frac{1}{4} \\ \frac{3}{8} & -\frac{1}{16} & -\frac{1}{4} \\ 8 & -\frac{3}{16} & \frac{1}{4} \end{bmatrix}$ transforms $[T(\mathbf{v})]_{\mathcal{E}_3}$ into $[T(\mathbf{v})]_{B_2}$. Figure 6.11 may be helpful in illustrating this point.

This "commutative diagram" indicates that there are two equivalent ways to get from $[\mathbf{v}]_{\mathcal{E}_2}$ to $[T(\mathbf{v})]_{B_2}$: $M_2[T]_{\mathcal{E}_2}^{\mathcal{E}_3}[\mathbf{v}]_{\mathcal{E}_2} = [T]_{B_1}^{B_2} M_1[\mathbf{v}]_{\mathcal{E}_2}$ for every vector $\mathbf{v} \in \mathbb{R}^2$. This equation implies $M_2[T]_{\mathcal{E}_2}^{\mathcal{E}_3} = [T]_{B_1}^{B_2} M_1$, or $M_2[T]_{\mathcal{E}_2}^{\mathcal{E}_3} M_1^{-1} = [T]_{B_1}^{B_2}$.

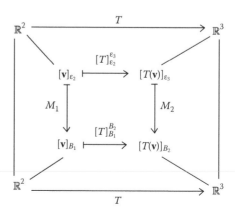

Figure 6.11 $M_2[T]_{\mathcal{E}_2}^{\mathcal{E}_3}[\mathbf{v}]_{\mathcal{E}_2} = [T]_{B_1}^{B_2} M_1[\mathbf{v}]_{\mathcal{E}_2}$.

A   simple   calculation   reveals   that $M_2[T]^{\mathcal{E}_3}_{\mathcal{E}_2}M_1^{-1} =$

$$\begin{bmatrix} -\frac{1}{8} & \frac{3}{16} & \frac{1}{4} \\ \frac{3}{8} & -\frac{1}{16} & -\frac{1}{4} \\ \frac{1}{8} & -\frac{3}{16} & \frac{1}{4} \end{bmatrix} \begin{bmatrix} 1 & 1 \\ 1 & -1 \\ 1 & 2 \end{bmatrix} \begin{bmatrix} 1 & 1 \\ -1 & 1 \end{bmatrix} = \begin{bmatrix} \frac{1}{8} & \frac{1}{2} \\ \frac{1}{8} & 0 \\ -\frac{5}{8} & 1 \end{bmatrix} = [T]^{B_2}_{B_1},$$

as expected.

Now we consider the most general case. We want to understand the relationship between two matrices, each of which represents the same linear transformation, $T:V \to W$, with respect to independent choices of bases for V and W.

Suppose that $T:V \to W$ is a linear transformation, where V has the basis B_V and W has the basis B_W. Then, with respect to these bases, T has a matrix representation $[T]^{B_W}_{B_V}$, and we can write

$$[T(\mathbf{v})]_{B_W} = [T]^{B_W}_{B_V}[\mathbf{v}]_{B_V}. \tag{6.7.1}$$

Similarly, if we change the basis of V to $\hat{B}_V$ and the basis of W to $\hat{B}_W$, then the matrix representing T changes, and we have the relation

$$[T(\mathbf{v})]_{\hat{B}_W} = [T]^{\hat{B}_W}_{\hat{B}_V}[\mathbf{v}]_{\hat{B}_V}. \tag{6.7.2}$$

If we let M_1 and M_2 be the invertible matrices that represent the changes of bases $B_V \to \hat{B}_V$ and $B_W \to \hat{B}_W$, respectively, then for any vectors $\mathbf{v} \in V$ and $\mathbf{w} \in W$, we have (see Equation 6.7.1)

$$[\mathbf{v}]_{\hat{B}_V} = M_1[\mathbf{v}]_{B_V}, \quad \text{or} \quad M_1^{-1}[\mathbf{v}]_{\hat{B}_V} = [\mathbf{v}]_{B_V}, \tag{6.7.3}$$

and

$$[\mathbf{v}]_{\hat{B}_W} = M_2[\mathbf{v}]_{B_W}. \tag{6.7.4}$$

If we substitute $T(\mathbf{v})$ for $\mathbf{v}$ in Equation 6.7.4 and combine these last equations with Equation 6.7.1, we obtain

$$[T(\mathbf{v})]_{\hat{B}_W} \overset{\text{Equation 6.7.4}}{=} M_2[T(\mathbf{v})]_{B_W} \overset{\text{Equation 6.7.1}}{=} M_2[T]^{B_W}_{B_V}[\mathbf{v}]_{B_V} \overset{\text{Equation 6.7.3}}{=}$$
$$M_2[T]^{B_W}_{B_V}M_1^{-1}[\mathbf{v}]_{\hat{B}_V}.$$

But, comparing this outcome to Equation 6.7.2, because relations (Equation 6.7.1 through 6.7.4) are valid for *every* $\mathbf{v} \in V$, we conclude that $M_2[T]^{B_W}_{B_V}M_1^{-1} = [T]^{\hat{B}_W}_{\hat{B}_V}$, that is, the matrix $M_2[T]^{B_W}_{B_V}M_1^{-1}$ describes the linear transformation T with respect to the bases $\hat{B}_V$ and $\hat{B}_W$ of V and W.

Once again, a diagram (Figure 6.12) provides a roadmap for our manipulations.

We can summarize our analysis as follows, simplifying our notation in order to show an important relationship in a clutter-free way.

Theorem 6.7.2

Let V and W be the (nonzero) finite-dimensional vector spaces and let $B_V, \hat{B}_V$ be ordered bases of V and $B_W, \hat{B}_W$ be ordered bases of W.

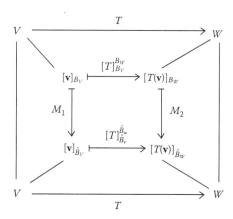

Figure 6.12 $M_2[T]^{B_W}_{B_V} M_1^{-1} = [T]^{\hat{B}_W}_{\hat{B}_V}$.

Furthermore, suppose that the invertible matrices M_1 and M_2 describe the change of bases $B_V \rightarrow \hat{B}_V$ and $B_W \rightarrow \hat{B}_W$, respectively. If the linear transformation $T : V \rightarrow W$ is represented by the matrix A_T with respect to B_V and B_W, and by the matrix $\hat{A}_T$ with respect to $\hat{B}_V$ and $\hat{B}_W$, then $\hat{A}_T = M_2 A_T M_1^{-1}$.

Matrices A and B are said to be **equivalent** if $B = PAQ$ for suitable invertible matrices P and Q. Thus, **two matrix representations of the same linear transformation $T : V \rightarrow W$ with respect to different ordered bases $B_V, \hat{B}_V$ and $B_W, \hat{B}_W$ for V and W, respectively, are equivalent**. This is the same relationship between matrices we saw in Section 3.6, where we discussed premultiplication and postmultiplication of a matrix by elementary matrices. Recall that a square matrix is invertible if and only if it can be written as a product of elementary matrices (Theorem 3.6.4). (Also see Definition 3.6.2 and Problem B11 of Exercises 3.6.)

An important special case of Theorem 6.7.2 is that of a linear operator $T : V \rightarrow V$, when the same ordered basis B is used for both the domain and the codomain.

Corollary 6.7.1

Let B_1 and B_2 be two ordered bases of a finite-dimensional vector space V, and let T be a linear operator on V. If T is represented by the matrices A_T and $\hat{A}_T$ with respect to B_1 and B_2, respectively, then $\hat{A}_T = MA_TM^{-1}$, where M is the matrix representing the change of basis $B_1 \rightarrow B_2$.

Note that if $\dim(V) = n$ in Corollary 6.7.1, then the change-of-basis matrix M is also $n \times n$. We recognize the conclusion of Corollary 6.7.1 as saying that matrix A_T is *similar* to matrix $\hat{A}_T$ (see Definition 4.5.1). Because of the intimate relationship between matrices and linear transformations, we can extend the concept of similarity to operators.

Definition 6.7.1

Two operators S and T on the same vector space V are **similar** if there exists an invertible operator R such that $RSR^{-1} = T$.

This completes the discussion begun in Section 4.5 in the context of matrix algebra. We can consider operators and the matrices that represent them and come to a significant conclusion.

Two $n \times n$ matrices A and B are similar if and only if A and B represent the same operator on an n-dimensional vector space V, with respect to two possibly different bases.

Example 6.7.3: Similar Matrices

Let us consider the operator $T: \mathbb{R}^3 \to \mathbb{R}^3$ defined by

$$T\left(\begin{bmatrix} x \\ y \\ z \end{bmatrix}\right) = \begin{bmatrix} z \\ y \\ x \end{bmatrix}.$$ Suppose that we have two bases for $\mathbb{R}^3$:

$$B_1 = \{\mathbf{v}_i\} = \left\{ \begin{bmatrix} 1 \\ 0 \\ 0 \end{bmatrix}, \begin{bmatrix} 1 \\ 1 \\ 0 \end{bmatrix}, \begin{bmatrix} 1 \\ 1 \\ 1 \end{bmatrix} \right\}, B_2 = \{\hat{\mathbf{v}}_i\} = \left\{ \begin{bmatrix} 1 \\ 0 \\ 1 \end{bmatrix}, \begin{bmatrix} 0 \\ 1 \\ 0 \end{bmatrix}, \begin{bmatrix} 1 \\ 0 \\ 0 \end{bmatrix} \right\}.$$

First, let us calculate the change-of-basis matrix M from B_1 to B_2. We know from previous discussions that $M = \left[[\mathbf{v}_1]_{B_2} \;\; [\mathbf{v}_2]_{B_2} \;\; [\mathbf{v}_3]_{B_2} \right]$. Leaving the details as an exercise (A15), we find that

$$M = \begin{bmatrix} 0 & 0 & 1 \\ 0 & 1 & 1 \\ 1 & 1 & 0 \end{bmatrix} \quad \text{and} \quad M^{-1} = \begin{bmatrix} 1 & -1 & 1 \\ -1 & 1 & 0 \\ 1 & 0 & 0 \end{bmatrix}.$$

Next, we determine $[T]_{B_1} = \left[[T(\mathbf{v}_1)]_{B_1} \;\; [T(\mathbf{v}_2)]_{B_1} \;\; [T(\mathbf{v}_3)]_{B_1} \right]$ and $[T]_{B_2} = \left[[T(\hat{\mathbf{v}}_1)]_{B_2} \;\; [T(\hat{\mathbf{v}}_2)]_{B_2} \;\; [T(\hat{\mathbf{v}}_3)]_{B_2} \right]$:

$$[T]_{B_1} = \begin{bmatrix} 0 & -1 & 0 \\ -1 & 0 & 0 \\ 1 & 1 & 1 \end{bmatrix} \quad \text{and} \quad [T]_{B_2} = \begin{bmatrix} 1 & 0 & 1 \\ 0 & 1 & 0 \\ 0 & 0 & -1 \end{bmatrix}.$$

Then

$$M[T]_{B_1}M^{-1} = \begin{bmatrix} 0 & 0 & 1 \\ 0 & 1 & 1 \\ 1 & 1 & 0 \end{bmatrix}\begin{bmatrix} 0 & -1 & 0 \\ -1 & 0 & 0 \\ 1 & 1 & 1 \end{bmatrix}\begin{bmatrix} 1 & -1 & 1 \\ -1 & 1 & 0 \\ 1 & 0 & 0 \end{bmatrix}$$

$$= \begin{bmatrix} 0 & 0 & 1 \\ 0 & 1 & 1 \\ 1 & 1 & 0 \end{bmatrix}\begin{bmatrix} 1 & -1 & 0 \\ -1 & 1 & -1 \\ 1 & 0 & 1 \end{bmatrix}$$

$$= \begin{bmatrix} 1 & 0 & 1 \\ 0 & 1 & 0 \\ 0 & 0 & -1 \end{bmatrix} = [T]_{B_2},$$

so matrices $[T]_{B_1}$ and $[T]_{B_2}$, representing the same operator T with respect to two different bases, are similar.

In the next section, we will see how the concepts of determinant, characteristic polynomial, and eigenvalue/eigenvector can be applied to a linear transformation. Similar matrices will play a significant role in that discussion.

Exercises 6.7

A.

1. Let $\mathbf{x} = \begin{bmatrix} 2 \\ 0 \\ 2 \\ 4 \end{bmatrix}$. Find the coordinates of $\mathbf{x}$ with respect to the following ordered bases for $\mathbb{R}^4$.

    (a) $\left\{ \begin{bmatrix} 1 \\ 1 \\ 0 \\ 0 \end{bmatrix}, \begin{bmatrix} 1 \\ -1 \\ 0 \\ 0 \end{bmatrix}, \begin{bmatrix} 0 \\ 0 \\ -1 \\ 1 \end{bmatrix}, \begin{bmatrix} 0 \\ 0 \\ 1 \\ 1 \end{bmatrix} \right\}$

    (b) $\left\{ \begin{bmatrix} 0 \\ 1 \\ 1 \\ 1 \end{bmatrix}, \begin{bmatrix} 1 \\ 0 \\ 1 \\ -1 \end{bmatrix}, \begin{bmatrix} 1 \\ -1 \\ 0 \\ 1 \end{bmatrix}, \begin{bmatrix} 1 \\ 1 \\ -1 \\ 0 \end{bmatrix} \right\}$

2. Let $\mathcal{E}_3$ denote the standard basis for $\mathbb{R}^3$ and let B be the basis
    $$\left\{ \begin{bmatrix} 2 \\ 0 \\ 0 \end{bmatrix}, \begin{bmatrix} -1 \\ 2 \\ 0 \end{bmatrix}, \begin{bmatrix} 1 \\ 1 \\ 1 \end{bmatrix} \right\}.$$
    Find the matrices that represent the basis changes $\mathcal{E}_3 \to B$ and $B \to \mathcal{E}_3$.

3. Find the change-of-basis matrix for changing from $B_1 = \{1+x, x, 1-x^2\}$ to $B_2 = \{x, 1-x, 1+x^2\}$, two bases for $P_2[x]$.

4. For each of the following ordered bases, find the transition matrix from the basis B to the standard basis and find the coordinates of the vectors listed relative to B.

   (a) $B = \left\{ \begin{bmatrix} 1 \\ 2 \end{bmatrix}, \begin{bmatrix} 2 \\ 1 \end{bmatrix} \right\}$ for $\mathbb{R}^2$; vectors: $\begin{bmatrix} 2 \\ 4 \end{bmatrix}, \begin{bmatrix} 4 \\ 2 \end{bmatrix}, \begin{bmatrix} 3 \\ 3 \end{bmatrix}$.

   (b) $B = \left\{ \begin{bmatrix} 1 \\ 1 \end{bmatrix}, \begin{bmatrix} 0 \\ 1 \end{bmatrix} \right\}$ for $\mathbb{R}^2$; vectors: $\begin{bmatrix} 17 \\ 4 \end{bmatrix}, \begin{bmatrix} 3 \\ 1 \end{bmatrix}, \begin{bmatrix} 0 \\ 0 \end{bmatrix}$.

   (c) $B = \left\{ \begin{bmatrix} 1 \\ 1 \\ 1 \end{bmatrix}, \begin{bmatrix} 0 \\ 1 \\ 1 \end{bmatrix}, \begin{bmatrix} 0 \\ 0 \\ 1 \end{bmatrix} \right\}$ for $\mathbb{R}^3$; vectors: $\begin{bmatrix} 3 \\ 1 \\ 1 \end{bmatrix}, \begin{bmatrix} 1 \\ 0 \\ 0 \end{bmatrix}$.

5. Let V denote the solution space of the homogeneous linear differential equation $y''' - 3y'' + y' - 3y = 0$. It can be shown that $B = \{e^{3x}, \cos x, \sin x\}$ is a basis for V.

   (a) Verify that $f_1(x) = 2e^{3x}$, $f_2(x) = \cos x + \sin x$, and $f_3(x) = e^{3x} - \sin x$ are all contained in V.

   (b) Show that $\hat{B} = \{f_1, f_2, f_3\}$ is a basis for V.

   (c) Find the B to $\hat{B}$ change-of-basis matrix and use it to find the coordinates of $g(x) = 5e^{3x} - 2\sin x + 4\cos x$ with respect to the basis $\hat{B}$.

6. Suppose $T : \mathbb{R}^3 \to \mathbb{R}^3$ is the linear transformation with matrix $\begin{bmatrix} 1 & 2 & 1 \\ 1 & 3 & 1 \\ 0 & 1 & 0 \end{bmatrix}$ relative to the standard basis. Find the matrix for T relative to the basis B in each case.

   (a) $B = \left\{ \begin{bmatrix} 1 \\ 2 \\ 0 \end{bmatrix}, \begin{bmatrix} 2 \\ 1 \\ 0 \end{bmatrix}, \begin{bmatrix} 0 \\ 0 \\ 1 \end{bmatrix} \right\}$ (b) $B = \left\{ \begin{bmatrix} 1 \\ 1 \\ 0 \end{bmatrix}, \begin{bmatrix} 0 \\ 1 \\ 1 \end{bmatrix}, \begin{bmatrix} 1 \\ 0 \\ 1 \end{bmatrix} \right\}$

   (c) $B = \left\{ \begin{bmatrix} 1 \\ 1 \\ 1 \end{bmatrix}, \begin{bmatrix} 1 \\ 1 \\ 0 \end{bmatrix}, \begin{bmatrix} 0 \\ 0 \\ 1 \end{bmatrix} \right\}$

7. Consider two ordered bases for $P_2[x]$: $B_1 = \{1, x, x^2\}$ and $B_2 = \{1, 2x, 4x^2 - 2\}$.

   (a) Find the matrix M that describes the change of basis $B_2 \to B_1$.

   (b) Find the matrix that describes the change of basis $B_1 \to B_2$.

   (c) Using the result of part (b), express $p(x) = a + bx + cx^2$ in terms of the basis B_2. [You can check your answer by simply solving for the scalars r, s, t in the equation $a + bx + cx^2 = r(1) + s(2x) + t(4x^2 - 2)$.]

8. If $A = \begin{bmatrix} 2 & 1 \\ 1 & 2 \end{bmatrix}$ is the matrix representation of a linear operator T on $\mathbb{R}^2$ relative to the standard basis, find the matrix of T relative to the basis $\left\{ \begin{bmatrix} 1 \\ 1 \end{bmatrix}, \begin{bmatrix} 1 \\ 2 \end{bmatrix} \right\}$.

9. If $A = \begin{bmatrix} 3 & 2 \\ 1 & 4 \end{bmatrix}$ is the matrix representation of a linear operator T on $\mathbb{R}^2$ relative to the standard basis, find the matrix of T relative to the basis $\left\{ \begin{bmatrix} 2 \\ 1 \end{bmatrix}, \begin{bmatrix} 1 \\ 1 \end{bmatrix} \right\}$.

10. Suppose B_1 and B_2 are the ordered bases for a three-dimensional space V, and the transition matrix from B_1 to B_2 is $M = \begin{bmatrix} 4 & 3 & 2 \\ 3 & 5 & 2 \\ 2 & 2 & 1 \end{bmatrix}$. Let T be the linear operator on V whose matrix representation relative to B_1 is $\begin{bmatrix} -1 & 2 & 0 \\ 1 & -2 & -1 \\ 1 & 3 & 2 \end{bmatrix}$. Find the matrix representing T relative to B_2.

11. If $T: \mathbb{R}^3 \to \mathbb{R}^2$ is a linear transformation with matrix $\begin{bmatrix} 2 & -1 & 3 \\ 1 & 1 & -2 \end{bmatrix}$ relative to the standard bases of both spaces, find the matrix of T relative to the bases

$$B_1 = \left\{ \begin{bmatrix} 1 \\ 1 \\ 1 \end{bmatrix}, \begin{bmatrix} 0 \\ -2 \\ 3 \end{bmatrix}, \begin{bmatrix} 2 \\ 0 \\ 1 \end{bmatrix} \right\} \quad \text{and} \quad B_2 = \left\{ \begin{bmatrix} 2 \\ 1 \end{bmatrix}, \begin{bmatrix} 3 \\ -2 \end{bmatrix} \right\}.$$

12. If the matrix of a linear operator T on a two-dimensional space is $\begin{bmatrix} 2 & 1 \\ 1 & 4 \end{bmatrix}$ relative to a basis $\{\mathbf{v}_1, \mathbf{v}_2\}$, determine the matrix of T relative to the basis $\{\hat{\mathbf{v}}_1, \hat{\mathbf{v}}_2\}$, where
 (a) $\hat{\mathbf{v}}_1 = 3\mathbf{v}_1 + \mathbf{v}_2$, $\hat{\mathbf{v}}_2 = 5\mathbf{v}_1 + 2\mathbf{v}_2$.
 (b) $\hat{\mathbf{v}}_1 = 2\mathbf{v}_1 - 3\mathbf{v}_2$, $\hat{\mathbf{v}}_2 = \mathbf{v}_1 + \mathbf{v}_2$.
 Now show that $T(\mathbf{v}_1 + 2\mathbf{v}_2)$ is independent of the matrix representation of T that is used to describe it.

13. Let $T: P_1[x] \to P_1[x]$ be defined by $T(a + bx) = 2a - bx$. Let $B_1 = \{1, x\}$ and $B_2 = \{-1, 1 + x\}$.
 (a) Find the matrix $[T]_{B_1}$ for T with respect to the basis B_1.
 (b) Find the matrix $[T]_{B_2}$ for T with respect to the basis B_2.
 (c) Find the transition matrix M from B_2 to B_1.
 (d) Verify that $[T]_{B_2} = M^{-1}[T]_{B_1} M$.

14. Let $T: \mathbb{R}^2 \to \mathbb{R}^2$ be the linear transformation defined by $T\left(\begin{bmatrix} x \\ y \end{bmatrix} \right) = \begin{bmatrix} 2x \\ x - y \end{bmatrix}$. Let $\mathcal{E}_2$ be the standard basis for $\mathbb{R}^2$ and let $B = \left\{ \begin{bmatrix} 1 \\ -1 \end{bmatrix}, \begin{bmatrix} 0 \\ 1 \end{bmatrix} \right\}$.
 (a) Find the matrix $[T]_{\mathcal{E}_2}$ for T with respect to the standard basis, $\mathcal{E}_2$.
 (b) Find the matrix $[T]_B$ for T with respect to the basis B.
 (c) Find the transition matrix M from B to $\mathcal{E}_2$.
 (d) Verify that $[T]_B = M^{-1}[T]_{\mathcal{E}_2} M$.

15. In Example 6.7.3, verify the calculations for $M, M^{-1}, [T]_{B_1}$, and $[T]_{B_2}$.

16. Suppose $D : P_2[x] \to P_1[x]$ is the derivative transformation. Let B_1 be the standard basis for $P_2[x]$, $B_2 = \{x^2 + 1, x^2 - 1, x\}$ be another basis for $P_2[x]$, and $C = \{2x + 1, x - 2\}$ be an ordered basis for $P_1[x]$.
   (a) Find the matrix A_D representing D relative to B_1 and C.
   (b) Find the transition matrix M from B_2 to B_1.
   (c) Using the matrices A_D and M, find the matrix $\hat{A}_D$ representing D relative to B_2 and C.

B.

1. Let $B_1 = \{1, t, t^2, \ldots, t^n\}$ and $B_2 = \{1, t-1, (t-1)^2, \ldots, (t-1)^n\}$ be bases for $P_n[t]$. Find the change-of-basis matrix M from B_1 to B_2 and find M^{-1}.

2. If A is a fixed 2×2 matrix, consider the operator T on M_2 defined by $T_A(B) = AB + BA$ for all $B \in M_2$. Letting $A = \begin{bmatrix} a & b \\ c & d \end{bmatrix}$ and taking $S = \left\{ \begin{bmatrix} 1 & 0 \\ 0 & 0 \end{bmatrix}, \begin{bmatrix} 0 & 0 \\ 1 & 0 \end{bmatrix}, \begin{bmatrix} 0 & 1 \\ 0 & 0 \end{bmatrix}, \begin{bmatrix} 0 & 0 \\ 0 & 1 \end{bmatrix} \right\}$ as the ordered basis for M_2, find $[T_A]_S$.

3. Let $B_1 = \{\mathbf{v}_1, \mathbf{v}_2, \mathbf{v}_3\}$ and $B_2 = \{\hat{\mathbf{v}}_1, \hat{\mathbf{v}}_2, \hat{\mathbf{v}}_3\}$ be bases for $\mathbb{R}^3$. It is known that $\mathbf{v}_1 = \begin{bmatrix} -3 \\ 5 \\ 2 \end{bmatrix}, \mathbf{v}_2 = \begin{bmatrix} 4 \\ 1 \\ 1 \end{bmatrix}$, and $\hat{\mathbf{v}}_2 = \begin{bmatrix} 4 \\ 0 \\ -7 \end{bmatrix}$, and that the B_1 to B_2 change-of-basis matrix is $M = \begin{bmatrix} -3 & 1 & 5 \\ 1 & -1 & 2 \\ 2 & -1 & -1 \end{bmatrix}$. Use this information to find $\mathbf{v}_3, \hat{\mathbf{v}}_1$, and $\hat{\mathbf{v}}_3$.

4. Prove that if A and B are similar $n \times n$ matrices, then $\text{trace}(A) = \text{trace}(B)$.

5. Show that all matrices of the form $\begin{bmatrix} \cos\theta & \sin\theta \\ \sin\theta & -\cos\theta \end{bmatrix}$ are similar.

6. Consider the bases $B = \left\{ \begin{bmatrix} 1 \\ 1 \\ 0 \end{bmatrix}, \begin{bmatrix} 1 \\ 0 \\ 1 \end{bmatrix}, \begin{bmatrix} 0 \\ 1 \\ 1 \end{bmatrix} \right\}$ and $\hat{B} = \left\{ \begin{bmatrix} 1 \\ -1 \\ 0 \end{bmatrix}, \begin{bmatrix} 1 \\ 1 \\ 1 \end{bmatrix}, \begin{bmatrix} 1 \\ 1 \\ -2 \end{bmatrix} \right\}$ of $\mathbb{R}^3$.
   (a) Find the transition matrix M from B to $\hat{B}$.

(b) Find a formula for the linear transformation T that takes the vectors of B onto the vectors of $\hat{B}$ in the same order. Also find a formula for T^{-1}.

(c) Find the matrix of T with respect to B, as well as the matrix of T with respect to $\hat{B}$.

(d) Find the matrices of T^{-1} with respect to B and $\hat{B}$. How are these matrices related to the matrices found in part (c)?

7. Let $P = QR$, where all the matrices are $n \times n$. Suppose that P is nonsingular.

(a) Show that the columns of P and Q are bases for $\mathbb{R}^n$.

(b) Find the change-of-basis matrix from the Q columns to the P columns.

8. Let B and $\hat{B}$ be the bases of a vector space V, and let P be the transition matrix from B to $\hat{B}$. Let π be the linear transformation taking the vectors of B to the vectors of $\hat{B}$ in the same order.

(a) Show that P is the matrix representation of π with respect to B, as well as the matrix representation of π with respect to $\hat{B}$.

(b) Show that P^{-1} is the matrix representation of π^{-1} with respect to B, as well as the matrix representation of π^{-1} with respect to $\hat{B}$.

C.

1. Let $\{v_i\}_{i=1}^n$ and $\{\hat{v}_i\}_{i=1}^n$ be the bases for a vector space V. Show that the change-of-basis matrix (in either direction) is upper triangular if and only if $span(\{v_1, v_2, \ldots, v_k\}) = span(\{\hat{v}_1, \hat{v}_2, \ldots, \hat{v}_k\})$ for $k = 1, 2, \ldots, n$.

2. Let V be an n-dimensional vector space. A linear operator T on V is said to be *nilpotent* if $T^p = O_V$ for some positive integer p. The smallest such integer p is called the *index of nilpotency* of T.

(a) Now suppose that T is nilpotent of index p. If $v \in V$ is such that $T^{p-1}(v) \neq 0_V$, prove that $\{v, T(v), T^2(v), \ldots, T^{p-1}(v)\}$ is linearly independent.

(b) Show that T is nilpotent of index n if and only if there is an ordered basis $B = \{v_1, v_2, \ldots, v_n\}$ of V, such that the matrix of T relative to B is of the form

$$\begin{bmatrix} 0 & 0 & 0 & \cdots & 0 & 0 \\ 1 & 0 & 0 & \cdots & 0 & 0 \\ 0 & 1 & 0 & \cdots & 0 & 0 \\ 0 & 0 & 1 & \cdots & 0 & 0 \\ \vdots & \vdots & \vdots & & \vdots & \vdots \\ 0 & 0 & 0 & \cdots & 1 & 0 \end{bmatrix}.$$

(c) Show that an $n \times n$ matrix M is such that $M^n = O_n$ and $M^{n-1} \neq O_n$ if and only if M is similar to a matrix of the form given in part (b).

6.8 Similarity Invariants of Operators

The intimate connection between linear transformations and matrices exposed in this chapter enables us to expand our view of linear algebra considerably. On a finite-dimensional space, matrices and linear transformations are interchangeable—we can use whichever form seems more convenient, more likely to shed light on the mathematical problem we are facing. In this section, we will, for the most part, restrict our attention to operators on a finite-dimensional space—that is, to linear transformations $T: V \to V$, where $\dim(V) = n < \infty$.

For example, all the important concepts developed in Chapter 4 have their counterparts in the theory of linear transformations. By the **determinant** of an operator T on a finite-dimensional vector space V, we mean the determinant of any matrix representation $[T]_B$ of T with respect to some ordered basis B of V. The fact that the determinant of a matrix is a *similarity invariant* (Theorem 4.5.2) means that this definition yields a unique scalar which will be denoted $\det(T)$: If $[T]_{B*}$ is another matrix representation of T, then there exists a nonsingular matrix M such that $[T]_{B*} = M[T]_B M^{-1}$ (Corollary 6.7.1), and so $\det([T]_{B*}) = \det(M[T]_B M^{-1}) = \det(M) \det([T]_B) \det(M^{-1}) = \det(M) \det(M^{-1}) \det([T]_B) = \det(I_n) \det([T]_B) = \det([T]_B)$. **The determinant of an operator has all the properties of the determinant of its representative matrix**. In particular, if S and T are operators on V, then $\det(ST) = \det(TS) = \det(S) \det(T)$.

We say that a nonzero vector $\mathbf{v}$ is an **eigenvector** of the operator T on a vector space V if $T(\mathbf{v}) = \lambda\mathbf{v}$ for some scalar λ. We say that the scalar λ is the **eigenvalue** of T, associated with the eigenvector $\mathbf{v}$. This definition is valid even if the space V is infinite-dimensional (see Example 6.8.1(b)). For a finite-dimensional space, as we saw in Section 4.4, eigenvalues are the roots of the **characteristic equation**, $\det(\lambda I_V - T) = 0$.

Example 6.8.1: Eigenvalues and Eigenvectors
of Operators

(a) Let us determine eigenvalues and eigenvectors for the operator on $\mathbb{R}^3$ defined in Example 6.7.3:
$$T\left(\begin{bmatrix} x \\ y \\ z \end{bmatrix}\right) = \begin{bmatrix} z \\ y \\ x \end{bmatrix}.$$ We want to find scalars λ and nonzero vectors $\mathbf{v} \in \mathbb{R}^3$ such that $T(\mathbf{v}) = \lambda\mathbf{v}$, or
$$T\left(\begin{bmatrix} x \\ y \\ z \end{bmatrix}\right) = \begin{bmatrix} z \\ y \\ x \end{bmatrix} = \lambda \begin{bmatrix} x \\ y \\ z \end{bmatrix}.$$ This equality is equivalent to the equations (1) $z = \lambda x$, (2) $y = \lambda y$, and (3) $x = \lambda z$.

We see that $\lambda = 0$ is not an eigenvalue because this forces $x = y = z = 0$, and the zero vector cannot be an eigenvector. If $\lambda = 1$, then equations (1)–(3) become (1)* $z = x$, (2)* $y = y$, and (3)* $x = z$. This says that $\lambda = 1$ is an eigenvalue and that any vector of the form $\begin{bmatrix} x \\ y \\ x \end{bmatrix}$, where $x \neq 0$ or $y \neq 0$, is an eigenvector associated with $\lambda = 1$. (Recall from Section 4.4 that this set of all eigenvectors corresponding to $\lambda = 1$ is a subspace of $\mathbb{R}^3$ called an *eigenspace*. In this case, because $\left\{ \begin{bmatrix} x \\ y \\ x \end{bmatrix} : x, y \in \mathbb{R}; x \neq 0 \text{ or } y \neq 0 \right\} = \left\{ x \begin{bmatrix} 1 \\ 0 \\ 1 \end{bmatrix} + y \begin{bmatrix} 0 \\ 1 \\ 0 \end{bmatrix} : x, y \in \mathbb{R}; x \neq 0 \text{ or } y \neq 0 \right\}$, we have a two-dimensional subspace, a plane through the origin in $\mathbb{R}^3$.)

If $\lambda \neq 0, 1$, then equation (2) implies that $y = 0$. Furthermore, if $\lambda \neq 0, 1$, it follows from (1) and (3) that $z = \lambda^2 z$ and $\lambda = -1$ (or else $z = 0$, implying that $x = 0$). Thus $\lambda = -1$ is the only other eigenvalue, and all corresponding eigenvectors have the form $\begin{bmatrix} x \\ 0 \\ -x \end{bmatrix} = x \begin{bmatrix} 1 \\ 0 \\ -1 \end{bmatrix}$. In this case, the eigenspace associated with $\lambda = -1$ is a one-dimensional subspace of $\mathbb{R}^3$, a line through the origin.

We note that the eigenvalues of any matrix representing the operator T are the same as the eigenvalues of T (Exercise B1). However, the eigenvectors of an operator may not be the same as those of its matrix representations (see Problem B9 of Exercises 4.5.).

(b) Suppose we let $V = C^\infty(-\infty, \infty)$, an infinite-dimensional space, and $T = \frac{d}{dx}$, the differentiation operator. Then $f(x) = ce^{\lambda x} \in V$ for any real numbers c and λ, and $(T(f))(x) = \lambda ce^{\lambda x} = \lambda(ce^{\lambda x}) = \lambda f(x)$. Therefore, $f(x) = ce^{\lambda x}$ is an eigenvector of the operator T, with eigenvalue λ. Because of the nature of the vector space, such an eigenvector is usually called an *eigenfunction*.

Similarly, if $V = C^\infty(-\infty, \infty)$ and $T_2 = \frac{d^2}{dx^2}$, then

$(T_2(\sin))(x) = \left(\frac{d^2}{dx^2}(\sin) \right)(x) = -\sin x = ((-1)\sin)(x)$ and

$(T_2(\cos))(x) = \left(\frac{d^2}{dx^2}(\cos) \right)(x) = -\cos x = ((-1)\cos)(x)$, so the

sine and cosine are eigenfunctions of T_2 on V, with eigenvalue -1 in each case.

There are other fundamental concepts we discussed in the context of matrix algebra that remain meaningful for operators on a finite-dimensional space. For example, we call an operator, T, **diagonalizable** if any matrix, $[T]_B$, for T is diagonalizable. We can state the operator version of Theorem 4.5.3 as follows: Any operator T on a space V, where $\dim(V) = n$, is diagonalizable if and only if T has n linearly independent eigenvectors, $\mathbf{v}_1, \mathbf{v}_2, \ldots, \mathbf{v}_n$. (Furthermore, if T is diagonalizable, then $[T]_B$ is similar to a diagonal matrix whose elements are the corresponding eigenvalues, $\lambda_1, \lambda_2, \ldots, \lambda_n$, possibly repeated.)

This is where we address a problem we detoured around in Chapter 4. *The characteristic equation of an operator or matrix may have complex roots, so that these complex numbers do not qualify as "scalars," and therefore as eigenvalues, if we are dealing with real vector spaces (spaces whose scalars come from $\mathbb{R}$).* In theory, for finite-dimensional vector spaces, the determination of eigenvalues depends on the **Fundamental Theorem of Algebra**, which states that, for $n \geq 1$, any nth degree polynomial with complex coefficients has n complex zeros (counting multiplicity).* As we have stated before, the practical difficulty is that there is no general solution formula for the roots of polynomial equations of a degree greater than or equal to 5. Calculators and computers use various approximation methods. Furthermore, a complex eigenvalue may lead to a *complex eigenvector*, which is an element of $\mathbb{C}^n = \left\{ [z_1 \quad z_2 \quad \cdots \quad z_n]^T : z_i \in \mathbb{C} \text{ for } 1 \leq i \leq n \right\}$ rather than $\mathbb{R}^n$. This, in turn, changes the nature of eigenspaces associated with a matrix or transformation.

Example 6.8.2: A Matrix with No Real Eigenvalues/Eigenvectors

The characteristic equation of the matrix $A = \begin{bmatrix} 0 & 1 \\ -1 & 0 \end{bmatrix}$ is $\lambda^2 + 1 = 0$, so the eigenvalues of A are i and $-i$, where $i = \sqrt{-1}$. If we consider this as a problem in a real vector space, then A has no eigenvalues. Geometrically, the operator $T_A = A\mathbf{x}$ rotates any vector $\mathbf{x}$ in $\mathbb{R}^2$ by $270°$ ($3\pi/2$ rad) in a counterclockwise direction (see Example 6.1.2(c)) or, equivalently, by $90°$ ($\pi/2$ rad) in a clockwise direction. Therefore, there can be no nonzero vector in $\mathbb{R}^2$ that is transformed into a scalar (real number) multiple of itself by T_A—that is, there is no real eigenvector.

However, for each complex eigenvalue, we can solve the equation $T_A(\mathbf{x}) = A\mathbf{x} = \lambda\mathbf{x}$ for a complex eigenvector: $\begin{bmatrix} 0 & 1 \\ -1 & 0 \end{bmatrix}\begin{bmatrix} x \\ y \end{bmatrix} = i\begin{bmatrix} x \\ y \end{bmatrix}$ is equivalent to the system $\{y = ix,$ $-x = iy\}$, with the solution $\begin{bmatrix} x \\ ix \end{bmatrix} = x\begin{bmatrix} 1 \\ i \end{bmatrix}$. Thus, the representative eigenvector corresponding to the eigenvalue i is

---

* Even the so-called elementary proofs of this important result can be intricate. See, for example, Section 4.4 of E.J. Barbeau, *Polynomials* (New York: Springer-Verlag, 1989).

$\begin{bmatrix} 1 \\ i \end{bmatrix} \in \mathbb{C}^2$. Similarly, the eigenvector corresponding to $-i$ is $\begin{bmatrix} -1 \\ i \end{bmatrix} \in \mathbb{C}^2$.

To enjoy the matrix-transformation duality fully, we will have to invite the field of complex numbers to our party. There are two basic reasons for us to consider expanding the set of scalars to be all of $\mathbb{C}$, the complex number system: (1) We can generalize many of the key results we have already seen and (2) we can solve problems whose solutions are not obtained easily by sticking with real numbers. Jacques Hadamard (1865–1963) has said that "the shortest path between two truths in the real domain passes through the complex domain." Although he may have had a different mathematical area in mind, we can borrow his philosophy. Paradoxically (and punningly), we can simplify our linear algebraic lives by making things more complex. Beginning in the next chapter, we will consider vector spaces whose field of scalars is $\mathbb{C}$.

Exercises 6.8

A.

1. Find all eigenvalues and eigenvectors (real or complex) of the following matrices, each regarded as a matrix with complex entries. (Any real number r is a complex number with its imaginary part equal to 0: $r = r + 0 \cdot i$.)

(a) $\begin{bmatrix} 1 & 1 \\ -1 & 1 \end{bmatrix}$ (b) $\begin{bmatrix} 1 & 3 \\ -2 & 1 \end{bmatrix}$ (c) $\begin{bmatrix} 0 & -1 & 1 \\ 0 & 1 & 0 \\ -1 & 0 & 0 \end{bmatrix}$

(d) $\begin{bmatrix} 2 & -2 & 1 \\ 2 & 1 & -2 \\ 1 & 2 & 2 \end{bmatrix}$

2. (a) Show that $\begin{bmatrix} 1 \\ i \end{bmatrix}$ and $\begin{bmatrix} i \\ 1 \end{bmatrix}$ are eigenvectors of
$A = \begin{bmatrix} \cos\theta & \sin\theta \\ -\sin\theta & \cos\theta \end{bmatrix}$.

(b) If $P = \begin{bmatrix} 1 & i \\ i & 1 \end{bmatrix}$, compute the product $P^{-1}AP$.

3. Calculate the determinant of the operator T in Example 6.8.1.

B.

1. Prove that the eigenvalues of an operator T on an n-dimensional space V are the same as the eigenvalues of any matrix representing the operator on $\mathbb{R}^n$.

2. Find an inequality that must be satisfied by the real numbers a, b, c, and d if the matrix $\begin{bmatrix} a & b \\ c & d \end{bmatrix}$ is to have only real eigenvalues.

3. Suppose T is an invertible operator on a vector space V.
   (a) Prove that all eigenvalues of T are nonzero.
   (b) Show that λ is an eigenvalue of T if and only if λ^{-1} is an eigenvalue of T^{-1}.

4. Show that the operator on $P_n[x]$ defined by $T(f) = xf'$, where f' is the derivative of f, is diagonalizable.

C.

1. Let $\mathcal{E}_4 = \{\mathbf{e}_i\}_{i=1}^{4}$ be the standard basis for $\mathbb{R}^4$, and let T be the operator on $\mathbb{R}^4$ defined by $T(\mathbf{e}_i) = \mathbf{e}_{5-i}$ for $i = 1, 2, 3, 4$.
   (a) Find the matrix representation of T with respect to $\mathcal{E}_4$.
   (b) Find the eigenvalues of T. [*Hint*: There are only two distinct eigenvalues, each of multiplicity 2.]
   (c) Show that $\mathbf{e}_1 + \mathbf{e}_4, \mathbf{e}_2 + \mathbf{e}_3$, and $\mathbf{e}_3 + \mathbf{e}_2$ are linearly independent eigenvectors corresponding to the positive eigenvalue found in part (b).
   (d) Show that $\mathbf{e}_1 - \mathbf{e}_4, \mathbf{e}_2 - \mathbf{e}_3$, and $\mathbf{e}_3 - \mathbf{e}_2$ are linearly independent eigenvectors corresponding to the negative eigenvalue found in part (b).
   (e) Show that T is diagonalizable.

6.9 Summary

If V and W are vector spaces, a **transformation** (**mapping** or **function**) T from V into W, denoted by $T : V \to W$, is a rule that assigns to each vector $\mathbf{v} \in V$ one and only one element $T(\mathbf{v}) \in W$, called the **image of v under T**. The vector spaces V and W are called the **domain** and the **codomain** of T, respectively. The set of all images of vectors of V under the transformation T is called the **range** of T (or the **image** of T), written range T. A transformation T from a vector space V to a vector space W is called a **linear transformation** if $T(a\mathbf{v}_1 + b\mathbf{v}_2) = aT(\mathbf{v}_1) + bT(\mathbf{v}_2)$ for all $\mathbf{v}_1, \mathbf{v}_2 \in V$ and $a, b \in \mathbb{R}$. In general, if T is a linear transformation, $T(c_1\mathbf{v}_1 + c_2\mathbf{v}_2 + \cdots + c_k\mathbf{v}_k) = c_1 T(\mathbf{v}_1) + c_2 T(\mathbf{v}_2) + \cdots + c_k T(\mathbf{v}_k)$ for $\mathbf{v}_1, \mathbf{v}_2, \ldots, \mathbf{v}_k \in V$ and any scalars $c_1, c_2, \ldots, c_k$. **If $T : V \to W$ is linear, then T is completely determined by its effect on any basis of V.**

The **null space** (or **kernel**) of a linear transformation $T, T : V \to W$, denoted by null space T, is the set of all vectors $\mathbf{v}$ in V such that $T(\mathbf{v}) = \mathbf{0}_W$. The **range** of T (also called the **image** of T), denoted by range T, is the set $T(V) = \{T(\mathbf{v}) | \mathbf{v} \in V\}$, a subset of W. The set null space T is a subspace of V, whereas range T is a subspace of W. The **rank** of a linear transformation T, denoted by rank T, is the dimension of its range, whereas the **nullity** of a linear transformation T, denoted by nullity T, is the dimension of its null space. **If T is a linear transformation from V to W, where $\dim(V) = n$, then rank T + nullity $T = n$.** Also, if A is an $m \times n$ matrix and $T_A : \mathbb{R}^n \to \mathbb{R}^m$ is the linear transformation defined by $T_A(\mathbf{v}) = A\mathbf{v}$ for every $\mathbf{v} \in \mathbb{R}^n$, then rank $T_A = \text{rank}(A)$.

Given linear transformations T and S from V to W, we can define the **sum**, **difference**, and the **scalar multiple** transformations in an obvious way. With respect to the operations just mentioned, **the set $L(V, W)$ of all linear transformations from a vector space V to a vector space W is itself a vector space**.

Given vector spaces U, V, W and linear transformations T from U to V and S from V to W, we can define the **composite** (or **product**) **transformation** ST as follows: $(ST)(\mathbf{u}) = S(T(\mathbf{u}))$ for $\mathbf{u} \in U$. We can extend this definition to the composition of finitely many linear transformations; also, if $T \in L(V) = L(V, V)$, we can define **powers** of T recursively. Theorem 6.3.1 provides some useful algebraic properties of linear transformations.

If B_1 and B_2 are two ordered bases of a finite-dimensional vector space V, and M is defined to be the $n \times n$ matrix whose ith column is the coordinate vector of the ith vector of B_1 with respect to the basis B_2, then M is invertible and $[\mathbf{v}]_{B_2} = M[\mathbf{v}]_{B_1}$ and $[\mathbf{v}]_{B_1} = M^{-1}[\mathbf{v}]_{B_2}$ for every $\mathbf{v} \in V$. In this case, M is called the **transition matrix (change-of-basis matrix** or **change-of-coordinates matrix)** from B_1 to B_2.

Now let V and W be nontrivial, finite-dimensional vector spaces and let $B_V, \hat{B}_V$ be ordered bases of V and $B_W, \hat{B}_W$ be ordered bases of W. Furthermore, suppose that (invertible) matrices M_1 and M_2 describe the change of bases $B_V \to \hat{B}_V$ and $B_W \to \hat{B}_W$, respectively. If the linear transformation $T : V \to W$ is represented by the matrix A_T with respect to B_V and B_W and by the matrix $\hat{A}_T$ with respect to $\hat{B}_V$ and $\hat{B}_W$, then $\hat{A}_T = M_2 A_T M_1^{-1}$. Matrices A and B are said to be **equivalent** if $B = PAQ$ for suitable nonsingular matrices P and Q. Thus, two matrix representations of the same linear transformation, $T : V \to W$, with respect to different ordered bases, $B_V, \hat{B}_V$ and $B_W, \hat{B}_W$, for V and W, respectively, are equivalent. Let B_1 and B_2 be two ordered bases of a finite-dimensional vector space V, and let T be a linear operator on V. If T is represented by matrices A_T and $\hat{A}_T$ with respect to B_1 and B_2, respectively, then $\hat{A}_T = MA_TM^{-1}$, where M is the matrix representing the change of basis $B_1 \to B_2$. Saying that **two $n \times n$ matrices A and B are similar means that A and B represent the same linear transformation on an n-dimensional vector space V, with respect to two possibly different ordered bases**.

If we have linear transformations $S, T : V \to W$ and $R : W \to X$, then the fundamental connection between the algebra of linear transformations and the algebra of matrices can be expressed as (1)

$[(T \pm S)(\mathbf{v})]_{B_W} = \left([T]^{B_W}_{B_V} \pm [S]^{B_W}_{B_V} \right) [\mathbf{v}]_{B_V}$, (2) $[(cT)(\mathbf{v})]_{B_W} = \left(c[T]^{B_W}_{B_V} \right) [\mathbf{v}]_{B_V}$, and (3) $[(RT)(\mathbf{u})]_{B_X} = \left([R]^{B_X}_{B_W} [T]^{B_W}_{B_V} \right) [\mathbf{u}]_{B_V}$.

A linear transformation T from V to W is said to be **invertible** if there exists a mapping $\hat{T}$ from W to V such that $\hat{T}T = I_V$ and $T\hat{T} = I_W$, where I_V is the identity transformation on V and I_W is the identity transformation on W. Otherwise, the linear transformation is called **noninvertible**. Such a mapping $\hat{T}$ is called an **inverse** of T. If it exists, the inverse map $\hat{T}$ must be linear and unique, and we use the symbol T^{-1}. **A linear transformation is invertible if and only if it is both one-to-one and onto**. If T is nonsingular, then (1) T^{-1} is nonsingular; (2) $(T^{-1})^{-1} = T$; (3) $(cT)^{-1} = c^{-1}T^{-1}$ if $c \neq 0$. Furthermore, if T is a linear transformation from V to W and S is a linear transformation from W to Y, then (1) ST is nonsingular if and only if both T and S are nonsingular; (2) if ST is nonsingular, then $(ST)^{-1} = T^{-1}S^{-1}$. We can express the relationship between an invertible transformation and its matrix representation as $[T^{-1}(\mathbf{w})]_{B_V} = \left([T]_{B_W} \right)^{-1} [\mathbf{w}]_{B_W}$.

An invertible linear transformation, $T : V \to W$, is called an **isomorphism**. In this case, we say that V and W are **isomorphic. Two finite-dimensional vector spaces V and W with the same set of scalars are isomorphic if and only if $\dim(V) = \dim(W)$**. In particular, **any n-dimensional vector space V (with scalars $\mathbb{R}$) is isomorphic to the Euclidean space $\mathbb{R}^n$.**

On a finite-dimensional space, matrices and linear transformations are interchangeable. Familiar concepts from matrix theory can be extended to linear transformations. For example, the **determinant** of an operator, T, on a finite-dimensional vector space V is the determinant of any matrix representation, $[T]_B$, of T with respect to some ordered basis B of V. The fact that the determinant of a matrix is a *similarity invariant* (Theorem 4.5.2) means that this definition yields a unique scalar, which will be denoted $\det(T)$. If S and T are operators on V, then $\det(ST) = \det(TS) = \det(S) \det(T)$. Similarly, we say that a nonzero vector $\mathbf{v}$ is an **eigenvector** of the operator T on a vector space V if $T(\mathbf{v}) = \lambda \mathbf{v}$ for some scalar λ, called the **eigenvalue** of T associated with the eigenvector $\mathbf{v}$. For operators on a finite-dimensional space, eigenvalues are the roots of the **characteristic equation**, $\det(\lambda I_V - T) = 0$. An operator T is **diagonalizable** if any matrix, A_T, for T is diagonalizable. There is an operator version of Theorem 4.5.3: Any operator T on a space V, where $\dim(V) = n$, is diagonalizable if and only if T has n linearly independent eigenvectors, $\mathbf{v}_1, \mathbf{v}_2, \ldots, \mathbf{v}_n$. Furthermore, if T is diagonalizable, then $[T]_B$ is similar to a diagonal matrix whose elements are the corresponding eigenvalues $\lambda_1, \lambda_2, \ldots, \lambda_n$, possibly repeated.

The characteristic equation of a matrix or an operator may have complex roots, so the eigenvalues may be complex, possibly leading to eigenvectors with complex components. Thus, we broaden our scope and consider the possibility that the "scalars" in the definition of a vector space V could refer to the complex number system, $\mathbb{R}$.

7

Inner Product Spaces

In Chapter 5, we saw the extension and abstraction of many concepts introduced in Chapter 1, in particular the notions of *linear independence/ dependence*, *basis*, *dimension*, and *subspace*. In Chapters 1 through 5, there were occasional references in the text and in the exercises to the *dot product*, *length*, *angle*, and *orthogonality* applied to vectors in $\mathbb{R}^n$. In this chapter, we return for a deeper, wider look at questions of length, distance, and orthogonality.

7.1 Complex Vector Spaces

From now on, we will assume, unless stated otherwise, that the "scalars" in the definition of a vector space V refer to the complex number system $\mathbb{C}$. In this case, we refer to V as **a vector space over** $\mathbb{C}$, or **a vector space over the complex numbers**. Most of the time, we will describe this algebraic structure simply as a **complex vector space**. In a complex vector space, we define the fundamental concepts of *linear combination, independence/dependence*, *basis*, *dimension*, and *subspace* in the usual ways, remembering that multiplication of vectors by complex numbers is allowed. Considering complex numbers as scalars changes nothing in the theory of systems of linear equations and in the basic algebra of matrices/linear transformations. As Example 6.8.1 indicated, working with complex numbers gives us more room in which to solve problems. One thing to remember as we go forward is that the real numbers constitute a subset of the complex numbers—that is, **any real number can be regarded as a complex number with imaginary part equal to 0**: $r = r + 0 \cdot i$, where $r \in \mathbb{R}$ and $i = \sqrt{-1}$. (See Appendix D for some background in working with complex numbers.) **Any result we prove about complex vector spaces is true for vector spaces over** $\mathbb{R}$.

Example 7.1.1: Some Complex Vector Spaces

(a) The set of complex numbers $\mathbb{C}$ is itself a complex vector space if we interpret vector addition and scalar multiplication as the usual addition and multiplication of complex numbers. One basis for this space is $\{1\}$: Any vector $a + bi$ can be written as $(a + bi)1$, where $a + bi$ is taken as the scalar and 1 is regarded as the basis vector. Similarly, $\{i\}$ is a basis because $a + bi = (b - ai)i$ for any $a + bi \in \mathbb{C}$. Thus, we have dim $(\mathbb{C}) = 1$.

Note that $\mathbb{R} \subset \mathbb{C}$ and in fact $\mathbb{R}$ is a *subspace* of $\mathbb{C}$ if $\mathbb{C}$ is considered a *real* vector space (a vector space over $\mathbb{R}$). A basis for $\mathbb{C}$ in this situation is $\{1, i\}$, so dim $(\mathbb{C}) = 2$. In the usual two-dimensional representation of the complex plane (see Appendix D), $\mathbb{R}$ is the "real axis," a one-dimensional subspace of $\mathbb{C}$ over $\mathbb{R}$.

(b) We can define **n-dimensional complex coordinate space:**

$$\mathbb{C}^n = \left\{ \begin{bmatrix} c_1 \\ c_2 \\ \vdots \\ c_n \end{bmatrix} : c_i \in \mathbb{C} \quad \text{for all } i \right\}$$

$$= \left\{ \begin{bmatrix} a_1 \\ a_2 \\ \vdots \\ a_n \end{bmatrix} + i \begin{bmatrix} b_1 \\ b_2 \\ \vdots \\ b_n \end{bmatrix} : a_i, b_i \in \mathbb{R} \quad \text{for all } i \right\},$$

in which vector addition and multiplication of a vector by complex numbers are defined in the usual component-by-component fashion. The standard basis $\mathcal{E}_n = \{\mathbf{e}_i\}_{i=1}^n$ for $\mathbb{R}^n$ will be considered the standard basis for $\mathbb{C}^n$:

$$\begin{bmatrix} c_1 \\ c_2 \\ \vdots \\ c_n \end{bmatrix} = \sum_{i=1}^n c_i \mathbf{e}_i.$$ Therefore, dim $(\mathbb{C}^n) = n$. (In Example

6.6.1(a), we saw that dim $(\mathbb{C}) = 2$ and dim $(\mathbb{C}^n) = 2n$ if we consider $\mathbb{C}$ and $\mathbb{C}^n$ as *real* vector spaces.)

In $\mathbb{C}^3$, for instance, we have

$$\begin{bmatrix} i \\ 5 \\ 2 - 3i \end{bmatrix} = i\mathbf{e}_1 + 5\mathbf{e}_2 + (2 - 3i)\mathbf{e}_3.$$

(c) The set $M_{mn}(\mathbb{C})$ of all $m \times n$ matrices with complex entries is a complex vector space. The standard basis for the real vector space M_{mn} serves as the standard basis for $M_{mn}(\mathbb{C})$:

$$\begin{bmatrix} c_{11} & c_{12} & \cdots & c_{1n} \\ c_{21} & c_{22} & \cdots & c_{2n} \\ \vdots & \vdots & \cdots & \vdots \\ c_{m1} & c_{m2} & \cdots & c_{mn} \end{bmatrix} = \sum_{i=1}^m \sum_{j=1}^n c_{ij} E_{ij},$$

where E_{ij} is the matrix with 1 in row i, column j and 0's elsewhere. This says that dim$(M_{mn}(\mathbb{C})) = mn$.

In $M_{23}(\mathbb{C})$, for example, we can write

$$\begin{bmatrix} 3i & -2 & 0 \\ -i & 3+5i & 7 \end{bmatrix} = 3i \begin{bmatrix} 1 & 0 & 0 \\ 0 & 0 & 0 \end{bmatrix} - 2 \begin{bmatrix} 0 & 1 & 0 \\ 0 & 0 & 0 \end{bmatrix} - i \begin{bmatrix} 0 & 0 & 0 \\ 1 & 0 & 0 \end{bmatrix}$$
$$+ (3+5i) \begin{bmatrix} 0 & 0 & 0 \\ 0 & 1 & 0 \end{bmatrix} + 7 \begin{bmatrix} 0 & 0 & 0 \\ 0 & 0 & 1 \end{bmatrix}.$$

Because we will be working with complex matrices in the remaining chapters, we need to establish basic notation and some fundamental properties. Most of the properties are obvious extensions of properties of complex numbers (see Appendix D).

First of all, because each column of an $m \times n$ complex matrix C is a vector in $\mathbb{C}^m$ and can be split as in Example 7.1.1(b), we can decompose C into real and imaginary parts. If $C = [c_{ij}]$, where $c_{ij} = a_{ij} + b_{ij}i$ and $a_{ij}, b_{ij} \in \mathbb{R}$, then

$$\begin{bmatrix} c_{11} & c_{12} & \cdots & c_{1n} \\ c_{21} & c_{22} & \cdots & c_{2n} \\ \vdots & \vdots & \cdots & \vdots \\ c_{m1} & c_{m2} & \cdots & c_{mn} \end{bmatrix} = \begin{bmatrix} a_{11} & a_{12} & \cdots & a_{1n} \\ a_{21} & a_{22} & \cdots & a_{2n} \\ \vdots & \vdots & \cdots & \vdots \\ a_{m1} & a_{m2} & \cdots & a_{mn} \end{bmatrix} + i \begin{bmatrix} b_{11} & b_{12} & \cdots & b_{1n} \\ b_{21} & b_{22} & \cdots & b_{2n} \\ \vdots & \vdots & \cdots & \vdots \\ b_{m1} & b_{m2} & \cdots & b_{mn} \end{bmatrix}$$
$$= A + iB,$$

where A and B are real matrices.

The (**complex**) **conjugate** of a complex matrix $C = [c_{ij}]$ is denoted by $\bar{C}$ and is defined as $\bar{C} = [\bar{c}_{ij}] = \begin{bmatrix} \bar{c}_{11} & \bar{c}_{12} & \cdots & \bar{c}_{1n} \\ \bar{c}_{21} & \bar{c}_{22} & \cdots & \bar{c}_{2n} \\ \vdots & \vdots & \cdots & \vdots \\ \bar{c}_{m1} & \bar{c}_{m2} & \cdots & \bar{c}_{mn} \end{bmatrix}$. If a complex matrix C is written as $A + iB$, where A and B are real matrices, then $\bar{C} = A - iB$.

Theorem 7.1.1: *Properties of the Conjugate*

If $C, C_1,$ and C_2 are conformable complex matrices and α is a complex number, then

(1) $\overline{(\bar{C})} = C.$
(2) $\overline{(\alpha C)} = \bar{\alpha}\bar{C}.$
(3) $\overline{(C_1 + C_2)} = \bar{C}_1 + \bar{C}_2.$
(4) $\overline{(C_1 C_2)} = \bar{C}_1 \bar{C}_2.$

Proof Exercise A6.

Example 7.1.2: The Conjugate of a Complex Matrix

Let $A = \begin{bmatrix} i & 2-i \\ 3 & -i \end{bmatrix}$ and $B = \begin{bmatrix} 0 & 2i \\ 1+i & 2 \end{bmatrix}$. Then $\bar{A} = \begin{bmatrix} -i & 2+i \\ 3 & i \end{bmatrix}$,

$\bar{B} = \begin{bmatrix} 0 & -2i \\ 1-i & 2 \end{bmatrix}$,   and $\overline{(A+B)} = \begin{bmatrix} i & 2+i \\ 4+i & 2-i \end{bmatrix} =$

$\begin{bmatrix} -i & 2-i \\ 4-i & 2+i \end{bmatrix} = \begin{bmatrix} -i & 2+i \\ 3 & i \end{bmatrix} + \begin{bmatrix} 0 & -2i \\ 1-i & 2 \end{bmatrix} = \bar{A} + \bar{B}$.

Also, $\overline{(iAB)} = -i\bar{A}\bar{B} = -i\begin{bmatrix} -i & 2+i \\ 3 & i \end{bmatrix}\begin{bmatrix} 0 & -2i \\ 1-i & 2 \end{bmatrix} =$

$-i\begin{bmatrix} 3-i & 2+2i \\ 1+i & -4i \end{bmatrix} = \begin{bmatrix} -1-3i & 2-2i \\ 1-i & -4 \end{bmatrix}$.

An important transformation of a complex matrix C is the **conjugate transpose** $C \to C^*$ defined by $C^* = (\bar{C})^{\mathrm{T}}$. The properties of the conjugate transpose will be particularly useful in this chapter and in Chapter 8.

Theorem 7.1.2: Properties of the Conjugate Transpose

If C, C_1, and C_2 are conformable complex matrices and α is a complex number, then

(1) $C^* = \overline{(C^{\mathrm{T}})}$.
(2) $(C^*)^* = C$.
(3) $(\alpha C)^* = \bar{\alpha}C^*$.
(4) $(C_1 + C_2)^* = C_1^* + C_2^*$.
(5) $(C_1 C_2)^* = C_2^* C_1^*$.

Proof Exercises B8 through B12.

Example 7.1.3: The Conjugate Transpose
of a Complex Matrix

Let $A = \begin{bmatrix} 2-i & 3 & 2+2i \\ i & 4i & 1 \end{bmatrix}$ and $B = \begin{bmatrix} 0 & i \\ -i & 2i \\ 1 & 1+i \end{bmatrix}$. Then,

$A^* = \begin{bmatrix} 2+i & -i \\ 3 & -4i \\ 2-2i & 1 \end{bmatrix}$, $B^* = \begin{bmatrix} 0 & i & 1 \\ -i & -2i & 1-i \end{bmatrix}$,   and

$$(AB)^* = \left(\begin{bmatrix} 2-i & 1+12i \\ 5 & -8+i \end{bmatrix}\right)^* = \left(\begin{bmatrix} 2+i & 1-12i \\ 5 & -8-i \end{bmatrix}\right)^{\mathsf{T}} =$$

$$\begin{bmatrix} 2+i & 5 \\ 1-12i & -8-i \end{bmatrix} = \begin{bmatrix} 0 & i & 1 \\ -i & -2i & 1-i \end{bmatrix} \begin{bmatrix} 2+i & -i \\ 3 & -4i \\ 2-2i & 1 \end{bmatrix} = B^*A^*.$$

Also, $\overline{(A^{\mathsf{T}})} = \overline{\left(\begin{bmatrix} 2-i & i \\ 3 & 4i \\ 2+2i & 1 \end{bmatrix}\right)} = \begin{bmatrix} 2+i & -i \\ 3 & -4i \\ 2-2i & 1 \end{bmatrix} = A^*.$

Exercises 7.1

A.

1. (a) Prove that the set $\left\{ \begin{bmatrix} 3-i \\ 2+2i \\ 4 \end{bmatrix}, \begin{bmatrix} 2 \\ 2+4i \\ 3 \end{bmatrix}, \begin{bmatrix} 1-i \\ -2i \\ -1 \end{bmatrix} \right\}$ is a basis for the complex vector space $\mathbb{C}^3$.

   (b) Express each of the vectors $\begin{bmatrix} 1 \\ 0 \\ 0 \end{bmatrix}, \begin{bmatrix} 0 \\ 1 \\ 0 \end{bmatrix}, \begin{bmatrix} 0 \\ 0 \\ 1 \end{bmatrix}$ as a linear combination of the basis vectors in part (a).

2. (a) Prove that $\{1, i\}$ is a basis for $\mathbb{C}$, regarded as a real vector space.
   (b) Show that $T: \mathbb{C} \to \mathbb{C}$ defined by $T(z) = \bar{z}$ is linear and find the matrix of T with respect to the basis $\{1, i\}$.
   (c) Show that T is not linear when $\mathbb{C}$ is regarded as a complex vector space.

3. Show that $\{z\}$, where z is any nonzero complex number, is a basis for $\mathbb{C}$ over $\mathbb{C}$.

4. Solve the system:
$$(1-i)x + y = 0$$
$$-2x - (1+i)y = 0$$

5. Solve the system:
$$ix + y + (2+2i)z = 1$$
$$y + (1-2i)z = -1+i$$
$$ix + 2y + 3z = i$$

6. Prove Theorem 7.1.1.

7. Prove Theorem 7.1.2.

8. Let $A = \begin{bmatrix} 2+i & 0 & 3i \\ 1 & -i & 1-i \end{bmatrix}$ and $B = \begin{bmatrix} -i & 2 & 1 \\ i & 1+i & -1 \end{bmatrix}$. Find

(a) A^* (b) B^* (c) $(A+B)^*$.

9. Find the eigenvalues and eigenvectors of

$$A = \begin{bmatrix} -2 & -2 & -9 \\ -1 & 1 & -3 \\ 1 & 1 & 4 \end{bmatrix}.$$

10. Find the eigenvalues and eigenvectors of

$$A = \begin{bmatrix} 3 & 3 & 6 & 9 \\ 1 & 4 & 3 & 7 \\ 2 & -5 & 8 & 3 \\ 2 & -9 & 7 & 4 \end{bmatrix}.$$

11. Suppose that $A = \begin{bmatrix} i \\ 2+2i \end{bmatrix}$ and $B = [\, 1-i \quad 2i\,]$. Verify that $(AB)^* = B^*A^*$.

12. Let $A = \begin{bmatrix} 2 & 1-i \\ 1+i & 3 \end{bmatrix}$ and $B = \begin{bmatrix} -1 & 2+i \\ 2-i & 1 \end{bmatrix}$. A square matrix M is called *Hermitian* if $M^* = M$.

(a) Show that $A^* = A$ and $B^* = B$.
(b) Compute AB and $(AB)^*$.
(c) Is the product of Hermitian matrices necessarily Hermitian?

13. Show that the following matrices are Hermitian (see the previous exercise) and find the eigenvalues of each matrix:

(a) $\begin{bmatrix} 1 & -i \\ i & 1 \end{bmatrix}$ (b) $\begin{bmatrix} 1 & i & 0 \\ -i & -1 & 1 \\ 0 & 1 & 1 \end{bmatrix}$

14. Find V^{-1} if

$$V = \begin{bmatrix} 1 & 1 & 1 & 1 \\ 1 & -i & -1 & -i \\ 1 & -1 & 1 & -1 \\ 1 & -i & -1 & i \end{bmatrix}.$$

15. Define $T : \mathbb{C} \to \mathbb{C}$ by $T(z) = \bar{z}$, where $\mathbb{C}$ is regarded as a *real* vector space. What is T^2? What is T^{-1}?

B.

1. If $\theta \in \mathbb{R}$, prove that the following complex matrices are similar:

$$\begin{bmatrix} \cos\theta & -\sin\theta \\ \sin\theta & \cos\theta \end{bmatrix}, \quad \begin{bmatrix} e^{i\theta} & 0 \\ 0 & e^{-i\theta} \end{bmatrix}.$$

(Recall that $e^{i\theta} = \cos\theta + i\sin\theta$.)

2. Prove that if λ is a (possibly complex) eigenvalue of an $n \times n$ complex matrix A with corresponding eigenvector $\mathbf{v}$, then $\bar{\lambda}$ is an eigenvalue of $\bar{A}$ with eigenvector $\bar{\mathbf{v}}$.

3. Find three eigenvalues and corresponding eigenvectors for the complex matrix:

$$\begin{bmatrix} 1 & -\sqrt{6} & -i\sqrt{2} \\ \sqrt{6} & 0 & i\sqrt{3} \\ i\sqrt{2} & i\sqrt{3} & 2 \end{bmatrix}.$$

Now find a matrix U such that U^*AU is diagonal. (See Section 4.5.)

4. Prove that $(\bar{A})^* = A^{\mathsf{T}}$.

5. If A is an $n \times n$ complex matrix, show that $\text{trace}(\bar{A}) = \overline{\text{trace}(A)}$.

6. If A is an $n \times n$ complex matrix, show that $\det(\bar{A}) = \overline{\det(A)}$.

7. If A is nonsingular, show that $(\bar{A})^{-1} = \overline{(A^{-1})}$.

8. Prove that A is a real matrix if and only if $A^* = A^{\mathsf{T}}$.

9. Prove that $\text{rank}(A^*) = \text{rank}(A)$.

10. If A, B, C, and D are complex matrices of size $m \times n$, $m \times p$, $q \times n$, and $q \times p$, respectively, show that

$$\overline{\begin{bmatrix} A & B \\ C & D \end{bmatrix}} = \begin{bmatrix} \bar{A} & \bar{B} \\ \bar{C} & \bar{D} \end{bmatrix} \quad \text{and} \quad \begin{bmatrix} A & B \\ C & D \end{bmatrix}^* = \begin{bmatrix} A^* & C^* \\ B^* & D^* \end{bmatrix}.$$

11. The *adjoint* of a linear operator $T: \mathbb{C}^n \to \mathbb{C}^n$ is the linear operator $T^*: \mathbb{C}^n \to \mathbb{C}^n$ whose matrix with respect to the standard basis for $\mathbb{C}^n$ is the conjugate transpose of the matrix of T—that is, $[T^*]_{\mathcal{E}_n} = \left([T]_{\mathcal{E}_n}\right)^*$. Find the adjoint of $T: \mathbb{C}^2 \to \mathbb{C}^2$ defined by $T\left(\begin{bmatrix} z_1 \\ z_2 \end{bmatrix}\right) = \begin{bmatrix} z_1 \\ iz_2 \end{bmatrix}$.

12. A nonsingular complex matrix A is called *unitary* if $A^* = A^{-1}$. Verify that

$$U = \begin{bmatrix} -i/\sqrt{2} & 0 & i/\sqrt{2} \\ 0 & i & 0 \\ i/\sqrt{2} & 0 & i/\sqrt{2} \end{bmatrix}$$

is a unitary matrix.

13. An $n \times n$ real matrix is called *orthogonal* if $A^{\mathsf{T}} = A^{-1}$. If A is an orthogonal matrix, show that $U = \left((1+i)/\sqrt{2}\right)A$ is a unitary matrix (see the previous exercise).

14. Show that the diagonal entries of a Hermitian matrix (see Exercise A12) are real numbers.

15. Show that the transpose of a Hermitian matrix is Hermitian (see Exercise A12). Is the conjugate transpose of a Hermitian matrix Hermitian?

C.

1. Consider the mapping of $\mathbb{C}$ into itself given by $Az = az + b\bar{z}$, where $a, b \in \mathbb{C}$.
   Prove that this mapping is invertible if and only if $|a| \neq |b|$.

2. Let C be an $n \times n$ complex matrix.
   (a) Prove that C can be written as $C = A + iB$, where A and B are Hermitian (see Exercise A12).
   (b) Prove that $C^*C = CC^*$ if and only if $AB = BA$, where A and B are the Hermitian matrices in part (a).

7.2 Inner Products

We start with a generalization of the *dot product* (Euclidean inner product), a function that takes two vectors in $\mathbb{R}^n$ and produces a unique real number from them.

Definition 7.2.1

Let V be a complex vector space. An **inner product** on V is a function that assigns to each pair of vectors $\mathbf{u}, \mathbf{v} \in V$ a scalar $\langle \mathbf{u}, \mathbf{v} \rangle$ such that the following properties hold:

1. $\langle a\mathbf{u} + b\mathbf{v}, \mathbf{w} \rangle = a\langle \mathbf{u}, \mathbf{w} \rangle + b\langle \mathbf{v}, \mathbf{w} \rangle$ for all $\mathbf{u}, \mathbf{v}, \mathbf{w} \in V$ and a, b scalars.
2. $\langle \mathbf{u}, \mathbf{v} \rangle = \overline{\langle \mathbf{v}, \mathbf{u} \rangle}$.
3. $\langle \mathbf{u}, \mathbf{u} \rangle \geq 0$ with equality if and only if $\mathbf{u} = \mathbf{0}$.

In more sophisticated mathematical terms, we can describe the inner product as a function from the cross product $V \times V$ to the set $\mathbb{R}$ or $\mathbb{C}$, depending on whether V is a real or a complex vector space. (See Appendix A for a description of $V \times V$.)

A vector space together with an inner product is called an **inner product space**. A finite-dimensional real inner product space is often called a **Euclidean space**, whereas a finite-dimensional complex inner product space is often referred to as a **unitary space**. Any subspace of an inner product space is also an inner product space.

Note that if V is a vector space over $\mathbb{R}$, the scalar $\langle \mathbf{u}, \mathbf{v} \rangle$ is a real number. The bar in property 2 denotes the complex conjugate, so if V is a real vector space we have $\langle \mathbf{u}, \mathbf{v} \rangle = \overline{\langle \mathbf{v}, \mathbf{u} \rangle} = \langle \mathbf{v}, \mathbf{u} \rangle$. We see that property 2 implies that $\langle \mathbf{u}, \mathbf{u} \rangle$ is always a real number (even if $\mathbf{u}$ comes from a complex vector space) because $\langle \mathbf{u}, \mathbf{u} \rangle$ equals its complex conjugate.

Example 7.2.1: Inner Product Spaces

(a) If we define the function $\mathbb{R}^3 \times \mathbb{R}^3 \to \mathbb{R}$ by $\langle \mathbf{u}, \mathbf{v} \rangle = u_1 v_1 + 2u_2 v_2 + 3u_3 v_3$, where $\mathbf{u} = [\, u_1 \quad u_2 \quad u_3 \,]^T$ and $\mathbf{v} = [\, v_1 \quad v_2 \quad v_3 \,]^T$, then we can show that this function is an inner product.

First, suppose that $\mathbf{u}, \mathbf{v}$, and $\mathbf{w}$ are vectors in $\mathbb{R}^3$ and a, b are real numbers. Then, $a\mathbf{u} + b\mathbf{v} = [\, au_1 + bv_1 \quad au_2 + bv_2 \quad au_3 + bv_3 \,]^T = [\, au_1 + bv_1 \quad au_2 + bv_2 \quad au_3 + bv_3]^T$ and

$$\langle a\mathbf{u} + b\mathbf{v}, \mathbf{w} \rangle = \left\langle [\, au_1 + bv_1 \quad au_2 + bv_2 \quad au_3 + bv_3 \,]^T, [\, w_1 \quad w_2 \quad w_3 \,]^T \right\rangle$$

$$= (au_1 + bv_1)w_1 + 2(au_2 + bv_2)w_2 + 3(au_3 + bv_3)w_3$$

$$= (au_1 w_1 + 2au_2 w_2 + 3au_3 w_3) + (bv_1 w_1 + 2bv_2 w_2 + 3bv_3 w_3)$$

$$= a(u_1 w_1 + 2u_2 w_2 + 3u_3 w_3) + b(v_1 w_1 + 2v_2 w_2 + 3v_3 w_3)$$

$$= a\langle \mathbf{u}, \mathbf{w} \rangle + b\langle \mathbf{v}, \mathbf{w} \rangle.$$

Next, we see that

$$\langle \mathbf{u}, \mathbf{v} \rangle = u_1 v_1 + 2u_2 v_2 + 3u_3 v_3 = v_1 u_1 + 2v_2 u_2 + 3v_3 u_3$$

$$= \langle \mathbf{v}, \mathbf{u} \rangle = \overline{\langle \mathbf{v}, \mathbf{u} \rangle}$$

because $\langle \mathbf{v}, \mathbf{u} \rangle$ is a real number.

Finally, $\langle \mathbf{u}, \mathbf{u} \rangle = u_1 u_1 + 2u_2 u_2 + 3u_3 u_3 = u_1^2 + 2u_2^2 + 3u_3^2 \geq 0$.

Clearly, $\langle \mathbf{u}, \mathbf{u} \rangle = u_1^2 + 2u_2^2 + 3u_3^2 = 0$ if and only if $u_1 = u_2 = u_2 = 0$—that is, if and only if $\mathbf{u} = \mathbf{0}$.

(b) Let us consider $\mathbb{C}^2$ as a complex vector space and define a function by $\langle \mathbf{u}, \mathbf{v} \rangle = u_1 \bar{v}_1 + 3u_2 \bar{v}_2$ for all vectors $\mathbf{u} = \begin{bmatrix} u_1 \\ u_2 \end{bmatrix}$ and $\mathbf{v} = \begin{bmatrix} v_1 \\ v_2 \end{bmatrix}$ in $\mathbb{C}^2$. This function is an inner product on $\mathbb{C}^2$. To see this, first note that $\langle \mathbf{u}, \mathbf{v} \rangle \in \mathbb{C}$.

Now for $a, b \in \mathbb{C}$, we have $a\mathbf{u} + b\mathbf{v} = \begin{bmatrix} au_1 + bv_1 \\ au_2 + bv_2 \end{bmatrix}$, so

$$\langle a\mathbf{u} + b\mathbf{v}, \mathbf{w} \rangle = \left\langle \begin{bmatrix} au_1 + bv_1 \\ au_2 + bv_2 \end{bmatrix}, \begin{bmatrix} w_1 \\ w_2 \end{bmatrix} \right\rangle$$

$$= (au_1 + bv_1)\bar{w}_1 + 3(au_2 + bv_2)\bar{w}_2$$

$$= (au_1 \bar{w}_1 + 3au_2 \bar{w}_2) + (bv_1 \bar{w}_1 + 3bv_2 \bar{w}_2)$$

$$= a(u_1 \bar{w}_1 + 3u_2 \bar{w}_2) + b(v_1 \bar{w}_1 + 3v_2 \bar{w}_2)$$

$$= a\langle \mathbf{u}, \mathbf{w} \rangle + b\langle \mathbf{v}, \mathbf{w} \rangle.$$

Remembering that $\bar{\bar{z}} = z$ for any complex number z, we calculate

$$\langle \mathbf{u}, \mathbf{v} \rangle = u_1\bar{v}_1 + 3u_2\bar{v}_2 = \bar{\bar{u}}_1\bar{v}_1 + 3\bar{\bar{u}}_2\bar{v}_2 = \overline{(\bar{u}_1 v_1)} + 3\overline{(\bar{u}_2 v_2)}$$
$$= \overline{(\bar{u}_1 v_1 + 3\bar{u}_2 v_2)} = \overline{(v_1\bar{u}_1 + 3v_2\bar{u}_2)} = \overline{\langle \mathbf{v}, \mathbf{u} \rangle}.$$

Finally, $\langle \mathbf{u}, \mathbf{u} \rangle = u_1\bar{u}_1 + 3u_2\bar{u}_2 = |u_1|^2 + 3|u_2|^2 \geq 0$ as the sum of two squares. We see that $\langle \mathbf{u}, \mathbf{u} \rangle = 0$ if and only if $|u_1|^2 + 3|u_2|^2 = 0$—that is, if and only if $u_1 = u_2 = 0$, so that $\mathbf{u} = \mathbf{0}$.

Some important examples follow. We will use these inner product spaces or related spaces to illustrate various concepts in this chapter and in the next. Verification of the properties of an inner product space is left as an exercise in each case (see Problems B1 through B4).

Example 7.2.2: Inner Product Spaces

(a) The vector space $\mathbb{R}^n$ together with the inner product given by the dot product $\langle \mathbf{x}, \mathbf{y} \rangle = \mathbf{x} \cdot \mathbf{y} = x_1 y_1 + x_2 y_2 + \cdots + x_n y_n$ is a real inner product space. Here, $\mathbf{x} = [x_1 \quad x_2 \quad \ldots \quad x_n]^\mathsf{T}$ and $\mathbf{y} = [y_1 \quad y_2 \quad \ldots \quad y_n]^\mathsf{T}$, where $x_i, y_i \in \mathbb{R}$. In this inner product space, we can interpret $\langle \mathbf{x}, \mathbf{y} \rangle$ as the product of a row matrix and a column matrix: $\langle \mathbf{x}, \mathbf{y} \rangle = \mathbf{x}^\mathsf{T}\mathbf{y}$.

(b) If $\mathbf{x} = [x_1 \quad x_2 \quad \ldots \quad x_n]^\mathsf{T}$ and $\mathbf{y} = [y_1 \quad y_2 \quad \ldots \quad y_n]^\mathsf{T}$, where $x_i, y_i \in \mathbb{C}$, the space $\mathbb{C}^n$ with the inner product

$$\langle \mathbf{x}, \mathbf{y} \rangle = \mathbf{x} \cdot \bar{\mathbf{y}} = x_1\bar{y}_1 + x_2\bar{y}_2 + \cdots + x_n\bar{y}_n$$

is a complex inner product space (a unitary space). We can also write the inner product in the matrix form $\langle \mathbf{x}, \mathbf{y} \rangle = \mathbf{x}^\mathsf{T}\bar{\mathbf{y}}$. As a final, equivalent way of expressing this inner product, we have $\langle \mathbf{x}, \mathbf{y} \rangle = \overline{\langle \mathbf{y}, \mathbf{x} \rangle} = \overline{(\mathbf{y}^\mathsf{T}\bar{\mathbf{x}})} = \bar{\mathbf{y}}^\mathsf{T}\bar{\bar{\mathbf{x}}} = \mathbf{y}^*\mathbf{x}$, where $\mathbf{y}^*$ is the complex conjugate of $\mathbf{y}$, $\mathbf{y}$ being viewed as an $n \times 1$ matrix.

(c) The vector space $M_{mn}(\mathbb{C})$ of all $m \times n$ matrices with complex entries is an inner product space if we define $\langle A, B \rangle = \sum_{i=1}^m \sum_{j=1}^n a_{ij}\bar{b}_{ij} = \text{trace}\,(A^\mathsf{T}\bar{B})$ for $A = [a_{ij}]$ and $B = [b_{ij}]$. It is easily proved that $\text{trace}\,(A^\mathsf{T}) = \text{trace}\,(A)$, so $(A^\mathsf{T}\bar{B})^\mathsf{T} = (\bar{B})^\mathsf{T}(A^\mathsf{T})^\mathsf{T} = B^*A$. Thus, an equivalent definition of $\langle A, B \rangle$ is $\text{trace}\,(B^*A)$.

By removing the conjugate bars in this definition, we obtain an inner product for $M_{mn} = M_{mn}(\mathbb{R})$: $\langle A, B \rangle = \text{trace}(A^\mathsf{T}B) = \text{trace}\,(B^\mathsf{T}A)$. (Recall that $\text{trace}\,(M^\mathsf{T}) = \text{trace}\,(M)$ for any square matrix M.)

(d) The space $\mathbb{P}_n[x] = \{c_0 + c_1 x + \cdots + c_n x^n \mid c_i \in \mathbb{C}$ for $i = 0, 1, \ldots, n.\}$ with the inner product defined by

$$\langle a_0 + a_1 x + \cdots + a_n x^n, b_0 + b_1 x + \cdots + b_n x^n \rangle$$
$$= \bar{a}_0 b_0 + \bar{a}_1 b_1 + \cdots + \bar{a}_n b_n$$

is a unitary space.

(e) The space $C[a, b]$ of real-valued continuous functions on the interval $[a, b]$ together with the inner product $\langle f, g \rangle = \int_a^b f(x)g(x)dx$ is an infinite-dimensional real inner product space.

(f) Suppose we consider the infinite-dimensional vector space ℓ^2 consisting of all infinite sequences $\mathbf{a} = \{a_i\}_{i=1}^{\infty}$ of real numbers such that $\sum_{i=1}^{\infty} a_i^2$ converges. If we define $\langle \mathbf{a}, \mathbf{b} \rangle = \langle \{a_i\}, \{b_i\} \rangle = \sum_{i=1}^{\infty} a_i b_i$ for $\mathbf{a}, \mathbf{b} \in \ell^2$, then this last series converges and defines an inner product on ℓ^2: we have $\left| \sum_{k=1}^{\infty} x_k y_k \right| \leq \left(\sum_{k=1}^{\infty} x_k^2 \right)^{1/2} \left(\sum_{k=1}^{\infty} y_k^2 \right)^{1/2}$ by using the Cauchy–Schwarz inequality (Theorem 7.2.1(2)) and taking the limit as $n \to \infty$. (See the comments after Theorem 7.2.1.) The remaining properties of an inner product follow from standard manipulations with infinite sequences.

*In each part of the preceding examples, the inner product defined will be taken as the **standard (or canonical) inner product** for the associated vector space in all future discussions.*

Property (3) of Definition 7.2.1 indicates that $\langle \mathbf{u}, \mathbf{u} \rangle \geq 0$ for any vector $\mathbf{u}$ in an inner product space. Therefore, the positive square root of $\langle \mathbf{u}, \mathbf{u} \rangle$ is a real number and we can define the **norm** (or **length**) of $\mathbf{u}$:

$$\|\mathbf{u}\| = \sqrt{\langle \mathbf{u}, \mathbf{u} \rangle}.$$

This is sometimes referred to as "the norm induced by the inner product." A useful form of this definition is $\|\mathbf{u}\|^2 = \langle \mathbf{u}, \mathbf{u} \rangle$.

There are several basic and important properties of the norm in any inner product space. Property (3) of Definition 7.2.1 is equivalent to $\|\mathbf{u}\| > 0$ for $\mathbf{u} \neq \mathbf{0}$ We have seen other properties in Chapter 1, in the context of $\mathbb{R}^n$. In the next theorem, keep in mind that if c is a complex number, then $c\bar{c} = |c|^2$, whereas if c is real, we have $c\bar{c} = c \cdot c = c^2$.

Theorem 7.2.1: Properties of the Norm

If V is an inner product space, then for any vectors $\mathbf{u}, \mathbf{v}$ in V and any scalar c:

1. $\|c\,\mathbf{u}\| = |c|\|\mathbf{u}\|$.
2. $|\langle \mathbf{u}, \mathbf{v} \rangle| \leq \|\mathbf{u}\|\|\mathbf{v}\|$. [*Cauchy–Schwarz Inequality*]*
3. $\|\mathbf{u} + \mathbf{v}\| \leq \|\mathbf{u}\| + \|\mathbf{v}\|$. [*Triangle Inequality*]

Proof   (1) $\|c\mathbf{u}\| = \sqrt{\langle c\mathbf{u}, c\mathbf{u} \rangle} = \sqrt{c\langle \mathbf{u}, c\mathbf{u} \rangle} = \sqrt{c\langle c\mathbf{u}, \mathbf{u} \rangle}$
$= \sqrt{c\bar{c} < \mathbf{u}, \mathbf{u} >} = |c|\|\mathbf{u}\|$.

---

* This inequality is also referred to as the *Schwarz inequality* or the *Cauchy–Schwarz–Bunyakovsky inequality*. See J.S. Steele, *The Cauchy–Schwarz Master Class: An Introduction to the Art of Inequalities* (New York and Washington, DC: Cambridge University Press/Mathematical Association of America, 2004).

(2) If either $\mathbf{u}$ or $\mathbf{v}$ is the zero vector, it is easy to see that $|\langle \mathbf{u}, \mathbf{v} \rangle| = \|\mathbf{u}\|\|\mathbf{v}\|$. Now, we assume neither $\mathbf{u}$ nor $\mathbf{v}$ is $\mathbf{0}$ and we consider two cases. First, if $\mathbf{u}$ and $\mathbf{v}$ are linearly dependent, then one vector is a scalar multiple of the other, say $\mathbf{u} = c\mathbf{v}$ for some nonzero scalar c. Then,

$$\langle \mathbf{u}, \mathbf{v} \rangle^2 = \langle c\mathbf{v}, \mathbf{v} \rangle^2 = c^2 \langle \mathbf{v}, \mathbf{v} \rangle^2 = c^2\|\mathbf{v}\|^4 = \frac{c^2}{|c|^2}\|\mathbf{u}\|^2\|\mathbf{v}\|^2,$$

so we have $|\langle \mathbf{u}, \mathbf{v} \rangle| = \|\mathbf{u}\|\|\mathbf{v}\|$.

If $\mathbf{u}$ and $\mathbf{v}$ are linearly independent, then $t\mathbf{u} + \mathbf{v}$ cannot be the zero vector for any real scalar t. It follows that the quadratic polynomial in t defined by

$$\|t\mathbf{u} + \mathbf{v}\|^2 = \langle t\mathbf{u} + \mathbf{v}, t\mathbf{u} + \mathbf{v} \rangle = \|\mathbf{u}\|^2 t^2 + 2\langle \mathbf{u}, \mathbf{v} \rangle t + \|\mathbf{v}\|^2$$

must be *positive* for every real value of t—that is, it has no real zeros. This tells us that the *discriminant* of the quadratic polynomial is negative:

$$4\langle \mathbf{u}, \mathbf{v} \rangle^2 - 4\|\mathbf{u}\|^2\|\mathbf{v}\|^2 < 0,$$

or

$$\langle \mathbf{u}, \mathbf{v} \rangle^2 < \|\mathbf{u}\|^2\|\mathbf{v}\|^2,$$

so $|\langle \mathbf{u}, \mathbf{v} \rangle| < \|\mathbf{u}\|\|\mathbf{v}\|$.

To summarize, we find that $|\langle \mathbf{u}, \mathbf{v} \rangle| \leq \|\mathbf{u}\|\|\mathbf{v}\|$, with equality if and only if one vector is a scalar multiple of the other.

(3) In what follows, we need the fact that $\text{Re}(z) \leq |z|$ for any $z \in \mathbb{C}$: if $z = a + bi = \text{Re}(z) + \text{Im}(z)i$, then $\text{Re}(z) = a \leq |a| = \sqrt{a^2} \leq \sqrt{a^2 + b^2} = |z|$. Now, we see that

$$\|\mathbf{u} + \mathbf{v}\|^2 = \langle \mathbf{u} + \mathbf{v}, \mathbf{u} + \mathbf{v} \rangle = \langle \mathbf{u}, \mathbf{u} + \mathbf{v} \rangle + \langle \mathbf{v}, \mathbf{u} + \mathbf{v} \rangle$$
$$= \overline{\langle \mathbf{u} + \mathbf{v}, \mathbf{u} \rangle} + \overline{\langle \mathbf{u} + \mathbf{v}, \mathbf{v} \rangle} = \overline{\langle \mathbf{u}, \mathbf{u} \rangle} + \overline{\langle \mathbf{v}, \mathbf{u} \rangle} + \overline{\langle \mathbf{u}, \mathbf{v} \rangle} + \overline{\langle \mathbf{v}, \mathbf{v} \rangle}$$
$$= \|\mathbf{u}\|^2 + 2\text{Re}\langle \mathbf{u}, \mathbf{v} \rangle + \|\mathbf{v}\|^2$$
$$\leq \|\mathbf{u}\|^2 + 2|\langle \mathbf{u}, \mathbf{v} \rangle| + \|\mathbf{v}\|^2$$
$$\leq \|\mathbf{u}\|^2 + 2\|\mathbf{u}\|\|\mathbf{v}\| + \|\mathbf{v}\|^2 \quad \text{[by the Cauchy–Schwarz inequality]}$$
$$= (\|\mathbf{u}\| + \|\mathbf{v}\|)^2.$$

Taking the square root of each side gives us the desired inequality.

One consequence of the Cauchy–Schwarz inequality in any *real* inner product space (Euclidean space) is that

$$-1 \leq \frac{\langle \mathbf{u}, \mathbf{v} \rangle}{\|\mathbf{u}\|\|\mathbf{v}\|} \leq 1,$$

provided that $\mathbf{u} \neq \mathbf{0}$ and $\mathbf{v} \neq \mathbf{0}$. Therefore, there is a unique number $\theta \in [0, \pi]$ such that $\cos\theta = \langle \mathbf{u}, \mathbf{v} \rangle/(\|\mathbf{u}\|\|\mathbf{v}\|)$. This real number θ is called the **angle** between the vectors **u** and **v**. (Recall Definition 1.2.3.) We add that the angle between two nonzero vectors $\mathbf{u}, \mathbf{v} \in \mathbb{C}^n$ can be defined by setting $\cos \theta = |\langle \mathbf{u}, \mathbf{v} \rangle| / \|\mathbf{u}\| \|\mathbf{v}\|$ for $0 \leq \theta \leq \pi/2$. Although $\langle \mathbf{u}, \mathbf{v} \rangle$ may be a complex number, $|\langle \mathbf{u}, \mathbf{v} \rangle|$ is always a nonnegative real number.

We should note here that, in general, *the calculation of the angle between two vectors depends on the inner product that is being used.* However, in a real vector space, if one vector is a scalar multiple of the other, then the angle will be computed as 0 or π—depending on whether the scalar is positive or negative, respectively—no matter what inner product is used in the calculation.

The Cauchy–Schwarz inequality has great applicability to real and complex function theory, including applications in statistics. For example, applying this result to $\mathbb{R}^n$ with the standard inner product given in Example 7.1.1(a), we have $\left(\sum_{k=1}^{n} x_k y_k \right)^2 \leq \left(\sum_{k=1}^{n} x_k^2 \right) \left(\sum_{k=1}^{n} y_k^2 \right)$, or $\left| \sum_{k=1}^{n} x_k y_k \right| \leq \left(\sum_{k=1}^{n} x_k^2 \right)^{1/2} \left(\sum_{k=1}^{n} y_k^2 \right)^{1/2}$ for finite ordered sets of real numbers $x_1, x_2, \ldots, x_n$ and $y_1, y_2, \ldots, y_n$.* If the infinite sequences $\{x_k\}, \{y_k\}$ are *square-summable*—that is, if both $\sum_{k=1}^{\infty} x_k^2$ and $\sum_{k=1}^{\infty} y_k^2$ converge—then we can take limits as $n \to \infty$ in the last stated inequality to conclude that $\left| \sum_{k=1}^{\infty} x_k y_k \right| \leq \left(\sum_{k=1}^{\infty} x_k^2 \right)^{1/2} \left(\sum_{k=1}^{\infty} y_k^2 \right)^{1/2}$, so the series $\sum_{k=1}^{\infty} x_k y_k$ must converge. Similarly, if we consider the inner product space $C[a, b]$ of real-valued continuous functions on an interval $[a, b]$ (Example 7.2.2(e)), we can apply the Cauchy–Schwarz inequality to conclude

$$\left| \int_a^b f(x)g(x)dx \right| \leq \left[\int_a^b f^2(x)dx \right]^{1/2} \left[\int_a^b g^2(x)dx \right]^{1/2} \dagger .$$

Example 7.2.3

(a) Let us consider $C[-1, 1]$ with inner product defined by $\langle f, g \rangle = \int_{-1}^{1} f(x)g(x)dx$ for $f, g \in C[-1, 1]$. Then, for example, $\langle x, x^3 \rangle = \int_{-1}^{1} x(x^3)dx = \int_{-1}^{1} x^4 \, dx = 2/5$. Also, we can calculate $\|x\| = \sqrt{\langle x, x \rangle} = \sqrt{\int_{-1}^{1} x^2 \, dx} = \sqrt{2/3} = \sqrt{6}/3$ and $\|x^3\| = \sqrt{\langle x^3, x^3 \rangle} = \sqrt{\int_{-1}^{1} x^6 \, dx} = \sqrt{2/7} = \sqrt{14}/7$.

Now we can compute the cosine of the angle θ between these vectors in $C[-1, 1]$ as

$$\frac{\langle x, x^3 \rangle}{\|x\| \|x^3\|} = \frac{2/5}{(\sqrt{6}/3)(\sqrt{14}/7)} = \frac{\sqrt{21}}{5},$$

so $\theta = \cos^{-1}(\sqrt{21}/5) \approx 0.4115$ rad $\approx 23.6°$.

---

* This formula was given by Cauchy in 1821.
† This form of the inequality appeared in work by B. Bunyakovsky in 1859. A double integral version was proved by H.A. Schwarz in 1885.

If we take $P(x) = (3x^2 - 1)/2$ and $Q(x) = (5x^3 - 3x)/2$, then $\langle P, Q \rangle = \int_{-1}^{1} ((3x^2 - 1)/2) \cdot ((5x^3 - 3x)/2)dx = 0$, so the angle between vectors P and Q is $\pi/2$ rad, or $90°$.

(b) Now consider $M_2(\mathbb{C})$, the vector space of all 2×2 matrices with complex entries, with the inner product defined by $\langle A, B \rangle = \text{trace}(A^{\mathsf{T}}\bar{B})$ (Example 7.2.2(c)). If we take $A = \begin{bmatrix} 1 & 2+i \\ 3 & i \end{bmatrix}$ and $B = \begin{bmatrix} 1+i & 0 \\ i & -i \end{bmatrix}$ in $M_2(\mathbb{C})$, then $A^{\mathsf{T}} = \begin{bmatrix} 1 & 3 \\ 2+i & i \end{bmatrix}$, $\bar{B} = \begin{bmatrix} 1-i & 0 \\ -i & i \end{bmatrix}$, and

$$A^{\mathsf{T}}\bar{B} = \begin{bmatrix} 1 & 3 \\ 2+i & i \end{bmatrix} \begin{bmatrix} 1-i & 0 \\ -i & i \end{bmatrix} = \begin{bmatrix} 1-4i & 3i \\ 4-i & -1 \end{bmatrix},$$ so

$\langle A, B \rangle = \text{trace}(A^{\mathsf{T}}\bar{B}) = (1-4i) + (-1) = -4i.$

We can also calculate the norm of A and the norm of B:

$$\|A\| = \sqrt{\langle A, A \rangle} = \sqrt{\text{trace}(A^{\mathsf{T}}\bar{A})}$$

$$= \sqrt{\text{trace}\left(\begin{bmatrix} 1 & 3 \\ 2+i & i \end{bmatrix} \begin{bmatrix} 1 & 2-i \\ 3 & -i \end{bmatrix} \right)}$$

$$= \sqrt{\text{trace}\left(\begin{bmatrix} 10 & 2-4i \\ 2+4i & 6 \end{bmatrix} \right)} = 4,$$

$$\|B\| = \sqrt{\langle B, B \rangle} = \sqrt{\text{trace}(B^{\mathsf{T}}\bar{B})}$$

$$= \sqrt{\text{trace}\left(\begin{bmatrix} 1+i & i \\ 0 & -i \end{bmatrix} \begin{bmatrix} 1-i & 0 \\ -i & i \end{bmatrix} \right)}$$

$$= \sqrt{\text{trace}\left(\begin{bmatrix} 3 & -1 \\ -1 & 1 \end{bmatrix} \right)} = 2.$$

Finally, $|\langle A, B \rangle| = |-4i| = 4 < 8 = \|A\|\|B\|$, so the Cauchy–Schwarz inequality holds in this example.

Exercises 7.2

A.

1. If $\mathbf{x} = [1 \quad -1 \quad 1 \quad -1]^{\mathsf{T}}$, calculate $\mathbf{x}^{\mathsf{T}}\mathbf{x}$ and $\mathbf{x}\mathbf{x}^{\mathsf{T}}$.

2. Define a function on $\mathbb{R}^3$ by $\langle \mathbf{u}, \mathbf{v} \rangle = u_1v_1 - u_2v_2 + 3u_3v_3$ for every $\mathbf{u} = [u_1 \quad u_2 \quad u_3]^{\mathsf{T}}$ and $\mathbf{v} = [v_1 \quad v_2 \quad v_3]^{\mathsf{T}}$ in $\mathbb{R}^3$. Show that this function is not an inner product on $\mathbb{R}^3$.

3. In the space $P_2[x]$, with the inner product defined by

$$\langle a_0 + a_1x + a_2x^2, b_0 + b_1x + b_2x^2 \rangle = a_0b_0 + a_1b_1 + a_2b_2.$$

Calculate each of the inner products $\langle f, g \rangle$, where
(a) $f(x) = -1 + 2x + 3x^2$ and $g(x) = -4x + 2x^2$.
(b) $f(x) = -1 + 2x - x^2$ and $g(x) = -8 - 3x + 2x^2$.
(c) $f(x) = 2 + 4x + 6x^2$ and $g(x) = 2 + 4x + 6x^2$.

4. In $\mathbb{C}^4$, with the inner product defined in Example 7.2.2(b), compute the inner products $\langle \mathbf{u}, \mathbf{v} \rangle$ and $\langle \mathbf{v}, \mathbf{u} \rangle$, where
(a) $\mathbf{u} = \begin{bmatrix} 1 & i & 0 & 2i \end{bmatrix}^T$ and $\mathbf{v} = \begin{bmatrix} 1+i & 2 & 2-3i & 0 \end{bmatrix}^T$.
(b) $\mathbf{u} = \begin{bmatrix} i & 2i & 3i & 4i \end{bmatrix}^T$ and $\mathbf{v} = \begin{bmatrix} 4i & 3i & 2i & i \end{bmatrix}^T$.
(c) $\mathbf{u} = \begin{bmatrix} 1 & i & -i & 2i \end{bmatrix}^T$ and $\mathbf{v} = \begin{bmatrix} 1 & i & -i & 2i \end{bmatrix}^T$.

5. By using the appropriate inner product from Example 7.2.2(c), calculate each of the inner products $\langle A, B \rangle$, where
(a) $A = \begin{bmatrix} 1 & 2 \\ 3 & 4 \end{bmatrix}$ and $B = \begin{bmatrix} -2 & 0 \\ 4 & -3 \end{bmatrix}$.
(b) $A = \begin{bmatrix} -1 & 5 \\ 2 & 4 \end{bmatrix}$ and $B = \begin{bmatrix} 3 & -1 \\ 4 & 0 \end{bmatrix}$.
(c) $A = \begin{bmatrix} i & 2 \\ 0 & -i \end{bmatrix}$ and $B = \begin{bmatrix} 2+3i & -2i \\ 3 & 5 \end{bmatrix}$.
(d) $A = \begin{bmatrix} i & 2i \\ 3i & -1 \end{bmatrix}$ and $B = \begin{bmatrix} 1 & -i \\ -2i & -i-8 \end{bmatrix}$.

6. Suppose $V = C[-1, 1]$. By using the inner product defined in Example 7.2.2(e), calculate $\langle f, g \rangle$, where
(a) $f(x) = 1 + x + x^2$ and $g(x) = -x + 2x^2$.
(b) $f(x) = e^{-x}$ and $g(x) = e^x$.
(c) $f(x) = \sin x$ and $g(x) = \cos x$.

B.

1. Prove that the scalar-valued function $\langle \mathbf{x}, \mathbf{y} \rangle$ defined in Example 7.2.2(b) is an inner product for $\mathbb{C}^n$.

2. Prove that the scalar-valued function $\langle A, B \rangle$ defined in Example 7.2.2(c) is an inner product for $M_{23}(\mathbb{C})$.

3. Prove that the scalar-valued function $\langle f, g \rangle$ defined in Example 7.2.2(d) is an inner product for $\mathbb{P}_2[x]$.

4. Prove that the scalar-valued function $\langle f, g \rangle$ defined in Example 7.2.2(e) is an inner product for $C[a, b]$.

5. Suppose that V is an inner product space, $\mathbf{u}, \mathbf{v}, \mathbf{w} \in \mathbf{v}$, and a, b are scalars.

(a) Prove that $\langle \mathbf{u}, a\mathbf{v} + b\mathbf{w} \rangle = a\langle \mathbf{u}, \mathbf{v} \rangle + b\langle \mathbf{u}, \mathbf{w} \rangle$ if V is a real vector space.

(b) Prove that $\langle \mathbf{u}, a\mathbf{v} + b\mathbf{w} \rangle = \bar{a}\langle \mathbf{u}, \mathbf{v} \rangle + \bar{b}\langle \mathbf{u}, \mathbf{w} \rangle$ if V is a complex vector space.

6. Let V be a real or complex vector space.
    (a) Show that the sum of two inner products on V is an inner product on V. (That is, if $\langle \, , \rangle_1$ and $\langle \, , \rangle_2$ are inner products on V, then $\langle \mathbf{u}, \mathbf{v} \rangle = \langle \mathbf{u}, \mathbf{v} \rangle_1 + \langle \mathbf{u}, \mathbf{v} \rangle_2$ is an inner product on V.)
    (b) Is the difference of two inner products an inner product?
    (c) Show that a positive multiple of an inner product is an inner product.

7. Let V be an inner product space, and let $\boldsymbol{\alpha}$ and $\boldsymbol{\beta}$ be vectors in V. Show that $\boldsymbol{\alpha} = \boldsymbol{\beta}$ if and only if $\langle \boldsymbol{\alpha}, \boldsymbol{\gamma} \rangle = \langle \boldsymbol{\beta}, \boldsymbol{\gamma} \rangle$ for every $\boldsymbol{\gamma} \in V$.

8. Find an inner product on $\mathbb{R}^2$ such that $\langle \mathbf{e}_1, \mathbf{e}_2 \rangle = 2$, where $\mathbf{e}_1$ and $\mathbf{e}_2$ are the standard basis vectors for $\mathbb{R}^2$.

9. Prove that

$$\left(\sum_{j=1}^{n} a_j b_j \right)^2 \leq \left(\sum_{j=1}^{n} j a_j^2 \right) \left(\sum_{j=1}^{n} \frac{b_j^2}{j} \right)$$

for n a positive integer and all real numbers $a_1, \ldots, a_n$ and $b_1, \ldots, b_n$.

C.

1. Given four complex numbers $\alpha, \beta, \gamma,$ and δ, try to define an inner product in $\mathbb{C}^2$ of the form

$$\langle \mathbf{u}, \mathbf{v} \rangle = \alpha \xi_1 \bar{\eta}_1 + \beta \xi_2 \bar{\eta}_1 + \gamma \xi_1 \bar{\eta}_2 + \delta \xi_2 \bar{\eta}_2,$$

where $\mathbf{u} = \begin{bmatrix} \xi_1 \\ \xi_2 \end{bmatrix}$ and $\mathbf{v} = \begin{bmatrix} \eta_1 \\ \eta_2 \end{bmatrix}$. Under what conditions on $\alpha, \beta, \gamma,$ and δ does this equation define an inner product?

2. If V is a unitary space, prove that

$$4\langle \mathbf{u}, \mathbf{v} \rangle = \|\mathbf{u} + \mathbf{v}\|^2 - \|\mathbf{u} - \mathbf{v}\|^2 + \|\mathbf{u} + i\mathbf{v}\|^2 i - \|\mathbf{u} - i\mathbf{v}\|^2 i$$

for all $\mathbf{u}, \mathbf{v} \in V$.

3. Suppose T is a linear transformation on a unitary space V. The *numerical range* (or *field of values*) of T is defined as the set of complex numbers:

$$W(T) = \{ \langle T(\mathbf{x}), \mathbf{x} \rangle \, | \, \mathbf{x} \in V, \|\mathbf{x}\| = 1 \}.^*$$

---

* The letter "W" stands for *Wertvorrat*, the original term used in Otto Toeplitz's 1918 paper defining the concept.

(a) Show that $W(T + cI) = W(T) + c$ for every $c \in \mathbb{C}$.

(b) Show that $W(cT) = cW(T)$ for every $c \in \mathbb{C}$.

(c) If B is a basis for V and $[T]_B = A$, show that the diagonal entries of A are contained in $W(T)$.

(d) Show that the eigenvalues of T are contained in $W(T)$.

(e) Determine $W(T)$ if $T\left(\begin{bmatrix} c_1 \\ c_2 \end{bmatrix}\right) = \begin{bmatrix} c_1 \\ (1+i)c_2 \end{bmatrix}$, where $c_1, c_2 \in \mathbb{C}$.

7.3 Orthogonality and Orthonormal Bases

Now we extend the concept of orthogonality in $\mathbb{R}^n$ (Section 1.2) to an arbitrary inner product space. This extension is important in many theoretical and applied areas of mathematics and science.

Definition 7.3.1

Two vectors **u** and **v** in an inner product space are called **orthogonal** if $\langle \mathbf{u}, \mathbf{v} \rangle = 0$. (We also say "**u** is orthogonal to **v**" or "**v** is orthogonal to **u**.") If this is the case, we write $\mathbf{u} \perp \mathbf{v}$ (pronounced "**u** perp **v**").

We recognize that the vectors in the standard basis for $\mathbb{R}^n$ are **mutually orthogonal**: $\langle \mathbf{e}_i, \mathbf{e}_j \rangle = 0$ if $i \neq j$. A set of vectors in any inner product space is called **orthogonal** if the vectors in the set are mutually orthogonal: $\langle \mathbf{u}, \mathbf{v} \rangle = 0$ if $\mathbf{u} \neq \mathbf{v}$. If an inner product space has a basis consisting of mutually orthogonal vectors, this basis is called an **orthogonal basis**.

Orthogonal subsets of an inner product space V are useful, as the next theorem shows.

Theorem 7.3.1

Let V be an inner product space. Then, any orthogonal subset of V consisting of nonzero vectors is linearly independent.

Proof   Suppose that $\{\mathbf{v}_1, \mathbf{v}_2, \ldots, \mathbf{v}_k\}$ is an orthogonal subset of V and assume that there is a linear relation of the form $a_1\mathbf{v}_1 + a_2\mathbf{v}_2 + \cdots + a_k\mathbf{v}_k = \mathbf{0}$.

Then, forming the inner product of both sides of this relation with any $\mathbf{v}_j$, we get $0 = \langle \sum_{i=1}^{k} a_i\mathbf{v}_i, \mathbf{v}_j \rangle = \sum_{i=1}^{k} a_i \langle \mathbf{v}_i, \mathbf{v}_j \rangle = a_j \langle \mathbf{v}_j, \mathbf{v}_j \rangle = a_j \|\mathbf{v}_j\|^2$ because $\langle \mathbf{v}_i, \mathbf{v}_j \rangle = 0$ if $i \neq j$. We know that $\|\mathbf{v}_j\| \neq 0$, so we conclude that $a_j = 0$. Because $\mathbf{v}_j$ is arbitrary, we conclude that $a_j = 0$ for $j = 1, 2, \ldots, k$—that is, $\{\mathbf{v}_i\}_{i=1}^{k}$ is linearly independent.

It follows that if the dimension of V is n and an orthogonal subset S of V consists of n nonzero vectors, then S is a basis for V (Corollary 5.4.2).

In $\mathbb{R}^n$, we see that $\|\mathbf{e}_i\| = 1$, $i = 1, 2, \ldots, n$. In any inner product space, if $\|\mathbf{u}\| = 1$ or, equivalently, if $\langle \mathbf{u}, \mathbf{u} \rangle = 1$, we call **u** a **unit vector**. Any nonzero vector **u** can be multiplied by the reciprocal of its norm to get the unit vector $\hat{\mathbf{u}} = \mathbf{u}/\|\mathbf{u}\|$. This process is called **normalizing u**.

Definition 7.3.2

A set of vectors that are mutually orthogonal and of unit length is called an **orthonormal set**. In particular, a *basis* consisting of vectors that are mutually orthogonal and of unit length is called an **orthonormal basis**.

Example 7.3.1

If we consider the inner product space $C[0, 2\pi]$ with the standard inner product, then the vectors $f(x) = \sin x$ and $g(x) = \cos x$ are orthogonal:

$\langle f, g \rangle = \int_0^{2\pi} \sin x \cos x \, dx = \frac{1}{2} \int_0^{2\pi} \sin 2x \, dx = 0$. We calculate $\|f\| = \|g\| = \left[\int_0^{2\pi} \cos^2 x \, dx \right]^{1/2} = 1/\sqrt{\pi}$. Thus, we can normalize f and g by calculating $f/\|f\|$ and $g/\|g\|$ to get the orthonormal set $\{\hat{f}, \hat{g}\} = \{\sqrt{\pi} \sin x, \sqrt{\pi} \cos x\}$.

Example 7.3.2

If we consider the vector space $P_2[x]$ restricted to the closed interval $[-1, 1]$ and define $\langle p, q \rangle = \int_{-1}^1 p(x)q(x) \, dx$ for $p, q \in P_2[x]$, we get an inner product space. The set $\{1, x, x^2\}$ is a basis for $P_2[x]$. An *orthogonal* basis for $P_2[x]$ is $\{1, x, x^2 - \frac{1}{3}\}$. In the original basis, for example, we have $\langle 1, x^2 \rangle = \int_{-1}^1 1 \cdot x^2 \, dx = 2/3$, but in the new basis $\langle 1, x^2 - \frac{1}{3} \rangle = \int_{-1}^1 x^2 - \frac{1}{3} \, dx = 0$. Finally, $\{1/\sqrt{2}, \sqrt{6}x/2, (3\sqrt{10}/4)(x^2 - \frac{1}{3})\}$ is an *orthonormal basis* for $P_2[x]$.

Even though any inner product space (indeed, any vector space) may have many bases, it is sometimes possible to find a particular basis so that problems in the vector space are easier to solve. For example, working with analytic geometry in $\mathbb{R}^2$ or $\mathbb{R}^3$ is usually much easier if the axes (basis vectors) are mutually orthogonal. The next three theorems indicate how useful orthogonal sets, especially orthonormal sets and orthonormal bases, are.

Theorem 7.3.2: The Pythagorean Theorem

If $\mathbf{u}$ and $\mathbf{v}$ are orthogonal vectors in V, then $\|\mathbf{u} + \mathbf{v}\|^2 = \|\mathbf{u}\|^2 + \|\mathbf{v}\|^2$.

Proof Exercise B1.

If V is a Euclidean space, then $\|\mathbf{u}+\mathbf{v}\|^2=\|\mathbf{u}\|^2+\|\mathbf{v}\|^2$ is both a necessary and sufficient condition for vectors $\mathbf{u}$ and $\mathbf{v}$ to be orthogonal. However, in a unitary space $\|\mathbf{u}+\mathbf{v}\|^2=\|\mathbf{u}\|^2+\|\mathbf{v}\|^2$ does not imply that $\mathbf{u}$ and $\mathbf{v}$ are orthogonal. For example, if we take $\mathbf{u}=\begin{bmatrix}1\\-i\end{bmatrix}$ and $\mathbf{v}=\begin{bmatrix}i\\1\end{bmatrix}$ in $\mathbb{C}^2$ with the standard inner product, then $\|\mathbf{u}+\mathbf{v}\|^2=4=\|\mathbf{u}\|^2+\|\mathbf{v}\|^2$, whereas $\langle\mathbf{u},\mathbf{v}\rangle=-2i\neq 0$ (Exercise A6).

Theorem 7.3.3

If $\{\mathbf{v}_1,\mathbf{v}_2,\ldots,\mathbf{v}_k\}$ is an orthogonal set of vectors in V, then

$$\|a_1\mathbf{v}_1+a_2\mathbf{v}_2+\cdots+a_k\mathbf{v}_k\|^2=|a_1|^2\|\mathbf{v}_1\|^2+|a_2|^2\|\mathbf{v}_2\|^2+\cdots+|a_k|^2\|\mathbf{v}_k\|^2$$

for all scalars $a_1,a_2,\ldots,a_k$.

Proof   The theorem follows from the repeated use of the Pythagorean Theorem (Theorem 7.3.2). To do this, we first establish that for any $j<k$, we have $\langle a_1\mathbf{v}_1+a_2\mathbf{v}_2+\cdots+a_{j-1}\mathbf{v}_{j-1},a_j\mathbf{v}_j\rangle=\sum_{i=1}^{j-1}\langle a_i\mathbf{v}_i, a_j\mathbf{v}_j\rangle=\sum_{i=1}^{j-1}a_i\bar{a}_j\langle\mathbf{v}_i,\mathbf{v}_j\rangle=0$. Then

$$\left\|\sum_{i=1}^{k}a_i\mathbf{v}_i\right\|^2=\|(a_1\mathbf{v}_1+a_2\mathbf{v}_2+\cdots+a_{k-1}\mathbf{v}_{k-1})+a_k\mathbf{v}_k\|^2$$
$$=\|(a_1\mathbf{v}_1+a_2\mathbf{v}_2+\cdots+a_{k-1}\mathbf{v}_{k-1})\|^2+\|a_k\mathbf{v}_k\|^2$$
$$=\|(a_1\mathbf{v}_1+a_2\mathbf{v}_2+\cdots+a_{k-1}\mathbf{v}_{k-1})\|^2+|a_k|^2\|\mathbf{v}_k\|^2$$
$$=\|(a_1\mathbf{v}_1+a_2\mathbf{v}_2+\cdots+a_{k-2}\mathbf{v}_{k-2})+a_{k-1}\mathbf{v}_{k-1}\|^2+|a_k|^2\|\mathbf{v}_k\|^2$$
$$=\|(a_1\mathbf{v}_1+a_2\mathbf{v}_2+\cdots+a_{k-2}\mathbf{v}_{k-2})\|^2+\|a_{k-1}\mathbf{v}_{k-1}\|^2+|a_k|^2\|\mathbf{v}_k\|^2$$
$$=\|(a_1\mathbf{v}_1+a_2\mathbf{v}_2+\cdots+a_{k-2}\mathbf{v}_{k-2})\|^2+|a_{k-1}|^2\|\mathbf{v}_{k-1}\|^2+|a_k|^2\|\mathbf{v}_k\|^2$$
$$=\ldots\ldots\ldots\ldots\ldots\ldots\ldots\ldots\ldots\ldots\ldots\ldots\ldots\ldots\ldots\ldots\ldots\ldots\ldots$$
$$=\|a_1\mathbf{v}_1\|^2+|a_2|^2\|\mathbf{v}_2\|^2+\cdots+|a_k|^2\|\mathbf{v}_k\|^2$$
$$=|a_1|^2\|\mathbf{v}_1\|^2+|a_2|^2\|\mathbf{v}_2\|^2+\cdots+|a_k|^2\|\mathbf{v}_k\|^2.$$

If the orthogonal set of vectors $\{\mathbf{v}_1,\mathbf{v}_2,\ldots,\mathbf{v}_k\}$ in Theorem 7.3.3 is in fact orthonormal, then the next result follows immediately.

Corollary 7.3.1

If $\{\mathbf{v}_1,\mathbf{v}_2,\ldots,\mathbf{v}_k\}$ is an orthonormal set of vectors in V, then

$$\|a_1\mathbf{v}_1+a_2\mathbf{v}_2+\cdots+a_k\mathbf{v}_k\|^2=|a_1|^2+|a_2|^2+\cdots+|a_k|^2$$

for all scalars $a_1,a_2,\ldots,a_k$.

One of the advantages of using an orthonormal basis for an inner product space V is that it is very easy to compute the coordinates of a vector with respect to such a basis.

Theorem 7.3.4

If $\{\mathbf{v}_1, \mathbf{v}_2, \ldots, \mathbf{v}_n\}$ is an orthogonal basis for V, then

(1) $\mathbf{v} = \frac{\langle \mathbf{v}, \mathbf{v}_1 \rangle}{\|\mathbf{v}_1\|^2} \mathbf{v}_1 + \frac{\langle \mathbf{v}, \mathbf{v}_2 \rangle}{\|\mathbf{v}_2\|^2} \mathbf{v}_2 + \cdots + \frac{\langle \mathbf{v}, \mathbf{v}_n \rangle}{\|\mathbf{v}_n\|^2} \mathbf{v}_n$

and

(2) $\|\mathbf{v}\|^2 = \frac{|\langle \mathbf{v}, \mathbf{v}_1 \rangle|^2}{\|\mathbf{v}_1\|^2} + \frac{|\langle \mathbf{v}, \mathbf{v}_2 \rangle|^2}{\|\mathbf{v}_2\|^2} + \cdots + \frac{|\langle \mathbf{v}, \mathbf{v}_n \rangle|^2}{\|\mathbf{v}_n\|^2}$

for every $\mathbf{v} \in V$.

Proof Let $\mathbf{v} \in V$. Because $\{\mathbf{v}_1, \mathbf{v}_2, \ldots, \mathbf{v}_n\}$ is a basis for V, there exist unique scalars $a_1, a_2, \ldots, a_n$ such that

$$\mathbf{v} = a_1 \mathbf{v}_1 + a_2 \mathbf{v}_2 + \cdots + a_n \mathbf{v}_n.$$

Then, for any $\mathbf{v}_j$, we have

$$\langle \mathbf{v}, \mathbf{v}_j \rangle = \langle a_1 \mathbf{v}_1 + a_2 \mathbf{v}_2 + \cdots + a_n \mathbf{v}_n, \mathbf{v}_j \rangle = \sum_{i=1}^{n} a_i \langle \mathbf{v}_i, \mathbf{v}_j \rangle = a_j \|\mathbf{v}_j\|^2,$$

so statement (1) is true. Finally, statement (2) follows from statement (1) and Theorem 7.3.3.

When the basis in Theorem 7.3.4 is an orthonormal set, we get a particularly neat result.

Corollary 7.3.2

If $\{\mathbf{v}_1, \mathbf{v}_2, \ldots, \mathbf{v}_n\}$ is an orthonormal basis for V, then

(1) $\mathbf{v} = \langle \mathbf{v}, \mathbf{v}_1 \rangle \mathbf{v}_1 + \langle \mathbf{v}, \mathbf{v}_2 \rangle \mathbf{v}_2 + \cdots + \langle \mathbf{v}, \mathbf{v}_n \rangle \mathbf{v}_n.$
(2) $\|\mathbf{v}\|^2 = |\langle \mathbf{v}, \mathbf{v}_1 \rangle|^2 + |\langle \mathbf{v}, \mathbf{v}_2 \rangle|^2 + \cdots + |\langle \mathbf{v}, \mathbf{v}_n \rangle|^2$
for every $\mathbf{v} \in V$.

In conclusion (1) of Corollary 7.3.2, the unique coordinates of $\mathbf{v}$ relative to the orthonormal basis $\{\mathbf{v}_1, \mathbf{v}_2, \ldots, \mathbf{v}_n\}$ are called the **Fourier coefficients** of $\mathbf{v}$ with respect to the basis.* The relation shown in conclusion (2) is called **Parseval's identity**.

Exercises 7.3

A.

1. Consider the space $\mathbb{R}^3$ with the standard inner product. Determine which of the following bases are orthogonal, and then determine which are orthonormal:

---

* (Jean-Baptiste) Joseph Fourier (1768–1830) was a French mathematician and engineer who made contributions to various areas of mathematics, as well as to mechanics and the theory of heat transfer.

(a) $B = \left\{ \begin{bmatrix} 1 \\ 2 \\ 3 \end{bmatrix}, \begin{bmatrix} 1 \\ 0 \\ 2 \end{bmatrix}, \begin{bmatrix} -2 \\ -1 \\ 3 \end{bmatrix} \right\}$

(b) $B = \left\{ \begin{bmatrix} \frac{1}{2}\sqrt{2} \\ 0 \\ -\frac{1}{2}\sqrt{2} \end{bmatrix}, \begin{bmatrix} 0 \\ 1 \\ 0 \end{bmatrix}, \begin{bmatrix} \frac{1}{2}\sqrt{2} \\ 0 \\ \frac{1}{2}\sqrt{2} \end{bmatrix} \right\}$

(c) $B = \left\{ \begin{bmatrix} 1 \\ -1 \\ 2 \end{bmatrix}, \begin{bmatrix} -2 \\ 0 \\ 1 \end{bmatrix}, \begin{bmatrix} 1 \\ 5 \\ 2 \end{bmatrix} \right\}$

(d) $B = \left\{ \begin{bmatrix} 1 \\ 1 \\ 0 \end{bmatrix}, \begin{bmatrix} 1 \\ -1 \\ 1 \end{bmatrix}, \begin{bmatrix} -1 \\ 1 \\ 2 \end{bmatrix} \right\}$

2. Consider the unitary space $\mathbb{C}^2$ with the standard inner product. Determine which of the following bases are orthogonal, and then determine which are orthonormal:

(a) $B = \left\{ \begin{bmatrix} 1 \\ 0 \end{bmatrix}, \begin{bmatrix} 0 \\ i \end{bmatrix} \right\}$ (b) $B = \left\{ \begin{bmatrix} i \\ 1 \end{bmatrix}, \begin{bmatrix} 2 \\ i \end{bmatrix} \right\}$

(c) $B = \left\{ \begin{bmatrix} 2 \\ i \end{bmatrix}, \begin{bmatrix} -1 \\ 2i \end{bmatrix} \right\}$

3. Assuming the standard inner product for $\mathbb{C}^2$, for what values of a is the basis $\left\{ \begin{bmatrix} a \\ -a \end{bmatrix}, \begin{bmatrix} i/\sqrt{2} \\ i/\sqrt{2} \end{bmatrix} \right\}$ an orthonormal basis for $\mathbb{C}^2$?

4. Verify that $\left\{ \begin{bmatrix} 1 & 0 \\ 0 & 0 \end{bmatrix}, \begin{bmatrix} 0 & 1 \\ 0 & 0 \end{bmatrix}, \begin{bmatrix} 0 & 0 \\ 1 & 0 \end{bmatrix}, \begin{bmatrix} 0 & 0 \\ 0 & 1 \end{bmatrix} \right\}$ is an orthonormal basis for M_2 with the standard inner product.

5. In the space of real integrable functions on $\mathbb{R}$ with inner product defined by $\langle f, g \rangle = \int_{-1}^{1} f(x)g(x)dx$, find a polynomial function of degree 2 that is orthogonal to both f and g, where $f(x) = 2$ and $g(x) = x + 1$ for all $x \in \mathbb{R}$.

6. If $\mathbf{u} = \begin{bmatrix} 1 \\ -i \end{bmatrix}$ and $\mathbf{v} = \begin{bmatrix} i \\ 1 \end{bmatrix}$ in $\mathbb{C}^2$ with the standard inner product, show that $\|\mathbf{u} + \mathbf{v}\|^2 = 4 = \|\mathbf{u}\|^2 + \|\mathbf{v}\|^2$ and $\langle \mathbf{u}, \mathbf{v} \rangle = -2i$. (See the comment after Theorem 7.3.2.)

7. Verify that the set $\{1/\sqrt{2}, \sqrt{6}x/2, (3\sqrt{10}/4)(x^2 - \frac{1}{3})\}$ given in Example 7.3.3 is an orthonormal basis for the inner product space $P_2[x]$ described there.

B.

1. Prove Theorem 7.3.2.

2. Prove Theorem 7.3.1 by using the result of Theorem 7.3.4.

3. Assume that $\{v_1, v_2, \ldots, v_n\}$ is an arbitrary linearly independent set in a complex vector space V. Define a function $\langle \, , \, \rangle \colon V \times V \to \mathbb{C}$ by

    $$\left\langle \sum_{i=1}^{n} a_i v_i, \sum_{j=1}^{n} b_j v_j \right\rangle = \sum_{k=1}^{n} a_k \bar{b}_k.$$

    Show that this function defines an inner product with respect to which $\{v_1, v_2, \ldots, v_n\}$ is orthonormal.

4. A real symmetric $n \times n$ matrix A is called *positive-definite* if $x^T A x > 0$ for all nonzero $x \in \mathbb{R}^n$ (see Section 3.2 and Problem C3 in Exercises 3.2).

    Suppose $B = \{v_1, v_2, \ldots, v_n\}$ is any ordered basis for $\mathbb{R}^n$. Prove that in $\mathbb{R}^n$, a function $\mathbb{R}^n \times \mathbb{R}^n \to \mathbb{R}$ is an inner product $\langle x, y \rangle$ if and only if there exists an $n \times n$ positive-definite symmetric matrix A such that $[x]_B^T A [y]_B = \langle x, y \rangle$ for every $x = \sum_{i=1}^{n} x_i v_i, y = \sum_{j=1}^{n} y_j v_j \in \mathbb{R}^n$. [*Hint*: Consider $A = [a_{ij}]$, where $a_{ij} = \langle v_i, v_j \rangle$ for $i, j = 1, 2, \ldots, n$.]

5. Consider the vector space $P_2[x]$ with inner product defined by $\langle f, g \rangle = \int_0^1 f(x)g(x)\,dx$ for $f, g \in P_2[x]$. Use the result of the preceding exercise, including the hint given, to represent this inner product on $P_2[x]$ in matrix form.

C.

1. Consider the set $S = \{e_k(t) = e^{ikt} : k = 0, \pm 1, \pm 2 \ldots\}$, where $e^{ikt} = \cos(kt) + i\sin(kt)$ by *Euler's formula* (see Appendix D).

    (a) Show that S is orthogonal with respect to the inner product:

    $$\langle f, g \rangle = \int_0^{2\pi} f(t)\overline{g(t)}\,dt.$$

    (b) Show that $\|e_k\|^2 = 2\pi$ for all k.
    (c) Show that $\hat{S} = \{f_k(t)\} = \{1, \cos(kt), \sin(kt) : k = 1, 2, 3 \ldots\}$ is orthogonal with respect to the inner product defined in part (a).
    (d) Show that $\|f_k\|^2 = \pi$ for all members f_k of the set $\hat{S}$ except $f_0 = 1$. What is $\|f_0\|^2$?

2. Let $b > 0$ and c be given, and let $f_n(x) = \sin(\alpha_n x)$, where α_n is the nth positive root of $\tan(\alpha b) = c\alpha$. Show that $\{f_k\}_{k=1}^{\infty}$ is orthogonal with respect to the inner product $\langle f, g \rangle = \int_0^b f(x)g(x)\,dx$. [*Hint*: Notice that $f_n'' = -\alpha_n^2 f_n$. Then, consider $\langle f_n'', f_m \rangle$ and integrate by parts twice.]

3. Define

$$f_k(x) = \begin{cases} 1 & \text{if } k/n \le x < (k+1)/n \\ 0 & \text{otherwise} \end{cases} , \quad k = 0, 1, 2, \ldots, n-1.$$

Show that $\{f_k\}_{k=0}^{n-1}$ is an orthogonal set with respect to the inner product $\langle f, g \rangle = \int_0^1 f(x)g(x)\,dx$.

7.4 The Gram–Schmidt Process

Even though we may be impressed by the usefulness of orthonormal bases in the examples we have seen, the question of *existence* remains: Can we always find an orthonormal basis for an inner product space? The answer to this burning question is positive: **Every finite-dimensional inner product space has an orthonormal basis**.

The proof of this important result provides a practical method, the **Gram–Schmidt process**,* for constructing orthonormal bases (see Example 1.5.2 for an early glimpse at this method). The algorithm works by turning a linearly independent set into an orthonormal set, one vector at a time. Specifically, the procedure starts with an arbitrary basis $B = \{\mathbf{v}_1, \mathbf{v}_2, \ldots, \mathbf{v}_n\}$ for the n-dimensional inner product space V and uses B to construct an orthonormal basis $\hat{B} = \{\hat{\mathbf{v}}_1, \hat{\mathbf{v}}_2, \ldots, \hat{\mathbf{v}}_n\}$ *sequentially* in such a way that $\hat{B}_k = \{\hat{\mathbf{v}}_1, \hat{\mathbf{v}}_2, \ldots, \hat{\mathbf{v}}_k\}$ is an orthonormal basis for $V_k = span\{\mathbf{v}_1, \mathbf{v}_2, \ldots, \mathbf{v}_k\}$, $k = 1, 2, \ldots, n$. Before getting into the details of the algorithm, we pause to understand the geometry that is at the heart of the process; and this is best done in $\mathbb{R}^n$, particularly in $\mathbb{R}^2$ and $\mathbb{R}^3$.

If $\mathbf{u}$ and $\mathbf{v}$ are vectors in $\mathbb{R}^n$ and $\mathbf{u} \ne \mathbf{0}$, then the **orthogonal projection of v onto u** is the vector $\mathbf{proj}_{\mathbf{u}}(\mathbf{v})$ defined by $\mathbf{p} = \mathbf{proj}_{\mathbf{u}}(\mathbf{v}) = (\mathbf{u} \bullet \mathbf{v}/\mathbf{u} \bullet \mathbf{u})\mathbf{u}$. Here is an illustration of the orthogonal projection of a vector $\mathbf{v}$ onto another vector $\mathbf{u}$ in $\mathbb{R}^2$ (Figure 7.1).

Loosely speaking, this projection indicates "how much" of vector $\mathbf{v}$ is pointing in the direction of vector $\mathbf{u}$. The expression **component of v**

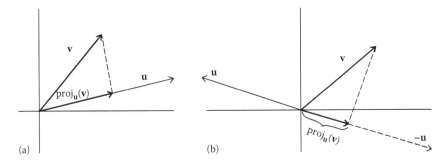

(a)   (b)

Figure 7.1 Orthogonal projections

---

* The algorithm was named for Jorgen P. Gram, a Danish actuary who gave an *implicit* version of the procedure in 1883, and Erhard Schmidt, a German mathematician who used the process explicitly in 1907 in his work on integral equations.

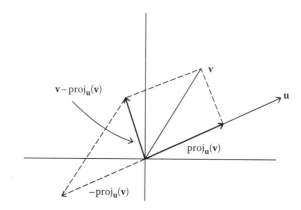

Figure 7.2 $\mathbf{u}$ is orthogonal to $\mathbf{v} - \text{proj}_u(\mathbf{v})$.

in the direction of u is also used for $\text{proj}_u(\mathbf{v})$. To understand the orthonormalization algorithm (Theorem 7.4.1), it is important to realize that **u is orthogonal to $\mathbf{v} - \text{proj}_u(\mathbf{v})$ for all vectors u and v in $\mathbb{R}^n$,** where $\mathbf{u} \neq \mathbf{0}$ (Figure 7.2). (Also see Exercise B1.)

Thus, starting with two vectors **u** and **v**, we keep **u** and use **v** to produce a vector $\mathbf{u}'$ orthogonal to **u**. Finally, we normalize by taking $\mathbf{u}/\|\mathbf{u}\|$ and $\mathbf{u}'/\|\mathbf{u}'\|$. The resulting set $\{\mathbf{u}/\|\mathbf{u}\|, \mathbf{u}'/\|\mathbf{u}'\|\}$ is an orthonormal set. An example in $\mathbb{R}^3$ will serve as a model and aid our understanding of the process.

Example 7.4.1: Orthonormalizing Vectors in $\mathbb{R}^3$

The vectors $\mathbf{v}_1 = \begin{bmatrix} 1 \\ 1 \\ 1 \end{bmatrix}, \mathbf{v}_2 = \begin{bmatrix} 0 \\ 1 \\ 0 \end{bmatrix}$, and $\mathbf{v}_3 = \begin{bmatrix} -1 \\ -1 \\ 1 \end{bmatrix}$ form a basis for $\mathbb{R}^3$. With respect to the standard inner product on $\mathbb{R}^3$, this is not an orthogonal set, nor are the vectors $\mathbf{v}_1$ and $\mathbf{v}_3$ of unit length. We will construct an orthogonal set of vectors from $\mathbf{v}_1, \mathbf{v}_2$, and $\mathbf{v}_3$ and normalize each vector at the end.

We start with $\mathbf{v}_1' = \mathbf{v}_1$. Now

$$\mathbf{v}_2' = \mathbf{v}_2 - \text{proj}_{\mathbf{v}_1'}(\mathbf{v}_2) = \mathbf{v}_2 - \left(\frac{\mathbf{v}_1' \cdot \mathbf{v}_2}{\mathbf{v}_1' \cdot \mathbf{v}_1'} \right) \cdot \mathbf{v}_1'$$

$$= \begin{bmatrix} 0 \\ 1 \\ 0 \end{bmatrix} - \frac{\begin{bmatrix} 1 \\ 1 \\ 1 \end{bmatrix} \cdot \begin{bmatrix} 0 \\ 1 \\ 0 \end{bmatrix}}{\begin{bmatrix} 1 \\ 1 \\ 1 \end{bmatrix} \cdot \begin{bmatrix} 1 \\ 1 \\ 1 \end{bmatrix}} \begin{bmatrix} 1 \\ 1 \\ 1 \end{bmatrix} = \begin{bmatrix} -1/3 \\ 2/3 \\ -1/3 \end{bmatrix}.$$

Note that $\mathbf{v}_2' \perp \mathbf{v}_1'$. Now we must ensure that the third vector $\mathbf{v}_3'$ we construct is orthogonal to both $\mathbf{v}_1'$ and $\mathbf{v}_2'$. To do this, we subtract the projection of $\mathbf{v}_3$ on $\mathbf{v}_1'$ and the projection of $\mathbf{v}_3$ on $\mathbf{v}_2'$ from $\mathbf{v}_3$:

$$\mathbf{v}_3' = \mathbf{v}_3 - \mathbf{proj}_{\mathbf{v}_1'}(\mathbf{v}_3) - \mathbf{proj}_{\mathbf{v}_2'}(\mathbf{v}_3)$$

$$= \mathbf{v}_3 - \left(\frac{\mathbf{v}_1' \bullet \mathbf{v}_3}{\mathbf{v}_1' \bullet \mathbf{v}_1'}\right) \cdot \mathbf{v}_1' - \left(\frac{\mathbf{v}_2' \bullet \mathbf{v}_3}{\mathbf{v}_2' \bullet \mathbf{v}_2'}\right) \cdot \mathbf{v}_2'$$

$$= \begin{bmatrix} -1 \\ -1 \\ 1 \end{bmatrix} - \frac{\begin{bmatrix} 1 \\ 1 \\ 1 \end{bmatrix} \bullet \begin{bmatrix} -1 \\ -1 \\ 1 \end{bmatrix}}{\begin{bmatrix} 1 \\ 1 \\ 1 \end{bmatrix} \bullet \begin{bmatrix} 1 \\ 1 \\ 1 \end{bmatrix}} \begin{bmatrix} 1 \\ 1 \\ 1 \end{bmatrix} - \frac{\begin{bmatrix} -1/3 \\ 2/3 \\ -1/3 \end{bmatrix} \bullet \begin{bmatrix} -1 \\ -1 \\ 1 \end{bmatrix}}{\begin{bmatrix} -1/3 \\ 2/3 \\ -1/3 \end{bmatrix} \bullet \begin{bmatrix} -1/2 \\ 2/3 \\ -1/3 \end{bmatrix}}$$

$$\times \begin{bmatrix} -1/3 \\ 2/3 \\ -1/3 \end{bmatrix} = \begin{bmatrix} -1 \\ 0 \\ 1 \end{bmatrix}.$$

Finally, we normalize $\mathbf{v}_1'$, $\mathbf{v}_2'$, and $\mathbf{v}_3'$ to get our orthonormal basis:

$$\hat{\mathbf{v}}_1 = \frac{\mathbf{v}_1'}{\|\mathbf{v}_1'\|} = \frac{1}{\sqrt{3}}\begin{bmatrix} 1 \\ 1 \\ 1 \end{bmatrix}, \quad \hat{\mathbf{v}}_2 = \frac{\mathbf{v}_2'}{\|\mathbf{v}_2'\|} = \frac{1}{\sqrt{6}}\begin{bmatrix} -1 \\ 2 \\ -1 \end{bmatrix} \quad \text{and}$$

$$\hat{\mathbf{v}}_3 = \frac{\mathbf{v}_3'}{\|\mathbf{v}_3'\|} = \frac{1}{\sqrt{2}}\begin{bmatrix} -1 \\ 0 \\ 1 \end{bmatrix}.$$

Even though the "official" statement of the Gram–Schmidt process given below starts with a basis for a subspace, we can start with any linearly independent set of vectors in an inner product space. By Corollary 5.4.1, we can extend such a set to form a basis. In fact, we can apply the Gram–Schmidt process to any finite set of vectors spanning a subspace S of an inner product space, provided that we take care to remove any occurrences of the zero vector in the process.

Theorem 7.4.1: The Gram–Schmidt Orthonormalization Process

Given a basis $\{\mathbf{v}_1, \mathbf{v}_2, \ldots, \mathbf{v}_k\}$ for a subspace S of an inner product space, define

$$\mathbf{v}_1' = \mathbf{v}_1$$

$$\mathbf{v}_2' = \mathbf{v}_2 - \frac{\langle \mathbf{v}_2, \mathbf{v}_1' \rangle}{\|\mathbf{v}_1'\|^2}\mathbf{v}_1'$$

$$\mathbf{v}_3' = \mathbf{v}_3 - \frac{\langle \mathbf{v}_3, \mathbf{v}_1' \rangle}{\|\mathbf{v}_1'\|^2}\mathbf{v}_1' - \frac{\langle \mathbf{v}_3, \mathbf{v}_2' \rangle}{\|\mathbf{v}_2'\|^2}\mathbf{v}_2'$$

$$\vdots$$

$$\mathbf{v}_k' = \mathbf{v}_k - \frac{\langle \mathbf{v}_k, \mathbf{v}_1' \rangle}{\|\mathbf{v}_1'\|^2}\mathbf{v}_1' - \frac{\langle \mathbf{v}_k, \mathbf{v}_2' \rangle}{\|\mathbf{v}_2'\|^2}\mathbf{v}_2' - \cdots - \frac{\langle \mathbf{v}_k, \mathbf{v}_{k-1}' \rangle}{\|\mathbf{v}_{k-1}'\|^2}\mathbf{v}_{k-1}'.$$

Then, $\{\mathbf{v}'_1/\|\mathbf{v}'_1\|, \mathbf{v}'_2/\|\mathbf{v}'_2\|, \ldots, \mathbf{v}'_k/\|\mathbf{v}'_k\|\}$ is an orthonormal basis for S. In addition, $span\{\mathbf{v}'_1, \mathbf{v}'_2, \ldots, \mathbf{v}'_j\} = span\{\mathbf{v}_1, \mathbf{v}_2, \ldots, \mathbf{v}_j\}$ for $1 \leq j \leq k$.

Proof The proof is by mathematical induction on j, the dimension of the subspace $S_j = span\{\mathbf{v}_1, \mathbf{v}_2, \ldots, \mathbf{v}_j\}$, $1 \leq j \leq k$.

Because $\mathbf{v}'_1 = \mathbf{v}_1$, we have $span\{\mathbf{v}'_1\} = span\{\mathbf{v}_1\}$. Now, assume that the set $\{\mathbf{v}'_1, \mathbf{v}'_2, \ldots, \mathbf{v}'_j\}$ has been constructed by the formulas given here to be an orthogonal basis for S_j. We show that the set $\{\mathbf{v}'_1, \mathbf{v}'_2, \ldots, \mathbf{v}'_j, \mathbf{v}'_{j+1}\}$ is an orthogonal basis for $S_{j+1} = span\{\mathbf{v}_1, \mathbf{v}_2, \ldots, \mathbf{v}_j, \mathbf{v}_{j+1}\}$. First of all, $\mathbf{v}'_{j+1} \neq \mathbf{0}$, because otherwise $\mathbf{v}_{j+1}$ would be a linear combination of the vectors $\{\mathbf{v}'_1, \mathbf{v}'_2, \ldots, \mathbf{v}'_j\}$ and hence a combination of $\mathbf{v}_1, \mathbf{v}_2, \ldots, \mathbf{v}_j$, contradicting the assumption that the set $\{\mathbf{v}_1, \mathbf{v}_2, \ldots, \mathbf{v}_k\}$ is linearly independent. Next, if $1 \leq i \leq j$, then

$$\langle \mathbf{v}'_{j+1}, \mathbf{v}'_i \rangle = \left\langle \mathbf{v}_{j+1} - \sum_{p=1}^{j} \frac{\langle \mathbf{v}_{j+1}, \mathbf{v}'_p \rangle}{\|\mathbf{v}'_p\|^2} \mathbf{v}'_p, \mathbf{v}'_i \right\rangle$$

$$= \langle \mathbf{v}_{j+1}, \mathbf{v}'_i \rangle - \sum_{p=1}^{j} \frac{\langle \mathbf{v}_{j+1}, \mathbf{v}'_p \rangle}{\|\mathbf{v}'_p\|^2} \langle \mathbf{v}'_p, \mathbf{v}'_i \rangle$$

$$= \langle \mathbf{v}_{j+1}, \mathbf{v}'_i \rangle - \langle \mathbf{v}_{j+1}, \mathbf{v}'_i \rangle = 0.$$

Therefore, $\{\mathbf{v}'_1, \mathbf{v}'_2, \ldots, \mathbf{v}'_j, \mathbf{v}'_{j+1}\}$ is an *orthogonal* set consisting of $j + 1$ nonzero vectors in the subspace spanned by $\{\mathbf{v}_1, \mathbf{v}_2, \ldots, \mathbf{v}_j, \mathbf{v}_{j+1}\}$. By Theorem 7.2.1, the set $\{\mathbf{v}'_1, \mathbf{v}'_2, \ldots, \mathbf{v}'_j, \mathbf{v}'_{j+1}\}$ is a basis for this subspace.

Thus, the vectors $\mathbf{v}'_1, \mathbf{v}'_2, \ldots, \mathbf{v}'_k$ may be constructed sequentially as indicated in the statement of the theorem. Normalizing the orthogonal basis, we have constructed is the last step in the process.

The next three examples illustrate the Gram–Schmidt process in different inner product spaces.

Example 7.4.2: Orthonormalizing a Basis in $\mathbb{C}^3$

Consider the unitary space $\mathbb{C}^3$ with the standard inner product (Example 7.2.1(b)). The vectors $\mathbf{v}_1 = \begin{bmatrix} 1 \\ 0 \\ 1 \end{bmatrix}, \mathbf{v}_2 = \begin{bmatrix} 1 \\ 1 \\ 0 \end{bmatrix}$, and

$\mathbf{v}_3 = \begin{bmatrix} 0 \\ i \\ 1+i \end{bmatrix}$ form a basis for $\mathbb{C}^3$, and we will apply the

Gram–Schmidt process to find an *orthonormal* basis for $\mathbb{C}^3$. Define

$$\mathbf{v}_1' = \mathbf{v}_1 = \begin{bmatrix} 1 \\ 0 \\ 1 \end{bmatrix}$$

$$\mathbf{v}_2' = \mathbf{v}_2 - \frac{\langle \mathbf{v}_2, \mathbf{v}_1' \rangle}{\|\mathbf{v}_1'\|^2} \mathbf{v}_1' = \begin{bmatrix} 1 \\ 1 \\ 0 \end{bmatrix} - \frac{1}{2} \begin{bmatrix} 1 \\ 0 \\ 1 \end{bmatrix} = \begin{bmatrix} 1/2 \\ 1 \\ -1/2 \end{bmatrix}$$

$$\mathbf{v}_3' = \mathbf{v}_3 - \frac{\langle \mathbf{v}_3, \mathbf{v}_1' \rangle}{\|\mathbf{v}_1'\|^2} \mathbf{v}_1' - \frac{\langle \mathbf{v}_3, \mathbf{v}_2' \rangle}{\|\mathbf{v}_2'\|^2} \mathbf{v}_2'$$

$$= \begin{bmatrix} 0 \\ i \\ 1+i \end{bmatrix} - \frac{1}{3}(i-1) \begin{bmatrix} 1/2 \\ 1 \\ -1/2 \end{bmatrix} - \frac{1}{2}(i+1) \begin{bmatrix} 1 \\ 0 \\ 1 \end{bmatrix} = \frac{1}{3} \begin{bmatrix} -1-2i \\ 1+2i \\ 1+2i \end{bmatrix}.$$

Now, all we have to do is normalize $\mathbf{v}_1'$, $\mathbf{v}_2'$, and $\mathbf{v}_3'$:

$$\mathbf{v}_1'/\|\mathbf{v}_1'\| = \sqrt{2}/2 \begin{bmatrix} 1 \\ 0 \\ 1 \end{bmatrix}, \quad \mathbf{v}_2'/\|\mathbf{v}_2'\| = \sqrt{6}/3 \begin{bmatrix} 1/2 \\ 1 \\ -1/2 \end{bmatrix}, \quad \text{and}$$

$\mathbf{v}_3'/\|\mathbf{v}_3'\| = \sqrt{15}/15 \begin{bmatrix} -1-2i \\ 1+2i \\ 1+2i \end{bmatrix}$. It is easy to confirm that these

vectors are orthonormal. (Note that we could have avoided unnecessary fractions if we had taken the mutually orthogonal vectors $\mathbf{v}_1'$, $2\mathbf{v}_2'$, and $3\mathbf{v}_3'$ and normalized these.)

Example 7.4.3: Orthonormalizing a Subset in M_2

The span of the vectors $\mathbf{v}_1 = \begin{bmatrix} 3 & 5 \\ -1 & 1 \end{bmatrix}$, $\mathbf{v}_2 = \begin{bmatrix} -1 & 9 \\ 5 & -1 \end{bmatrix}$,

and $\mathbf{v}_3 = \begin{bmatrix} 7 & -17 \\ 2 & -6 \end{bmatrix}$ is a (three-dimensional) subspace of M_2.

We will find an orthonormal basis for this subspace with respect to the inner product defined by $\langle A, B \rangle = \sum_{i=1}^{2} \sum_{j=1}^{2} a_{ij} b_{ij} = \text{trace}(A^\mathsf{T} B)$.

Using Theorem 7.4.1, we define

$$\mathbf{v}_1' = \mathbf{v}_1 = \begin{bmatrix} 3 & 5 \\ -1 & 1 \end{bmatrix},$$

$$\mathbf{v}_2' = \mathbf{v}_2 - \frac{\langle \mathbf{v}_2, \mathbf{v}_1' \rangle}{\|\mathbf{v}_1'\|^2} \mathbf{v}_1' = \begin{bmatrix} -1 & 9 \\ 5 & -1 \end{bmatrix} - \frac{36}{36} \begin{bmatrix} 3 & 5 \\ -1 & 1 \end{bmatrix} = \begin{bmatrix} -4 & 4 \\ 6 & -2 \end{bmatrix},$$

$$\mathbf{v}_3' = \mathbf{v}_3 - \frac{\langle \mathbf{v}_3, \mathbf{v}_1' \rangle}{\|\mathbf{v}_1'\|^2} \mathbf{v}_1' - \frac{\langle \mathbf{v}_3, \mathbf{v}_2' \rangle}{\|\mathbf{v}_2'\|^2} \mathbf{v}_2'$$

$$= \begin{bmatrix} 7 & -17 \\ 2 & -6 \end{bmatrix} - \frac{(-72)}{36} \begin{bmatrix} 3 & 5 \\ -1 & 1 \end{bmatrix} - \frac{(-72)}{72} \begin{bmatrix} -4 & 4 \\ 6 & -2 \end{bmatrix} = \begin{bmatrix} 9 & -3 \\ 6 & -6 \end{bmatrix}.$$

Normalizing $\mathbf{v}_1'$, $\mathbf{v}_2'$, and $\mathbf{v}_3'$, we get the following orthonormal vectors as a basis for $span\{\mathbf{v}_1, \mathbf{v}_2, \mathbf{v}_3\}$:

$$\frac{\mathbf{v}_1'}{\|\mathbf{v}_1'\|} = \frac{1}{6}\begin{bmatrix} 3 & 5 \\ -1 & 1 \end{bmatrix}, \quad \frac{\mathbf{v}_2'}{\|\mathbf{v}_2'\|} = \frac{1}{6\sqrt{2}}\begin{bmatrix} -1 & 9 \\ 5 & -1 \end{bmatrix}, \quad \text{and} \quad \frac{\mathbf{v}_3'}{\|\mathbf{v}_3'\|} = \frac{1}{9\sqrt{2}}\begin{bmatrix} 7 & -17 \\ 2 & -6 \end{bmatrix}.$$

(Note, for example, that $\|\mathbf{v}_1'\|^2 = \langle \mathbf{v}_1', \mathbf{v}_1' \rangle = \text{trace}\,(\mathbf{v}_1'^T, \mathbf{v}_1') = $
$\text{trace}\left(\begin{bmatrix} 3 & -1 \\ 5 & 1 \end{bmatrix}\begin{bmatrix} 3 & 5 \\ -1 & 1 \end{bmatrix}\right) = \text{trace}\left(\begin{bmatrix} 10 & 14 \\ 14 & 26 \end{bmatrix}\right) = 36,$ so
$\|\mathbf{v}_1'\| = \sqrt{36} = 6.$)

Example 7.4.4: Orthogonalizing a Subset in $C[-1, 1]$

An important inner product on $C[-1, 1]$ is defined by
$\langle f, g \rangle = \frac{2}{\pi} \int_{-1}^{1} \frac{f(x)g(x)}{\sqrt{1-x^2}} \, dx.$
This is an example of an **inner product with respect to a weight function**. In this case, the weight function $(2/\pi)/\sqrt{1-x^2}$ emphasizes the values of x for which $|x|$ is near 1.

The set $\{\mathbf{v}_0, \mathbf{v}_1, \mathbf{v}_2, \ldots, \mathbf{v}_n \ldots\} = \{1, x, x^2, \ldots, x^n \ldots\}$ is a basis for $\mathbb{R}[x]$, the space of all polynomials with real coefficients, which is a subspace of $C[-1, 1]$. We will use the Gram–Schmidt process on the first four vectors of this set to find polynomials $T_0(x), T_1(x), T_2(x),$ and $T_3(x)$ such that $\langle T_i, T_j \rangle = 0$ when $i \neq j$. (We will do without normalization in this case.) These functions are the first four *Chebyshev polynomials of the first kind* and are useful in approximating functions by polynomials:*

$\mathbf{v}_0' = \mathbf{v}_0 = 1$

$$\mathbf{v}_1' = \mathbf{v}_1 - \frac{\langle \mathbf{v}_1, \mathbf{v}_0' \rangle}{\|\mathbf{v}_0'\|^2}\mathbf{v}_0' = x - \frac{\frac{2}{\pi}\int_{-1}^{1}\frac{x \cdot 1}{\sqrt{1-x^2}}dx}{\frac{2}{\pi}\int_{-1}^{1}\frac{1 \cdot 1}{\sqrt{1-x^2}}dx} \cdot 1 = x$$

$$\mathbf{v}_2' = \mathbf{v}_2 - \frac{\langle \mathbf{v}_2, \mathbf{v}_0' \rangle}{\|\mathbf{v}_0'\|^2}\mathbf{v}_0' - \frac{\langle \mathbf{v}_2, \mathbf{v}_1' \rangle}{\|\mathbf{v}_1'\|^2}\mathbf{v}_1'$$

$$= x^2 - \frac{\frac{2}{\pi}\int_{-1}^{1}\frac{x^2 \cdot 1}{\sqrt{1-x^2}}dx}{\frac{2}{\pi}\int_{-1}^{1}\frac{1 \cdot 1}{\sqrt{1-x^2}}dx} \cdot 1 - \frac{\frac{2}{\pi}\int_{-1}^{1}\frac{x^2 \cdot x}{\sqrt{1-x^2}}dx}{\frac{2}{\pi}\int_{-1}^{1}\frac{x \cdot x}{\sqrt{1-x^2}}dx} \cdot x = \frac{2x^2 - 1}{2}$$

$$\mathbf{v}_2' = \mathbf{v}_3 - \frac{\langle \mathbf{v}_3, \mathbf{v}_0' \rangle}{\|\mathbf{v}_0'\|^2}\mathbf{v}_0' - \frac{\langle \mathbf{v}_3, \mathbf{v}_1' \rangle}{\|\mathbf{v}_1'\|^2}\mathbf{v}_1' - \frac{\langle \mathbf{v}_3, \mathbf{v}_2' \rangle}{\|\mathbf{v}_2'\|^2}\mathbf{v}_2'$$

$$= x^3 - \frac{\frac{2}{\pi}\int_{-1}^{1}\frac{x^3 \cdot 1}{\sqrt{1-x^2}}dx}{\frac{2}{\pi}\int_{-1}^{1}\frac{1 \cdot 1}{\sqrt{1-x^2}}dx} \cdot 1 - \frac{\frac{2}{\pi}\int_{-1}^{1}\frac{x^3 \cdot x}{\sqrt{1-x^2}}dx}{\frac{2}{\pi}\int_{-1}^{1}\frac{x \cdot x}{\sqrt{1-x^2}}dx} \cdot x - \frac{\frac{2}{\pi}\int_{-1}^{1}\frac{x^3 \cdot x^2}{\sqrt{1-x^2}}dx}{\frac{2}{\pi}\int_{-1}^{1}\frac{x^2 \cdot x^2}{\sqrt{1-x^2}}dx} \cdot x^2$$

$$= x^3 - \frac{3}{4}x = \frac{4x^3 - 3x}{4}.$$

---

* See, for example, R.L. Burden and J.D. Faires, *Numerical Analysis* (7th edn.) (Pacific Grove, CA: Brooks/Cole, 2001), Section 8.3.

Thus, $T_0(x) = 1$, $T_1(x) = x$, $T_2(x) = \frac{1}{2}(2x^2 - 1)$,   and $T_3(x) = \frac{1}{4}(4x^3 - 3x)$. (The integrals in the inner product calculations can be evaluated by tables or by using technology. It helps to recognize that $\int_{-a}^{a} f(x)dx = 0$ if f is an *odd* function, which is true when the numerators of the integrals above are odd powers of x.)

The next example will appear again in Section 7.6, where an important matrix factorization is introduced and applied to significant problems.

Example 7.4.5: An Orthonormal Basis
for a Column Space

Suppose we want to find an orthonormal basis for the column space of the matrix:

$$A = \begin{bmatrix} 1 & -1 & 1 \\ 0 & 0 & 1 \\ 1 & -1 & 0 \\ 0 & 1 & -1 \end{bmatrix} = [\,\mathbf{a}_1 \quad \mathbf{a}_2 \quad \mathbf{a}_3\,].$$

First, we use the Gram–Schmidt procedure to orthogonalize the linearly independent columns $\mathbf{a}_1, \mathbf{a}_2$, and $\mathbf{a}_3$:

$$\mathbf{a}_1' = \mathbf{a}_1 = [\,1 \quad 0 \quad 1 \quad 0\,]^\mathsf{T},$$

$$\mathbf{a}_2' = \mathbf{a}_2 - \frac{\langle \mathbf{a}_2, \mathbf{a}_1' \rangle}{\|\mathbf{a}_1'\|^2} \mathbf{a}_1' = \mathbf{a}_2 + \mathbf{a}_1 = [\,0 \quad 0 \quad 0 \quad 1\,]^\mathsf{T},$$

$$\mathbf{a}_3' = \mathbf{a}_3 - \frac{\langle \mathbf{a}_3, \mathbf{a}_1' \rangle}{\|\mathbf{a}_1'\|^2} \mathbf{a}_1' - \frac{\langle \mathbf{a}_3, \mathbf{a}_2' \rangle}{\|\mathbf{a}_2'\|^2} \mathbf{a}_2' = \mathbf{a}_3 + \tfrac{1}{2}\mathbf{a}_1 + \mathbf{a}_2$$

$$= [\,1/2 \quad 1 \quad -1/2 \quad 0\,]^\mathsf{T}.$$

Now, we normalize the orthogonal vectors:

$$\frac{\mathbf{a}_1'}{\|\mathbf{a}_1'\|} = [\,\sqrt{2}/2 \quad 0 \quad \sqrt{2}/2 \quad 0\,]^\mathsf{T},$$

$$\frac{\mathbf{a}_2'}{\|\mathbf{a}_2'\|} = [\,0 \quad 0 \quad 0 \quad 1\,]^\mathsf{T},$$

$$\frac{\mathbf{a}_3'}{\|\mathbf{a}_3'\|} = [\,\sqrt{6}/6 \quad \sqrt{6}/3 \quad -\sqrt{6}/6 \quad 0\,]^\mathsf{T}.$$

Thus, an orthonormal basis for the column space of A is

$$\left\{ [\,\sqrt{2}/2 \quad 0 \quad \sqrt{2}/2 \quad 0\,]^\mathsf{T}, [\,0 \quad 0 \quad 0 \quad 1\,]^\mathsf{T}, [\,\sqrt{6}/6 \quad \sqrt{6}/3 \quad -\sqrt{6}/6 \quad 0\,]^\mathsf{T} \right\}.$$

Exercises 7.4

A.

1. In $\mathbb{R}^2$, define the operator $T_\mathbf{v}$ by $T_\mathbf{v}(\mathbf{x}) = \mathbf{proj_v}(\mathbf{x})$ for a fixed $\mathbf{v} \in V$ and every $\mathbf{x} \in \mathbb{R}^2$.
    (a) Determine the matrices corresponding to $T_{\mathbf{v}_1}$ and $T_{\mathbf{v}_2}$, where
    $$\mathbf{v}_1 = \begin{bmatrix} 1 \\ 1 \end{bmatrix} \quad \text{and} \quad \mathbf{v}_2 = \begin{bmatrix} 1 \\ -1 \end{bmatrix}.$$
    (b) Use your answer to part (a) to find $T_{\mathbf{v}_1}(\mathbf{x})$, where
    $$\mathbf{x} = \begin{bmatrix} 3 \\ 1 \end{bmatrix}.$$

2. In $\mathbb{C}^2$, with the standard inner product, define the operator $T_\mathbf{v}$ by $T_\mathbf{v}(\mathbf{x}) = \mathbf{proj_v}(\mathbf{x})$ for every $\mathbf{x} \in \mathbb{C}^2$.
    (a) Determine the matrices corresponding to $T_{\mathbf{v}_1}$ and $T_{\mathbf{v}_2}$, where
    $$\mathbf{v}_1 = \begin{bmatrix} 1 \\ i \end{bmatrix} \quad \text{and} \quad \mathbf{v}_2 = \begin{bmatrix} 1 \\ -i \end{bmatrix}.$$
    (b) What is $T_{\mathbf{v}_1} + T_{\mathbf{v}_2}$? Explain.

3. Verify that the vectors $\mathbf{v}_1', \mathbf{v}_2'$, and $\mathbf{v}_3'$ in Example 7.4.2 are orthonormal.

4. Use the Gram–Schmidt process to transform $\left\{ \begin{bmatrix} -1 \\ 2 \end{bmatrix}, \begin{bmatrix} 1 \\ 1 \end{bmatrix} \right\}$ into an orthonormal basis for $\mathbb{R}^2$, assuming the standard inner product for this space.

5. In $\mathbb{R}^3$, with the standard inner product, orthonormalize the vectors $\begin{bmatrix} 1 \\ 1 \\ 0 \end{bmatrix}, \begin{bmatrix} 3 \\ 1 \\ 1 \end{bmatrix}$, and $\begin{bmatrix} 1 \\ 1 \\ 3 \end{bmatrix}$.

6. Use the Gram–Schmidt process to transform $\left\{ \begin{bmatrix} 1 \\ 0 \\ 1 \end{bmatrix}, \begin{bmatrix} 1 \\ 2 \\ -2 \end{bmatrix}, \begin{bmatrix} 2 \\ -1 \\ 1 \end{bmatrix} \right\}$ into an orthogonal basis for $\mathbb{R}^3$, assuming the standard inner product for this space.

7. Let $B = \left\{ \begin{bmatrix} 1 \\ 2 \\ 0 \end{bmatrix}, \begin{bmatrix} 10 \\ 0 \\ 4 \end{bmatrix}, \begin{bmatrix} 0 \\ 1 \\ 1 \end{bmatrix} \right\}$ be a basis for $\mathbb{R}^3$ with the standard inner product. Orthonormalize B.

8. Find an orthonormal basis for $span\left\{ \begin{bmatrix} 1 \\ 1 \\ 1 \end{bmatrix}, \begin{bmatrix} 3 \\ 2 \\ 1 \end{bmatrix}, \begin{bmatrix} 1 \\ -3 \\ -1 \end{bmatrix} \right\}$ in $\mathbb{R}^3$ with the standard inner product.

9.   Find   an   orthonormal   basis   for   the   subspace
     $U = \{[x_1 \ x_2 \ x_3 \ x_4]^T : x_1 + x_2 + x_3 + x_4 = 0\}$ of $\mathbb{R}^4$, assuming
     the standard inner product.

10. Find a basis for the subspace of all vectors in $\mathbb{R}^4$ that are
     orthogonal to $[1 \ \ 1 \ \ 2 \ \ 1]^T$ and $[2 \ \ 0 \ \ 1 \ \ 0]^T$.

11. Let $A = \begin{bmatrix} 1 & 1 & 2 \\ 1 & 2 & 2 \\ 1 & 0 & 4 \\ 1 & 1 & 0 \end{bmatrix}$. Find an orthonormal basis for the
     column space of A.

12. Find an orthonormal basis for the row space of the matrix A in
     the preceding exercise.

13. Let $V = C[0, 1]$ with the inner product $\langle f, g \rangle = \int_0^1 f(x)g(x)dx$
     for any f, g in V. Find an orthonormal basis for the subspace
     spanned by $1, x$, and x^2.

14. Let $V = \mathbb{R}^2$ with the inner product defined by

$$\langle \mathbf{x}, \mathbf{y} \rangle = \left\langle \begin{bmatrix} x_1 \\ x_2 \end{bmatrix}, \begin{bmatrix} y_1 \\ y_2 \end{bmatrix} \right\rangle = \frac{1}{2}x_1y_1 + \frac{1}{3}x_2y_2.$$

Convert the basis $\left\{ \begin{bmatrix} 1 \\ 1 \end{bmatrix}, \begin{bmatrix} 1 \\ -1 \end{bmatrix} \right\}$ into an orthonormal basis
using this inner product.

15. Let $V = \mathbb{R}^3$ with the inner product defined by

$$\langle \mathbf{x}, \mathbf{y} \rangle = \left\langle \begin{bmatrix} x_1 \\ x_2 \\ x_3 \end{bmatrix}, \begin{bmatrix} y_1 \\ y_2 \\ y_3 \end{bmatrix} \right\rangle = x_1y_1 + 2x_2y_2 + \frac{1}{3}x_3y_3.$$

Convert the basis $\left\{ \begin{bmatrix} 1 \\ 0 \\ 1 \end{bmatrix}, \begin{bmatrix} 1 \\ 1 \\ 0 \end{bmatrix}, \begin{bmatrix} 0 \\ 1 \\ 1 \end{bmatrix} \right\}$ into an orthonormal
basis using this inner product.

16. Using the inner product for M_2 defined by $\langle A, B \rangle =$
     trace $(A^T B)$, construct an orthonormal basis for M_2 starting
     with the basis:

$$B = \left\{ \begin{bmatrix} 1 & 1 \\ 1 & 0 \end{bmatrix}, \begin{bmatrix} 0 & 1 \\ 1 & 1 \end{bmatrix}, \begin{bmatrix} 1 & 0 \\ 1 & 1 \end{bmatrix}, \begin{bmatrix} 1 & 1 \\ 0 & 1 \end{bmatrix} \right\}.$$

17. Let $V = \{p \in P_2[x] : p(0) = 0\}$ and $\langle f, g \rangle = \int_0^1 f(x)g(x)\,dx$.
     Find an orthonormal basis for V.

18. Let $V = C[0, \pi]$ with the inner product $\langle f, g \rangle = \int_0^1 f(x)g(x)\,dx$
     for any f, g in V. Find an orthonormal basis for the subspace
     spanned by $1, \sin x$, and $\sin^2 x$.

B.

1. Prove that $\mathbf{u}$ is orthogonal to $\mathbf{v} - \mathbf{proj}_{\mathbf{u}}(\mathbf{v})$ for all vectors $\mathbf{u}$ and $\mathbf{v}$ in $\mathbb{R}^n$, where $\mathbf{u} \neq \mathbf{0}$.

2. Find an orthonormal basis of $\mathbb{R}^3$ that contains the vector
$$\mathbf{v} = \frac{1}{\sqrt{5}}\begin{bmatrix} 1 \\ 0 \\ -2 \end{bmatrix}.$$

3. Extend $\left\{\begin{bmatrix} 2 \\ 3 \\ -1 \end{bmatrix}, \begin{bmatrix} 1 \\ -2 \\ -4 \end{bmatrix}\right\}$ to an orthogonal basis of $\mathbb{R}^3$ with integer components.

4. In Example 7.4.4, calculate $\|T_n(x)\|$ for $n = 1, 2, 3$. Does there seem to be a formula for $\|T_n(x)\|, n \geq 1$?

5. Suppose $abc \neq 0$. Show that the Gram–Schmidt method leads to the same orthonormal basis for $span\{\mathbf{u}_1, \mathbf{u}_2, \mathbf{u}_3\}$ and $span\{a\mathbf{u}_1, b\mathbf{u}_2, c\mathbf{u}_3\}$.

6. Explain why the Gram–Schmidt process leads to different orthonormal bases for $span\{\mathbf{u}_1, \mathbf{u}_2, \mathbf{u}_3\}$ and $span\{\mathbf{u}_1, \mathbf{u}_2 + \mathbf{u}_3, \mathbf{u}_3\}$. (Compare this with the result of the preceding exercise.)

C.

1. (a) Apply the Gram–Schmidt process to the *dependent* set of vectors $\left\{\begin{bmatrix} 1 \\ -1 \\ 1 \end{bmatrix}, \begin{bmatrix} -1 \\ -1 \\ 2 \end{bmatrix}, \begin{bmatrix} -2 \\ 0 \\ 1 \end{bmatrix}\right\}$ in $\mathbb{R}^3$ with the usual inner product.

   (b) Use your experience in part (a) to explain what happens if the Gram–Schmidt procedure is applied to a set of vectors that is not linearly independent. Prove that your explanation is correct.

2. If $\mathbf{v}_1, \mathbf{v}_2, \ldots, \mathbf{v}_k$ are arbitrary vectors of an inner product space V, we define the *Gram matrix* (or a *Grammian*) G (named for Jorgen P. Gram of Gram–Schmidt fame) to be the $k \times k$ matrix $[a_{ij}]$, where $a_{ij} = \langle \mathbf{v}_i, \mathbf{v}_j \rangle, i, j = 1, 2, \ldots, k$.

   (a) Let $\mathbf{v}_1 = \begin{bmatrix} 1 \\ 2 \end{bmatrix}$ and $\mathbf{v}_2 = \begin{bmatrix} 2 \\ 3 \end{bmatrix}$. Compute the Grammian corresponding to these vectors using the standard inner product in $\mathbb{R}^2$.

   (b) Prove that $\mathbf{v}_1, \mathbf{v}_2, \ldots, \mathbf{v}_k$ are linearly dependent if and only if $\det(G) = 0$.

   (c) If $\dim(V) = n$, select an orthonormal basis for V and use it to represent G as the product of a $k \times n$ and an $n \times k$ matrix. Then, verify that $\det(G) \geq 0$.

3. Let $V = \mathbb{R}[x]$, the space of all polynomials in x with real coefficients, and define the inner product $\langle f, g \rangle = \frac{1}{\sqrt{\pi}} \int_{-\infty}^{\infty} f(x)g(x)e^{-x^2} dx$. This inner product is related to the probability density function for the normal distribution and is useful in quantum mechanics.

    Apply the Gram–Schmidt process to orthogonalize the vectors $v_1(x) = 1$, $v_2(x) = x$, $v_3(x) = x^2$, and $v_4(x) = x^3$. (The orthogonal vectors are called the first four *Hermite polynomials*.)

7.5 Unitary Matrices and Orthogonal Matrices

Suppose that $B_1 = \{v_k\}_{k=1}^n$ is an orthonormal basis for a complex inner product space V. We want to investigate the conditions under which a new basis $B_2 = \{\tilde{v}_j\}_{j=1}^n$, where $\tilde{v}_j = \sum_{i=1}^n p_{ij}v_i$, is also orthonormal.

As we saw in Section 6.7, in this situation the change of coordinates can be written as $[v]_{B_1} = P[v]_{B_2}$, with $P = [p_{ij}]$. Because B_1 is orthonormal, Corollary 7.3.2 tells us that the entries of P are the Fourier coefficients $p_{ij} = \langle \tilde{v}_j, v_i \rangle$. Then,

$$\langle \tilde{v}_i, \tilde{v}_i \rangle = \left\langle \sum_{k=1}^n p_{ki}v_k, \sum_{r=1}^n p_{rj}v_r \right\rangle = \sum_{r=1}^n \langle p_{ri}v_r, p_{rj}v_r \rangle$$

$$= \sum_{r=1}^n p_{ri}\langle v_r, p_{rj}v_r \rangle = \sum_{r=1}^n p_{ri}\overline{\langle p_{rj}v_r, v_r \rangle}$$

$$= \sum_{r=1}^n p_{ri}\overline{p_{rj}}\langle v_r, v_r \rangle = \sum_{r=1}^n p_{ri}\overline{p_{rj}}\|v_r\|^2$$

$$= \sum_{r=1}^n p_{ri}\overline{p_{rj}} = \text{ the } (j,i) \text{ entry of } \bar{P}^T P = P^*P.$$

Thus, $B_2 = \{\tilde{v}_j\}_{j=1}^n$ is orthonormal if and only if entry (j,i) of P^*P is 1 if $j = i$ and is 0 if $j \neq i$—that is, if and only if $P^*P = I_n$.

This last condition is significant and deserves a definition.

Definition 7.5.1

An $n \times n$ complex matrix A such that $A^*A = I_n = AA^*$ is called a **unitary matrix**. A real unitary matrix ($A^T A = I_n = AA^T$) is called an **orthogonal matrix**.

Notice that this definition implies that if A is unitary, then A is nonsingular and $A^{-1} = A^*$. A unitary matrix has important properties and applications that we will examine further in the rest of this section.

The discussion that began this section established that if $B_1 = \{v_k\}_{k=1}^n$ is an orthonormal basis for V, then $B_2 = \{\tilde{v}_j\}_{j=1}^n$ is also an orthonormal basis for V if and only if the change of basis matrix $[I_n]_{B_2}^{B_1}$ is unitary.

Example 7.5.1: Unitary and Orthogonal Matrices

(a) The matrix

$$A = \frac{1}{2}\begin{bmatrix} 1-i & 1+i \\ 1-i & -1-i \end{bmatrix}$$

is unitary: $\bar{A} = (1/2)\begin{bmatrix} 1+i & 1-i \\ 1+i & -1+i \end{bmatrix}$ and $A^* = \bar{A}^T = (1/2)$

$\times \begin{bmatrix} 1+i & 1+i \\ 1-i & -1+i \end{bmatrix}$, so $A^*A = (1/2)\begin{bmatrix} 1+i & 1+i \\ 1-i & -1+i \end{bmatrix} \cdot (1/2)$

$\times \begin{bmatrix} 1-i & 1+i \\ 1-i & -1-i \end{bmatrix} = \begin{bmatrix} 1 & 0 \\ 0 & 1 \end{bmatrix} = I_2.$

(b) The matrices

$$A = \begin{bmatrix} 3/5 & -4/5 & 0 \\ 4/5 & 3/5 & 0 \\ 0 & 0 & 0 \end{bmatrix} \quad \text{and} \quad B = \begin{bmatrix} 0 & 0 & 0 & 1 \\ 0 & 0 & 1 & 0 \\ 1 & 0 & 0 & 0 \\ 0 & 1 & 0 & 0 \end{bmatrix}$$

are orthogonal matrices: $A^T A = I_3$ and $B^T B = I_4$.

(c) Any matrix of the form:

$$A_\theta = \begin{bmatrix} \cos\theta & -\sin\theta \\ \sin\theta & \cos\theta \end{bmatrix} \quad \text{or} \quad B_\theta = \begin{bmatrix} \cos\theta & \sin\theta \\ \sin\theta & -\cos\theta \end{bmatrix}$$

is orthogonal (see Exercise A9). (In fact, *any* 2×2 orthogonal matrix must be of the form A_θ or B_θ for some value of θ—see Exercise B1.)

The nature of unitary and orthogonal matrices leads to some useful properties. The following result offers a computationally simpler way to determine if a matrix A is unitary. We do not have to calculate A^*A. We just have to perform some simple calculations on the rows or columns of the matrix.

Theorem 7.5.1

Let A be a complex $n \times n$ matrix. The following statements are equivalent.

  (1) A is a unitary matrix.
  (2) The column vectors of A form an orthonormal set of vectors in $\mathbb{C}^n$ under the standard inner product for $\mathbb{C}^n$.
  (3) The row vectors of A form an orthonormal set of vectors in $\mathbb{C}^n$ under the standard inner product for $\mathbb{C}^n$.

Proof We will establish the implications (1) ⇔ (2) and (1) ⇔ (3).
(1) ⇒ (2): Suppose that A is an $n \times n$ unitary matrix. If $\mathbf{e}_i$ is an element of the standard (orthonormal) basis for $\mathbb{C}^n$, then $A\mathbf{e}_i = A \operatorname{col}_i(I_n) = \operatorname{col}_i (AI_n) = \operatorname{col}_i (A)$ by Theorem 3.3.1. Thus the standard inner product of column i and column j of A is given by $< A\mathbf{e}_i, A\mathbf{e}_j > = (A\mathbf{e}_j)^*(A\mathbf{e}_i) = \mathbf{e}_j^*(A^*A)\mathbf{e}_i = \mathbf{e}_j^*\mathbf{e}_i = < \mathbf{e}_i, \mathbf{e}_j >$, so the columns of A are orthonormal.

(2) ⇒ (1): For the converse, we use the easily established fact that the (i,j) entry of a matrix M is $\mathbf{e}_i^T M\mathbf{e}_j$. If we assume that the columns of a matrix A are orthonormal, then $\mathbf{e}_i^T(A^*A)\mathbf{e}_j = (A\mathbf{e}_i)^*(A\mathbf{e}_j) = <$ column j of A, column i of $A >=$
$$\begin{cases} 1 & \text{if } i=j \\ 0 & \text{if } i \neq j \end{cases}$$—that is, $A^*A = I_n$.

(1) ⇔ (3): The equivalence of statements (1) and (3) follows from the fact that A is unitary if and only if A^T is unitary. To see the truth of this last statement, notice that $(A^T)^* A^T = (A^*)^T A^T = (AA^*)^T$. If A is unitary, then $(AA^*)^T = I_n$ (because $I_n^T = I_n$), so $(A^T)^* A^T = I_n$—that is, A^T is unitary.

Now suppose A is unitary. Then A^T is unitary and the equivalence of statements (1) and (2) implies that the columns of A^T are orthonormal. But the columns of A^T are the rows of A, so the rows of A are orthonormal. Conversely, if the rows of A are orthonormal, then the columns of A^T are orthonormal. Again, by the equivalence of statements (1) and (2), we conclude that A^T is unitary. This, in turn, implies that A is unitary.

The proof of Theorem 7.5.1 can be represented schematically as follows:

A is unitary ⇔ the columns of A are orthonormal ⇔ the rows of A^T are orthonormal

⇕

A^T is unitary ⇔ the columns of A^T are orthonormal ⇔ the rows of A are orthonormal.

Example 7.5.2: Unitary and Orthogonal Matrices

(a) The matrix

$$A = \begin{bmatrix} 0 & 0 & 1/\sqrt{2} & i/\sqrt{2} \\ 0 & 0 & i/\sqrt{2} & 1/\sqrt{2} \\ 1/\sqrt{2} & i/\sqrt{2} & 0 & 0 \\ i/\sqrt{2} & 1/\sqrt{2} & 0 & 0 \end{bmatrix}$$

is unitary. As vectors in $\mathbb{C}^4$, the columns are clearly linearly independent. It is also obvious that both columns 1 and 2 are orthogonal to columns 3 and 4.

Simple calculations demonstrate that columns 1 and 2 are orthogonal to each other, as are columns 3 and 4. We can also verify that the rows and columns of A are unit vectors in $\mathbb{C}^4$.

Now,

$$A^* = \overline{\left(\begin{bmatrix} 0 & 0 & 1/\sqrt{2} & i/\sqrt{2} \\ 0 & 0 & i/\sqrt{2} & 1/\sqrt{2} \\ 1/\sqrt{2} & i/\sqrt{2} & 0 & 0 \\ i/\sqrt{2} & 1/\sqrt{2} & 0 & 0 \end{bmatrix}\right)^T}$$

$$= \begin{bmatrix} 0 & 0 & 1/\sqrt{2} & -i/\sqrt{2} \\ 0 & 0 & -i/\sqrt{2} & 1/\sqrt{2} \\ 1/\sqrt{2} & -i/\sqrt{2} & 0 & 0 \\ -i/\sqrt{2} & 1/\sqrt{2} & 0 & 0 \end{bmatrix}$$

and $A^*A = I_4 = AA^*$. Finally, we can verify that A^* is also unitary.

(b) The matrix

$$B = \frac{1}{2} \begin{bmatrix} 1 & -1 & -1 & -1 \\ 1 & -1 & 1 & 1 \\ 1 & 1 & -1 & 1 \\ 1 & 1 & 1 & -1 \end{bmatrix}$$

is orthogonal. It is easy to check that both the rows and columns of B constitute orthonormal bases of $\mathbb{R}^4$. It is trivial to confirm that

$$B^T B = I_4 = BB^T.$$

The orthogonal matrix $A_\theta = \begin{bmatrix} \cos\theta & -\sin\theta \\ \sin\theta & \cos\theta \end{bmatrix}$ of Example 7.5.1(c) represents a linear transformation $T : \mathbb{R}^2 \to \mathbb{R}^2$ that rotates a vector through the angle θ in a counterclockwise direction (see Example 6.1.2(c)). Geometrically, the orthogonal matrix $B_\theta = \begin{bmatrix} \cos\theta & \sin\theta \\ \sin\theta & -\cos\theta \end{bmatrix}$ of Example 7.5.1(c) represents the reflection across the line passing through the origin that forms an angle $\theta/2$ with the positive x-axis. Equivalently,

$$B_\theta = \begin{bmatrix} \cos\theta & \sin\theta \\ \sin\theta & -\cos\theta \end{bmatrix} = \begin{bmatrix} \cos\theta & -\sin\theta \\ \sin\theta & \cos\theta \end{bmatrix} \begin{bmatrix} 1 & 0 \\ 0 & -1 \end{bmatrix},$$

so B_θ represents a reflection across the x-axis followed by a rotation through an angle θ in a counterclockwise direction. Figure 7.3 shows what happens to $\mathbf{v} = \begin{bmatrix} 1 \\ 1 \end{bmatrix}$ if $\theta = 120°$. In part (a) $\mathbf{v}$ is reflected across the

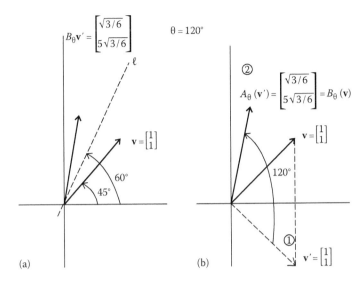

Figure 7.3 The effects of reflections and rotations.

line through the origin forming an angle of 60° with the positive x-axis, whereas in part (b) of Figure 7.3 **v** is first reflected across the x-axis and then rotated 120° in a counterclockwise direction. In any case, **v** is transformed into the vector $\begin{bmatrix} \sqrt{3}/6 \\ 5\sqrt{3}/6 \end{bmatrix}$.

Exercise B1 asks for a proof that **any orthogonal 2×2 matrix represents either a rotation transformation or a reflection across the x-axis followed by a rotation transformation.**

Thinking geometrically, we should realize that the rotation given by $T(\mathbf{v}) = A_\theta \mathbf{v}$ and the reflection defined by $S(\mathbf{v}) = B_\theta \mathbf{v}$ preserve both the lengths of vectors and the angles between two vectors. **In an inner product space V, any unitary matrix A preserves the lengths of vectors**:

$$\|A\mathbf{v}\| = \sqrt{\langle A\mathbf{v}, A\mathbf{v}\rangle} = \sqrt{(A\mathbf{v})^\mathrm{T}\overline{A\mathbf{v}}} = \sqrt{(\mathbf{v}^\mathrm{T}A^\mathrm{T})(\bar{A}\bar{\mathbf{v}})} = \sqrt{\mathbf{v}^\mathrm{T}(A^\mathrm{T}\bar{A})\bar{\mathbf{v}}}$$
$$= \sqrt{\mathbf{v}^\mathrm{T}\,\bar{\mathbf{v}}} = \sqrt{\langle \mathbf{v}, \mathbf{v}\rangle} = \|\mathbf{v}\|.$$

(Because A is unitary, $A^* = (\bar{A})^\mathrm{T} = A^{-1}$, or $\bar{A} = (A^{-1})^\mathrm{T} = (A^\mathrm{T})^{-1}$ by Theorem 3.5.3(4).)

This property leads to a general definition in any inner product space. In this situation, we have to realize that different inner products generally lead to different norms: If U and V are inner product spaces with inner products $\langle,\rangle_U$ and $\langle,\rangle_V$, respectively, then $\|\mathbf{u}\|_U = \sqrt{\langle \mathbf{u}, \mathbf{u}\rangle_U}$ and $\|\mathbf{v}\|_V = \sqrt{\langle \mathbf{v}, \mathbf{v}\rangle_V}$, where $\mathbf{u} \in U$ and $\mathbf{v} \in V$.

> ### Definition 7.5.2
>
> (a) Let U and V be inner product spaces, with norms $\|\cdot\|_U$ and $\|\cdot\|_V$, respectively. A linear transformation $T: U \to V$ is called an **isometry**,* or an **orthogonal transformation**, if it preserves the lengths of vectors: $\|T(\mathbf{u})\|_V = \|\mathbf{u}\|_U$ for every $\mathbf{u} \in U$.
>
> (b) An isometry on a real inner product space is called an **orthogonal operator**, whereas an isometry on a complex inner product space is called a **unitary operator**.

The term *unitary* we have been using comes from the fact that unit vectors are preserved by an isometry: $\|T(\mathbf{u})\|_V = \|\mathbf{u}\|_U = 1$ for every unit vector $\mathbf{u} \in U$.

Imitating the calculation shown just before Definition 7.5.2, we see that any unitary matrix A gives rise to an isometry $T_A(\mathbf{v}) = A\mathbf{v}$. The converse of this statement is also true, as we will see later (Theorem 7.5.3).

Although we have defined an isometry in terms of preserving lengths, we can also characterize an isometry in terms of inner products.

Theorem 7.5.2

Let U and V be inner product spaces with inner products $\langle, \rangle_U$ and $\langle, \rangle_V$, respectively, and let $T: U \to V$ be a linear transformation. Then, T is an isometry if and only if T preserves inner products—that is, $\langle T(\mathbf{x}), T(\mathbf{y}) \rangle_V = \langle \mathbf{x}, \mathbf{y} \rangle_U$ for all vectors $\mathbf{x}, \mathbf{y} \in U$.

Proof First suppose that T is an isometry. Then, $\|T(\mathbf{x})\|_V^2 = \|\mathbf{x}\|_U^2$ for every $\mathbf{x} \in U$. Hence,

$$\langle T(\mathbf{x}+\mathbf{y}), T(\mathbf{x}+\mathbf{y}) \rangle_V = \|T(\mathbf{x}+\mathbf{y})\|_V^2 = \|\mathbf{x}+\mathbf{y}\|_U^2 = \langle \mathbf{x}+\mathbf{y}, \mathbf{x}+\mathbf{y} \rangle_U$$

for all $\mathbf{x}, \mathbf{y} \in U$. Now, we can consider the alternative expansion:

$$\langle T(\mathbf{x}+\mathbf{y}), T(\mathbf{x}+\mathbf{y}) \rangle_V = \langle T(\mathbf{x}), T(\mathbf{x}) \rangle_V + 2\langle T(\mathbf{x}), T(\mathbf{y}) \rangle_V + \langle T(\mathbf{y}), T(\mathbf{y}) \rangle_V$$
$$= \|T(\mathbf{x})\|_V^2 + 2\langle T(\mathbf{x}), T(\mathbf{y}) \rangle_V + \|T(\mathbf{y})\|_V^2$$
$$= \|\mathbf{x}\|_U^2 + 2\langle T(\mathbf{x}), T(\mathbf{y}) \rangle_V + \|\mathbf{y}\|_U^2 \qquad (*)$$

and

$$\langle \mathbf{x}+\mathbf{y}, \mathbf{x}+\mathbf{y} \rangle_U = \langle \mathbf{x}, \mathbf{x} \rangle_U + 2\langle \mathbf{x}, \mathbf{y} \rangle_U + \langle \mathbf{y}, \mathbf{y} \rangle_U$$
$$= \|\mathbf{x}\|_U^2 + 2\langle \mathbf{x}, \mathbf{y} \rangle_U + \|\mathbf{y}\|_U^2. \qquad (**)$$

Equating (*) and (**), we get $\langle T(\mathbf{x}), T(\mathbf{y}) \rangle_V = \langle \mathbf{x}, \mathbf{y} \rangle_U$.

---

* **Isometry** comes from two Greek words meaning "equal" and "measure."

To prove the converse statement, assume that $\langle T(\mathbf{x}), T(\mathbf{y}) \rangle_V = \langle \mathbf{x}, \mathbf{y} \rangle_U$ and let $\mathbf{y} = \mathbf{x}$. This yields $\langle T(\mathbf{x}), T(\mathbf{x}) \rangle_V = \langle \mathbf{x}, \mathbf{x} \rangle_U$, or $\|T(\mathbf{x})\|_V^2 = \|\mathbf{x}\|_U^2$, from which it follows that $\|T(\mathbf{x})\|_V = \|\mathbf{x}\|_U$ for every $\mathbf{x} \in U$.

Corollary 7.5.1

Let A be a real $n \times n$ matrix. Then, A is an orthogonal matrix if and only if $T_A : \mathbb{R}^n \to \mathbb{R}^n$ defined by $T_A(\mathbf{x}) = A\mathbf{x}$ preserves the Euclidean inner product:

$$T_A(\mathbf{x}) \cdot T_A(\mathbf{y}) = A\mathbf{x} \cdot A\mathbf{y} = \mathbf{x} \cdot \mathbf{y} \quad \text{for all vectors } \mathbf{x}, \mathbf{y} \in \mathbb{R}^n.$$

Proof   If A is an orthogonal matrix, then

$$A\mathbf{x} \cdot A\mathbf{y} = (A\mathbf{x})^{\mathrm{T}}(A\mathbf{y}) = \mathbf{x}^{\mathrm{T}} A^{\mathrm{T}} A\mathbf{y} = \mathbf{x}^{\mathrm{T}}\mathbf{y} = \mathbf{x} \cdot \mathbf{y}.$$

Now, suppose that A preserves the dot product. Then, for all vectors $\mathbf{x}, \mathbf{y} \in \mathbb{R}^n, A\mathbf{x} \cdot A\mathbf{y} = \mathbf{x} \cdot \mathbf{y}$. Take $\mathbf{x} = \mathbf{e}_i$ and $\mathbf{y} = \mathbf{e}_j$, standard basis vectors for $\mathbb{R}^n$, in this last equation to get

$$\overbrace{A\mathbf{e}_i}^{\text{col. } i \text{ of } A} \cdot \overbrace{A\mathbf{e}_j}^{\text{col. } j \text{ of } A} = \mathbf{e}_i \cdot \mathbf{e}_j = \begin{cases} 1 & \text{if } i = j \\ 0 & \text{if } i \neq j \end{cases}.$$

Because i and j are arbitrary, $1 \leq i, j \leq n$, the columns of A are orthonormal and A is orthogonal by Theorem 7.5.1.

If θ is the angle between two nonzero vectors $\mathbf{x}$ and $\mathbf{y}$ in a real inner product space U, then for any isometry $T : U \to V$, where V is also a real inner product space:

$$\cos \theta = \frac{\langle \mathbf{x}, \mathbf{y} \rangle_U}{\|\mathbf{x}\|_U \|\mathbf{y}\|_U} = \frac{\langle T(\mathbf{x}), T(\mathbf{y}) \rangle_V}{\|T(\mathbf{x})\|_V \|T(\mathbf{y})\|_V}.$$

Because $\langle T(\mathbf{x}), T(\mathbf{y}) \rangle_V / \|T(\mathbf{x})\|_V \|T(\mathbf{y})\|_V$ equals the cosine of the angle between the vectors $T(\mathbf{x})$ and $T(\mathbf{y})$, we recognize another consequence of Theorem 7.5.2.

Corollary 7.5.2

An isometry on a real inner product space preserves angles between vectors.

However, as the next example shows, the converse of Corollary 7.5.2 is not true in general.

Example 7.5.3: *T* Preserves Angles, But Is Not an Isometry

On any space $\mathbb{R}^n$, define the operator T as follows: $T(\mathbf{v}) = A\mathbf{v}$, where

$$A = 3I_n = \begin{bmatrix} 3 & 0 & 0 & \cdots & 0 \\ 0 & 3 & 0 & \cdots & 0 \\ 0 & 0 & 3 & \cdots & 0 \\ \vdots & \vdots & \vdots & \vdots & \vdots \\ 0 & 0 & 0 & \cdots & 3 \end{bmatrix}.$$

Then, for any nonzero $\mathbf{x}, \mathbf{y} \in \mathbb{R}^n$, $\cos\theta = \mathbf{x} \cdot \mathbf{y}/\|\mathbf{x}\|\,\|\mathbf{y}\|$ and

$$\frac{\langle T(\mathbf{x}), T(\mathbf{y})\rangle}{\|T(\mathbf{x})\|\,\|T(\mathbf{y})\|} = \frac{(A\mathbf{x})\cdot(A\mathbf{y})}{\|A\mathbf{x}\|\,\|A\mathbf{y}\|} = \frac{9\mathbf{x}\cdot\mathbf{y}}{3\|\mathbf{x}\|\,3\|\mathbf{y}\|} = \frac{\mathbf{x}\cdot\mathbf{y}}{\|\mathbf{x}\|\,\|\mathbf{y}\|},$$

so the angle between $\mathbf{x}$ and $\mathbf{y}$ is preserved by T.

But, as we observed in the last calculation:

$$\|T(\mathbf{x})\| = \|A\mathbf{x}\| = \|3I_n\mathbf{x}\| = 3\|\mathbf{x}\| \neq \|\mathbf{x}\|.$$

(A linear transformation such as T for which $\|T(\mathbf{x})\| = r\|\mathbf{x}\|$ with $r > 1$ is called a *dilation*.)

We have seen that any unitary (or orthogonal) matrix A gives rise to an isometry T defined by $T(\mathbf{x}) = A\mathbf{x}$. Now, we make good on our earlier claim that the converse holds.

Theorem 7.5.3

Let U and V be inner product spaces of the same dimension and let $T:U \to V$ be an isometry. Let $B_U = \{\mathbf{u}_1, \mathbf{u}_2, \ldots, \mathbf{u}_n\}$ and $B_V = \{\mathbf{v}_1, \mathbf{v}_2, \ldots, \mathbf{v}_n\}$ be orthonormal bases for U and V, respectively. Then, the matrix representation $[T]_{B_U}^{B_V}$ of T with respect to the bases B_U and B_V, respectively, is a unitary (or orthogonal) matrix.

Proof First we note that the kth column vector of $[T]_{B_U}^{B_V}$ is just $[T(\mathbf{u}_k)]_{B_V}$ (see the proof of Theorem 6.4.3).

Next, we establish that

$$\langle T(\mathbf{u}_k), T(\mathbf{u}_r)\rangle_V = \left\langle \sum_{i=1}^{n} a_i\mathbf{v}_i, \sum_{j=1}^{n} b_j\mathbf{v}_j \right\rangle_V = \sum_{i=1}^{n} a_i \left\langle \mathbf{v}_i, \sum_{j=1}^{n} b_j\mathbf{v}_j \right\rangle_V$$

$$= \sum_{i=1}^{n} a_i \left(\sum_{j=1}^{n} \bar{b}_j \langle \mathbf{v}_i, \mathbf{v}_j\rangle_V \right) = \sum_{i=1}^{n} a_i\bar{b}_i$$

$$= \langle [T(\mathbf{u}_k)]_{B_V}, [T(\mathbf{u}_r)]_{B_V}\rangle.$$

Because T preserves inner products (Theorem 7.5.2) and B_U is orthonormal:

$$\langle \text{column } k \text{ of } [T]_{B_U}^{B_V}, \text{ column } r \text{ of } [T]_{B_U}^{B_V} \rangle = \langle [T(\mathbf{u}_k)]_{B_V}, [T(\mathbf{u}_r)]_{B_V} \rangle$$

$$= \langle T(\mathbf{u}_k), T(\mathbf{u}_r) \rangle_V = \langle \mathbf{u}_k, \mathbf{u}_r \rangle_U$$

$$= \begin{cases} 1 & \text{if } k = r \\ 0 & \text{if } k \neq r \end{cases},$$

which shows that the column vectors of $[T]_{B_U}^{B_V}$ are orthonormal.

Therefore, **an operator $T : V \rightarrow V$ is an isometry if and only if $[T]_{B_V}$ is a unitary (or orthogonal) matrix with respect to an orthonormal basis B_V.** The emphasis on an orthonormal basis in producing $[T]_{B_V}$ is important. Furthermore, **a real $n \times n$ matrix A preserves the dot product if and only if it preserves the length of vectors.**

Exercises 7.5

A.

1. Determine which of the following matrices is (are) orthogonal:

   (a) $\begin{bmatrix} 1 & 0 \\ 0 & 1 \end{bmatrix}$ (b) $\begin{bmatrix} 0 & 1 \\ 1 & 0 \end{bmatrix}$ (c) $\begin{bmatrix} 1 & 1 \\ -1 & 1 \end{bmatrix}$

   (d) $\begin{bmatrix} \cos\theta & \sin\theta & 0 \\ -\sin\theta & \cos\theta & 0 \\ 0 & 0 & 1 \end{bmatrix}$

2. Which of the following matrices is (are) unitary?

   (a) $\begin{bmatrix} 0 & 1 \\ i & 0 \end{bmatrix}$ (b) $\begin{bmatrix} 0 & i \\ -i & 0 \end{bmatrix}$ (c) $\frac{1}{\sqrt{2}} \begin{bmatrix} 1 & i \\ i & 1 \end{bmatrix}$

   (d) $\begin{bmatrix} \frac{1+i}{\sqrt{3}} & \frac{1+i}{\sqrt{6}} \\ \frac{i}{\sqrt{3}} & -\frac{2i}{\sqrt{6}} \end{bmatrix}$

3. Find values for a, b, and c for which the following matrix will be orthogonal:

   $$A = \begin{bmatrix} 0 & -\frac{2}{3} & a \\ \frac{1}{\sqrt{5}} & \frac{2}{3} & b \\ -\frac{2}{\sqrt{5}} & \frac{1}{3} & c \end{bmatrix}.$$

4. Find a 2×2 orthogonal matrix whose first column is $\begin{bmatrix} 1/\sqrt{2} \\ 1/\sqrt{2} \end{bmatrix}$.

5. Find a 3×3 orthogonal matrix whose first column is $[1/3 \quad 2/3 \quad 2/3]^T$.

6. Find a 3×3 symmetric orthogonal matrix A whose first column is $\begin{bmatrix} 1/3 & 2/3 & 2/3 \end{bmatrix}^{\mathsf{T}}$. Compute A^2.

7. Fill in the entries so that

$$A = \begin{bmatrix} 1/\sqrt{3} & 0 & * & * \\ 1/\sqrt{3} & 1/\sqrt{2} & * & * \\ 1/\sqrt{3} & -1/\sqrt{2} & * & * \\ 0 & 0 & * & * \end{bmatrix}$$

becomes an orthogonal 4×4 matrix.

8. Which of the following operators is (are) unitary or orthogonal?

    (a) $T: \mathbb{R}^2 \to \mathbb{R}^2$ defined by $T\left(\begin{bmatrix} x \\ y \end{bmatrix}\right) = \begin{bmatrix} y \\ x \end{bmatrix}$.

    (b) $T: \mathbb{C}^2 \to \mathbb{C}^2$ defined by $T\left(\begin{bmatrix} \alpha \\ \beta \end{bmatrix}\right) = \begin{bmatrix} -i\beta \\ i\alpha \end{bmatrix}$.

    (c) $T: \mathbb{R}^3 \to \mathbb{R}^3$ defined by $T(\mathbf{v}) = A\mathbf{v}$, where $\mathbf{v} \in \mathbb{R}^3$ and

    $$A = \begin{bmatrix} 2 & 1 & -2 \\ 1 & 2 & 2 \\ 2 & -2 & 1 \end{bmatrix}.$$

    (d) $T: C[-\pi, \pi] \to C[-\pi, \pi]$ defined by $T(f) = -f$, where $\langle f, g \rangle = \int_{-\pi}^{\pi} f(x)g(x)\, dx$.

9. Show that any matrix of the form

$$A_\theta = \begin{bmatrix} \cos\theta & -\sin\theta \\ \sin\theta & \cos\theta \end{bmatrix} \quad \text{or} \quad B_\theta = \begin{bmatrix} \cos\theta & \sin\theta \\ \sin\theta & -\cos\theta \end{bmatrix}$$

is orthogonal.

10. Verify that the eigenvalues and determinant of the unitary matrix $\begin{bmatrix} 0 & i \\ i & 0 \end{bmatrix}$ have absolute value (modulus) equal to 1.

11. Show that the matrix

$$A_b = \begin{bmatrix} \sqrt{1+b^2} & bi \\ -bi & \sqrt{1+b^2} \end{bmatrix} \in M_2(\mathbb{C})$$

satisfies $A_b A_b^{\mathsf{T}} = I_2$ for any real number b, but except for $b = 0$, it is not unitary.

B.

1. Show that any 2×2 orthogonal matrix must be of the form:

$$A_\theta = \begin{bmatrix} \cos\theta & -\sin\theta \\ \sin\theta & \cos\theta \end{bmatrix} \quad \text{or} \quad B_\theta = \begin{bmatrix} \cos\theta & \sin\theta \\ \sin\theta & -\cos\theta \end{bmatrix}$$

for some value of θ (cf. Exercise A9).

2. Prove that every eigenvalue of an isometry $T: U \rightarrow V$ has absolute value 1.
   (Equivalently, *the spectrum of T lies on the unit circle in* $\mathbb{C}$.)

3. Prove that any $n \times n$ permutation matrix (see Section 3.7) is an orthogonal matrix.

4. If $A \in M_n(\mathbb{C})$ is such that $A^2 = I_n$, show that the following statements are equivalent:
   (a) $A = A^*$.
   (b) $AA^* = A^*A$.
   (c) A is a unitary matrix.

5. Suppose V is a complex inner product space and $\{v_1, v_2, \ldots, v_n\}$ is an orthonormal basis for V. Prove that a linear transformation $T: V \rightarrow V$ is an isometry if and only if $\{T(v_1), T(v_2), \ldots, T(v_n)\}$ is an orthonormal basis for V.

6. If $U \in M_n(\mathbb{C})$ is a unitary matrix, show that $|\det(U)| = 1$. Give an example of a complex matrix A, which shows that $|\det(A)| = 1$ does not imply A is unitary.

7. Let $U \in M_n(\mathbb{C})$ be a unitary matrix. Prove that U^{T} and $\overline{U}$ are unitary.

8. Let $U \in M_n(\mathbb{C})$ be a unitary matrix. Prove that UV is unitary for every $n \times n$ unitary matrix V.

9. Prove that U^k is unitary if U is unitary and k is a positive integer.

10. Show that the sum of unitary operators (matrices) is not necessarily unitary.

11. Let $U \in M_n(\mathbb{C})$ be a unitary matrix. If λ is an eigenvalue of U, show that $1/\lambda$ is an eigenvalue of U^*. (Exercise B2 implies that $\lambda \neq 0$.)

12. If U is an $n \times n$ unitary matrix, show that $|x^*Ux| \leq 1$ for every unit vector $x \in \mathbb{C}^n$.

13. If $U = [u_{ij}]$ is an $n \times n$ unitary matrix, show that each row and column sum of the matrix $[|u_{ij}|^2]$ equals 1. (This is the *Hadamard* (or *Schur*) *product* U^*U defined in problem C3 of Exercises 3.3.)

14. Suppose U is an $n \times n$ unitary matrix. If x and y are eigenvectors of U belonging to distinct eigenvalues, show that $x^*y = 0$.

15. How many 3×3 matrices are both diagonal and orthogonal? List all of them.

16. Under what conditions on the real numbers a and b will

$$A = \begin{bmatrix} a+b & b-a \\ a-b & b+a \end{bmatrix}$$ be an orthogonal matrix?

17. If A is a real $n \times n$ matrix such that $A^{\mathrm{T}} = -A$ (that is, A is skew-symmetric), show
    (a) The matrix $I_n - A$ is nonsingular.
    (b) The matrix $(I_n - A)^{-1}(I_n + A)$ is orthogonal.
    (The matrix in part (b) is the *Cayley transform* of A.)

18. If U is a unitary matrix, show that $\begin{bmatrix} I & O \\ O & U \end{bmatrix}$ is also unitary.

19. Suppose that $\mathbf{v}$ is a fixed unit vector in $\mathbb{C}^n$, and define the matrix $A = I_n - 2\mathbf{v}\mathbf{v}^*$.*
    (a) Show that $A^* = A$.
    (b) Show that A is a unitary matrix.

20. Let A be the matrix defined in the preceding exercise. If $A\mathbf{x} = \mathbf{x}$ and $n > 1$, prove that $\mathbf{x}$ must be orthogonal to $\mathbf{v}$.

21. The orthogonal operator on $\mathbb{R}^2$ defined by $T\left(\begin{bmatrix} x \\ y \end{bmatrix}\right) = \begin{bmatrix} x \\ -y \end{bmatrix}$
    has matrices $A = \begin{bmatrix} 0 & -1 \\ -1 & 0 \end{bmatrix}$ and $B = \begin{bmatrix} 1 & -2 \\ 0 & -1 \end{bmatrix}$ with
    respect to the bases $\left\{\begin{bmatrix} 1 \\ 1 \end{bmatrix}, \begin{bmatrix} -1 \\ 1 \end{bmatrix}\right\}$ and $\left\{\begin{bmatrix} 1 \\ 0 \end{bmatrix}, \begin{bmatrix} -1 \\ 1 \end{bmatrix}\right\}$,
    respectively. Show that A is orthogonal and B is not. Explain how this can happen.

22. Describe all the unitary operators on a one-dimensional inner product space.

C.

1. Find conditions on the real scalars α and β so that

$$A = \begin{bmatrix} 0 & 0 & \beta & \alpha i \\ 0 & 0 & \alpha i & \beta \\ \beta & \alpha i & 0 & 0 \\ \alpha i & \beta & 0 & 0 \end{bmatrix}$$

is a unitary matrix.

---

* Such a matrix is called a **Householder matrix** or a **Householder transformation**, named for A. S. Householder (1904–1993), a leading figure in numerical analysis, especially matrix computations.

2. Show that if $Q = Q_1 + iQ_2$ is unitary, with $Q_1, Q_2 \in M_n$, then the $2n \times 2n$ real matrix:

$$Z = \begin{bmatrix} Q_1 & -Q_2 \\ Q_2 & Q_1 \end{bmatrix}$$

is orthogonal.

3. An operator T on $P_2[x]$ has the matrix:

$$\begin{bmatrix} 3 & -2 & -2 \\ 2 & -1 & -2 \\ 2 & -2 & -1 \end{bmatrix}$$

with respect to the basis 1, x, and x^2. Define an inner product on $P_2[x]$ so that T becomes an orthogonal operator.

4. Let T be an operator on a complex inner product space V. Prove that if T preserves the orthogonality of any two vectors—that is, if $\langle \mathbf{x}, \mathbf{y} \rangle = 0$ implies $\langle T(\mathbf{x}), T(\mathbf{y}) \rangle = 0$ for $\mathbf{x}, \mathbf{y} \in V$—then $T = \alpha U$, where α is a scalar and U is a unitary operator.

7.6 Schur Factorization and the Cayley–Hamilton Theorem

A crucial result that sheds some light on the structure of various classes of matrices and their associated operators is due to Issai Schur.* The proof is a bit daunting, but it amounts to an algorithm for the **Schur decomposition** (or **Schur factorization**) of an $n \times n$ complex matrix.

Theorem 7.6.1a: Schur's Theorem

Let A be an $n \times n$ complex matrix. Then, there is a unitary matrix U such that U^*AU is upper triangular.

Proof We will proceed by induction on n. Let A be an $n \times n$ complex matrix. If $n = 1$, then A is already upper triangular. Now suppose the theorem is true for $n = k - 1$, where $k > 2$. We will show that the result holds true for $n = k$. We start with the fact that the $k \times k$ matrix A has an eigenvector $\mathbf{x}_1$, which we can choose to be a unit vector in $\mathbb{C}^k$, with associated eigenvalue λ_1. By Theorem 7.4.1 (Gram–Schmidt), we can extend the set $\{\mathbf{x}_1\}$ to an orthonormal basis $\{\mathbf{x}_1, \mathbf{x}_2, \ldots, \mathbf{x}_k\}$ for $\mathbb{C}^k$.

Let $U_0 = [\mathbf{x}_1 \ \mathbf{x}_2 \ \cdots \ \mathbf{x}_k]$, so $U_0^* = [\mathbf{x}_1^* \ \mathbf{x}_2^* \ \cdots \ \mathbf{x}_k^*]^T$. Then U_0 is unitary because its columns are orthonormal. Now $U_0^*A\mathbf{x}_1 = U_0^*(\lambda_1\mathbf{x}_1) = \lambda_1\left(U_0^*\mathbf{x}_1\right)$. Also, if $i > 1$, $0 = \langle \mathbf{x}_i, \mathbf{x}_1 \rangle = \mathbf{x}_i^T\bar{\mathbf{x}}_1 = (\mathbf{x}_i^T\bar{\mathbf{x}}_1) = \mathbf{x}_i^*\mathbf{x}_1$, whereas $1 = \langle \mathbf{x}_1, \mathbf{x}_1 \rangle = \mathbf{x}_1^T\bar{\mathbf{x}}_1 = (\mathbf{x}_1^T\bar{\mathbf{x}}_1) = \mathbf{x}_1^*\mathbf{x}_1$.

---

* Issai Schur (1875–1941) did fundamental work in group theory and the theory of matrices. He also did research in number theory and analysis. Theorem 7.6.1a was proved in 1909.

Therefore,

$$U_0^* A \mathbf{x}_1 = \lambda_1 \left(U_0^* \mathbf{x}_1 \right) = \lambda_1 \begin{bmatrix} \mathbf{x}_1^* \mathbf{x}_1 \\ \mathbf{x}_2^* \mathbf{x}_1 \\ \vdots \\ \mathbf{x}_k^* \mathbf{x}_1 \end{bmatrix} = \lambda_1 \begin{bmatrix} 1 \\ 0 \\ \vdots \\ 0 \end{bmatrix} = \begin{bmatrix} \lambda_1 \\ 0 \\ \vdots \\ 0 \end{bmatrix}.$$

Because $U_0^* A U_0 = U_0^* A [\mathbf{x}_1 \ \mathbf{x}_2 \ \ldots \ \mathbf{x}_k] = [U_0^* A \mathbf{x}_1 \ U_0^* A \mathbf{x}_2 \ \ldots \ U_0^* A \mathbf{x}_k]$, we see that

$$U_0^* A U_0 = \begin{bmatrix} \lambda_1 & B \\ \mathbf{0} & A_1 \end{bmatrix},$$

where A_1 is an $(k-1) \times (k-1)$ matrix, B is a $1 \times (k-1)$ row matrix, and $\mathbf{0}$ is a $(k-1) \times 1$ column matrix.

By the induction hypothesis applied to A_1, there is a $k \times k$ unitary matrix U_1 such that $U_1^* A_1 U_1 = T_1$ is upper triangular. We write

$$U_2 = \begin{bmatrix} 1 & \mathbf{0}^T \\ \mathbf{0} & U_1 \end{bmatrix},$$

which is a $(k+1) \times (k+1)$ unitary matrix because the columns are orthonormal. Now let $U = U_0 U_2$, which is also unitary because $U^*U = (U_0 U_2)^*(U_0 U_2) = (U_2^* U_0^*)(U_0 U_2) = U_2^* (U_0^* U_0) U_2 = U_2^* U_2 = I_k$. Finally, $U^* A U = (U_0 U_2)^* A (U_0 U_2) = U_2^* (U_0^* A U_0)$ $U_2 = U_2^* \begin{bmatrix} \lambda_1 & B \\ \mathbf{0} & A_1 \end{bmatrix} U_2 = \begin{bmatrix} \lambda_1 & B U_1 \\ \mathbf{0} & U_1^* A_1 U_1 \end{bmatrix}$, which shows that $U^* A U = \begin{bmatrix} \lambda_1 & B U_1 \\ \mathbf{0} & T_1 \end{bmatrix}$, an upper triangular matrix, as required.

If A and B are square complex matrices such that $B = U^* A U$ for a unitary matrix U, we say that A is **unitarily similar** to B (because $U^* = U^{-1}$).

If $U^* A U = T$, where T is upper triangular, then $A U = U T$ and $A = U T U^*$. This last expression is called the **Schur decomposition,** or **Schur factorization**, of A. Because similar matrices have the same eigenvalues (Theorem 4.5.2) and the eigenvalues of a triangular matrix lie along the main diagonal (Theorem 4.4.7)—results which are easily shown to be true for complex matrices—we see that **the diagonal elements of the upper triangular matrix $U^* A U$ are the eigenvalues of A**, which can be made to appear in any order. Therefore, the Schur decomposition $U^* A U$ is not unique. Also, the proof of Theorem 7.6.1a is easily modified to show that any square complex matrix A is unitarily similar to a *lower* triangular matrix.

Example 7.6.1: Schur Factorizations

Consider the complex matrix $A = \begin{bmatrix} 3 & i \\ -i & 3 \end{bmatrix}$. The eigenvalues of A are 2 and 4, with corresponding orthogonal eigenvectors $\begin{bmatrix} -i \\ 1 \end{bmatrix}$ and $\begin{bmatrix} i \\ 1 \end{bmatrix}$.

Normalizing these eigenvectors, we can form the unitary matrix U having the vectors as its columns:

$$U = \frac{1}{\sqrt{2}} \begin{bmatrix} -i & i \\ 1 & 1 \end{bmatrix}.$$

Then,

$$U^*AU = \frac{1}{\sqrt{2}} \begin{bmatrix} i & 1 \\ -i & 1 \end{bmatrix} \begin{bmatrix} 3 & i \\ -i & 3 \end{bmatrix} \frac{1}{\sqrt{2}} \begin{bmatrix} -i & i \\ 1 & 1 \end{bmatrix}$$

$$= \begin{bmatrix} 2 & 0 \\ 0 & 4 \end{bmatrix},$$

an upper triangular matrix with the eigenvalues of A as its main diagonal entries. We can write the result of Schur's Theorem as

$$A = U \begin{bmatrix} 2 & 0 \\ 0 & 4 \end{bmatrix} U^* = \frac{1}{\sqrt{2}} \begin{bmatrix} -i & i \\ 1 & 1 \end{bmatrix} \begin{bmatrix} 2 & 0 \\ 0 & 4 \end{bmatrix} \frac{1}{\sqrt{2}} \begin{bmatrix} i & 1 \\ -i & 1 \end{bmatrix}$$

$$= \frac{1}{2} \begin{bmatrix} -i & i \\ 1 & 1 \end{bmatrix} \begin{bmatrix} 2 & 0 \\ 0 & 4 \end{bmatrix} \begin{bmatrix} i & 1 \\ -i & 1 \end{bmatrix}.$$

However, this factorization is not unique. If we interchange the columns of the matrix U given above, we get another unitary matrix:

$$\hat{U} = \frac{1}{\sqrt{2}} \begin{bmatrix} i & -i \\ 1 & 1 \end{bmatrix},$$

and $\hat{U}^*A\hat{U} = \begin{bmatrix} 4 & 0 \\ 0 & 2 \end{bmatrix}$ or $A = \hat{U} \begin{bmatrix} 4 & 0 \\ 0 & 2 \end{bmatrix} \hat{U}^*$

$$= \frac{1}{\sqrt{2}} \begin{bmatrix} i & -i \\ 1 & 1 \end{bmatrix} \begin{bmatrix} 4 & 0 \\ 0 & 2 \end{bmatrix} \frac{1}{\sqrt{2}} \begin{bmatrix} -i & 1 \\ i & 1 \end{bmatrix}$$

$$= \frac{1}{2} \begin{bmatrix} i & -i \\ 1 & 1 \end{bmatrix} \begin{bmatrix} 4 & 0 \\ 0 & 2 \end{bmatrix} \begin{bmatrix} -i & 1 \\ i & 1 \end{bmatrix}.$$

The last example demonstrated that the diagonal entries of the triangular matrix in Schur's Theorem may appear in any order. However, the upper triangular matrices guaranteed by the theorem may also appear very different above the diagonal.

Example 7.6.2: Schur Factorizations

The upper triangular matrices

$$T_1 = \begin{bmatrix} 1 & 1 & 4 \\ 0 & 2 & 2 \\ 0 & 0 & 3 \end{bmatrix} \quad \text{and} \quad T_2 = \begin{bmatrix} 2 & -1 & 3\sqrt{2} \\ 0 & 1 & \sqrt{2} \\ 0 & 0 & 3 \end{bmatrix}$$

are unitarily similar via the unitary matrix

$$U = \frac{1}{\sqrt{2}} \begin{bmatrix} 1 & 1 & 0 \\ 1 & -1 & 0 \\ 0 & 0 & \sqrt{2} \end{bmatrix}.$$

This means that $T_1 = U^* T_2 U$ and $T_2 = U^* T_1 U$. Now if A is a complex 3×3 matrix and $T_1 = U_1^* A U_1$ for some unitary matrix U_1 by Schur's Theorem, then

$$U^* T_2 U = U_1^* A U_1, \quad \text{so} \quad T_2 = \overbrace{(U U_1^*)}^{\text{unitary}} A \overbrace{(U_1 U^*)}^{\text{unitary}} = U^* A U.$$

Thus, A is unitarily similar to both T_1 and T_2.

Alternatively, in the language of transformations, Schur's Theorem states that **any linear operator T on a finite-dimensional inner product space V has an upper triangular matrix representation with respect to a suitable orthonormal basis**.

Theorem 7.6.1b: Schur's Theorem (Operator Form)

Let V be an inner product space, where $\dim(V)$ is finite and nonzero, and let $T : V \rightarrow V$ be linear. Then there exits an orthonormal basis $\{\mathbf{v}_1, \mathbf{v}_2, \dots, \mathbf{v}_n\}$ for V such that $T(\mathbf{v}_j) = \sum_{i=1}^{j} t_{ij} \mathbf{v}_i$, $j = 1, 2, \dots, n$, and t_{ii} is an eigenvalue of T for each $i = 1, 2, \dots, n$.

If $V = \mathbb{C}^n$ and $T(\mathbf{x}) = A\mathbf{x}$, then Theorem 7.6.1b asserts the existence of a unitary matrix U and an upper triangular matrix T such that $A = UTU^*$. Furthermore, the diagonal entries of T are the eigenvalues of A. We will not prove this operator version of Schur's Theorem.*

Another consequence of Schur's Theorem is that we can prove the general form of the Cayley–Hamilton Theorem, not just the version for A diagonalizable (Theorem 4.8.1). However, before we get to this proof, let us examine a concrete example that illustrates the strategy.

Example 7.6.3: A Matrix Satisfies Its Characteristic Equation

Consider the matrix $A = \begin{bmatrix} 2 & -1 & 1 \\ 0 & -1 & 3 \\ 0 & 0 & 1 \end{bmatrix}$. The characteristic polynomial of A is $c_A(\lambda) = \det(\lambda I_3 - A) = \lambda^3 - 2\lambda^2 - \lambda + 2 = (\lambda - 2)(\lambda + 1)(\lambda - 1)$. If we substitute A for λ, we get $c_A(A) = A^3 - 2A^2 - A + 2I = (A - 2I)(A + I)(A - I)$, which we will examine multiplication by multiplication in the following way:

---

* See J.T. Scheick, *Linear Algebra with Applications* (New York: McGraw-Hill, 1997), pp. 389–390.

$$A - 2I = \begin{bmatrix} 0 & -1 & 1 \\ 0 & -3 & 3 \\ 0 & 0 & -1 \end{bmatrix}$$

$$(A - 2I)(A + I) = \begin{bmatrix} 0 & -1 & 1 \\ 0 & -3 & 3 \\ 0 & 0 & -1 \end{bmatrix}\begin{bmatrix} 3 & -1 & 1 \\ 0 & 0 & 3 \\ 0 & 0 & 2 \end{bmatrix} = \begin{bmatrix} 0 & 0 & -1 \\ 0 & 0 & -3 \\ 0 & 0 & -2 \end{bmatrix}$$

$$(A - 2I)(A + I)(A - I) = \begin{bmatrix} 0 & 0 & -1 \\ 0 & 0 & -3 \\ 0 & 0 & -2 \end{bmatrix}\begin{bmatrix} 1 & -1 & 1 \\ 0 & -2 & 3 \\ 0 & 0 & 0 \end{bmatrix} = \begin{bmatrix} 0 & 0 & 0 \\ 0 & 0 & 0 \\ 0 & 0 & 0 \end{bmatrix}.$$

Therefore, $c_A(A) = (A - 2I)(A + I)(A - I) = O_3$—that is, A satisfies its characteristic equation. Notice the pattern as we continue to postmultiply.

The number of zero columns increases with each multiplication until we find ourselves with the zero matrix.

Thanks to Schur's Theorem, we need prove the Cayley–Hamilton Theorem only for upper triangular matrices, as the next lemma shows.

Lemma 7.6.1

Suppose A is any complex $n \times n$ matrix, and $T = U^{-1}AU$, where U is unitary and T is upper triangular. If there is a polynomial $f(x) = a_n x^n + a_{n-1}x^{n-1} + \cdots + a_1 x + a_0$ such that $f(T) = O_n$, then $f(A) = O_n$.

Proof If $T = U^{-1}AU$, it is easy to show that $T^k = (U^{-1}AU)^k = U^{-1}A^k U$ for any positive integer k (see problem B12, Exercises 4.5). Then,

$$O_n = f(T) = a_n T^n + a_{n-1}T^{n-1} + \cdots + a_1 T + a_0 I$$

$$= a_n[U^{-1}A^n U] + a_{n-1}[U^{-1}A^{n-1}U] + \cdots + a_1[U^{-1}AU] + a_0 I$$

$$= \sum_{k=0}^{n} a_{n-k}[U^{-1}A^{n-k}U] = U^{-1}\left(\sum_{k=0}^{n} a_{n-k}A^{n-k}\right)U,$$

which implies that $O_n = UO_n U^{-1} = \left(\sum_{k=0}^{n} a_{n-k}A^{n-k}\right) = f(A)$.

Now for the main event.

Theorem 7.6.2: The Cayley–Hamilton Theorem

Every $n \times n$ complex matrix satisfies its own characteristic equation— that is, if matrix A has the characteristic polynomial $p_A(\lambda)$, then $p_A(A) = O_n$.

Proof By Schur's Theorem, there exists an invertible matrix U such that $T = U^{-1}AU$ is an upper triangular matrix. Moreover, from Lemma 7.6.1, if there is a polynomial $f(x) = a_n x^n + a_{n-1}x^{n-1} + \cdots + a_1 x + a_0$ such that $f(T) = O_n$, then $f(A) = O_n$ Therefore, it is enough to prove that any upper triangular matrix satisfies its characteristic equation.

Theorem 4.4.7 states that the eigenvalues of a triangular matrix are the entries on its main diagonal. Now, let $\lambda_1, \lambda_2, \ldots, \lambda_n$ be eigenvalues of T, so we can express the characteristic polynomial of T in the form $c_T(\lambda) = \det(\lambda I - T) = (\lambda - \lambda_1)(\lambda - \lambda_2)(\lambda - \lambda_3) \cdots (\lambda - \lambda_n)$. Substituting T, we get $c_T(T) = (T - \lambda_1 I)(T - \lambda_2 I)(T - \lambda_3 I) \cdots (T - \lambda_n I)$.

Then, we can write

$$T = \begin{bmatrix} \lambda_1 & X \\ \mathbf{0} & B \end{bmatrix},$$

as in the proof of Schur's Theorem, where X, $\mathbf{0}$, and B are matrix blocks of sizes $1 \times (n-1), (n-1) \times 1$, and $(n-1) \times (n-1)$, respectively.

Now

$$(T - \lambda_1 I)(T - \lambda_2 I) = \begin{bmatrix} 0 & X \\ \mathbf{0} & B - \lambda_1 I \end{bmatrix} \begin{bmatrix} \lambda_1 - \lambda_2 & X \\ \mathbf{0} & B - \lambda_2 I \end{bmatrix}$$

$$= \begin{bmatrix} 0 & X(B - \lambda_2 I) \\ \mathbf{0} & (B - \lambda_1 I)(B - \lambda_2 I) \end{bmatrix}$$

Next,

$$(T - \lambda_1 I)(T - \lambda_2 I)(T - \lambda_3 I) = \begin{bmatrix} 0 & X(B - \lambda_2 I) \\ \mathbf{0} & (B - \lambda_1 I)(B - \lambda_2 I) \end{bmatrix} \begin{bmatrix} \lambda_1 - \lambda_3 & X \\ \mathbf{0} & (B - \lambda_3 I) \end{bmatrix}$$

$$= \begin{bmatrix} 0 & X(B - \lambda_2 I)(B - \lambda_3 I) \\ \mathbf{0} & (B - \lambda_1 I)(B - \lambda_2 I)(B - \lambda_3 I) \end{bmatrix}.$$

In general,

$$(T - \lambda_1 I)(T - \lambda_2 I) \cdots (T - \lambda_n I) = \begin{bmatrix} 0 & X(B - \lambda_2 I)(B - \lambda_3 I) \cdots (B - \lambda_n I) \\ \mathbf{0} & (B - \lambda_1 I)(B - \lambda_2 I) \cdots (B - \lambda_n I) \end{bmatrix}.$$

$$(*)$$

To finish the proof, we use induction on n. Assume that the Cayley–Hamilton is true for all $k \times k$ matrices, where $k < n$. Then, $(B - \lambda_2 I)(B - \lambda_3 I) \cdots (B - \lambda_n I) = O_{n-1}$. Therefore, by equation $(*)$, $(T - \lambda_1 I)(T - \lambda_2 I) \cdots (T - \lambda_n I) = O_n$, and the proof is complete.

Section 4.8 describes some of the consequences and applications of the Cayley–Hamilton Theorem.

Exercises 7.6

A.

1. Let $A = \begin{bmatrix} 2 & 1-i \\ 1+i & 1 \end{bmatrix}$. Find a Schur factorization of A.

2. Find a Schur factorization of $A = \begin{bmatrix} 4 & 4 & 1 \\ -1 & 0 & 0 \\ 0 & 0 & 2 \end{bmatrix}$.

3. Use Schur factorization to find eigenvalues of the matrix:
$$\begin{bmatrix} 5 & 6 & 18 \\ 11 & 6 & 24 \\ -4 & -2 & -8 \end{bmatrix}.$$

4. Verify the Cayley–Hamilton Theorem for the following matrices:

   (a) $\begin{bmatrix} 0 & 1 \\ 2 & 1 \end{bmatrix}$ (b) $\begin{bmatrix} 1 & 1 \\ 2 & 1 \end{bmatrix}$ (c) $\begin{bmatrix} 1 & 1 \\ 1 & 1 \end{bmatrix}$

   (d) $\begin{bmatrix} 1 & i \\ -2 & 1+i \end{bmatrix}$ (e) $\begin{bmatrix} 2 & 1 & 1 \\ 0 & 1 & 2 \\ 0 & 0 & 1 \end{bmatrix}$

   (f) $\begin{bmatrix} 3 & 1 & 1 \\ 2 & 4 & 2 \\ 1 & 1 & 3 \end{bmatrix}$ (g) $\begin{bmatrix} 1 & 2 & 2 \\ 1 & 2 & -1 \\ -1 & 1 & 4 \end{bmatrix}$

   (h) $\begin{bmatrix} 1+i & 0 & 3 \\ -1 & 4 & i \\ 2 & 2i & 0 \end{bmatrix}$

5. Verify the Cayley–Hamilton Theorem for the matrix:
$$A = \begin{bmatrix} 1 & 1 & 1 & 1 & 1 \\ 1 & 1 & 1 & 1 & 1 \\ 1 & 1 & 1 & 1 & 1 \\ 1 & 1 & 1 & 1 & 1 \\ 1 & 1 & 1 & 1 & 1 \end{bmatrix}.$$

6. Give a direct computational proof of the Cayley–Hamilton Theorem for any 2×2 matrix.

B.

1. Suppose $\lambda_1, \lambda_2, \ldots, \lambda_n$ are the eigenvalues (including multiplicities) of an $n \times n$ complex matrix A. Use Schur's Theorem and Theorem 4.5.2 to show that

   $$\det(A) = \lambda_1, \lambda_2, \ldots, \lambda_n \quad \text{and} \quad \text{trace}(A) = \lambda_1 + \lambda_2 + \cdots + \lambda_n.$$

2. If $A = \begin{bmatrix} 0 & -1 \\ 1 & 0 \end{bmatrix}$, show that $U^{-1}AU$ is not upper triangular for any invertible *real* matrix U.

3. Suppose $A = \begin{bmatrix} 1 & 1 \\ 0 & 1 \end{bmatrix}$. Show that $U^{-1}AU$ is diagonal for *no* invertible complex matrix U.

4. Show that the matrix $A = \begin{bmatrix} 0 & -1 \\ 1 & 0 \end{bmatrix}$ is unitarily diagonalizable.

5. Let $A = \begin{bmatrix} 4 & 3-i \\ 3+i & 2 \end{bmatrix}$. Find a unitary matrix U such that $U*AU$ is diagonal.

6. If $A = \begin{bmatrix} 1 & 2 \\ 2 & 1 \end{bmatrix}$, use the Cayley–Hamilton Theorem to calculate $A^4 - 7A^3 - 3A^2 + A + 4I_2$. [*Hint:* Recall the Division Algorithm for polynomials.]

C.

1. (a) If A is an $n \times n$ nonsingular matrix and $c_A(\lambda) = \lambda^n + a_{n-1}\lambda^{n-1} + \cdots + a_1\lambda + a_0$ is the characteristic polynomial of A, show that

$$A^{-1} = -\frac{1}{a_0}\left(A^{n-1} + a_{n-1}A^{n-2} + \cdots + a_1 I\right).$$

   (b) Use the result of part (a) to find the inverse of the nonsingular matrix:

$$A = \begin{bmatrix} 4 & 2 & -2 \\ -5 & 3 & 2 \\ -2 & 4 & 1 \end{bmatrix}.$$

7.7 The QR Factorization and Applications

The Gram–Schmidt process described and illustrated in Section 7.4 is a very useful technique, both in theory and in practice. The matrix version of this algorithm is equally powerful and has important applications.

In Example 7.4.5, we found an orthonormal basis for the column space of

$$A = \begin{bmatrix} 1 & -1 & 1 \\ 0 & 0 & 1 \\ 1 & -1 & 0 \\ 0 & 1 & -1 \end{bmatrix} = \begin{bmatrix} \mathbf{a}_1 & \mathbf{a}_2 & \mathbf{a}_3 \end{bmatrix}.$$

Now, we reexamine this example and focus on the end result of the Gram–Schmidt process in terms of column operations on A:

$$\mathbf{q}_1 = \frac{\mathbf{a}_1'}{\|\mathbf{a}_1'\|} = \begin{bmatrix} \sqrt{2}/2 & 0 & \sqrt{2}/2 & 0 \end{bmatrix}^T = \frac{\sqrt{2}}{2}\mathbf{a}_1,$$

$$\mathbf{q}_2 = \frac{\mathbf{a}_2'}{\|\mathbf{a}_2'\|} = \begin{bmatrix} 0 & 0 & 0 & 1 \end{bmatrix}^T = \mathbf{a}_1 + \mathbf{a}_2, \qquad (*)$$

$$\mathbf{q}_3 = \frac{\mathbf{a}_3'}{\|\mathbf{a}_3'\|} = \begin{bmatrix} \sqrt{6}/6 & \sqrt{6}/3 & -\sqrt{6}/6 & 0 \end{bmatrix}^T = \frac{\sqrt{6}}{6}\mathbf{a}_1 + \frac{\sqrt{6}}{3}\mathbf{a}_2 + \frac{\sqrt{6}}{3}\mathbf{a}_3.$$

We see that the orthonormalization process (*) can be interpreted as postmultiplication (multiplication on the right) of A by an upper triangular matrix:

$$Q = [\mathbf{q}_1 \ \ \mathbf{q}_2 \ \ \mathbf{q}_3] = \left[\frac{\mathbf{a}_1'}{\|\mathbf{a}_1'\|} \ \ \frac{\mathbf{a}_2'}{\|\mathbf{a}_2'\|} \ \ \frac{\mathbf{a}_3'}{\|\mathbf{a}_3'\|} \right] = [\mathbf{a}_1 \ \ \mathbf{a}_2 \ \ \mathbf{a}_3] \begin{bmatrix} \sqrt{2}/2 & 1 & \sqrt{6}/6 \\ 0 & 1 & \sqrt{6}/3 \\ 0 & 0 & \sqrt{6}/3 \end{bmatrix}$$

$$= \begin{bmatrix} 1 & -1 & 1 \\ 0 & 0 & 1 \\ 1 & -1 & 0 \\ 0 & 1 & -1 \end{bmatrix} \begin{bmatrix} \sqrt{2}/2 & 1 & \sqrt{6}/6 \\ 0 & 1 & \sqrt{6}/3 \\ 0 & 0 & \sqrt{6}/3 \end{bmatrix} = \begin{bmatrix} \sqrt{2}/2 & 0 & \sqrt{6}/6 \\ 0 & 0 & \sqrt{6}/3 \\ \sqrt{2}/2 & 0 & -\sqrt{6}/6 \\ 0 & 1 & 0 \end{bmatrix},$$

or

$$A \cdot \begin{bmatrix} \sqrt{2}/2 & 1 & \sqrt{6}/6 \\ 0 & 1 & \sqrt{6}/3 \\ 0 & 0 & \sqrt{6}/3 \end{bmatrix} = Q.$$

It is important to recognize that the columns of the matrix Q are the orthonormal vectors resulting from applying the Gram–Schmidt procedure to the columns of A. Inverting the 3×3 matrix in the last equation and multiplying on the right by this inverse, we get the relation

$$A = \begin{bmatrix} 1 & -1 & 1 \\ 0 & 0 & 1 \\ 1 & -1 & 0 \\ 0 & 1 & -1 \end{bmatrix} = \begin{bmatrix} \sqrt{2}/2 & 0 & \sqrt{6}/6 \\ 0 & 0 & \sqrt{6}/3 \\ \sqrt{2}/2 & 0 & -\sqrt{6}/6 \\ 0 & 1 & 0 \end{bmatrix} \begin{bmatrix} \sqrt{2} & -\sqrt{2} & \sqrt{2}/2 \\ 0 & 1 & -1 \\ 0 & 0 & \sqrt{6}/2 \end{bmatrix}.$$

This factorization of A is known as a **QR factorization**, and we label the factors according to the decomposition $A = QR$, where the matrix Q has orthonormal columns and the square matrix R is upper triangular.

Alternatively, we can solve system (*) for vectors $\mathbf{a}_1$, $\mathbf{a}_2$, and $\mathbf{a}_3$ in terms of $\mathbf{q}_1$, $\mathbf{q}_2$, and $\mathbf{q}_3$:

$$\begin{aligned} \mathbf{a}_1 &= \sqrt{2}\mathbf{q}_1, \\ \mathbf{a}_2 &= -\sqrt{2}\mathbf{q}_1 + \mathbf{q}_2, \\ \mathbf{a}_3 &= \frac{\sqrt{2}}{2}\mathbf{q}_1 - \mathbf{q}_2 + \frac{\sqrt{6}}{2}\mathbf{q}_3. \end{aligned} \tag{**}$$

Writing the system (**) in matrix form, we have

$$A = [\mathbf{a}_1 \ \ \mathbf{a}_2 \ \ \mathbf{a}_3] = [\mathbf{q}_1 \ \ \mathbf{q}_2 \ \ \mathbf{q}_3] \begin{bmatrix} \sqrt{2} & -\sqrt{2} & \sqrt{2}/2 \\ 0 & 1 & -1 \\ 0 & 0 & \sqrt{6}/2 \end{bmatrix}$$

$$= \begin{bmatrix} \sqrt{2}/2 & 0 & \sqrt{6}/6 \\ 0 & 0 & \sqrt{6}/3 \\ \sqrt{2}/2 & 0 & -\sqrt{6}/6 \\ 0 & 1 & 0 \end{bmatrix} \begin{bmatrix} \sqrt{2} & -\sqrt{2} & \sqrt{2}/2 \\ 0 & 1 & -1 \\ 0 & 0 & \sqrt{6}/2 \end{bmatrix},$$

so once again we find that

$$A = \begin{bmatrix} 1 & -1 & 1 \\ 0 & 0 & 1 \\ 1 & -1 & 0 \\ 0 & 1 & -1 \end{bmatrix} = \begin{bmatrix} \sqrt{2}/2 & 0 & \sqrt{6}/6 \\ 0 & 0 & \sqrt{6}/3 \\ \sqrt{2}/2 & 0 & -\sqrt{6}/6 \\ 0 & 1 & 0 \end{bmatrix} \begin{bmatrix} \sqrt{2} & -\sqrt{2} & \sqrt{2}/2 \\ 0 & 1 & -1 \\ 0 & 0 & \sqrt{6}/2 \end{bmatrix} = QR.$$

Essentially, *this factorization is a description of the Gram–Schmidt process in matrix notation*. Noting that the Gram–Schmidt process (Theorem 7.4.1) starts with a *linearly* independent set of vectors, we conclude that the Gram–Schmidt algorithm shows that **any $m \times n$ matrix of rank n has a QR factorization**. However, we will state and prove the result only for a *real* matrix A.

In the example we just completed, note that because $\{\mathbf{q}_1, \mathbf{q}_2, \mathbf{q}_3\}$ is an orthonormal basis for $\mathbb{R}^3$, the nonzero entries of R are just the Fourier coefficients of the columns of A (Corollary 7.3.2(1)): $r_{ji} = \langle \mathbf{a}_i, \mathbf{q}_j \rangle$ for $i, j = 1, 2, 3$; or $r_{ij} = \langle \mathbf{q}_j, \mathbf{a}_i \rangle$.

Theorem 7.7.1: The QR Factorization

Let A be a real $m \times n$ matrix with rank n. Then, A can be written as a product QR, where Q is a real $m \times n$ matrix whose columns form an orthonormal set and R is a real $n \times n$ upper triangular matrix with positive entries on its main diagonal.

Proof   First write A in column form: $[\mathbf{a}_1 \quad \mathbf{a}_2 \quad \ldots \quad \mathbf{a}_n]$. Using the standard inner product in $\mathbb{R}^m$, apply the Gram–Schmidt process (Theorem 7.4.1) to the columns $\mathbf{a}_1, \mathbf{a}_2, \ldots, \mathbf{a}_n$ to get the orthonormal vectors $\mathbf{q}_1, \mathbf{q}_2, \ldots, \mathbf{q}_n$ in $\mathbb{R}^m$. One consequence of the Gram–Schmidt process is that $span\{\mathbf{q}_1, \mathbf{q}_2, \ldots, \mathbf{q}_k\} = span\{\mathbf{a}_1, \mathbf{a}_2, \ldots, \mathbf{a}_k\}$ for $k = 1, 2, \ldots, n$.

Thus, there exist real numbers r_{ik} such that

$$\mathbf{a}_1 = r_{11}\mathbf{q}_1$$
$$\mathbf{a}_2 = r_{12}\mathbf{q}_1 + r_{22}\mathbf{q}_2$$
$$\vdots \qquad \vdots$$
$$\mathbf{a}_n = r_{1n}\mathbf{q}_1 + r_{2n}\mathbf{q}_2 + \cdots + r_{nn}\mathbf{q}_n$$

Now define $r_{ik} = 0$ if $i > k$, and form the matrix $R = [r_{ik}] = [\mathbf{r}_1 \quad \mathbf{r}_2 \quad \ldots \quad \mathbf{r}_n]$ in column form. If some diagonal entry r_{kk} is negative, just replace r_{kk} by $-r_{kk}$ and $\mathbf{q}_k$ by $-\mathbf{q}_k$. Now, we can write, for $1 \leq k \leq n$:

$$\mathbf{a}_k = \sum_{i=1}^{k} r_{ik}\mathbf{q}_i = [\mathbf{q}_1 \quad \mathbf{q}_2 \quad \ldots \quad \mathbf{q}_n] \begin{bmatrix} r_{1k} \\ r_{2k} \\ \vdots \\ r_{kk} \\ \cdots \\ \mathbf{0} \end{bmatrix} = Q\mathbf{r}_k,$$

and so $A = [\mathbf{a}_1 \quad \mathbf{a}_2 \quad \ldots \quad \mathbf{a}_n] = [Q\mathbf{r}_1 \quad Q\mathbf{r}_2 \quad \ldots \quad Q\mathbf{r}_n] = QR.$

So far, we have guaranteed that $r_{kk} \geq 0$ for $k = 1, 2, \ldots, n$.

If any main diagonal entry of R is zero, then R (as an upper triangular matrix) is noninvertible. But then the system $R\mathbf{x} = \mathbf{0}$ and hence $QR\mathbf{x} = A\mathbf{x} = \mathbf{0}$ has nontrivial solutions—contradicting Theorem 2.8.1, which states that $A\mathbf{x} = \mathbf{0}$ has no nonzero solution if rank$(A) = n$.

Corollary 7.7.1

If A is a real $m \times n$ matrix with rank n that has a QR factorization as described in Theorem 7.7.1, then the QR factorization is unique.

Proof Suppose that A has two QR decompositions: $A = Q_1 R_1 = Q_2 R_2$. Let $U = Q_2^T Q_1 = R_2 R_1^{-1}$. The matrix $R_2 R_1^{-1}$ is upper triangular with positive diagonal entries,* whereas $Q_2^T Q_1$ is an orthogonal matrix by Theorem 7.7.1:
$$\left(Q_2^T Q_1\right)^T \left(Q_2^T Q_1\right) = Q_1^T Q_2 Q_2^T Q_1 = Q_1^T Q_1 = I_n.$$ Therefore, U is an upper triangular matrix whose columns are mutually orthogonal and whose diagonal entries are positive.

If $U = [\,\mathbf{u}_1 \quad \mathbf{u}_2 \quad \ldots \quad \mathbf{u}_n\,]$, then $\|\mathbf{u}_1\| = \left\|[\,u_{11} \quad 0 \quad \ldots \quad 0\,]^T\right\| = 1$ implies $u_{11} = \pm 1$; but $u_{11} > 0$ tells us that $u_{11} = 1$, so $\mathbf{u}_1 = \mathbf{e}_1$.

A similar argument, together with the fact that the columns of U are orthogonal, produces $\mathbf{u}_1^T \mathbf{u}_2 = 0$, which implies that $u_{12} = 0$, so $u_{22} = 1$ and $\mathbf{u}_2 = \mathbf{e}_2$. Using induction, we establish that $\mathbf{u}_k = \mathbf{e}_k$ for $k = 1, 2, \ldots, n$—that is, $U = I_n$. Thus, $Q_1 = Q_2$ and $R_1 = R_2$.

Example 7.7.1: A QR Factorization

Returning to Example 7.4.1, we consider the matrix

$$A = \begin{bmatrix} 1 & 0 & -1 \\ 1 & 1 & -1 \\ 1 & 0 & 1 \end{bmatrix} = [\,\mathbf{a}_1 \quad \mathbf{a}_2 \quad \mathbf{a}_3\,]$$

and see that the rank of A is three.

The calculations of that example give us the orthonormal form of the column vectors $\mathbf{a}_i$ of A:

$$\mathbf{q}_1 = \frac{\sqrt{3}}{3}\begin{bmatrix} 1 \\ 1 \\ 1 \end{bmatrix}, \quad \mathbf{q}_2 = \frac{\sqrt{6}}{6}\begin{bmatrix} -1 \\ 2 \\ -1 \end{bmatrix}, \quad \mathbf{q}_3 = \frac{\sqrt{2}}{2}\begin{bmatrix} -1 \\ 0 \\ 1 \end{bmatrix}.$$

---

* The fact that the diagonal entries are positive comes from the (easily provable) fact that the diagonal elements of the product of two upper triangular matrices U_1 and U_2 are the products of the corresponding diagonal elements of U_1 and U_2.

Thus,

$$Q = [\mathbf{q}_1 \quad \mathbf{q}_2 \quad \mathbf{q}_3] = \begin{bmatrix} \sqrt{3}/3 & -\sqrt{6}/6 & -\sqrt{2}/2 \\ \sqrt{3}/3 & \sqrt{6}/3 & 0 \\ \sqrt{3}/3 & -\sqrt{6}/6 & \sqrt{2}/2 \end{bmatrix}.$$

To find the upper triangular matrix R, we use the observation made just before the statement of Theorem 7.7.1: $R = [r_{ij}]$, where $r_{ij} = \langle \mathbf{q}_i, \mathbf{a}_j \rangle$. Thus,

$$r_{11} = \langle \mathbf{q}_1, \mathbf{a}_1 \rangle = \begin{bmatrix} \sqrt{3}/3 \\ \sqrt{3}/3 \\ \sqrt{3}/3 \end{bmatrix} \cdot \begin{bmatrix} 1 \\ 1 \\ 1 \end{bmatrix} = \sqrt{3}, \quad r_{12} = \langle \mathbf{q}_1, \mathbf{a}_2 \rangle \begin{bmatrix} \sqrt{3}/3 \\ \sqrt{3}/3 \\ \sqrt{3}/3 \end{bmatrix} \cdot \begin{bmatrix} 0 \\ 1 \\ 0 \end{bmatrix} = \sqrt{3}/3,$$

$$r_{13} = \langle \mathbf{q}_1, \mathbf{a}_3 \rangle = \begin{bmatrix} \sqrt{3}/3 \\ \sqrt{3}/3 \\ \sqrt{3}/3 \end{bmatrix} \cdot \begin{bmatrix} -1 \\ -1 \\ 1 \end{bmatrix} = -\sqrt{3}/3, \quad r_{21} = \langle \mathbf{q}_2, \mathbf{a}_1 \rangle = \begin{bmatrix} -\sqrt{6}/6 \\ \sqrt{6}/3 \\ -\sqrt{6}/6 \end{bmatrix} \cdot \begin{bmatrix} 1 \\ 1 \\ 1 \end{bmatrix} = 0,$$

$$r_{22} = \langle \mathbf{q}_2, \mathbf{a}_2 \rangle = \begin{bmatrix} -\sqrt{6}/6 \\ \sqrt{6}/3 \\ -\sqrt{6}/6 \end{bmatrix} \cdot \begin{bmatrix} 0 \\ 1 \\ 0 \end{bmatrix} = \sqrt{6}/3, \quad r_{23} = \langle \mathbf{q}_2, \mathbf{a}_3 \rangle = \begin{bmatrix} -\sqrt{6}/6 \\ \sqrt{6}/3 \\ -\sqrt{6}/6 \end{bmatrix} \cdot \begin{bmatrix} -1 \\ -1 \\ 1 \end{bmatrix} = -\sqrt{6}/3,$$

$$r_{31} = \langle \mathbf{q}_3, \mathbf{a}_1 \rangle = \begin{bmatrix} -\sqrt{2}/2 \\ 0 \\ \sqrt{2}/2 \end{bmatrix} \cdot \begin{bmatrix} 1 \\ 1 \\ 1 \end{bmatrix} = 0, \quad r_{32} = \langle \mathbf{q}_3, \mathbf{a}_2 \rangle = \begin{bmatrix} -\sqrt{2}/2 \\ 0 \\ \sqrt{2}/2 \end{bmatrix} \cdot \begin{bmatrix} 0 \\ 1 \\ 0 \end{bmatrix} = 0,$$

$$r_{33} = \langle \mathbf{q}_3, \mathbf{a}_3 \rangle = \begin{bmatrix} -\sqrt{2}/2 \\ 0 \\ \sqrt{2}/2 \end{bmatrix} \cdot \begin{bmatrix} -1 \\ -1 \\ 1 \end{bmatrix} = \sqrt{2}.$$

Therefore,

$$R = \begin{bmatrix} \sqrt{3} & \sqrt{3}/3 & -\sqrt{3}/3 \\ 0 & \sqrt{6}/3 & -\sqrt{6}/3 \\ 0 & 0 & \sqrt{2} \end{bmatrix}$$

and we see that

$$A = \begin{bmatrix} 1 & 0 & -1 \\ 1 & 1 & -1 \\ 1 & 0 & 1 \end{bmatrix} = \begin{bmatrix} \sqrt{3}/3 & -\sqrt{6}/6 & -\sqrt{2}/2 \\ \sqrt{3}/3 & \sqrt{6}/3 & 0 \\ \sqrt{3}/3 & -\sqrt{6}/6 & \sqrt{2}/2 \end{bmatrix}$$

$$\times \begin{bmatrix} \sqrt{3} & \sqrt{3}/3 & -\sqrt{3}/3 \\ 0 & \sqrt{6}/3 & -\sqrt{6}/3 \\ 0 & 0 & \sqrt{2} \end{bmatrix} = QR.$$

Theorem 7.7.1 is the matrix version of the Gram–Schmidt process and as such assumes that we start with an $m \times n$ matrix of rank n. However, for the record, we state (without proof) a more general result:*

---

* See R.A. Horn and C.R. Johnson, *Matrix Analysis* (New York: Cambridge University Press, 1990), Section 2.6.1.

Suppose $A \in M_{mn}(\mathbb{C})$. If $m \geq n$, there is a matrix $Q \in M_{mn}(\mathbb{C})$ with ortho-normal columns and an upper triangular matrix $R \in M_n(\mathbb{C})$ such that $A = QR$. If $m = n$, Q is unitary. If, in addition, A is nonsingular (i.e., rank $(A) = n$), then R may be chosen so that all its diagonal entries are positive; and in this event, the factors Q and R are both unique. If $A \in M_{mn}(\mathbb{R})$ (i.e., A is real), then both Q and R may be taken to be real.

The QR factorization can be used to solve a consistent $m \times n$ system of linear equations $A\mathbf{x} = \mathbf{b}$ in a way that is similar to the use of the LU decomposition discussed in Section 3.7. This assumes that rank$(A) = n$. By replacing A by QR, we have the convenience of working with an upper triangular matrix as well as with an orthonormal matrix:

$$A\mathbf{x} = \mathbf{b}$$
$$A^T A\mathbf{x} = A^T \mathbf{b}$$
$$(QR)^T(QR)\mathbf{x} = (QR)^T\mathbf{b}$$
$$R^T Q^T QR\mathbf{x} = R^T Q^T\mathbf{b}$$
$$R^T R\mathbf{x} = R^T Q^T\mathbf{b} \quad [Q \text{ is orthonormal, so } Q^T Q = I]$$
$$R\mathbf{x} = Q^T\mathbf{b}$$

This series of implications is reversible, so we have the result that $A\mathbf{x} = \mathbf{b}$ **if and only if** $R\mathbf{x} = Q^T\mathbf{b}$. The advantage of the system $R\mathbf{x} = Q^T\mathbf{b}$ is that the matrix R is upper triangular, so the system can be solved by back substitution. Note that if A is an $m \times n$ matrix of rank n, then the factors Q and R are unique, and this unique solution is given by $\mathbf{x} = R^{-1}Q^T\mathbf{b}$. Even though R is invertible and $\mathbf{x} = R^{-1}Q^T\mathbf{b}$, it is easier to use back substitution.

Example 7.7.2: Solving $A\mathbf{x} = \mathbf{b}$ via the QR Factorization

Let us use the QR decomposition to solve the system $A\mathbf{x} = \mathbf{b}$, where

$$A = \begin{bmatrix} 1 & 0 & 0 \\ 0 & 1 & 1 \\ 0 & 0 & 1 \\ 1 & 1 & 0 \end{bmatrix}, \quad \mathbf{x} = \begin{bmatrix} x \\ y \\ z \end{bmatrix}, \quad \text{and} \quad \mathbf{b} = \begin{bmatrix} 1 \\ 5 \\ 3 \\ 3 \end{bmatrix}.$$

We can write

$$A = \frac{1}{6}\overbrace{\begin{bmatrix} 3\sqrt{2} & -\sqrt{6} & \sqrt{3} \\ 0 & 2\sqrt{6} & \sqrt{3} \\ 0 & 0 & 3\sqrt{3} \\ 3\sqrt{2} & \sqrt{6} & -\sqrt{3} \end{bmatrix}}^{Q} \frac{1}{6}\overbrace{\begin{bmatrix} 6\sqrt{2} & 3\sqrt{2} & 0 \\ 0 & 3\sqrt{6} & 2\sqrt{6} \\ 0 & 0 & 4\sqrt{3} \end{bmatrix}}^{R},$$

so $R\mathbf{x} = Q^T\mathbf{b}$ is equivalent to

$$\begin{bmatrix} 6\sqrt{2} & 3\sqrt{2} & 0 \\ 0 & 3\sqrt{6} & 2\sqrt{6} \\ 0 & 0 & 4\sqrt{3} \end{bmatrix} \begin{bmatrix} x \\ y \\ z \end{bmatrix} = \begin{bmatrix} 3\sqrt{2} & 0 & 0 & 3\sqrt{2} \\ -\sqrt{6} & 2\sqrt{6} & 0 & \sqrt{6} \\ \sqrt{3} & \sqrt{3} & 3\sqrt{3} & -\sqrt{3} \end{bmatrix}$$

$$\times \begin{bmatrix} 1 \\ 5 \\ 3 \\ 3 \end{bmatrix} = \begin{bmatrix} 12\sqrt{2} \\ 12\sqrt{6} \\ 12\sqrt{3} \end{bmatrix}.$$

Solving by back substitution—first for z, then for y, and finally for x—we find that $x = 1$, $y = 2$, and $z = 3$.

Exercises 7.7

A.

1. Find the QR factorization of each of the following matrices:

    (a) $\begin{bmatrix} 1 & 1 \\ 3 & -1 \end{bmatrix}$ (b) $\begin{bmatrix} 1 & 0 \\ -1 & 1 \end{bmatrix}$ (c) $\begin{bmatrix} 1 & 1 \\ 1 & 2 \end{bmatrix}$

    (d) $\begin{bmatrix} 0 & 1 \\ 1 & -1 \end{bmatrix}$

2. Find the QR factorization of the matrix $A = \begin{bmatrix} 1 & 10 \\ 1 & 15 \\ 1 & 20 \end{bmatrix}$.

3. Determine the QR factorization of $A = \begin{bmatrix} 1 & 1 \\ 0 & 1 \\ 1 & 1 \end{bmatrix}$.

4. Determine the QR factorization of $A = \begin{bmatrix} 1 & 0 & 2 \\ 0 & 1 & 1 \\ 1 & 2 & 0 \end{bmatrix}$.

5. Determine the QR factorization of $A = \begin{bmatrix} 0 & 3 & 2 \\ 3 & 5 & 5 \\ 4 & 0 & 5 \end{bmatrix}$.

6. Find a QR factorization of the matrix $A = \begin{bmatrix} 1 & 1 & 1 \\ 1 & 0 & 0 \\ 1 & 1 & 0 \\ 1 & 0 & 1 \end{bmatrix}$.

7. Find the QR factorization of $A = \begin{bmatrix} 1 & 2 & 2 \\ 1 & 1 & 2 \\ 0 & 1 & 3 \\ 1 & 0 & -1 \end{bmatrix}$.

8. Use the QR factorization to find the exact solution of the system:

$$a + b + 2c = 1$$
$$a + 2b + 3c = 2$$
$$a + 2b + c = 1$$

9. Use the QR factorization to solve the system:

$$x + 2y + 2z = 1$$
$$x + 2z = 11$$
$$y + z = -1$$

10. If A is a diagonal matrix, what is its QR factorization?

11. What is the QR factorization of an orthogonal matrix?

12. Calculate the inverse of the matrix $A = \begin{bmatrix} 1 & 0 & 2 \\ 2 & 4 & 2 \\ 1 & 2 & 6 \end{bmatrix}$ by first finding the QR decomposition of A and then using $A^{-1} = (QR)^{-1} = R^{-1}Q^{-1} = R^{-1}Q^{T}$. [Note that the inverse of the upper triangular matrix R is found easily by back substitution.]

B.

Unless otherwise stated, all matrices in the following problems are real matrices.

1. Compute the QR factorization of A, where

$$A = \begin{bmatrix} 1 & 1 & 1 \\ 1 & 2 & 4 \\ 1 & 3 & 9 \\ 1 & 4 & 16 \\ 1 & 5 & 25 \\ 1 & 6 & 36 \end{bmatrix}.$$

Calculate Q and R to five decimal places.

2. Consider an upper triangular $n \times n$ matrix A whose diagonal entries are negative real numbers. What does the Q factor in the QR factorization of A look like?
[*Hint*: Use the uniqueness of the QR factorization.]

3. Let $A = QR$ be a QR factorization of A. Find QQ^*A. Is it true that $QQ^* = I$?

4. Suppose $A = PT$, where P has orthogonal columns $\mathbf{p}_i$ and T is a nonsingular upper triangular matrix. Find the QR factorization of A.

5. If the real matrix A has the QR factorization $A = QR$, show that $A^{T}A = R^{T}R$.

6. Suppose an $m \times n$ matrix A has rank n and let $A = QR$ be a QR factorization.

   Show that $A\mathbf{x} = \mathbf{b}$ is consistent if and only if $Q\mathbf{y} = \mathbf{b}$ is consistent. Why does this result say that A and Q have the same column space?

7. Let A be a matrix with nonzero columns that form an orthogonal set of vectors. If $A = QR$ is a QR factorization, show that R is a diagonal matrix.

8. Use the ideas in the proof of Theorem 7.7.1 to determine the QR factorization of $\begin{bmatrix} -i & i \\ 1+i & 2 \end{bmatrix}$? (Also see the generalization of Theorem 7.7.1.)

C.

1. Modify the QR factorization method to include the case in which the columns of A are not necessarily linearly independent.

2. Find the QR factorization of the $n \times n$ matrix U, where

$$U = \begin{bmatrix} 1 & 1 & 1 & \cdots & 1 \\ 0 & 1 & 1 & \cdots & 1 \\ 0 & 0 & 1 & \cdots & 1 \\ \vdots & \vdots & \vdots & \cdots & \vdots \\ 0 & 0 & 0 & \cdots & 1 \end{bmatrix}.$$

   [*Hint*: Experiment with the $2 \times 2, 3 \times 3,$ and 4×4 cases to see the pattern.]

3. Show that any matrix $B \in M_n(\mathbb{C})$ of the form $B = A^*A$, $A \in M_n(\mathbb{C})$, may be written as $B = LL^*$, where $L \in M_n(\mathbb{C})$ is lower triangular with nonnegative diagonal entries. Also, show that this factorization is unique if A is nonsingular.

   (This is called the *Cholesky factorization* of B.) [*Hint*: Write $A = QR$.]

7.8 Orthogonal Complements

Suppose that A is a real $m \times n$ matrix. Now let $\mathbf{z}$ be a vector in the null space of A, meaning that $\mathbf{z} \in \mathbb{R}^n$ and $A\mathbf{z} = \mathbf{0}$. Then the row view of matrix multiplication (Theorem 3.3.2) yields the fact that $0 = \text{row}_i(A\mathbf{z}) = \text{row}_i(A)\mathbf{z}$, so that **every row of A has a zero dot product with z**. Another way to say this is that **every vector in the null space of A is orthogonal to every row of A**. Similarly, we can show that **every vector v in the null space of A^T is orthogonal to every column of A**. These facts about the rows and columns of a matrix first appeared as problem B28 of Exercises 3.3; and we will prove them as Theorem 7.8.1. The special orthogonality relationship just described can be generalized to any inner product space and formalized.

Definition 7.8.1

If a vector $\mathbf{z}$ is orthogonal to every vector in a nonempty subset S of an inner product space V, then $\mathbf{z}$ is said to be **orthogonal to S** The set of all vectors $\mathbf{z}$ that are orthogonal to S is called the **orthogonal complement** of S and is denoted by $S^{\perp}$ (pronounced "S perp"):

$$S^{\perp} = \{\mathbf{z} \in V \mid \langle \mathbf{z}, \mathbf{v} \rangle = 0 \text{ for all } \mathbf{v} \in S\}.$$

The use of the word "complement" will be justified by Theorem 7.8.2.

Example 7.8.1: An Orthogonal Complement in $\mathbb{R}^3$

In $\mathbb{R}^3$, let S consist of the vector $\mathbf{v} = [1 \quad -1 \quad 2]^{\mathsf{T}}$. A vector $\mathbf{w} = [x \quad y \quad z]^{\mathsf{T}}$ is orthogonal to $\mathbf{v}$ if $\mathbf{w} \bullet \mathbf{v} = (1)x + (-1)y + (2)z = x - y + 2z = 0$. Thus $S^{\perp} = \{[x \quad y \quad z]^{\mathsf{T}} : x - y + 2z = 0\}$, a plane through the origin:

The next two examples depict the concept in more general spaces.

Example 7.8.2: An Orthogonal Complement in M_2

Consider the space M_2 with its standard inner product:

$$\langle A, B \rangle = \sum_{i=1}^{2} \sum_{j=1}^{2} a_{ij} b_{ij} = \text{trace}(A^{\mathsf{T}} B).$$

Let S be the subspace of M_2 consisting of the diagonal matrices. We want to determine $S^{\perp}$.

If $A = \begin{bmatrix} a_{11} & a_{12} \\ a_{21} & a_{22} \end{bmatrix}$ is in $S^{\perp}$, then for any element $D = \begin{bmatrix} d_1 & 0 \\ 0 & d_2 \end{bmatrix}$ in S,

$$0 = \langle A, D \rangle = \text{trace}(A^{\mathsf{T}} D) = \text{trace}\left(\begin{bmatrix} a_{11} & a_{21} \\ a_{12} & a_{22} \end{bmatrix} \begin{bmatrix} d_1 & 0 \\ 0 & d_2 \end{bmatrix} \right)$$

$$= \text{trace} \begin{bmatrix} a_{11} d_1 & a_{21} d_2 \\ a_{12} d_1 & a_{22} d_2 \end{bmatrix} = a_{11} d_1 + a_{22} d_2.$$

Because this equality must be true for *all real values of* d_1 and d_2, we can choose $d_1 = 1$ and $d_2 = 0$, which forces $a_{11} = 0$. Similarly, choosing $d_1 = 0$ and $d_2 = 1$, we see that $a_{22} = 0$. So far, we have proved that $S^\perp \subseteq \left\{ \begin{bmatrix} 0 & r \\ s & 0 \end{bmatrix} : r, s \in \mathbb{R} \right\}$. The strategy is to show that $\left\{ \begin{bmatrix} 0 & r \\ s & 0 \end{bmatrix} : r, s \in \mathbb{R} \right\} \subseteq S^\perp$ as well, so $S^\perp$ must equal $\left\{ \begin{bmatrix} 0 & r \\ s & 0 \end{bmatrix} : r, s \in \mathbb{R} \right\}$.

Now if $A = \begin{bmatrix} 0 & r \\ s & 0 \end{bmatrix}$ and $D = \begin{bmatrix} d_1 & 0 \\ 0 & d_2 \end{bmatrix}$ is any element of S, we have $\langle A, D \rangle = \text{trace}(A^\mathsf{T} D) = \text{trace} \left(\begin{bmatrix} 0 & s \\ r & 0 \end{bmatrix} \begin{bmatrix} d_1 & 0 \\ 0 & d_2 \end{bmatrix} \right) = \text{trace} \begin{bmatrix} 0 & sd_2 \\ rd_1 & 0 \end{bmatrix} = 0$, showing that $\left\{ \begin{bmatrix} 0 & r \\ s & 0 \end{bmatrix} : r, s \in \mathbb{R} \right\} \subseteq S^\perp$. Therefore, we conclude that $S^\perp = \left\{ \begin{bmatrix} 0 & r \\ s & 0 \end{bmatrix} : r, s \in \mathbb{R} \right\}$.

Example 7.8.3: An Orthogonal Complement in $C[-1, 1]$

Consider the vector space $V = C[-1, 1]$ with its standard inner product, $\langle f, g \rangle = \int_{-1}^{1} f(x)g(x)dx$. Let W be the subspace of *odd* functions—that is, functions $f \in V$ satisfying $f(-x) = -f(x)$ for all $x \in [-1, 1]$. For example, $f(x) = \sin x$ and $f(x) = x^5$ belong to W. We want to find the orthogonal complement of W:

$$W^\perp = \left\{ g \in V \,\middle|\, \int_{-1}^{1} f(x)g(x)dx = 0 \text{ for all } f \in W \right\}.$$

If g is any element of V, then

$$g(x) = \frac{g(x) + g(-x)}{2} + \frac{g(x) - g(-x)}{2}$$

for $x \in [-1, 1]$. Let $E(x) = (g(x) + g(-x))/2$ and $O(x) = (g(x) - g(-x))/2$. Then E is an *even* function—that is, $E(-x) = E(x)$—whereas O is odd.* Now suppose that $g \in W^\perp$. Then,

$$0 = \int_{-1}^{1} O(x)g(x)dx = \int_{-1}^{1} O(x)[E(x) + O(x)]dx$$

$$= \int_{-1}^{1} O(x)E(x)dx + \int_{-1}^{1} O^2(x)dx = \text{Int}_1 + \text{Int}_2.$$

---

* A useful fact in analysis is that every function defined on an interval I can be written (uniquely) as the sum of an even function and an odd function, provided that $-x \in I$ whenever $x \in I$.

Because $O(x)E(x)$ is odd, $\text{Int}_1 = 0$ (Exercise A2), so Int_2 must equal 0 as well. Hence, $O^2(x), O(x)$, and consequently $g(x) - g(-x)$ are each equal to 0 for all $x \in [-1, 1]$, which implies that $g(x)$ must be an *even* function. Therefore,

$W^\perp \subseteq \{\text{even functions in } V\}$. Conversely, we can see that if g is an even function in V, then the product of g and any odd function is odd, so $g \in W^\perp$—that is, $\{\text{even functions in } V\} \subseteq W^\perp$. Hence, $W^\perp$ equals the set of all even functions in V.

An important application of the concept of orthogonal complements is to the row and column spaces of any $m \times n$ matrix.

For the rest of this section, the word "real" may be replaced by the word "complex" and $(\cdot)^T$ may be replaced by $(\cdot)^*$.

Theorem 7.8.1

If A is a real $m \times n$ matrix, then

> (1) (Row space of $A)^\perp = $ (Null space of A)
>
> (2) (Column space of $A)^\perp = $ (Null space of A^T)

Proof (1): In the beginning of this section, we showed that (Null space of $A) \subseteq$ (Row space of $A)^\perp$. Now suppose that $\mathbf{z} \in$ (Row *space of* $A)^\perp$. Then $0 = \text{row}_i(A)\mathbf{z} = \text{row}_i(A\mathbf{z})$ for $i = 1, 2, \ldots, m$, which indicates that every entry in the vector $A\mathbf{z}$ is zero—that is, $A\mathbf{z} = \mathbf{0}$ and $\mathbf{z} \in$ (Null space of A). Thus, (Row space of $A)^\perp \subseteq$ (Null space of A) and so (Row space of $A)^\perp = $ (Null space of A).

Assertion (2) follows by replacing A by A^T in statement (1) and recalling that the row space of A^T is the column space of A (see Section 2.7).

There are some basic properties of orthogonal complements whose proofs will be left as exercises (B3(a)–(d)). Note that **we assume that V is a finite-dimensional space**.

Theorem 7.8.2

If S a nonempty subset of a finite-dimensional inner product space V, then

(a) $S \cap S^\perp = \{\mathbf{0}\}$.

(b) If S a nonempty subset of a finite-dimensional inner product space V, then $S^\perp$ is a subspace of V.

(c) $S^\perp = (span(S))^\perp$—that is, a vector $\mathbf{z}$ belongs to the orthogonal complement of a nonempty set S if and only if $\mathbf{z}$ is orthogonal to every linear combination of vectors from S.

(d) $(S^\perp)^\perp = S$.

Orthogonal subspaces can be used to illuminate the structure of finite-dimensional inner product spaces. The following theorem describes the decomposition of any such space into simpler components. (Although it may not be obvious, Theorem 2.8.5 is an analogous result.)

Theorem 7.8.3: The Orthogonal Decomposition Theorem

If U is a subspace of a finite-dimensional inner product space V, then any vector $\mathbf{v}$ in V can be written uniquely in the form $\mathbf{v} = \mathbf{u}_1 + \mathbf{u}_2$, where $\mathbf{u}_1$ is in U and $\mathbf{u}_2$ is in $U^\perp$—that is,

$$V = U \oplus U^\perp.$$

Proof Suppose $\dim(U) = k \leq n = \dim(V)$. Let $\{\mathbf{v}_1, \mathbf{v}_2, \ldots, \mathbf{v}_k\}$ be an orthonormal basis for U and let $\{\mathbf{v}_1, \mathbf{v}_2, \ldots, \mathbf{v}_k, \mathbf{v}_{k+1}, \ldots, \mathbf{v}_n\}$ be an extension of this basis to an orthonormal basis for V. (We extend the orthonormal basis for U to a basis for V by Corollary 5.4.1 and then orthonormalize the added vectors by the Gram–Schmidt process.)

If $\mathbf{v}$ is any vector in V, then

$$\mathbf{v} = \sum_{i=1}^{n} a_i \mathbf{v}_i = \sum_{i=1}^{k} a_i \mathbf{v}_i + \sum_{i=k+1}^{n} a_i \mathbf{v}_i = \mathbf{u} + \mathbf{z}, \quad \text{with } \mathbf{u} = \sum_{i=1}^{k} a_i \mathbf{v}_i \in U.$$

Now, we show that $\mathbf{z} \in U^\perp$: If $\mathbf{x} \in U$, then $\mathbf{x} = \sum_{j=1}^{k} b_j \mathbf{v}_j$ and

$$
\begin{aligned}
\langle \mathbf{x}, \mathbf{z} \rangle &= \left\langle \sum_{j=1}^{k} b_j \mathbf{v}_j, \sum_{i=k+1}^{n} a_i \mathbf{v}_i \right\rangle = \sum_{j=1}^{k} b_j \left\langle \mathbf{v}_j, \sum_{i=k+1}^{n} a_i \mathbf{v}_i \right\rangle \\
&= \sum_{j=1}^{k} b_j \left(\sum_{i=k+1}^{n} \bar{a}_i \langle \mathbf{v}_j, \mathbf{v}_i \rangle \right) \\
&= 0
\end{aligned}
$$

because $\{\mathbf{v}_i\}_{i=1}^{n}$ is an orthonormal set. Thus, $V = U + U^\perp$. If $\mathbf{y} \in U \cap U^\perp$, then $0 = \langle \mathbf{y}, \mathbf{y} \rangle = \|\mathbf{y}\|^2$, which implies $\mathbf{y} = \mathbf{0}$. Thus, $U \cap U^\perp = \{\mathbf{0}\}$ and $V = U \oplus U^\perp$ by Theorem 5.2.3.

If we combine Theorem 7.8.3 with Corollary 5.4.3 on the dimension of a direct sum of subspaces—and Theorem 7.8.2(b) states that the orthogonal complement of a nonempty subset of an inner product space is a subspace—we can derive the next result.

Corollary 7.8.1

If U is a subspace of an n-dimensional inner product space V, then $\dim(U) + \dim(U^\perp) = n$.

As an example of the insight provided by Corollary 7.8.1, we see that the restriction $\dim(U) + \dim(U^\perp) = n$ forces $\mathbb{R}^3$ to have only two types of orthogonal complements: (1) U is a plane through the origin [$\dim(U) = 2$]

and $U^\perp$ is the line through the origin perpendicular to the plane $[\dim(U^\perp) = 1]$ or (2) $U = \mathbb{R}^3$ and $U^\perp = \{\mathbf{0}\}$. See Example 7.8.1 and Example 7.8.4(a).

Theorems 7.8.1 and 7.8.3 together have the following important consequence.

Corollary 7.8.2

If A is a real $m \times n$ matrix, then

> (1) $\mathbb{R}^m = $ (Column space of A) $\oplus$ (Column space of A)$^\perp$
> $= $ (Column space of A) $\oplus$ (Null space of A^T)
>
> (2) $\mathbb{R}^n = $ (Null space of A) $\oplus$ (Null space of A)$^\perp$
> $= $ (Null space of A) $\oplus$ (Column space of A^T).

Corollary 7.8.2 emphasizes the importance of what are sometimes called the **four Fundamental Subspaces** of an $m \times n$ matrix A, two belonging to $\mathbb{R}^n$ and two belonging to $\mathbb{R}^m$:

The **column space** of $A^T = $ the **row space of** $A = \{A^T\mathbf{y} \mid \mathbf{y} \in \mathbb{R}^m\} \subseteq \mathbb{R}^n$

The **column space of** $A = \{A\mathbf{x} \mid \mathbf{x} \in \mathbb{R}^n\} \subseteq \mathbb{R}^m$

The **null space of** $A = \{\mathbf{x} \mid A\mathbf{x} = \mathbf{0}\} \subseteq \mathbb{R}^n$

The **null space of** $A^T = \{\mathbf{y} \mid A^T\mathbf{y} = \mathbf{0}\} \subseteq \mathbb{R}^m$

Theorem 7.8.1 and Corollary 7.8.2 make up part of what can be called the "Fundamental Theorem of Linear Algebra."*

Example 7.8.4: The Four Fundamental Subspaces

Suppose $A = \begin{bmatrix} 1 & 1 & -1 \\ 0 & 2 & 1 \\ 1 & 2 & 0 \end{bmatrix}$. We will determine the four fundamental subspaces of A.

To find **null space A**, we calculate the reduced row echelon form of A and solve the simplified equation rref(A)$\mathbf{x} = \mathbf{0}$:

$$\begin{bmatrix} 1 & 0 & -3/2 \\ 0 & 1 & 1/2 \\ 0 & 0 & 0 \end{bmatrix} \begin{bmatrix} x \\ y \\ z \end{bmatrix} = \begin{bmatrix} 0 \\ 0 \\ 0 \end{bmatrix} \Rightarrow y + \frac{1}{2}z = 0 \text{ and } x - \frac{3}{2}z = 0$$

$$\Rightarrow \text{null space } A = \left\{ \begin{bmatrix} \frac{3}{2}z \\ -\frac{1}{2}z \\ z \end{bmatrix} : z \in \mathbb{R} \right\} = span\left\{ \begin{bmatrix} 3 \\ -1 \\ 2 \end{bmatrix} \right\}.$$

---

* Gilbert Strang, The fundamental theorem of linear algebra, *Amer. Math. Monthly* **100** (1993): 848–855; reprinted in *Resources for Teaching Linear Algebra*, D. Carlson et al., eds. (Washington, DC: Mathematical Association of America, 1997).

Similarly, **null space** A^T can be found by solving rref$(A^T)\mathbf{x} = \mathbf{0}$:

$$\begin{bmatrix} 1 & 0 & 1 \\ 0 & 1 & 1 \\ 0 & 0 & 0 \end{bmatrix} \begin{bmatrix} x \\ y \\ z \end{bmatrix} = \begin{bmatrix} 0 \\ 0 \\ 0 \end{bmatrix} \Rightarrow x + z = 0 \quad \text{and} \quad y + z = 0$$

$$\Rightarrow \text{null space } A = \left\{ \begin{bmatrix} \frac{3}{2}z \\ -\frac{1}{2}z \\ z \end{bmatrix} : z \in \mathbb{R} \right\} = span \left\{ \begin{bmatrix} 3 \\ -1 \\ 2 \end{bmatrix} \right\}$$

$$\Rightarrow \text{null space } A^T = \left\{ \begin{bmatrix} -z \\ -z \\ z \end{bmatrix} : z \in \mathbb{R} \right\} = span \left\{ \begin{bmatrix} -1 \\ -1 \\ 1 \end{bmatrix} \right\}$$

To determine the **column space of A**, we use Theorem 2.7.3, which states that the pivot columns of rref(A) identify the columns of A that serve as basis vectors for the column space:

$$\text{rref}(A) = \begin{bmatrix} 1 & 0 & -3/2 \\ 0 & 1 & 1/2 \\ 0 & 0 & 0 \end{bmatrix} \Rightarrow \text{Columns 1 and 2 are pivot columns}$$

$\Rightarrow$ columns 1 and 2 of A are basis vectors for the column space of A.

Thus, the column space of $A = A = span\left\{ \begin{bmatrix} 1 \\ 0 \\ 1 \end{bmatrix}, \begin{bmatrix} 1 \\ 2 \\ 3 \end{bmatrix} \right\}$.

Finally, the **column space of A^T** can be found by examining the pivot columns of rref(A^T):

$$\text{rref}(A^T) = \begin{bmatrix} 1 & 0 & 1 \\ 0 & 1 & 1 \\ 0 & 0 & 0 \end{bmatrix} \Rightarrow \text{Columns 1 and 2 are pivot columns}$$

$\Rightarrow$ columns 1 and 2 of A^T are basis vectors for the column space of A^T.

Thus, the column space of A^T is $span\left\{ \begin{bmatrix} 1 \\ 1 \\ -1 \end{bmatrix}, \begin{bmatrix} 0 \\ 2 \\ 1 \end{bmatrix} \right\}$.

Example 7.8.5: $V = U \oplus U^\perp$

(a) From Example 7.8.1 and Theorem 7.8.2(c), we deduce that the subspace $U = span\left\{ [1 \ -1 \ 2]^T \right\} = \left\{ c[1 \ -1 \ 2]^T : c \in \mathbb{R} \right\}$ has $\left\{ [x \ y \ z]^T : x - y + 2z = 0 \right\}$ as its orthogonal complement. Thus, $\mathbb{R}^3 = U \oplus U^\perp$, meaning that every vector in $\mathbb{R}^3$ can be written uniquely as the sum of a scalar multiple of $[1 \ -1 \ 2]^T$ and a vector $[x \ y \ z]^T$ such that $x - y + 2z = 0$. For example,

$$[1 \quad 2 \quad 3]^{\mathsf{T}} = \tfrac{5}{6}[1 \quad -1 \quad 2]^{\mathsf{T}} + \tfrac{1}{3}[1/2 \quad 17/2 \quad 4]^{\mathsf{T}}.$$

(b) We see that any 2×2 real matrix $\begin{bmatrix} a & b \\ c & d \end{bmatrix}$ can be

written as $\begin{bmatrix} a & 0 \\ 0 & d \end{bmatrix} + \begin{bmatrix} 0 & b \\ c & 0 \end{bmatrix}$. The information in Example 7.8.2 and Theorem 7.8.3 guarantee that this representation as a sum is unique.

(c) Example 7.8.3 and Theorem 7.8.3 imply that any function continuous on the interval $[-1, 1]$ can be written uniquely as the sum of an even function and an odd function, each from $C[-1, 1]$.

The conclusion of Theorem 7.8.3 is not true for infinite-dimensional inner product spaces, as the next example shows. The span of an infinite set of vectors is generally understood to mean the set of all *finite* linear combinations of the set of vectors.

Example 7.8.6: $V \neq U \oplus U^{\perp}$

Consider the inner product space ℓ^2 of "square-summable" real sequences (Example 7.2.2(f)). Let $\mathbf{e}_k$ be the sequence whose kth entry is 1 and all of whose other terms are 0, and let $B = \{\mathbf{e}_k | k = 1, 2 \ldots\}$. Then, B is an orthonormal subset of ℓ^2 (Exercise A11). However, B is *not* a basis for ℓ^2. The sequence $\{a_n\}$ with $a_n = 1/2^{n/2}$ for $n = 1, 2 \ldots$ is in ℓ^2 but is not in *span* (B): The series $\sum_{n=1}^{\infty} a_n^2 = \sum_{n=1}^{\infty} 1/2^n$ is a geometric series converging to 1, but the proper subspace $span(B)$ is the space of all sequences in ℓ^2, which have only finitely many nonzero entries. Furthermore, if $U = span(B)$, then clearly $U^{\perp} = \{\mathbf{0}\}$. Therefore, for $V = \ell^2$, $V \neq U \oplus U^{\perp}$.

Exercises 7.8

A.

1. Consider the inner product space M_2 and subspace S of Example 7.8.2.
   (a) If $A \in S$, show that (Row space of $A)^{\perp} =$ (Null space of A) by determining these spaces explicitly.
   (b) If $A \in S$, show that (Column space of $A)^{\perp} =$ (Null space of A^{T}) by determining these spaces explicitly.

2. If $O(x)$ is an odd function on $[-1, 1]$ and $E(x)$ is an even function on $[-1, 1]$, prove that $\langle O(x), E(x) \rangle = \int_{-1}^{1} O(x)E(x)\, dx = 0$. (See Example 7.8.3.)

3. In each of the following cases, assume the standard inner product for the appropriate space $\mathbb{R}^n$. Find $U^\perp$ and determine a basis for $U^\perp$ (if possible). Also, describe $U^\perp$ geometrically.

   (a) $U = \left\{ \begin{bmatrix} x \\ y \end{bmatrix} \in \mathbb{R}^2 : x + y = 1 \right\}$.

   (b) U is the z-axis in $\mathbb{R}^3$.

   (c) $U = span \left\{ \begin{bmatrix} 1 \\ -1 \end{bmatrix} \right\}$ in $\mathbb{R}^2$.

   (d) $U = \left\{ \begin{bmatrix} x \\ y \\ z \end{bmatrix} \in \mathbb{R}^3 : x + y + z = 0 \right\}$.

4. Suppose U consists of the single vector $\begin{bmatrix} 1 \\ -2 \\ 5 \end{bmatrix}$ in $\mathbb{R}^3$. What is the orthogonal complement of U? Describe $U^\perp$ geometrically.

5. Let $V = P_2[x]$ with $\langle f, g \rangle = \int_0^1 f(x)g(x)dx$.
   (a) Find the orthogonal complement of the subspace $span\{1, x\}$.
   (b) Now change the definition of the inner product to $\langle f, g \rangle = \int_{-1}^1 f(x)g(x)dx$ and find the orthogonal complement of the subspace $span\{1, x\}$.

6. Let $V = P_2[x]$ with $\langle f, g \rangle = \int_0^1 f(x)g(x)dx$, and let $S = span\{1 - x^2, 2 - x + x^2\}$. Find a basis for the orthogonal complement of S.

7. Consider $\mathbb{R}^3$ with the inner product defined by $\langle \mathbf{u}, \mathbf{v} \rangle = u_1 v_1 + 2u_2 v_2 + 3u_3 v_3$ for $\mathbf{u} = \begin{bmatrix} u_1 \\ u_2 \\ u_3 \end{bmatrix}, \mathbf{v} = \begin{bmatrix} v_1 \\ v_2 \\ v_3 \end{bmatrix} \in \mathbb{R}^3$. Find the orthogonal complement of the subspace spanned by $\begin{bmatrix} 3 \\ -1 \\ 1 \end{bmatrix}$.

8. Let $S = span\{f_1, f_2\}$, where $f_1(x) = \cos x$ and $f_2(x) = \sin x$, the inner product being defined by $\langle f, g \rangle = \int_{-\pi}^{\pi} f(x)g(x)\, dx$. If $U = span\{f_1\}$, what is $U^\perp$?

9. Find an orthogonal complement of the subspace of $M_2(\mathbb{C})$ consisting of all 2×2 diagonal matrices having complex entries, with the inner product given by $\langle A, B \rangle = \text{trace}(A\bar{B}^{\mathsf{T}})$.

10. Consider the subspaces $U = \{[x_1 \ x_2 \ x_3 \ x_4]^{\mathsf{T}} : x_1 = x_2 = x_3\}$ and $W = \{[x_1 \ x_2 \ x_3 \ x_4]^{\mathsf{T}} : x_1 = x_2$ and $x_4 = 0\}$ of $\mathbb{R}^4$.
    (a) Find $U + W$, $U^\perp$, and $W^\perp$.
    (b) Verify that $(U + W)^\perp = U^\perp \cap W^\perp$. (See Exercise B2 below.)

11. In Example 7.8.6, prove B is an orthonormal subset of ℓ^2.

12.   Find bases for the four fundamental subspaces of the matrix:

$$A = \begin{bmatrix} 1 & 2 & 3 \\ 0 & 1 & 0 \end{bmatrix}.$$

13.   Find bases for the four fundamental subspaces of the matrix:

$$A = \begin{bmatrix} 1 & 0 & 0 & 1 \\ 0 & 1 & 1 & 1 \\ 1 & 1 & 1 & 2 \\ 1 & 2 & 2 & 3 \end{bmatrix}.$$

B.

1.   Let U and W be subspaces of an inner product space V. If $U \subseteq W$, show that $W^{\perp} \subseteq U^{\perp}$.

2.   Let U and W be subspaces of an inner product space V. Show that
     (a) $(U + W)^{\perp} = U^{\perp} \cap W^{\perp}$.
     (b) $(U \cap W)^{\perp} = U^{\perp} + W^{\perp}$.

3.   Let S be a nonempty set of vectors in a finite-dimensional inner product space V.
     (a) Show that $S \cap S^{\perp} = \{\mathbf{0}\}$.
     (b) $S^{\perp}$ is a subspace of V.
     (c) Show that $S^{\perp} = [span(S)]^{\perp}$—that is, a vector $\mathbf{z}$ belongs to the orthogonal complement of a nonempty set S if and only if $\mathbf{z}$ is orthogonal to every linear combination of vectors from S.
     (d) Show that $(S^{\perp})^{\perp} = S$.

4.   Let $U = span\{[\,1 \quad 1 \quad 0 \quad 0\,]^{\mathsf{T}}, [\,1 \quad 0 \quad 1 \quad 0\,]^{\mathsf{T}}\}$ and $W = span\{[\,0 \quad 1 \quad 0 \quad 1\,]^{\mathsf{T}}, [\,0 \quad 0 \quad 1 \quad 1\,]^{\mathsf{T}}\}$. Find a basis for each of the following subspaces.
     (a) $U + W$ (b) $U^{\perp}$ (c) $U^{\perp} + W^{\perp}$ (d) $U \cap W$.

5.   Let $V = P_5[x]$ with the inner product $\langle p, q \rangle = \int_{-1}^{1} p(x)q(x)\mathrm{d}x$. Find $U^{\perp}$ if $U = \{p \in P_5[x] | p(0) = p'(0) = p''(0) = 0\}$.

6.   Find an orthogonal complement of the subspace of $M_n(\mathbb{C})$ consisting of all diagonal matrices having complex entries, with the inner product given by $\langle A, B \rangle = \mathrm{trace}(A\bar{B}^{\mathsf{T}})$. (This is a generalization of Problem A9.)

C.

1.   Let W be a subspace of an inner product space V and let S be a nonempty subset of V.
     (a) Show that $(S^{\perp} + W)^{\perp} = (S^{\perp})^{\perp} \cap W^{\perp}$.
     (b) Show that $(S^{\perp} \cap W)^{\perp} = (S^{\perp})^{\perp} + W^{\perp}$.

7.9 Projections

Let V be a vector space and suppose U and W are subspaces of V such that $V = U \oplus W$. Then, any $\mathbf{v} \in V$ can be written uniquely as $\mathbf{v} = \mathbf{u} + \mathbf{w}$ for $\mathbf{u} \in U$ and $\mathbf{w} \in W$. Now, define the function $P_U \colon V \to V$ by $P_U(\mathbf{v}) = P_U(\mathbf{u} + \mathbf{w}) = \mathbf{u}$. This transformation is called the **projection of V (or of $\mathbf{v}$) on the subspace U**, or the **projection of V (or of $\mathbf{v}$) on U along W**.* Because $U \oplus W = W \oplus U$, we can also define the projection of V on W along U: $P_W(\mathbf{v}) = P_W(\mathbf{u} + \mathbf{w}) = \mathbf{w}$. It is easy to show that (1) P_U is a linear transformation; (2) the range of P_U is U: range $P_U = P_U (V) = U$; and (3) the null space of P_U is W: null space $P_U = W$ (Exercise B1).

Furthermore, if $U_1, U_2, \ldots, U_m$ are subspaces of a vector space V such that $V = U_1 \oplus U_2 \oplus \cdots \oplus U_m$, then we can define the **projection P_{U_k} of V on U_k** as $P_{U_k}(\mathbf{u}_1 + \mathbf{u}_2 + \cdots + \mathbf{u}_m) = \mathbf{u}_k$, where $\mathbf{u}_i \in U_i, i = 1, 2, \ldots, m$.

If $V = U \oplus U^{\perp}$, then the projection of $\mathbf{v} \in V$ onto U is called the **orthogonal projection of V on U along $U^{\perp}$**. Then $P_U(\mathbf{v}) \in U$ and $\mathbf{v} - P_U(\mathbf{v}) \in U^{\perp}$ by the unique representation of any $\mathbf{v} \in V$ as $\mathbf{u} + \mathbf{u}'$, where $\mathbf{u} \in U$ and $\mathbf{u}' \in U^{\perp}$ (Theorem 7.8.3) (see Figure 7.4).

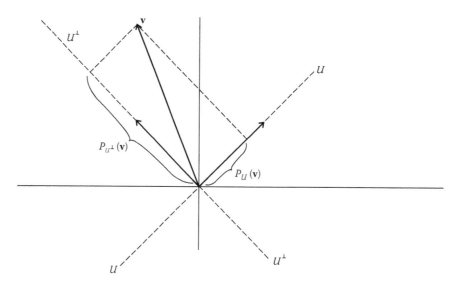

Figure 7.4 Orthogonal projections.

---

* Sometimes, this is described as "the projection of V on U *parallel* to W."

Example 7.9.1: Projections in $\mathbb{R}^2$

In $\mathbb{R}^2$, let $U = span\left\{\begin{bmatrix} 1 \\ 2 \end{bmatrix}\right\}$ and $W_1 = span\left\{\begin{bmatrix} 1 \\ 0 \end{bmatrix}\right\}$. Then, we have $\mathbb{R}^2 = U \oplus W_1$, so any vector $\mathbf{v} = \begin{bmatrix} x \\ y \end{bmatrix}$ can be expressed uniquely as $\mathbf{v} = \begin{bmatrix} y/2 \\ y \end{bmatrix} + \begin{bmatrix} x - y/2 \\ 0 \end{bmatrix}$. Now, we can define the projection P_U of $\mathbb{R}^2$ on U by $P_U(\mathbf{v}) = \begin{bmatrix} y/2 \\ y \end{bmatrix}$ and the projection P_{W_1} of $\mathbb{R}^2$ on W_1 by $P_{W_1}(\mathbf{v}) = \begin{bmatrix} x - y/2 \\ 0 \end{bmatrix}$. Figure 7.5 shows the projections of the vector $\begin{bmatrix} 1 \\ 1 \end{bmatrix}$ on both U and W_1.

Alternatively, we could let $W_2 = span\left\{\begin{bmatrix} -2 \\ 3 \end{bmatrix}\right\}$, so $\mathbb{R}^2 = U \oplus W_2$ and any vector $\mathbf{v} = \begin{bmatrix} x \\ y \end{bmatrix}$ can be expressed as $\mathbf{v} = \begin{bmatrix} (3x + 2y)/7 \\ (6x + 4y)/7 \end{bmatrix} + \begin{bmatrix} 2(2x - y)/7 \\ -3(2x - y)/7 \end{bmatrix}$. In this case, $P_U(\mathbf{v}) = \begin{bmatrix} (3x + 2y)/7 \\ (6x + 4y)/7 \end{bmatrix}$ and $P_{W_2}(\mathbf{v}) = \begin{bmatrix} 2(2x - y)/7 \\ -3(2x - y)/7 \end{bmatrix}$.

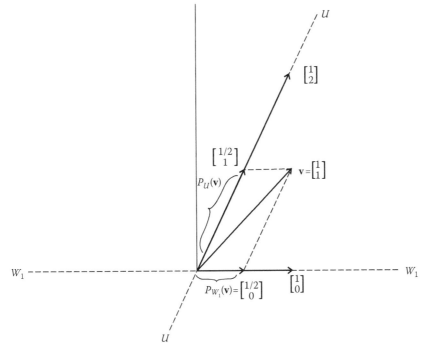

Figure 7.5 Projections of $\begin{bmatrix} 1 \\ 1 \end{bmatrix}$ on U and W_1.

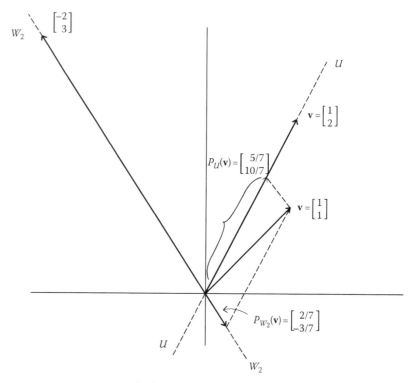

Figure 7.6 Projections of $\begin{bmatrix} 1 \\ 1 \end{bmatrix}$ on U and W_2.

Figure 7.6 shows the projections of the vector $\begin{bmatrix} 1 \\ 1 \end{bmatrix}$ on both U and W_2.

Note that we are not dealing with orthogonal projections in this example because neither W_1 nor W_2 is the orthogonal complement of U.

The point of the last example is that the projections of the same vector $\mathbf{v}$ onto a subspace U are different along different complements. Thus, there are many projections for a given subspace U, depending on the complementary subspace W chosen. However, for fixed complementary subspaces U and W, the projections P_U and P_W are uniquely determined.

Example 7.9.2: A Projection in $P_4[x]$

In the vector space $P_4[x]$, the subspace U of even polynomials is a complement of the subspace W of odd polynomials. (In fact, U and V are *orthogonal* complements. See Example 7.8.3.) Let us determine P_U and P_W, the (orthogonal) projections of $3 - x + 2x^2 + x^3 - 5x^4$ onto the subspaces U and W, respectively.

First of all, we can write

$$\mathbf{v} = 3 - x + 2x^2 + x^3 - 5x^4 = (3 + 2x^2 - 5x^4) + (-x + x^3),$$

where $3 + 2x^2 - 5x^4 \in U$ and $-x + x^3 \in W$. Consequently,
$P_U(\mathbf{v}) = 3 + 2x^2 - 5x^4$ and $P_W(\mathbf{v}) = -x + x^3$.

Example 7.9.3: A Projection on a Column Space

Consider the vector $\mathbf{v} = \begin{bmatrix} 1 \\ -2 \\ 3 \end{bmatrix}$ and the matrix $A = \begin{bmatrix} 1 & 0 \\ 2 & -4 \\ 3 & 5 \end{bmatrix}$.

We will find the orthogonal projection of $\mathbf{v}$ on the column space of A.

Let U denote the column space of A. Because the columns of A are linearly independent, they form a basis for U. We have to find a vector $\mathbf{w}$ in U such that $\mathbf{v} - \mathbf{w}$ is orthogonal to both columns of A. Then, $\mathbf{v} - \mathbf{w} \in U^\perp$ and $\mathbf{w}$ will be the orthogonal projection of $\mathbf{v}$ on U.

Now, $\mathbf{w}$ must have the form:

$$\mathbf{w} = x \begin{bmatrix} 1 \\ 2 \\ 3 \end{bmatrix} + y \begin{bmatrix} 0 \\ -4 \\ 5 \end{bmatrix} = \begin{bmatrix} x \\ 2x - 4y \\ 3x + 5y \end{bmatrix}$$

for some scalars x and y. Denoting the columns of A by $\mathbf{a}_1$ and $\mathbf{a}_2$, the conditions for $\mathbf{v} - \mathbf{w}$ to belong to $U^\perp$ are

$$\langle \mathbf{v} - \mathbf{w}, \mathbf{a}_1 \rangle = (1 - x) + 2(-2 - 2x + 4y) + 3(3 - 3x - 5y)$$
$$= 0$$

and

$$\langle \mathbf{v} - \mathbf{w}, \mathbf{a}_2 \rangle = -4(-2 - 2x + 4y) + 5(3 - 3x - 5y) = 0.$$

The solutions of this system are $x = 17/105$ and $y = 8/15$, so the orthogonal projection of $\mathbf{v}$ on the column space of A is

$$\mathbf{w} = \begin{bmatrix} x \\ 2x - 4y \\ 3x + 5y \end{bmatrix} = \frac{1}{105} \begin{bmatrix} 17 \\ -190 \\ 331 \end{bmatrix}.$$

A finite-dimensional subspace U of a unitary space V has an orthonormal basis, and this makes the calculation of the orthogonal projections on U and $U^\perp$ easy.

Theorem 7.9.1

Let $\{\mathbf{u}_1, \mathbf{u}_2, \ldots, \mathbf{u}_k\}$ be an orthonormal basis for the subspace U of the unitary space V and let $\mathbf{v}$ be any element of V. Then, the orthogonal projection of $\mathbf{v}$ on U is $\mathbf{v}_1 = \sum_{i=1}^{k} \langle \mathbf{v}, \mathbf{u}_i \rangle \mathbf{u}_i$ and the orthogonal projection of $\mathbf{v}$ on $U^\perp$ is

$$\mathbf{v}_2 = \mathbf{v} - \mathbf{v}_1.$$

The proof of Theorem 7.9.1 is buried in the proof of Theorem 7.8.3, which is also called the **Projection Theorem**.

Exercises 7.9

A.

1. Let $\quad U = \left\{ \begin{bmatrix} x \\ y \\ z \end{bmatrix} \in \mathbb{R}^3 : x - 2y + z = 0 \right\} \quad$ and suppose that

   $\mathbf{v} = \begin{bmatrix} 1 \\ -1 \\ 4 \end{bmatrix}.$

   (a) Find the orthogonal projection of $\mathbf{v}$ on U.
   (b) Find the orthogonal projection of $\mathbf{v}$ on $U^\perp$.
   (c) What is the distance of $\mathbf{v}$ to the plane U?

2. (a) Find the orthogonal projection of $\mathbf{v} = \begin{bmatrix} 4 \\ -1 \\ 1 \end{bmatrix}$ on the plane $x - y - z = 0$ in $\mathbb{R}^3$.
   (b) What is the component of $\mathbf{v}$ that is perpendicular to this plane?
   (c) What is the distance from $\mathbf{v}$ to the plane?

3. Find the projection of the vector $\mathbf{v} = \begin{bmatrix} 1 \\ 1 \\ 1 \end{bmatrix}$ on the column space of the matrix $A = \begin{bmatrix} 1 & 3 \\ 2 & -1 \\ 1 & 4 \end{bmatrix}$.

B.

1. Let P_U be the projection of a vector space $V = U \oplus W$ on the subspace U.
   (a) Prove that P_U is a linear transformation on V.
   (b) Prove that the range of P_U is U.
   (c) Prove that the null space of P_U is W.

2. Find the orthogonal projection of vector $\mathbf{v}$ on the subspace U of $\mathbb{R}^4$ if
$$\mathbf{v} = [\,1 \quad 2 \quad 2 \quad 9\,]^T \quad \text{and} \quad U = span\{[\,2 \quad 1 \quad -1 \quad 2\,]^T,$$
$$[\,1 \quad 2 \quad 0 \quad 1\,]^T\}.$$

3. Let $V = P_3[t]$ with the inner product defined by $\langle f, g \rangle = \int_0^1 f(t)g(t)dt$ for any $f, g \in V$. Let W be the subspace of V spanned by $\{1, x\}$, and define $f(x) = x^2$.
   Find the orthogonal projection P_W of f on W.

4. Let $V = P_3[t]$ with the inner product defined by $\langle x, y \rangle = \int_{-1}^1 x(t)y(t)dt$. Find the orthogonal complement in V of $S = span\{1, x\}$. What is the vector in S closest to $x(t) = t + t^3$?

5. If U is the subspace of $\mathbb{C}^3$ spanned by
$$\mathbf{v}_1 = \begin{bmatrix} i \\ 1 \\ 0 \end{bmatrix} \text{ and } \mathbf{v}_2 = \begin{bmatrix} -1 \\ 1+i \\ 1 \end{bmatrix}, \text{ what is the orthogonal projec-}$$
tion of $\mathbf{v} = \begin{bmatrix} 2-i \\ 1 \\ i \end{bmatrix}$ on U?

C.

1. Let U be an isometry on a finite-dimensional inner product space and let S be the subspace of all solutions of $U\mathbf{x} = \mathbf{x}$. Define the sequence of operators $\{V_n\}$ by
$$V_n = \frac{1}{n}(I + U + U^2 + \cdots + U^{n-1}).$$

   (a) If $\mathbf{x} = \mathbf{y} - U\mathbf{y} \in \mathbb{R}_{I-U}$, prove that $V_n\mathbf{x} = \frac{1}{n}(\mathbf{y} - U^n\mathbf{y})$ and $\|V_n\mathbf{x}\| \leq \frac{2}{n}\|\mathbf{y}\|$.
   (b) If $\mathbf{x} \in S$, show that $V_n\mathbf{x} = \mathbf{x}$.
   (c) Prove that $R_{I-U}^\perp = S$.
   (d) Use the results of parts (a)-(c) to prove that $\lim_{n \to \infty} V_n = P_S$, the orthogonal projection on S. (This is called the *Ergodic Theorem*.)

7.10 Summary

The properties of complex vectors and matrices are obvious extensions of properties of complex numbers. An important pattern is established by the definition of **n-dimensional complex coordinate space**, $\mathbb{C}^n$. If $C = [c_{ij}] = A + iB$, where A and B are real matrices, the **complex conjugate** of C is denoted $\bar{C}$ and is defined as $\bar{C} = [\bar{c}_{ij}] = A - iB$. Properties of the conjugate of a complex matrix are given in Theorem 7.1.1. The **conjugate transpose** of a complex matrix C is denoted by C^* and defined as $C^* = (\bar{C})^{\mathrm{T}}$. Properties of the conjugate transpose are provided by Theorem 7.1.2.

An **inner product** is a rule that assigns to each pair of vectors $\mathbf{u}, \mathbf{v}$ in a real or complex vector space V a scalar $\langle \mathbf{u}, \mathbf{v} \rangle$—a real number if V is real and a complex number if V is complex—such that the following properties hold:

1. $\langle a\mathbf{u} + b\mathbf{v}, \mathbf{w} \rangle = a\langle \mathbf{u}, \mathbf{w} \rangle + b\langle \mathbf{v}, \mathbf{w} \rangle$ for all $\mathbf{u}, \mathbf{v}, \mathbf{w} \in V$ and a, b scalars;
2. $\langle \mathbf{u}, \mathbf{v} \rangle = \overline{\langle \mathbf{v}, \mathbf{u} \rangle}$;
3. $\langle \mathbf{u}, \mathbf{u} \rangle \geq 0$ with equality if and only if $\mathbf{u} = \mathbf{0}$. A vector space together with an inner product is called an **inner product space**—a **Euclidean space** if the vector space is real and a **unitary space** if we are dealing with a complex vector space. Example 7.1.1 defines **standard inner products** for several important Euclidean and unitary spaces.

In an inner product space, we define the **norm** (or **length**) of any vector: $\|\mathbf{u}\| = \sqrt{\langle \mathbf{u}, \mathbf{u} \rangle}$. If V is an inner product space, then for any vectors $\mathbf{u}, \mathbf{v}$ in V and any scalar c, we have the following properties: 1. $\|c\mathbf{u}\| = |c| \|\mathbf{u}\|$; 2. $|\langle \mathbf{u}, \mathbf{v} \rangle| \leq \|\mathbf{u}\| \|\mathbf{v}\|$ [**Cauchy–Schwarz Inequality**]; 3. $\|\mathbf{u} + \mathbf{v}\| \leq \|\mathbf{u}\| + \|\mathbf{v}\|$ [**Triangle Inequality**]. In a real inner product space, if $\mathbf{u} \neq \mathbf{0}$ and $\mathbf{v} \neq \mathbf{0}$, there is a unique number $\theta \in [0, \pi]$ such that $\cos \theta = \langle \mathbf{u}, \mathbf{v} \rangle / \|\mathbf{u}\| \|\mathbf{v}\|$. This real number θ is called the **angle** between the vectors $\mathbf{u}$ and $\mathbf{v}$.

Vectors $\mathbf{u}$ and $\mathbf{v}$ in an inner product space are called **orthogonal**, written $\mathbf{u} \perp \mathbf{v}$, if $\langle \mathbf{u}, \mathbf{v} \rangle = 0$. A set of vectors in an inner product space is called **orthogonal** if the vectors in the set are mutually orthogonal. If an inner product space has a basis consisting of mutually orthogonal vectors, this basis is called an **orthogonal basis**. If V is an inner product space, then **any orthogonal subset of V consisting of nonzero vectors is linearly independent**. Any nonzero vector $\mathbf{u}$ can be **normalized**: $\hat{\mathbf{u}} = \mathbf{u}/\|\mathbf{u}\|$, with $\|\hat{\mathbf{u}}\| = 1$. A set of vectors that are mutually orthogonal and of unit length is called an **orthonormal set**. In particular, a *basis* that is an orthonormal set is called an **orthonormal basis**. In an inner product space V, we have the **Pythagorean Theorem**: **If $\mathbf{u}$ and $\mathbf{v}$ are orthogonal vectors in V, then $\|\mathbf{u} + \mathbf{v}\|^2 = \|\mathbf{u}\|^2 + \|\mathbf{v}\|^2$.** Also, if $\{\mathbf{v}_1, \mathbf{v}_2, \ldots, \mathbf{v}_k\}$ is an orthogonal set of vectors in V, then

$$\|a_1\mathbf{v}_1 + a_2\mathbf{v}_2 + \cdots + a_k\mathbf{v}_k\|^2 = |a_1|^2\|\mathbf{v}_1\|^2 + |a_2|^2\|\mathbf{v}_2\|^2 + \cdots + |a_k|^2\|\mathbf{v}_k\|^2$$

for all scalars $a_1, a_2, \ldots, a_k$. In particular, if $\{\mathbf{v}_1, \mathbf{v}_2, \ldots, \mathbf{v}_k\}$ is an orthonormal set of vectors in V, then $\|a_1\mathbf{v}_1 + a_2\mathbf{v}_2 + \cdots + a_k\mathbf{v}_k\|^2 = |a_1|^2 + |a_2|^2 + \cdots + |a_k|^2$. Furthermore, if $\{\mathbf{v}_1, \mathbf{v}_2, \ldots, \mathbf{v}_n\}$ is an orthogonal basis for V, then (1) $\mathbf{v} = \frac{\langle \mathbf{v}, \mathbf{v}_1 \rangle}{\|\mathbf{v}_1\|^2}\mathbf{v}_1 + \frac{\langle \mathbf{v}, \mathbf{v}_2 \rangle}{\|\mathbf{v}_2\|^2}\mathbf{v}_2 + \cdots + \frac{\langle \mathbf{v}, \mathbf{v}_n \rangle}{\|\mathbf{v}_n\|^2}\mathbf{v}_n$ and (2) $\|\mathbf{v}\|^2 = \frac{|\langle \mathbf{v}, \mathbf{v}_1 \rangle|^2}{\|\mathbf{v}_1\|^2} + \frac{|\langle \mathbf{v}, \mathbf{v}_2 \rangle|^2}{\|\mathbf{v}_2\|^2} + \cdots + \frac{|\langle \mathbf{v}, \mathbf{v}_n \rangle|^2}{\|\mathbf{v}_n\|^2}$ for every $\mathbf{v} \in V$. These results imply that if $\{\mathbf{v}_1, \mathbf{v}_2, \ldots, \mathbf{v}_n\}$ is an orthonormal basis for V, then (1) $\mathbf{v} = \langle \mathbf{v}, \mathbf{v}_1 \rangle \mathbf{v}_1 + \langle \mathbf{v}, \mathbf{v}_2 \rangle \mathbf{v}_2 + \cdots + \langle \mathbf{v}, \mathbf{v}_n \rangle \mathbf{v}_n$ and (2) $\|\mathbf{v}\|^2 = |\langle \mathbf{v}, \mathbf{v}_1 \rangle|^2 + |\langle \mathbf{v}, \mathbf{v}_2 \rangle|^2 + \cdots + |\langle \mathbf{v}, \mathbf{v}_n \rangle|^2$ for every $\mathbf{v} \in V$. **Every finite-dimensional inner product space has an orthonormal basis.** The **Gram–Schmidt process** (Theorem 7.4.1) provides a practical algorithm for constructing orthonormal bases.

An $n \times n$ complex matrix whose columns form an orthonormal basis for $\mathbb{C}^n$ is called a **unitary matrix**. A real matrix whose columns form an orthonormal basis for $\mathbb{R}^n$ is called an **orthogonal matrix**. If A is a complex $n \times n$ matrix, the following statements are equivalent: (1) A is a unitary matrix; (2) A is nonsingular and $A^* = A^{-1}$; (3) the row vectors of A are orthonormal—that is, A^* is a unitary matrix.

Let U and V be inner product spaces, with norms $\|\cdot\|_U$ and $\|\cdot\|_V$, respectively. A linear transformation $T: U \to V$ is called an **isometry**, or an **orthogonal transformation**, if it preserves the lengths of vectors: $\|T(\mathbf{u})\|_V = \|\mathbf{u}\|_U$ for every $\mathbf{u} \in U$. An isometry on a real inner product space is called an **orthogonal operator**, whereas an isometry on a complex inner product space is called a **unitary operator**. **The linear transformation T is an isometry if and only if T preserves inner products—that is, $\langle T(\mathbf{x}), T(\mathbf{y}) \rangle_V = \langle \mathbf{x}, \mathbf{y} \rangle_U$ for all vectors $\mathbf{x}, \mathbf{y} \in U$.** Consequently, a real $n \times n$ matrix A is an orthogonal matrix if and only if $T: \mathbb{R}^n \to \mathbb{R}^n$ defined by $T(\mathbf{x}) = A\mathbf{x}$ preserves the dot product: $A\mathbf{x} \cdot A\mathbf{y} = \mathbf{x} \cdot \mathbf{y}$ for all vectors $\mathbf{x}, \mathbf{y} \in \mathbb{R}^n$. Furthermore, an isometry on a real inner product space preserves angles between vectors.

Let $T: U \to V$ be an isometry between inner product spaces of the same dimension and let B_U and B_V be orthonormal bases for U and V, respectively. Then, the matrix representation $[T]_{B_U}^{B_V}$ of T with respect to the bases B_U and B_V, respectively, is a unitary (or orthogonal) matrix. In particular, an operator $T: V \to V$ is an isometry if and only if $[T]_{B_V}$ is a unitary (or orthogonal) matrix *with respect to an orthonormal basis B_V*.

An important result that sheds some light on the structure of complex matrices is **Schur's Theorem: Let A be an $n \times n$ complex matrix. Then, there is a unitary matrix U such that U^*AU is upper triangular—that is, any square complex matrix A is *unitarily similar* to an upper triangular matrix.** The expression U^*AU in Schur's Theorem is called the **Schur decomposition** or **Schur factorization** of A. The diagonal elements of the upper triangular matrix U^*AU are the eigenvalues of A. Alternatively, Schur's Theorem states that **any linear operator T on a finite-dimensional inner product space V has an upper triangular matrix representation with respect to a suitable orthonormal basis**. One consequence of Schur's Theorem is a proof of the general form of the **Cayley–Hamilton Theorem**: Every $n \times n$ complex matrix satisfies its own characteristic equation.

The matrix version of the Gram–Schmidt orthonormalization process is known as a **QR factorization: Let A be a real $m \times n$ matrix with rank n. Then, A can be written as a product QR, where Q is a real $m \times n$ matrix whose columns form an orthonormal set and R is a real $n \times n$ upper triangular matrix with positive entries on its main diagonal.** The QR factorization is unique and can be used to solve a consistent $m \times n$ system of linear equations $A\mathbf{x} = \mathbf{b}$.

If a vector $\mathbf{z}$ is orthogonal to every vector in a nonempty subset S of an inner product space V, then $\mathbf{z}$ is said to be **orthogonal to S**. The set of all vectors $\mathbf{z}$ that are orthogonal to S is called the **orthogonal complement** of S and is denoted by $S^{\perp}$ (pronounced "S perp"). The **Orthogonal Decomposition Theorem** states that if U is a subspace of a finite-dimensional inner product space V, then any vector $\mathbf{v}$ in V can be written uniquely in the form $\mathbf{v} = \mathbf{u}_1 + \mathbf{u}_2$, where $\mathbf{u}_1$ is in U and $\mathbf{u}_2$ is in $U^\perp$—that is, $V = U \oplus U^\perp$.

Let V be a vector space and suppose U and W are subspaces of V such that $V = U \oplus W$. Then, any $\mathbf{v} \in V$ can be written uniquely as $\mathbf{v} = \mathbf{u} + \mathbf{w}$ for $\mathbf{u} \in U$ and $\mathbf{w} \in W$. Now, define the function $P_U: V \to V$ by $P_U(\mathbf{v}) = P_U(\mathbf{u} + \mathbf{w}) = \mathbf{u}$. This transformation is called the **projection of V** (or of $\mathbf{v}$) **on the subspace U**, or the **projection of V** (or of $\mathbf{v}$) **on U along W**. Because $U \oplus W = W \oplus U$, we can also define the projection of V on W along $U: P_W(\mathbf{v}) = P_W(\mathbf{u} + \mathbf{w}) = \mathbf{w}$. We know that (1) P_U is a linear transformation; (2) the range of P_U is U: range $P_U = P_U(V) = U$; and (3) the null space of P_U is W: null space $P_U = W$. Furthermore, if $U_1, U_2, \ldots, U_m$ are subspaces of a vector space V such that $V = U_1 \oplus U_2 \oplus \cdots \oplus U_m$, then we can define the **projection P_{U_k} of V on U_k** as $P_{U_k}(\mathbf{u}_1 + \mathbf{u}_2 + \cdots + \mathbf{u}_m) = \mathbf{u}_k$, where $\mathbf{u}_i \in U_i, i = 1, 2, \ldots, m$. If $V = U \oplus U^\perp$, then the projection of $\mathbf{v} \in V$ onto U is called the **orthogonal projection of V on U along $U^\perp$**. Consequently, $P_U(\mathbf{v}) \in U$ and $\mathbf{v} - P_U(\mathbf{v}) \in U^\perp$ by the unique representation of any $\mathbf{v} \in V$ as $\mathbf{u} + \mathbf{u}'$, where $\mathbf{u} \in U$ and $\mathbf{u}' \in U^\perp$.

Hermitian Matrices and Quadratic Forms

We have considered symmetric matrices ($A^T = A$) previously; also, in Chapters 4 and 6, we proved results about similar matrices and the diagonalization of matrices. In this chapter, we extend our understanding of these concepts, stating and proving theorems of great theoretical and practical importance. Most of our results will take place in the complex domain.

8.1 Linear Functionals and the Adjoint of an Operator

In Chapter 7, we associated certain operators (isometries) with unitary or orthogonal matrices, depending on whether the inner product space V was complex or real. The conjugate transpose of a matrix A, $A^* = \bar{A}^T$, played an important role in the discussion. In this section, we will define an operator T^* on V whose matrix representation with respect to any orthonormal basis B for V is $([T]_B)^*$.

First, we define a **linear functional** φ on a vector space V as a linear transformation from V to its set of scalars. In our case, $\varphi : V \to \mathbb{R}$ or $\varphi : V \to \mathbb{C}$. For example, if $V = C[0, 2\pi]$ and g is a fixed function in V, then $G(f) = \frac{1}{2\pi} \int_0^{2\pi} f(t)g(t)\mathrm{d}t$ for all $f \in C[0, 2\pi]$ is a linear functional on V. (When $g(t) = \sin nt$ or $g(t) = \cos nt$, $G(f)$ is called the nth *Fourier coefficient* of f.) In general, if V is an inner product space and $\mathbf{y}$ is a fixed vector in V, then the mapping defined by $g(\mathbf{x}) = \langle \mathbf{x}, \mathbf{y} \rangle$ for all $\mathbf{x} \in V$ is clearly a linear functional. The interesting point here is that if V is finite-dimensional, then *every* linear functional from V to $\mathbb{R}$ (or $\mathbb{C}$) has this form.

Theorem 8.1.1*

Suppose φ is a linear functional on an inner product space V, where $\dim(V) = n$. Then there is a unique vector $\mathbf{v} \in V$ such that $\varphi(\mathbf{u}) = \langle \mathbf{u}, \mathbf{v} \rangle$ for every $\mathbf{u} \in V$.

Proof First we produce a vector $\mathbf{v} \in V$ such that $\varphi(\mathbf{u}) = \langle \mathbf{u}, \mathbf{v} \rangle$ for every $\mathbf{u} \in V$.

Let $\{\mathbf{v}_i\}_{i=1}^n$ be an orthonormal basis for V and suppose that $\mathbf{u} \in V$. Then, by

Corollary 7.3.2, $\mathbf{u} = \sum_{i=1}^n \langle \mathbf{u}, \mathbf{v}_i \rangle \mathbf{v}_i$, so

$$
\begin{aligned}
\varphi(\mathbf{u}) &= \varphi\left(\sum_{i=1}^n \langle \mathbf{u}, \mathbf{v}_i \rangle \right) = \sum_{i=1}^n \langle \mathbf{u}, \mathbf{v}_i \rangle \varphi(\mathbf{v}_i) \\
&= \left\langle \mathbf{u}, \overline{\varphi(\mathbf{v}_1)}\mathbf{v}_1 + \overline{\varphi(\mathbf{v}_2)}\mathbf{v}_2 + \cdots + \overline{\varphi(\mathbf{v}_n)}\mathbf{v}_n \right\rangle \\
&= \left\langle \mathbf{u}, \sum_{i=1}^n \overline{\varphi(\mathbf{v}_i)}\mathbf{v}_i \right\rangle
\end{aligned}
$$

for every $\mathbf{u} \in V$. If we set $\mathbf{v} = \sum_{i=1}^n \overline{\varphi(\mathbf{v}_i)}\mathbf{v}_i$, then we have $\varphi(\mathbf{u}) = \langle \mathbf{u}, \mathbf{v} \rangle$ for every $\mathbf{u} \in V$, as promised.

To prove the uniqueness of this vector $\mathbf{v}$ we have produced, suppose that $\mathbf{v}_1, \mathbf{v}_2 \in V$ are such that $\varphi(\mathbf{u}) = \langle \mathbf{u}, \mathbf{v}_1 \rangle = \langle \mathbf{u}, \mathbf{v}_2 \rangle$ for every $\mathbf{u} \in V$. Then $0 = \langle \mathbf{u}, \mathbf{v}_1 \rangle - \langle \mathbf{u}, \mathbf{v}_2 \rangle = \langle \mathbf{u}, \mathbf{v}_1 - \mathbf{v}_2 \rangle$ for every $\mathbf{u} \in V$. If we take $\mathbf{u} = \mathbf{v}_1 - \mathbf{v}_2$, we see that $0 = \langle \mathbf{v}_1 - \mathbf{v}_2, \mathbf{v}_1 - \mathbf{v}_2 \rangle = |\mathbf{v}_1 - \mathbf{v}_2|^2$, so $\mathbf{v}_1 - \mathbf{v}_2 = \mathbf{0}$. This proves that the vector $\mathbf{v}$ we constructed in the first part of this proof is unique.

Now we are ready to produce an operator that corresponds to the complex conjugate of a matrix.

Theorem 8.1.2

Let V be a finite-dimensional inner product space, and let T be a linear operator on V. Then there exists a unique linear transformation $T^* : V \rightarrow V$ such that $\langle T(\mathbf{x}), \mathbf{y} \rangle = \langle \mathbf{x}, T^*(\mathbf{y}) \rangle$ for all $\mathbf{x}, \mathbf{y} \in V$.

Proof Let $\mathbf{y}$ be a fixed vector in V. Define $g : V \rightarrow \mathbb{R}$ (or $g : V \rightarrow \mathbb{C}$) by $g(\mathbf{x}) = \langle T(\mathbf{x}), \mathbf{y} \rangle$ for all $\mathbf{x} \in V$. It is easy to show that g is linear (Exercise A7).

Now we apply Theorem 8.1.1 to obtain a unique vector $\hat{\mathbf{y}}$ such that $g(\mathbf{x}) = \langle \mathbf{x}, \hat{\mathbf{y}} \rangle$, that is, $\langle T(\mathbf{x}), \mathbf{y} \rangle = \langle \mathbf{x}, \hat{\mathbf{y}} \rangle$ for all $\mathbf{x} \in V$. If we define $T^* : V \rightarrow V$ by $T^*(\mathbf{y}) = \hat{\mathbf{y}}$, we have $\langle T(\mathbf{x}), \mathbf{y} \rangle = \langle \mathbf{x}, T^*(\mathbf{y}) \rangle$. The linearity of T^* is left as an exercise (A8).

Finally, we show that T^* is unique. Suppose that $F : V \rightarrow V$ is linear and that $\langle T(\mathbf{x}), \mathbf{y} \rangle = \langle \mathbf{x}, F(\mathbf{y}) \rangle$ for all $\mathbf{x}, \mathbf{y} \in V$. Then $\langle \mathbf{x}, T^*(\mathbf{y}) \rangle = \langle \mathbf{x}, F(\mathbf{y}) \rangle$ for all $\mathbf{x}, \mathbf{y} \in V$, so $T^* = F$.

---

* This is the **Riesz Representation Theorem**, named after the Hungarian mathematician Frigyes Riesz (1880–1956). Riesz was one of the founders of functional analysis and his work has many important applications in physics.

The linear operator T^* (pronounced "T star"), described in Theorem 8.1.2, is called the **adjoint** of the operator T. The term "adjoint" is sometimes used to describe a matrix associated with an inverse matrix and is often distinguished by the term "classical adjoint." For our purposes, the adjoint is the unique operator on V satisfying $\langle T(\mathbf{x}), \mathbf{y} \rangle = \langle \mathbf{x}, T^*(\mathbf{y}) \rangle$ for all $\mathbf{x}, \mathbf{y} \in V$.

It follows that

$$\langle \mathbf{x}, T(\mathbf{y}) \rangle = \overline{\langle T(\mathbf{y}), \mathbf{x} \rangle} = \overline{\langle \mathbf{y}, T^*(\mathbf{x}) \rangle} = \langle T^*(\mathbf{x}), \mathbf{y} \rangle,$$

so $\langle \mathbf{x}, T(\mathbf{y}) \rangle = \langle T^*(\mathbf{x}), \mathbf{y} \rangle$ for all $\mathbf{x}, \mathbf{y} \in V$. In practical terms, this says that when we are calculating with the adjoint and move the "T" within the inner product brackets, we must "star" it, with the understanding that $(T^*)^* = T$ (Exercise A9).

It should be noted that the adjoint may not exist on an infinite-dimensional inner product space (see Exercise C1), hence the importance of having finite dimensionality in the hypothesis of Theorem 8.1.2.

A useful method to construct T^* from T when the adjoint exists is to start with $\langle T(\mathbf{x}), \mathbf{y} \rangle (= \langle \mathbf{x}, T^*(\mathbf{y}) \rangle)$ and use properties of the adjoint to get an expression of the form $\langle \mathbf{x}, \mathbf{z} \rangle$. By the uniqueness of the operator T^*, we conclude that $\mathbf{z} = T^*(\mathbf{y})$.

Example 8.1.1: Adjoints of Operators

(a) Let $V = \mathbb{R}^2$, with the inner product defined by $\langle \mathbf{x}, \mathbf{y} \rangle = 3x_1 y_1 + x_2 y_2$, where $\mathbf{x} = \begin{bmatrix} x_1 \\ x_2 \end{bmatrix}$ and $\mathbf{y} = \begin{bmatrix} y_1 \\ y_2 \end{bmatrix}$. Define $T: \mathbb{R}^2 \to \mathbb{R}^2$ by $T\left(\begin{bmatrix} x_1 \\ x_2 \end{bmatrix}\right) = \begin{bmatrix} 2x_1 + x_2 \\ 4x_1 + x_2 \end{bmatrix}$. Then for any $\mathbf{y} = \begin{bmatrix} y_1 \\ y_2 \end{bmatrix}$,

$$\left\langle T\left(\begin{bmatrix} x_1 \\ x_2 \end{bmatrix}\right), \begin{bmatrix} y_1 \\ y_2 \end{bmatrix} \right\rangle = \left\langle \begin{bmatrix} 2x_1 + x_2 \\ 4x_1 + x_2 \end{bmatrix}, \begin{bmatrix} y_1 \\ y_2 \end{bmatrix} \right\rangle$$
$$= 3(2x_1 + x_2)y_1 + (4x_1 + x_2)y_2$$
$$= (6y_1 + 4y_2)x_1 + (3y_1 + y_2)x_2$$
$$= 3(2y_1 + \tfrac{4}{3}y_2)x_1 + (3y_1 + y_2)x_2$$
$$= \left\langle \begin{bmatrix} x_1 \\ x_2 \end{bmatrix}, \begin{bmatrix} 2y_1 + \tfrac{4}{3}y_2 \\ 3y_1 + y_2 \end{bmatrix} \right\rangle,$$

so $T^*\left(\begin{bmatrix} y_1 \\ y_2 \end{bmatrix}\right) = \begin{bmatrix} 2y_1 + \tfrac{4}{3}y_2 \\ 3y_1 + y_2 \end{bmatrix}$.

(b) Let $V = \mathbb{R}^2$, with the standard inner product, and define $T: \mathbb{R}^2 \to \mathbb{R}^2$ by $T\left(\begin{bmatrix} z_1 \\ z_2 \end{bmatrix}\right) = \begin{bmatrix} 2z_1 + iz_2 \\ (1-i)z_1 \end{bmatrix}$. Choose any $\begin{bmatrix} \alpha \\ \beta \end{bmatrix} \in \mathbb{R}^2$. Then,

$$\left\langle T\left(\begin{bmatrix} z_1 \\ z_2 \end{bmatrix}\right), \begin{bmatrix} \alpha \\ \beta \end{bmatrix}\right\rangle = \left\langle \begin{bmatrix} 2z_1 + iz_2 \\ (1-i)z_1 \end{bmatrix}, \begin{bmatrix} \alpha \\ \beta \end{bmatrix}\right\rangle = (2z_1 + iz_2)\bar{\alpha} + (1-i)z_1\bar{\beta}$$
$$= z_1(2\bar{\alpha} + \bar{\beta} - i\bar{\beta}) + z_2(i\bar{\alpha})$$
$$= z_1\overline{(2\alpha + \beta + i\beta)} + z_2\overline{(-i\alpha)}$$
$$= \left\langle \begin{bmatrix} z_1 \\ z_2 \end{bmatrix}, \begin{bmatrix} 2\alpha + (1+i)\beta \\ -i\alpha \end{bmatrix}\right\rangle,$$

so $T^*\left(\begin{bmatrix} \alpha \\ \beta \end{bmatrix}\right) = \begin{bmatrix} 2\alpha + (1+i)\beta \\ -i\alpha \end{bmatrix}.$

Example 8.1.2: The Adjoint of an Operator on an Infinite-Dimensional Space

Let V be the vector space of all real-valued functions that are differentiable on a closed interval $[a, b]$. This is an infinite-dimensional space. Define an inner product on V by setting $\langle f, g \rangle = \int_a^b f(x)g(x)dx$ for all $f, g \in V$.

Now let T be an operator on V satisfying $T(f)(x) = \int_a^b e^{-tx}f(t)dt$. (Because the variable of integration is t, the integral is a function of the parameter x.) Then it is an exercise in iterated integration to show that $\langle T(f), g \rangle = \langle f, T(g) \rangle$ for all $f, g \in V$. Thus T^* exists and equals T.

Let us consider the familiar linear transformation, $T_A(\mathbf{v}) = A\mathbf{v}$, where A is a given $n \times n$ complex matrix and $\mathbf{v}$ is any vector in $\mathbb{R}^n$. If we assume the standard basis, $\mathcal{E}_n$, (which is orthonormal) and use the standard inner product on $\mathbb{R}^n$, $\langle \mathbf{x}, \mathbf{y} \rangle = \mathbf{y}^*\mathbf{x}$, then

$$\langle \mathbf{x}, T_A{}^*(\mathbf{y}) \rangle = \langle T_A(\mathbf{x}), \mathbf{y} \rangle = \langle A\mathbf{x}, \mathbf{y} \rangle$$
$$= \mathbf{y}^*A\mathbf{x} = \mathbf{y}^*A^{**}\mathbf{x} = (A^*\mathbf{y})^*\mathbf{x}$$
$$= \langle \mathbf{x}, A^*\mathbf{y} \rangle.$$

Thus $T_A^*(\mathbf{y}) = A^*\mathbf{y}$ for all $\mathbf{y} \in \mathbb{R}^n$, which says that $\left[T_A^*\right]_{\mathcal{E}_n} = A^* = \bar{A}^{\mathrm{T}}$.

This analysis of T_A and its adjoint serves as a model for the relationship between a matrix representation of an operator and a matrix corresponding to its adjoint (Theorem 8.1.3). First we state and prove an auxiliary result.

Lemma 8.1.1

Let V be a finite-dimensional inner product space and let $B = \{\mathbf{v}_i\}_{i=1}^n$ be an (ordered) orthonormal basis for V. Let T be a linear operator on V and suppose that $A = [a_{ij}]$ is the matrix of T with respect to the basis B. Then $a_{ij} = \langle T(\mathbf{v}_j), \mathbf{v}_i \rangle$.

Proof Since B is an orthonormal basis, we have $\mathbf{v} = \sum_{i=1}^n \langle \mathbf{v}, \mathbf{v}_i \rangle \mathbf{v}_i$ for every vector $\mathbf{v} \in V$. The matrix A is defined by $T(\mathbf{v}_j) = \sum_{i=1}^n a_{ij}\mathbf{v}_i$. Because $T(\mathbf{v}_j) = \sum_{i=1}^n \langle T(\mathbf{v}_j), \mathbf{v}_i \rangle \mathbf{v}_i$, we have $a_{ij} = \langle T(\mathbf{v}_j), \mathbf{v}_i \rangle$ by the uniqueness of the representation of a vector as a linear combination of basis vectors.

Theorem 8.1.3

Let V be a finite-dimensional inner product space and let B be an orthonormal basis for V. If T is a linear operator on V, then $[T^*]_B = ([T]_B)^*$—that is, the matrix of the adjoint operator of T with respect to B is the complex conjugate of the matrix of T with respect to B.

Proof Let $B = \{v_i\}_{i=1}^n$ be an orthonormal basis for V, and let $A = [a_{ij}] = [T]_B$ and $B = [b_{ij}] = [T^*]_B$. Then Lemma 8.1.1 states that $a_{ij} = \langle T(v_j), v_i \rangle$ and $b_{ij} = \langle T^*(v_j), v_i \rangle$.

By the definition of T^*, we have

$$b_{ij} = \langle T^*(v_j), v_i \rangle$$
$$= \overline{\langle v_i, T^*(v_j) \rangle} = \overline{\langle T(v_i), v_j \rangle}$$
$$= \bar{a}_{ji}.$$

Thus $B = \bar{A}^T = A^*$, or $[T^*]_B = ([T]_B)^*$.

Before we continue, we should note three things about the adjoint operator:

1. The adjoint of a linear transformation on V does not depend on the choice of basis for V, but its matrix representation does. (See the Kaplansky quote at the end of Section 6.6.)
2. The adjoint of T depends not only on the definition of T but also on the inner product as well (see Example 8.1.3).
3. The relation between $[T]_B$ and $[T^*]_B$ is generally more complicated than that given in Theorem 8.1.3 when the basis B is not orthonormal (see Exercise C2).

Example 8.1.3: $[T^*]_B$ and $[T]_B^*$

(a) In Example 8.1.1(a), the standard basis, $\mathcal{E}_2$, for $\mathbb{R}^2$ is assumed, but the inner product is not the standard inner product. We see that $[T]_{\mathcal{E}_2} = \begin{bmatrix} 2 & 1 \\ 4 & 1 \end{bmatrix}$ and $[T^*]_{\mathcal{E}_2} = \begin{bmatrix} 2 & 4/3 \\ 3 & 1 \end{bmatrix}$, so $[T^*]_{\mathcal{E}_2} \neq [T]_{\mathcal{E}_2}^*$.

(b) In Example 8.1.1(b), we used the standard basis, B, and the standard inner product for $\mathbb{C}^2$. Here, we find that $[T]_B = \begin{bmatrix} 2 & i \\ 1-i & 0 \end{bmatrix}$ and $[T^*]_B = \begin{bmatrix} 2 & 1+i \\ -i & 0 \end{bmatrix} = [T]_B^*$.

The following theorem is the operator analog of Theorem 7.1.1. In effect, we are considering the properties of the transformation $T \to T^*$.

Theorem 8.1.4

Let V be a finite-dimensional inner product space. If T and U are linear operators on V and c is a scalar, then

(1) $(T + U)^* = T^* + U^*$,

(2) $(cT)^* = \bar{c}T^*$,

(3) $(T^*)^* = T$,

(4) $(TU)^* = U^*T^*$,

(5) $I^* = I$, where I is the identity transformation on V,

(6) If one of the inverses T^{-1} or $(T^*)^{-1}$ exists, then so does the other, and $(T^{-1})^* = (T^*)^{-1}$.

Proof Exercises A10 through A14 and B1.

Before ending our preliminary discussion of adjoints, we note another analogy between the transformation $T \to T^*$ and complex conjugation. A complex number z is a real number if and only if $\bar{z} = z$. In the next section, we will consider *self-adjoint* operators—operators T such that $T^* = T$—and see to what extent they behave like real numbers. For example, the linear transformation, $T_A(\mathbf{v}) = A\mathbf{v}$, where A is a given $n \times n$ real matrix and $\mathbf{v}$ is any vector in $\mathbb{R}^n$, is self-adjoint (Exercise A15).

Exercises 8.1

A.

1. Define $f : \mathbb{R}^2 \to \mathbb{R}$ by $f\left(\begin{bmatrix} x \\ y \end{bmatrix}\right) = 2x + y$. Show that f is a linear functional.

2. If $V = \mathbb{R}^3$ with the standard inner product, and $f : \mathbb{R}^3 \to \mathbb{R}$ is defined by $f\left(\begin{bmatrix} x \\ y \\ z \end{bmatrix}\right) = x - 2y + 4z$, find a vector $\mathbf{y} \in \mathbb{R}^3$ such that $f(\mathbf{x}) = \langle \mathbf{x}, \mathbf{y} \rangle$ for all $\mathbf{x} \in \mathbb{R}^3$.

3. Suppose that $V = P_2[x]$, with the inner product defined by $\langle f, g \rangle = \int_0^1 f(t)g(t)\mathrm{d}t$.

   Define a linear functional $H : P_2[x] \to \mathbb{R}$ by $H(f) = f(0) + f'(1)$ for all $f \in P_2[x]$. Find a vector $\hat{g} \in P_2[x]$ such that $H(f) = \langle f, \hat{g} \rangle$ for all $f \in P_2[x]$.

4. Suppose that $V = \mathbb{R}^2$ with the standard inner product. Let $T : V \to V$ be defined by $T\left(\begin{bmatrix} x \\ y \end{bmatrix}\right) = \begin{bmatrix} 2x + y \\ x - 3y \end{bmatrix}$. Find T^* and calculate $T^*\left(\begin{bmatrix} 3 \\ 5 \end{bmatrix}\right)$.

5. Suppose that $V = P_1[x]$ with $\langle f, g \rangle = \int_{-1}^1 f(t)g(t)\mathrm{d}t$. Let T be the operator on V defined by $T(f) = f' + 3f$ for all $f \in P_1[x]$. Find T^* and calculate $T^*(4 - 2x)$.

6. Let $V = \mathbb{R}^2$ with the standard inner product. Let T be the linear operator on $\mathbb{C}^2$ defined by $T\left(\begin{bmatrix} 1 \\ 0 \end{bmatrix}\right) = \begin{bmatrix} 1 \\ -2 \end{bmatrix}$ and $T\left(\begin{bmatrix} 0 \\ 1 \end{bmatrix}\right) = \begin{bmatrix} i \\ -1 \end{bmatrix}$. If $\mathbf{v} = \begin{bmatrix} x \\ y \end{bmatrix} \in \mathbb{R}^2$, find $T^*(\mathbf{v})$.

7. Suppose T is a linear operator on an inner product space V and $\mathbf{y}$ is a fixed vector in V. If $g : V \to \mathbb{C}$ is defined by $g(\mathbf{x}) = \langle T(\mathbf{x}), \mathbf{y} \rangle$ for all $\mathbf{x} \in V$, show that g is a linear transformation.

8. If T is a linear operator on the inner product space V and T^* is defined by the relation $\langle T(\mathbf{x}), \mathbf{y} \rangle = \langle \mathbf{x}, T^*(\mathbf{y}) \rangle$ for all $\mathbf{x}, \mathbf{y} \in V$, show that T^* is a linear operator on V.

9. If T is an operator on a finite-dimensional inner product space V, prove that $(T^*)^* = T$.

10. Prove Theorem 8.1.4(1).

11. Prove Theorem 8.1.4(2).

12. Prove Theorem 8.1.4(3).

13. Prove Theorem 8.1.4(4).

14. Prove Theorem 8.1.4(5).

15. If A is an $n \times n$ real matrix and $T_A : \mathbb{R}^n \to \mathbb{R}^n$ is defined by $T_A(\mathbf{v}) = A\mathbf{v}$ for every $\mathbf{v} \in \mathbb{R}^n$, show that $(T_A)^* = T_{A^\mathsf{T}}$.

16. Let T be a linear operator on an inner product space V. Let $U_1 = T + T^*$ and $U_2 = TT^*$. Show that $U_1^* = U_1$ and $U_2^* = U_2$.

17. If $T^* = T$ and $U^* = U$, show that $(TU)^* = TU$ if and only if $TU = UT$.

B.

1. Prove Theorem 8.1.4(6).

2. Let $A = \begin{bmatrix} 0 & 1 & 0 \\ 1 & 0 & 0 \\ 0 & 0 & 1 \end{bmatrix}$. Define $T_A : \mathbb{R}^3 \to \mathbb{R}^3$ by $T_A(\mathbf{x}) = A\mathbf{x}$ for every $\mathbf{x} \in \mathbb{R}^3$. Determine T^* relative to the standard inner product on $\mathbb{R}^3$.

3. Prove that if $V = W \oplus W^\perp$ and T is the projection on W along $W^\perp$, then $T = T^*$. [*Hint*: null space $T = W^\perp$ (see Section 7.8)]

4. Let V be an inner product space and let $\mathbf{y}, \mathbf{z} \in V$. Define $T: V \to V$ by $T(\mathbf{x}) = \langle \mathbf{x}, \mathbf{y} \rangle \mathbf{z}$ for all $\mathbf{x} \in V$.
   (a) Prove that T is linear.
   (b) Show that T^* exists and find an explicit expression for it.

5. Suppose that T is a linear operator on an inner product space V and U is a subspace of V. Prove that $T(U) = \{T(\mathbf{u}): \mathbf{u} \in U\} \subseteq U$ if and only if $T^*(U^\perp) \subseteq U^\perp$.

6. Let $V = P_3[x]$ with $\langle f, g \rangle = \int_0^1 f(t)g(t)dt$ for all $f, g \in V$. Let D be the differentiation operator on V, $D(f) = f'$ for all $f \in V$. Find D^*.

7. Suppose T^* is the adjoint of a linear transformation T on a finite-dimensional complex inner product space.
   (a) Show that null space $T^* = (\text{range } T)^\perp$.
   (b) Show that range $T^* = (\text{null space } T)^\perp$.
   (c) Using the results of parts (a) and (b), show that $V = (\text{range } T) \oplus (\text{null space } T^*) = (\text{range } T^*) \oplus (\text{null space } T)$.

8. Suppose n is a positive integer. Define $S: \mathbb{R}^n \to \mathbb{R}^n$ by $S([z_1 \quad z_2 \quad \cdots \quad z_n]^{\mathsf{T}}) = [0 \quad z_1 \quad z_2 \quad \cdots \quad z_{n-1}]^{\mathsf{T}}$. Find a formula for S^*.

9. Let $V = M_n(\mathbb{R})$ with inner product $\langle A, B \rangle = \text{trace}(AB^*)$. For any $A \in V$, define $T_A: V \to V$ by $T_A(M) = AM$ for all $M \in V$. Prove that $T_A^* = T_A$, that is, $T_A^*(M) = A^*M$.

C.

1. Let V be the vector space of all infinite sequences, $\mathbf{a} = \{a_i\}_{i=1}^\infty$, of real numbers such that $a_i \neq 0$ for only finitely many positive integers i. (The space V becomes an inner product space if we define $\langle \mathbf{a}, \mathbf{b} \rangle = \sum_{n=1}^\infty \mathbf{a}(n)\mathbf{b}(n)$ for all $\mathbf{a}, \mathbf{b} \in V$, where $\boldsymbol{\sigma}(k)$ denotes the kth term in the sequence $\boldsymbol{\sigma}$.)

   For each positive integer n, let $\mathbf{e}_n$ be an infinite sequence of real numbers such that $\mathbf{e}_n(k) = 0$ if $n \neq k$, $\mathbf{e}_n(k) = 1$ if $n = k$. Then, $\{\mathbf{e}_n\}_{n=1}^\infty$ is an orthonormal basis for V, indicating that V is infinite-dimensional. Now define $T: V \to V$ by $T(\mathbf{a})(k) = \sum_{i=k}^\infty \mathbf{a}(i)$. (Because $\mathbf{a}(i) \neq 0$ for only finitely many values of i, the infinite series in the definition of T converges.)
   (a) Prove that T is a linear operator on V.
   (b) Prove that for any positive integer n, $T(\mathbf{e}_n) = \sum_{i=1}^n \mathbf{e}_i$.
   (c) Prove that T has no adjoint. [*Hint*: Suppose that T^* exists. Show that for any positive integer n, $T^*(\mathbf{e}_n)(k) \neq 0$ for infinitely many values of k, and note the contradiction.]

2. Suppose that $S = \{\mathbf{v}_i\}_{i=1}^n$ is any basis for a vector space with an inner product $\langle \cdot, \cdot \rangle$, and let T be any linear operator on V.

The *Gram matrix*, $G = [g_{ij}]$, of $\{\mathbf{v}_i\}_{i=1}^n$ with respect to the given inner product is defined by $g_{ij} = \langle \mathbf{v}_j, \mathbf{v}_i \rangle$ for $i, j = 1, 2, \ldots, n$. (Note that $G^* = G$.) Now let $[T]_S = A$ and $[T^*]_S = B$.

(a) Prove that $\langle \mathbf{x}, \mathbf{y} \rangle = [\mathbf{y}]_S^* G [\mathbf{x}]_S$ for all $\mathbf{x}, \mathbf{y} \in V$.

(b) Show that

$$\langle T(\mathbf{x}), \mathbf{y} \rangle = [\mathbf{y}]_S^* G A [\mathbf{x}]_S \quad \text{for all } \mathbf{x}, \mathbf{y} \in V$$

and

$$\langle \mathbf{x}, T^*(\mathbf{y}) \rangle = [\mathbf{y}]_S^* B^* G [\mathbf{x}]_S \text{ for all } \mathbf{x}, \mathbf{y} \in V.$$

(c) Use the result of part (b) to show that $B = G^{-1} A^* G$—that is, $[T^*]_S = G^{-1} [T]_S^* G$.

(d) If the basis S is orthonormal, show that $[T^*]_S = [T]_S^*$.

8.2 Hermitian Matrices

A **Hermitian*** (sometimes **hermitian**) **matrix** A is a square matrix with complex entries such that $A^* = A$, where $A^* = \bar{A}^T$. Matrix A is called **skew-Hermitian** if $A^* = -A$. If we recall that a complex number z is real if and only if $\bar{z} = z$, we realize that a Hermitian matrix is analogous to a real number. (In fact, a 1×1 Hermitian matrix can be interpreted as a real number.) Such a matrix is the complex analogue of a symmetric matrix. More to the point, every symmetric matrix is a Hermitian matrix, which implies that every result that is true for a Hermitian matrix is true for a symmetric matrix. Three fundamental properties of Hermitian matrices are provided by the next theorem.

Theorem 8.2.1

Let A be a Hermitian matrix. Then,

(1) $\langle A\mathbf{x}, \mathbf{y} \rangle = \langle \mathbf{x}, A\mathbf{y} \rangle = \langle \mathbf{x}, A^*\mathbf{y} \rangle$ for every $\mathbf{x}, \mathbf{y} \in \mathbb{C}^n$.

(2) All eigenvalues of A are real.

(3) Eigenvectors of A associated with distinct eigenvalues are orthogonal.

Proof (1) If A is Hermitian, $\bar{A} = \overline{A^*} = \overline{(\bar{A}^T)} = (\overline{\bar{A}})^T = A^T$. Thus, $\langle A\mathbf{x}, \mathbf{y} \rangle = (A\mathbf{x})^T \bar{\mathbf{y}} = \mathbf{x}^T A^T \bar{\mathbf{y}} = \mathbf{x}^T \bar{A} \bar{\mathbf{y}} = \mathbf{x}^T \overline{(A\mathbf{y})} = \langle \mathbf{x}, A\mathbf{y} \rangle = \langle \mathbf{x}, A^*\mathbf{y} \rangle$.

(2) Let λ be an eigenvalue of A associated with an eigenvector $\mathbf{x}$. Then $A\mathbf{x} = \lambda\mathbf{x}$ and $\mathbf{x}^* A\mathbf{x} = \mathbf{x}^* \lambda\mathbf{x} = \lambda\mathbf{x}^*\mathbf{x} = \lambda\langle \mathbf{x}, \mathbf{x} \rangle = \lambda\|\mathbf{x}\|^2$.

The left side of this last equation is real because $\mathbf{x}^* A\mathbf{x}$ is a 1×1 matrix, which implies $\overline{(\mathbf{x}^* A\mathbf{x})} = (\mathbf{x}^* A\mathbf{x})^* = \mathbf{x}^* A^* \mathbf{x}^{**} = \mathbf{x}^* A\mathbf{x}$. But on the far

---

* Named for Charles Hermite (1822–1901), a French mathematician who made contributions in various areas of number theory, algebra, and analysis. In 1873, he published the first proof that e is not an algebraic number, i.e., e is not a zero of a polynomial with integer coefficients. Hermitian matrices are also called **self-adjoint** matrices.

right side of the equation, $\|\mathbf{x}\|^2$ is real (and positive) because $\mathbf{x} \neq \mathbf{0}$. Therefore, λ must be real.

(3) Let $\mathbf{x}$ and $\mathbf{y}$ be eigenvectors of A associated with eigenvalues λ and μ, respectively, with $\lambda \neq \mu$. Because $A = A^*$ and λ, μ are real, we have

$$\lambda \langle \mathbf{x}, \mathbf{y} \rangle = \langle \lambda \mathbf{x}, \mathbf{y} \rangle = \langle A\mathbf{x}, \mathbf{y} \rangle \overset{(1)}{=} \langle \mathbf{x}, A\mathbf{y} \rangle = \langle \mathbf{x}, \mu \mathbf{y} \rangle = \mu \langle \mathbf{x}, \mathbf{y} \rangle,$$

or $(\lambda - \mu)\langle \mathbf{x}, \mathbf{y} \rangle = 0$. Because $\lambda \neq \mu$, we have $\langle \mathbf{x}, \mathbf{y} \rangle = 0$, so $\mathbf{x}$ is orthogonal to $\mathbf{y}$.

Example 8.2.1: The Eigenpairs of a Hermitian Matrix

Consider the matrix

$$A = \begin{bmatrix} 2 & 1 - i \\ 1 + i & 1 \end{bmatrix},$$

which is Hermitian:

$$A^* = \bar{A}^\mathsf{T} = \begin{bmatrix} 2 & 1 + i \\ 1 - i & 1 \end{bmatrix}^\mathsf{T} = \begin{bmatrix} 2 & 1 - i \\ 1 + i & 1 \end{bmatrix} = A.$$

The eigenvalues of A are $\lambda_1 = 0$ and $\lambda_2 = 3$, distinct real numbers, with representative eigenvectors $\mathbf{x}_1 = \begin{bmatrix} -1 + i \\ 2 \end{bmatrix}$ and $\mathbf{x}_2 = \begin{bmatrix} 1 - i \\ 1 \end{bmatrix}$, respectively. Finally, $\langle \mathbf{x}_1, \mathbf{x}_2 \rangle = \mathbf{x}_1^\mathsf{T} \bar{\mathbf{x}}_2 = [-1 + i \quad 2] \begin{bmatrix} 1 + i \\ 1 \end{bmatrix} = 0$, so the eigenvectors are orthogonal.

It is easy to see that if A is Hermitian, then iA is skew-Hermitian; and if A is skew-Hermitian, then iA is Hermitian. Consequently, skew-Hermitian matrices have properties analogous to those given in Theorem 8.2.1 (see Exercise B5).

Schur's Theorem (Theorem 7.6.1) is a key result that sheds some light on the structure of Hermitian matrices. The following important theorem is one of several results called *spectral theorems*. The terminology emphasizes the fact that the eigenvalues of a Hermitian matrix A, elements of the spectrum of A, are the entries of the diagonal matrices involved in the result. Note that matrix A in Example 8.2.2 is Hermitian, and the upper triangular matrix guaranteed by Schur's Theorem turns out to be a diagonal matrix.

Theorem 8.2.2: Spectral Theorem for Hermitian Matrices

Let A be a Hermitian matrix. Then there is a unitary matrix U such that U^*AU is diagonal. Furthermore, if A is a real symmetric matrix, then U can be chosen to be real and orthogonal.

Proof By Schur's Theorem, there is a unitary matrix U such that $U^*AU = T$ is upper triangular. Then $T^* = (U^*AU)^* = U^*A^*U = U^*AU = T$, so T is Hermitian. But T is upper triangular and T^* is lower triangular, so the only way that T and T^* can be equal is if all the entries of T off the main diagonal are zero—that is, T is a diagonal matrix. The proof for a real symmetric matrix A is essentially the same.

Note that in Theorem 8.2.2 the elements of the diagonal matrix T are the eigenvalues of A. Example 8.2.2 can serve as an illustration of this spectral theorem.

Corollary 8.2.1

If A is an $n \times n$ Hermitian matrix, there is an orthonormal basis for $\mathbb{C}^n$ which entirely consists eigenvectors of A. If A is real, there is an orthonormal basis for $\mathbb{R}^n$ consisting of eigenvectors of A.

Proof Theorem 8.2.2 guarantees the existence of a unitary matrix U such that $U^*AU = D$ is diagonal, say with diagonal entries $d_1, d_2, \ldots, d_n$. If $\mathbf{u}_1, \mathbf{u}_2, \ldots, \mathbf{u}_n$ are the columns of U, then the equation $AU = (U^*)^{-1}D = UD$ is equivalent to $\begin{bmatrix} A\mathbf{u}_1 & A\mathbf{u}_2 & \ldots & A\mathbf{u}_n \end{bmatrix} = \begin{bmatrix} d_1\mathbf{u}_1 & d_2\mathbf{u}_2 & \ldots & d_n\mathbf{u}_n \end{bmatrix}$, which implies that $A\mathbf{u}_i = d_i\mathbf{u}_i$ for $i = 1, 2, \ldots, n$. Therefore, $\mathbf{u}_i$ are eigenvectors of A, and, because U is unitary, these vectors form an orthonormal basis for $\mathbb{C}^n$. The argument in the real case is similar.

If A is a Hermitian matrix, the Spectral Theorem allows us to write $A = UDU^*$, where U is unitary and D is diagonal. Furthermore, the eigenvalues, $\lambda_1, \lambda_2, \ldots, \lambda_n$, of A are the diagonal entries of D. Letting $U = \begin{bmatrix} \mathbf{u}_1 & \mathbf{u}_2 & \ldots & \mathbf{u}_n \end{bmatrix}$ and $U^* = \begin{bmatrix} \mathbf{u}_1^* & \mathbf{u}_2^* & \ldots & \mathbf{u}_n^* \end{bmatrix}^{\mathrm{T}}$, we write the spectral form of A as follows:

$$A = UDU^* = \begin{bmatrix} \mathbf{u}_1 & \mathbf{u}_2 & \ldots & \mathbf{u}_n \end{bmatrix} \begin{bmatrix} \lambda_1 & 0 & \ldots & 0 \\ 0 & \lambda_2 & \ldots & 0 \\ \vdots & \vdots & \ldots & \vdots \\ 0 & 0 & \ldots & \lambda_n \end{bmatrix} \begin{bmatrix} \mathbf{u}_1^* \\ \mathbf{u}_2^* \\ \vdots \\ \mathbf{u}_n^* \end{bmatrix}$$

$$= \lambda_1 \mathbf{u}_1 \mathbf{u}_1^* + \lambda_2 \mathbf{u}_2 \mathbf{u}_2^* + \cdots + \lambda_n \mathbf{u}_n \mathbf{u}_n^*.$$

This representation of a Hermitian matrix, in terms of its eigenvalues and corresponding orthonormal eigenvectors,

$$\boxed{A = \lambda_1 \mathbf{u}_1 \mathbf{u}_1^* + \lambda_2 \mathbf{u}_2 \mathbf{u}_2^* + \cdots + \lambda_n \mathbf{u}_n \mathbf{u}_n^*,} \tag{S}$$

is an alternative form of the Spectral Theorem, usually called the **spectral decomposition** of A (see the remarks before Lemma 4.8.1). Note that each summand $\lambda_i \mathbf{u}_i \mathbf{u}_i^*$ is an $n \times n$ Hermitian matrix of rank 1 because every column of $\lambda_i \mathbf{u}_i \mathbf{u}_i^*$ is a multiple of $\mathbf{u}_i$. Furthermore, $\mathbf{u}_i \mathbf{u}_i^*$ is a *projection matrix* in the sense that for each $\mathbf{x} \in \mathbb{R}^n$, $\left(\mathbf{u}_i \mathbf{u}_i^* \right) \mathbf{x}$ is the orthogonal projection of $\mathbf{x}$ onto the subspace spanned by $\mathbf{u}_i$ (Exercise B14).

Corollary 8.2.2

Let A be an $m \times n$ complex matrix. Then the matrix AA^* (or A^*A) is Hermitian and its eigenvalues are nonnegative.

Proof Clearly AA^* is an $m \times m$ matrix and $(AA^*)^* = (A^*)^*A^* = AA^*$. Because AA^* is Hermitian, there exists an orthonormal set, $\mathbf{v}_1, \mathbf{v}_2, \ldots, \mathbf{v}_m$, of eigenvectors corresponding to its eigenvalues, $\lambda_1, \lambda_2, \ldots, \lambda_m$, respectively. Now $AA^*\mathbf{v}_i = \lambda_i \mathbf{v}_i$ implies $\mathbf{v}_i^*AA^*\mathbf{v}_i = \lambda_i \mathbf{v}_i^*\mathbf{v}_i$, or $(A^*\mathbf{v}_i)^*A^*\mathbf{v}_i = \lambda_i \mathbf{v}_i^*\mathbf{v}_i$, so $\|A^*\mathbf{v}_i\|^2 = \lambda_i \|\mathbf{v}_i\|^2$. Thus $\lambda_i \geq 0$. Similarly, the eigenvalues of A^*A are nonnegative.

In the proof of the next result, $\hat{\sigma}(M) = \sigma(M)\backslash\{0\}$, the set of all nonzero eigenvalues of matrix M. (See Section 4.4 for the basic notation.)

Corollary 8.2.3

Let A be an $m \times n$ matrix. Then the matrices AA^* and A^*A have the same nonzero eigenvalues.

Proof The strategy in this proof is to show that both $\hat{\sigma}(A^*A) \subseteq \hat{\sigma}(AA^*)$ and $\hat{\sigma}(AA^*) \subseteq \hat{\sigma}(A^*A)$ are true.

Corollary 8.2.2 states that AA^* and A^*A have nonnegative eigenvalues. Suppose that $\lambda_1, \lambda_2, \ldots, \lambda_n$ are the nonnegative eigenvalues of A^*A, with corresponding eigenvectors, $\mathbf{v}_1, \mathbf{v}_2, \ldots, \mathbf{v}_n$. Furthermore, we can assume that $\{\mathbf{v}_1, \mathbf{v}_2, \ldots, \mathbf{v}_n\}$ is an orthonormal basis for $\mathbb{C}^n$. Then $\langle A^*A\mathbf{v}_i, \mathbf{v}_j \rangle = \langle \lambda_i \mathbf{v}_i, \mathbf{v}_j \rangle = \lambda_i \langle \mathbf{v}_i, \mathbf{v}_j \rangle = \lambda_i$ if $i = j$ and 0 if $i \neq j$, $1 \leq i, j \leq n$. On the other hand, $\langle A^*A\mathbf{v}_i, \mathbf{v}_j \rangle = \langle A\mathbf{v}_i, A\mathbf{v}_j \rangle$, and a comparison with the previous calculation shows that $\|A\mathbf{v}_i\|^2 = \langle A\mathbf{v}_i, A\mathbf{v}_i \rangle = \lambda_i$, $i = 1, 2, \ldots, n$. Thus $A\mathbf{v}_i = 0$ if and only if $\lambda_i = 0$ ($1 \leq i \leq n$).

Because $AA^*(A\mathbf{v}_i) = A(A^*A\mathbf{v}_i) = A(\lambda_i \mathbf{v}_i) = \lambda_i(A\mathbf{v}_i)$, $1 \leq i \leq n$, the conclusion of the last paragraph shows that for $\lambda_i \neq 0$, the vector $A\mathbf{v}_i$ is an eigenvector of AA^*. Hence, if a nonzero eigenvalue λ_i of A^*A is associated with the eigenvector $\mathbf{v}_i$, then $\lambda_i \in \hat{\sigma}(AA^*)$ and is associated with the eigenvector $A\mathbf{v}_i$. In particular, we have shown that $\hat{\sigma}(A^*A) \subseteq \hat{\sigma}(AA^*)$.

We can prove $\hat{\sigma}(AA^*) \subseteq \hat{\sigma}(A^*A)$ by exchanging the roles of A and A^* in the argument we have just completed. Then $\hat{\sigma}(A^*A) \subseteq \hat{\sigma}(AA^*)$ and $\hat{\sigma}(AA^*) \subseteq \hat{\sigma}(A^*A)$ imply that $\hat{\sigma}(AA^*) = \hat{\sigma}(A^*A)$—that is, AA^* and A^*A have the same nonzero eigenvalues.

Example 8.2.2: Eigenvalues of AA^* and A^*A

Let $A = [1 \quad 2 \quad 3]$. Then $A^* = A^\mathsf{T} = \begin{bmatrix} 1 \\ 2 \\ 3 \end{bmatrix}$ and $AA^* = [14]$,

with the eigenvalue equal to 14. It is easy to calculate

$A^*A = \begin{bmatrix} 1 & 2 & 3 \\ 2 & 4 & 6 \\ 3 & 6 & 9 \end{bmatrix}$, with eigenvalues $\lambda_1 = \lambda_2 = 0$ and $\lambda_3 = 14$.

Thus AA^* and A^*A have the same nonzero eigenvalues.

Similarly, if $A = \begin{bmatrix} i & 1 \\ -2 & 1+i \end{bmatrix}$ and $A^* = \begin{bmatrix} -i & -2 \\ 1 & 1-i \end{bmatrix}$, then

$A^*A = \begin{bmatrix} 5 & -2-3i \\ -2+3i & 3 \end{bmatrix}$, with eigenvalues $4 \pm \sqrt{14}$, and

$AA^* = \begin{bmatrix} 2 & 1-3i \\ 1+3i & 6 \end{bmatrix}$, with eigenvalues $4 \pm \sqrt{14}$.

Theorem 8.2.2 should be compared with Theorem 4.5.3 and Corollary 4.5.1, which dealt with real matrices. In the situation where the matrix A is real and symmetric, Theorem 8.2.2 is sometimes referred to as a form of the **Principal Axes Theorem**, although we will reserve this name for a theorem we will see later in this chapter when we explore quadratic forms.

Exercises 8.2

A.

1. Identify each of the following matrices as Hermitian, skew-Hermitian, symmetric, skew-symmetric, or none of these.

   (a) $\begin{bmatrix} 1 & 1+i \\ 1+i & 2 \end{bmatrix}$ (b) $\begin{bmatrix} 1 & 1 \\ -1 & 0 \end{bmatrix}$ (c) $\begin{bmatrix} 0 & 1 \\ -1 & 0 \end{bmatrix}$

   (d) $\begin{bmatrix} 4 & 3-5i \\ 3+5i & -7 \end{bmatrix}$ (e) $\begin{bmatrix} 0 & 5-7i \\ 4+3i & 0 \end{bmatrix}$

   (f) $\begin{bmatrix} 0 & 1+i \\ -1+i & i \end{bmatrix}$ (g) $\begin{bmatrix} 3 & 1-2i & 4+7i \\ 1+2i & -4 & -2i \\ 4-7i & 2i & 2 \end{bmatrix}$

   (h) $\begin{bmatrix} 6 & 2 & -5 \\ 2 & -1 & 3 \\ -5 & 3 & 2 \end{bmatrix}$

2. A matrix A is said to be *involutory* (or an *involution*) if $A^2 = I$. Verify that the following *Pauli spin matrices* (which are

important in quantum mechanics) are involutory, unitary (or orthogonal), and Hermitian (or symmetric):

$$\sigma_x = \begin{bmatrix} 0 & 1 \\ 1 & 0 \end{bmatrix}, \quad \sigma_y = \begin{bmatrix} 0 & -i \\ i & 0 \end{bmatrix}, \quad \sigma_z = \begin{bmatrix} 1 & 0 \\ 0 & -1 \end{bmatrix}.$$

3. The matrices

$$D_1 = \begin{bmatrix} 0 & 0 & 0 & 1 \\ 0 & 0 & 1 & 0 \\ 0 & 1 & 0 & 0 \\ 1 & 0 & 0 & 0 \end{bmatrix} \quad D_2 = \begin{bmatrix} 0 & 0 & 0 & i \\ 0 & 0 & -i & 0 \\ 0 & i & 0 & 0 \\ -i & 0 & 0 & 0 \end{bmatrix}$$

$$D_3 = \begin{bmatrix} 0 & 0 & 1 & 0 \\ 0 & 0 & 0 & -1 \\ 1 & 0 & 0 & 0 \\ 0 & -1 & 0 & 0 \end{bmatrix} \quad D_4 = \begin{bmatrix} 1 & 0 & 0 & 0 \\ 0 & 1 & 0 & 0 \\ 0 & 0 & -1 & 0 \\ 0 & 0 & 0 & -1 \end{bmatrix}$$

occur in quantum mechanics and are called the *Dirac spin matrices*.

(a) Show that D_i is Hermitian, $i = 1, 2, 3, 4$.

(b) Show that $D_i D_j + D_j D_i = 2I_4$ if $i = j$ and O_4 if $i \neq j$ for $i, j = 1, 2, 3, 4$.

4. Let A be a Hermitian matrix. For what scalars k is kA Hermitian?

5. Let A be a skew-Hermitian matrix. For what scalars k is kA skew-Hermitian?

B.

1. Show that the determinant of any Hermitian matrix is real.

2. Describe all 3×3 matrices that are simultaneously Hermitian, unitary, and diagonal. How many are there?

3. If A and B are Hermitian matrices, what assumptions, if any, are required to show that AB is Hermitian?

4. If $p(x)$ is a polynomial with real coefficients and A is Hermitian, show that $p(A)$ is Hermitian.

5. Let A be a skew-Hermitian matrix. Prove that
(a) $\mathbf{x}^*A\mathbf{x}$ is purely imaginary for any complex vector $\mathbf{x} \neq \mathbf{0}$.
(b) every eigenvalue of A purely imaginary.
(c) the eigenvectors of A associated with distinct eigenvalues are orthogonal.

6. Show that any complex matrix A can be written uniquely as $B + C$, where B is Hermitian and C is skew-Hermitian. (See Exercises 3.1, C1 for the corresponding result for real matrices.)

7. Prove that if A is skew-Hermitian, then $I - A$ and $I + A$ are nonsingular.

8. Let A be a real $n \times n$ skew-symmetric matrix.
   (a) If n is odd, show that $\det(A) = 0$.
   (b) If n is even, show that $\det(A) \geq 0$.
   (c) For any positive integer n, show that $\det(I + A) \geq 1$.

9. Show that $\text{trace}(A^*A) = 0$ if and only if the eigenvalues of A^*A are zero.

10. If A is an $n \times n$ matrix, show that $\text{trace}(A^*A) = 0$ implies $A = O_n$. [*Hint*: Use the result of the previous exercise and the fact that A^*A is Hermitian.]

11. Construct a real symmetric matrix with eigenvectors
$$\begin{bmatrix} 1 \\ -1 \\ -1 \end{bmatrix}, \begin{bmatrix} 2 \\ 1 \\ 1 \end{bmatrix}, \begin{bmatrix} 0 \\ -1 \\ 1 \end{bmatrix} \text{ and corresponding eigenvalues}$$

   (a) $0, 1, 2$ (b) $1, 1, 2$ (c) $\lambda_1 = \lambda_2 = \lambda_3$
   (d) $\lambda, \lambda + 1, \lambda - 1$

12. Find an orthogonal matrix U and a diagonal matrix D such that $A = U^T D U$, if
$$A = \begin{bmatrix} 1 & 0 & -4 \\ 0 & 5 & 4 \\ -4 & 4 & 3 \end{bmatrix}.$$

13. Show that for any $A \in M_{mn}(\mathbb{R})$, nullspace $A^*A = $ nullspace A and range $A^*A = $ range A^*.

14. Consider the spectral decomposition of an $n \times n$ Hermitian matrix A,
$$A = \lambda_1 \mathbf{u}_1 \mathbf{u}_1^* + \lambda_2 \mathbf{u}_2 \mathbf{u}_2^* + \cdots + \lambda_n \mathbf{u}_n \mathbf{u}_n^*,$$

   where $\mathbf{u}_i$, $1 \leq i \leq n$, is an orthonormal eigenvector of A. Show that for each $\mathbf{x} \in \mathbb{R}^n$, $(\mathbf{u}_i \mathbf{u}_i^*)\mathbf{x}$ is the orthogonal projection of $\mathbf{x}$ onto the subspace spanned by $\mathbf{u}_i$.

C.

1. Suppose that $\{\mathbf{u}_1, \mathbf{u}_2, \ldots, \mathbf{u}_n\}$ is an orthonormal set of vectors in $\mathbb{R}^n$ and that the scalars, $\lambda_1, \lambda_2, \ldots, \lambda_n$, are real.
   (a) Show that there exists a matrix H such that $H\mathbf{u}_i = \lambda_i \mathbf{u}_i$ for each i.

(b) Show that every matrix H with the property $H\mathbf{u}_i = \lambda_i \mathbf{u}_i$ is Hermitian. [*Hint*: Consider $U = [\,\mathbf{u}_1 \quad \mathbf{u}_2 \quad \ldots \quad \mathbf{u}_n\,]$ and $D = \text{diag}(\lambda_1, \lambda_2, \ldots, \lambda_n).]$

A Hermitian matrix, A, is said to be **positive definite** *if all its eigenvalues are positive. A Hermitian matrix, A, is said to be* **positive semidefinite** *if all its eigenvalues are nonnegative. We write $A > 0$ and $A \geq 0$ for positive definite and positive semidefinite matrices, respectively. Use these concepts for Exercises 2 through 5.*

2. Given a Hermitian matrix $A \in M_n(\mathbb{R})$, prove that $A > 0$ if and only if $\langle A\mathbf{x}, \mathbf{x} \rangle > 0$ for all nonzero $\mathbf{x} \in \mathbb{C}^n$.

3. Deduce from Exercise B13 that the matrix A^*A, where $A \in M_{mn}(\mathbb{R})$ and $m \geq n$, is positive definite if and only if $\text{rank}(A) = n$.

4. Show that if A is positive semidefinite and $\text{rank}(A) = r$, then A has exactly r positive eigenvalues.

5. Let A be Hermitian and let B be a positive definite matrix of the same size. Show that AB has only real eigenvalues.

8.3 Normal Matrices

Corollary 8.2.1 highlights an important property of any Hermitian matrix A: There is an orthonormal basis for $\mathbb{C}^n$, which consists entirely of eigenvectors of A—thus, A can be diagonalized with a unitary matrix (Theorem 4.5.3 for a complex matrix). However, there are non-Hermitian matrices that also have this important characteristic, for example, skew-Hermitian (and therefore skew-symmetric) matrices. We want to describe *all* matrices that have this useful property.

Now, suppose that there is an orthonormal basis for $\mathbb{C}^n$, which consists entirely of eigenvectors of a matrix A. Then there is a unitary matrix, U, such that $U^*AU = D$ is diagonal: Let U be the matrix whose columns are orthonormal basis vectors. Then, $A = UDU^*$ because $U^* = U^{-1}$. Now we perform the calculations

$$AA^* = (UDU^*)(UD^*U^*) = UDD^*U^*,$$

and

$$A^*A = (UD^*U^*)(UDU^*) = UD^*DU^*.$$

Because diagonal matrices commute, we have $DD^* = D^*D$. It follows that $AA^* = A^*A$. Hence, a *necessary* condition that there be an orthonormal basis for $\mathbb{C}^n$, which consists entirely of eigenvectors of a matrix A, is that A and A^* commute. Matrices that behave this way are important.

Definition 8.3.1

(a) A matrix A is said to be **normal** if $AA^* = A^*A$. If A is a real matrix, then A is normal if $AA^\mathrm{T} = A^\mathrm{T}A$.

(b) A linear operator on a finite-dimensional inner product space is called **normal** if $TT^* = T^*T$.

It follows immediately from Theorem 8.1.3 that an operator T is normal if and only if $[T]_B$ is normal, where B is an orthonormal basis.

Our next objective is to show that **if A is normal, then there is an orthonormal basis for $\mathbb{C}^n$, which consists entirely of eigenvectors of A.**

Suppose that A is normal. Schur's Theorem (Theorem 7.6.1) states that there is a unitary matrix U such that $U^*AU = T$ is upper triangular. Furthermore, T is normal:

$$T^*T = (U^*AU)^*(U^*AU) = (U^*A^*U)(U^*AU) = U^*(A^*A)U,$$

and

$$TT^* = (U^*AU)(U^*AU)^* = (U^*AU)(U^*A^*U) = U^*(AA^*)U.$$

But $A^*A = AA^*$, so $T^*T = TT^*$—that is, T is normal.

Now let $T = [t_{ij}]$ and $T^* = [\hat{t}_{ij}]$, where $\hat{t}_{ij} = \bar{t}_{ji}$. Because T^*T and TT^* are equal, we can, in particular, equate the first row, first column entries of T^*T and TT^*. If we let $T^*T = [a_{ij}]$ and $TT^* = [b_{ij}]$, then

$$a_{11} = \sum_{k=1}^{n} \hat{t}_{1k}t_{k1} = \sum_{k=1}^{n} \bar{t}_{k1}t_{k1} = \sum_{k=1}^{n} |t_{k1}|^2 = |t_{11}|^2,$$

because T^*T upper triangular implies $t_{k1} = 0$ for $k \geq 2$. Also,

$$b_{11} = \sum_{k=1}^{n} t_{1k}\hat{t}_{k1} = \sum_{k=1}^{n} t_{1k}\bar{t}_{1k} = \sum_{k=1}^{n} |t_{1k}|^2.$$

Equating entries gives us the equation

$$|t_{11}|^2 = |t_{11}|^2 + |t_{12}|^2 + \cdots + |t_{1n}|^2,$$

which implies that $t_{12} = t_{13} = \cdots = t_{1n} = 0$. Similarly, by looking at the $(2,2), (3,3), \ldots, (n,n)$ entries of T^*T and TT^*, we see that all the other off-diagonal entries of T vanish. Thus, T is actually a diagonal matrix.

Finally, because $AU = UT$ and T is diagonal, the columns of U are eigenvectors of A: If $T = \mathrm{diag}(d_1, d_2, \ldots, d_n)$, then $AU = [A\mathbf{u}_1 \ \ A\mathbf{u}_2 \ \ \ldots \ \ A\mathbf{u}_n]$ and $UT = [d_1\mathbf{u}_1 \ \ d_2\mathbf{u}_2 \ \ \ldots \ \ d_n\mathbf{u}_n]$, so $A\mathbf{u}_i = d_i\mathbf{u}_i$ for $i = 1, 2, \ldots, n$. Furthermore, the columns form an orthonormal basis of $\mathbb{C}^n$ because U is unitary. This concludes the proof:

There is an orthonormal basis for $\mathbb{C}^n$, which consists entirely of eigenvectors of a matrix A if and only if A is normal.

We note that the set of all normal matrices includes Hermitian and therefore real symmetric matrices; skew-Hermitian and so real skew-symmetric matrices; unitary and, therefore, orthogonal matrices. It follows that *all these types of matrices are diagonalizable.*

Example 8.3.1: A Normal Matrix That Is Not Hermitian, Skew-Hermitian, or Unitary

Consider the matrix $A = \begin{bmatrix} 1 & 3+4i \\ 4+3i & 1 \end{bmatrix}$. We see that

$A^* = \begin{bmatrix} 1 & 4-3i \\ 3-4i & 1 \end{bmatrix}$. Then $A^*A = \begin{bmatrix} 1 & 3+4i \\ 4+3i & 1 \end{bmatrix}$

$\begin{bmatrix} 1 & 4-3i \\ 3-4i & 1 \end{bmatrix} = \begin{bmatrix} 26 & 7+i \\ 7-i & 26 \end{bmatrix}$ and $AA^* = \begin{bmatrix} 1 & 4-3i \\ 3-4i & 1 \end{bmatrix}$

$\begin{bmatrix} 1 & 3+4i \\ 4+3i & 1 \end{bmatrix} = \begin{bmatrix} 26 & 7+i \\ 7-i & 26 \end{bmatrix}$, showing that A is normal.

However, $A \neq A^*$ and $A^* \neq -A$, so A is neither Hermitian nor

skew-Hermitian. Finally, because $A^*A = \begin{bmatrix} 26 & 7+i \\ 7-i & 26 \end{bmatrix} \neq I_2$,

we see that A is not unitary.

We can summarize the information in this section in the next two theorems.

Theorem 8.3.1(a): A Spectral Theorem for Normal Matrices

If $A \in M_n(\mathbb{C})$, then there exists a unitary $n \times n$ matrix, U, such that $U^*AU = D$, a diagonal matrix, if and only if A is normal.

Example 8.3.2: Diagonalizing a Normal Matrix

The matrix $A = \begin{bmatrix} 1 & i \\ i & 1 \end{bmatrix}$, like the matrix in Example 8.3.1, is not Hermitian, skew-Hermitian, or unitary, but it is unitarily similar to a diagonal matrix by Theorem 8.3.1(a):

The eigenvalues of A are $\lambda_1 = 1 + i$ and $\lambda_2 = 1 - i$, with corresponding eigenvectors, $v_1 = \begin{bmatrix} 1 \\ 1 \end{bmatrix}$ and $v_2 = \begin{bmatrix} -1 \\ 1 \end{bmatrix}$, respectively. Forming the matrix $[v_1 \quad v_2]$, and then orthonormalizing the columns, we get $U = \frac{1}{\sqrt{2}} \begin{bmatrix} 1 & -1 \\ 1 & 1 \end{bmatrix}$, a unitary

matrix: $U^* = \frac{1}{\sqrt{2}} \begin{bmatrix} 1 & 1 \\ -1 & 1 \end{bmatrix} = U^{-1}$. Then

$$U^*AU = \frac{1}{\sqrt{2}} \begin{bmatrix} 1 & 1 \\ -1 & 1 \end{bmatrix} \begin{bmatrix} 1 & i \\ i & 1 \end{bmatrix} \frac{1}{\sqrt{2}} \begin{bmatrix} 1 & -1 \\ 1 & 1 \end{bmatrix}$$

$$= \begin{bmatrix} 1+i & 0 \\ 0 & 1-i \end{bmatrix}.$$

We notice that the diagonal elements of this last matrix are the eigenvalues of A.

In the spirit of the spectral decomposition of a Hermitian matrix, expression (**S**) in Section 8.2, we have the following alternative form of a spectral theorem for normal matrices. (Compare this with Lemma 4.8.1.)

Theorem 8.3.1(b): A Spectral Theorem for Normal Matrices

Let $\lambda_1, \lambda_2, \ldots, \lambda_r$ be the distinct eigenvalues of $A \in M_n(\mathbb{C})$, with multi-plicities $m_1, m_2, \ldots, m_r$. Then, A is normal if and only if there exist r pairwise orthogonal, orthogonal projections, $P_1, P_2, \ldots, P_r$, such that $\sum_{i=1}^r P_i = I_n$, rank$(P_i) = m_i$, and $A = \sum_{i=1}^r \lambda_i P_i$. [Two orthogonal projections, P and Q, are *pairwise orthogonal* if and only if range P and range Q are orthogonal subspaces.]

Other properties of normal operators and normal matrices can be found in the exercises.

Exercises 8.3

A.

1. Determine which of the following matrices, if any, is (are) normal.

(a) $\begin{bmatrix} 1 & i \\ i & 1 \end{bmatrix}$ (b) $\begin{bmatrix} 6 & 2i \\ 2i & 1 \end{bmatrix}$ (c) $\begin{bmatrix} 1 & i \\ 0 & 1 \end{bmatrix}$

(d) $\begin{bmatrix} i & -i & 1 \\ 1 & i & -i \\ -i & 1 & i \end{bmatrix}$ (e) $\begin{bmatrix} 1 & 2 & 0 \\ 0 & 1 & 2 \\ 2 & 0 & 1 \end{bmatrix}$

(f) $\begin{bmatrix} 2-i & -1 & 0 \\ -1 & 1-i & 1 \\ 0 & 1 & 2-i \end{bmatrix}$ (g) $\begin{bmatrix} 1 & 1 & 1 & 1 \\ 1 & 1 & -1 & -1 \\ -1 & 1 & -1 & 1 \\ -1 & 1 & 1 & -1 \end{bmatrix}$

2. Diagonalize the normal matrix $A = \begin{bmatrix} 2 & i \\ i & 2 \end{bmatrix}$.

3. Let $A \in M_n(\mathbb{R})$. Show that if $A + A^*$ and $A - A^*$ commute, then A is normal.

4. Show that A^*A and AA^* are normal matrices for every square matrix A.

5. Show that $A \in M_n(\mathbb{R})$ is normal if and only if every matrix that is unitarily similar to A is normal.

6. Prove that $A = B + iC$, where B and C are real matrices, is normal if and only if $BC = CB$.

7. Prove that an idempotent operator, P (i.e., $P^2 = P$), is normal if and only if $P = P^*$.

8. Show that a normal matrix is skew-Hermitian if and only if its eigenvalues are purely imaginary—numbers of the form ci, where c is a real number and $i = \sqrt{-1}$.

9. Show that A is normal if and only if every eigenvector of A is also an eigenvector of A^*.

10. Show that a complex $n \times n$ matrix, A, is normal if and only if A commutes with A^*A.

11. Show that if A is normal, then the eigenvectors of A corresponding to distinct eigenvalues are orthogonal.

12. Show that a triangular matrix is normal if and only if it is diagonal.

13. If A is a normal matrix, prove that the following matrices are also normal:
    (a) kA, where k is a scalar
    (b) A^m, where m is a positive integer
    (c) A^*
    (d) U^*AU, where U is unitary

B.

1. Show that if an $n \times n$ matrix, A, is normal, then so is $A - \lambda I_n$, where λ is any scalar.

2. If $p(x)$ is a polynomial with real coefficients and A is normal, show that $p(A)$ is normal.

3. Let $A = B + iC$, where B and C are real matrices. Show that if A is normal, then so is $\begin{bmatrix} B & -C \\ C & B \end{bmatrix}$. (See Exercise A6.)

4. Show that a matrix $A \in M_n(\mathbb{R})$ is normal if and only if $\|A\mathbf{v}\| = \|A^*\mathbf{v}\|$ for all $\mathbf{v} \in \mathbb{R}^n$, where $\|\cdot\|$ denotes the standard norm in $\mathbb{R}^n$.

5. Show that if A is normal with real eigenvalues, then A is Hermitian.

6. Show that every 2×2 *real* normal matrix is symmetric or skew-symmetric. (Compare this statement with Example 8.2.5.)

7. Show that a normal matrix is unitary if and only if every eigenvalue has an absolute value equal to 1.

8. Prove or find a counterexample: Every complex symmetric matrix is normal.

9. Show that the matrix representation of the operator $T: \mathbb{C}^2 \to \mathbb{C}^2$ defined by $T\left(\begin{bmatrix} \alpha \\ \beta \end{bmatrix}\right) = \begin{bmatrix} 2\alpha + i\beta \\ \alpha + 2\beta \end{bmatrix}$ is normal.

10. Show that $A \in M_n(\mathbb{R})$ is normal if and only if $\text{trace}(AA^*)^2 = \text{trace}[A^2(A^2)^*]$.

11. Show that the following statements are equivalent.
    (1) A is normal.
    (2) A commutes with $A + A^*$.
    (3) A commutes with $A - A^*$.
    (4) A commutes with $AA^* - A^*A$.

12. Show that if $A = [a_{ij}] \in M_n(\mathbb{R})$ has eigenvalues $\lambda_1, \lambda_2, \ldots, \lambda_n$ (counting multiplicities), then the following statements are equivalent.
    (1) A is normal.
    (2) $\sum_{i,j=1}^{n} |a_{ij}|^2 = \sum_{i=1}^{n} |\lambda_i|^2$.

13. Let A be normal with the spectral representation $A = \sum_{i=1}^{r} \lambda_i P_i$ (as in Theorem 8.3.1(b)). Show that $A^2 = \sum_{i=1}^{r} \lambda_i^2 P_i$.

C.

1. If $\mathbf{u}, \mathbf{v} \in \mathbb{R}^n$, under what circumstances is the rank-one matrix $\mathbf{u}\mathbf{v}^*$ normal?

2. Show that if A is a normal matrix, then A can be expressed as $A = B + iC$, where B and C are Hermitian matrices and $BC = CB$. [*Hint*: $(A + A^*)/2$ is Hermitian.]

3. If $\lambda_1, \lambda_2, \ldots, \lambda_r$ are the distinct eigenvalues of a normal matrix, $A \in M_n(\mathbb{R})$, then the eigenspaces, $S_{\lambda_i}(A)$ (see Section 4.4), are pairwise orthogonal, and

$$\mathbb{R}^n = S_{\lambda_1}(A) \oplus S_{\lambda_2}(A) \oplus \cdots \oplus S_{\lambda_r}(A).$$

*8.4 Quadratic Forms

In multivariable calculus, the **Taylor expansion** of a function f of two variables at a point (a, b), or in a neighborhood of (a, b), is expressed in terms of the partial derivatives of f by $f(a + h, b + k) = f(a, b) + hf_x(a, b) + kf_y(a, b) + \frac{1}{2}[h^2 f_{xx}(a, b) + 2hk f_{xy}(a, b) + k^2 f_{yy}(a, b)] + \cdots$, provided that $f_{xy} = \frac{\partial}{\partial y}\left(\frac{\partial f}{\partial x}\right) = \frac{\partial}{\partial x}\left(\frac{\partial f}{\partial y}\right) = f_{yx}$. In this expansion, h and k are real numbers. In determining the points (if any) at which the function f attains local maximum or local minimum values, it is important to locate the function's **critical points**—points $(\hat{a}, \hat{b})$ such that $f_x(\hat{a}, \hat{b}) = 0 = f_y(\hat{a}, \hat{b})$. As we will see, the expression

$$h^2 f_{xx}(\hat{a}, \hat{b}) + 2hk f_{xy}(\hat{a}, \hat{b}) + k^2 f_{yy}(\hat{a}, \hat{b}) \tag{*}$$

plays a key role in determining the nature of the critical points. For now, we note that the expression (*) can be written in the matrix form:

$$\begin{bmatrix} h & k \end{bmatrix} \begin{bmatrix} f_{xx}(\hat{a}, \hat{b}) & f_{xy}(\hat{a}, \hat{b}) \\ f_{xy}(\hat{a}, \hat{b}) & f_{yy}(\hat{a}, \hat{b}) \end{bmatrix} \begin{bmatrix} h \\ k \end{bmatrix} = \mathbf{x}^\mathrm{T} A \mathbf{x},$$

with $\mathbf{x} = \begin{bmatrix} h \\ k \end{bmatrix} \in \mathbb{R}^2$ and A symmetric. This important expression is an example of a **quadratic form** (see Example 3.2.3).

Definition 8.4.1

An expression of the form

$$Q(\mathbf{x}) = \mathbf{x}^\mathrm{T} A \mathbf{x} = \begin{bmatrix} x_1 & x_2 & \cdots & x_n \end{bmatrix} \begin{bmatrix} a_{11} & a_{12} & \cdots & a_{1n} \\ a_{21} & a_{22} & \cdots & a_{2n} \\ \vdots & \vdots & \vdots & \vdots \\ a_{n1} & a_{n2} & \cdots & a_{nn} \end{bmatrix} \begin{bmatrix} x_1 \\ x_2 \\ \vdots \\ x_n \end{bmatrix}$$

$$= \sum_{i=1}^{n} \sum_{j=1}^{n} a_{ij} x_i x_j,$$

where $A = [a_{ij}]$ is an $n \times n$ symmetric matrix and $\mathbf{x} \in \mathbb{R}^n$, is called a **(real) quadratic form** (in $\mathbb{R}^n$, or in n variables), and A is called the **matrix of the quadratic form**. (Note that Q is a scalar-valued function, $Q: \mathbb{R}^n \to \mathbb{R}$.)

Some treatments of quadratic forms state the definition without requiring that the matrix of the quadratic form be symmetric. In this case, there will be many ways to choose A so that $Q(\mathbf{x}) = \mathbf{x}^T A \mathbf{x}$. However, it turns out that there will always be a *unique* symmetric matrix that works (see Exercise B1), hence we can say "the" matrix of the quadratic form.

Example 8.4.1: Matrices of Quadratic Forms

(a) The quadratic form $(2x + 3y)^2 = 4x^2 + 12xy + 9y^2$ can be written as $\begin{bmatrix} x & y \end{bmatrix} \begin{bmatrix} 4 & 6 \\ 6 & 9 \end{bmatrix} \begin{bmatrix} x \\ y \end{bmatrix}$, so the matrix of the quadratic form is $A = \begin{bmatrix} 4 & 6 \\ 6 & 9 \end{bmatrix}$.

(b) The quadratic form $x^2 + 2z^2 - xy$ can be written as
$$\begin{bmatrix} x & y & z \end{bmatrix} \begin{bmatrix} 1 & -1/2 & 0 \\ -1/2 & 0 & 0 \\ 0 & 0 & 2 \end{bmatrix} \begin{bmatrix} x \\ y \\ z \end{bmatrix} = \mathbf{x}^T A \mathbf{x}, \qquad \text{where}$$
$\mathbf{x} \in \mathbb{R}^3$.

(c) In classical mechanics, the kinetic energy of a particle of mass m having n degrees of freedom is given by

$$KE = \frac{m}{2} \sum_{i=1}^{n} \left(\frac{dx_i}{dt} \right)^2,$$

where x_i are the position coordinates of the particle. The formula can be expressed as

$$\begin{bmatrix} \frac{dx_1}{dt} & \frac{dx_2}{dt} & \cdots & \frac{dx_n}{dt} \end{bmatrix} \begin{bmatrix} m/2 & 0 & \cdots & 0 \\ 0 & m/2 & \cdots & 0 \\ \vdots & \vdots & \cdots & \vdots \\ 0 & 0 & \cdots & m/2 \end{bmatrix} \begin{bmatrix} \frac{dx_1}{dt} \\ \frac{dx_2}{dt} \\ \vdots \\ \frac{dx_n}{dt} \end{bmatrix},$$

where the matrix of the quadratic form is diagonal.

(d) Suppose we want to analyze data from an experiment or a survey. If n observations are represented by the vector $\mathbf{x} = \begin{bmatrix} x_1 & x_2 & \cdots & x_n \end{bmatrix}^T$, then the *mean* of the observations is $\bar{\mathbf{x}} = (x_1 + x_2 + \cdots + x_n)/n$ and the *sample variance* is

$$SV = (x_1 - \bar{\mathbf{x}})^2 + (x_2 - \bar{\mathbf{x}})^2 + \cdots + (x_n - \bar{\mathbf{x}})^2 = \sum_{i=1}^{n} x_i^2 - n\bar{\mathbf{x}}^2.$$

In matrix notation,

$$\sum_{i=1}^{n} x_i^2 = \begin{bmatrix} x_1 & x_2 & \cdots & x_n \end{bmatrix} \begin{bmatrix} x_1 \\ x_2 \\ \vdots \\ x_n \end{bmatrix} = \mathbf{x}^T \mathbf{x}$$

and

$$n\bar{\mathbf{x}}^2 = \bar{\mathbf{x}} \sum_{i=1}^{n} x_i = \begin{bmatrix} \bar{\mathbf{x}} & \bar{\mathbf{x}} & \cdots & \bar{\mathbf{x}} \end{bmatrix} \begin{bmatrix} x_1 \\ x_2 \\ \vdots \\ x_n \end{bmatrix}$$

$$= \mathbf{x}^T \begin{bmatrix} 1/n & 1/n & \cdots & 1/n \\ 1/n & 1/n & \cdots & 1/n \\ \vdots & \vdots & \cdots & \vdots \\ 1/n & 1/n & \cdots & 1/n \end{bmatrix} \mathbf{x}.$$

Hence, $SV = \mathbf{x}^T\mathbf{x} - \mathbf{x}^T U_n\mathbf{x} = \mathbf{x}^T(I - U_n)\mathbf{x}$, where U_n is the $n \times n$ matrix, every entry of which equals $1/n$, and where $(I - U_n)$ is an $n \times n$ symmetric matrix, with every diagonal element equal to $(n-1)/n$ and all off-diagonal entries equal to $-1/n$.

Quadratic forms were discussed in a limited way in Section 3.2, but now we will use our more sophisticated knowledge of matrices to deepen our understanding of this useful topic.

An essential role is played by a real symmetric matrix in our definition of a quadratic form and, accordingly, we can use Theorem 8.2.3 to express an arbitrary quadratic form in a simple way as a sum of squares.* This process is referred to as the **diagonalization** of the quadratic form. In elementary terms, the next theorem is a generalization of the method of completing the square in elementary algebra. Geometrically, it is a counterpart of the spectral theorems we have seen and has applications in statistics and physics. As we will see, the result involves a change of basis, with respect to which a quadratic form has a particularly simple appearance.

Theorem 8.4.1: The Principal Axes Theorem

Let $Q(\mathbf{x}) = \mathbf{x}^T A \mathbf{x}$ be an arbitrary quadratic form. Then, there is a real orthogonal matrix, S, such that $Q(\mathbf{x}) = \lambda_1\tilde{x}_1^2 + \lambda_2\tilde{x}_2^2 + \cdots + \lambda_n\tilde{x}_n^2$, where $\tilde{x}_1, \tilde{x}_2, \ldots, \tilde{x}_n$ are the entries of $\tilde{\mathbf{x}} = S^T\mathbf{x}$ and $\lambda_1, \lambda_2, \ldots, \lambda_n$ are the eigenvalues of the matrix A. (Note that there are no "cross terms" in the new version of $Q(\mathbf{x})$.)

Proof By Theorem 8.2.3, there is a real orthogonal matrix, S, such that $S^T A S = D$ is diagonal, say with diagonal entries $\lambda_1, \lambda_2, \ldots, \lambda_n$. Define $\tilde{\mathbf{x}} = S^T\mathbf{x}$, so $\mathbf{x} = S\tilde{\mathbf{x}}$. Substituting for $\mathbf{x}$, we find that

$$Q(\mathbf{x}) = \mathbf{x}^T A \mathbf{x} = (S\tilde{\mathbf{x}})^T A (S\tilde{\mathbf{x}}) = (\tilde{\mathbf{x}})^T (S^T A S)\tilde{\mathbf{x}} = (\tilde{\mathbf{x}})^T D\tilde{\mathbf{x}}.$$

---

* A famous number theoretical result of Lagrange (1736–1813) is that *every positive integer n can be expressed as the sum of four squares*, $n = x_1^2 + x_2^2 + x_3^2 + x_4^2$, *where x_i are nonnegative integers*. See, for example, I. Niven, H.S. Zuckerman, and H.L. Montgomery, *An Introduction to the Theory of Numbers*, 5th edn. (New York: Wiley, 1991).

Multiplying out the last matrix product $(\tilde{\mathbf{x}})^T D \tilde{\mathbf{x}}$, we find that

$$Q(\mathbf{x}) = \lambda_1 \tilde{x}_1^2 + \lambda_2 \tilde{x}_2^2 + \cdots + \lambda_n \tilde{x}_n^2.$$

In Example 3.2.3, we saw that a quadratic form in two variables may be a component of the equation of a conic section. Diagonalizing such a quadratic form makes it easier for us to recognize the specific conic section.

Example 8.4.2: Diagonalizing a Quadratic Form

Consider the second degree equation, $7x^2 + 2\sqrt{3}xy + 5y^2 = 1$. This second degree equation represents an ellipse with the major axis along the line $y = -\sqrt{3}x$ and the minor axis along $y = (\sqrt{3}/3)x$ (Figure 8.1).
   We can write this equation in the form $[x \ \ y] \begin{bmatrix} 7 & \sqrt{3} \\ \sqrt{3} & 5 \end{bmatrix} \begin{bmatrix} x \\ y \end{bmatrix} = 1$, or $\mathbf{x}^T A \mathbf{x} = 1$. The eigenvalues of A are 4 and 8, so according to Theorem 8.4.1, we can express our original quadratic form as $4\tilde{x}^2 + 8\tilde{y}^2$ and the entire second degree equation as $4\tilde{x}^2 + 8\tilde{y}^2 = 1$. In the standard form, this is $\frac{\tilde{x}^2}{(\sqrt{2})^2} + \frac{\tilde{y}^2}{1^2} = \left(\frac{\sqrt{2}}{4}\right)^2$, the equation of an ellipse centered at the origin, with axes along the standard axes (Figure 8.2).
   To see what has happened, first we note that the eigenvectors of A are $\begin{bmatrix} -\sqrt{3}/3 \\ 1 \end{bmatrix}$ and $\begin{bmatrix} \sqrt{3} \\ 1 \end{bmatrix}$, corresponding to the eigenvalues 4 and 8, respectively. The major and minor axes of the original ellipse lie along these eigenvectors (and their

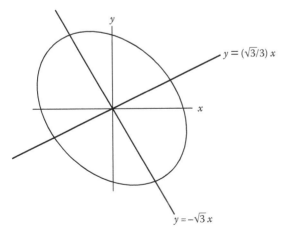

Figure 8.1 The ellipse $7x^2 + 2\sqrt{3}xy + 5y^2 = 1$.

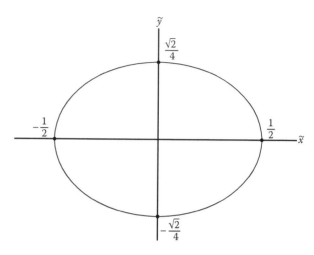

Figure 8.2 The ellipse $4\tilde{x}^2 + 8\tilde{y}^2$.

negatives). Furthermore, if we form the orthogonal matrix, S, with columns from the normalized eigenvectors, we find that A is diagonalized by $S = \frac{1}{2}\begin{bmatrix} -1 & \sqrt{3} \\ \sqrt{3} & 1 \end{bmatrix}$. Now let $\tilde{\mathbf{x}} = \begin{bmatrix} \tilde{x} \\ \tilde{y} \end{bmatrix} = S^{\mathsf{T}}\mathbf{x}$, so that $\mathbf{x} = S\tilde{\mathbf{x}} = \frac{1}{2}\begin{bmatrix} -1 & \sqrt{3} \\ \sqrt{3} & 1 \end{bmatrix}\begin{bmatrix} \tilde{x} \\ \tilde{y} \end{bmatrix} = \begin{bmatrix} \frac{1}{2}(-\tilde{x} + \sqrt{3}\tilde{y}) \\ \frac{1}{2}(\sqrt{3}\tilde{x} + \tilde{y}) \end{bmatrix}$, which is equivalent to

$$\tilde{x} = \tfrac{1}{2}(-\tilde{x} + \sqrt{3}\tilde{y}) \quad \text{and} \quad \tilde{y} = \tfrac{1}{2}(\sqrt{3}\tilde{x} + \tilde{y}).$$

These equations describe a *change of coordinates* resulting from the standard coordinate axes being rotated in a clockwise direction, whereas the transformed second degree equation represents the ellipse after it has had its axes rotated 60° ($\pi/3$ rad) in a counterclockwise direction. This change of coordinates has the effect of eliminating the cross-product term, $2\sqrt{3}xy$, from the original quadratic form.

Before we get back to the maximum/minimum discussion at the beginning of this section, we must distinguish among certain types of quadratic forms.

Definition 8.4.2

A quadratic form $Q : \mathbb{R}^n \to \mathbb{R}$ is called **positive definite** if $Q(\mathbf{x}) > 0$ for all nonzero $\mathbf{x} \in \mathbb{R}^n$ and **negative definite** if $Q(\mathbf{x}) < 0$ for all nonzero $\mathbf{x} \in \mathbb{R}^n$. If the quadratic form takes both positive and negative values, it is called **indefinite**.

By extension, we call a matrix A **positive definite** (**negative definite**) if A is a symmetric matrix such that the quadratic form, $Q(\mathbf{x}) = \mathbf{x}^T A \mathbf{x}$, is positive definite (negative definite). (See the "C" exercises after Section 8.2.)

The theory of multivariable calculus* tells us how to determine the nature of a critical point $(\hat{a}, \hat{b})$ of a function f of two variables. If $f_x(\hat{a}, \hat{b}) = 0 = f_y(\hat{a}, \hat{b})$, then $(\hat{a}, \hat{b})$ yields

(a) A local *minimum* of the function, f, if and only if the quadratic form $h^2 f_{xx}(\hat{a}, \hat{b}) + 2hk f_{xy}(\hat{a}, \hat{b}) + k^2 f_{yy}(\hat{a}, \hat{b})$ is *positive definite*.

(b) A local *maximum* of f if and only if $h^2 f_{xx}(\hat{a}, \hat{b}) + 2hk f_{xy}(\hat{a}, \hat{b}) + k^2 f_{yy}(\hat{a}, \hat{b})$ is *negative definite*.

(c) A *saddle point* (i.e., neither a maximum nor a minimum) if and only if $h^2 f_{xx}(\hat{a}, \hat{b}) + 2hk f_{xy}(\hat{a}, \hat{b}) + k^2 f_{yy}(\hat{a}, \hat{b})$ is *indefinite*.

Of course, it would be helpful if we had criteria for deciding whether a quadratic form is positive definite, negative definite, or indefinite. Here is where orthogonal diagonalization comes to our assistance.

Theorem 8.4.2

A quadratic form $Q: \mathbb{R}^n \to \mathbb{R}$ given by $Q(\mathbf{x}) = \mathbf{x}^T A \mathbf{x}$ is positive definite if and only if all eigenvalues of A are positive, and is negative definite if and only if all eigenvalues of A are negative. The quadratic form is indefinite if and only if A has both positive and negative eigenvalues.

Proof Suppose that in the (principal axes) form

$$Q(\mathbf{x}) = \lambda_1 \tilde{x}_1^2 + \lambda_2 \tilde{x}_2^2 + \cdots + \lambda_n \tilde{x}_n^2,$$

all the eigenvalues, $\lambda_1, \lambda_2, \ldots, \lambda_n$, of A are positive. Clearly, Q is positive definite because $Q(\mathbf{x}) > 0$ for all nonzero $\mathbf{x} \in \mathbb{R}^n$: If $\mathbf{x} \neq \mathbf{0}$, at least one of its coordinates, $\tilde{x}_1, \tilde{x}_2, \ldots, \tilde{x}_n$, is nonzero.

On the other hand, suppose that $Q(\mathbf{x}) > 0$ for all nonzero $\mathbf{x} \in \mathbb{R}^n$. Then, we can calculate the value of Q on the ith vector, $\mathbf{v}_i$, of the orthogonal basis $B: \lambda_i = Q(\mathbf{v}_i) > 0$, where $[\mathbf{v}_i]_B = \mathbf{e}_i$, the ith vector of the standard basis for $\mathbb{R}^n$.

For a negative definite quadratic form, the proof is similar (see Exercise B7).

Now we have all the information needed to tackle certain maximum/minimum problems in the calculus of several variables.

---

* See, for example, Section 3.6 of J.H. Hubbard and B.B. Hubbard, *Vector Calculus, Linear Algebra, and Differential Forms: A Unified Approach*, 2nd edn. (Upper Saddle River, NJ: Prentice Hall, 2002).

Example 8.4.3: Analyzing Critical Points

Suppose that we want to find the local maximum and local minimum values (if any) of $f(x, y) = x^3 + y^3 - 3x - 12y + 20$. First, we calculate $f_x = 3x^2 - 3$ and $f_y = 3y^2 - 12$, so $f_x = 0$ when $x = \pm 1$ and $f_y = 0$ when $y = \pm 2$. Thus, the critical points of f are $(1, 2)$, $(-1, 2)$, $(1, -2)$, and $(-1, -2)$. Next, we find $f_{xx} = 6x$, $f_{xy} = 0$, and $f_{yy} = 6y$.

Now we examine the matrix of the quadratic form

$$\begin{bmatrix} h & k \end{bmatrix} \begin{bmatrix} f_{xx}(\hat{a}, \hat{b}) & f_{xy}(\hat{a}, \hat{b}) \\ f_{xy}(\hat{a}, \hat{b}) & f_{yy}(\hat{a}, \hat{b}) \end{bmatrix} \begin{bmatrix} h \\ k \end{bmatrix}$$

at each of the critical points:

At $(1, 2)$, $A = \begin{bmatrix} f_{xx}(1, 2) & f_{xy}(1, 2) \\ f_{xy}(1, 2) & f_{yy}(1, 2) \end{bmatrix} = \begin{bmatrix} 6 & 0 \\ 0 & 12 \end{bmatrix}$. Because A is diagonal, its eigenvalues are 6 and 12. Hence, by Theorem 8.3.2, Q is positive definite, and $(1, 2)$ yields a *local minimum* value, $f(1, 2) = 2$.

At $(-1, 2)$, $A = \begin{bmatrix} f_{xx}(-1, 2) & f_{xy}(-1, 2) \\ f_{xy}(-1, 2) & f_{yy}(-1, 2) \end{bmatrix} = \begin{bmatrix} -6 & 0 \\ 0 & 12 \end{bmatrix}$. The eigenvalues of A are -6 and 12, so Q is indefinite, that is, $(-1, 2)$ is a *saddle point*.

At $(1, -2)$, $A = \begin{bmatrix} f_{xx}(1, -2) & f_{xy}(1, -2) \\ f_{xy}(1, -2) & f_{yy}(1, -2) \end{bmatrix} = \begin{bmatrix} 6 & 0 \\ 0 & -12 \end{bmatrix}$. Because the eigenvalues are positive and negative, $(1, -2)$ is a *saddle point*.

Finally, at $(-1, -2)$, $A = \begin{bmatrix} f_{xx}(-1, -2) & f_{xy}(-1, -2) \\ f_{xy}(-1, -2) & f_{yy}(-1, -2) \end{bmatrix} = \begin{bmatrix} -6 & 0 \\ 0 & -12 \end{bmatrix}$, so Q is negative definite, and f has a *local maximum* value at $(-1, -2)$, $f(-1, -2) = 38$ (Figure 8.3).

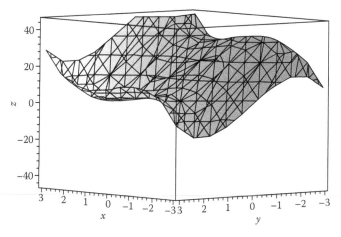

Figure 8.3 $f(x, y) = x^3 + y^3 - 3x - 12y + 20$.

Exercises 8.4

A.

1. Confirm that the symmetric matrix
$$A = \begin{bmatrix} 1 & 0 & 1/2 \\ 0 & 0 & -1/2 \\ 1/2 & -1/2 & -1 \end{bmatrix} \text{ represents the quadratic form}$$
$x^2 + xz - yz - z^2$.

2. Find the (symmetric) matrices of the following quadratic forms:
   (a) $9x^2 - y^2 + 4z^2 + 6xy - 8xz + 2yz$ (b) $xy + xz + yz$
   (c) $(x + 2y - z)(3x - y)$ (d) $2xy + z^2$ (e) $x^2 + 4xy - 2y^2 + z^2$
   (f) $x^2 + y^2 - z^2 - w^2 + 2xy - 10xw + 4zw$

3. Show that the dot product in $\mathbb{R}^n$ defines a quadratic form Q by
   $Q(\mathbf{x}) = \mathbf{x} \cdot \mathbf{x}$, or $Q(\mathbf{x}) = \|\mathbf{x}\|^2$. What is the associated matrix?

4. If $p \in P_2[x]$, show that the integral $Q(p) = \int_0^1 [p(x)]^2 dx$ defines
   a positive definite quadratic form.

5. Diagonalize each of the following quadratic forms:
   (a) $2x^2 + 5y^2 + 2z^2 - 4xy - 2xz + 4yz$
   (b) $-3y^2 + 4xy + 10xz - 4yz$
   (c) $-x^2 + y^2 - 5z^2 + 6xz + 4yz$ (d) $2xw + 6yz$

6. Diagonalize the quadratic form $2x^2 + 4(y - x)^2 + 3(y - z)^2$.

7. Diagonalize the quadratic form
   $$x^2 + 4y^2 + z^2 + 4w^2 + 4xy + 2xz + 4xw + 4yz + 8yw + 4zw.$$

8. Identify each of the following conics and sketch a graph of
   each.
   (a) $x^2 + 6xy + y^2 = 4$ (b) $x^2 + 2xy + 4y^2 = 6$
   (c) $4x^2 - xy + 4y^2 = 12$

9. Determine the conic section given by the quadratic equation,
   $3x^2 + 2xy + 3y^2 - 8 = 0$.

10. Determine if each of the following matrices is positive defin-
    ite, negative definite, or indefinite.
    (a) $\begin{bmatrix} 1 & -1 & -1 \\ -1 & 2 & 4 \\ -1 & 4 & 6 \end{bmatrix}$ (b) $\begin{bmatrix} 4 & 2 & -2 \\ 2 & 4 & 2 \\ -2 & 2 & 4 \end{bmatrix}$
    (c) $\begin{bmatrix} 1 & 2 & 0 \\ 2 & 1 & 3 \\ 0 & 3 & -3 \end{bmatrix}$ (d) $\begin{bmatrix} 2 & 1 & 1 \\ 1 & 2 & 1 \\ 1 & 1 & 2/3 \end{bmatrix}$

11.   Find the critical points of the function

$$f(x, y, z) = x + \frac{y^2}{4x} + \frac{z^2}{y} + \frac{2}{z}, \quad x > 0, \quad y > 0, \quad z > 0$$

and classify them—that is, determine if each point yields a maximum value, a minimum value, or neither.

12.   Find and identify all critical points of the function

$$f(x, y) = y^3 + 3x^2y - 3x^2 - 3y^2 + 2.$$

B.

1.   Suppose that we had defined a quadratic form as an expression of the form $\mathbf{x}^T A \mathbf{x}$, where $\mathbf{x} \in \mathbb{R}^n$ and A is *any* square matrix.
   (a) Use the result of problem C1, Exercises 3.1 to show that any square matrix A can be written uniquely as the sum of a symmetric matrix B and a skew-symmetric matrix C.
   (b) If C is the skew-symmetric matrix found in part (b), show that $\mathbf{x}^T C \mathbf{x} = -\mathbf{x}^T C \mathbf{x}$ and conclude that, as a real number, $\mathbf{x}^T C \mathbf{x} = 0$.
   (c) Using the results of parts (a) and (b), conclude that $\mathbf{x}^T A \mathbf{x} = \mathbf{x}^T B \mathbf{x}$, where B is symmetric.

2.   Rewrite $(a_1 x_1 + a_2 x_2 + \cdots + a_n x_n)^2$ in the form $\mathbf{x}^T A \mathbf{x}$, where A is a symmetric matrix.

3.   If A is the matrix of the quadratic form $Q(\mathbf{x})$ and B is a real symmetric matrix such that $Q(\mathbf{x}) = \mathbf{x}^T B \mathbf{x}$ for all $\mathbf{x} \in \mathbb{R}^n$ show that $A = B$. [*Hint*: Consider $\mathbf{x}^T(A - B)\mathbf{x}$ and the orthogonal matrix, S, that diagonalizes $A - B$; also use the fact that a real symmetric matrix with all its eigenvalues zero is the zero matrix.]

4.   Suppose A is a Hermitian matrix. Restate and prove Theorem 8.4.1 with the real orthogonal matrix, S, replaced by a unitary matrix, U.

5.   Show that the quadratic form defined by

$$Q(\mathbf{x}) = x_1^2 + (x_1 + x_2)^2 + \cdots + (x_1 + x_2 + \cdots + x_n)^2,$$

where $\mathbf{x} = \begin{bmatrix} x_1 & x_2 & \cdots & x_n \end{bmatrix}^T$, is positive definite.

6.   Determine what values of the scalar x make the following matrix positive definite, negative definite, or indefinite.

$$\begin{bmatrix} x & 1 & 0 \\ 1 & x & 1 \\ 0 & 1 & x \end{bmatrix}$$

7.   Prove that a quadratic form $Q: \mathbb{R}^n \to \mathbb{R}$, given by $Q(\mathbf{x}) = \mathbf{x}^T A \mathbf{x}$, is negative definite if and only if all eigenvalues of A are negative. (See the proof of Theorem 8.4.2.)

8. Show that a real symmetric matrix, A, is positive definite if and only if A^{-1} exists and is positive definite and symmetric.

C.

1. Under what conditions on x and a is the real $n \times n$ matrix,

$$A = \begin{bmatrix} x & a & a & \dots & a \\ a & x & a & \dots & a \\ a & a & x & \dots & a \\ \vdots & \vdots & \vdots & \dots & \vdots \\ a & a & a & \dots & x \end{bmatrix},$$

positive definite? [*Hint*: Use technology to make a conjecture. Then prove you are right.]

2. When a quadratic form, Q, is written as a sum of squares,

$$Q(\mathbf{x}) = \lambda_1 \tilde{x}_1^2 + \lambda_2 \tilde{x}_2^2 + \cdots + \lambda_n \tilde{x}_n^2,$$

let k be the number of positive eigenvalues λ_i. Show that the number k is the largest dimension of a subspace of $\mathbb{R}^n$ on which Q is positive definite.

*8.5 Singular Value Decomposition

We have seen some important matrix factorizations, such as the LU and QR decompositions. In Section 8.2, we discussed a number of structure theorems for special classes of matrices. In particular, we saw how any real symmetric matrix A can be diagonalized by a real orthogonal matrix, U (Theorem 8.2.3): $D = U^{\mathrm{T}}AU$. What the *singular value decomposition* (SVD) does is provide for *all* matrices a form of diagonalization that has far-reaching consequences, both theoretical and applied.[†]

Our strategy is to work with the Hermitian matrix, A^*A (or AA^*), which has real, nonnegative eigenvalues (see Theorem 8.2.1 and Corollary 8.2.2). Even if A is nonsquare—say, an $m \times n$ matrix—the matrices AA^* and A^*A have the same nonzero eigenvalues, counting multiplicity (Corollary 8.2.3). The **singular values** of A are the square roots of the positive eigenvalues of A^*A (or of AA^*), listed with their multiplicities.

---

* The development in this section was inspired by D. Kalman, A singularly valuable decomposition: The SVD of a matrix, *Coll. Math. J.* **27** (1996): 2–23 and by C. Mulcahy and J. Rossi, A fresh approach to the singular value decomposition, *Coll. Math. J.* **29** (1998): 199–207.

[†] The early development of the SVD is generally credited to Eugenio Beltrami (1835–1899) and M.E. Camille Jordan (1838–1922). See G.W. Stewart, On the early history of the singular value decomposition, *SIAM Rev.* **35** (1993): 551–566 or R.A. Horn and C.R. Johnson *Topics in Matrix Analysis*, Chap. 3 (New York: Cambridge University Press, 1991).

Example 8.5.1: Singular Values

(a) Suppose $A = \begin{bmatrix} 2 & 0 \\ 0 & 3 \end{bmatrix}$. Then $A^* = \begin{bmatrix} 2 & 0 \\ 0 & 3 \end{bmatrix}$, and $A^*A = \begin{bmatrix} 4 & 0 \\ 0 & 9 \end{bmatrix} =$ AA^* has eigenvalues 4 and 9, so the singular values are (in descending order) $\sigma_1 = \sqrt{9} = 3$ and $\sigma_2 = \sqrt{4} = 2$.

(b) If $B = \begin{bmatrix} 3 \\ 4 \end{bmatrix}$ and $B^* = \begin{bmatrix} 3 & 4 \end{bmatrix}$, then $A^*A = [25]$. Thus, the only eigenvalue is 25, and the sole singular value is $\sigma_1 = \sqrt{25} = 5$. On the other hand, $AA^* = \begin{bmatrix} 9 & 12 \\ 12 & 16 \end{bmatrix}$, with eigenvalues 0 and 25. Once again, the singular value is $\sigma_1 = \sqrt{25} = 5$.

(c) Finally, if $A = \begin{bmatrix} i & 1 \\ -2 & 1+i \end{bmatrix}$ and $A^* = \begin{bmatrix} -i & -2 \\ 1 & 1-i \end{bmatrix}$, then $A^*A = \begin{bmatrix} 5 & -2-3i \\ -2+3i & 3 \end{bmatrix}$, with eigenvalues $4 \pm \sqrt{14}$; and $AA^* = \begin{bmatrix} 2 & 1-3i \\ 1+3i & 6 \end{bmatrix}$, with eigenvalues $4 \pm \sqrt{14}$ (see Example 8.1.4). Therefore, the singular values are $\sigma_1 = \sqrt{4+\sqrt{14}}$ and $\sigma_2 = \sqrt{4-\sqrt{14}}$.

These singular values of a matrix allow us to factor an arbitrary $m \times n$ matrix A in a way that illuminates the structure of A and provides a valuable tool for work in numerical analysis, statistics, and many other pure and applied areas. This should be repeated: **Any matrix, square or not, invertible or singular, has a singular value decomposition, as described in the next theorem.** In the statement of the theorem, "quasidiagonal" means that the rectangular matrix, Σ, has the general block form

$$\Sigma = \begin{bmatrix} D_r & O \\ O & O \end{bmatrix},$$

where the important point is that D_r is a particular $r \times r$ diagonal matrix with positive entries, and all other entries of Σ are zeros. For example,

$$\begin{bmatrix} 1 & 0 & 0 \\ 0 & 5 & 0 \end{bmatrix}, \quad \begin{bmatrix} 1 & 0 & 0 & 0 & 0 \\ 0 & 2 & 0 & 0 & 0 \\ 0 & 0 & 3 & 0 & 0 \\ 0 & 0 & 0 & 0 & 0 \\ 0 & 0 & 0 & 0 & 0 \\ 0 & 0 & 0 & 0 & 0 \end{bmatrix}, \quad \begin{bmatrix} 4 & 0 & 0 \\ 0 & 7 & 0 \\ 0 & 0 & 0 \end{bmatrix}$$

are quasidiagonal matrices with diagonal blocks D_r, where $r = 2, 3$, and 2, respectively.

Theorem 8.5.1: SVD

Given any $m \times n$ matrix A, there exists a real $m \times n$ "quasidiagonal" matrix, Σ, together with unitary matrices $U(m \times m)$ and $V(n \times n)$, such that $A = U\Sigma V^*$.

Proof We will start with the assumption $m \geq n$ for convenience. (If $m < n$, we prove the existence of an SVD for A^T. Then, if $A^\mathrm{T} = U\Sigma V^*$, the SVD of A is $(U\Sigma V^*)^\mathrm{T} = \bar{V}\Sigma^\mathrm{T}U^\mathrm{T}$.)

As we noted above (Corollary 8.2.3), AA^* and A^*A have the same nonzero eigenvalues. Now arrange all the eigenvalues of A^*A in the descending order:

$$\lambda_1 \geq \lambda_2 \geq \cdots \geq \lambda_n \geq 0.$$

Because A^*A is Hermitian, Theorem 8.2.3 guarantees that there is a unitary matrix, V, such that

$$V^*(A^*A)V = D, \tag{8.5.1}$$

where D is a diagonal matrix whose diagonal elements are λ_1, $\lambda_2, \ldots, \lambda_n$ in descending order. Assume that exactly r of the λ_i's are nonzero, and define

$$\sigma_i = \sqrt{\lambda_i}$$

for $i = 1, 2, \ldots, r$. Now let

$$\Sigma = \begin{bmatrix} \sigma_1 & 0 & 0 & \cdots & 0 & \cdots & \cdots & 0 \\ 0 & \sigma_2 & 0 & \cdots & 0 & \cdots & \cdots & 0 \\ \vdots & \vdots & \vdots & \ddots & \vdots & \ddots & \ddots & \vdots \\ 0 & 0 & 0 & \cdots & \sigma_r & \cdots & \cdots & 0 \\ \vdots & \vdots & \vdots & \ddots & \vdots & \ddots & \ddots & \vdots \\ 0 & 0 & 0 & \cdots & 0 & \cdots & \cdots & 0 \end{bmatrix} = \begin{bmatrix} D_r & O \\ O & O \end{bmatrix}, \tag{8.5.2}$$

where

Σ is an $m \times n$ matrix whose diagonal elements are $\sigma_1, \sigma_2, \ldots, \sigma_r, 0, 0, \ldots, 0$

D_r is an $r \times r$ diagonal matrix with diagonal elements $\sigma_1, \sigma_2, \ldots, \sigma_r$

Because $\Sigma^\mathrm{T}\Sigma = D$, Equation 8.5.1 reduces to

$$V^*(A^*A)V = (V^*A^*)(AV) = (AV)^*(AV) = \Sigma^\mathrm{T}\Sigma. \tag{8.5.3}$$

Let $\mathbf{v}_1, \mathbf{v}_2, \ldots, \mathbf{v}_r$ be the column vectors of V corresponding to the eigenvalues $\lambda_1 = \sigma_1^2, \lambda_2 = \sigma_2^2, \ldots, \lambda_r = \sigma_r^2$, and let $\mathbf{v}_{r+1}, \mathbf{v}_{r+2}, \ldots, \mathbf{v}_n$ be the column vectors of V corresponding to the eigenvalue 0. Thus, we can write the orthogonal matrix V as $V = \begin{bmatrix} \mathbf{v}_1 & \mathbf{v}_2 & \cdots & \mathbf{v}_r & \cdots & \mathbf{v}_n \end{bmatrix}$.

Let $W = AV$. Equation 8.5.3 reduces to

$$W^*W = \Sigma^\mathrm{T}\Sigma. \tag{8.5.4}$$

Write the matrix W in a column form: $W = \begin{bmatrix} \mathbf{w}_1 & \mathbf{w}_2 & \cdots & \mathbf{w}_r & \cdots & \mathbf{w}_n \end{bmatrix}$, where $\mathbf{w}_i \in \mathbb{C}^m$. From Equation 8.5.4, we have

$$\mathbf{w}_i{}^*\mathbf{w}_i = \begin{cases} \sigma_i^2 & \text{if } 1 \le i \le r \\ 0 & \text{if } i > r \end{cases} \tag{8.5.5}$$

and

$$\mathbf{w}_i^*\mathbf{w}_j = 0 \quad \text{if } i \ne j. \tag{8.5.6}$$

Because $\mathbf{w}_i = \mathbf{0}$ for $i > r$ from Equations 8.5.5 and 8.5.6, the first r columns of W are linearly independent. Hence $r \le m$.

Finally, define

$$\mathbf{u}_i = \frac{1}{\sigma_i} \mathbf{w}_i = \frac{1}{\sigma_i} A\mathbf{v}_i \tag{8.5.7}$$

for $i = 1, 2, \ldots, r$. From Equations 8.5.5 and 8.5.6, Equation 8.5.7 defines an orthonormal set, $\{\mathbf{u}_i\}$. If $r < m$, then choose $\mathbf{u}_{r+1}, \mathbf{u}_{r+2}, \ldots, \mathbf{u}_m$, so that $U = \begin{bmatrix} \mathbf{u}_1 & \mathbf{u}_2 & \cdots & \mathbf{u}_m \end{bmatrix}$ is a unitary matrix.

From Equation 8.5.7, it is clear that $U\Sigma = AV$. Hence, $A = U\Sigma V^*$, as the theorem claims.

A careful analysis of the proof of Theorem 8.5.1 reveals that the matrix Σ *is unique because it is determined by the singular values arranged in descending order of magnitude.* However, we could have chosen the matrix V differently, thereby yielding a different unitary matrix U. Therefore, given an arbitrary rectangular matrix A, we should not refer to "the" SVD of A, but rather to "a" singular value decomposition of A.

Also, because multiplication of a matrix A by a nonsingular matrix does not change the rank of A (Problem B9 of Exercises 3.5), we know that r, the number of nonsingular values of A, is the rank of A: $\text{rank}(A) = \text{rank}(U\Sigma V^*) = \text{rank}(\Sigma V^*) = \text{rank}(\Sigma) = r$.

Example 8.5.2: A Singular Value Decomposition

Let us find a singular value decomposition of the real matrix

$$A = \begin{bmatrix} 0 & -2 & 1 \\ -1 & 1 & 0 \end{bmatrix}.$$

First we form the key symmetric matrix

$$A^\mathsf{T}A = \begin{bmatrix} 1 & -1 & 0 \\ -1 & 5 & -2 \\ 0 & -2 & 1 \end{bmatrix},$$

and arrange its eigenvalues, 0, 6, and 1, in descending order, with corresponding orthogonal and orthonormal eigenvectors:

$$\lambda_1 = 6 \quad > \quad \lambda_2 = 1 \quad > \quad \lambda_3 = 0$$

$$\updownarrow \qquad\qquad \updownarrow \qquad\qquad \updownarrow$$

$$\begin{bmatrix} 1 \\ -5 \\ 2 \end{bmatrix} \qquad \begin{bmatrix} -2 \\ 0 \\ 1 \end{bmatrix} \qquad \begin{bmatrix} 1 \\ 1 \\ 2 \end{bmatrix}$$

$$\updownarrow \qquad\qquad \updownarrow \qquad\qquad \updownarrow$$

$$\mathbf{v}_1 = \frac{1}{\sqrt{30}} \begin{bmatrix} 1 \\ -5 \\ 2 \end{bmatrix}, \quad \mathbf{v}_2 = \frac{1}{\sqrt{5}} \begin{bmatrix} -2 \\ 0 \\ 1 \end{bmatrix}, \quad \mathbf{v}_3 = \frac{1}{\sqrt{6}} \begin{bmatrix} 1 \\ 1 \\ 2 \end{bmatrix}.$$

Just as in the proof of Theorem 8.5.1, we form the matrix V from the normalized eigenvectors of $A^T A$:

$$V = [\mathbf{v}_1 \quad \mathbf{v}_2 \quad \mathbf{v}_3] = \begin{bmatrix} 1/\sqrt{30} & -2/\sqrt{5} & 1/\sqrt{6} \\ -5/\sqrt{30} & 0 & 1/\sqrt{6} \\ 2/\sqrt{30} & 1/\sqrt{5} & 2/\sqrt{6} \end{bmatrix}.$$

$$\text{Now } V^T(A^T A)V = D = \begin{bmatrix} 6 & 0 & 0 \\ 0 & 1 & 0 \\ 0 & 0 & 0 \end{bmatrix}.$$

Next, we let $\mu_1 = \sqrt{\lambda_1} = \sqrt{6}$ and $\mu_1 = \sqrt{\lambda_2} = 1$, and use these values to define the matrix Σ, which must have the same size as A:

$$\Sigma = \begin{bmatrix} \sqrt{6} & 0 & 0 \\ 0 & 1 & 0 \end{bmatrix} = [D_2 \quad \mathbf{0}], \quad \text{where } D_2 = \begin{bmatrix} \sqrt{6} & 0 \\ 0 & 1 \end{bmatrix}.$$

Now we calculate

$$W = AV = \begin{bmatrix} 12/\sqrt{30} & 1/\sqrt{5} & 0 \\ -6/\sqrt{30} & 2/\sqrt{5} & 0 \end{bmatrix} = [\mathbf{w}_1 \quad \mathbf{w}_2 \quad \mathbf{w}_3]$$

and use the column vectors of W to define the column vectors of U:

$$\mathbf{u}_1 = \frac{1}{\mu_1}\mathbf{w}_1 = \frac{1}{\sqrt{6}}\begin{bmatrix} 12/\sqrt{30} \\ -6/\sqrt{30} \end{bmatrix} \quad \text{and}$$

$$\mathbf{u}_2 = \frac{1}{\mu_2}\mathbf{w}_2 = \begin{bmatrix} 1/\sqrt{5} \\ 2/\sqrt{5} \end{bmatrix}.$$

So we have, after some algebraic simplification,

$$U = \begin{bmatrix} 2/\sqrt{5} & 1/\sqrt{5} \\ -1/\sqrt{5} & 2/\sqrt{5} \end{bmatrix},$$

which is clearly an orthogonal matrix. Finally, we can put the pieces together and write

$$A = U\Sigma V^{\mathsf{T}} = \begin{bmatrix} 2/\sqrt{5} & 1/\sqrt{5} \\ -1/\sqrt{5} & 2/\sqrt{5} \end{bmatrix} \begin{bmatrix} \sqrt{6} & 0 & 0 \\ 0 & 1 & 0 \end{bmatrix}$$

$$\times \begin{bmatrix} 1/\sqrt{30} & -5/\sqrt{30} & 2/\sqrt{30} \\ -2/\sqrt{5} & 0 & 1/\sqrt{5} \\ 1/\sqrt{6} & 1/\sqrt{6} & 2/\sqrt{6} \end{bmatrix}.$$

For the matrix A in the last example, we could have found, for instance, the (slightly) different SVD:

$$A = \begin{bmatrix} -2/\sqrt{5} & 1/\sqrt{5} \\ -1/\sqrt{5} & 2/\sqrt{5} \end{bmatrix} \begin{bmatrix} \sqrt{6} & 0 & 0 \\ 0 & 1 & 0 \end{bmatrix}$$

$$\times \begin{bmatrix} -1/\sqrt{30} & 5/\sqrt{30} & -2/\sqrt{30} \\ -2/\sqrt{5} & 0 & 1/\sqrt{5} \\ 1/\sqrt{6} & 1/\sqrt{6} & 2/\sqrt{6} \end{bmatrix}.$$

Note that only the matrix Σ is the same.

The SVD has an important connection to the four "Fundamental Subspaces" we discussed in Section 7.8.

Corollary 8.5.1

If $A = U\Sigma V^{\mathsf{T}}$ is the SVD of a real $m \times n$ matrix A, where $\sigma_1 \geq \sigma_2 \geq \cdots \geq \sigma_r > 0$ are the singular values of A, $U = [\mathbf{u}_1 \ \ldots \ \mathbf{u}_r \ \ldots \ \mathbf{u}_m]$, and $V = [\mathbf{v}_1 \ \ldots \ \mathbf{v}_r \ \ldots \ \mathbf{v}_n]$, then

(a) range$(A) = span\{\mathbf{u}_1, \mathbf{u}_2, \ldots, \mathbf{u}_r\}$.
(b) null space$(A) = span\{\mathbf{v}_{r+1}, \mathbf{v}_{r+2}, \ldots, \mathbf{v}_n\}$.
(c) range$(A^{\mathsf{T}}) = span\{\mathbf{v}_1, \mathbf{v}_2, \ldots, \mathbf{v}_r\}$.
(d) null space$(A^{\mathsf{T}}) = span\{\mathbf{u}_{r+1}, \mathbf{u}_{r+2}, \ldots, \mathbf{u}_m\}$.

Proof Equations 8.5.5 through 8.5.7 of the proof of Theorem 8.5.1 indicate that $A\mathbf{v}_i = \sigma_i\mathbf{u}_i$ for $i = 1, 2, \ldots, r$ and $A\mathbf{v}_i = \mathbf{w}_i = 0$ for $i = r+1, r+2, \ldots, n$. Thus, range$(A) = span\{\mathbf{u}_1, \mathbf{u}_2, \ldots, \mathbf{u}_r\}$.

Similarly, taking the SVD of A^{T}, $A^{\mathsf{T}} = V\Sigma^{\mathsf{T}}U^{\mathsf{T}}$, we have $A^{\mathsf{T}}\mathbf{u}_i = \sigma_i\mathbf{v}_i$ for $i = 1, 2, \ldots, r$ and $A^{\mathsf{T}}\mathbf{u}_i = 0$ for $i = r+1$, $r = 2, \ldots, m$. Hence, range$(A^{\mathsf{T}}) = span\{\mathbf{v}_1, \mathbf{v}_2, \ldots, \mathbf{v}_r\}$.

Because $\mathbb{R}^n = $ null space$(A) \oplus$ range(A^{T}) and $\mathbb{R}^m = $ range$(A) \oplus$ null space(A^{T}) (Corollary 7.8.2), we can easily prove statements (b) and (d).

In problems involving the computer storage of large matrices, it is important to be able to store the matrix so that an excessive portion of the total available memory is not required for this data. If A is an $m \times n$ matrix with r nonzero singular values, then the following corollary shows that A

can be written as the sum of r matrices, each of rank 1. The important point is that an $m \times n$ rank-one matrix can be stored using only $m + n$ storage locations, rather than the mn locations, $a_{11}, \ldots, a_{1n}, a_{21}, \ldots, a_{2n}, \ldots, a_{m1}, a_{m2}, \ldots, a_{mn}$, needed for an arbitrary $m \times n$ matrix, A (Exercise B4). If r is considerably smaller than n, the result is a very substantial saving in storage.

In what follows, we assume that $\mathbf{u}_i$ and $\mathbf{v}_i$ are the ith columns of U and V, respectively, in the decomposition, $A = U\Sigma V^*$.

Corollary 8.5.2

If A is an $m \times n$ matrix with $\sigma_1 \geq \sigma_2 \geq \cdots \geq \sigma_r > 0$ as its r singular values, then

$$A = \sum_{j=1}^{r} \sigma_j \mathbf{u}_j \mathbf{v}_j^*,$$

the sum of rank-one matrices.

Proof Letting $U_1 = [\mathbf{u}_1 \ \mathbf{u}_2 \ \ldots \ \mathbf{u}_r]$, $U_2 = [\mathbf{u}_{r+1} \ \mathbf{u}_{r+2} \ \ldots \ \mathbf{u}_m]$,

$$V_1 = \begin{bmatrix} \mathbf{v}_1^* \\ \mathbf{v}_2^* \\ \vdots \\ \mathbf{v}_r^* \end{bmatrix}, \text{ and } V_2 = \begin{bmatrix} \mathbf{v}_{r+1}^* \\ \mathbf{v}_{r+2}^* \\ \vdots \\ \mathbf{v}_n^* \end{bmatrix},$$ we write the SVD in partitioned form:

$$A = U\Sigma V^* = \begin{bmatrix} U_1 & U_2 \end{bmatrix} \begin{bmatrix} \sigma_1 & 0 & \ldots & 0 & \vdots & 0 \\ 0 & \sigma_2 & \ldots & 0 & \vdots & 0 \\ \vdots & \vdots & \ddots & \vdots & \vdots & 0 \\ 0 & 0 & \ldots & \sigma_r & \vdots & 0 \\ \ldots & \ldots & \ldots & \ldots & \vdots & \ldots \\ 0 & 0 & 0 & 0 & \vdots & 0 \end{bmatrix} \begin{bmatrix} V_1 \\ \ldots \\ V_2 \end{bmatrix}$$

$$= \begin{bmatrix} U_1 & U_2 \end{bmatrix} \begin{bmatrix} \sigma_1 \mathbf{v}_1^* \\ \vdots \\ \sigma_r \mathbf{v}_r^* \\ \ldots \\ 0 \\ \vdots \\ 0 \end{bmatrix} = \begin{bmatrix} \mathbf{u}_1 & \ldots & \mathbf{u}_r & \vdots & U_2 \end{bmatrix} \begin{bmatrix} \sigma_1 \mathbf{v}_1^* \\ \vdots \\ \sigma_r \mathbf{v}_r^* \\ \ldots \\ 0 \\ \vdots \\ 0 \end{bmatrix}$$

$$= \mathbf{u}_1 (\sigma_1 \mathbf{v}_1^*) + \mathbf{u}_2 (\sigma_2 \mathbf{v}_2^*) + \cdots + \mathbf{u}_r (\sigma_r \mathbf{v}_r^*) = \sum_{j=1}^{r} \sigma_j \mathbf{u}_j \mathbf{v}_j^*.$$

The fact that each $m \times n$ matrix, $\mathbf{u}_j \mathbf{v}_j^*$, has rank 1 is easy to prove: With the row view of multiplication (Equation 3.3.2), we have

$$\mathbf{u}_j\mathbf{v}_j^* = \begin{bmatrix} \text{row}_1(\mathbf{u}_j) \\ \text{row}_2(\mathbf{u}_j) \\ \vdots \\ \text{row}_m(\mathbf{u}_j) \end{bmatrix} \cdot \mathbf{v}_j^* = \begin{bmatrix} \text{row}_1(\mathbf{u}_j) \cdot \mathbf{v}_j^* \\ \text{row}_2(\mathbf{u}_j) \cdot \mathbf{v}_j^* \\ \vdots \\ \text{row}_m(\mathbf{u}_j) \cdot \mathbf{v}_j^* \end{bmatrix} = \begin{bmatrix} u_{1j}\mathbf{v}_j^* \\ u_{2j}\mathbf{v}_j^* \\ \vdots \\ u_{mj}\mathbf{v}_j^* \end{bmatrix},$$

so each row of $\mathbf{u}_j\mathbf{v}_j^*$ is a scalar multiple of (row vector) $\mathbf{v}_j^*$, that is, rank $(\mathbf{u}_j\mathbf{v}_j^*) = $ the dimension of the row space of $\mathbf{u}_j\mathbf{v}_j^* = 1$.

The formula $A = \sum_{j=1}^r \sigma_j\mathbf{u}_j\mathbf{v}_j^*$ is called the **outer product** form of the SVD. Each term $\mathbf{u}_j\mathbf{v}_j^*$, is called an **outer product** of the matrices $\mathbf{u}_j$ and $\mathbf{v}_j^*$.

Example 8.5.3: A Rank-One Decomposition

In Example 8.4.2, we saw that the matrix $A = \begin{bmatrix} 0 & -2 & 1 \\ -1 & 1 & 0 \end{bmatrix}$ has a singular value decomposition

$$A = U\Sigma V^{\mathsf{T}} = \begin{bmatrix} 2/\sqrt{5} & 1/\sqrt{5} \\ -1/\sqrt{5} & 2/\sqrt{5} \end{bmatrix} \begin{bmatrix} \sqrt{6} & 0 & 0 \\ 0 & 1 & 0 \end{bmatrix}$$
$$\begin{bmatrix} 1/\sqrt{30} & -5/\sqrt{30} & 2/\sqrt{30} \\ -2/\sqrt{5} & 0 & 1/\sqrt{5} \\ 1/\sqrt{6} & 1/\sqrt{6} & 2/\sqrt{6} \end{bmatrix}.$$

We use the displayed singular values and the columns of U and V to write

$$A = \sqrt{6} \begin{bmatrix} 2/\sqrt{5} \\ -1/\sqrt{5} \end{bmatrix} \begin{bmatrix} 1/\sqrt{30} & -5/\sqrt{30} & 2/\sqrt{30} \end{bmatrix}$$
$$+ \begin{bmatrix} 1/\sqrt{5} \\ 2/\sqrt{5} \end{bmatrix} \begin{bmatrix} -2/\sqrt{5} & 0 & 1/\sqrt{5} \end{bmatrix}$$
$$= \begin{bmatrix} 2/5 & -2 & 4/5 \\ -1/5 & 1 & -2/5 \end{bmatrix} + \begin{bmatrix} -2/5 & 0 & 1/5 \\ -4/5 & 0 & 2/5 \end{bmatrix}.$$

(To write A in this form, it is important to realize that the last matrix in the SVD of A is the *transpose* of V.)

Each of the two matrices forming A clearly has rank equal to 1.

If we compare Theorem 8.2.3 and the representation of a Hermitian matrix as a sum of matrices with Theorem 8.5.1 and Corollary 8.5.2, we can see that the SVD is a generalization of the Spectral Theorem for Hermitian matrices.

Exercises 8.5

A.

1. Find the singular values of each of the following matrices:

    (a) $\begin{bmatrix} 1 & 1 \\ 1 & 0 \\ 0 & 1 \end{bmatrix}$ (b) $\begin{bmatrix} 1 & 1 & 0 \\ 0 & 0 & 1 \end{bmatrix}$ (c) $\begin{bmatrix} 2 & 0 & 1 \\ 0 & 2 & 0 \end{bmatrix}$

    (d) $\begin{bmatrix} 0 & 2 & 0 \\ 0 & 0 & -3 \\ 1 & 0 & 0 \end{bmatrix}$ (e) $\begin{bmatrix} 2 & 0 & 0 \\ 0 & 1 & 0 \\ 0 & 0 & -2 \end{bmatrix}$

2. Find an SVD for matrix (a) in Exercise 1.

3. Find an SVD for matrix (b) in Exercise 1.

4. Find an SVD for $A = \begin{bmatrix} 2 & -1 & 2 \end{bmatrix}$.

5. Find an SVD for $A = \begin{bmatrix} 1 & 1 \\ 1 & 1 \\ -2 & 1 \end{bmatrix}$.

6. Find an SVD for $A = \begin{bmatrix} 0 & 4 & 3 \\ 4 & 0 & 0 \\ 3 & 0 & 0 \end{bmatrix}$.

7. Let $A = \begin{bmatrix} 1 & 1 & 1 \\ 1 & 1 & 1 \end{bmatrix}$.
    (a) Find an SVD for A.
    (b) Express A as a sum of matrices of rank 1.

8. Let $A = \begin{bmatrix} 1 & 1 \\ -2 & 4 \end{bmatrix}$.
    (a) Find an SVD for A.
    (b) Describe the image of the unit circle in $\mathbb{R}^2$, $\{\mathbf{x} \in \mathbb{R}^2 : \|\mathbf{x}\| = 1\}$, under A.

9. Let $A = \begin{bmatrix} 1 & 1 & 1 \\ -1 & 0 & -2 \\ 1 & 2 & 0 \end{bmatrix}$.
    (a) Find an SVD for A.
    (b) Express A as a sum of matrices of rank 1.

10. Find an SVD for $\mathbf{x}^T$, where $\mathbf{x} \neq \mathbf{0} \in \mathbb{R}^n$.

B.

1. If A is invertible, what is the relationship between the singular values of A and those of A^{-1}?

2. Given the singular values $\sigma_1 \geq \sigma_2 \geq \cdots \geq \sigma_r > 0$ of a complex matrix A, find the singular values of
    (a) A^*,
    (b) αA, where $\alpha \in \mathbb{R}$.

3. Prove that if A is symmetric, then the singular values of A are the absolute values of the nonzero eigenvalues of A.

4. Prove that an $m \times n$ rank-one matrix can be stored using only $m + n$ storage locations.

5. Prove part (b) of Corollary 8.5.1.

6. Prove part (d) of Corollary 8.5.1.

7. Find an SVD for

    $$A = \begin{bmatrix} 0 & 1 & 1 & 1 \\ 1 & 0 & 1 & 1 \\ 1 & 1 & 0 & 1 \\ 1 & 1 & 1 & 0 \end{bmatrix}.$$

    [*Hint*: The eigenvalues of A are $3,\ -1,\ -1,\ -1$.]

8. Find an SVD for $\mathbf{uv}^\mathrm{T}$, where $\mathbf{u}, \mathbf{v} \in \mathbb{R}^n$, $\langle \mathbf{u}, \mathbf{v} \rangle \neq 0$. [*Hint*: If you have done Exercise A10, you may want to imitate your proof of that problem.]

9. Prove Corollary 8.5.2 by defining $B = \sum_{j=1}^r \sigma_j \mathbf{u}_j \mathbf{v}_j^*$ and showing that $B = A$.
    [*Hint*: Show that $B\mathbf{v}_i = A\mathbf{v}_i$ for $i = 1, 2, \ldots, n$, and then use the matrix version of Theorem 6.1.2.]

10. Use the SVD in Example 8.5.2 to find an orthonormal basis for null space(A) and one for range(A) if $A = \begin{bmatrix} 0 & -2 & 1 \\ -1 & 1 & 0 \end{bmatrix}$.

11. Let $A = \dfrac{1}{\sqrt{30}} U_0 \Sigma V_0^\mathrm{T} = \dfrac{1}{\sqrt{30}} \begin{bmatrix} \sqrt{2} & -1 & -\sqrt{3} \\ \sqrt{2} & 2 & 0 \\ \sqrt{2} & -1 & \sqrt{3} \end{bmatrix} \begin{bmatrix} 3 & 0 \\ 0 & 2 \\ 0 & 0 \end{bmatrix}$

    $\begin{bmatrix} 1 & -2 \\ 2 & 1 \end{bmatrix}$. Note that this is not an SVD for A.

    (a) Find the singular values of A.
    (b) Write out an SVD for A.

12. If $\sigma_1 \geq \sigma_2 \geq \cdots \geq \sigma_r > 0$ are the singular values of an $m \times n$ matrix A with $m \geq n$, show that $\sigma_r \|\mathbf{x}\| \leq \|A\mathbf{x}\| \leq \sigma_1 \|\mathbf{x}\|$ for any $\mathbf{x} \in \mathbb{R}^n$.

13. Prove that all singular values of an orthogonal matrix A are equal to one.

C.

1. Let $V = P_n[x]$ with the inner product $\langle f, g \rangle = \sum_{i=1}^n a_i b_i$, where $f(x) = a_0 + a_1 x + \cdots + a_n x^n$ and $g(x) = b_0 + b_1 x + \cdots + b_n x^n$. Define $T = \frac{d}{dx}$, the differential operator. Determine the singular values of $T : V \to V$.

2. Find the singular values of an $m \times n$ matrix $A = [a_{ij}]$ of rank 1.

3. Let A be any $m \times n$ complex matrix. Define the *Euclidean* (or *Frobenius*) *norm* of A as

$$\|A\| = [\text{trace}(A^*A)]^{1/2} = \left[\sum_{i=1}^m \sum_{j=1}^n |a_{ij}|^2 \right]^{1/2}.$$

   (a) Prove that $\|UA\| = \|A\|$ for any unitary matrix U.
   (b) As a result of part (a), show that $\|UAV\| = \|A\|$ for unitary matrices U and V.
   (c) Use the result of part (b) to prove that if $A = U\Sigma V^*$ is an SVD for A, then

$$\|A\|^2 = \sigma_1^2 + \sigma_2^2 + \cdots + \sigma_r^2,$$

   where σ_i are the singular values of A.

4. If $\sigma_1, \sigma_2, \ldots, \sigma_r$ are the singular values of a complex $n \times n$ matrix A, prove that $\sigma_1 + \sigma_2 + \cdots + \sigma_r = \max_W |\text{trace}(AW)| = \max_W \text{Re}\{\text{trace}(AW)\}$, where W ranges over the whole set of $n \times n$ unitary matrices.

8.6 The Polar Decomposition

Another useful consequence of the singular value decomposition is the **polar decomposition** of any $n \times n$ matrix. This is the matrix version of the *polar representation* of a complex number z: $z = re^{i\theta}$, where r is a nonnegative real number and $|e^{i\theta}| = 1$. Geometrically, the representation $re^{i\theta}$ (or $|z|e^{i\theta}$) decomposes a complex number z into a magnification factor r and a rotation factor $e^{i\theta}$. (See Appendix D for details of complex geometry.)

Expanding Definition 8.4.2, we say that a Hermitian matrix M is **positive semidefinite** (or **nonnegative definite**) if $Q(\mathbf{x}) = \mathbf{x}^*M\mathbf{x} \geq 0$ for all $\mathbf{x} \in \mathbb{R}^n$. Equivalently (see Theorem 8.4.2), M is positive semidefinite if all eigenvalues of M are nonnegative.

Theorem 8.6.1: The Polar Decomposition

Any $n \times n$ matrix A can be written as $A = PU$, where P is a Hermitian positive semidefinite matrix and U is a unitary matrix.

Proof Suppose A is an $n \times n$ matrix. Then we can write A as $U\Sigma V^*$ (Theorem 8.5.1), where U, Σ, and V are also $n \times n$. Because U is unitary, $U^*U = I_n$, and we have $A = U\Sigma V^* = U\Sigma(U^*U)V^* = (U\Sigma U^*)(UV^*)$. Let $P = U\Sigma U^*$ and $Q = UV^*$. We see that $P^* = (U\Sigma U^*)^* = U\Sigma^T U^* = U\Sigma U^* = P$, so P is Hermitian. Now let $\mathbf{y} = U^*\mathbf{x} = \begin{bmatrix} y_1 & y_2 & \cdots & y_n \end{bmatrix}^T$. Then, for $\mathbf{x} \neq \mathbf{0}$, $\mathbf{x}^*P\mathbf{x} = \mathbf{x}^*U\Sigma U^*\mathbf{x} = (U^*\mathbf{x})^*\Sigma(U^*\mathbf{x}) = \mathbf{y}^*\Sigma\mathbf{y} = \begin{bmatrix} y_1^* & y_2^* & \cdots & y_n^* \end{bmatrix} \begin{bmatrix} \sigma_1 y_1 & \sigma_2 y_2 & \cdots & \sigma_n y_n \end{bmatrix}^T = \sum_{i=1}^{n} \sigma_i |y_i|^2 \geq 0$.

Thus P is a Hermitian positive semidefinite matrix. Furthermore, $QQ^* = UV^*VU^* = UU^* = I_n$, so Q is unitary.

It can be shown that the matrix P in this theorem is unique. Also, if A is invertible, then the polar decomposition is unique, that is, $P_1U_1 = P_2U_2$ implies that $P_1 = P_2$ and $U_1 = U_2$. Various polar decompositions exist for general $m \times n$ complex matrices.*

Example 8.6.1: A Polar Decomposition

We will find a polar decomposition of

$$A = \begin{bmatrix} 1 & 0 & 2 \\ 0 & 1 & 0 \\ 0 & 0 & 0 \end{bmatrix}.$$

First, using the method shown in Example 8.5.2, we find a singular value decomposition of A (Exercise A1):

$$A = U\Sigma V^T = \begin{bmatrix} -1 & 0 & 0 \\ 0 & 1 & 0 \\ 0 & 0 & 1 \end{bmatrix} \begin{bmatrix} \sqrt{5} & 0 & 0 \\ 0 & 1 & 0 \\ 0 & 0 & 0 \end{bmatrix} \begin{bmatrix} -1/\sqrt{5} & 0 & -2/\sqrt{5} \\ 0 & 1 & 0 \\ -2/\sqrt{5} & 0 & 1/\sqrt{5} \end{bmatrix}.$$

As in the proof of Theorem 8.6.1, let $P = U\Sigma U^T$ and $Q = UV^T$:

$$P = \begin{bmatrix} -1 & 0 & 0 \\ 0 & 1 & 0 \\ 0 & 0 & 1 \end{bmatrix} \begin{bmatrix} \sqrt{5} & 0 & 0 \\ 0 & 1 & 0 \\ 0 & 0 & 0 \end{bmatrix} \begin{bmatrix} -1 & 0 & 0 \\ 0 & 1 & 0 \\ 0 & 0 & 1 \end{bmatrix} = \begin{bmatrix} \sqrt{5} & 0 & 0 \\ 0 & 1 & 0 \\ 0 & 0 & 0 \end{bmatrix},$$

---

* See, for example, Section 7.3 of R.A. Horn and C.R. Johnson, *Matrix Analysis* (New York: Cambridge University Press, 1990).

$$Q = UV^\mathsf{T} = \begin{bmatrix} -1 & 0 & 0 \\ 0 & 1 & 0 \\ 0 & 0 & 1 \end{bmatrix} \begin{bmatrix} -1/\sqrt{5} & 0 & -2/\sqrt{5} \\ 0 & 1 & 0 \\ -2/\sqrt{5} & 0 & 1/\sqrt{5} \end{bmatrix}$$

$$= \begin{bmatrix} 1/\sqrt{5} & 0 & 2/\sqrt{5} \\ 0 & 1 & 0 \\ -2/\sqrt{5} & 0 & 1/\sqrt{5} \end{bmatrix}.$$

Then the corresponding polar decomposition is

$$A = PQ = \begin{bmatrix} \sqrt{5} & 0 & 0 \\ 0 & 1 & 0 \\ 0 & 0 & 0 \end{bmatrix} \begin{bmatrix} 1/\sqrt{5} & 0 & 2/\sqrt{5} \\ 0 & 1 & 0 \\ -2/\sqrt{5} & 0 & 1/\sqrt{5} \end{bmatrix}$$

$$= \frac{1}{5} \begin{bmatrix} 5 & 0 & 0 \\ 0 & \sqrt{5} & 0 \\ 0 & 0 & 0 \end{bmatrix} \begin{bmatrix} 1 & 0 & 2 \\ 0 & \sqrt{5} & 0 \\ -2 & 0 & 1 \end{bmatrix}.$$

If A is a 2×2 or 3×3 matrix, with $T = A\mathbf{x}$ interpreted as a geometric transformation on $\mathbb{R}^2$ or $\mathbb{R}^3$, the polar decomposition, $A = PU$, shows that the effect of A on a geometrical object can be seen as a rigid rotation (or a rotation and reflection) induced by U, followed by stretching and/or compression corresponding to P. (See Section 7.5 and Exercise B1 there. Also, look at the discussion of determinants and geometry in Section 4.2.)

Example 8.6.2: Geometric Interpretation of Polar Decomposition

The matrix $A = \begin{bmatrix} 1 & 1 \\ 0 & 1 \end{bmatrix}$ has a polar decomposition of the form

$$A = \begin{bmatrix} \frac{3}{\sqrt{5}} & \frac{1}{\sqrt{5}} \\ \frac{1}{\sqrt{5}} & \frac{2}{\sqrt{5}} \end{bmatrix} \begin{bmatrix} \frac{2}{\sqrt{5}} & \frac{1}{\sqrt{5}} \\ -\frac{1}{\sqrt{5}} & \frac{2}{\sqrt{5}} \end{bmatrix} = PQ.$$

This factorization lets us see the effect of A in two stages. For example, we can determine how the transformation represented by A changes a unit square, $\left\{ \begin{bmatrix} x \\ y \end{bmatrix} : 0 \leq x \leq 1 \text{ and } 0 \leq y \leq 1 \right\}$.

According to the polar decomposition of A, each point, $\begin{bmatrix} x \\ y \end{bmatrix}$, in the unit square is first transformed by a rigid rotation represented by the matrix Q, a clockwise rotation of about $26.5°$ (Figure 8.4). (The angle of counterclockwise rotation is θ rad, where $\cos\theta = 2/\sqrt{5}$ and $\sin\theta = -1/\sqrt{5}$.)

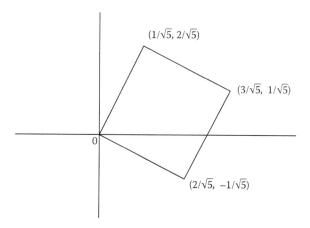

Figure 8.4 The effect of rotation θ.

Next, the rotated square is subjected to both a stretching and a compression (Figure 8.5).

Because $\det(P) = \det(Q) = \det(A) = 1$, at no stage of the transformation is the area of the figure changed (see Section 4.2)—the area is always 1.

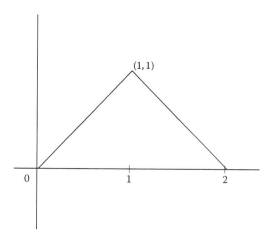

Figure 8.5 The effect of P on the rotated unit square.

Exercises 8.6

A.

1. Prove that the matrix A in Example 8.6.1 has the SVD shown.

2. Consider the matrix A given in Example 8.6.1 and find a polar decomposition of the form $A = UP$, where P is a symmetric positive semidefinite matrix and U is an orthogonal matrix. Is your polar form the same as that given in Example 8.5.1?

3. In Example 8.6.1, verify that P is a symmetric positive semi-definite matrix and that U is an orthogonal matrix.

4. If A is the 3×3 matrix given in Problem A9 of Exercises 8.5, use the SVD provided at the back of the book and find a corresponding polar decomposition of A.

5. If $A = \begin{bmatrix} 11 & -5 \\ -2 & 10 \end{bmatrix}$ has a singular value decomposition

$$A = \begin{bmatrix} 4/5 & 3/5 \\ -3/5 & 4/5 \end{bmatrix} \begin{bmatrix} 10\sqrt{2} & 0 \\ 0 & 5\sqrt{2} \end{bmatrix} \begin{bmatrix} 1/\sqrt{2} & 1/\sqrt{2} \\ -1/\sqrt{2} & 1/\sqrt{2} \end{bmatrix}^*,$$

find a polar decomposition of A.

6. Determine a polar decomposition for $A = \begin{bmatrix} 6 & 2 \\ 0 & 5 \end{bmatrix}$.

7. Find a polar representation for $A = \begin{bmatrix} 1 & 1 \\ 0 & 0 \end{bmatrix}$.

8. Find a polar decomposition for $\begin{bmatrix} 1 & 2 \\ -3 & -1 \end{bmatrix}$.

9. What are the possible polar representations of a positive semidefinite matrix A?

10. Find a polar representation of the matrix

$$\begin{bmatrix} 0 & 0 & 0 & 2 \\ 1 & 0 & 0 & 0 \\ 0 & 1 & 0 & 0 \\ 0 & 0 & 1 & 0 \end{bmatrix}.$$

11. If $A = \begin{bmatrix} 1+i & 1 \\ 1-i & -i \end{bmatrix}$ has a singular value decomposition

$$A = \begin{bmatrix} (1+i)/2 & (1+i)/2 \\ (1-i)/2 & (-1+i)/2 \end{bmatrix} \begin{bmatrix} \sqrt{6} & 0 \\ 0 & 0 \end{bmatrix} \begin{bmatrix} 2/\sqrt{6} & (1-i)/\sqrt{6} \\ (1+i)/\sqrt{6} & -2/\sqrt{6} \end{bmatrix}^*,$$

find a polar decomposition of A.

B.

1. If an $m \times n$ real matrix, A, is positive semidefinite, show that
    (a) A is symmetric and nonsingular.
    (b) A^{-1} is positive semidefinite as well.

2. Show that every positive semidefinite symmetric matrix, A, can be written as $A = X^T X$ for a nonsingular matrix X.

3. Show that if A is positive semidefinite, then all diagonal entries a_{ii} of A are nonnegative.

4. Let A be a square complex matrix with a polar decomposition $A = PU$.
    (a) Prove that A is normal if and only if $UP^2 = P^2 W$.
    (b) Use the result of part (a) to prove that A is normal if and only if $PU = UP$.

5. Let $A = PU$ be a polar representation of a matrix A. Prove that the matrix U transforms the orthonormal basis containing the eigenvectors of A^*A into a similar basis for AA^*.

6. Prove that any matrix can be represented in the form $A = UDW$, where U and W are unitary matrices and D is a diagonal matrix.

C.

1. If $A = PU$ is a polar decomposition of the $n \times n$ matrix A, show that $P = \sqrt{AA^*}$, that is, $P^2 = AA^*$.

2. Let A be a nonzero normal matrix in $M_n(\mathbb{R})$, and let its eigenvalues, $\lambda_1, \lambda_2, \ldots, \lambda_n$, be given in polar form, $\lambda_1 = r_1 e^{i\theta_1}, \lambda_2 = r_2 e^{i\theta_2}, \ldots, \lambda_n = r_n e^{i\theta_n}$. Prove that the matrices P and U in the polar decomposition of A have the eigenvalues $r_1, r_2, \ldots, r_n$ and $e^{i\theta_1}, e^{i\theta_2}, \ldots, e^{i\theta_n}$, respectively.

8.7 Summary

The **adjoint** of a linear operator on a finite-dimensional inner product space, V, is the unique linear transformation such that $\langle T(\mathbf{x}), \mathbf{y} \rangle = \langle \mathbf{x}, T^*(\mathbf{y}) \rangle$ for every $\mathbf{x}$, $\mathbf{y} \in V$. If B is an orthonormal basis for V, then $[T^*]_B = [T]_B^*$. The properties of an adjoint operator are similar to the properties of the conjugate of a complex number.

A **Hermitian** (sometimes **hermitian**) **matrix** A is a square matrix with complex entries such that $A = A^*$, where $A^* = \bar{A}^T$. If A is a Hermitian matrix, then (1) $\langle A\mathbf{x}, \mathbf{y} \rangle = \langle \mathbf{x}, A\mathbf{y} \rangle = \langle \mathbf{x}, A^*\mathbf{y} \rangle$ for every $\mathbf{x}, \mathbf{y} \in \mathbb{R}^n$; (2) all eigenvalues of A are real; (3) eigenvectors of A associated with distinct eigenvalues are orthogonal.

Another important result is the **Spectral Theorem for Hermitian Matrices**: **Let A be a Hermitian matrix. Then there is a unitary matrix, U, such that U^*AU is diagonal. Furthermore, if A is a real symmetric matrix, then U may be chosen to be real and orthogonal.** The Spectral Theorem has several useful consequences: (1) If A is an $n \times n$ Hermitian matrix, there is an orthonormal basis for $\mathbb{C}^n$, which consists entirely of eigenvectors of A. If A is real, there is an orthonormal basis for $\mathbb{R}^n$, consisting of eigenvectors of A. (2) If A is an $m \times n$ matrix, the matrix AA^* (or A^*A) is Hermitian and its eigenvalues are nonnegative. (3) If A is an $m \times n$ matrix, the matrices AA^* and A^*A have the same nonzero eigenvalues.

A matrix A is said to be **normal** if $AA^* = A^*A$. If A is a real matrix, then A is normal if $AA^T = A^TA$. **There is an orthonormal basis for $\mathbb{C}^n$, which consists entirely of eigenvectors of a matrix A if and only if A is normal**.

An expression of the form

$$Q(\mathbf{x}) = \mathbf{x}^T A \mathbf{x} = \begin{bmatrix} x_1 & x_2 & \cdots & x_n \end{bmatrix} \begin{bmatrix} a_{11} & a_{12} & \cdots & a_{1n} \\ a_{21} & a_{22} & \cdots & a_{2n} \\ \vdots & \vdots & \vdots & \vdots \\ a_{n1} & a_{n2} & \cdots & a_{nn} \end{bmatrix} \begin{bmatrix} x_1 \\ x_2 \\ \vdots \\ x_n \end{bmatrix}$$

$$= \sum_{i=1}^n \sum_{j=1}^n a_{ij} x_i x_j,$$

where $A = [a_{ij}]$ is an $n \times n$ symmetric matrix and $\mathbf{x} \in \mathbb{R}^n$, is called a **(real) quadratic form** (in $\mathbb{R}^n$, or in n variables), and A is called the **matrix of the quadratic form**. (Note that Q is a scalar-valued function, $Q: \mathbb{R}^n \to \mathbb{R}$.) **The Principal Axes Theorem** is fundamental in the theory and applications of quadratic forms: **Let $Q(\mathbf{x}) = \mathbf{x}^T A \mathbf{x}$ be an arbitrary quadratic form. Then there is a real orthogonal matrix, S, such that $Q(\mathbf{x}) = \lambda_1 \tilde{x}_1^2 + \lambda_2 \tilde{x}_2^2 + \cdots + \lambda_n \tilde{x}_n^2$, where $\tilde{x}_1, \tilde{x}_2, \ldots, \tilde{x}_n$ are the entries of $\tilde{\mathbf{x}} = S^T \mathbf{x}$ and $\lambda_1, \lambda_2, \ldots, \lambda_n$ are the eigenvalues of the matrix A.** Expressing an arbitrary quadratic form as a sum of squares is referred to as **diagonalizing** the quadratic form. A quadratic form $Q : \mathbb{R}^n \to \mathbb{R}$ is called **positive definite** if $Q(\mathbf{x}) > 0$ for all nonzero $\mathbf{x} \in \mathbb{R}^n$ and **negative definite** if $Q(\mathbf{x}) < 0$ for all nonzero $\mathbf{x} \in \mathbb{R}^n$. If the quadratic form takes both positive and negative values, it is called **indefinite**. By extension, we call a matrix A **positive definite** (**negative definite**) if A is a symmetric matrix such that the quadratic form $Q(\mathbf{x}) = \mathbf{x}^T A \mathbf{x}$ is positive definite (negative definite). **A quadratic form** $Q : \mathbb{R}^n \to \mathbb{R}$, **given by** $Q(\mathbf{x}) = \mathbf{x}^T A \mathbf{x}$, **is positive definite if and only if all eigenvalues of A are positive, and is negative definite if and only if all eigenvalues of A are negative. The quadratic form is indefinite if and only if A has both positive and negative eigenvalues.**

The **singular values** of an $m \times n$ matrix A are the square roots of the positive eigenvalues of A^*A (or of AA^*), listed with their multiplicities. A **quasidiagonal matrix**, Σ, is a rectangular matrix having the general block form

$$\Sigma = \begin{bmatrix} D_r & O \\ O & O \end{bmatrix},$$

where the important point is that D_r is a particular $r \times r$ diagonal matrix with positive entries, and all other entries of Σ are zeros. These notions lead to the **SVD**: **Given any** $m \times n$ **matrix A, there exists a real** $m \times n$ **quasidiagonal matrix**, Σ, **together with unitary matrices,** U **($m \times m$) and** V **($n \times n$), such that** $A = U\Sigma V^*$. In an SVD, the matrix Σ is unique, although U and V are not. The **outer product** form of the SVD is important for applications: **If A is an** $m \times n$ **matrix with** $\sigma_1 \geq \sigma_2 \geq \cdots \geq \sigma_r > 0$ **as its r singular values, then** $A = \sum_{j=1}^{r} \sigma_j \mathbf{u}_j \mathbf{v}_j^*$, **the sum of rank-one matrices.** The SVD has an important connection to the **Four Fundamental Subspaces** discussed in Section 5.5: If $A = U\Sigma V^T$ is the SVD of a real $m \times n$ matrix A, where $\sigma_1 \geq \sigma_2 \geq \cdots \geq \sigma_r > 0$ are the singular values of A, $U = [\mathbf{u}_1 \ \ldots \ \mathbf{u}_r \ \ldots \ \mathbf{u}_m]$, and $V = [\mathbf{v}_1 \ \ldots \ \mathbf{v}_r \ \ldots \ \mathbf{v}_n]$, then (a) range$(A) = span\{\mathbf{u}_1, \mathbf{u}_2, \ldots, \mathbf{u}_r\}$; (b) null space $(A) = span\{\mathbf{v}_{r+1}, \mathbf{v}_{r+2}, \ldots, \mathbf{v}_n\}$; (c) range$(A^T) = span\{\mathbf{v}_1, \mathbf{v}_2, \ldots, \mathbf{v}_r\}$; (d) null space$(A^T) = span\{\mathbf{u}_{r+1}, \mathbf{u}_{r+2}, \ldots, \mathbf{u}_m\}$.

A Hermitian matrix, M, is **positive semidefinite** (or **nonnegative definite**) if $Q(\mathbf{x}) = \mathbf{x}^* M \mathbf{x} \geq 0$ for all $\mathbf{x} \in \mathbb{R}^n$. Equivalently, M is positive semidefinite if all eigenvalues of M are nonnegative. Another useful consequence of the singular value decomposition is the **polar decomposition** of any $n \times n$ matrix: Any $n \times n$ real matrix A can be written as $A = PU$, where P is a Hermitian positive semidefinite matrix and U is a unitary matrix.

Appendix A: Basics of Set Theory

A.1 Basic Terminology

A **set** is a collection of objects, called **elements**, usually represented as enclosed by braces { }. If an element x belongs to a set S, we write $x \in S$. Similarly, if element x does *not* belong to S, we write $x \notin S$. If the contents of a set are obvious, a listing is used: $S = \{0, 1, 2, \ldots, n, \ldots\}$. Otherwise, some description is included: $R = \{n : n = 2k + 1, k$ a positive integer$\}$ or $R = \{n \mid n = 2k + 1, k$ a positive integer$\}$. If every element of set R is also an element of set S, we say that R is a **subset** of S, denoted $R \subseteq S$. Note that the **empty set** (or **null set**), usually denoted by $\emptyset$, is a subset of *every* set. We call R a **proper subset** of S, denoted $R \subset S$ or $R \subsetneq S$, if R is a subset of S but there are elements of S that are not contained in R. Looking at the relation $R \subseteq S$ from a different perspective, we say that S is a **superset** of R, denoted $S \supseteq R$. Sets are usually considered part of a **universe** or **universal set**, which we will denote U, some master set that contains all sets in a particular discussion.

The **union** of two sets R and S, denoted $R \cup S$ is the set of all elements in R or in S, possibly both: $R \cup S = \{x : x \in R$ or $x \in S\}$. Extending this concept, we can speak of the union of any finite number of sets S_k:

$$S = \bigcup_{k=1}^{n} S_k = \{x : x \in S_k \text{ for at least one value of } k\}.$$

In fact, we can consider unions over any (finite or infinite) set of indices (subscripts): $X = \bigcup_{\alpha \in A} X_\alpha$. The **intersection** of R and S, denoted $R \cap S$, is the set of elements contained in *both* R and S: $R \cap S = \{x : x \in R$ and $x \in S\}$. Note that $R \cap S \subseteq \left\{ \begin{matrix} R \\ S \end{matrix} \right\} \subseteq R \cup S$. If $R \cap S = \emptyset$, we say that R and S are **disjoint**. Just as we did for unions of sets, we can consider intersections of more than two sets:

$$X = \bigcap_{\alpha \in A} X_\alpha = \{x : x \in X_\alpha \text{ for at least one value of } \alpha\}.$$

Note that $\bigcap_{\alpha \in A} X_\alpha \subseteq X_\alpha$, for any $\alpha \in A$, $\subseteq \bigcup_{\alpha \in A} X_\alpha$.

Example A.1

Suppose n is a positive integer and $X_n = [-1/n, 1/n] = \{x \in \mathbb{R} : -1/n \leq x \leq 1/n\}$. Then $\bigcup_{n=1}^{\infty} X_i = [-1, 1]$ and $\bigcap_{n=1}^{\infty} X_i = \{0\}$.

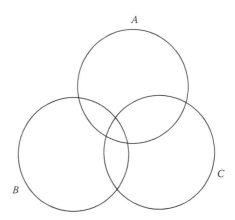

Figure A.1 A Venn diagram.

The **complement** of a set R is the set of all elements in the universe that are *not* in R. Common notations for this are $U\backslash R$, R^c, and R'. More generally, if R and S are two sets, we define the **complement of R with respect to S** as $R\backslash S = \{x : x \in R \text{ but } x \notin S\}$. We will use the notation R' for $U\backslash R$.

Set–theoretic relationships are often shown via **Venn diagrams** (Figure A.1).

A.2 Properties of Set Union and Intersection

Much of linear algebra is concerned with sets—sets of vectors, matrices, functions.... In working with sets, there are certain properties that make calculations easier. We can refer to these properties as making up the *algebra* of sets. It is important to know how to *prove* statements about sets. For example, to prove an inclusion $R \subseteq S$, we start with an arbitrary member of R and show that it satisfies the requirements for membership in S. To show that two sets are equal, $R = S$, we show that $R \subseteq S$ *and* $S \subseteq R$. This last technique is analogous to demonstrating that two numbers a and b are equal by showing $a \leq b$ and $b \leq a$.

Properties of Union, Intersection, and Complement

If A, B, and C are sets and U is the universe, then

(1) $\emptyset \cup A = A = A \cup \emptyset$ and $\emptyset \cap A = \emptyset = A \cap \emptyset$,
    $U \cup A = U = A \cup U$ and $U \cap A = A = A \cap U$
(2) $A \cup B = B \cup A$ and $A \cap B = B \cap A$ [*Commutativity*]
(3) $A \cup (B \cup C) = (A \cup B) \cup C$ and $A \cap (B \cap C) = (A \cap B) \cap C$
    [*Associativity*]
(4) $A \cap (B \cup C) = (A \cap B) \cup (A \cap C)$ and $A \cup (B \cap C) = (A \cup B) \cap$
    $(A \cup C)$ [*Distributivity*]
(5) $A \subseteq B$ and $B \subseteq C$ imply $A \subseteq C$ [*Transitivity*]

(6) $A \cap A' = \emptyset$ and $A \cup A' = U$

(7) $(A \cup B)' = A' \cap B'$ and $(A \cap B)' = A' \cup B'$ [*DeMorgan's Laws*]

(8) $(A')' = A$.

As an illustration of proof techniques in set theory, we will prove the first parts of property 4 and property 6.

$$A \cap (B \cup C) = (A \cap B) \cup (A \cap C)$$

The strategy is to prove that the left-hand set (LHS) is a subset of the right-hand set (RHS) and that the RHS is contained in the LHS.

First we prove that $A \cap (B \cup C) \subseteq (A \cap B) \cup (A \cap C)$. If $x \in A \cap (B \cup C)$, then $x \in A$ and $x \in (B \cup C)$. Thus we have $x \in A$ and $x \in B$ OR $x \in A$ and $x \in C$. But this last statement means that $x \in (A \cap B)$ OR $x \in (A \cap C)$. Thus $x \in (A \cap B) \cup (A \cap C)$, and the inclusion is proved.

On the other hand, if $x \in (A \cap B) \cup (A \cap C)$, then $x \in (A \cap B)$ OR $x \in (A \cap C)$—that is, $x \in A$ and $x \in B$ OR $x \in A$ and $x \in C$. This last statement means that $x \in A \cap (B \cup C)$, so $(A \cap B) \cup (A \cap C) \subseteq A \cap (B \cup C)$ and we are finished. ∎

$$(A \cup B)' = A' \cap B'$$

If $x \in (A \cup B)'$, then x is not an element of A and x is not an element of B. Thus $x \in A'$ and $x \in B'$, so $x \in A' \cap B'$. On the other hand, if $x \in A' \cap B'$, then $x \in A'$ and $x \in B'$, so x is neither in A nor in B. This says that x cannot be in $A \cup B$—that is, $x \in (A \cup B)'$. ∎

A.3 Cartesian Products of Sets

If R and S are two sets, the **Cartesian product** of R and S is denoted by $R \times S$ and consists of all ordered pairs (x, y) with $x \in R$ and $y \in S$:

$$R \times S = \{(x, y) : x \in R \text{ and } y \in S\}$$

Two ordered pairs (x, y) and (u, v) are **equal** if $x = u$ in R and $y = v$ in S.

We can define the Cartesian product of any finite number of sets $S_1, S_2, \ldots, S_n$ as the set of all ordered n-tuples $(x_1, x_2, \ldots, x_n)$, where $x_i \in S_i$ for $i = 1, 2, \ldots, n$:

$$S_1 \times S_2 \times \cdots \times S_n = \{(x_1, x_2, \ldots, x_n) : x_i \in S_i, 1 \leq i \leq n\}$$

If $S_i = S$ for $i = 1, 2, \ldots, n$, the Cartesian product is usually expressed as S^n. In the context of linear algebra or multivariable calculus, for example, the Cartesian product $\mathbb{R}^n$, **Euclidean n-space**, is important.

A.4 Mappings between Sets

A **function** (**map, mapping, transformation**) f from a set R to a set S, denoted $f : R \to S$, is a rule that assigns to each element r of R exactly one element s of S. The element s is called the **image** of r or the **value** of the function f at r. We usually write this relation as $s = f(r)$. The set R is called the **domain** of the function f and the set S is called the **codomain of the function** f. The set of all images of elements of R under f, denoted $f(R)$ or range f, is called the **range** of f: $f(R) = \{f(r) : r \in R\} = \mathcal{R}_f$. In general, $f(R) \subseteq S$. Two functions f and g are **equal** if their domains are equal and $f(x) = g(x)$ for every element x in the common domain of f and g.

If $R = S$, then the map $i : R \to R$ defined by $i(r) = r$ for every r in R is called the **identity map** on R. If it is necessary to distinguish identity maps on different sets, the map may be labeled—for example, as i_R.

If $f : R \to S$ and $\hat{S} \subseteq S$, we can investigate the set of elements of R that map into $\hat{S}$. The **preimage** of $\hat{S}$ under f is defined by $f^{-1}(\hat{S}) = \{r \in R : f(r) \in \hat{S}\}$. If $f : R \to S$ is a function, then the preimage $f^{-1} : S \to R$ is not necessarily a function. A necessary and sufficient condition for f^{-1} to be a function will be stated and proved after we consider a few more definitions.

A map $f : R \to S$ is said to be **one-to-one** (**1-1** or **injective**) if f maps distinct elements of R to distinct elements of S. Alternatively, a map f is one-to-one if $r_1 \neq r_2$ in R implies $f(r_1) \neq f(r_2)$ in S. Another useful way to describe a one-to-one map is to say that $f(r_1) = f(r_2)$ in S implies $r_1 = r_2$ in R. The mathematical implications in the last two sentences are *contrapositives* of each other, hence logically equivalent.

A map $f : R \to S$ is called **onto** (or **surjective**, from the French word *sur*, meaning "on" or "onto") if given any $s \in S$, there is at least one $r \in R$ such that $f(r) = s$. In other words, any element in S is the image of at least one element in R. It is easy to prove that $f : R \to S$ is onto if and only if $f(R) = S$. For instance, given a Cartesian product $\mathcal{S} = S_1 \times S_2 \times \cdots \times S_n = \{(x_1, x_2, \ldots, x_n) : x_i \in S_i,\ 1 \leq i \leq n\}$, we can define a **projection** $pr_i : \mathcal{S} \to S_i$ by x_i, where x_i denotes the ith component of $(x_1, x_2, \ldots, x_n)$. This projection mapping is onto, although not one-to-one.

The concepts "one-to-one" and "onto" are independent of each other. One property of a map does not imply the other. In particular, a map may have one, both, or neither of these properties.

A map $f : R \to S$ is **bijective** (**one-to-one onto**, or a **bijection**) if f is both injective and surjective.

Now we are ready to provide a necessary and sufficient condition for f^{-1} to be a function: f^{-1} **is a function from** S **to** R **if and only if** f **is bijective**. To prove this, first assume that f^{-1} is a function from S to R. To show that f is 1-1, assume that $f(r_1) = f(r_2) = s \in S$. Because f^{-1} is a function from S to R, every element of S has a unique image under f^{-1}. Thus, in particular, s has a unique image under f^{-1}. Because $f^{-1}(s) = r_1$ and $f^{-1}(s) = r_2$, it follows that $r_1 = r_2$, and so f is 1-1. To show that f is onto, let $s \in S$. Because f^{-1} is a function from S to R, there exists a unique element $r \in R$ such that $f^{-1}(s) = r$, implying $f(r) = s$. Therefore, f is onto. For the converse, assume that the function $f : R \to S$ is bijective. We show that f^{-1} is a function from S to R. Let $s \in S$. Because f is onto, there exists $r \in R$ such that $f(r) = s$. Hence $f^{-1}(s) = r$. Now we have to show that r is *unique*. So assume $f^{-1}(s) = \hat{r}$ as well. Then $f(r) = s$ and $f(\hat{r}) = s$. Because f is 1-1, we must have $r = \hat{r}$. Therefore, we have shown that for every $s \in S$, there exists a unique element $r \in R$ such that $f^{-1}(s) = r$—that is, f^{-1} is a function from S to R. ∎

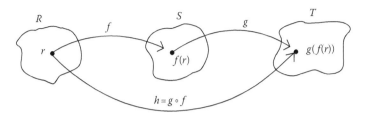

Figure A.2 The composition of mappings.

A.5 The Composition of Maps

A very important operation on mappings is that of *composition*. If $f: R \to S$ and $g: S \to T$, the **composition** of f and g is the mapping $h: R \to T$ defined by $h(r) = g(f(r))$ for all $r \in R$ (Figure A.2). We write $h = g \circ f$.

The properties of being injective, surjective, and bijective are preserved by composition.

Properties of the Composition

Let $f: R \to S$, $g: S \to T$, and $h: T \to V$ be three mappings, where R, S, T, and V are nonempty sets. Then

(1) If f and g are injective, so is $g \circ f$.
(2) If f and g are surjective, so is $g \circ f$.
(3) If f and g are bijective, so is $g \circ f$.
(4) $(h \circ g) \circ f = h \circ (g \circ f)$.

Proof (1) Suppose $f: R \to S$ and $g: S \to T$ are injective functions. Assume that $(g \circ f)(r_1) = (g \circ f)(r_2)$. By definition, $g(f(r_1)) = g(f(r_2))$. Because g is injective, it follows that $f(r_1) = f(r_2)$. However, because f is injective, we must have $r_1 = r_2$, proving that $g \circ f$ is injective.

(2): Now let $f: R \to S$ and $g: S \to T$ be surjective functions, and let $t \in T$. Because g is surjective, there exists $s \in S$ such that $g(s) = t$. On the other hand, because f is surjective, there must exist $r \in R$ such that $f(r) = s$. Hence $(g \circ f)(r) = g(f(r)) = g(s) = t$, implying that $g \circ f$ is also surjective.

(3): The fact that $g \circ f$ is bijective follows immediately from properties (1) and (2).

(4): Let $r \in R$. Also, let $f(r) = s$, $g(s) = t$, and $h(t) = v$. Then

$$((h \circ g) \circ f)(r) = (h \circ g)(f(r)) = (h \circ g)(s) = h(g(s)) = h(t) = v,$$

whereas

$$(h \circ (g \circ f))(r) = h((g \circ f)(r)) = h(g(f(r))) = h(g(s)) = h(t) = v.$$

Thus $(h \circ g) \circ f = h \circ (g \circ f)$.   ∎

If f is a function from R to S, and f^{-1} is a function from S to R, then $f^{-1} \circ f = i_R$ **and** $f \circ f^{-1} = i_S$. To prove this, let r be any element of R. Let $s = f(r) \in S$. Then $f^{-1}(s) = r$ because f^{-1} is a $1 - 1$ function. Thus $(f^{-1} \circ f)(r) = f^{-1}(f(r)) = f^{-1}(s) = r = i_R(r)$. Because r was an arbitrary element of R, we have just shown that $f^{-1} \circ f = i_R$. The proof of the second half of the theorem is similar.

Appendix B: Summation and Product Notation

B.1 Basic Notation

The Greek letter **sigma**, Σ, is used to denote the sum (finite or infinite) of a set of objects of a set, provided that these objects can be added in a meaningful way. Sigma is the Greek capital "s," which represents "summation." In particular, this notation is used for the addition of numbers, vectors, matrices, and functions. For example, if $x_1, x_2, \ldots, x_n$ are n numbers, we write $\sum_{i=1}^{n} x_i$ to mean the **finite series** $x_1 + x_2 + \cdots + x_n$. Each number i in the symbol x_i is called a **subscript**, which is just a label, a tag used to distinguish one element from another and order the elements in a particular way. In general, we can refer to these subscripts as **dummy variables** because in a mathematical discussion we can use other letters as subscripts.

If $\mathbf{x} = [x_1 \quad x_2 \quad \ldots \quad x_n]^T$ and $\mathbf{y} = [y_1 \quad y_2 \quad \ldots \quad y_n]^T$ are vectors in $\mathbb{R}^n$, we can express their **dot product** as $\mathbf{x} \cdot \mathbf{y} = \sum_{i=1}^{n} x_i y_i$ (Definition 1.2.1).

If $A = [a_{ij}]$ is an $n \times n$ matrix, for example, then the **trace** of A is the sum of the main diagonal entries of A: $\text{trace}(A) = \sum_{i=1}^{n} a_{ii}$. Alternatively, if $\lambda_1, \lambda_2, \ldots, \lambda_n$ are the eigenvalues of a matrix A, then $\text{trace}(A) = \sum_{i=1}^{n} \lambda_i$ (Theorem 4.4.3(2)).

If we have an infinite sequence of numbers $\{x_i\}_{i=1}^{\infty}$, we can denote the sum of this sequence as $\sum_{i=1}^{\infty} x_i$. Such an **infinite series** raises questions of **convergence** or **divergence**. We say that the infinite series **converges** if the sequence of **partial sums** $\left\{ \sum_{i=1}^{n} x_i \right\}_{n=1}^{\infty} = \{x_1, x_1 + x_2, \ldots, x_1 + x_2 + \cdots + x_n, \ldots\}$ converges. Otherwise, we say the series **diverges**.

There are some basic properties of sigma notation that simplify dealing with finite series.

Theorem B.1

Suppose $\{x_1, x_2, \ldots, x_n\}$, $\{y_1, y_2, \ldots, y_n\}$, and $\{z_1, z_2, \ldots, z_n\}$ are finite sets of mathematical objects with the same defined addition and k is a constant (a *scalar*, in the language of linear algebra). If, furthermore, this "addition" is commutative and associative, then

(1) $\sum_{i=1}^{n} (x_i + y_i + z_i) = \sum_{i=1}^{n} x_i + \sum_{i=1}^{n} y_i + \sum_{i=1}^{n} z_i.$

(2) $\sum_{i=1}^{n} k x_i = k \sum_{i=1}^{n} x_i.$

(3) $\sum_{i=1}^{n} k = nk.$

Proof (1) By using associativity and commutativity, we can regroup so that "like terms" appear together:

$$\sum_{i=1}^{n} (x_i + y_i + z_i) = (x_1 + y_1 + z_1) + (x_2 + y_2 + z_2) + \cdots$$

$$+ (x_n + y_n + z_n)$$
$$= (x_1 + x_2 + \cdots + x_n) + (y_1 + y_2 + \cdots + y_n)$$
$$+ (z_1 + z_2 + \cdots + z_n)$$
$$= \sum_{i=1}^{n} x_i + \sum_{i=1}^{n} y_i + \sum_{i=1}^{n} z_i.$$

(2): $$\sum_{i=1}^{n} kx_i = (kx_1 + kx_2 + \cdots + kx_n)$$

$$= k(x_1 + x_2 + \cdots + x_n) = k \sum_{i=1}^{n} x_i.$$

(3): $$\sum_{i=1}^{n} k = \sum_{i=1}^{n} k \cdot 1 \overset{(2)}{=} k \sum_{i=1}^{n} 1 = k(\overbrace{1 + 1 + \cdots + 1}^{n \text{ times}}) = nk.$$

The same properties are valid for infinite series, provided the series in question converge.

Example B.1

In statistics, an important measure of the variation present in a finite set of discrete measurements makes use of a sum of the form $S = \sum_{i=1}^{n} (X_i - \overline{X})^2$, where $\overline{X}$ is the *arithmetic mean* $\overline{X} = \left(\sum_{i=1}^{n} X_i \right) / n$. We want to use the properties of summation to find a form of S, that is, better suited to calculation. We have

$$S = \sum_{i=1}^{n} (X_i - \overline{X})^2 = \sum_{i=1}^{n} \left(X_i^2 - 2X_i \overline{X} + \overline{X}^2 \right)$$

$$= \sum_{i=1}^{n} X_i^2 - 2\overline{X} \sum_{i=1}^{n} X_i + \sum_{i=1}^{n} \overline{X}^2$$

$$= \sum_{i=1}^{n} X_i^2 - 2\left(\frac{\sum_{i=1}^{n} X_i}{n} \right) \sum_{i=1}^{n} X_i + n\left(\frac{\sum_{i=1}^{n} X_i}{n} \right)^2$$

$$= \sum_{i=1}^{n} X_i^2 - 2\frac{\left(\sum_{i=1}^{n} X_i \right)^2}{n} + \frac{\left(\sum_{i=1}^{n} X_i \right)^2}{n}$$

$$= \sum_{i=1}^{n} X_i^2 - \frac{\left(\sum_{i=1}^{n} X_i \right)^2}{n}.$$

The last expression is the form of S that is used most frequently because it is easy to calculate.

Related to sigma notation is the *pi notation* for finite and infinite products: $\prod_{i=1}^{n} a_i = a_1 a_2 \cdots a_n$ or $\prod_{i=1}^{\infty} a_i = a_1 a_2 \cdots a_n \cdots$. $\prod$ is the Greek capital "p," which represents "product." For example, the *factorial* of a positive integer n can be expressed as $n! = \prod_{k=0}^{n-1} (n-k)$, or as $n! = \prod_{k=1}^{n} k$.

If $a_1, a_2, \ldots, a_n$ are nonnegative real numbers, then the famous **Arithmetic-Geometric Mean (AGM) inequality**

$$\sqrt[n]{a_1 a_2 \cdots a_n} \leq \frac{a_1 + a_2 + \cdots + a_n}{n}$$

can be written succinctly as

$$\left(\prod_{k=1}^{n} a_k \right)^{\frac{1}{n}} \leq \frac{1}{n} \sum_{k=1}^{n} a_k.$$

Definition 4.1.1 describes the **determinant** of a square matrix A as a product. Equivalently, if $\lambda_1, \lambda_2, \ldots, \lambda_n$ are the eigenvalues of a matrix A, then Theorem 4.4.3(1) states that $\det(A) = \prod_{k=1}^{n} \lambda_k$.

B.2 Double Series

A **double series** or **double sum** is a series having terms depending on two subscripts (indices): $\sum_{i,j} a_{ij}$, $\sum_{i=1}^{m} \sum_{j=1}^{n} a_{ij}$, or $\sum_{1 \leq i,j \leq n} a_{ij}$. If $A = [a_{ij}]$ is a 3×4 matrix, for example, then

$$S = \sum_{i=1}^{3} \sum_{j=1}^{4} a_{ij} = \sum_{i=1}^{3} (a_{i1} + a_{i2} + a_{i3} + a_{i4})$$

$$= (a_{11} + a_{12} + a_{13} + a_{14}) + (a_{21} + a_{22} + a_{23} + a_{24})$$
$$+ (a_{31} + a_{32} + a_{33} + a_{34}) + (a_{41} + a_{42} + a_{43} + a_{44})$$

gives the sum of all the elements of A. The same sum can be calculated by *reversing the order of summation*:

$$S = \sum_{j=1}^{4} \sum_{i=1}^{3} a_{ij} = \sum_{j=1}^{4} (a_{1j} + a_{2j} + a_{3j} + a_{4j})$$

$$= (a_{11} + a_{21} + a_{31} + a_{41}) + (a_{12} + a_{22} + a_{32} + a_{42})$$
$$+ (a_{13} + a_{23} + a_{33} + a_{43}) + (a_{14} + a_{24} + a_{34} + a_{44})$$

In the first calculation of S, the sums in the parentheses are *row* sums RSi, whereas in the second calculation the sums in parentheses are *column* sums CSj:

a_{11}	a_{12}	$\cdots$	a_{1n}	$\rightarrow$	RS1
a_{21}	a_{22}	$\cdots$	a_{2n}	$\rightarrow$	RS2
$\vdots$	$\vdots$	$\cdots$	$\vdots$	$\vdots$	$\vdots$
a_{m1}	a_{m2}	$\cdots$	a_{mn}	$\rightarrow$	RSm
$\downarrow$	$\downarrow$	$\cdots$	$\downarrow$		$\downarrow$
CS1	CS2	$\cdots$	CSn	$\rightarrow$	S

Some finite double series can be written as a product of series:

$$\sum_{i=1}^{m}\sum_{j=1}^{n}a_i b_j = (a_1 b_1 + a_1 b_2 + \cdots + a_1 b_n) + (a_2 b_1 + a_2 b_2 + \cdots + a_2 b_n)$$

$$+ \cdots + (a_m b_1 + a_m b_2 + \cdots + a_m b_n)$$
$$= (a_1 + a_2 + \cdots + a_m) b_1 + (a_1 + a_2 + \cdots + a_m) b_2$$
$$+ \cdots + (a_1 + a_2 + \cdots + a_m) b_n$$
$$= \left(\sum_{i=1}^{m} a_i\right)(b_1 + b_2 + \cdots + b_n)$$
$$= \left(\sum_{i=1}^{m} a_i\right)\left(\sum_{j=1}^{n} b_i\right)$$

Appendix C:
Mathematical Induction

Mathematical Induction is a very useful method of proving statements (often formulas) involving natural numbers. It is claimed that the Italian scientist Francesco Maurolico was the first European to use mathematical induction to provide rigorous proofs, whereas the English logician Augustus De Morgan (1806–1873) coined the phrase "mathematical induction" in the early nineteenth century.*

Fundamentally, we can think of induction in this way: If we can climb onto some rung of a ladder (rung K) and if, whenever we are on rung n, we can always climb to rung $(n+1)$, then we can climb the entire ladder, beginning with rung K. Alternatively, the idea has been expressed in terms of a row of dominos standing on edge: If we can knock over domino K and if, whenever we knock over domino n, domino $(n+1)$ is knocked over, then we can topple the entire row of dominos, beginning with domino K.

Many of the results in the text—and even more exercises—depend upon one of the forms of induction described in this appendix.

C.1 Weak Induction

The Principle of Mathematical Induction (or the **Weak Principle of Mathematical Induction**):

Let $P(n)$ represent a statement relative to a positive integer n. If

(1) $P(K)$ is true, where K is the smallest integer for which the statement can be made.
and
(2) whenever $P(n)$ is true, it follows that $P(n+1)$ must also be true,
then $P(n)$ is true for *all $n \geq K$*.

In mathematical induction, the assumption in (2) that $P(n)$ is true is called the **inductive hypothesis**. To be able to use induction effectively, we must be able to express the statement $P(n+1)$ in terms of the statement $P(n)$ in a fairly straightforward way. If this cannot be done neatly, then we will probably have to try another method of proof.

---

* See, for example, W.H. Bussey, The origin of mathematical induction, *Amer. Math. Monthly* **24** (1917): 199–207 and D.E. Knuth, *The Art of Computer Programming*, 3rd edn., vol. 1, p. 17 (Boston, MA: Addison-Wesley, 1997).

Example C.1

We will use induction to prove the statement $P(n)$: A set of n elements has 2^n subsets. First we prove that $P(1)$ is true: $S = \{a\}$ has $2^1 = 2$ subsets, the empty set and S itself. Now we suppose that $P(n)$ is true and show that $P(n+1)$ must be true—that is, assuming that any set of n elements has 2^n subsets, we demonstrate that a set of $n+1$ elements has 2^{n+1} subsets.

Let $S = \{a_1, a_2, \ldots, a_n, a_{n+1}\}$ be any set of $n+1$ elements. The key observation is that every one of the subsets of the n-element set $\hat{S} = \{a_1, a_2, \ldots, a_n\}$ is automatically a subset of S, so this accounts for 2^n subsets of S by the inductive hypothesis. Furthermore, if R is any subset of $\hat{S}$, then $R \cup \{a_{n+1}\}$ is a subset of S. This gives us another 2^n subsets of S, for a total of $2^n + 2^n = 2^{n+1}$ subsets so far. In fact, we now have *all* the subsets of S because every subset of S either contains a_{n+1} or it does not. Thus $P(n+1)$ is true, so $P(n)$ is true for all positive integers n. (Note that we could have begun our induction with $n = 0$.)

Example C.2

Let $A = \begin{bmatrix} 1 & 0 \\ 1 & 1 \end{bmatrix}$. We want to prove that the statement $P(n)$:

$A^n = \begin{bmatrix} 1 & 0 \\ n & 1 \end{bmatrix}$ is true for all integers $n \geq 2$. (See Section 3.3 for powers of square matrices.)

We start by verifying $P(2)$: $A^2 = \begin{bmatrix} 1 & 0 \\ 1 & 1 \end{bmatrix}\begin{bmatrix} 1 & 0 \\ 1 & 1 \end{bmatrix} = \begin{bmatrix} 1 & 0 \\ 2 & 1 \end{bmatrix}$.

Now we assume that $A^n = \begin{bmatrix} 1 & 0 \\ n & 1 \end{bmatrix}$ and demonstrate that

$A^{n+1} = \begin{bmatrix} 1 & 0 \\ n+1 & 1 \end{bmatrix}$. We have, for $n \geq 2$,

$$A^{n+1} = A \cdot A^n = \begin{bmatrix} 1 & 0 \\ 1 & 1 \end{bmatrix} \cdot \overbrace{\begin{bmatrix} 1 & 0 \\ n & 1 \end{bmatrix}}^{\text{By the inductive hypothesis}} = \begin{bmatrix} 1 & 0 \\ n+1 & 1 \end{bmatrix}.$$

Thus $P(n+1)$ is true, and so $P(n)$ must be true for all $n \geq 2$.

The proof of Theorem 1.4.2 uses a limited form of induction, showing that a statement is true for a finite set of positive integer values. Theorem 1.2.2 and Schur's Theorem (Theorem 7.6.1a) use the usual Principle of Mathematical Induction.

C.2 Strong Induction

Sometimes assuming the truth of $P(n)$ is not strong enough to prove the truth of $P(n+1)$. An alternative version of induction, called the **Principle of Complete Induction**, replaces statement (2) in the earlier definition by a stronger hypothesis. We can also call this form of induction the **Principle of Strong Mathematical Induction**.

The **Principle of Complete Induction** (or the **Principle of Strong Mathematical Induction**):

Let $P(n)$ represent a statement relative to a positive integer n. If

(1) $P(K)$ is true, where K is the smallest integer for which the statement can be made.
(2) whenever $P(m)$ is true for all numbers m with $m = K, K+1, \ldots, n$, it follows that $P(n+1)$ must also be true, then $P(n)$ is true for *all* $n \geq K$.

Notice that the Principle of Strong Mathematical Induction implies the Principle of Mathematical Induction. In fact, the two principles are equivalent, although we will not prove this. Which method of induction we use (if any) is sometimes a matter of taste and usually depends on the nature of the problem.

Example C.3

Suppose there is a country where only two types of coins are minted—a 3-cent coin and a 5-cent coin. We want to prove that any amount of money greater than 7 cents can be paid in coins. In formal terms, we want to prove that the statement $P(n)$: "Any amount of money $n > 7$ can be paid using 3-cent coins and 5-cent coins."

We start by verifying $P(8): 8 = 3 + 5$. Now let k be greater than or equal to 8 and assume that any amount of money greater than 7 and less than or equal to k can be paid in coins. We consider two cases:

(a) If $k - 5 \geq 8$, then by the inductive hypothesis $k - 5$ can be paid using the two kinds of coins. But then so can $k + 1$: $k + 1 = (k - 5) + 3 + 3$.
(b) If $k - 5 < 8$, then $k + 1 < 14$, and we have to check just a few values, namely $k + 1 = 9$, 10, 11, 12, and 13. But $9 = 3 + 3 + 3$, $10 = 5 + 5$, $11 = 3 + 3 + 5$, $12 = 3 + 3 + 3 + 3$, and $13 = 3 + 5 + 5$.

Therefore we have proved that in either case $k + 1$ can be paid in coins. Thus, any amount of money greater than 7 cents can be paid in coins.

Example C.4

Suppose we define a sequence $\{a_n\}$ as follows:

$$a_1 = 2, \quad a_2 = 4, \quad \text{and} \quad a_{n+2} = 5a_{n+1} - 6a_n \quad \text{for all } n \geq 3.$$

If we use the *recursion formula* to calculate a few terms, we see that the sequence starts out $2, 4, 8, 16, 32, \ldots$. Therefore we *conjecture* (make an intelligent guess) that the statement $P(n): ``a_n = 2^{n"}$ is true for all natural numbers n. Now we prove the conjecture.

As usual, first we establish $P(1)$: Clearly $a_1 = 2^1 = 2$. Because the formula we are given for the $(n+2)$nd term does not depend only on the previous term, but on the previous *two* terms, we use strong induction with the hypothesis that the formula for a_n is valid for all integers $1, 2, \ldots, n+1$. Then, using the inductive hypothesis, we have $a_{n+2} = 5a_{n+1} - 6a_n = 5(2^{n+1}) - 6(2^n) = 2^n[(5 \cdot 2) - 6] = 2^n \cdot 2^2 = 2^{n+2}$, that is, $P(n+2)$ is true. This proves that $P(n)$ is true for all $n \geq 1$.

The proof that the reduced row echelon form of a matrix is unique (Theorem 2.5.1) uses complete induction. For a more complex use of the Principle of Strong Mathematical Induction, see the proof of Theorem 7.6.2 (the Cayley–Hamilton Theorem).

Appendix D:
Complex Numbers

D.1 The Algebraic View

Historically, the need for complex numbers arose when people tried to solve equations such as $x^2 + 1 = 0$ and realized that there was no real number satisfying this equation. The basic element in the expansion of the number system is the **imaginary unit**, $i = \sqrt{-1}$. A **complex number** is any expression of the form $x + yi$, where x and y are real numbers. In the complex number $z = x + yi$, x is called the **real part**—denoted Re(z)—and y is called the **imaginary part**—denoted Im(z)—of the complex number. Note that despite its name, y is a real number. In particular, any real number x is a member of the family of complex numbers because it can be written $x + 0 \cdot i$. Any complex number of the form yi $(= 0 + yi)$, where y is real, is called a **pure imaginary number**.

We add and subtract complex numbers by combining real and imaginary parts as follows:

$$(a + bi) + (c + di) = (a + c) + (b + d)i,$$

and

$$(a + bi) - (c + di) = (a - c) + (b - d)i.$$

We can also multiply complex numbers as we would multiply any binomials in algebra, remembering to replace i^2 whenever it occurs by -1:

$$(a + bi) \cdot (c + di) = ac + adi + bci + bdi^2 = (ac - bd) + (ad + bc)i.$$

Division of complex numbers is not as obvious. If $z = x + yi$, then the **complex conjugate** of z, denoted $\bar{z}$, is defined as follows: $\bar{z} = x - yi$. The complex conjugate is important in division because $z \cdot \bar{z} = x^2 + y^2$, a real number. In dividing complex numbers, the complex conjugate plays much the same role as the conjugate used in algebra to "rationalize the denominator." In dividing two complex numbers, we use the complex conjugate of the divisor to get the quotient to look like a complex number. For example,

$$\frac{-1 + 2i}{2 + 3i} = \frac{-1 + 2i}{2 + 3i} \cdot \frac{2 - 3i}{2 - 3i} = \frac{4 + 7i}{4 + 9} = \frac{4}{13} + \frac{7}{13}i.$$

In general, if $z = a + bi$ and $w = c + di$, then

$$\frac{z}{w} = \frac{a+bi}{c+di} = \frac{a+bi}{c+di} \cdot \frac{c-di}{c-di} = \frac{ac+bd}{c^2+d^2} + \frac{bc-ad}{c^2+d^2} i.$$

If z and w are complex numbers, then the following basic properties are easy to prove: (1) $\bar{\bar{z}} = z$; (2) $\overline{(z+w)} = \bar{z} + \bar{w}$; (3) $\overline{z \cdot w} = \bar{z} \cdot \bar{w}$; and (4) $\overline{\left(\frac{z}{w}\right)} = \frac{\bar{z}}{\bar{w}}$ for $w \neq 0$. Also, $\text{Re}(z) = \frac{z+\bar{z}}{2}$ and $\text{Im}(z) = \frac{z-\bar{z}}{2i}$.

Given the quadratic equation $ax^2 + bx + c = 0$, where a, b, and c are real numbers with $a \neq 0$, the quadratic formula yields a complex conjugate pair of solutions

$$x = \frac{-b + \sqrt{q}i}{2a} \quad \text{and} \quad x = \frac{-b - \sqrt{q}i}{2a},$$

when the discriminant $b^2 - 4ac$ equals $-q$ for q a positive real number.

The important algebraic properties of commutativity, associativity, and distributivity are valid for complex numbers. The complex numbers form an algebraic structure called a **field**. In more general treatments of linear algebra, the scalars are usually elements of a (nameless) field. Furthermore, all the properties in this appendix extend to vectors and matrices with complex entries. For example, if $\mathbf{v} = [c_1 \quad c_2 \quad \ldots \quad c_n]^{\text{T}}$ is a vector with complex components (i.e., $\mathbf{v} \in \mathbb{C}^n$), then $\bar{\mathbf{v}} = [\bar{c}_1 \quad \bar{c}_2 \quad \ldots \quad \bar{c}_n]^{\text{T}}$. If $A = [a_{ij}]$ is a matrix with complex entries, then $\bar{A} = \overline{[a_{ij}]} = [\bar{a}_{ij}]$. Theorems 7.1.1 and 7.1.2 in the text capture the most important algebraic properties of complex matrices.

D.2 The Geometric View

We can represent a complex number using the familiar Cartesian coordinate system, making the horizontal axis the **real axis** and the vertical axis the **imaginary axis**. Such a system is called the **complex plane**. What this does is establish a special correspondence (an **isomorphism**, in the language of Section 6.6) between complex numbers $z = x + yi$ and vectors $\begin{bmatrix} x \\ y \end{bmatrix}$ in $\mathbb{R}^2$ (or the point with coordinates (x, y) in the Cartesian plane). Figure D.1 illustrates this.

In particular, the addition/subtraction of complex numbers corresponds to the **Parallelogram Law** of vector algebra (Figure D.2).

The **modulus**, or **absolute value**, of the complex number $z = x + yi$, denoted $|z|$, is the nonnegative real number defined as $|z| = \sqrt{x^2 + y^2}$. The number $|z|$ represents the *length* of the complex number (vector) z. Note that $|z|^2 = z \cdot \bar{z}$.

D.3 Euler's Formula

Around 1740, while studying certain differential equations, the great Swiss mathematician Leonhard Euler (pronounced "oiler") discovered his famous formula for complex exponentials:

$$e^{iy} = \cos y + i \sin y.$$

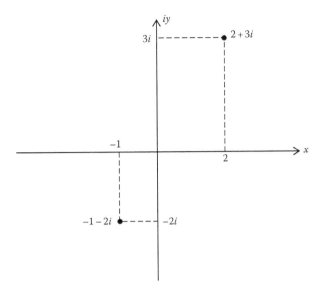

Figure D.1 The complex plane.

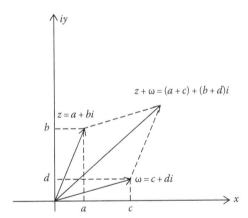

Figure D.2 The Parallelogram Law.

If $z = x + yi$, then we can write

$$e^z = e^{x+iy} = e^x e^{iy} = e^x(\cos y + i \sin y).$$

Letting $y = \pi$, we get the famous formula connecting five of the most basic constants in mathematics: $e^{i\pi} + 1 = 0$. For a marvelous account of this formula and its consequences, see *Dr. Euler's Fabulous Formula: Cures Many Mathematical Ills* by Paul J. Nahin (Princeton, NJ: Princeton University Press, 2006).

Instead of using coordinates to label a number in the complex plane, we can identify a complex number z uniquely by its modulus r and the angle z makes with the positive real axis. Using Euler's result, we can write any

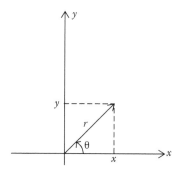

Figure D.3 The polar form of a complex number.

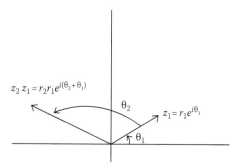

Figure D.4 The multiplication of complex numbers in polar form.

complex number $z = x + yi$ in **polar form**, $z = re^{i\theta}$, where $r = |z| = \sqrt{x^2 + y^2}$ and $\theta = \arctan(y/x)$ (Figure D.3).

If $z_1 = r_1 e^{i\theta_1}$ and $z_2 = r_2 e^{i\theta_2}$, then multiplication of these complex numbers has a nice geometric interpretation: $z_2 z_1 = (r_2 e^{i\theta_2})(r_1 e^{i\theta_1}) = r_2 r_1 e^{i(\theta_2 + \theta_1)}$, so multiplication of z_1 by z_2 magnifies the length of z_1 by a factor of $|z_2| = r_2$ and rotates the vector representing z_1 counterclockwise through an angle of θ_2 (Figure D.4).

Answers/Hints
to Odd-Numbered Problems

Exercises 1.1

A1. a. $[8 \quad 7 \quad 2 \quad 0]^T$ b. $[30 \quad 30 \quad -30 \quad -30]^T$
     c. $[1 \quad -6 \quad -6 \quad -6]^T$ d. $[15 \quad -44 \quad 50 \quad 30]^T$
     e. $[9/2 \quad 15/4 \quad 3 \quad 3/2]^T$ f. $[-7 \quad 0 \quad 12 \quad 12]^T$
     g. $[-605 \quad -617 \quad 590 \quad 588]^T$.

A3. $\mathbf{u} = [1 \quad 1/2 \quad -5/4]^T$, $\mathbf{v} = [2 \quad 1 \quad -5/2]^T$,
     $\mathbf{w} = [3 \quad 3/2 \quad -15/4]^T$.

A5. No, unless this vector will be used to calculate some sort of average (see A6).

B1. We can let b be an arbitrary nonzero real number (a "free variable") and then choose $a = -2b$ and $c = b$.

B5. (c) No.

C3. No. If $\mathbf{y} \neq \mathbf{0}$ is a vector in the complement of $\{\mathbf{x}\}$, then $\mathbf{u} = \mathbf{x} + \mathbf{y}$ and $\mathbf{v} = \mathbf{x} - \mathbf{y}$ are in the complement, but $\frac{1}{2}\mathbf{u} + \frac{1}{2}\mathbf{v}$ is not.

Exercises 1.2

A1. a. -38 b. -198 c. -35 d. -23 e. 18 f. -638.

A3. a. $\sqrt{34}$ b. $5\sqrt{2}$ c. 4 d. $\sqrt{19}$ e. $\sqrt{91}$ f. $3\sqrt{5}$.

A5. a. $\theta \approx 0.2487 \text{ rad} \approx 14.25°$ b. $\theta = \pi/2 \text{ rad} = 90°$
     c. $\theta = \pi/2 \text{ rad} = 90°$ d. $\theta \approx 1.4624 \text{ rad} \approx 83.79°$
     e. $\theta \approx 1.7609 \text{ rad} \approx 100.89°$ f. $\theta = \pi/2 \text{ radians} = 90°$
     g. $\theta \approx 1.3353 \text{ rad} \approx 76.51°$ h. $\theta \approx 1.7672 \text{ rad} \approx 101.26°$.

A7. Use properties (c) and (d) of Theorem 1.2.1.

A9. $\alpha = \mathbf{u} \cdot \mathbf{v}$.

A11.

X	i	j	k
i	0	k	j
j	−k	0	i
k	−j	i	0

B5. $a = 0, -2, 2$.

B7.   Position side $\mathbf{w}$ along the positive x-axis and calculate $\cos\theta_1 = ((\mathbf{v}+\mathbf{w})\bullet\mathbf{w})/(\|\mathbf{v}+\mathbf{w}\|\|\mathbf{w}\|)$ and $\cos\theta_2 = ((\mathbf{v}+\mathbf{w})\bullet\mathbf{v})/(\|\mathbf{v}+\mathbf{w}\|\|\mathbf{v}\|)$, remembering that we have unit vectors.

B9. $\theta \approx 0.9553$ rad $\approx 54.7°$.

B13.   For example, if $\mathbf{u} = \begin{bmatrix} 1 \\ 0 \end{bmatrix}$, $\mathbf{v} = \begin{bmatrix} 0 \\ 1 \end{bmatrix}$, and $\mathbf{w} = \begin{bmatrix} 0 \\ 5 \end{bmatrix}$ in $\mathbb{R}^2$, then $\mathbf{u} \bullet \mathbf{v} = 0 = \mathbf{u} \bullet \mathbf{w}$, but $\begin{bmatrix} 0 \\ 1 \end{bmatrix} \neq \begin{bmatrix} 0 \\ 5 \end{bmatrix}$.

C1.   a. 64 multiplications and 48 additions/subtractions.
      b. 48 multiplications and 120 additions/subtractions.

Exercises 1.3

A1.   The span is a straight line through the origin, in the same direction as the vector $[\,1 \quad 1 \quad 1\,]^T$.

A3.   There are *no* values of t for which $\mathbf{u}$ and $\mathbf{v}$ span $\mathbb{R}^3$.

A5.   Notice that $[\,2 \quad -1 \quad 6\,]^T + [\,-3 \quad 4 \quad 1\,]^T = [\,-1 \quad 3 \quad 7\,]^T$ and $[\,2 \quad -1 \quad 6\,]^T - 2[\,-3 \quad 4 \quad 1\,]^T = [\,8 \quad -9 \quad 4\,]^T$. Now show that the first set is spanned by the second set.

A7.   Show that $\mathbf{u}_1$ and $\mathbf{u}_2$ are linear combinations of $\mathbf{v}_1$ and $\mathbf{v}_2$.

A9.   The span is $\{[a+2b \quad 4a \quad 4b+4c \quad c]^T : a, b, c \in \mathbb{R}\}$, a three-dimensional subset of $\mathbb{R}^4$.

C1.   b. Each $\mathbf{v}_i \in span\{\mathbf{w}_1, \mathbf{w}_2, \ldots, \mathbf{w}_r\}$ and each $\mathbf{w}_i \in span\{\mathbf{v}_1, \mathbf{v}_2, \ldots, \mathbf{v}_k\}$.

Exercises 1.4

A1.   a. linearly dependent   b. linearly independent
       c. linearly dependent   d. linearly independent
       e. linearly independent   f. linearly independent
       g. linearly dependent   h. linearly dependent
       i. linearly dependent   j. linearly independent
       k. linearly dependent   l. linearly independent.

A7.   The intersection represents the intersection of two planes through the origin in $\mathbb{R}^3$. This could be a single point, a line, or a plane.

A9.   There are *no* values of t for which **u** and **v** are linearly dependent.

B7.   The vectors are linearly dependent if $a = 0$ or $a = \pm\sqrt{2}$. The vectors are linearly independent for all other real values of a.

B9.   For example, take $[1 \ \ 0 \ \ 0]^{\mathrm{T}}, [0 \ \ 1 \ \ 0]^{\mathrm{T}}$, and $[1 \ \ 1 \ \ 0]^{\mathrm{T}}$.

C1. $1 - \beta + \alpha\beta \neq 0$.

Exercises 1.5

A1.   a. No. Any basis for $\mathbb{R}^2$ must consist of two vectors.
       b. Yes. The vectors are linearly independent and span $\mathbb{R}^2$.
       c. No. Any three vectors in $\mathbb{R}^2$ must be linearly dependent.
       d. No. Any set containing the zero vector must be linearly dependent.
       e. Yes.
       f. No. Any four vectors in $\mathbb{R}^3$ must be linearly dependent.
       g. Yes.
       h. No. Any four vectors in $\mathbb{R}^3$ must be linearly dependent.
       i. No. The vectors are not linearly independent.
       j. No. Any basis for $\mathbb{R}^4$ must have exactly four vectors.

A3.   a. Add any vector that is not in the span of the two given vectors, for example, $[1 \ \ 1 \ \ 3]^{\mathrm{T}}$.
       b. First find a vector that is not a scalar multiple of the given vector. Now find the span of these two vectors and choose a third vector not in this span.

A5.   No one of the colors red, green, and blue can be obtained by combining the remaining colors.

A7. $B = \left\{ [\,-4 \quad -7\,]^T, \, \left[\tfrac{7}{2} \quad 6\right]^T \right\}$.

A9.   b. $\mathbf{e}_1 = \tfrac{7}{10}\mathbf{v}_1 + \tfrac{3}{10}\mathbf{v}_2 + \tfrac{1}{5}\mathbf{v}_3, \quad \mathbf{e}_2 = -\tfrac{1}{5}\mathbf{v}_1 + \tfrac{1}{5}\mathbf{v}_2 - \tfrac{1}{5}\mathbf{v}_3,$
        $\mathbf{e}_3 = -\tfrac{3}{10}\mathbf{v}_1 + \tfrac{3}{10}\mathbf{v}_2 + \tfrac{1}{5}\mathbf{v}_3.$

A11. $[\mathbf{e}_1]_B = [\,1 \quad -1 \quad 0 \quad 0\,]^T, \quad [\mathbf{e}_2]_B = [\,0 \quad 1 \quad -1 \quad 0\,]^T,$
        $[\mathbf{e}_3]_B = [\,0 \quad 0 \quad 1 \quad -1\,]^T, \quad [\mathbf{e}_4]_B = [\,0 \quad 0 \quad 0 \quad 1\,]^T$
        $[\mathbf{v}_1]_B = [\,1 \quad 1 \quad 1 \quad 1\,]^T, \quad [\mathbf{v}_2]_B = [\,0 \quad 1 \quad 1 \quad 1\,]^T,$
        $[\mathbf{v}_3]_B = [\,0 \quad 0 \quad 1 \quad 1\,]^T, \quad [\mathbf{v}_4]_B = [\,0 \quad 0 \quad 0 \quad 1\,]^T.$

B3.   For example, $\mathbf{v}$ could be $[\,2 \quad -11 \quad 8\,]^T$.

B5.   b. $\mathbf{x} = -\tfrac{2}{3\sqrt{2}}\mathbf{u}_1 + \tfrac{5}{3}\mathbf{u}_2 + 0 \cdot \mathbf{u}_3 = -\tfrac{\sqrt{2}}{3}\mathbf{u}_1 + \tfrac{5}{3}\mathbf{u}_2.$

B7. $c_1 = 4, \quad c_2 = 0, \quad c_3 = \pm 3.$

C1.   All real values of a except 0, -1, and 1.

C3.   b. If $B = \{\varepsilon_i\}_{i=1}^n$, then $[\mathbf{v}]_B = [\,-1 \quad -1 \quad \cdots \quad -1 \quad n\,]^T.$

Exercises 1.6

A3. $[\mathbf{w}_1]_{B_1} = \begin{bmatrix} 1 \\ 1 \end{bmatrix}, \quad [\mathbf{w}_2]_{B_1} = \begin{bmatrix} 1 \\ -1 \end{bmatrix}.$

A5. W is a line with slope 5 passing through the origin in $\mathbb{R}^2$. The dimension of W is 1.

A7.   No. The set is not closed under vector addition.

A9.   No. The set is not closed under scalar multiplication.

B1.   The set V is not closed under scalar multiplication (cf. Exercise A9).

B3.   Expand to a basis for the entire space. The new vectors (4 of them) make up the basis for $V^\perp$, for example, a basis is
        $\{[\,-4 \quad 0 \quad 0 \quad 0 \quad 1\,]^T, [\,-3 \quad 0 \quad 0 \quad 1 \quad 0\,]^T,$
        $[\,2 \quad 0 \quad 1 \quad 0 \quad 0\,]^T, [\,0 \quad 1 \quad 0 \quad 0 \quad 0\,]^T\}.$

B5.   Let U be the x-axis in $\mathbb{R}^2$ and let V be the y-axis. Then $\dim(U) = \dim(V) = 1$, but $U \neq V$.

Exercises 2.1

A1. $x = -1$, $y = 2$.

A3.   There is no solution. The lines are parallel.

A5.   There are infinitely many solutions. There is only one line.

A9.   All ordered pairs (x, y) satisfying $-12x + 2y = -3$.

A11.   Good crop: 9.25 dou/sheaf; mediocre crop: 4.25 dou/sheaf; bad crop: 2.75 dou/sheaf.

A13.   a. $y = 14x + 19$.

A15.   No. The solution set may be a line or a plane (if the two equations represent the same plane).

A17.

A19.   For arbitrary x, take $y = x$ and $z = 0$. The solution set is the span of $[\,1 \quad 1 \quad 0\,]^{\mathrm{T}}$, a line in three-dimensional space.

A21. $I_A = 27/38$, $I_B = -15/38$, $I_C = -6/19$.

B5. $k = 1$.

B7. $c - 2a - b = 0$.

C1. $x_1 = 1/2 = x_2 = x_3$; $x_4 = -1/2$.

C3. $x = y = z = 1/(\lambda + 2)$ if $\lambda \neq -2$. There is no solution if $\lambda = -2$.

Exercises 2.2

A1. $\begin{bmatrix} -3 & 4 \\ 1 & -2 \end{bmatrix} \begin{bmatrix} x \\ y \end{bmatrix} = \begin{bmatrix} 2 \\ -1 \end{bmatrix}$.

A3. $\begin{bmatrix} 1 & 1 & 1 \\ 1 & 2 & 3 \\ 2 & 3 & 4 \end{bmatrix} \begin{bmatrix} x \\ y \\ z \end{bmatrix} = \begin{bmatrix} 1 \\ 2 \\ 3 \end{bmatrix}$.

A5. $\begin{bmatrix} 1 & -1 & 1 & -1 \\ 2 & 3 & -1 & 4 \\ 3 & -2 & 4 & -1 \\ -2 & 1 & 2 & 1 \end{bmatrix} \begin{bmatrix} a \\ b \\ c \\ d \end{bmatrix} = \begin{bmatrix} 0 \\ -2 \\ 1 \\ -3 \end{bmatrix}$.

A7. $\begin{bmatrix} 1 & 1 & 1 \\ 3 & -4 & -3 \end{bmatrix} \begin{bmatrix} x \\ y \\ z \end{bmatrix} = \begin{bmatrix} -2 \\ 4 \end{bmatrix}$.

A9.   a. $\begin{bmatrix} 4 & 3 & 2 & 0 \\ 5 & -1 & 4 & 2 \\ 0 & 1 & -2 & 3 \end{bmatrix}$ b. $\begin{bmatrix} 0 & 1 & -2 & 3 \\ 5 & -1 & 4 & 2 \\ 16 & 12 & 8 & 0 \end{bmatrix}$

   c. $\begin{bmatrix} 0 & 1 & -2 & 3 \\ 5 & -1 & 4 & 2 \\ -11 & 6 & -10 & -6 \end{bmatrix}$.

A11.   For example, $\begin{bmatrix} 2 & -1 & 3 & 4 \\ 2 & 0 & 5 & 3 \\ 5 & 2 & -3 & 4 \end{bmatrix}$, $\begin{bmatrix} 2 & -1 & 3 & 4 \\ 0 & 1 & 2 & -1 \\ 5 & 3 & -1 & 3 \end{bmatrix}$,

   and $\begin{bmatrix} 2 & -1 & 3 & 4 \\ 0 & 1 & 2 & -1 \\ 9 & 0 & 3 & 12 \end{bmatrix}$.

A13. $x = 2$, $y = -1$.

A15. $a = 1/3$, $b = 5/3$.

A17.   The system has no solution. It is inconsistent.

B1. $span\{[-1 \quad -2 \quad 0 \quad 1]^T, [-1 \quad 1 \quad 1 \quad 0]^T\}$.

B3.   If $y = 2$, $a = b = c = d = e = k$, where k is any real number; if $y \neq 2$ and $y^2 + y - 1 \neq 0$, then $a = b = c = d = e = 0$; if $y \neq 2$ and $y^2 + y - 1 = 0$, then $a = s$, $b = t$, $c = yt - s$, $d = (y^2 - 1)t - ys$, $e = ys - t$, where s, t are arbitrary and y is a solution of $y^2 + y - 1 = 0$.

B5.   a. If x_i is the number initially in chamber i, then
       $x_1 = 10$, $x_2 = 20$, $x_3 = 30$, and $x_4 = 40$.
       b. 16, 22, 22, 40.

B9.   A basis for the column space is $\left\{ \begin{bmatrix} 1 \\ 0 \\ -1 \end{bmatrix}, \begin{bmatrix} 1 \\ 1 \\ 0 \end{bmatrix} \right\}$.

B11.   a. 3 b. 3 c. The answers are equal.

B13. $\begin{bmatrix} 1 & * & * \\ 0 & 1 & * \\ 0 & 0 & 1 \end{bmatrix}, \begin{bmatrix} 1 & * & * \\ 0 & 1 & * \\ 0 & 0 & 0 \end{bmatrix}, \begin{bmatrix} 1 & * & * \\ 0 & 0 & 1 \\ 0 & 0 & 0 \end{bmatrix}, \begin{bmatrix} 1 & * & * \\ 0 & 0 & 0 \\ 0 & 0 & 0 \end{bmatrix},$

       $\begin{bmatrix} 0 & 1 & * \\ 0 & 0 & 1 \\ 0 & 0 & 0 \end{bmatrix}, \begin{bmatrix} 0 & 1 & * \\ 0 & 0 & 0 \\ 0 & 0 & 0 \end{bmatrix}, \begin{bmatrix} 0 & 0 & 1 \\ 0 & 0 & 0 \\ 0 & 0 & 0 \end{bmatrix}.$

C3. $a_n = c_n/k!$

Exercises 2.3

A1. $x_1 = 3$, $x_2 = 1$, $x_3 = 1$.

A3. $x = 2$, $y = -2$, $z = 3$.

A5. $x = -1$, $y = -1$, $z = 0$, $w = 1$.

A7. $x = -2/33$, $y = 35/33$, $z = -42/11$, $w = 245/66$.

A9.   a. 4: There is at most one pivot per row.
       b. 4: There is at most one pivot per column.

B1.   a. $\{y[-2 \quad 1 \quad 0 \quad 0]^T + w[3 \quad 0 \quad -1 \quad 1]^T : y, w \in \mathbb{R}\}$.
       b. Theorem 2.2.2.
       c. Any vector in $\mathbb{R}^3$ whose last component is nonzero.

B3. $x = 2, y = 4, z = 5; x = -4, y = -7, z = -8; x = 3, y = 4, z = 4$.

B5. $\alpha = \pi/2$, $\beta = 3\pi/2$, $\gamma = 0$.

C5.   Suppose $A\mathbf{x} = \mathbf{b}$ is a consistent system with solution
       $\mathbf{x} = [x_1 \quad x_2 \quad \dots \quad x_n]^T$.
       a. If both j and k are not equal to $n+1$, the result is $x_j \leftrightarrow x_k$.
          If column j is switched with the last column, $\mathbf{b} = $ column
          $n+1$, then $x_j \rightarrow 1/x_j$ for $x_j \neq 0$ and each of the remaining
          x_i's is multiplied by $-1/j$.

b. If column j is not the last column, then $x_j \rightarrow x_j/c$ and the remaining x_i's are unchanged. If column **b** is multiplied by c, then each x_j is multiplied by c.

c. If a nonzero multiple of column k is added to column j, then solution x_k will change.

Exercises 2.4

A1.
$$\begin{bmatrix} 1 & 0 & 0 & 0 & 1 \\ 0 & 1 & 0 & 0 & 2 \\ 0 & 0 & 1 & 0 & 4 \\ 0 & 0 & 0 & 1 & 8 \\ 0 & 0 & 0 & 0 & 16 \end{bmatrix}.$$

A3.   a./b. $x = 1$, $y = 1$, $z = 1$.

B1.   a. $x = 0.999$, $y = -0.999$.

b. $\{0.001x - y = 0.999999, \ x + y = 0\}$.

c. $x = 1$, $y = -1$.

d. $\{x - y = 2, \ x + y = 0\}$.

e. $x = 0.\overline{999000}$, $y = -0.\overline{999000}$: This solution equals neither the solution in (a) nor the solution in (c).

f. $x = 0.999$, $y = -0.999$: This equals the solution in (a), but not in (c).

Exercises 2.5

A1.   a. Yes   b. No—(2) is violated   c. No—(2) is violated

d. No—the first nonzero entry in row 1 is 2, not 1

e. No—(3) is violated

f. No—(2) is violated   g. Yes

h. the first nonzero entry in row 3 is 3, not 1

i. No—(3) is violated.

A3.
$$\begin{bmatrix} 1 & 0 & 0 \\ 0 & 1 & 0 \\ 0 & 0 & 1 \end{bmatrix}.$$

A5.   Any $k \neq -2$.

A7.   a. $a = 5$, $b = 4$

b. $a = 5$, $b \neq 4$.

A9. $x^2 + y^2 - x + 3y - 10 = 0$,   or $(x - \frac{1}{2})^2 + (y + \frac{3}{2})^2 = (5\sqrt{2}/2)^2$.

A11.   200 shares of Gillette implies that there are 300 shares of IBM
       and 100 shares of Coca Cola.

A13. $a = 6, b = 6, c = 6, d = 1$.

A15. $a = 15, b = 44, c = 22, d = 88, e = 5, f = 90$.

B1. $A = \begin{bmatrix} 1 & 0 & 2 & 1 & 4 \\ -1 & -1 & 3 & -2 & -7 \\ 3 & 1 & 1 & 0 & -9 \end{bmatrix}$.

B3. $b_1 + b_2 - b_3 = 0$.

B5. $\begin{bmatrix} 1 & 0 \\ 0 & 1 \end{bmatrix}, \begin{bmatrix} 1 & 0 \\ 0 & 0 \end{bmatrix}, \begin{bmatrix} 0 & 1 \\ 0 & 0 \end{bmatrix}, \begin{bmatrix} 0 & 0 \\ 0 & 0 \end{bmatrix}$.

B7.   GOOD JOB.

C1. $\begin{bmatrix} 1 & 0 & * \\ 0 & 1 & * \end{bmatrix}, \begin{bmatrix} 1 & * & 0 \\ 0 & 0 & 1 \end{bmatrix}, \begin{bmatrix} 1 & * & * \\ 0 & 0 & 0 \end{bmatrix}, \begin{bmatrix} 0 & 1 & 0 \\ 0 & 0 & 1 \end{bmatrix}, \begin{bmatrix} 0 & 1 & * \\ 0 & 0 & 0 \end{bmatrix},$

$\begin{bmatrix} 0 & 0 & 1 \\ 0 & 0 & 0 \end{bmatrix}, \begin{bmatrix} 0 & 0 & 0 \\ 0 & 0 & 0 \end{bmatrix}$.

C3.   There are 26:

$\begin{bmatrix} 1 & 0 & 0 & * & * \\ 0 & 1 & 0 & * & * \\ 0 & 0 & 1 & * & * \end{bmatrix}, \begin{bmatrix} 1 & 0 & * & 0 & * \\ 0 & 1 & * & 0 & * \\ 0 & 0 & 0 & 1 & * \end{bmatrix}, \begin{bmatrix} 1 & 0 & * & * & 0 \\ 0 & 1 & * & * & 0 \\ 0 & 0 & 0 & 0 & 1 \end{bmatrix},$

$\begin{bmatrix} 1 & 0 & * & * & * \\ 0 & 1 & * & * & * \\ 0 & 0 & 0 & 0 & 0 \end{bmatrix}, \begin{bmatrix} 1 & * & 0 & 0 & * \\ 0 & 0 & 1 & 0 & * \\ 0 & 0 & 0 & 1 & 0 \end{bmatrix},$

$\begin{bmatrix} 1 & * & 0 & * & 0 \\ 0 & 0 & 1 & * & 0 \\ 0 & 0 & 0 & 0 & 1 \end{bmatrix}, \begin{bmatrix} 1 & * & 0 & * & * \\ 0 & 0 & 1 & * & * \\ 0 & 0 & 0 & 0 & 0 \end{bmatrix}, \begin{bmatrix} 1 & * & * & 0 & 0 \\ 0 & 0 & 0 & 1 & 0 \\ 0 & 0 & 0 & 0 & 1 \end{bmatrix},$

$\begin{bmatrix} 1 & * & * & 0 & * \\ 0 & 0 & 0 & 1 & * \\ 0 & 0 & 0 & 0 & 0 \end{bmatrix}, \begin{bmatrix} 1 & * & * & * & 0 \\ 0 & 0 & 0 & 0 & 1 \\ 0 & 0 & 0 & 0 & 0 \end{bmatrix},$

$\begin{bmatrix} 0 & 1 & 0 & 0 & * \\ 0 & 0 & 1 & 0 & * \\ 0 & 0 & 0 & 1 & * \end{bmatrix}, \begin{bmatrix} 0 & 1 & 0 & * & 0 \\ 0 & 0 & 1 & * & 0 \\ 0 & 0 & 0 & 0 & 1 \end{bmatrix}, \begin{bmatrix} 0 & 1 & 0 & * & * \\ 0 & 0 & 1 & * & * \\ 0 & 0 & 0 & 0 & 0 \end{bmatrix},$

$\begin{bmatrix} 0 & 1 & * & 0 & 0 \\ 0 & 0 & 0 & 1 & 0 \\ 0 & 0 & 0 & 0 & 1 \end{bmatrix}, \begin{bmatrix} 0 & 1 & * & 0 & * \\ 0 & 0 & 0 & 1 & * \\ 0 & 0 & 0 & 0 & 0 \end{bmatrix}, \begin{bmatrix} 0 & 1 & * & * & 0 \\ 0 & 0 & 0 & 0 & 1 \\ 0 & 0 & 0 & 0 & 0 \end{bmatrix},$

$\begin{bmatrix} 0 & 1 & * & * & * \\ 0 & 0 & 0 & 0 & 0 \\ 0 & 0 & 0 & 0 & 0 \end{bmatrix}, \begin{bmatrix} 0 & 0 & 1 & 0 & 0 \\ 0 & 0 & 0 & 1 & 0 \\ 0 & 0 & 0 & 0 & 1 \end{bmatrix}, \begin{bmatrix} 0 & 0 & 1 & 0 & * \\ 0 & 0 & 0 & 1 & * \\ 0 & 0 & 0 & 0 & 0 \end{bmatrix},$

$\begin{bmatrix} 0 & 0 & 1 & * & 0 \\ 0 & 0 & 0 & 0 & 1 \\ 0 & 0 & 0 & 0 & 0 \end{bmatrix}, \begin{bmatrix} 0 & 0 & 1 & * & * \\ 0 & 0 & 0 & 0 & 0 \\ 0 & 0 & 0 & 0 & 0 \end{bmatrix},$

$$
\begin{bmatrix} 0 & 0 & 0 & 1 & 0 \\ 0 & 0 & 0 & 0 & 1 \\ 0 & 0 & 0 & 0 & 0 \end{bmatrix}, \quad
\begin{bmatrix} 0 & 0 & 0 & 1 & * \\ 0 & 0 & 0 & 0 & 0 \\ 0 & 0 & 0 & 0 & 0 \end{bmatrix}, \quad
\begin{bmatrix} 0 & 0 & 0 & 0 & 1 \\ 0 & 0 & 0 & 0 & 0 \\ 0 & 0 & 0 & 0 & 0 \end{bmatrix},
$$

$$
\begin{bmatrix} 0 & 0 & 0 & 0 & 0 \\ 0 & 0 & 0 & 0 & 0 \\ 0 & 0 & 0 & 0 & 0 \end{bmatrix}, \quad
\begin{bmatrix} 1 & * & * & * & * \\ 0 & 0 & 0 & 0 & 0 \\ 0 & 0 & 0 & 0 & 0 \end{bmatrix}.
$$

C5.   a. There is no solution.

b. If $r < \min(m, n)$, there could be infinitely many solutions or a unique solution.

Exercises 2.6

A1.   a. $x = 10$, $y = -10$.

b. The right-hand sides are now 29 and 19—close.

c. Now $x = 35$, $y = -38$. This is a relatively large change.

A3.   a. $x_1 = 1 = x_2$.

b. $x_1 = 20.97$, $x_2 = -18.99$: A small change in the output vector results in a large change to the solution.

B1.   The first system has solution $x = -98$, $y = 102$, whereas the second system has solution $x = -16$, $y = 20$. A slight change in the coefficient of x results in a large change in the solution.

B3.   a. $x = 3$, $y = 2$.

b. $x = 3 + h/3$, $y = 2 - h/3$: The absolute error is $|h|/3$ for both x and y. This is small for small values of $|h|$.

c. $\|\Delta \mathbf{b}\| / \|\mathbf{b}\| = \sqrt{26}|h|/26$ and $\|\Delta \mathbf{x}\| / \|\mathbf{x}\| = \sqrt{26}|h|/39$. The system is well-conditioned.

C1.   a. $x = 1$, $y = 1$.

b. $\mathbf{r} = \begin{bmatrix} 0 \\ 0.00002 \end{bmatrix}$.

c. $\mathbf{x} - \tilde{\mathbf{x}} = \begin{bmatrix} -2 \\ 1 \end{bmatrix}$: A small residual vector does not guarantee the accuracy of an approximate solution.

C3.   a. $x = 2$, $y = 1$.

b. $x = 2 - 100,000h$, $y = 1 + 100,000h$: If $|h| \ll 10^{-5}$, the change in the output vector does not change the solution much in terms of absolute error.

c. $\|\Delta \mathbf{b}\| / \|\mathbf{b}\| \approx |h|/3.46$ and $\|\Delta \mathbf{x}\| / \|\mathbf{x}\| \approx 63,246|h|$: The system is ill-conditioned.

Exercises 2.7

A1. The row space is $\mathbb{R}^3$, so the row rank is 3.

A3. The row rank is 2 and the column rank is 2.

A5. For example, $\{[2\ 0\ 11\ -1\ -4]^T, [0\ 2\ 5\ -1\ 0]^T\}$.

A7. For example, $\{[1\ 1\ -1]^T, [0\ 1\ 1]^T\}$.

A9. For example, $\{[1\ 0\ 0\ 2]^T, [2\ 1\ 1\ 0]^T, [4\ -5\ 3\ -1]^T, [0\ -1\ -1\ -1]^T\}$.

A11. 6

A13. $\{[a+2b\ \ b-2a\ \ a]^T : a, b \in \mathbb{R}\}$.

A15. Rank$(A) = 3$, nullity$(A) = 0$: The unique solution is $\mathbf{0}$.

B1. If $b \neq 3$, a basis for the row space is $\{[1\ 0\ 1\ 0]^T, [0\ 1\ 1\ 0]^T, [0\ 0\ 0\ 1]^T\}$. A basis for the column space is $\{[0\ 1\ 1]^T, [1\ 0\ -1]^T, [b\ 4\ 1]^T\}$.

B3. b. A basis for the null space is $\{[-a-b-c\ \ 1\ \ 1]^T\}$.

Exercises 2.8

A1. 3

A3. 1

A5. 2

A7. 1

A9. 3

A11. 3

A13. Rank$(A) = 2$. A basis for the row space is $\{[3\ 0\ -1\ -3\ 5\ 7]^T, [0\ 6\ -2\ 3\ 7\ 2]^T\}$ and a basis for the column space is $\{[1\ 2\ -1\ 0]^T, [2\ -2\ -8\ -6]^T\}$.

A15.   a. $x - y + 2z + 3w = 4, \quad 2x + y - z + 2w = 0,$
           $-x + 2y + z + w = 3, \quad x + 5y - 8z - 5w = -12,$
           $3x - 7y + 8z + 9w = 13.$
       b. $x = \frac{9}{14} - \frac{9w}{7}, y = \frac{11}{14} - \frac{4w}{7}, z = \frac{29}{14} - \frac{8w}{7}$, w is arbitrary.

A17.   a.   2.
       b.   5.

A19.   a.   Yes, by Theorem 2.8.1 because rank$(A) = 2 < 3$.
       b.   No, if **b** is not in the two-dimensional column space of A, a
           subspace of $\mathbb{R}^3$.

A21. $\begin{bmatrix} \frac{17}{12} & \frac{2}{3} & \frac{3}{4} & 0 \end{bmatrix}^T + w \begin{bmatrix} -25 & 14 & -9 & 6 \end{bmatrix}^T.$

A23. $x = 1, \quad y = -1, \quad z = 1.$

B3.   a.   Linearly dependent: We have six vectors in $\mathbb{R}^4$ (see
           Theorem 1.4.3).
       b.   Linearly independent: The (column) rank is 4.

B5.   For example, consider

$$\begin{bmatrix} 1 & 0 & 0 \\ 0 & 1 & 0 \\ 0 & 0 & 1 \end{bmatrix} \begin{bmatrix} x \\ y \\ z \end{bmatrix} = \begin{bmatrix} 1 \\ 2 \\ 3 \end{bmatrix}$$

with obvious solution $x = 1$, $y = 2$, $z = 3$. Now perform the
column operation $C_1 \rightarrow C_1 + C_2$. The new system

$$\begin{bmatrix} 1 & 0 & 0 \\ 1 & 1 & 0 \\ 0 & 0 & 1 \end{bmatrix} \begin{bmatrix} x \\ y \\ z \end{bmatrix} = \begin{bmatrix} 1 \\ 2 \\ 3 \end{bmatrix}$$

has the solution $x = 1$, $y = 1$, $z = 3$.

B7.   Rank $= 3$ if $\lambda \neq 0$; Rank $= 2$ if $\lambda = 0$.

B11. $t = 2$ or $t = -3$.

B13.   b. $\gamma_1 = [-7 \quad 4 \quad 9 \quad -2]^T, \quad \gamma_2 = [2 \quad 0 \quad 1 \quad 1]^T,$
           $\gamma_3 = [-5 \quad 1 \quad -2 \quad 3]^T.$
       c.   The system $A\mathbf{x} = \boldsymbol{\beta}$ is inconsistent since $\boldsymbol{\beta}$ cannot be
           expressed as a linear combination of the γ's.

B15.   For example, $\{x + 2y + 3w = 1, z - 2w = 1, 2y + 3w = 0\}$.

Exercises 3.1

A1.   a. A and B are not the same size
       b. $\begin{bmatrix} 1 & 4 & 8 \\ 11 & 14 & 7 \end{bmatrix}$ c. $\begin{bmatrix} 2 & 1 \\ 2 & 1 \\ 1 & 11 \end{bmatrix}$ d. $\begin{bmatrix} 1 & 3 \\ 2 & 4 \end{bmatrix}$

e. B^T and C^T are not the same size.

f. $\begin{bmatrix} -4 & -8 \\ -12 & -16 \end{bmatrix}$ g. $\begin{bmatrix} -4 & 5 \\ -2 & 7 \\ 3 & -21 \end{bmatrix}$ h. $\begin{bmatrix} 2 & 5 \\ 5 & 8 \end{bmatrix}$

i. C and C^T are not the same size.

A3.   For example, $B = \frac{1}{3}A = \begin{bmatrix} 1 & 2 & -1/3 \\ 0 & -2/3 & 4/3 \end{bmatrix}$.

A7.   a. $[-2 \quad 0 \quad 4 \quad 7]$ b. $\begin{bmatrix} 0 \\ 2 \\ 4 \end{bmatrix}$ c. $\begin{bmatrix} -3 & 1 \\ 0 & 2 \end{bmatrix}$ d. $\begin{bmatrix} 1 & 0 & -3 \\ 2 & 3 & 0 \end{bmatrix}$

e. $\begin{bmatrix} 5 & -2 \\ -2 & 7 \end{bmatrix}$—symmetric   f. $\begin{bmatrix} -1 & 0 & 1 \\ 2 & 4 & -5 \\ -3 & 5 & 0 \end{bmatrix}$

g. $\begin{bmatrix} 1 & 5 & 9 \\ 5 & 2 & -3 \\ 9 & -3 & 0 \end{bmatrix}$—symmetric.

B7.   a. For example, $A = \begin{bmatrix} 0 & 1 & 2 & 3 \\ -1 & 0 & -1 & 5 \\ -2 & 1 & 0 & 3 \\ -3 & -5 & -3 & 0 \end{bmatrix}$.

b. The diagonal entries are all equal to zero. If A is an $n \times n$ skew-symmetric matrix, all its diagonal entries must equal zero.

Exercises 3.2

A1. $\begin{bmatrix} -5 \\ -13 \end{bmatrix}$ A3. The number of columns of A does not equal the number of rows of $\mathbf{x}$.

A5. $[10 \quad -1 \quad -5 \quad 26]^T$ A7. $[0 \quad -19 \quad -26 \quad 6 \quad -25]^T$

A9. $[36]$ A11. $[38]$ A13. $[-30]$

A15. $[x \quad y] \begin{bmatrix} 4 & 0 \\ 0 & 9 \end{bmatrix} \begin{bmatrix} x \\ y \end{bmatrix}$ A17. $[x \quad y] \begin{bmatrix} 13 & -5 \\ -5 & 13 \end{bmatrix} \begin{bmatrix} x \\ y \end{bmatrix}$

A19. $[x_1 \quad x_2 \quad x_3] \begin{bmatrix} 2 & -2 & -5/2 \\ -2 & 3 & 0 \\ -5/2 & 0 & 10 \end{bmatrix} \begin{bmatrix} x_1 \\ x_2 \\ x_3 \end{bmatrix}$

A21.   There is a counterclockwise rotation and the original vector has been lengthened.

A23.   The vector has been rotated $45°$ counterclockwise, with no change in length.

A25.   There has been a $180°$ counterclockwise rotation and the length has been multiplied by 4.

A27.   a. $\mathbf{x}_1$ is an eigenvector with eigenvalue 6; $\mathbf{x}_3$ is an eigenvector with eigenvalue 6.
       b. $\mathbf{x}_2$ is an eigenvector with eigenvalue 4; $\mathbf{x}_3$ is an eigenvector with eigenvalue 4.
       c. $\mathbf{x}_2$ is an eigenvector with eigenvalue 1; $\mathbf{x}_3$ is an eigenvector with eigenvalue 2; $\mathbf{x}_4$ is an eigenvector with eigenvalue -1.

B1.   For example, $\begin{bmatrix} 1 & 0 & 5 \\ 7 & 0 & -3 \end{bmatrix} \begin{bmatrix} 0 \\ 1 \\ 0 \end{bmatrix} = \begin{bmatrix} 0 \\ 0 \end{bmatrix}$.

B3.   If $A = \operatorname{diag}(d_1, d_2, \ldots, d_n)$ and $\mathbf{x} = [\,x_1 \quad x_2 \quad \ldots \quad x_n\,]^{\mathrm{T}}$, then $A\mathbf{x} = [\,d_1 x_1 \quad d_2 x_2 \quad \ldots \quad d_n x_n\,]^{\mathrm{T}}$.

B5.   a. $\mathbf{X} = \begin{bmatrix} \frac{1}{8} & \frac{1}{3} & \frac{1}{4} \\ \frac{1}{2} & \frac{1}{6} & \frac{1}{4} \\ \frac{1}{4} & \frac{1}{6} & \frac{1}{4} \end{bmatrix} \mathbf{X} + \mathbf{Y}$

      b. $\begin{bmatrix} x_1 \\ x_2 \\ x_3 \end{bmatrix} = \begin{bmatrix} 70 \\ \frac{1395}{14} \\ \frac{505}{7} \end{bmatrix}$ c. $\begin{bmatrix} x_1 \\ x_2 \\ x_3 \end{bmatrix} = \begin{bmatrix} 160 \\ \frac{1620}{7} \\ \frac{1200}{7} \end{bmatrix}$

C3.   a./b. Positive definite

Exercises 3.3

A1.   a. 2×7 b. Not defined   c. 7×3 d. 7×3 e. 2×2
      f. Not defined   g. 2×2 h. Not defined

A3.   a. $\begin{bmatrix} -14 & 1 \\ -12 & 4 \end{bmatrix}$ b. $\begin{bmatrix} 13 & -8 & 10 \\ 21 & -16 & 26 \end{bmatrix}$ c. $\begin{bmatrix} -39 & -4 & -81 \\ 24 & 4 & 51 \\ -9 & -4 & -21 \end{bmatrix}$

      d. $\begin{bmatrix} -20 & 10 \\ -8 & 40 \\ 1 & 123 \end{bmatrix}$ e. $\begin{bmatrix} -9 & 60 & 46 & 111 & -19 \\ -68 & 22 & -81 & -16 & 17 \\ 80 & -58 & 60 & 6 & 83 \\ 4 & 2 & 16 & 4 & 11 \\ 10 & -42 & -80 & -2 & 8 \end{bmatrix}$

B1.   a. For example, take $\mathbf{x} = \begin{bmatrix} 1 \\ 1 \end{bmatrix}$ or any multiple of this vector.

      b. For example, take $\mathbf{y} = [\,3 \quad 1\,]$ or any multiple of this row vector.

B3.   For example, let $A = \begin{bmatrix} 1 & 0 & 1 \\ -2 & 1 & 0 \end{bmatrix}$ and $B = \begin{bmatrix} 1 & 2 \\ 2 & 5 \\ 0 & -2 \end{bmatrix}$.

B9.   a. For example, let $A = \begin{bmatrix} 1 & -2 \\ -2 & 3 \end{bmatrix}$ and $B = \begin{bmatrix} 0 & 1 \\ 1 & 0 \end{bmatrix}$.

    b. For example, let $A = \begin{bmatrix} 1 & -2 \\ -2 & 3 \end{bmatrix}$ and $B = \begin{bmatrix} 1 & 0 \\ 0 & 1 \end{bmatrix}$.

B15.   a. O_3

B17.   a. 140   b. 512   c. 196   d. 2000   e. 64

B31.   For example, let $A = \begin{bmatrix} 0 & 0 \\ 1 & 0 \end{bmatrix}$ and $B = \begin{bmatrix} 0 & 0 \\ 0 & 1 \end{bmatrix}$.

B33.   b. $m \times n$

    c. $AB = \begin{bmatrix} 1 & -2 & 3 & -4 \\ 17 & -18 & 19 & -20 \\ 25 & -30 & 35 & -40 \end{bmatrix}$.

Exercises 3.4

A1.   a. $\begin{bmatrix} 42 & 45 \\ -9 & 3 \end{bmatrix}$ b. $\begin{bmatrix} 8 & 7 & -4 & 11 \\ 17 & 29 & 2 & 28 \\ 20 & 25 & -5 & 30 \end{bmatrix}$ c. $\begin{bmatrix} 4 & 6 & 9 \\ 2 & 5 & 6 \\ 1 & 2 & 3 \end{bmatrix}$

    d. $\begin{bmatrix} 7 & 6 & 0 & 0 \\ 17 & 10 & 0 & 0 \\ 27 & 14 & 0 & 0 \\ 0 & 0 & -1 & 9 \\ 0 & 0 & 7 & -5 \end{bmatrix}$ e. $\begin{bmatrix} -6 & 11 & -5 \\ 16 & 7 & 13 \\ 31 & -17 & 15 \end{bmatrix}$

A3. $\begin{bmatrix} \times & \vdots & \times & \times & \times \\ \times & \vdots & \times & \times & \times \\ \times & \vdots & \times & \times & \times \\ \cdots & \vdots & \cdots & \cdots & \cdots \\ \times & \vdots & \times & \times & \times \end{bmatrix} \begin{bmatrix} \times & \vdots & \times & \times & \times \\ \cdots & \vdots & \cdots & \cdots & \cdots \\ \times & \vdots & \times & \times & \times \\ \times & \vdots & \times & \times & \times \\ \times & \vdots & \times & \times & \times \end{bmatrix} = \begin{bmatrix} \times & \vdots & \times & \times & \times \\ \times & \vdots & \times & \times & \times \\ \times & \vdots & \times & \times & \times \\ \cdots & \vdots & \cdots & \cdots & \cdots \\ \times & \vdots & \times & \times & \times \end{bmatrix}$

A5. $B^2 = \begin{bmatrix} A^2 + I_n & 2A & I_n \\ 2A & A^2 + 2I_n & 2A \\ I_n & 2A & A^2 + I_n \end{bmatrix}$,

    $B^3 = \begin{bmatrix} A^3 + 3A & 2A^2 + A + 2I_n & 3A \\ 3A^2 + 2I_n & A^3 + 6A & 3A^2 + 2I_n \\ 3A & 2A^2 + A + 2I_n & A^3 + 3A \end{bmatrix}$

B1.   There are 18 ways.

B3. $A^T = \begin{bmatrix} A_{11}^T & A_{21}^T \\ A_{12}^T & A_{22}^T \end{bmatrix}$. In general, if $A = \begin{bmatrix} A_{11} & A_{12} & \cdots & A_{1n} \\ A_{21} & A_{21} & \cdots & A_{2n} \\ \vdots & \vdots & \vdots & \vdots \\ A_{m1} & A_{m1} & \cdots & A_{m1} \end{bmatrix}$,

then $A^T = \begin{bmatrix} A_{11}^T & A_{21}^T & \cdots & A_{m1}^T \\ A_{12}^T & A_{22}^T & \cdots & A_{m2}^T \\ \vdots & \vdots & \vdots & \vdots \\ A_{1n}^T & A_{2n}^T & \cdots & A_{mn}^T \end{bmatrix}$.

B7.   b. For example, let $A = \begin{bmatrix} 1 & 0 \\ 0 & 0 \end{bmatrix}$, $B = \begin{bmatrix} 0 & 0 \\ 0 & 1 \end{bmatrix}$, and $C = \begin{bmatrix} 0 & 0 \\ 1 & 0 \end{bmatrix}$.

Then rank $\begin{bmatrix} A & B \\ O & C \end{bmatrix} = 3$ and $\text{rank}(A) + \text{rank}(B) = 2$.

c. Write any upper triangular matrix M in block form as $\begin{bmatrix} A & B \\ O & C \end{bmatrix}$ and consider where the diagonal elements of M occur in A and C. Then use the result of part (a).

C1.   b. 465.

Exercises 3.5

A5.   a. $M^n = \begin{bmatrix} k^n & k^{n-1} + k^{n-2} + \cdots + 1 \\ 0 & 1 \end{bmatrix} = \begin{bmatrix} k^n & \frac{k^n - 1}{k-1} \\ 0 & 1 \end{bmatrix}$.

b. $M^{-n} = \frac{1}{k^n} \begin{bmatrix} 1 & -(k^{n-1} + k^{n-2} + \cdots + 1) \\ 0 & k^n \end{bmatrix} = \frac{1}{k^n} \begin{bmatrix} 1 & -\left(\frac{k^n - 1}{k-1}\right) \\ 0 & k^n \end{bmatrix}$.

A7. $\begin{bmatrix} a & -\sqrt{1 - a^2} \\ \sqrt{1 - a^2} & a \end{bmatrix}$ for $|a| < 1$.

B7.   Yes, $\lambda = \left(-5 + \sqrt{17}\right)/4$ and $\lambda = \left(-5 - \sqrt{17}\right)/4$.

Exercises 3.6

A1.   a. No   b. Yes   c. Yes   d. No   e. No   f. Yes   g. No
h. No

A3.   a. $B = \begin{bmatrix} 1 & 0 & 0 \\ 2 & 1 & 0 \\ 0 & 0 & 1 \end{bmatrix} \begin{bmatrix} 1 & 0 & 0 \\ 0 & 1 & 0 \\ 0 & 1 & 1 \end{bmatrix} \begin{bmatrix} 1 & 0 & 0 \\ 0 & 1 & 0 \\ 0 & 0 & 5 \end{bmatrix}$
$\begin{bmatrix} 1 & 0 & 0 \\ 0 & 1 & -4 \\ 0 & 0 & 1 \end{bmatrix} \begin{bmatrix} 1 & 0 & 2 \\ 0 & 1 & 0 \\ 0 & 0 & 1 \end{bmatrix}$.

b. $B^{-1} = \begin{bmatrix} 1 & 0 & -2 \\ 0 & 1 & 0 \\ 0 & 0 & 1 \end{bmatrix} \begin{bmatrix} 1 & 0 & 0 \\ 0 & 1 & 4 \\ 0 & 0 & 1 \end{bmatrix} \begin{bmatrix} 1 & 0 & 0 \\ 0 & 1 & 0 \\ 0 & 0 & \frac{1}{5} \end{bmatrix}$

$\begin{bmatrix} 1 & 0 & 0 \\ 0 & 1 & 0 \\ 0 & -1 & 1 \end{bmatrix} \begin{bmatrix} 1 & 0 & 0 \\ -2 & 1 & 0 \\ 0 & 0 & 1 \end{bmatrix}.$

A5. $A = \begin{bmatrix} 1 & 0 & 0 \\ 2 & 1 & 0 \\ 0 & 0 & 1 \end{bmatrix} \begin{bmatrix} 1 & 0 & 0 \\ 0 & 1 & 0 \\ -1 & 0 & 1 \end{bmatrix} \begin{bmatrix} 1 & 0 & 3 \\ 0 & -1 & 0 \\ 0 & 0 & 1 \end{bmatrix} \begin{bmatrix} 1 & 0 & 0 \\ 0 & 1 & -4 \\ 0 & 0 & 1 \end{bmatrix}$

$\begin{bmatrix} 1 & 0 & 3 \\ 0 & 1 & 0 \\ 0 & 0 & 1 \end{bmatrix} \begin{bmatrix} 1 & -2 & 0 \\ 0 & 1 & 0 \\ 0 & 0 & 1 \end{bmatrix},$

$A^{-1} = \begin{bmatrix} 1 & 2 & 0 \\ 0 & 1 & 0 \\ 0 & 0 & 1 \end{bmatrix} \begin{bmatrix} 1 & 0 & -3 \\ 0 & 1 & 0 \\ 0 & 0 & 1 \end{bmatrix} \begin{bmatrix} 1 & 0 & 0 \\ 0 & 1 & 4 \\ 0 & 0 & 1 \end{bmatrix} \begin{bmatrix} 1 & 0 & 3 \\ 0 & -1 & 0 \\ 0 & 0 & 1 \end{bmatrix}$

$\begin{bmatrix} 1 & 0 & 0 \\ 0 & 1 & 0 \\ 1 & 0 & 1 \end{bmatrix} \begin{bmatrix} 1 & 0 & 0 \\ -2 & 1 & 0 \\ 0 & 0 & 1 \end{bmatrix}.$

A9.   6

B9.   a. No. The order of the row exchanges is important because each such exchange changes the labels of the rows involved in successive interchanges.
       b. Yes. Such matrices are diagonal and diagonal matrices commute.
       c. No

Exercises 3.7

A1.   a. $\begin{bmatrix} 1 & 0 \\ -3 & 1 \end{bmatrix} \begin{bmatrix} 1 & 2 \\ 0 & 5 \end{bmatrix}$ b. $\begin{bmatrix} 1 & 0 \\ \frac{3}{2} & 1 \end{bmatrix} \begin{bmatrix} 2 & -4 \\ 0 & 7 \end{bmatrix}$

       c. $\begin{bmatrix} 1 & 0 & 0 \\ 2 & 1 & 0 \\ 2 & -1 & 1 \end{bmatrix} \begin{bmatrix} 1 & 2 & 3 \\ 0 & 2 & 1 \\ 0 & 0 & -1 \end{bmatrix}$

       d. $\begin{bmatrix} 1 & 0 & 0 \\ 2 & 1 & 0 \\ 3 & 4 & 1 \end{bmatrix} \begin{bmatrix} 2 & 2 & 2 \\ 0 & 3 & 3 \\ 0 & 0 & 4 \end{bmatrix}$ e. $\begin{bmatrix} 1 & 0 & 0 \\ 4 & 1 & 0 \\ 3 & 2 & 1 \end{bmatrix} \begin{bmatrix} 1 & 4 & 5 \\ 0 & 2 & 6 \\ 0 & 0 & 3 \end{bmatrix}$

       f. $\begin{bmatrix} 1 & 0 & 0 \\ 2 & 1 & 0 \\ \frac{3}{2} & -\frac{1}{4} & 1 \end{bmatrix} \begin{bmatrix} 2 & 2 & -1 \\ 0 & -4 & 6 \\ 0 & 0 & 7 \end{bmatrix}$

       g. $\begin{bmatrix} 1 & 0 & 0 & 0 \\ -3 & 1 & 0 & 0 \\ 1 & 2 & 1 & 0 \\ -1 & 8 & -5 & 1 \end{bmatrix} \begin{bmatrix} 1 & 2 & -1 & 4 \\ 0 & 1 & 3 & 7 \\ 0 & 0 & 1 & 2 \\ 0 & 0 & 0 & 1 \end{bmatrix}$

A3.

$$\begin{bmatrix} 1 & 0 & 0 & 0 & 0 & 0 \\ -7 & 1 & 0 & 0 & 0 & 0 \\ -1 & \frac{1}{6} & 1 & 0 & 0 & 0 \\ 2 & -\frac{3}{2} & -\frac{279}{13} & 1 & 0 & 0 \\ 5 & -2 & -\frac{252}{13} & \frac{316}{335} & 1 & 0 \\ 0 & -\frac{1}{6} & -\frac{25}{13} & \frac{4}{67} & -\frac{700}{129} & 1 \end{bmatrix} \begin{bmatrix} -1 & -2 & 4 & 2 & 3 & 1 \\ 0 & -6 & 31 & 13 & 21 & -2 \\ 0 & 0 & -\frac{13}{6} & \frac{17}{6} & \frac{1}{2} & \frac{13}{3} \\ 0 & 0 & 0 & \frac{1005}{13} & \frac{458}{13} & 91 \\ 0 & 0 & 0 & 0 & -\frac{516}{335} & -\frac{1621}{335} \\ 0 & 0 & 0 & 0 & 0 & -\frac{2798}{129} \end{bmatrix}$$

A5. $a = -\frac{23}{22}, b = \frac{9}{11}, c = -\frac{5}{11}$

A7. $x = 4, y = -\frac{5}{2}$

A9. $x = -\frac{3}{2}, y = -2, z = 2$

A11. $x = \frac{1}{16}, y = -\frac{1}{8}, z = \frac{1}{16}, w = \frac{1}{4}$

A13. $a = -\frac{1}{6}, b = -\frac{5}{6}, c = -\frac{5}{6}$

B11. $A = \begin{bmatrix} 1 & 0 & 0 & 0 \\ -3 & 1 & 0 & 0 \\ 2 & 4 & 1 & 0 \\ -1 & -2 & 0 & 1 \end{bmatrix} \begin{bmatrix} 2 & 3 & -1 \\ 0 & 3 & 2 \\ 0 & 0 & 0 \\ 0 & 0 & 0 \end{bmatrix}$

C1. $x = 0, \pm\sqrt{2}$

C3. $A = \begin{bmatrix} 1 & 0 & 0 \\ 2 & 1 & 0 \\ -1 & 3 & 1 \end{bmatrix} \begin{bmatrix} 2 & 0 & 0 \\ 0 & 1 & 0 \\ 0 & 0 & -11 \end{bmatrix} \begin{bmatrix} 1 & 1 & -\frac{1}{2} \\ 0 & 1 & 4 \\ 0 & 0 & 1 \end{bmatrix}$

Exercises 4.1

A1.   a. 1   b. 0   c. 1   d. 1   e. 2   f. 2   g. −35,000   h. 0

A3.   a. $-2(3^n)$ b. $-5^n/2$ c. −8   d. $(-3)^n(-2)$

A7.   −590

A9.   420

A13. $a^2(a - b)$

A15. $\det(AB) = 0$; $\det(BA) = 144$

A17.   a. 0   b. −3

A19. $-\sqrt{2}/2$

A21. -8

A23.   0 for all positive integers n

A25.   12

A27. $x = -6.$

B9.   No. For example, $\det \begin{bmatrix} 0 & 1 & 1 \\ 1 & 0 & 1 \\ 1 & 1 & 0 \end{bmatrix} = 2.$

B15. $x = -15/17, \quad y = 21/17, \quad z = 14/17.$

C1.   e. $A^{-1} = \begin{bmatrix} 1 & 0 & -2 & 3 \\ 0 & 1 & -1 & 1 \\ 0 & 0 & 1 & -3 \\ 0 & 0 & 0 & 1 \end{bmatrix}$

C7.   For example, let $P = O_2$, $Q = I_2$, $R = I_2$, and $S = O_2$. Then
$\det \begin{bmatrix} P & Q \\ R & S \end{bmatrix} = 1 \neq -1 = \det(P)\det(S) - \det(Q)\det(R).$

Exercises 4.2

A1.   a. 3/2   b. 34   c. 11/2

A3.   a. -7 b. 0   c. 1   d. -1 e. 0

Exercises 4.3

A1. $-35,000$

A3.   420

A5.   (a) 0   (b) -408

A7.   45

A9.   (a) $2x - 10$ (b) $x \neq 5$ (c) $1/(2x - 10)$

B1. -6

Exercises 4.4

A1.   a. **y** is an eigenvector with eigenvalue 1; **u** is an eigenvector with eigenvalue 4.
   b. **x** is an eigenvector with eigenvalue 6; **y** is an eigenvector with eigenvalue 9; **v** is an eigenvector with eigenvalue 6.
   c. **x** is an eigenvector with eigenvalue -6; **u** is an eigenvector with eigenvalue 12; **v** is an eigenvector with eigenvalue -6.
   d. **y** is an eigenvector with eigenvalue 3; **v** is an eigenvector with eigenvalue -2.
   e. **y** is an eigenvector with eigenvalue 0; **u** is an eigenvector with eigenvalue 3; **v** is an eigenvector with eigenvalue 0.

A3.   a. $\lambda_1 = 5$, $\mathbf{v}_1 = \begin{bmatrix} 1 \\ 1 \end{bmatrix}$; $\lambda_2 = -1$, $\mathbf{v}_2 = \begin{bmatrix} -1 \\ 1 \end{bmatrix}$

   b. $\lambda_1 = -1$, $\mathbf{v}_1 = \begin{bmatrix} -1 \\ 1 \end{bmatrix}$; $\lambda_2 = 1$, $\mathbf{v}_2 = \begin{bmatrix} 1 \\ 1 \end{bmatrix}$

   c. $\lambda_1 = 1$, $\mathbf{v}_1 = \begin{bmatrix} 2 \\ 3 \end{bmatrix}$; $\lambda_2 = 2$, $\mathbf{v}_2 = \begin{bmatrix} 1 \\ 1 \end{bmatrix}$

   d. $\lambda_1 = 4$, $\mathbf{v}_1 = \begin{bmatrix} 1 \\ 1 \\ 0 \end{bmatrix}$; $\lambda_2 = -6$, $\mathbf{v}_2 = \begin{bmatrix} 1 \\ -1 \\ 2 \end{bmatrix}$; $\lambda_3 = 2$,

   $\mathbf{v}_3 = \begin{bmatrix} -1 \\ 1 \\ 2 \end{bmatrix}$

   e. $\lambda_1 = -8$, $\mathbf{v}_1 = \begin{bmatrix} -1 \\ 1 \\ 2 \end{bmatrix}$; $\lambda_2 = 0$, $\mathbf{v}_2 = \begin{bmatrix} 1 \\ 1 \\ 0 \end{bmatrix}$; $\lambda_3 = -4$,

   $\mathbf{v}_3 = \begin{bmatrix} 1 \\ -1 \\ 2 \end{bmatrix}$

   f. $\lambda_1 = -2$, $\mathbf{v}_1 = \begin{bmatrix} -1 \\ 0 \\ 1 \end{bmatrix}$; $\lambda_2 = -2$, $\mathbf{v}_2 = \begin{bmatrix} 1 \\ 1 \\ 0 \end{bmatrix}$; $\lambda_3 = -4$,

   $\mathbf{v}_3 = \begin{bmatrix} -1 \\ -2 \\ 1 \end{bmatrix}$

   g. $\lambda_1 = 2$, $\mathbf{v}_1 = \begin{bmatrix} -2 \\ 0 \\ 1 \end{bmatrix}$; $\lambda_2 = 2$; $\lambda_3 = 1$, $\mathbf{v}_3 = \begin{bmatrix} -1 \\ 1 \\ 0 \end{bmatrix}$

   There is no linearly independent eigenvector corresponding to the second occurrence of the eigenvalue 2.

   h. $\lambda_1 = 1$, $\mathbf{v}_1 = [1 \ 0 \ 0 \ 0]^T$; $\lambda_2 = 2$, $\mathbf{v}_2 = [3 \ 1 \ 0 \ 0]^T$;
   $\lambda_3 = 4$, $\mathbf{v}_3 = [-29 \ -21 \ -6 \ 6]^T$;
   $\lambda_4 = 3$, $\mathbf{v}_4 = [11 \ 6 \ 2 \ 0]^T$

   i. $\lambda_1 = 2$, $\mathbf{v}_1 = [2 \ 1 \ 0 \ 0]^T$; $\lambda_2 = 3$, $\mathbf{v}_2 = [0 \ 0 \ 1 \ 0]^T$;
   $\lambda_3 = 5$, $\mathbf{v}_3 = [0 \ 0 \ 0 \ 1]^T$;
   $\lambda_4 = 1$, $\mathbf{v}_4 = [0 \ 0 \ -1 \ 1 \ 0]^T$

A5.   a. $m_1 = 20$, $m_2 = 11$, $m_3 = 10$, $m_4 = 10$,
          $m_5 = 10$, $b_1 = 10$, $b_2 = 19$, $b_3 = 20$, $b_4 = 20$, $b_5 = 20$.
      b. $m_1 = 10 = m_2 = m_3 = m_4 = m_5$,
          $b_1 = 20 = b_2 = b_3 = b_4 = b_5$.

A7. $x_{n+1} = (3^n + (-1)^{n+1})/2$, $y_{n+1} = (3^n + (-1)^n)/2$.

A9.   36

B3.   The eigenvalue of $\mathbf{y}$ is 1; the eigenvalue of $\mathbf{z}$ is -1.

B5.   The eigenvalue corresponding to A^k is λ^k.

B7.   The eigenvalue corresponding to A^{-1} is $1/\lambda$.

B15. See Exercise A9.

B17. $[\overbrace{1 \quad 1 \quad \ldots \quad 1}^{n\ 1\text{'s}}]^{\mathsf{T}}$ or any multiple of this vector.

C1.   For M, $(n-1)$ of the eigenvalues are 0 and one is n. The
      eigenvector corresponding to $\lambda = n$ is $[\overbrace{1 \quad 1 \quad \ldots \quad 1}^{n\ 1\text{'s}}]^{\mathsf{T}}$.
      The other $(n-1)$ eigenvectors are the vectors of the form
      $[-1 \quad * \quad \ldots \quad *]^{\mathsf{T}}$, where the $(n-1)$ entries after the first
      are 0's and 1's, with exactly one entry equal to 1.

Exercises 4.5

A5.   a. $\lambda_1 = -1$, $\mathbf{v}_1 = \begin{bmatrix} 1 \\ 0 \\ 1 \end{bmatrix}$; $\lambda_2 = 2$, $\mathbf{v}_2 = \begin{bmatrix} 1 \\ 3 \\ 1 \end{bmatrix}$; $\lambda_3 = 1$,

          $\mathbf{v}_3 = \begin{bmatrix} 3 \\ 2 \\ 1 \end{bmatrix}$.

      b. Take $P = \begin{bmatrix} 1 & 1 & 3 \\ 0 & 3 & 2 \\ 1 & 1 & 1 \end{bmatrix}$, for example.

      c. $A^{-1} = \frac{1}{2} \begin{bmatrix} 3 & 1 & -5 \\ 1 & 1 & -1 \\ 1 & 1 & -3 \end{bmatrix}$.

A7.   Use $P = \begin{bmatrix} 1 & 1 \\ 2 & -1 \end{bmatrix}$ or $P = \begin{bmatrix} 1 & 1 \\ -1 & 2 \end{bmatrix}$ and note that
      $P^{-1}AP = D$, implies that $P^{-1}A^kP = D^k$.

A9. $x_{n+1} = \frac{6}{13}[(-7)^n - 6^n]$, $y_{n+1} = \frac{1}{13}[-4(6^n) - 9(-7)^n]$.

A11.   For example, take $P = \begin{bmatrix} -1 & 1 & 1 \\ 0 & 1 & 1 \\ 1 & 0 & 2 \end{bmatrix}$.

B7.   b. For example, let $A = \begin{bmatrix} 0 & -1 \\ 1 & 0 \end{bmatrix}$, which has no real eigenvalues/eigenvectors. However, $A^2 = \begin{bmatrix} -1 & 0 \\ 0 & -1 \end{bmatrix}$ has linearly independent (real) eigenvectors and so is diagonalizable.

B9.   b. $A: \begin{bmatrix} 0 \\ 1 \end{bmatrix}, \begin{bmatrix} 1 \\ 0 \end{bmatrix}$; $B: \begin{bmatrix} -5 \\ 2 \end{bmatrix}, \begin{bmatrix} -3 \\ 1 \end{bmatrix}$.

c. No.

C1. $\begin{bmatrix} \cos\left(\frac{t}{2}\right) & \sin\left(\frac{t}{2}\right) \\ -\sin\left(\frac{t}{2}\right) & \cos\left(\frac{t}{2}\right) \end{bmatrix} \begin{bmatrix} \cos t & \sin t \\ \sin t & -\cos t \end{bmatrix} \begin{bmatrix} \cos\left(\frac{t}{2}\right) & -\sin\left(\frac{t}{2}\right) \\ \sin\left(\frac{t}{2}\right) & \cos\left(\frac{t}{2}\right) \end{bmatrix}$
$= \begin{bmatrix} 1 & 0 \\ 0 & -1 \end{bmatrix}$.

Exercises 4.6

A1.   a. $\left\{ \begin{bmatrix} -1 \\ 1 \end{bmatrix} \right\}$.

b. The algebraic multiplicity is 1, which equals the geometric multiplicity.

A3.   a.   (i) $\lambda_1 = 4, \lambda_2 = 2$.
(ii) Both eigenvalues have geometric multiplicity 1.
(iii) The matrix is diagonalizable. For example, take
$P = \begin{bmatrix} -1 & 1 \\ 1 & 1 \end{bmatrix}$.

b.   (i) $\lambda_1 = 3, \lambda_2 = 6, \lambda_3 = 2$.
(ii) Both eigenvalues have geometric multiplicity 1.
(iii) The matrix is diagonalizable. For example, take
$P = \begin{bmatrix} 1 & 1 & -1 \\ 1 & -2 & 0 \\ 1 & 1 & 1 \end{bmatrix}$.

c.   (i) $\lambda_1 = 6, \lambda_2 = 10, \lambda_3 = 8$.
(ii) The eigenvalues have geometric multiplicity 1, which is their geometric multiplicity.
(iii) Take $P = \begin{bmatrix} 1 & 1 & 3 \\ 1 & 1 & 1 \\ 0 & 2 & 2 \end{bmatrix}$.

d.   (i) $\lambda_1 = 2 = \lambda_2, \lambda_3 = 1$.
(ii) The eigenvalues λ_1 and λ_2 have geometric multiplicity 1, as does λ_3.

(iii) Because the geometric multiplicity of λ_1 is 1, but the algebraic multiplicity is 2, the matrix is not diagonalizable.

e. (i) $\lambda_1 = \lambda_2 = \lambda_3 = 2$.
    (ii) The eigenvalue has geometric multiplicity 1.
    (iii) Because the geometric multiplicity of λ_1 is 1, but the algebraic multiplicity is 3, the matrix is not diagonalizable.

f. (i) $\lambda_1 = -1$, $\lambda_2 = -2$, $\lambda_3 = 1$.
    (ii) The geometric multiplicity of each eigenvalue is 1.
    (iii) Because the algebraic multiplicity of each eigenvalue is 1, the matrix is diagonalizable. For example, take
    $$P = \begin{bmatrix} 0 & 1 & 1 \\ 1 & 3 & 0 \\ 1 & 0 & 0 \end{bmatrix}.$$

A5. a. $\lambda^4 - 4\lambda^3 + 16\lambda - 16$.
     b. $\{[-1 \ 0 \ 0 \ 1]^T, [-1 \ 0 \ 1 \ 0]^T, [-1 \ 1 \ 0 \ 0]^T\}$.
     c. The algebraic multiplicity of 2 is 3. The geometric multiplicity of 2 is 3.

B1. a. $\lambda_1 = -2 = \lambda_2$, $\lambda_3 = 4$, $\lambda_4 = 2 = \lambda_5$.
     b. The algebraic multiplicity of λ_1 is 2. The algebraic multiplicity of λ_3 is 1. The algebraic multiplicity of λ_4 is 2.
     c. $S_{-2}(A): \{[-1 \ 0 \ 1 \ 0 \ 0]^T, [-1 \ 1 \ 0 \ 0 \ 0]^T\}$;
        $S_4(A): \{[1 \ 1 \ 1 \ 0 \ 3]^T\}$; $S_2(A): \{[0 \ 0 \ 0 \ 0 \ 1]^T\}$.
     d. The geometric multiplicity of -2 is. The geometric multiplicity of 4 is 1. The geometric multiplicity of 2 is 1.
     e. A is not diagonalizable because the geometric multiplicity of 2 does not equal its algebraic multiplicity.

C1. $B = \begin{bmatrix} 2.5 & -0.5 & 0.5 \\ -0.5 & 2.5 & -0.5 \\ 0 & 0 & 2 \end{bmatrix}$.

Exercises 4.7

A1. (a) $\lambda_1 = 1$, $\lambda_2 = 3$; $\mathbf{v}_1 = \begin{bmatrix} 1 \\ 1 \end{bmatrix}$, $\mathbf{v}_2 = \begin{bmatrix} -1 \\ 1 \end{bmatrix}$.

     (c) $\hat{\mathbf{v}}_1 = \begin{bmatrix} 1/\sqrt{2} \\ 1/\sqrt{2} \end{bmatrix}$, $\hat{\mathbf{v}}_2 = \begin{bmatrix} -1/\sqrt{2} \\ 1/\sqrt{2} \end{bmatrix}$.

     (d) $P = \begin{bmatrix} 1/\sqrt{2} & -1/\sqrt{2} \\ 1/\sqrt{2} & 1/\sqrt{2} \end{bmatrix}$.

     (e) $P^{-1}AP = \begin{bmatrix} 3 & 0 \\ 0 & 1 \end{bmatrix}$.

A3. $P = \begin{bmatrix} 1 & -1 \\ 1 & 1 \end{bmatrix}$, $D = \begin{bmatrix} -2 & 0 \\ 0 & 4 \end{bmatrix}$, $A^5 = \begin{bmatrix} 496 & -528 \\ -528 & 496 \end{bmatrix}$.

A5. $P = \begin{bmatrix} 0 & 1 & -1 \\ 0 & 1 & 1 \\ 1 & 0 & 0 \end{bmatrix}$, $P^{-1} = \begin{bmatrix} 0 & 0 & 1 \\ \frac{1}{2} & \frac{1}{2} & 0 \\ -\frac{1}{2} & \frac{1}{2} & 0 \end{bmatrix}$, $P^{-1}AP = \begin{bmatrix} 1 & 0 & 0 \\ 0 & 1 & 0 \\ 0 & 0 & -1 \end{bmatrix}$.

A7. $P = \begin{bmatrix} -1 & 1 & 1 \\ 1 & 0 & 1 \\ 1 & 1 & 0 \end{bmatrix}$, $P^{-1} = \frac{1}{3}\begin{bmatrix} -1 & 1 & 1 \\ 1 & -1 & 2 \\ 1 & 2 & -1 \end{bmatrix}$,

$P^{-1}AP = \begin{bmatrix} -3 & 0 & 0 \\ 0 & 3 & 0 \\ 0 & 0 & 3 \end{bmatrix}$.

A9. (b) $P = \frac{\sqrt{13}}{13}\begin{bmatrix} -3 & 2 \\ 2 & 3 \end{bmatrix}$, $D = \begin{bmatrix} 1 & 0 \\ 0 & -12 \end{bmatrix}$.

B5. $P = \frac{1}{\sqrt{2}}\begin{bmatrix} 1 & -1 \\ 1 & 1 \end{bmatrix}$, $D = \begin{bmatrix} b+a & 0 \\ 0 & -b+a \end{bmatrix}$.

C1. (a) $[1 \;\; 1 \;\; 1 \;\; 1 \;\; 1]^T, [-1 \;\; 0 \;\; 0 \;\; 0 \;\; 1]^T,$
$[-1 \;\; 0 \;\; 0 \;\; 1 \;\; 0]^T, [-1 \;\; 0 \;\; 1 \;\; 0 \;\; 0]^T,$
$[-1 \;\; 1 \;\; 0 \;\; 0 \;\; 0]^T.$

(b) $P = \begin{bmatrix} 1 & -1 & -1 & -1 & -1 \\ 1 & 0 & 0 & 0 & 1 \\ 1 & 0 & 0 & 1 & 0 \\ 1 & 0 & 1 & 0 & 0 \\ 1 & 1 & 0 & 0 & 0 \end{bmatrix}$.

(c) $P^{-1} = \frac{1}{5}\begin{bmatrix} 1 & 1 & 1 & 1 & 1 \\ -1 & -1 & -1 & -1 & 4 \\ -1 & -1 & -1 & 4 & -1 \\ -1 & -1 & 4 & -1 & -1 \\ -1 & 4 & -1 & -1 & -1 \end{bmatrix}$.

(d) $P^{-1}AP = \begin{bmatrix} 5 & 0 & 0 & 0 & 0 \\ 0 & 0 & 0 & 0 & 0 \\ 0 & 0 & 0 & 0 & 0 \\ 0 & 0 & 0 & 0 & 0 \\ 0 & 0 & 0 & 0 & 0 \end{bmatrix}$.

Exercises 4.8

A1. a. $(\lambda - 1)^2 = \lambda^2 - 2\lambda + 1$.
b. $\lambda^2 + 3\lambda - 10$.
c. $\lambda^3 - \lambda$.
d. $\lambda^3 - 3\lambda^2 + 2\lambda$.
e. $\lambda^3 - 6\lambda^2 + 11\lambda - 6$.
f. $\lambda^3 - 3\lambda^2 + \lambda - 3$.

A3. O_3.

A5. $A^{-1} = \frac{1}{3} \begin{bmatrix} 7 & -2 & -2 \\ 12 & -3 & -6 \\ 0 & 0 & 3 \end{bmatrix}$.

A7. $A^{593} = A = \begin{bmatrix} -3 & -4 \\ 2 & 3 \end{bmatrix}$.

A9. The characteristic equation of A is $\lambda^2 - 1.1\lambda + 0.1$. The remainder when x^k is divided by $x^2 - 1.1x + 0.1$ is $\left[1 + \frac{1}{9}\left(1 - (0.1)^{k-1}\right)\right]x - \frac{1}{9}\left(1 - (0.1)^{k-1}\right)$.

B1. What kind of mathematical object is $p_A(A)$? What kind of mathematical object is $\det(A - \lambda I)$?

B5. This result is not true for 4×4 nilpotent matrices.

C1. b. $\beta_1 = (\lambda_2\lambda_1^n - \lambda_1\lambda_2^n)/(\lambda_2 - \lambda_1)$, $\beta_2 = (\lambda_2^n - \lambda_1^n)/(\lambda_2 - \lambda_1)$

C3. a. $\lambda^3 - 5\lambda^2 + 7\lambda - 3$.
b. $\lambda^2 - 4\lambda + 3 = (\lambda - 3)(\lambda - 1)$.
c. $p_1(\lambda) = \lambda - 1$, $p_2(\lambda) = \lambda - 3$.
d. $E_1 = \frac{1}{2} \begin{bmatrix} 1 & 2 & -5 \\ 3 & 6 & -15 \\ 1 & 2 & -5 \end{bmatrix}$, $E_2 = -\frac{1}{2} \begin{bmatrix} -1 & 2 & -5 \\ 3 & 4 & -15 \\ 1 & 2 & -7 \end{bmatrix}$.

i. $S_3(A) = span \left\{ \begin{bmatrix} 1 \\ 3 \\ 1 \end{bmatrix} \right\}$, $S_1(A) = span \left\{ \begin{bmatrix} 5 \\ 0 \\ 1 \end{bmatrix}, \begin{bmatrix} -2 \\ 1 \\ 0 \end{bmatrix} \right\}$.

j. Let $P = \begin{bmatrix} 1 & 5 & -2 \\ 3 & 0 & 1 \\ 1 & 1 & 0 \end{bmatrix}$. Then $P^{-1}AP = \begin{bmatrix} 3 & 0 & 0 \\ 0 & 1 & 0 \\ 0 & 0 & 1 \end{bmatrix}$.

Exercises 5.1

B1. The smallest possible vector space is $\{\mathbf{0}\}$. By closure, if $\mathbf{v} \neq \mathbf{0} \in V$, then $c\mathbf{v} \in V$ for every real number c. (It is easy to see that if $\mathbf{v}$ is a nonzero vector and $c\mathbf{v} = k\mathbf{v}$ for real numbers c and k, then $c = k$.)

B3. No. For example, $\mathbf{u} = \begin{bmatrix} 1 \\ 0 \\ 0 \end{bmatrix}$ and $\mathbf{v} = \begin{bmatrix} 0 \\ 1 \\ 0 \end{bmatrix}$ are in S, but $\|\mathbf{u} + \mathbf{v}\| = \sqrt{2}$, so $\mathbf{u} + \mathbf{v} \notin S$.

B7. Yes.

C1.   No. For $n = 2$, $\begin{bmatrix} 1 & 2 \\ 3 & 4 \end{bmatrix}$ and $\begin{bmatrix} 2 & -1 \\ -2 & 0 \end{bmatrix}$ are nonsymmetric,

but their sum $\begin{bmatrix} 3 & 1 \\ 1 & 4 \end{bmatrix}$ is symmetric.

C3.   Yes.

Exercises 5.2

A1. $S = \mathbb{R}^2$, the two-dimensional Euclidean plane.

A7.   For example, if $U_2 = span\left\{ \begin{bmatrix} 1 \\ 0 \end{bmatrix} \right\}$ and $U_3 = span\left\{ \begin{bmatrix} 0 \\ 1 \end{bmatrix} \right\}$,
then $\mathbb{R}^2 = U \oplus U_2 = U \oplus U_3$.

A9.   No. The vectors $\begin{bmatrix} 0 & 1 \\ 5 & 6 \end{bmatrix}$ and $\begin{bmatrix} 1 & 2 \\ 3 & 0 \end{bmatrix}$ are in the set, but their
sum is $\begin{bmatrix} 1 & 3 \\ 8 & 6 \end{bmatrix}$, not in the set.

A11.   a. Yes.

b. No. For example, $\begin{bmatrix} 2 & 2 \\ 0 & 0 \end{bmatrix} + \begin{bmatrix} 1 & 3 \\ 0 & 0 \end{bmatrix} = \begin{bmatrix} 3 & 5 \\ 0 & 0 \end{bmatrix}$, which
is not in the set.

c. No. For example, $\begin{bmatrix} 2 & 1 \\ 0 & 2 \end{bmatrix} + \begin{bmatrix} 4 & 0 \\ 0 & 1 \end{bmatrix} = \begin{bmatrix} 6 & 1 \\ 0 & 3 \end{bmatrix}$, which
is not in the set.

d. No. For example, $\overbrace{\begin{bmatrix} 1 & 0 \\ 0 & 1 \end{bmatrix}}^{\text{rank}=2} + \overbrace{\begin{bmatrix} -1 & 0 \\ 0 & 1 \end{bmatrix}}^{\text{rank}=2} = \overbrace{\begin{bmatrix} 0 & 0 \\ 0 & 2 \end{bmatrix}}^{\text{rank}=1}$.

A17. $U + U = U$.

B13.   For example, let $U_1 = \left\{ \begin{bmatrix} x \\ y \\ 0 \end{bmatrix} : x, y \in \mathbb{R} \right\}$,

$U_2 = \left\{ \begin{bmatrix} 0 \\ 0 \\ z \end{bmatrix} : z \in \mathbb{R} \right\}$, and $U_3 = \left\{ \begin{bmatrix} 0 \\ y \\ y \end{bmatrix} : y \in \mathbb{R} \right\}$. We can

write $\begin{bmatrix} 0 \\ 0 \\ 0 \end{bmatrix}$ in two different ways as $\mathbf{u}_1 + \mathbf{u}_2 + \mathbf{u}_3$.

C1.   b. No.

Exercises 5.3

A1.   No. Rewrite the vectors as vectors in $\mathbb{R}^4$, form the 3×4 matrix
       A using these vectors as rows, and find that rref(A) has a
       row of zeros. Alternatively, show that some nontrivial linear
       combination of the given functions is the zero function.

A3.   b. For example, let $a=0$, $b=2$, $c=0$, and $d=1$ or $a=3$,
       $b=-4$, $c=3$, and $d=-2$.

A5. $P_2[x]$.

A11.   a. $\pi x^2 + 2x - 1 = \left(\frac{1-\pi}{2}\right)\mathbf{p}_1 + \left(\frac{\pi+3}{2}\right)\mathbf{p}_2 + \left(\frac{-3+\pi}{2}\right)\mathbf{p}_3$

B13.   For example, let $W_1 = span\left\{ [1\ \ 0\ \ 0\ \ 1]^T, [0\ \ 1\ \ 0\ \ 0]^T \right\}$
        and $W_2 = span\left\{ [0\ \ 0\ \ 0\ \ 1]^T, [0\ \ 0\ \ 1\ \ 0]^T \right\}$.

Exercises 5.4

A3.   First find a vector not in the span of the given vectors.
       Next, add a vector not in the span of the three vectors. For
       example, add the vectors $[1\ \ \ 0\ \ \ 0\ \ \ 1]^T, [1\ \ \ 1\ \ \ 1\ \ \ 1]^T$ to
       the original set.

A5.   Use Theorem 5.4.7 to conclude that $\dim(U \cap W) = 0$.

A7. U and W are planes such that $U \cap W$ is a line through the
       origin or $U = W$ (the two planes are the same).

A9.   The *least* possible value of $\dim(U + W)$ is 36 and the *greatest*
       possible value is 55.

B1.   For example, $\left\{ 1 - x - 2x^3,\ 1 + x^3,\ x^2 \right\}$.

B3.   For example, $\left\{ \begin{bmatrix} 1 & 0 \\ 0 & 0 \end{bmatrix}, \begin{bmatrix} 0 & 0 \\ 0 & 1 \end{bmatrix}, \begin{bmatrix} 1 & 0 \\ 2 & 0 \end{bmatrix}, \begin{bmatrix} 2 & 1 \\ -2 & -1 \end{bmatrix} \right\}$.

B9.   b. $\alpha = \mathbf{f}_1 - \mathbf{f}_2$.

B15. $\dim(M_{mn}) = m \cdot n$.

C5.   The result is not valid. As a counterexample in $\mathbb{R}^2$, let
       $U_1 = \left\{ \begin{bmatrix} x \\ 0 \end{bmatrix} : x \in \mathbb{R} \right\}, U_2 = \left\{ \begin{bmatrix} 0 \\ y \end{bmatrix} : y \in \mathbb{R} \right\}, U_3 = \left\{ \begin{bmatrix} x \\ x \end{bmatrix} : x \in \mathbb{R} \right\}$.

Exercises 6.1

A1.   (a) No   (b) Yes   (c) No   (d) No   (e) Yes   (f) No
(g) No   (h) Yes.

A9.   Yes.

A13.   No.

A15.   No. Linearity demands that $T\left(\begin{bmatrix} -6 \\ -2 \\ 0 \end{bmatrix}\right) = -2T\left(\begin{bmatrix} 3 \\ 1 \\ 0 \end{bmatrix}\right)$,

which is not true.

B3.   For example, $T\left(\begin{bmatrix} x \\ y \end{bmatrix}\right) = \sqrt[3]{x^3 + y^3}$.   (Why   doesn't

$T\left(\begin{bmatrix} x \\ y \end{bmatrix}\right) = \sqrt{x^2 + y^2}$ work?)

Exercises 6.2

A1.   Range $T = \mathbb{R}$. Null space $T = \{$all vectors in $\mathbb{R}^n$ whose ith
component is 0$\}$.

A3.   (a) Range $T = \mathbb{R}$, rank $T = 1$. The range is the entire real
number line.
(b) Range $T = \mathbb{R}^2$, rank $T = 2$. The range is Euclidean 2-space.
(c) Range $T = \mathbb{R}^2$, rank $T = 2$. The range is Euclidean 2-space.
(d) Range $T = span\left\{\begin{bmatrix} 2 \\ 0 \\ -1 \end{bmatrix}\right\}$, rank $T = 1$. The range is a
line in $\mathbb{R}^3$.

A5.   Range $T = \mathbb{R}$, null space $T = \{a \cos(x): a \in \mathbb{R}\}$.

A7.   Basis for the range: $\{-x^2 - 2, 3x^2 - 2x\}$; basis for the null
space: $\{2x^2 + 11x + 1\}$.

B3.   (b) Choose the scalars so that $c_1 + c_2 + c_3 = 1$.

B13.   Assume that there *is* such a transformation and get a contradiction.

C1.   (a) Yes   (b) Yes   (c) Yes.

Exercises 6.3

A3. (a) S represents a reflection of a vector in $\mathbb{R}^2$ across the line $y = x$. (Note that $S^2 = I$.) The operator T projects a vector in $\mathbb{R}^2$ onto the real axis (see the explanation in Exercises 1.2, between C8 and C9).

   (b) $(S+T)\left(\begin{bmatrix} x \\ y \end{bmatrix}\right) = \begin{bmatrix} x+y \\ x \end{bmatrix}$; $(TS)\left(\begin{bmatrix} x \\ y \end{bmatrix}\right) = \begin{bmatrix} y \\ 0 \end{bmatrix}$;

   $(ST)\left(\begin{bmatrix} x \\ y \end{bmatrix}\right) = \begin{bmatrix} 0 \\ x \end{bmatrix}$; $(S^2)\left(\begin{bmatrix} x \\ y \end{bmatrix}\right) = \begin{bmatrix} x \\ y \end{bmatrix}$;

   $(T^2)\left(\begin{bmatrix} x \\ y \end{bmatrix}\right) = \begin{bmatrix} x \\ 0 \end{bmatrix}$.

A5. (a) $(ST)\left(\begin{bmatrix} x \\ y \end{bmatrix}\right) = \begin{bmatrix} x-y \\ -y \end{bmatrix}$ (b) $(TS)\left(\begin{bmatrix} x \\ y \end{bmatrix}\right) = \begin{bmatrix} x+y \\ -y \end{bmatrix}$

   (c) $(S^2)\left(\begin{bmatrix} x \\ y \end{bmatrix}\right) = \begin{bmatrix} x+2y \\ y \end{bmatrix}$

   (d) $(S^3)\left(\begin{bmatrix} x \\ y \end{bmatrix}\right) = \begin{bmatrix} x+3y \\ y \end{bmatrix}$ (e) $((ST)U)\left(\begin{bmatrix} x \\ y \end{bmatrix}\right) = \begin{bmatrix} y-x \\ -x \end{bmatrix}$

   (f) $(S(TU))\left(\begin{bmatrix} x \\ y \end{bmatrix}\right) = \begin{bmatrix} y-x \\ -x \end{bmatrix}$ (g) $(ST-TS)\left(\begin{bmatrix} x \\ y \end{bmatrix}\right) = \begin{bmatrix} -2y \\ 0 \end{bmatrix}$

   (h) $(TU-UT)\left(\begin{bmatrix} x \\ y \end{bmatrix}\right) = \begin{bmatrix} 2y \\ -2x \end{bmatrix}$ (i) $(U^2-I)\left(\begin{bmatrix} x \\ y \end{bmatrix}\right) = \begin{bmatrix} 0 \\ 0 \end{bmatrix}$

   (j) $(S^2-2S+I)\left(\begin{bmatrix} x \\ y \end{bmatrix}\right) = \begin{bmatrix} 0 \\ 0 \end{bmatrix}$ (k) $(SU-US)\left(\begin{bmatrix} x \\ y \end{bmatrix}\right) = \begin{bmatrix} x \\ -y \end{bmatrix}$.

A7. (a) $(SU)(f) = f'' + f'''$ (b) $(US)(f) = f'' + f'''$.
   (c) $(UT)(f) = 5f'' + 5xf''' + x^2 f^{(4)}$.
   (d) $(TU)(f) = f'' + xf''' + x^2 f^{(4)}$.
   (e) $(TU-UT)(f) = -4f'' - 4xf'''$.
   (f) $(U^2)(f) = f^{(4)}$ (g) $(S^2)(f) = f + 2f' + f''$.
   (h) $(S^3)(f) = f + 3f' + 3f'' + f'''$.
   (i) $(TS)(f) = f + (1+x)f' + (x+x^2)f'' + x^2 f'''$.
   (j) $(ST)(f) = f + (x+2)f' + (x^2+3x)f'' + x^2 f'''$.

A11. 0

B5. For example, on $\mathbb{R}^2$ define S by $S\left(\begin{bmatrix} x \\ y \end{bmatrix}\right) = \begin{bmatrix} x \\ 0 \end{bmatrix}$ and T by $T\left(\begin{bmatrix} x \\ y \end{bmatrix}\right) = \begin{bmatrix} 0 \\ y \end{bmatrix}$.

Exercises 6.4

A1. (a) $\begin{bmatrix} 1 & -1 \\ -3 & 7 \end{bmatrix}$ (b) $\begin{bmatrix} 7 & 1 \\ -1 & -3 \end{bmatrix}$ (c) $\begin{bmatrix} -4 & -26 \\ 2 & 12 \end{bmatrix}$

(d) $\begin{bmatrix} 12 & 2 \\ -26 & -4 \end{bmatrix}$.

A3. $\begin{bmatrix} 2 & 2 & 1 \\ 0 & 3 & 0 \\ 0 & -4 & 1 \end{bmatrix}$.

A5. $\begin{bmatrix} \frac{1}{2} & -\frac{1}{2} \\ -\frac{1}{2} & \frac{1}{2} \end{bmatrix}$.

B1. (a) $\begin{bmatrix} -\frac{4}{3} \\ \frac{1}{2} \\ -\frac{1}{6} \end{bmatrix}$ (b) $a = -1/3$, $b = -1/2$, $c = 13/54$

(c) For part (a): $\begin{bmatrix} 1 & 1 & -\frac{4}{3} \\ 5 & 2 & \frac{1}{2} \\ 3 & 2 & -\frac{1}{6} \end{bmatrix}$; for part (b): $\begin{bmatrix} 1 & 1 & -1 \\ 5 & 2 & -2 \\ 3 & 2 & -3 \end{bmatrix}$.

Exercises 6.5

A1. (a) $T_A^{-1}(\mathbf{v}) = -\frac{1}{14} \begin{bmatrix} 4 & -2 \\ 1 & -3 \end{bmatrix} \mathbf{v}$ (b) Not invertible

(c) $T_A^{-1}(\mathbf{v}) = \begin{bmatrix} \cos\theta & \sin\theta \\ -\sin\theta & \cos\theta \end{bmatrix} \mathbf{v}$

(d) Not invertible (e) $T_A^{-1}(\mathbf{v}) = \frac{1}{24} \begin{bmatrix} 3 & -3 & 6 \\ 7 & 1 & 14 \\ -7 & -1 & 10 \end{bmatrix} \mathbf{v}$.

A3. (b) $\left(T_{a,b}^{-1}(f) \right)(x) = f\left(\frac{x-b}{a}\right)$, if $a \neq 0$.

A5. An antiderivative is given by $2(-6 + 6x - 3x^2 + x^3)e^x$.

B1. (b) Given $S: \mathbb{R}^m \to \mathbb{R}^n$ and $T: \mathbb{R}^n \to \mathbb{R}^m$, where $m > n$, prove that $TS \in L(\mathbb{R}^m)$ is not invertible.

C1. (a) domain $T = P[x]$, range $T = P_0[x]$; domain $D = P[x]$, range $D = P[x]$; domain $D_0 = P_0[x]$, range $D_0 = P[x]$; domain $DT = P[x]$, range $DT = P[x]$; domain $TD = P[x]$, range $DT = P_0[x]$; domain $D_0T = P[x]$, range $D_0T = P[x]$; domain $TD_0 = P_0[x]$, range $D_0T = P_0[x]$.
(b) DT, D_0T, TD_0.
(c) Only T, D_0, D_0T, and TD_0 are invertible.

Exercises 6.6

A3.   (a) No. T is not 1–1.   (b) No. T is not onto.
(c) Yes. T is both 1–1 and onto.

A5.   For example, $\begin{bmatrix} a_{11} & a_{12} \\ a_{21} & a_{22} \\ a_{31} & a_{32} \end{bmatrix} \leftrightarrow [\, a_{11} \quad a_{12} \quad a_{21} \quad a_{22} \quad a_{31} \quad a_{32} \,]^{\mathrm{T}}.$

Similarly, $\begin{pmatrix} a_{11} & \cdots & a_{1n} \\ \vdots & \ddots & \vdots \\ a_{m1} & \cdots & a_{mn} \end{pmatrix} \leftrightarrow$

$[\, a_{11} \quad \cdots \quad a_{1n} \quad a_{21} \quad \cdots \quad a_{2n} \quad \cdots \quad a_{m1} \quad \cdots \quad a_{mn} \,]^{\mathrm{T}}.$

A7. $a_0 + a_1 x + a_2 x^2 + a_3 x^3 + a_4 x^4 \leftrightarrow a_0 x + a_1 x^2 + a_2 x^3 + a_3 x^4 + a_4 x^5.$

A13. $\mathbb{R}^7 \cong P_6[x]$, $\mathbb{R}^{15} \cong M_{3,5}$, $\mathbb{R}^{15} \cong P_{14}[x]$, $M_3 \cong P_8[x]$, $P_{14}[x] \cong M_{3,5}.$

B1.   All real values of λ except $0, 1, \ldots, n$.

B3. $\begin{bmatrix} a_{11} & a_{12} & a_{13} & a_{14} \\ a_{21} & a_{22} & a_{23} & a_{24} \\ a_{31} & a_{32} & a_{33} & a_{34} \end{bmatrix} \leftrightarrow \begin{bmatrix} a_{11} & a_{12} & a_{13} & a_{14} & a_{21} & a_{22} \\ a_{23} & a_{24} & a_{31} & a_{32} & a_{33} & a_{34} \end{bmatrix}.$
The space of $k \times l$ matrices is isomorphic to the space of $m \times n$ matrices if and only if $k \cdot l = m \cdot n$.

B5. $n = 3$.

B9.   (b) The distance is $\sqrt{10}$.
(c) Yes, for lines passing through the origin; No, for lines not passing through the origin.

B11.   The game is tic-tac-toe (or *noughts and crosses, hugs and kisses*).

Exercises 6.7

A1.   (a) $[\mathbf{x}]_B = [\, 1 \quad 1 \quad 1 \quad 3 \,]^{\mathrm{T}}$ (b) $[\mathbf{x}]_B = [\, 2 \quad 0 \quad 2 \quad 0 \,]^{\mathrm{T}}.$

A3. $\begin{bmatrix} 2 & 1 & 2 \\ 1 & 0 & 2 \\ 0 & 0 & -1 \end{bmatrix}.$

A5. (c) $\begin{bmatrix} \frac{1}{2} & -\frac{1}{2} & \frac{1}{2} \\ 0 & 1 & 0 \\ 0 & 1 & -1 \end{bmatrix}$, $[g]_{\hat{B}} = \begin{bmatrix} -\frac{1}{2} \\ 4 \\ 6 \end{bmatrix}$.

A7. (a) $\begin{bmatrix} 1 & 0 & -2 \\ 0 & 2 & 0 \\ 0 & 0 & 4 \end{bmatrix}$ (b) $\begin{bmatrix} 1 & 0 & \frac{1}{2} \\ 0 & \frac{1}{2} & 0 \\ 0 & 0 & \frac{1}{4} \end{bmatrix}$ (c) $[p(x)]_{B_2} = \begin{bmatrix} a + \frac{1}{2}c \\ \frac{1}{2}b \\ \frac{1}{4}c \end{bmatrix}$.

A9. $\begin{bmatrix} 2 & 0 \\ 4 & 5 \end{bmatrix}$.

A11. $\frac{1}{7}\begin{bmatrix} 8 & -2 & 14 \\ 4 & 27 & 7 \end{bmatrix}$.

A13. (a) $\begin{bmatrix} 2 & 0 \\ 0 & -1 \end{bmatrix}$ (b) $\begin{bmatrix} 2 & -3 \\ 0 & -1 \end{bmatrix}$ (c) $\begin{bmatrix} -1 & 1 \\ 0 & 1 \end{bmatrix}$.

B1. $M = \begin{bmatrix} 1 & 1 & 1 & 1 \\ 0 & 1 & 2 & 3 \\ 0 & 0 & 1 & 3 \\ 0 & 0 & 0 & 1 \end{bmatrix}$, $M^{-1} = \begin{bmatrix} 1 & -1 & 1 & -1 \\ 0 & 1 & -2 & 3 \\ 0 & 0 & 1 & -3 \\ 0 & 0 & 0 & 1 \end{bmatrix}$.

B3. $\mathbf{v}_3 = \begin{bmatrix} -11 \\ 9 \\ 8 \end{bmatrix}$, $\hat{\mathbf{v}}_1 = \begin{bmatrix} -9 \\ -6 \\ 11 \end{bmatrix}$, $\hat{\mathbf{v}}_3 = \begin{bmatrix} 4 \\ 23 \\ 4 \end{bmatrix}$.

B5. (a) $A(v) = k_v \begin{bmatrix} 1 & -\frac{v}{c^2} \\ -v & 1 \end{bmatrix}$ (b) $A^{-1}(v) = A(-v) = k_v \begin{bmatrix} 1 & \frac{v}{c^2} \\ v & 1 \end{bmatrix}$

(c) $v_3 = \frac{(v_1+v_2)c^2}{v_1 v_2 + c^2}$.

B7. (b) R^{-1}.

Exercises 7.1

A5. The solution set is $\{[(-1 - 2i) - (4 - i)z \quad (-1 + i) - (1 - 2i)z \quad z]^T : z \in \mathbb{C}\}$

A9. $\lambda_1 = 1 + i, \mathbf{v}_1 = \begin{bmatrix} -5 + i \\ -1 + i \\ 1 \end{bmatrix}$; $\lambda_2 = 1 - i, \mathbf{v}_2 = \begin{bmatrix} -5 - i \\ -1 - i \\ 1 \end{bmatrix}$;

$\lambda_3 = 1, \mathbf{v}_3 = \begin{bmatrix} -3 \\ 0 \\ 1 \end{bmatrix}$.

A13. (a) $\lambda_1 = 0, \lambda_2 = 2$ (b) $\lambda_1 = 1, \lambda_2 = \sqrt{3}, \lambda_3 = -\sqrt{3}$.

A15. $T^2 = I$, $T^{-1} = T$.

B3. $\lambda_1 = 3i, \mathbf{v}_1 = \begin{bmatrix} \sqrt{2}i \\ \sqrt{3} \\ 1 \end{bmatrix}; \lambda_2 = -3i, \mathbf{v}_2 = \begin{bmatrix} \sqrt{2}i \\ -\sqrt{3} \\ 1 \end{bmatrix};$

$\lambda_3 = 3, \mathbf{v}_3 = \begin{bmatrix} -\sqrt{2}i \\ 0 \\ 2 \end{bmatrix}.$ Let $U = \begin{bmatrix} \sqrt{2}i & \sqrt{2}i & -\sqrt{2}i \\ \sqrt{3} & -\sqrt{3} & 0 \\ 1 & 1 & 2 \end{bmatrix}.$

Then $U^*AU = \mathrm{diag}(18i, -18i, 18)$.

B11. $T^*\left(\begin{bmatrix} z_1 \\ z_2 \end{bmatrix}\right) = \begin{bmatrix} z_1 \\ -iz_2 \end{bmatrix}.$

Exercises 7.2

A1. $\mathbf{x}^T\mathbf{x} = [4]$ and $\mathbf{x}\mathbf{x}^T = \begin{bmatrix} 1 & -1 & 1 & -1 \\ -1 & 1 & -1 & 1 \\ 1 & -1 & 1 & -1 \\ -1 & 1 & -1 & 1 \end{bmatrix}.$

A3.   (a) -2 (b) 0 (c) 56.

A5.   (a) -2 (b) 0 (c) $i+3$ (d) 0.

B9.   Use the Cauchy–Schwarz Inequality (Theorem 7.2.1(2)).

C1.   (e) The closed line segment joining $\begin{bmatrix} 1 \\ 0 \end{bmatrix}$ and $\begin{bmatrix} 1 \\ 1 \end{bmatrix}.$

Exercises 7.3

A1.   (a) not orthogonal (b) orthogonal and orthonormal
(c) orthogonal, but not orthonormal (d) orthogonal, but
not orthonormal.

A3. $a = \pm 1/\sqrt{2}, \pm i/\sqrt{2}.$

A5.   For example, $1 - 3x^2$ or $-\pi + 3\pi x^2$.

B5.   If we use the standard basis $B = \{1, x, x^2\}$ for $P_2[x]$, let

$A = \begin{bmatrix} 1 & \frac{1}{2} & \frac{1}{3} \\ \frac{1}{2} & \frac{1}{3} & \frac{1}{4} \\ \frac{1}{3} & \frac{1}{4} & \frac{1}{5} \end{bmatrix}$ and then $[f]_B^T A[g]_B = \langle f, g \rangle.$

C1.   (d) $\|f_0\|^2 = 1$.

Exercises 7.4

A1. (a) $[T_{v_1}]_{E_2} = \begin{bmatrix} \frac{1}{2} & \frac{1}{2} \\ \frac{1}{2} & \frac{1}{2} \end{bmatrix}$ and $[T_{v_2}]_{E_2} = \begin{bmatrix} \frac{1}{2} & -\frac{1}{2} \\ -\frac{1}{2} & \frac{1}{2} \end{bmatrix}$ (b) $\begin{bmatrix} 2 \\ 2 \end{bmatrix}$.

A5. $\begin{bmatrix} \sqrt{2}/2 \\ \sqrt{2}/2 \\ 0 \end{bmatrix}, \begin{bmatrix} 0 \\ 0 \\ 1 \end{bmatrix}, \begin{bmatrix} \sqrt{2}/2 \\ -\sqrt{2}/2 \\ 0 \end{bmatrix}$.

A7. $\left\{ \begin{bmatrix} \sqrt{5}/5 \\ 2\sqrt{5}/5 \\ 0 \end{bmatrix}, \begin{bmatrix} \sqrt{6}/3 \\ -\sqrt{6}/6 \\ \sqrt{6}/6 \end{bmatrix}, \begin{bmatrix} -30/\sqrt{15} \\ \sqrt{30}/30 \\ \sqrt{30}/6 \end{bmatrix} \right\}$.

A9. $[1 \quad -1 \quad 0 \quad 0]^T, [1 \quad 1 \quad -2 \quad 0]^T, [1 \quad 1 \quad 1 \quad -3]^T$.

A11. $\{[0 \quad \sqrt{2}/2 \quad 0 \quad -\sqrt{2}/2]^T, [0 \quad -\sqrt{6}/6 \quad \sqrt{6}/3 \quad -\sqrt{6}/6]^T,$
$[1/2 \quad 1/2 \quad 1/2 \quad 1/2]^T\}$.

A13. $\{1, \sqrt{3}(2x - 1), \sqrt{5}(6x^2 - 6x + 1)\}$.

A15. $\left\{ \begin{bmatrix} \sqrt{3}/2 \\ 0 \\ \sqrt{3}/2 \end{bmatrix}, \begin{bmatrix} 1/6 \\ 2/3 \\ -1/2 \end{bmatrix}, \begin{bmatrix} -\sqrt{2}/3 \\ \sqrt{2}/6 \\ \sqrt{2} \end{bmatrix} \right\}$.

A17. $\{x, x^2 - \frac{3}{4}x\}$.

A19. $\left\{ \begin{bmatrix} 2 \\ 3 \\ -1 \end{bmatrix}, \begin{bmatrix} 1 \\ -2 \\ -4 \end{bmatrix}, \begin{bmatrix} 2 \\ -1 \\ 1 \end{bmatrix} \right\}$, where $\begin{bmatrix} 1 \\ 0 \\ 0 \end{bmatrix}$ was chosen as the
original third basis vector.

B3. For example, add the vector $\begin{bmatrix} 2 \\ -1 \\ 1 \end{bmatrix}$.

B5. The first set is presumably linearly independent, whereas the
second set is linearly dependent.

C1. (a) $\left\{ \begin{bmatrix} -2\sqrt{5}/5 \\ 0 \\ \sqrt{5}/5 \end{bmatrix}, \begin{bmatrix} 3\sqrt{70}/70 \\ -\sqrt{70}/14 \\ 3\sqrt{70}/35 \end{bmatrix} \right\}$.
     (b) The usual Gram–Schmidt algorithm leads to a division
         by 0.

C3. $\{1, 2x, 4x^2 - 2, 8x^3 - 12x\}$.

Exercises 7.5

A1.   (a) Yes   (b) Yes   (c) Yes   (d) Yes.

A3.   For example, $a = 5$, $b = 4$, $c = 2$.

A5.   For example, $\begin{bmatrix} \frac{1}{3} & \frac{2}{3} & \frac{2}{3} \\ \frac{2}{3} & \frac{1}{3} & -\frac{2}{3} \\ \frac{2}{3} & -\frac{2}{3} & \frac{1}{3} \end{bmatrix}$.

A7.   The last two columns can be $[\,0 \ \ 0 \ \ 0 \ \ 1\,]^T$ and $[\,-2\sqrt{6} \ \ \sqrt{6} \ \ \sqrt{6} \ \ 0\,]^T$, for example.

B15.   There are eight: $\begin{bmatrix} \pm 1 & 0 & 0 \\ 0 & \pm 1 & 0 \\ 0 & 0 & \pm 1 \end{bmatrix}$.

B21.   The first basis is an orthogonal basis, whereas the second is not.

C1. $\alpha^2 + \beta^2 = 1$.

C3.   For example, for $f(x) = a_0 + a_1 x + a_2 x^2$ and $g(x) = b_0 + b_1 x + b_2 x^2$, we can define $\langle f, g \rangle = 3a_0 b_0 - 2a_0 b_1 - 2a_0 b_2 - 2a_1 b_0 + 2a_1 b_1 + a_1 b_2 - 2a_2 b_0 + a_2 b_1 + 2a_2 b_2$.

Exercises 7.6

A1. $A = \frac{1}{3} \begin{bmatrix} 1-i & -1 \\ 1 & 1+i \end{bmatrix} \begin{bmatrix} 3 & 0 \\ 0 & 0 \end{bmatrix} \begin{bmatrix} 1+i & 1 \\ -1 & 1-i \end{bmatrix}$.

A3. $\lambda_1 = 2$, $\lambda_2 = 3$, $\lambda_3 = -2$.

B5.   For example, $U = \dfrac{1}{\sqrt{2}} \begin{bmatrix} -i & i & 0 \\ 0 & 0 & \sqrt{2} \\ 1 & 1 & 0 \end{bmatrix}$.

C1.   (b) $\dfrac{1}{10} \begin{bmatrix} -5 & -10 & 10 \\ 1 & 0 & 2 \\ -14 & -20 & 22 \end{bmatrix}$.

Exercises 7.7

A1. (a) $Q = \dfrac{1}{\sqrt{10}}\begin{bmatrix} 1 & 3 \\ 3 & -1 \end{bmatrix}$, $R = \dfrac{1}{\sqrt{10}}\begin{bmatrix} 10 & -2 \\ 0 & 4 \end{bmatrix}$

  (b) $Q = \dfrac{1}{\sqrt{2}}\begin{bmatrix} 1 & 1 \\ -1 & 1 \end{bmatrix}$, $R = \dfrac{1}{\sqrt{2}}\begin{bmatrix} 2 & -1 \\ 0 & 1 \end{bmatrix}$

  (c) $Q = \dfrac{1}{\sqrt{2}}\begin{bmatrix} 1 & -1 \\ 1 & 1 \end{bmatrix}$, $R = \dfrac{1}{\sqrt{2}}\begin{bmatrix} 2 & 3 \\ 0 & 1 \end{bmatrix}$

  (d) $Q = \begin{bmatrix} 0 & 1 \\ 1 & 0 \end{bmatrix}$, $R = \begin{bmatrix} 1 & -1 \\ 0 & 1 \end{bmatrix}$.

A3. $Q = \begin{bmatrix} \sqrt{2}/2 & 0 \\ 0 & 1 \\ \sqrt{2}/2 & 0 \end{bmatrix}$, $R = \begin{bmatrix} \sqrt{2} & \sqrt{2} \\ 0 & 1 \end{bmatrix}$.

A5. $Q = \dfrac{1}{25}\begin{bmatrix} 0 & 15 & 20 \\ 15 & 16 & -12 \\ 20 & -12 & 9 \end{bmatrix}$, $R = \begin{bmatrix} 5 & 3 & 7 \\ 0 & 5 & 2 \\ 0 & 0 & 1 \end{bmatrix}$.

A7. $Q = \dfrac{\sqrt{3}}{3}\begin{bmatrix} 1 & 1 & -1 \\ 1 & 0 & 1 \\ 0 & 1 & 1 \\ 1 & -1 & 0 \end{bmatrix}$, $R = \sqrt{3}\begin{bmatrix} 1 & 1 & 1 \\ 0 & 1 & 2 \\ 0 & 0 & 1 \end{bmatrix}$.

A9. $Q = \dfrac{1}{6}\begin{bmatrix} 3\sqrt{2} & 2\sqrt{3} & -\sqrt{6} \\ 3\sqrt{2} & -2\sqrt{3} & \sqrt{6} \\ 0 & 2\sqrt{3} & \sqrt{6} \end{bmatrix}$,

  $R = \dfrac{1}{3}\begin{bmatrix} 3\sqrt{2} & 3\sqrt{2} & 6\sqrt{2} \\ 0 & 3\sqrt{3} & \sqrt{3} \\ 0 & 0 & \sqrt{6} \end{bmatrix}$; $x = 3,\ y = -5,\ z = 4$.

A11. If Q is orthogonal, then $Q = QI$ is the factorization.

B1. $Q = \begin{bmatrix} 0.40825 & -0.59761 & 0.54554 \\ 0.40825 & -1.79284 & -0.10911 \\ 0.40825 & -0.11952 & -0.43644 \\ 0.40825 & 0.11952 & -0.43644 \\ 0.40825 & 1.79284 & -0.10911 \\ 0.40825 & 0.59761 & 0.54554 \end{bmatrix}$,

  $R = \begin{bmatrix} 2.44949 & 8.57321 & 37.15059 \\ 0 & 4.18330 & 29.28310 \\ 0 & 0 & 6.11010 \end{bmatrix}$.

B3. No, unless A is an $n \times n$ matrix of rank n.

C1. Start from the leftmost column of A and ignore those columns that are linearly dependent on the previous ones—that is, choose only the pivot columns of A. The columns of the matrix Q are the outputs of the Gram–Schmidt process applied to pivot column vectors.

Exercises 7.8

A3. (a) $U^{\perp} = \{\mathbf{0}\}$; $\{\mathbf{0}\}$.

   (b) $U^{\perp} =$ the x-y plane; $\left\{ \begin{bmatrix} 1 \\ 0 \\ 0 \end{bmatrix}, \begin{bmatrix} 0 \\ 1 \\ 0 \end{bmatrix} \right\}$, for example.

   (c) $U^{\perp} =$ the line $y = x$; $\left\{ \begin{bmatrix} 1 \\ 1 \end{bmatrix} \right\}$, for example.

   (d) $U^{\perp} = span \left\{ \begin{bmatrix} 1 \\ 1 \\ 1 \end{bmatrix} \right\}$; $\left\{ \begin{bmatrix} 1 \\ 1 \\ 1 \end{bmatrix} \right\}$, for example.

A5. (a) $span\{x^2 - x + \frac{1}{6}\}$ (b) $span\{3x^2 - 1\}$.

A7. $span \left\{ \begin{bmatrix} \frac{2}{3} \\ 1 \\ 0 \end{bmatrix}, \begin{bmatrix} -1 \\ 0 \\ 1 \end{bmatrix} \right\}$.

A9. The orthogonal complement is $\left\{ \begin{bmatrix} 0 & a \\ b & 0 \end{bmatrix} : a, b \in \mathbb{R} \right\}$.

A13. (1) Column space of $A^{\mathrm{T}} =$ row space of A: Basis is
   $\left\{ \begin{bmatrix} 1 \\ 0 \\ 0 \\ 1 \end{bmatrix}, \begin{bmatrix} 0 \\ 1 \\ 1 \\ 1 \end{bmatrix} \right\}$.

   (2) Column space of A: Basis is $\left\{ \begin{bmatrix} 1 \\ 0 \\ 1 \\ 1 \end{bmatrix}, \begin{bmatrix} 0 \\ 1 \\ 1 \\ 2 \end{bmatrix} \right\}$.

   (3) Null space of A: Basis is $\left\{ \begin{bmatrix} -1 \\ -1 \\ 0 \\ 1 \end{bmatrix}, \begin{bmatrix} 0 \\ -1 \\ 1 \\ 0 \end{bmatrix} \right\}$.

   (4) Null space of A^{T}: Basis is $\left\{ \begin{bmatrix} -1 \\ -2 \\ 0 \\ 1 \end{bmatrix}, \begin{bmatrix} -1 \\ -1 \\ 1 \\ 0 \end{bmatrix} \right\}$.

B5. $U^{\mathrm{T}} = span\{1 - \frac{9}{5}x^4, \ x - \frac{18}{5}x^3 + \frac{99}{35}x^5, \ x^2 - \frac{9}{7}x^4\}$.

Exercises 7.9

A1.   (a) $\dfrac{1}{6}\begin{bmatrix} -1 \\ 8 \\ 17 \end{bmatrix}$ (b) $\dfrac{7}{6}\begin{bmatrix} 1 \\ -2 \\ 1 \end{bmatrix}$ (c) $7/\sqrt{6}$, or $7\sqrt{6}/6$.

A3. $\dfrac{1}{131}\begin{bmatrix} 122 \\ 132 \\ 138 \end{bmatrix}$.

B3. $-\dfrac{1}{6}+x$.

Exercises 8.1

A3. $\hat{g}(x) = 33 - 204x + 210x^2$.

A5. $T^*(4 - 2x) = 12 + 6x$.

Exercises 8.2

A1.   (a) Symmetric   (b) None of these   (c) Skew-symmetric
(d) Hermitian   (e) None of these   (f) Skew-Hermitian
(g) Hermitian   (h) Symmetric.

A5.   The scalar k must be a pure imaginary number.

B3. A and B must commute.

B11.   (a) $\dfrac{1}{10}\begin{bmatrix} 4 & 2 & 2 \\ 2 & 10 & -5 \\ 2 & -5 & 10 \end{bmatrix}$ (b) $\dfrac{1}{2}\begin{bmatrix} 2 & 0 & 0 \\ 0 & 3 & -1 \\ 0 & -1 & 3 \end{bmatrix}$

(c) $\lambda_1 I_3$ (d) $\dfrac{1}{3}\begin{bmatrix} 2+3\lambda & 1 & 1 \\ 1 & 3\lambda-1 & 2 \\ 1 & 2 & 3\lambda-1 \end{bmatrix}$.

Exercises 8.3

A1.   (a) Yes   (b) No   (c) No   (d) Yes   (e) Yes   (f) Yes
(g) Yes.

C1.   A sufficient condition is that $\mathbf{u}$ or $\mathbf{v}$ is the zero vector. If neither is the zero vector, then $\mathbf{u}\mathbf{v}^*$ is normal if and only if one of the vectors is a scalar multiple of the other.

Exercises 8.4

A3.   The associated matrix is I_n.

A5.   (a) $\tilde{x}_1^2 + 7\tilde{x}_2^2 + \tilde{x}_3^2$ (b) $-\tilde{x}_1^2 - 7\tilde{x}_2^2 + 5\tilde{x}_3^2$ (c) $-7\tilde{x}_1^2 + 2\tilde{x}_2^2$
      (d) $\tilde{x}_1^2 + 3\tilde{x}_2^2 - 3\tilde{x}_3^2 - \tilde{x}_4^2$.

A7. $10\tilde{x}_1^2$.

A9.   An ellipse.

A11. The only critical point is $(\frac{1}{2}, 1, 1)$, which yields a local minimum.

C1. $x > a$ if $a > 0$; $x > -(n-1)a$ if $a < 0$.

Exercises 8.5

A1.   (a) $\sqrt{3}, 1$ (b) $\sqrt{2}, 1$ (c) $\sqrt{5}, 2$ (d) $1, 2, 3$ (e) $1, 2, 2$.

A3. $A = \begin{bmatrix} 1 & 0 \\ 0 & 1 \end{bmatrix} \begin{bmatrix} \sqrt{2} & 0 & 0 \\ 0 & 1 & 0 \end{bmatrix} \begin{bmatrix} \sqrt{2}/2 & \sqrt{2}/2 & 0 \\ 0 & 0 & 1 \\ -\sqrt{2}/2 & \sqrt{2}/2 & 0 \end{bmatrix}$.

A5. $U = \begin{bmatrix} \sqrt{6}/6 & \sqrt{3}/3 & \sqrt{2}/2 \\ \sqrt{6}/6 & \sqrt{3}/3 & -\sqrt{2}/2 \\ -\sqrt{6}/3 & \sqrt{3}/3 & 0 \end{bmatrix}$, $\Sigma = \begin{bmatrix} \sqrt{6} & 0 \\ 0 & \sqrt{3} \\ 0 & 0 \end{bmatrix}$,
      $V = \begin{bmatrix} 1 & 0 \\ 0 & 1 \end{bmatrix}$.

A7.   (a) $A = \begin{bmatrix} \sqrt{2}/2 & -\sqrt{2}/2 \\ \sqrt{2}/2 & \sqrt{2}/2 \end{bmatrix} \begin{bmatrix} \sqrt{6} & 0 & 0 \\ 0 & 0 & 0 \end{bmatrix} \begin{bmatrix} \sqrt{3}/3 & \sqrt{2}/2 & \sqrt{6}/6 \\ \sqrt{3}/3 & -\sqrt{2}/2 & \sqrt{6}/6 \\ \sqrt{3}/3 & 0 & -\sqrt{6}/3 \end{bmatrix}$
      (b) $A = \sqrt{6} \begin{bmatrix} \sqrt{2}/2 \\ \sqrt{2}/2 \end{bmatrix} \begin{bmatrix} \sqrt{3}/3 & \sqrt{3}/3 & \sqrt{3}/3 \end{bmatrix}$.

A9. $U = \begin{bmatrix} \sqrt{3}/3 & 0 & -\sqrt{6}/3 \\ -\sqrt{3}/3 & -\sqrt{2}/2 & -\sqrt{6}/6 \\ \sqrt{3}/3 & -\sqrt{2}/2 & \sqrt{6}/6 \end{bmatrix}, \quad \Sigma = \begin{bmatrix} 3 & 0 & 0 \\ 0 & 2 & 0 \\ 0 & 0 & 0 \end{bmatrix},$

$V = \begin{bmatrix} \sqrt{3}/3 & 0 & -\sqrt{6}/3 \\ \sqrt{3}/3 & -\sqrt{2}/2 & \sqrt{6}/6 \\ \sqrt{3}/3 & \sqrt{2}/2 & \sqrt{6}/6 \end{bmatrix}.$

B1. The singular values of A^{-1} are the reciprocals of the singular values of A.

B7. $U = V = \begin{bmatrix} 1/2 & 0 & 0 & \sqrt{3}/2 \\ 1/2 & 0 & \sqrt{6}/3 & -\sqrt{3}/6 \\ 1/2 & -\sqrt{2}/2 & -\sqrt{6}/6 & -\sqrt{3}/6 \\ 1/2 & \sqrt{2}/2 & -\sqrt{6}/6 & -\sqrt{3}/6 \end{bmatrix},$

$\Sigma = \begin{bmatrix} 3 & 0 & 0 & 0 \\ 0 & 1 & 0 & 0 \\ 0 & 0 & 1 & 0 \\ 0 & 0 & 0 & 1 \end{bmatrix}.$

B11. (a) $12\sqrt{3}$ and $8\sqrt{3}$

(b) $A = \begin{bmatrix} \sqrt{3}/3 & -\sqrt{6}/6 & -\sqrt{2}/2 \\ \sqrt{3}/3 & \sqrt{6}/3 & 0 \\ \sqrt{3}/3 & -\sqrt{6}/6 & \sqrt{2}/2 \end{bmatrix} \begin{bmatrix} 3 & 0 \\ 0 & 2 \\ 0 & 0 \end{bmatrix} \begin{bmatrix} \sqrt{5}/5 & -2\sqrt{5}/5 \\ 2\sqrt{5}/5 & \sqrt{5}/5 \end{bmatrix}.$

C1. $n, n-1, \dots, 2, 1.$

Exercises 8.6

A5. $A = PQ,$ where $P = \begin{bmatrix} \frac{41\sqrt{2}}{5} & -\frac{12\sqrt{2}}{5} \\ -\frac{12\sqrt{2}}{5} & \frac{34\sqrt{2}}{5} \end{bmatrix}$ and $Q = \begin{bmatrix} \frac{7\sqrt{2}}{10} & -\frac{\sqrt{2}}{10} \\ \frac{\sqrt{2}}{10} & \frac{7\sqrt{2}}{10} \end{bmatrix}.$

A7. $A = PQ,$ where $P = \begin{bmatrix} \sqrt{2} & 0 \\ 0 & 0 \end{bmatrix}$ and $Q = \begin{bmatrix} \frac{\sqrt{2}}{2} & \frac{\sqrt{2}}{2} \\ -\frac{\sqrt{2}}{2} & \frac{\sqrt{2}}{2} \end{bmatrix}.$

A9. $A = AI.$

A11. $A = PQ,$ where $P = \begin{bmatrix} \frac{\sqrt{6}}{2} & \frac{\sqrt{6}}{2}i \\ -\frac{\sqrt{6}}{2}i & \frac{\sqrt{6}}{2} \end{bmatrix}$ and

$Q = \begin{bmatrix} \frac{\sqrt{6}}{6}(1+2i) & -\frac{\sqrt{6}}{6}i \\ -\frac{\sqrt{6}}{6}i & \frac{\sqrt{6}}{6}(1-2i) \end{bmatrix}.$

Index

645